Lecture Notes in Artificial Intelligence 13757

Subseries of Lecture Notes in Computer Science

More information about this subseries at https://link.springer.com/bookseries/1244

Ngoc Thanh Nguyen · Tien Khoa Tran ·
Ualsher Tukayev · Tzung-Pei Hong ·
Bogdan Trawiński · Edward Szczerbicki (Eds.)

Intelligent Information and Database Systems

14th Asian Conference, ACIIDS 2022
Ho Chi Minh City, Vietnam, November 28–30, 2022
Proceedings, Part I

 Springer

Editors
Ngoc Thanh Nguyen ⓘ
Wrocław University of Science
and Technology
Wrocław, Poland

Nguyen Tat Thanh University
Ho Chi Minh city, Vietnam

Ualsher Tukayev ⓘ
Al-Farabi Kazakh National University
Almaty, Kazakhstan

Bogdan Trawiński ⓘ
Wrocław University of Science
and Technology
Wrocław, Poland

Tien Khoa Tran ⓘ
Vietnam National University, Ho Chi Minh
City
Ho Chi Minh City, Vietnam

Tzung-Pei Hong ⓘ
National University of Kaohsiung
Kaohsiung, Taiwan

Edward Szczerbicki ⓘ
University of Newcastle
Newcastle, NSW, Australia

ISSN 0302-9743   ISSN 1611-3349 (electronic)
Lecture Notes in Artificial Intelligence
ISBN 978-3-031-21742-5   ISBN 978-3-031-21743-2 (eBook)
https://doi.org/10.1007/978-3-031-21743-2

LNCS Sublibrary: SL7 – Artificial Intelligence

This Springer imprint is published by the registered company Springer Nature Switzerland AG
The registered company address is: Gewerbestrasse 11, 6330 Cham, Switzerland

Preface

ACIIDS 2022 was the 14th event in a series of international scientific conferences on research and applications in the field of intelligent information and database systems. The aim of ACIIDS 2022 was to provide an international forum for research with scientific backgrounds in the technology of intelligent information and database systems and its various applications. The ACIIDS 2022 conference was co-organized by the International University - Vietnam National University HCMC (Vietnam) and the Wrocław University of Science and Technology (Poland) in cooperation with the IEEE SMC Technical Committee on Computational Collective Intelligence, the European Research Center for Information Systems (ERCIS), Al-Farabi Kazakh National University (Kazakhstan), the University of Newcastle (Australia), Yeungnam University (South Korea), Quang Binh University (Vietnam), Leiden University (The Netherlands), Universiti Teknologi Malaysia (Malaysia), Nguyen Tat Thanh University (Vietnam), BINUS University (Indonesia), the Committee on Informatics of the Polish Academy of Sciences (Poland) and Vietnam National University, Hanoi (Vietnam). ACIIDS 2022 was scheduled to be held in Almaty, Kazakhstan, during June 6–9, 2022. However, due to the unstable political situation, the conference was moved to Ho Chi Minh City, Vietnam, and was conducted as a hybrid event during 28–30 November 2022.

The ACIIDS conference series is already well established, having taken place at various locations throughout Asia (Vietnam, South Korea, Taiwan, Malaysia, Thailand, Indonesia, and Japan) since 2009. The 12th and 13th events were planned to take place in Phuket (Thailand). However, the global COVID-19 pandemic resulted in both editions of the conference being held online in virtual space. We were, therefore, pleased to be able to hold ACIIDS 2022 in person, whilst still providing an option for people participate online.

These two volumes contain 113 peer-reviewed papers selected for presentation from 406 submissions, with each submission receiving at least 3 reviews in a single-blind process. Papers included in this volume cover the following topics: data mining and machine learning methods, advanced data mining techniques and applications, intelligent and contextual systems, natural language processing, network systems and applications, computational imaging and vision, decision support and control systems, and data modeling and processing for Industry 4.0.

The accepted and presented papers focus on new trends and challenges facing the intelligent information and database systems community. The presenters at ACIIDS 2022 showed how research work could stimulate novel and innovative applications. We hope that you find these results useful and inspiring for your future research work.

We would like to express our sincere thanks to the honorary chairs for their support: Arkadiusz Wójs (Rector of Wroclaw University of Science and Technology, Poland) and Zhanseit Tuymebayev (Rector of Al-Farabi Kazakh National University, Kazakhstan). We would like to express our thanks to the keynote speakers for their world-class plenary speeches: Tzung-Pei Hong from the National University of Kaohsiung (Taiwan), Michał Woźniak from the Wrocław University of Science and Technology (Poland), Minh-Triet

Tran from the University of Science and the John von Neumann Institute, VNU-HCM (Vietnam), and Minh Le Nguyen from the Japan Advanced Institute of Science and Technology (Japan).

We cordially thank our main sponsors: International University - Vietnam National University HCMC, Hitachi Vantara Vietnam Co., Ltd, Polish Ministry of Education and Science, and Wrocław University of Science and Technology, as well as all of the aforementioned cooperating universities and organizations. Our special thanks are also due to Springer for publishing the proceedings and to all the other sponsors for their kind support.

We are grateful to the Special Session Chairs, Organizing Chairs, Publicity Chairs, Liaison Chairs, and Local Organizing Committee for their work towards the conference. We sincerely thank all the members of the international Program Committee for their valuable efforts in the review process, which helped us to select the highest quality papers for the conference. We cordially thank all the authors for their valuable contributions and the other conference participants. The conference would not have been possible without their support. Thanks are also due to the many experts who contributed to the event being a success.

November 2022

Ngoc Thanh Nguyen
Tien Khoa Tran
Ualsher Tukeyev
Tzung-Pei Hong
Bogdan Trawiński
Edward Szczerbicki

Organization

Honorary Chairs

Arkadiusz Wójs — Wrocław University of Science and Technology, Poland

Zhanseit Tuymebayev — Al-Farabi Kazakh National University, Kazakhstan

Conference Chairs

Tien Khoa Tran — International University - Vietnam National University HCMC, Vietnam

Ngoc Thanh Nguyen — Wrocław University of Science and Technology, Poland

Ualsher Tukeyev — Al-Farabi Kazakh National University, Kazakhstan

Program Chairs

Tzung-Pei Hong — National University of Kaohsiung, Taiwan

Edward Szczerbicki — University of Newcastle, Australia

Bogdan Trawiński — Wrocław University of Science and Technology, Poland

Steering Committee

Ngoc Thanh Nguyen (Chair) — Wrocław University of Science and Technology, Poland

Longbing Cao — University of Science and Technology Sydney, Australia

Suphamit Chittayasothorn — King Mongkut's Institute of Technology Ladkrabang, Thailand

Ford Lumban Gaol — Bina Nusantara University, Indonesia

Tu Bao Ho — Japan Advanced Institute of Science and Technology, Japan

Tzung-Pei Hong — National University of Kaohsiung, Taiwan

Dosam Hwang — Yeungnam University, South Korea

Bela Stantic — Griffith University, Australia

Geun-Sik Jo — Inha University, South Korea

Hoai An Le-Thi	University of Lorraine, France
Toyoaki Nishida	Kyoto University, Japan
Leszek Rutkowski	Częstochowa University of Technology, Poland
Ali Selamat	Universiti Teknologi Malaysia, Malaysia

Special Session Chairs

Van Sinh Nguyen	International University - Vietnam National University HCMC, Vietnam
Krystian Wojtkiewicz	Wroclaw University of Science and Technology, Poland
Bogumiła Hnatkowska	Wroclaw University of Science and Technology, Poland
Madina Mansurova	Al-Farabi Kazakh National University, Kazakhstan

Doctoral Track Chairs

Marek Krótkiewicz	Wrocław University of Science and Technology, Poland
Marcin Pietranik	Wrocław University of Science and Technology, Poland
Thi Thuy Loan Nguyen	International University - Vietnam National University HCMC, Vietnam
Paweł Sitek	Kielce University of Technology, Poland

Liaison Chairs

Ford Lumban Gaol	Bina Nusantara University, Indonesia
Quang-Thuy Ha	VNU-University of Engineering and Technology, Vietnam
Mong-Fong Horng	National Kaohsiung University of Applied Sciences, Taiwan
Dosam Hwang	Yeungnam University, South Korea
Le Minh Nguyen	Japan Advanced Institute of Science and Technology, Japan
Ali Selamat	Universiti Teknologi Malaysia, Malaysia

Organizing Chairs

| Van Sinh Nguyen | International University - Vietnam National University HCMC, Vietnam |
| Krystian Wojtkiewicz | Wrocław University of Science and Technology, Poland |

Publicity Chairs

Thanh Tung Tran	International University - Vietnam National University HCMC, Vietnam
Marek Kopel	Wrocław University of Science and Technology, Poland
Marek Krótkiewicz	Wrocław University of Science and Technology, Poland

Webmaster

Marek Kopel	Wroclaw University of Science and Technology, Poland

Local Organizing Committee

Le Van Canh	International University - Vietnam National University HCMC, Vietnam
Le Hai Duong	International University - Vietnam National University HCMC, Vietnam
Le Duy Tan	International University - Vietnam National University HCMC, Vietnam
Marcin Jodłowiec	Wrocław University of Science and Technology, Poland
Patient Zihisire Muke	Wrocław University of Science and Technology, Poland
Thanh-Ngo Nguyen	Wrocław University of Science and Technology, Poland
Rafał Palak	Wrocław University of Science and Technology, Poland

Keynote Speakers

Tzung-Pei Hong	National University of Kaohsiung, Taiwan
Michał Woźniak	Wrocław University of Science and Technology, Poland
Minh-Triet Tran	University of Science and John von Neumann Institute, VNU-HCM, Vietnam
Minh Le Nguyen	Japan Advanced Institute of Science and Technology, Japan

Special Sessions Organizers

ACMLT 2022: Special Session on Awareness Computing Based on Machine Learning

Yung-Fa Huang	Chaoyang University of Technology, Taiwan
Rung Ching Chen	Chaoyang University of Technology, Taiwan

ADMTA 2022: Special Session on Advanced Data Mining Techniques and Applications

Chun-Hao Chen	Tamkang University, Taiwan
Bay Vo	Ho Chi Minh City University of Technology, Vietnam
Tzung-Pei Hong	National University of Kaohsiung, Taiwan

AIIS 2022: Special Session on Artificial Intelligence in Information Security

Shynar Mussiraliyeva	Al-Farabi Kazakh National University, Kazakhstan
Batyrkhan Omarov	Al-Farabi Kazakh National University, Kazakhstan

BMLLC 2022: Special Session on Bio-modeling and Machine Learning in Prediction of Metastasis in Lung Cancer

Andrzej Swierniak	Silesian University of Technology, Poland
Rafal Suwinski	Institute of Oncology, Poland

BTAS 2022: Special Session on Blockchain Technology and Applications for Sustainability

Chien-wen Shen	National Central University, Taiwan
Ping-yu Hsu	National Central University, Taiwan

CIV 2022: Special Session on Computational Imaging and Vision

Manish Khare	Dhirubhai Ambani Institute of Information and Communication Technology, India
Prashant Srivastava	NIIT University, India
Om Prakash	HNB Garwal University, India
Jeonghwan Gwak	Korea National University of Transportation, South Korea

DMPI-APP 2022: Special Session on Data Modelling and Processing: Air Pollution Prevention

Marek Krótkiewicz	Wrocław University of Science and Technology, Poland
Krystian Wojtkiewicz	Wrocław University of Science and Technology, Poland
Hoai Phuong Ha	UiT The Arctic University of Norway, Norway
Jean-Marie Lepioufle	Norwegian Institute for Air Research, Norway

DMSN 2022: Special Session on Data Management in Sensor Networks

Khouloud Salameh	American University of Ras Al Khaimah, UAE
Yannis Manolopoulos	Open University of Cyprus, Cyprus
Richard Chbeir	Université de Pau et des Pays de l'Adour (UPPA), France

ICxS 2022: Special Session on Intelligent and Contextual Systems

Maciej Huk	Wroclaw University of Science and Technology, Poland
Keun Ho Ryu	Ton Duc Thang University, Vietnam
Rashmi Dutta Baruah	Indian Institute of Technology Guwahati, India
Tetsuji Kuboyama	Gekushuin University, Japan
Goutam Chakraborty	Iwate Prefectural University, Japan
Seo-Young Noh	Chungbuk National University, South Korea
Chao-Chun Chen	National Cheng Kung University, Taiwan

IPROSE 2022: Special Session on Intelligent Problem Solving for Smart Real World

Doina Logofătu	Frankfurt University of Applied Sciences, Germany
Costin Bădică	University of Craiova, Romania
Florin Leon	Gheorghe Asachi Technical University of Iaşi, Romania
Mirjana Ivanovic	University of Novi Sad, Serbia

ISCEC 2022: Special Session on Intelligent Supply Chains and e-Commerce

Arkadiusz Kawa — Łukasiewicz Research Network – The Institute of Logistics and Warehousing, Poland

Bartłomiej Pierański — Poznan University of Economics and Business, Poland

ISMSFuN 2022: Special Session on Intelligent Solutions for Management and Securing Future Networks

Grzegorz Kołaczek — Wrocław University of Science and Technology, Poland

Łukasz Falas — Wrocław University of Science and Technology, Poland

Patryk Schauer — Wrocław University of Science and Technology, Poland

Krzysztof Gierłowski — Gdańsk University of Technology, Poland

LPAIA 2022: Special Session on Learning Patterns/Methods in Current AI Applications

Urszula Boryczka — University of Silesia, Poland

Piotr Porwik — University of Silesia, Poland

LRLSTP 2022: Special Session on Low Resource Languages Speech and Text Processing

Ualsher Tukeyev — Al-Farabi Kazakh National University, Kazakhstan

Orken Mamyrbayev — Al-Farabi Kazakh National University, Kazakhstan

MISSI 2022: Satellite Workshop on Multimedia and Network Information Systems

Kazimierz Choroś — Wrocław University of Science and Technology, Poland

Marek Kopel — Wrocław University of Science and Technology, Poland

Mikołaj Leszczuk — AGH University of Science and Technology, Poland

Maria Trocan — Institut Supérieur d'Electronique de Paris, France

MLLSCP 2022: Special Session on Machine Learning in Large-Scale and Complex Problems

Jan Kozak	University of Economics in Katowice, Poland
Przemysław Juszczuk	University of Economics in Katowice, Poland
Barbara Probierz	University of Economics in Katowice, Poland

MLND 2022: Special Session on Machine Learning Prediction of Neurodegenerative Diseases

Andrzej W. Przybyszewski	Polish-Japanese Academy of Information Technology, Poland
Jerzy P. Nowacki	Polish-Japanese Academy of Information Technology, Poland

MMAML 2022: Special Session on Multiple Model Approach to Machine Learning

Tomasz Kajdanowicz	Wrocław University of Science and Technology, Poland
Edwin Lughofer	Johannes Kepler University Linz, Austria
Bogdan Trawiński	Wrocław University of Science and Technology, Poland

PPiBDA 2022: Special Session on Privacy Protection in Big Data Approaches

Abdul Razaque	International Information Technology University, Kazakhstan
Saleem Hariri	University of Arizona, USA
Munif Alotaibi	Shaqra University, Saudi Arabia
Fathi Amsaad	Eastern Michigan University, USA
Bandar Alotaibi	University of Tabuk, Saudi Arabia

SIOTBDTA 2022: Special Session on Smart IoT and Big Data Technologies and Applications

Octavian Postolache	ISCTE-University Institute of Lisbon, Portugal
Madina Mansurova	Al-Farabi Kazakh National University, Kazakhstan

Senior Program Committee

Ajith Abraham	Machine Intelligence Research Labs, USA
Jesus Alcala-Fdez	University of Granada, Spain
Lionel Amodeo	University of Technology of Troyes, France
Ahmad Taher Azar	Prince Sultan University, Saudi Arabia

Thomas Bäck	Leiden University, Netherlands
Costin Badica	University of Craiova, Romania
Ramazan Bayindir	Gazi University, Turkey
Abdelhamid Bouchachia	Bournemouth University, UK
David Camacho	Universidad Autonoma de Madrid, Spain
Leopoldo Eduardo Cardenas-Barron	Tecnologico de Monterrey, Mexico
Oscar Castillo	Tijuana Institute of Technology, Mexico
Nitesh Chawla	University of Notre Dame, USA
Rung-Ching Chen	Chaoyang University of Technology, Taiwan
Shyi-Ming Chen	National Taiwan University of Science and Technology, Taiwan
Simon Fong	University of Macau, Macau SAR
Hamido Fujita	Iwate Prefectural University, Japan
Mohamed Gaber	Birmingham City University, UK
Marina L. Gavrilova	University of Calgary, Canada
Daniela Godoy	ISISTAN Research Institute, Argentina
Fernando Gomide	University of Campinas, Brazil
Manuel Grana	University of the Basque Country, Spain
Claudio Gutierrez	Universidad de Chile, Chile
Francisco Herrera	University of Granada, Spain
Tzung-Pei Hong	National University of Kaohsiung, Taiwan
Dosam Hwang	Yeungnam University, South Korea
Mirjana Ivanovic	University of Novi Sad, Serbia
Janusz Jeżewski	Institute of Medical Technology and Equipment ITAM, Poland
Piotr Jedrzejowicz	Gdynia Maritime University, Poland
Kang-Hyun Jo	University of Ulsan, South Korea
Jason J. Jung	Chung-Ang University, South Korea
Janusz Kacprzyk	Systems Research Institute, Polish Academy of Sciences, Poland
Nikola Kasabov	Auckland University of Technology, New Zealand
Muhammad Khurram Khan	King Saud University, Saudi Arabia
Frank Klawonn	Ostfalia University of Applied Sciences, Germany
Joanna Kolodziej	Cracow University of Technology, Poland
Józef Korbicz	University of Zielona Gora, Poland
Ryszard Kowalczyk	Swinburne University of Technology, Australia
Bartosz Krawczyk	Virginia Commonwealth University, USA
Ondrej Krejcar	University of Hradec Králové, Czech Republic
Adam Krzyzak	Concordia University, Canada
Mark Last	Ben-Gurion University of the Negev, Israel
Hoai An Le Thi	University of Lorraine, France

Kun Chang Lee	Sungkyunkwan University, South Korea
Edwin Lughofer	Johannes Kepler University Linz, Austria
Nezam Mahdavi-Amiri	Sharif University of Technology, Iran
Yannis Manolopoulos	Open University of Cyprus, Cyprus
Klaus-Robert Müller	Technical University of Berlin, Germany
Saeid Nahavandi	Deakin University, Australia
Grzegorz J. Nalepa	AGH University of Science and Technology, Poland
Ngoc Thanh Nguyen	Wrocław University of Science and Technology, Poland
Dusit Niyato	Nanyang Technological University, Singapore
Yusuke Nojima	Osaka Prefecture University, Japan
Manuel Núñez	Universidad Complutense de Madrid, Spain
Jeng-Shyang Pan	Fujian University of Technology, China
Marcin Paprzycki	Systems Research Institute, Polish Academy of Sciences, Poland
Bernhard Pfahringer	University of Waikato, New Zealand
Hoang Pham	Rutgers University, USA
Tao Pham Dinh	INSA Rouen, France
Radu-Emil Precup	Politehnica University of Timisoara, Romania
Leszek Rutkowski	Częstochowa University of Technology, Poland
Juergen Schmidhuber	Swiss AI Lab IDSIA, Switzerland
Björn Schuller	University of Passau, Germany
Ali Selamat	Universiti Teknologi Malaysia, Malaysia
Andrzej Skowron	Warsaw University, Poland
Jerzy Stefanowski	Poznań University of Technology, Poland
Edward Szczerbicki	University of Newcastle, Australia
Ryszard Tadeusiewicz	AGH University of Science and Technology, Poland
Muhammad Atif Tahir	National University of Computing and Emerging Sciences, Pakistan
Bay Vo	Ho Chi Minh City University of Technology, Vietnam
Gottfried Vossen	University of Münster, Germany
Dinh Duc Anh Vu	Vietnam National University HCMC, Vietnam
Lipo Wang	Nanyang Technological University, Singapore
Junzo Watada	Waseda University, Japan
Michał Woźniak	Wrocław University of Science and Technology, Poland
Farouk Yalaoui	University of Technology of Troyes, France
Sławomir Zadrożny	Systems Research Institute, Polish Academy of Sciences, Poland
Zhi-Hua Zhou	Nanjing University, China

Program Committee

Muhammad Abulaish	South Asian University, India
Bashar Al-Shboul	University of Jordan, Jordan
Toni Anwar	Universiti Teknologi PETRONAS, Malaysia
Taha Arbaoui	University of Technology of Troyes, France
Mehmet Emin Aydin	University of the West of England, UK
Amelia Badica	University of Craiova, Romania
Kambiz Badie	ICT Research Institute, Iran
Hassan Badir	École Nationale des Sciences Appliquées de Tanger, Morocco
Zbigniew Banaszak	Warsaw University of Technology, Poland
Dariusz Barbucha	Gdynia Maritime University, Poland
Maumita Bhattacharya	Charles Sturt University, Australia
Leon Bobrowski	Białystok University of Technology, Poland
Bülent Bolat	Yildiz Technical University, Turkey
Mariusz Boryczka	University of Silesia, Poland
Urszula Boryczka	University of Silesia, Poland
Zouhaier Brahmia	University of Sfax, Tunisia
Stephane Bressan	National University of Singapore, Singapore
Peter Brida	University of Žilina, Slovakia
Piotr Bródka	Wroclaw University of Science and Technology, Poland
Grażyna Brzykcy	Poznan University of Technology, Poland
Robert Burduk	Wrocław University of Science and Technology, Poland
Aleksander Byrski	AGH University od Science and Technology, Poland
Dariusz Ceglarek	WSB University in Poznań, Poland
Somchai Chatvichienchai	University of Nagasaki, Japan
Chun-Hao Chen	Tamkang University, Taiwan
Leszek J. Chmielewski	Warsaw University of Life Sciences, Poland
Kazimierz Choroś	Wrocław University of Science and Technology, Poland
Kun-Ta Chuang	National Cheng Kung University, Taiwan
Dorian Cojocaru	University of Craiova, Romania
Jose Alfredo Ferreira Costa	Federal University of Rio Grande do Norte (UFRN), Brazil
Ireneusz Czarnowski	Gdynia Maritime University, Poland
Piotr Czekalski	Silesian University of Technology, Poland
Theophile Dagba	University of Abomey-Calavi, Benin
Tien V. Do	Budapest University of Technology and Economics, Hungary

Rafał Doroz	University of Silesia, Poland
El-Sayed M. El-Alfy	King Fahd University of Petroleum and Minerals, Saudi Arabia
Keiichi Endo	Ehime University, Japan
Sebastian Ernst	AGH University of Science and Technology, Poland
Nadia Essoussi	University of Carthage, Tunisia
Usef Faghihi	Université du Québec à Trois-Rivières, Canada
Dariusz Frejlichowski	West Pomeranian University of Technology, Szczecin, Poland
Blanka Frydrychova Klimova	University of Hradec Králové, Czech Republic
Janusz Getta	University of Wollongong, Australia
Daniela Gifu	Alexandru Ioan Cuza University of Iaşi, Romania
Gergo Gombos	Eötvös Loránd University, Hungary
Manuel Grana	University of the Basque Country, Spain
Janis Grundspenkis	Riga Technical University, Latvia
Dawit Haile	Addis Ababa University, Ethiopia
Marcin Hernes	Wrocław University of Business and Economics, Poland
Koichi Hirata	Kyushu Institute of Technology, Japan
Bogumiła Hnatkowska	Wrocław University of Science and Technology, Poland
Bao An Mai Hoang	Vietnam National University HCMC, Vietnam
Huu Hanh Hoang	Posts and Telecommunications Institute of Technology, Vietnam
Van-Dung Hoang	Quang Binh University, Vietnam
Jeongkyu Hong	Yeungnam University, South Korea
Yung-Fa Huang	Chaoyang University of Technology, Taiwan
Maciej Huk	Wrocław University of Science and Technology, Poland
Kha Tu Huynh	Vietnam National University HCMC, Vietnam
Sanjay Jain	National University of Singapore, Singapore
Khalid Jebari	LCS Rabat, Morocco
Joanna Jędrzejowicz	University of Gdańsk, Poland
Przemysław Juszczuk	University of Economics in Katowice, Poland
Krzysztof Juszczyszyn	Wroclaw University of Science and Technology, Poland
Mehmet Karaata	Kuwait University, Kuwait
Rafał Kern	Wroclaw University of Science and Technology, Poland
Zaheer Khan	University of the West of England, UK
Marek Kisiel-Dorohinicki	AGH University of Science and Technology, Poland

Attila Kiss	Eötvös Loránd University, Hungary
Shinya Kobayashi	Ehime University, Japan
Grzegorz Kołaczek	Wrocław University of Science and Technology, Poland
Marek Kopel	Wrocław University of Science and Technology, Poland
Jan Kozak	University of Economics in Katowice, Poland
Adrianna Kozierkiewicz	Wrocław University of Science and Technology, Poland
Dalia Kriksciuniene	Vilnius University, Lithuania
Dariusz Król	Wrocław University of Science and Technology, Poland
Marek Krótkiewicz	Wrocław University of Science and Technology, Poland
Marzena Kryszkiewicz	Warsaw University of Technology, Poland
Jan Kubicek	VSB -Technical University of Ostrava, Czech Republic
Tetsuji Kuboyama	Gakushuin University, Japan
Elżbieta Kukla	Wrocław University of Science and Technology, Poland
Marek Kulbacki	Polish-Japanese Academy of Information Technology, Poland
Kazuhiro Kuwabara	Ritsumeikan University, Japan
Annabel Latham	Manchester Metropolitan University, UK
Tu Nga Le	Vietnam National University HCMC, Vietnam
Yue-Shi Lee	Ming Chuan University, Taiwan
Florin Leon	Gheorghe Asachi Technical University of Iasi, Romania
Chunshien Li	National Central University, Taiwan
Horst Lichter	RWTH Aachen University, Germany
Tony Lindgren	Stockholm University, Sweden
Igor Litvinchev	Nuevo Leon State University, Mexico
Doina Logofatu	Frankfurt University of Applied Sciences, Germany
Lech Madeyski	Wrocław University of Science and Technology, Poland
Bernadetta Maleszka	Wrocław University of Science and Technology, Poland
Marcin Maleszka	Wrocław University of Science and Technology, Poland
Tamás Matuszka	Eötvös Loránd University, Hungary
Michael Mayo	University of Waikato, New Zealand
Héctor Menéndez	University College London, UK

Mercedes Merayo	Universidad Complutense de Madrid, Spain
Jacek Mercik	WSB University in Wrocław, Poland
Radosław Michalski	Wrocław University of Science and Technology, Poland
Peter Mikulecky	University of Hradec Králové, Czech Republic
Miroslava Mikusova	University of Žilina, Slovakia
Marek Milosz	Lublin University of Technology, Poland
Jolanta Mizera-Pietraszko	Opole University, Poland
Dariusz Mrozek	Silesian University of Technology, Poland
Leo Mrsic	IN2data Data Science Company, Croatia
Agnieszka Mykowiecka	Institute of Computer Science, Polish Academy of Sciences, Poland
Pawel Myszkowski	Wrocław University of Science and Technology, Poland
Huu-Tuan Nguyen	Vietnam Maritime University, Vietnam
Le Minh Nguyen	Japan Advanced Institute of Science and Technology, Japan
Loan T. T. Nguyen	Vietnam National University HCMC, Vietnam
Quang-Vu Nguyen	Korea-Vietnam Friendship Information Technology College, Vietnam
Thai-Nghe Nguyen	Cantho University, Vietnam
Thi Thanh Sang Nguyen	Vietnam National University HCMC, Vietnam
Van Sinh Nguyen	Vietnam National University HCMC, Vietnam
Agnieszka Nowak-Brzezińska	University of Silesia, Poland
Alberto Núñez	Universidad Complutense de Madrid, Spain
Tarkko Oksala	Aalto University, Finland
Mieczysław Owoc	Wrocław University of Business and Economics,
Panos Patros	University of Waikato, New Zealand
Maciej Piasecki	Wroclaw University of Science and Technology, Poland
Bartłomiej Pierański	Poznan University of Economics and Business, Poland
Dariusz Pierzchała	Military University of Technology, Poland
Marcin Pietranik	Wrocław University of Science and Technology, Poland
Elias Pimenidis	University of the West of England, UK
Jaroslav Pokorný	Charles University in Prague, Czech Republic
Nikolaos Polatidis	University of Brighton, UK
Elvira Popescu	University of Craiova, Romania
Piotr Porwik	University of Silesia in Katowice, Poland
Petra Poulova	University of Hradec Králové, Czech Republic
Małgorzata Przybyła-Kasperek	University of Silesia, Poland
Paulo Quaresma	Universidade de Evora, Portugal

David Ramsey	Wrocław University of Science and Technology, Poland
Mohammad Rashedur Rahman	North South University, Bangladesh
Ewa Ratajczak-Ropel	Gdynia Maritime University, Poland
Sebastian A. Rios	University of Chile, Chile
Keun Ho Ryu	Chungbuk National University, South Korea
Daniel Sanchez	University of Granada, Spain
Rafał Scherer	Częstochowa University of Technology, Poland
Donghwa Shin	Yeungnam University, South Korea
Andrzej Siemiński	Wrocław University of Science and Technology, Poland
Dragan Simic	University of Novi Sad, Serbia
Bharat Singh	Universiti Teknologi PETRONAS, Malaysia
Paweł Sitek	Kielce University of Technology, Poland
Krzysztof Slot	Łódź University of Technology, Poland
Adam Słowik	Koszalin University of Technology, Poland
Vladimir Sobeslav	University of Hradec Králové, Czech Republic
Kamran Soomro	University of the West of England, UK
Zenon A. Sosnowski	Białystok University of Technology, Poland
Bela Stantic	Griffith University, Australia
Stanimir Stoyanov	University of Plovdiv "Paisii Hilendarski", Bulgaria
Ja-Hwung Su	Cheng Shiu University, Taiwan
Libuse Svobodova	University of Hradec Králové, Czech Republic
Jerzy Świątek	Wrocław University of Science and Technology, Poland
Andrzej Swierniak	Silesian University of Technology, Poland
Julian Szymański	Gdańsk University of Technology, Poland
Yasufumi Takama	Tokyo Metropolitan University, Japan
Zbigniew Telec	Wrocław University of Science and Technology, Poland
Dilhan Thilakarathne	Vrije Universiteit Amsterdam, Netherlands
Satoshi Tojo	Japan Advanced Institute of Science and Technology, Japan
Diana Trandabat	Alexandru Ioan Cuza University of Iaşi, Romania
Bogdan Trawiński	Wrocław University of Science and Technology, Poland
Maria Trocan	Institut Superieur d'Electronique de Paris, France
Krzysztof Trojanowski	Cardinal Stefan Wyszyński University in Warsaw, Poland
Ualsher Tukeyev	Al-Farabi Kazakh National University, Kazakhstan

Olgierd Unold	Wrocław University of Science and Technology, Poland
Jørgen Villadsen	Technical University of Denmark, Denmark
Thi Luu Phuong Vo	Vietnam National University HCMC, Vietnam
Wahyono Wahyono	Universitas Gadjah Mada, Indonesia
Paweł Weichbroth	Gdańsk University of Technology, Poland
Izabela Wierzbowska	Gdynia Maritime University, Poland
Krystian Wojtkiewicz	Wrocław University of Science and Technology, Poland
Xin-She Yang	Middlesex University London, UK
Tulay Yildirim	Yildiz Technical University, Turkey
Drago Zagar	University of Osijek, Croatia
Danuta Zakrzewska	Łódź University of Technology, Poland
Constantin-Bala Zamfirescu	Lucian Blaga University of Sibiu, Romania
Katerina Zdravkova	Ss. Cyril and Methodius University in Skopje, Macedonia
Vesna Zeljkovic	Lincoln University, USA
Aleksander Zgrzywa	Wroclaw University of Science and Technology, Poland
Jianlei Zhang	Nankai University, China
Zhongwei Zhang	University of Southern Queensland, Australia
Adam Ziębiński	Silesian University of Technology, Poland

Contents – Part I

Internet of Things and Sensor Networks

Natural Language Processing

Social Networks and Recommender Systems

Contents – Part II

Innovations in Intelligent Systems

Advanced Data Mining Techniques and Applications

Textual One-Pass Stream Clustering with Automated Distance Threshold Adaption

Dennis Assenmacher[2]([✉])(iD) and Heike Trautmann[1](iD)

[1] University of Münster, Münster, Germany
[2] GESIS - Leibniz Institute for the Social Sciences, Cologne, Germany
dennis.assenmacher@gesis.org

Abstract. Stream clustering is a technique capable of identifying homogeneous groups of observations that continuously arrive in a digital stream. In this work, we inherently refine a TF-IDF-based text stream clustering algorithm by the introduction of an automated distance threshold adaption technique for document insertion and cluster merging, improving the performance during distributional changes in the data stream. By conducting a thorough evaluation study, we show that our new fast approach outperforms state-of-the-art one-pass and batch-based stream clustering algorithms on various existing benchmarking datasets as well as a newly introduced dataset that poses additional challenges to the community. Moreover, we find that current evaluation approaches in the field of textual stream clustering are not adequate for a sound clustering performance assessment of evolving distributions. We thus demand the utilization of time-based evaluation.

Keywords: Stream clustering · Text mining · Concept drift

1 Introduction

Nowadays, researchers are confronted with large amounts of potentially unbounded data that have to be processed to extract desired information as quickly as possible in (almost) real-time. Even worse, there is so much data available that algorithms can only iterate over the data once to ensure timely processing leading to a new sub-domain of unsupervised learning: stream clustering [13]. In this work, we compare the performance of state-of-the art text-based stream clustering algorithms on various challenging datasets. Moreover, we improve and refine the initial draft of a textual stream clustering algorithm that we proposed in [5] to analyze chats on the Twitch platform. We introduce a mechanism that automatically determines the critical distance threshold parameter that controls the creation of new clusters and develop a new strategy to merge existing clusters close to each other. To validate the performance of our new approach, we conduct a thorough comparison between this new algorithm which we call `textClust` and existing state-of-the-art competitors on multiple datasets and evaluation metrics.

© The Author(s), under exclusive license to Springer Nature Switzerland AG 2022
N. T. Nguyen et al. (Eds.): ACIIDS 2022, LNAI 13757, pp. 3–16, 2022.
https://doi.org/10.1007/978-3-031-21743-2_1

We show that we outperform existing one-pass stream clustering algorithms and even their computationally more complex batch-based alternatives. We identify shortcomings of current evaluation approaches that do not account for concept drift and propose time-based evaluation in upcoming studies.

Stream Clustering. Usually, stream clustering algorithms follow a widely accepted two-phase approach [1]. First, incoming data is aggregated on-the-fly, i.e., within an online phase. Here, not the complete dataset but only statistical summaries of dense areas in data-space (text-documents with high similarity) are kept for further analyses. These statistics are often referred to as *micro-clusters*. While the online phase is continuously running, micro-clusters may be re-clustered on demand and in an offline manner into so-called *macro-clusters*. In this offline step, traditional clustering approaches can be applied, and it is possible to iterate over the micro-clusters multiple times. An inherent property of data streams is that the underlying data distribution may change over time (also called concept drift). A comprehensive benchmark study between different stream clustering algorithms was conducted in [4].

1.1 Recent Work on Textual Stream Clustering

Textual stream clustering algorithms can be broadly classified according to *processing mode* and *representation*. The processing mode differentiates between one-pass and batch-based approaches. While an established assumption in the field is that streaming data must be processed and discarded afterward, some recent works relax this strict assumption and allow for batch-processing. In each batch, multiple iterations over the data are allowed [16]. Well-known representatives of this batch variant are the MStream algorithm presented by Yin et al. [16], and DP-BMM by Chen et al. [6]. One-pass clustering only allows that a new observation is processed once and has to be discarded after cluster assignment. OSDM, EStream, and the textClust algorithm are all representatives of this type of stream algorithms, although MStream and DP-BMM also offer a one-pass variant that can be utilized (for MStream this variant is called MStreamF).

Next to the processing mode, textual stream clustering algorithms are differentiated by their underlying representation of input documents. In *vector-space methods* each document and also the corresponding clusters are represented in a high-dimensional vector space using Term Frequency (TF) and Inverse Document Frequency (IDF) based representations [2,5,8]. The cosine similarity and the Euclidean distance are usually employed for calculating the proximity between clusters and new documents. Most recent research endeavors focus on *model-based approaches*, arguing that vector-space variants, in general, suffer from high dimensionality. Moreover, it is argued that selecting an appropriate distance threshold for deciding if a document is assigned to a cluster or not is infeasible. On the other hand, model-based approaches assume that topics (or clusters) emerge from an underlying generative model [6,9,17]. Those algorithms try to infer the distributional parameters by various means, such as Gibbs Sampling. Recently, Kumar et al. proposed a new stream clustering algorithm that explicitly targets short text content [9]. Their model-based OSDM algorithm follows the idea of several other approaches [6,9,17], which assume that (text)-documents are constructed by a generative Dirichlet multinomial mixture model.

While MStream utilizes unigrams for text representation, DP-BMM uses bigrams to overcome the data sparsity problem that often comes with short-text processing. The authors of OSDM extend these works by incorporating the semantic importance of words via co-occurrence information. Thereupon, Rakib et al. [11] improved the semantic clustering approach by utilizing pretrained document embeddings for outlier detection. Xu et al. improve the original MStream by considering topical correlations between different time steps (batches) with the introduction of DCSS [15]. In very recent work, Rakib et al. introduce the EStream online one-pass algorithm that dynamically computes similarity thresholds by utilizing distributional properties of common feature similarities [12]. In the following, we describe textClust in detail and moreover elaborate on significant improvements of the original work, such as an automated distance threshold selection.

2 Distance Based Clustering with Automatic Threshold Determination (textClust)

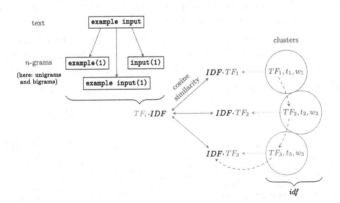

Fig. 1. Visualisation of the insertion strategy. A new incoming text is tokenized and n-grams and term frequencies created. Then, the IDF is computed from the term frequencies of all existing clusters. This allows to compute the TF-IDF vectors in order to measure the cosine similarity between the existing clusters and the new text.

textClust makes use of the common two-phase approach of stream clustering (see Sect. 1). First, an online component summarizes the stream in real-time, resulting in several micro-clusters that represent dense areas in the data space. In order to remove outdated clusters and merge similar clusters, a cleanup procedure is executed periodically. Finally, when the algorithm is supposed to output the final topics, these micro-clusters are reclustered into a final set of macro-clusters. textClust represents its micro-clusters mc as 3-tuples:

$$mc = (TF, t, w) \tag{1}$$

With $mc.TF$, the cluster holds a vector of token term frequencies as n-grams. The scalar $mc.t$ depicts the last time the micro-cluster was updated and the weight $mc.w$ how often a new observation was merged into that cluster (indirectly depicting cluster importance). Each time a new observation is merged into an existing cluster, the respective weight increases by 1. To account for distributional changes (concept drift), the cluster weights are exponentially faded each time a new observation appears in the data stream:

$$f(w) := w \cdot 2^{-\lambda(t_{now} - t)}, \lambda > 0, t_{now} > t \tag{2}$$

The parameter λ represents the fading (decay) factor, t_{now} the recent time stamp, and t the time of the last micro-cluster update. t_{gap} specifies the interval (number of new observations), after which a cleanup procedure and micro-cluster merging are triggered. This cleanup procedure removes all micro-clusters below a predefined weight threshold (i.e., clusters that were not recently updated) from the clustering result, ensuring that outdated micro-clusters are ultimately removed from the set of all existing clusters. The same fading method is applied to the token level (term fading). All tokens within a micro-cluster $(mc.TF)$ have an associated weight which is updated and increased by one when a new observation with the same token is merged into the cluster. To determine the distance between two micro-clusters or a micro-cluster and a new incoming text, the inverse document frequency of a term is calculated over all micro-clusters (which are assumed to represent all documents). To calculate the distance between two of these sparse TF-IDF vectors a and b, initially, the cosine similarity is calculated by using the dot product:

$$\cos(\theta) = \frac{a \cdot b}{\|a\|_2 \cdot \|b\|_2} \quad \text{with } \theta := \angle(a, b) \tag{3}$$

As the TF-IDF vectors contain positive frequencies, $cos(\omega) \in [0, 1]$, this leads to the *cosine distance*

$$D_{\cos}(a, b) = 1 - \cos(\theta). \tag{4}$$

When a new observation arrives at the stream, a new virtual micro-cluster mc_{new} is created (with $w = 1$), and its cosine distance to all other micro-clusters is calculated. The closest micro-cluster is selected to be merged when its distance falls below a predefined threshold tr. The merging process between both clusters is realized by fading their weight according to the current time t_{now} and taking the union of their term frequencies:

$$mc_1 + mc_2 = \left(tf_1 \cup tf_2, \quad t_{now}, \quad w_1 \cdot 2^{-\lambda(t_{now} - t_1)} + w_2 \cdot 2^{-\lambda(t_{now} - t_2)} \right). \tag{5}$$

If, however, the distance threshold tr is not reached, the new virtual micro-cluster is added to the set of all clusters. The hyperparameters tr, λ, and t_{gap} have to be set in advance and especially tr highly influences the final clustering result (see Sect. 3.4). textClust's insertion strategy is displayed in Fig. 1.

2.1   Automatic Tresholding During the Online Phase

One of the main limitations of similarity or distance-based stream clustering is the adequate threshold (tr) selection for micro-cluster assignments. In the preliminary concept of our algorithm, a fixed distance threshold was used for (a) deciding whether new observations are merged into the closest existing micro-cluster and (b) for determining when existing micro-clusters should be merged during the cleanup procedure. Experiments and several real-world applications of the algorithm revealed that the threshold parameter highly influences the final clustering result [3]. Intuitively, a distance threshold that is too large will result in a small number of micro-clusters that contain diverse topics. On the contrary, a threshold that is too small will result in multiple, small mixed-topic micro-clusters. Selecting an adequate threshold is therefore imperative for achieving good clustering results. However, as we are dealing with stream data that is subject to distributional changes, we are additionally confronted with the problem that a fixed threshold may not be optimal along the entire stream.

Until now, when a new observation arrives at the stream, a new virtual micro-cluster mc_{new} is created and textClust calculates the distance between mc_{new} and each existing micro-cluster. If this distance is smaller than the previously specified fixed threshold (tr), the algorithm merges mc_{new} into the closest cluster (mc_{cl}). Otherwise, mc_{new} is added to the set of all micro-clusters. To automate the threshold selection, we assume that we want to merge an observation into an existing micro-cluster if it is sufficiently close to it compared to all other currently existing micro-clusters. As the distance between mc_{new} and each existing micro-cluster is already calculated during the online phase, we can easily record the mean distance as:

$$\mu = \frac{1}{|MC| - 1} \sum_{mc \in MC \setminus \{mc_{cl}\}} D_{cos}(mc_{new}, mc) \tag{6}$$

where MC denotes the set of micro-clusters. The standard deviation σ of those distances results as:

$$\sigma = \sqrt{\left(\frac{1}{|MC| - 1} \sum_{mc \in MC \setminus \{mc_{cl}\}} D_{cos}(mc_{new}, mc)^2 \right) - \mu^2}. \tag{7}$$

We can calculate both μ and σ by storing the (square) sum of all distances leading to a new definition of the threshold tr as

$$tr = \mu - c \cdot \sigma. \tag{8}$$

Intuitively, tr is set such that the observation's distance to its closest micro-cluster is smaller (within c times sd) than the average distance to all other micro-clusters. Thus, tr is always adjusted w.r.t. existing micro-cluster distances. Hence, distributional changes in the stream are also reflected in a dynamic change of the threshold, instead of using a fixed distance for the complete stream. In the calculation of μ and σ, we explicitly exclude the closest micro-cluster mc_{cl} in order to avoid bias. Our experiments (see Sect. 3.4) indicate that a constant weight c of 0.5 for σ proved to produce very satisfying results for a variety of different datasets.

A distance threshold is also required during the micro-cluster merging process for cleaning purposes. Originally, textClust used the same fixed threshold as for cluster assignment. For dynamic adaption, we now keep track of the minimum distances D_{cos} w.r.t. merging observations into an existing micro-cluster:

$$D_{merge} = D_{merge} + D_{cos}(mc_{new}, mc_{cl}), \quad \text{if } D_{cos}(mc_{new}, mc_{cl}) < tr \quad (9)$$

Intuitively, we collect a sum of distances for each time window that lead to merging new observations rather than creating new micro-clusters. If two micro-clusters are closer than the average observation merging distance, we assume that they belong together and should be merged. The observation merging distance may vary over time due to concept drift and our dynamic threshold adaption. Therefore, we reset D_{merge} at the end of each cleanup procedure.

2.2   Algorithm Specification

The pseudo-code of our new online-component is shown in Algorithm 1. First, the algorithm reads a new text x from the stream (Line 3). As a preprocessing step, the text is then converted to lower case and split into individual tokens or words (Line 4). In contrast to the initial proposal of this algorithm, we also remove all stop words from the input.

The remaining list of tokens is then used to construct n-grams and count their frequency with n_{min} and n_{max} specifying the gram-range (Lines 5 – Line 6). From the results we create a new virtual micro-cluster mc_{new} (Line 7). Before computing the cosine similarity/distance, we also require the IDF vector across all documents which are computed on-the-fly from the set of all existing clusters. With the IDF vector we calculate the TF-IDF representations of each existing and the new virtual micro-cluster mc_{new}, allowing us to compute the cosine distance between them (Line 12). The closest micro-cluster becomes our candidate for merging (Line 15).

Subsequently we check whether the candidate is sufficiently close to mc_{new} such that both can be merged. Suppose its distance is not larger than tr, which we compute from the mean and standard deviation (μ, σ) of all other micro-cluster distances. In that case, we assume that the text fits into the cluster (μ and σ are computed as specified in Line 16). We then merge the components of mc_{new} into the candidate cluster (Line 21). In order to update the weight correctly, we first need to ensure that the cluster's weight has been decayed to the current time (Line 20). However, if the distance to the candidate is larger than tr, mc_{new} does not fit the cluster sufficiently well. In this case, we add the temporary micro-cluster mc_{new} to the set of all clusters (Line 23). Note that the set of micro-clusters MC will be empty initially. In this case, we can skip the candidate selection and initialize a new cluster for x directly. The cleanup

---

Algorithm 1. textClust

Require: $n_{\min}, n_{\max}, \lambda, t_{gap}, c$
Initialize: $MC = \emptyset, S = \emptyset$
1: **while** stream is active **do**
2:     $t_{now} \leftarrow$ current timestamp
3:     Read new text x from stream at time t_{now}
4:     Preprocess x (lowercase, tokenization and stopword removal)
5:     Build n-grams from tokens $\forall n \in \{n_{\min}, \ldots, n_{\max}\}$
6:     $tf \leftarrow$ Count frequency of n-grams
7:     $mc_{new} \leftarrow (tf, t_{now}, 1)$ $\triangleright$ cluster with weight one, Eq. (1)
8:     $\mu, \sigma, D_s, D_{ss} \leftarrow 0$
9:     **if** $|MC| > 2$ **then** $\triangleright$ The first two observations are considered as separate clusters
10:        $idf \leftarrow$ Calculate IDF from term-frequencies of micro-clusters in MC
11:        **for** each micro-cluster $mc \in MC$ **do**
12:            $dist \leftarrow D_{cos}(mc_{new}, mc)$ $\triangleright$ Cosine distance between TF-IDF vectors, Eq. (4)
13:            $D_s = D_s + dist$
14:            $D_{ss} = D_{ss} + dist^2$
15:        $mc_{cl} \leftarrow$ closest micro-cluster according to cosine distance D_{cos}
16:        $\mu \leftarrow \frac{D_s}{|MC|-1}, \sigma \leftarrow \sqrt{\frac{D_{ss}}{|MC|-1} - \mu^2}$
17:        $tr \leftarrow \mu - c \cdot \sigma$
18:        **if** $D_{cos}(mc_{new}, mc_{cl}) \lesssim tr$ **then**
19:            $D_{merge} \leftarrow D_{merge} + D_{cos}(mc_{new}, mc_{cl})$
20:            $mc_{cl}.w \leftarrow mc_{cl}.w \cdot 2^{-\lambda(t_{now}-mc_{cl}.t)}$ $\triangleright$ fade weights (last update $mc_{cl}.t$) Eq. (2)
21:            Merge mc_{new} into micro-cluster mc_{cl} $\triangleright$ Eq. (5)
22:        **else**
23:            Add mc_{new} to set of micro-clusters MC
24:     **else**
25:        Add mc_{new} to set of micro-clusters MC
26:     **if** $t_{now} \mod t_{gap} = 0$ **then**
27:        CLEANUP($\cdot$)
28:        MERGE(D_{merge})

---

procedure which is responsible for cluster- and token fading, as well as micro-cluster merging is triggered each t_{gap} timesteps.

3   Experiments

3.1   Benchmarking Datasets

The most frequently used and accepted text-based stream datasets are News-T and Tweets-T [16]. Recently, Rakib et al. combined both datasets to NT. Also, they proposed a new dataset containing question titles from Stackoverflow (SO-T), and combined it with NT into NTS. All datasets (except for SO-T) are sorted by topics since it is argued that in a real-world application, topics usually emerge and disappear after some time [16]. However, we see that there is still potential for additional evaluation sets that focus on different problems not captured yet by research. In this work, we generate a new Twitter-based dataset that is (a) significantly larger than the existing benchmarking data and (b) captures a low number of popular (trending) and intertwined topics on Twitter over time. In contrast to existing benchmarking datasets, the captured data is close to a real-world scenario, where raw social streams (including irrelevant posts) are monitored, and cluster algorithms should reveal groups of similar topics (conversations) in real-time. This allows us to investigate whether an algorithm works comparably well in a realistic setting and not only when, unrealistically, the text is already preprocessed and sorted. Additionally, we include timestamp information to enable clustering based on real-world throughput. We use the Twitter API to collect labelled data from the new dedicated Academic API and assume that a filtered search result for one hashtag/trend represents a single topic distribution (one class). We query Tweets of different trending hashtags/topics on the day they occur (we capture 24h of the day) and assign each tweet belonging to the same hashtag to the same class. Data was collected from 10 different topics/trends. Our new Trends-T dataset consists of 200.000 Tweets.

3.2 Experimental Setup

We evaluate textClust with $c = 0.5$ for automated threshold adaption on five different datasets frequently used in related literature for benchmarking purposes. Additionally, we conduct a comparison of the algorithms on our new Trends-T dataset. We compare the performance of the previously described state-of-the-art stream clustering one-pass algorithms (EStream, OSDM), as well as four batch-based alternatives (MStreamF, DP-BMM, MS-Rakib and DCSS) [6,10,11,16]. To establish a fair comparison, we use the deterministic one-pass variants of these algorithms. We did not shuffle the datasets randomly, as most of them are specifically constructed in a way such that topics only appear in specific time windows (similar to real-world conversation scenarios). Upon inspection, the previously introduced SO-T dataset exhibits unique characteristics. Topics are globally shuffled and do not reveal true time-dependency. Secondly, there are systematically more ground-truth clusters (10% of the total data stream vs. less than 1% for the other datasets). Choosing the correct evaluation procedure for a textual stream clustering scenario is not an easy task. We find that there does not seem to exist consent in the literature on the procedure to be applied. In practice, we observe the utilization of two different evaluation strategies. The first evaluation method we will refer to as *global evaluation*. Each time a new observation is assigned to a new cluster, the cluster's ID is stored, and then the clustering is updated. After the complete data-input stream has been clustered, the different metrics are calculated on the stored clustering IDs and compared to the ground-truth. This technique is quite accepted in literature, as it was used in several recent works [6,11]. Second, there is an *interval-based evaluation*. Here, the cluster results are split into equally sized batches (horizons), and all evaluation metrics are calculated for each horizon individually. This allows for tracking the performance of the algorithms over time. When applied, this approach is usually conducted in addition to the global evaluation [9,16]. We argue that all evaluation approaches should conduct such a time- or interval-based evaluation procedure as a sole global evaluation is insufficient for determining how algorithms can adapt to upcoming distributional changes (concept drift) and often overestimate the actual performance. If, however, prediction quality and robustness of the stream is focused, prequential [7] evaluation is recommended. To establish a fair comparison, we use the reference parameter settings, specified by the authors and do not change them for all other datasets. For both batch-based algorithms (DP-BMM and MStreamF) the standard settings are often optimized on one of the well-established baseline datasets (News-T or Tweets-T). For OSDM, we set $\alpha = 2e^{-3}$, $\beta = 4e^{-5}$ and $\lambda = 6e^{-6}$. MStreamF uses $\alpha = 3e^{-2}$, $\beta = 3e^{-2}$, $iter = 10$ (number of iterations) and the number of stored batches to 1. The EStream algorithm only consists of a single parameter which determines the delete interval (DI). We set DI to 500, as proposed in the original work [12]. For textClust we only need two parameters, as the threshold is now dynamically determined, i.e. we fix standard parameters $\lambda = 1e^{-2}$ and $t_{gap} = 200$.

3.3   Evaluation Metrics

In this work, we focus on external evaluation metrics, i.e., metrics that use ground-truth (*a-priori*) information. Moreover, we only evaluate micro-clusters and leave the macro-cluster perspective to a future endeavor as all of the competing algorithms are developed as pure online methods, only focusing on micro-level. While there are internal cluster evaluation metrics such as the Silhouette coefficient, the Dunn Index and other density-measures, we omit these in our experimental evaluation since those evaluation metrics are intrinsic measures that are usually biased towards certain cluster shapes (and thus favor those algorithms which optimize towards them). For micro-cluster evaluation, we use the homogeneity, completeness and Normalized Mutual Information (NMI) [14]. We employ the widely accepted normalized variant of Mutual Information (NMI) for evaluation which scales the results to a range between 0 and 1 [14].

3.4   Experimental Results

In Table 1 the performances of the algorithms for both *global* and *interval-based* evaluation are displayed for all datasets[1]. The horizon for the interval-based approach is heuristically set to 1000 and the mean result of all horizons is calculated. As textClust and EStream are deterministic, no repeated runs are necessary. For the other algorithms we report the mean of 5 runs. Clearly, the NMI results show that textClust outperforms existing algorithms on most of the examined datasets (News-T, Tweets-T, NT, NTS and Trends-T). An exception is the new SO-T dataset, where EStream produces slightly better results. Nonetheless, textClust can compete in contrast to other approaches which seem to deteriorate over time (see Fig. 2). Interestingly and in accordance with previous publications, we find that completeness scores tend to be lower than the homogeneity scores. This indicates that online clusters themselves are quite pure with low entropy, but the topics themselves are distributed over multiple clusters instead of a single one. We argue that this behavior does not necessarily exhibit wrong clustering behavior or a bad clustering at all. To achieve a perfect completeness score, each observation of a specific class must be assigned to the same cluster. However, because of the dynamic nature of stream clustering algorithms,

---

[1] All evaluation scripts and results are made publicly available on Github: https://github.com/Dennis1989/textClust-experiments.

12   D. Assenmacher and H. Trautmann

Table 1. Algorithm performance for global and interval-based mean evaluation. Top performances are highlighted in bold.

	Metric	Ns-T	Ts-T	NT	NTS	SO-T	Tr-T	Ns-T	Ts-T	NT	NTS	SO-T	Tr-T
		Global results						Mean results					
DP-BMM		0.548	0.395	0.506	0.453	0.118	0.039	0.875	0.913	0.948	0.895	0.402	0.229
MSTREAM		0.872	0.872	0.927	0.875	0.120	0.352	0.884	0.910	0.956	0.885	0.359	0.524
MS-RAKIB		0.866	0.948	**0.982**	0.965	0.632	0.342	0.839	0.956	**0.982**	0.931	0.652	0.509
DCSS		0.916	0.928	0.949	0.947	0.314	0.231	0.874	0.872	0.917	0.832	0.121	0.027
OSDM	Hom.	0.900	0.939	0.961	0.903	0.146	0.373	0.891	0.957	**0.982**	0.900	0.168	0.387
EStream		**0.956**	**0.963**	0.931	0.930	**0.720**	0.381	**0.949**	0.966	0.975	**0.967**	**0.687**	0.244
textClust-bigram		0.912	0.941	0.972	**0.973**	0.628	0.927	0.901	**0.959**	0.960	0.940	0.611	0.918
textClust-unigram		0.900	0.937	0.968	0.960	0.595	**0.929**	0.893	0.953	0.954	0.905	0.614	**0.904**
DP-BMM		0.718	0.561	0.769	0.641	0.351	0.070	0.791	0.765	0.691	0.537	0.953	0.375
MSTREAM		0.870	0.879	0.873	0.798	0.387	0.343	0.793	0.798	0.691	0.535	0.947	0.475
MS-RAKIB		0.785	0.689	0.741	0.674	0.705	0.397	0.684	0.570	0.492	0.370	0.963	0.540
DCSS		0.820	0.77	0.887	0.872	0.556	0.097	0.750	0.717	0.713	0.605	0.928	0.097
OSDM	Comp.	0.818	0.764	0.825	0.732	0.726	0.232	0.726	0.635	0.604	0.458	0.956	0.298
EStream		0.757	0.750	0.798	0.686	**0.730**	0.119	0.656	0.660	0.579	0.434	0.964	0.176
textClust-bigram		**0.871**	0.914	0.908	**0.843**	0.706	0.350	0.798	0.853	0.751	**0.588**	0.966	0.572
textClust-unigram		0.866	**0.920**	**0.912**	0.826	0.697	**0.412**	**0.800**	**0.864**	**0.758**	0.583	**0.967**	**0.585**
DP-BMM		0.622	0.463	0.610	0.531	0.177	0.050	0.829	0.828	0.781	0.640	0.565	0.281
MSTREAM		0.871	0.875	0.899	0.834	0.183	0.347	0.838	0.846	0.783	0.632	0.520	0.496
MS-RAKIB		0.823	0.798	0.845	0.794	0.666	0.368	0.750	0.709	0.638	0.501	0.777	0.521
DCSS		0.860	0.842	0.916	0.902	0.401	0.137	0.787	0.790	0.787	0.682	0.211	0.038
OSDM	NMI	0.857	0.842	0.888	0.809	0.243	0.286	0.796	0.759	0.730	0.573	0.283	0.335
EStream		0.845	0.843	0.859	0.789	**0.725**	0.182	0.773	0.780	0.709	0.557	**0.802**	0.203
textClust-bigram		**0.891**	0.927	0.939	**0.903**	0.665	0.509	**0.845**	0.898	0.823	**0.687**	0.748	0.703
textClust-unigram		0.883	**0.929**	**0.939**	0.887	0.601	**0.582**	0.842	**0.903**	**0.825**	0.668	0.751	**0.709**

they tend to produce more micro-clusters than classes, decreasing overall completeness. This specifically holds for more complex datasets with a labelling that captures broader topics like our new Trend-T dataset. An explanation for this is that the labeling for these datasets is not perfect and only captures "global" topics.

Figure 2 displays the algorithm's NMI performance in batches containing 1000 observations for all evaluation datasets. With this, we want to emphasize the necessity to capture the performance over time, as we can identify potential shortcomings of current techniques that lead to considerable performance dips (especially in the context of NT and NTS). Manual inspection revealed that these dips, which occur for each tested algorithm, are caused by batches that indicate a radical decrease in the number of contained topics. In extreme cases, a batch only consisted of one single topic resulting in a dip to 0 NMI. Interestingly, despite capturing more context and thus being more computationally demanding, the utilization of bigrams proved to be less efficient or only marginally better than unigrams. This indicates that good clustering can already be achieved on a single word level. Figure 2 again shows that textClust is always superior or en par with the top-performing competitors when evaluated for different horizons.

Within an additional sensitivity analysis, we tested the influence on the weight put on σ and how well our threshold adaption performed compared to

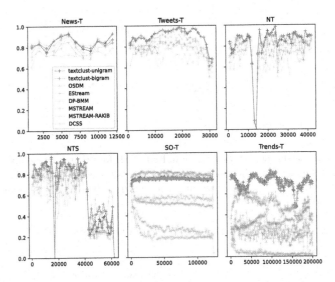

Fig. 2. Algorithm's NMI performance over time with a horizon of 1000.

the previously used fixed threshold selection. The left subplot indicates that calculating the new threshold by using 0.5 times the standard deviation, leads to good results. While 0.5 is not optimal for each of the tested datasets, we consider it as a suitable standard configuration and an acceptable trade-off for avoiding the manual selection of a hard-to-determine parameter. For the experiments shown in the right subplot we executed textClust on all datasets with a set of different fixed thresholds: $tr \in \{0.1, .., 1\}$. The resulting NMI score distributions are displayed as boxplots and the NMI performance with our automated threshold adaption ($0.5 \cdot \sigma$) is added in blue. It is evident that the performance variance when searching for the weighting factor for σ is significantly smaller than the performance variance during the threshold search, indicating that our new parameter is more stable. In general, our fully automated threshold approach leads to excellent clustering solutions that are clearly above the average score achieved with this threshold search. However, we also observe that the SO-T dataset behaves differently than the other evaluation datasets. While with increasing σ weight, the performance slowly decreases for all other data streams, it increases for SO-T. Further inspection of this dataset revealed that the number of clusters in SO-T is unusually high, compared to the other ones (10% of the number of observations). This leads to a situation, where good performance can be simply achieved by creating a lot of micro-clusters. High weight on σ leads to low tr. Consequently, with higher weights, more micro-clusters are created. However, it remains unclear whether the scenario of the SO-T dataset is representative for the respective domain (Fig. 3).

Last, we inspected the runtime of all algorithms on each individual dataset (Table 2). As described in the original paper, EStream was specifically designed for rapid clustering. It is not surprising that this approach outperforms all other

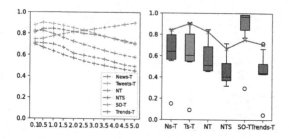

Fig. 3. Interval-based sensitivity evaluation. Left: NMI for different σ weighting factors. Right: NMI influence for different *tr* without automatic adjustments. The blue line indicates the performance of our new approach with automatic adjustments enabled. (Color figure online)

Table 2. Average algorithm runtime (in seconds)

Algorithm	Ns-T	Ts-T	NT	NTS	SO-T	Tr-T
DP-BMM	13.21	79.3	60.18	86.32	420.00	1901.20
MSTREAM	21.82	124.36	185.34	347.21	1588.14	4780.54
MS-RAKIB	32.37	102.29	143.00	210.60	711.34	10606.94
OSDM	20.09	122.77	94.38	138.96	754.41	5770.57
EStream	**2.22**	**10.05**	**14.57**	**25.77**	**147.13**	**601.62**
textClust-unigram	**9.06**	**30.45**	**37.14**	**60.75**	**165.18**	**605.01**
textClust-bigram	17.76	61.88	76.88	114.24	344.44	1625.06

algorithms, including `textClust`. However, for the larger datasets such as `SO-T` and `Trends-T`, `textClust` (especially the unigram variant) performs comparably fast and almost achieves the same speed as `EStream`. All other algorithms are at least two times slower than `textClust`. Since `textClust` outperforms `EStream` on all datasets except one, we argue that we are confronted with a trade-off between speed and clustering accuracy.

4   Discussion and Future Work

In this work, we significantly improved our draft of a textual stream clustering algorithm by tackling various shortcomings such as fixed distance threshold settings during the online clustering. We showed that our improved approach could not only compete but also outperform state-of-the-art textual clustering algorithms in the field. Moreover, we proposed a new benchmarking dataset that poses additional challenges and moves evaluation closer towards real-world settings. Our evaluation results indicate that only evaluating the global clustering result is not sufficient and does not properly show algorithm adoption of concept drift nor possible improvements. Moreover, we showed that global evaluation often overestimates the actual performance of the different algorithms. We stress that the textual stream clustering community needs a standardized evaluation procedure to enable fair comparison and an objective performance assessment of different algorithms. Future work will focus on finding suitable and robust parameter settings for `textClust` across a large variety of textual benchmark datasets based on sophisticated algorithm configuration procedures. Ideally, the configuration framework should also allow for dynamic adjustments.

References

1. Aggarwal, C.C., Han, J., Wang, J., Yu, P.S.: A framework for clustering evolving data streams. In: Proceedings of the 29th International Conference on Very Large Data Bases, Berlin, Germany, vol. 29, pp. 81–92 (2003)
2. Aggarwal, C.C., Yu, P.S.: On clustering massive text and categorical data streams. Knowl. Inf. Syst. **24**(2), 171–196 (2010)
3. Assenmacher, D., Adam, L., Trautmann, H., Grimme, C.: Towards real-time and unsupervised campaign detection in social media. In: Barták, R., Bell, E. (eds.) FLAIRS 2020 - Proceedings of the 33rd International Florida Artificial Intelligence Research Society Conference, pp. 303–306. AAAI Press (2020)
4. Carnein, M., Assenmacher, D., Trautmann, H.: An empirical comparison of stream clustering algorithms. In: Proceedings of the ACM International Conference on Computing Frontiers (CF 2017), pp. 361–365. ACM (2017)
5. Carnein, M., Assenmacher, D., Trautmann, H.: Stream clustering of chat messages with applications to twitch streams. In: de Cesare, S., Frank, U. (eds.) ER 2017. LNCS, vol. 10651, pp. 79–88. Springer, Cham (2017). https://doi.org/10.1007/978-3-319-70625-2_8
6. Chen, J., Gong, Z., Liu, W.: A dirichlet process biterm-based mixture model for short text stream clustering. Appl. Intell. **50**, 1609–1619 (2020)
7. Gama, J.A., Rodrigues, P.P., Sebastião, R.: Evaluating algorithms that learn from data streams. In: Proceedings of the 2009 ACM Symposium on Applied Computing. SAC 2009, New York, pp. 1496–1500. Association for Computing Machinery (2009)
8. Khalilian, M., Sulaiman, N.: Data stream clustering by divide and conquer approach based on vector model. J. Big Data **3**, 1–21 (2016)
9. Kumar, J., Shao, J., Uddin, S., Ali, W.: An online semantic-enhanced dirichlet model for short text stream clustering. In: Proceedings of the 58th Annual Meeting of the Association for Computational Linguistics, pp. 766–776 (2020)
10. Rakib, M.R.H., Asaduzzaman, M.: Fast clustering of short text streams using efficient cluster indexing and dynamic similarity thresholds (2021)
11. Rakib, M.R.H., Zeh, N., Milios, E.: Short text stream clustering via frequent word pairs and reassignment of outliers to clusters. In: Proceedings of the ACM Symposium on Document Engineering 2020. DocEng 2020, New York, NY, USA. Association for Computing Machinery (2020)
12. Rakib, M.R.H., Zeh, N., Milios, E.: Efficient clustering of short text streams using online-offline clustering. In: Proceedings of the 21st ACM Symposium on Document Engineering. DocEng 2021, New York, NY, USA. Association for Computing Machinery (2021). https://doi.org/10.1145/3469096.3469866
13. Silva, J.A., Faria, E.R., Barros, R.C., Hruschka, E.R., Carvalho, A.C.D., Gama, J.: Data stream clustering: a survey. ACM Comput. Surv. **46**(1), 1–31 (2013)
14. Vinh, N.X., Epps, J., Bailey, J.: Information theoretic measures for clusterings comparison: Variants, properties, normalization and correction for chance. J. Mach. Learn. Res. **11**(95), 2837–2854 (2010)

15. Xu, W., Li, Y., Qiang, J.: Dynamic clustering for short text stream based on dirichlet process. Appl. Intell. **52**, 4651–4662 (2021)
16. Yin, J., Chao, D., Liu, Z., Zhang, W., Yu, X., Wang, J.: Model-based clustering of short text streams. In: KDD 2018, New York, USA, pp. 2634–2642. Association for Computing Machinery (2018)
17. Yin, J., Wang, J.: A text clustering algorithm using an online clustering scheme for initialization. In: KDD 2016, New York, USA, pp. 1995–2004. Association for Computing Machinery (2016)

Using GPUs to Speed Up Genetic-Fuzzy Data Mining with Evaluation on All Large Itemsets

Chun-Hao Chen[1,2] ⓘ, Yu-Qi Huang[3], and Tzung-Pei Hong[3,4](✉) ⓘ

[1] Department of Information and Finance Management, National Taipei University of Technology, Taipei, Taiwan
[2] Department of Computer Science and Information Engineering, National Kaohsiung University of Science and Technology, Kaohsiung, Taiwan
[3] Department of Computer Science and Information Engineering, National University of Kaohsiung, Kaohsiung, Taiwan
tphong@nuk.edu.tw
[4] Department of Computer Science and Engineering, National Sun Yat-Sen University, Kaohsiung, Taiwan

Abstract. In fuzzy data mining, the membership function significantly influences exploration performance. Therefore, some scholars have proposed genetic-fuzzy mining to determine a set of good membership functions for effectively mining fuzzy association rules. Some scholars proposed evaluating the membership functions using both the number of large 1-itemsets and the suitability of chromosomes. They only considered large 1-itemsets instead of all large itemsets because of the time-consuming problem. We analyzed the time-consuming reason and found that there are many independent calculations in the mining process. Given this, we adopt the GPU devices and propose a GPU-based mining algorithm with evaluation on all large itemsets to improve obtained membership functions and reduce time cost. Experimental results also show the efficiency of using GPUs on genetic-fuzzy data mining.

Keywords: Data mining · Genetic algorithm · Genetic-fuzzy mining · Large itemsets · Graphics processing unit

1 Introduction

Data mining is often used to find the association rules in the database, such as Apriori [1] and other algorithms. This association rule will be expressed as $x \rightarrow y$, where x and y both represent an itemset. The association rule means that a customer who has bought all the products in x has a high probability of purchasing all the products in y simultaneously.

However, the Apriori algorithm is only suitable for binary data. Thus, many scholars have proposed various algorithms to mine fuzzy association rules by adopting the concept of fuzzy sets [6, 7, 12]. They used the membership functions to convert quantitative values into fuzzy expressions and combined the fuzzy expression to mine fuzzy association

© The Author(s), under exclusive license to Springer Nature Switzerland AG 2022
N. T. Nguyen et al. (Eds.): ACIIDS 2022, LNAI 13757, pp. 17–26, 2022.
https://doi.org/10.1007/978-3-031-21743-2_2

rules. For example, Hong et al. proposed an algorithm for mining fuzzy association rules from quantitative data [6].

In addition to the pre-defined membership functions used by previous scholars, some researchers have also proposed to evolve the membership function through the genetic algorithm [2, 8, 16, 20]. The time-consuming problems plague existing methods, so some scholars have proposed some strategies to accelerate the process of evolution and exploration. Hong et al. proposed a divide-and-conquer algorithm to speed up the mining process [8]. Hong et al. adopted MapReduce to solve the time-consuming issue [9].

With the rise of the Graphics Processing Unit (GPU), we propose an algorithm to fasten the evolution process using GPUs, named the genetic-fuzzy mining (GFM) using GPUs on large all itemsets (GFM-GPU-LAll). It first randomly generates the initial population, configures the transaction database and chromosomes into GPUs, converts all the transactions into fuzzy values, and calculates the fuzzy support. The fuzzy support of each item is compared with the minimum support to get the number of large 1-itemsets (L1). In our previous research [4], we used the GPU to accelerate the exploration process of L1 and got a good performance. Therefore, we want to complete the mining process to consider all large itemsets. After the L1 is obtained, all itemset candidates will be generated, calculated, and filtered. Finally, all frequent itemsets (LAll) are collected. When those operations are processed, better chromosomes will be selected into the next generation until the termination condition is met for the next iteration.

The massive and independent calculations are the reason that causes the time-consuming problem. Thus, we take those calculations into the GPUs and process these calculations through the massive parallel characteristic of the GPUs, thereby significantly reducing the execution time. Due to time constraints, we completed GFM-GPU-LAll within an acceptable time and successfully broke through all previously impossible items to explore. Through the experiments, the number of rules obtained by GFM-GPU-LAll is better than that by GFM-GPU-L1.

2   Related Work

In the literature, several works have been introduced to solve the time-consuming problem. Compute Unified Device Architecture (CUDA) is an integration technique developed by NVIDIA. CUDA is the first development environment, which allows the GPUs to be processed as a C program language compiler. It provides an abstract architecture for enabling the written code to reference the GPUs. The host in a computer represents the memory of the CPU, and the device means the memory of the GPU. A typical CUDA kernel function is executed as follows:

1. Allocate and initialize the memory of the host.
2. Allocate the device's memory and copy the data from the host to the device.
3. Call the CUDA kernel function to complete the specified operations on the device.
4. Copy the result of the operations on the device to the host.

The CUDA uses multiple threads to process each kernel function in parallel. The CUDA kernel functions on the GPU are used for highly independent and simple parallel calculations. Therefore, many studies are accelerating the evolution of GPU. Ohira et al. proposed a genetic algorithm (GA) that allows the island model to be diversified by migrating multiple independent populations [14]. The feature of independent computing is suitable for reducing the time costs with the GPUs architecture. Li et al. aimed to solve the model orientation problem and minimize the building time [13].

The performance of fuzzy data mining depends on the quality of the membership functions given in advance in many studies. The pre-given membership functions do not always apply to different data. Genetic-fuzzy mining is an evolutionary algorithm usually adopted to find the suitable membership functions for the data [3]. Chen et al. proposed a fuzzy data mining algorithm with a genetic algorithm to get type-2 membership functions and fuzzy association rules [5].

Although the evolutionary algorithm can dynamically find the better membership functions, it has the fatal disadvantage of being time-consuming. Hong et al. presented the mining algorithm with MapReduce to decrease the time cost [9]. Hong et al. proposed the master-slave architecture for Genetic-Fuzzy Mining to reduce the execution time [10]. Wu et al. proposed an HFUI-GA architecture to reduce the time cost of calculation [17]. It reduced the number of the chromosome calculations through the concept of the ant algorithm evaporation operator and used a threshold to end the iteration early to reduce time cost.

3   Components of the Proposed Algorithm

The idea of GFM-GPU-LAll and the component in the framework are described below.

3.1   Chromosome Representation

Our algorithm is crucial to how the GPUs can use the membership functions. Some methods have mentioned the encoding of membership functions [15, 18, 19]. Each set of membership functions is encoded into a chromosome. We apply two parameters to express the chromosome to avoid overly complex operations on the GPUs like Parodi and Bonelli do [15]. We assume that these membership functions belong to the isosceles triangular function. As shown in Fig. 1, where R_{jk} denotes the fuzzy region of each item, c_{jk} represents the center abscissa of the fuzzy region, and w_{jk} represents half the spread of the R_{jk}.

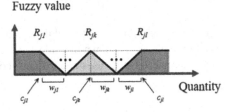

Fig. 1. Membership functions.

3.2   Population Initialization

The critical concept of the genetic algorithm is that given a set of initial feasible solutions, and through genetic operations to evolve these possible solutions. Each chromosome in the population is a set of multiple isosceles triangles, and these membership functions are all randomly generated.

3.3   Fitness Function and Selection

We adopt the fitness function designed by Hong et al. to select the better chromosome from the population of each generation [5]. The performance of membership function sets has feedback to the next generation in the genetic algorithm. There are two main factors in estimating the quality of the membership function set: *overlap_factor* and *coverage_factor*. The first factor is shown in Formula (1) below:

$$overlap_{factor(C_{pj})} = \Sigma_{k \neq 1} \left[max \left(\left(\frac{overlap(R_{jk}, R_{ji})}{min(w_{jk}, w_{ji})}, 1 \right), 1 \right) \right], \tag{1}$$

where $overlap(R_{jk}, R_{ji})$ is the overlap of R_{jk} and R_{ji}. The coverage factor of membership function sets for an item I_j is defined as the range to the maximum quantity of the item in the transactions. It is shown in Formula (2) below:

$$coverage(C_{pj}) = \frac{\frac{1}{range(R_{j1},...,R_{jl})}}{max(I_j)}, \tag{2}$$

where $range (R_{j1}, ..., R_{jl})$ is the coverage range of the membership functions, l is the number of membership functions for item I_j, and $max(I_j)$ is the maximum quantity of I_j in the transactions. The suitability with the two factors of the membership functions set, which is one chromosome C_p, is defined in Formula (3):

$$suitability(C_p) = \Sigma_{j=1}^{m} \left[overlap_{factor(C_{pj})} + coverage\_factor(C_{pj}) \right], \tag{3}$$

where m is the number of items in transactions. The fitness function of one chromosome is then denoted in Formula (4) below:

$$f(C_p) = \frac{|LAll|}{suitability(C_p)}, \tag{4}$$

where $|LAll|$ is the number of large all itemsets obtained by the chromosome C_p.

Assume that the transaction database is shown in Table 1 as an example. There are four candidate sets for large 1-itemsets. It thus uses a four-thread configuration in GPU.

Table 1. Transaction database.

Transaction ID	Quantities
1	(*bread, 1*)
2	(*milk, 1*), (*bread, 5*)
3	(*milk, 3*), (*candy, 3*)
4	(*milk, 1*), (*candy, 1*), (*bread, 3*)
5	(*milk, 2*), (*jam, 2*), (*bread, 2*)

Each item's three fuzzy support will be obtained as shown in Table 2.

Table 2. Result of fuzzy support.

	milk				bread		
Quantity	low	mid	high	Quantity	low	mid	high
1	1	0.33	0	1	1	0	0
3	0	1	0.67	5	0	0	1
1	1	0.33	0	3	0	1	0
2	0.5	0.67	0.33	2	0.5	0.5	0
Support	2.5/5 = 0.5	2.33/5 = 0.467	1/5 = 0.2	*Support*	*1.5/5 = 0.3*	*1.5/3 = 0.5*	*1/5 = 0.2*
	jam				*candy*		
Quantity	*low*	*mid*	*high*	*Quantity*	*low*	*mid*	*high*
1	1	0	0	3	0	1	0
				1	1	0	0
Support	*1/5 = 0.2*	*0/5 = 0*	*0/5 = 0*	*Support*	*1/5 = 0.2*	*1/5 = 0.2*	*0/5 = 0*

We calculate the fuzzy support of each itemset in parallel on the GPU by executing the CUDA kernel function that we designed as follows:

1. *GPU_fuzzy_table = cuda.mem_alloc(fuzzy_table.nbytes),*
2. *GPU_trans_data = cuda.mem_alloc(trans_data.nbytes),*
3. *GPU_fs = cuda.mem_alloc(GPU_fs, fuzzy_support),*
4. *cuda.memcpy_htod(GPU_fuzzy_table, fuzzy_table),*
5. *cuda.memcpy_htod(GPU_trans_data, trans_data),*
6. *cuda.memcpy_htod(GPU_fs, fuzzy_support),*
7. *fuzzy_support(GPU_trans_data, GPU_fuzzy_table, GPU_fs, block = (1, 1, 1)),*
8. *cuda.memcpy_dtoh(output_fuzzy_support, GPU_ofs).*

We generate the candidate set of L2 from L1. The approach then calculates them by the CUDA kernel function to generate L2 and the candidate set of L3 and so on. We use the minimum method to calculate the support larger than the large two itemsets. The example of calculation is shown in Table 3.

Table 3. Example of L2 fuzzy support calculation.

milk.low		bread.low			{ milk.low, bread.low}
Quantity	Fuzzy Value	Quantity	Fuzzy Value		Fuzzy Value
1	1	5	0	→	0
1	1	3	0	→	0
2	0.5	2	0.5	→	0.5
				Support	0.5 / 5 = 0.1
milk.low		bread.mid			{ milk.low, bread.mid}
Quantity	Fuzzy Value	Quantity	Fuzzy Value		Fuzzy Value
1	1	5	0	→	0
1	1	3	1	→	1
2	0.5	2	0.5	→	0.5
				Support	1.5 / 5 = 0.3
milk.mid		bread.low			{ milk.mid, bread.low}
Quantity	Fuzzy Value	Quantity	Fuzzy Value		Fuzzy Value
1	0.33	5	0	→	0
1	0.33	3	0	→	0
2	0.67	2	0.5	→	0.5
				Support	0.5 / 5 = 0.1
milk.mid		bread.mid			{ milk.mid, bread.mid}
Quantity	Fuzzy Value	Quantity	Fuzzy Value		Fuzzy Value
1	0.33	5	0	→	0
1	0.33	3	1	→	0.33
2	0.67	2	0.5	→	0.5
				Support	0.83 / 5 = 0.16

After obtaining the number of all large itemsets $|Lall|$, we could calculate the fitness value of each chromosome in parallel on the GPU. This process is executed in parallel based on chromosomes. If there are k chromosomes, there should be k threads configured.

3.4 Genetic Operators and Termination

In the proposed approach, when the termination condition is met, the algorithm outputs the best chromosome and LAll. Otherwise, it will execute a selection operator to select

P_{size} chromosomes for the next generation. As to crossover and mutation operators, we adopted are max-min-arithmetical (MMA) crossover proposed in [11] and one-point mutation. For MMA, it first selects two parent chromosomes. Four candidate chromosomes are then generated by applying the predefined operators on genes. The best two chromosomes are selected as the new offspring. The one-point mutation operation will generate a new membership function in a chromosome by changing a gene with a random value. In other words, the center point or the range of the fuzzy region will be changed.

4   The Proposed GFM-GPU-LAll Optimization Algorithm

IN this section, the pseudo code of the proposed GFM-GPU-LAll approach is illustrated in Table 4.

Table 4. Pseudo code of the proposed approach.

Input: A transaction data from IBM generator; population size (P_{size}); the number of generations; minimum support (MS); random number d.
Output: A set of fuzzy association rules; the best membership functions set from evaluation.
1   Generate the initial population randomly with the number of P_{size}.
2   Calculate the fitness value of each chromosome.
2.1   Transform the purchased quantity of item 1 to q into three fuzzy values of low, mid and high, where the q is the maximum quantity of the purchased item. Organize those fuzzy values into an index table, called fuzzy regions conversion table (FRCT).
2.2   Convert the purchased quantity of each item I_{ij} is in each transaction data, where j is the amount of item from 1 to n, and i is the transaction data of 1 to m, into the fuzzy set through the FRCT represented as: $$fuzzy\ support = FS(I_j) = \frac{\sum_{i=1}^{m} FRCT(I_{ij})}{m},$$
2.3   Compare the fuzzy support and minimum support to get the 1-item frequent itemset, call L1. $$L_1 = \{I_j \mid FS(I_j) \geq MS\},$$
2.4   Determine whether there are unprocessed chromosomes. If there are, go to Step 2-6; otherwise, go to Step 2-5.
2.5   Go to Step 2-12. Output the large itemsets and fuzzy support.
2.6   Determine whether there are unprocessed candidates. If there are, go to Step 2-7; otherwise, go to Step 2-4.
2.7   Generate candidate itemsets from the large itemset string, allocate and copy the data to the GPUs.
2.8   Scan the database and calculate the fuzzy support.
2.9   Select the large itemset by minimum support.
2.10   Copy the data to the CPU and concatenate it to the corresponding string.
2.11   Go to the Step 2-6.
2.12   Calculate the suitability and the fitness value of each chromosome (C_p).
3   Determine the termination condition is met or not. If not, continue the execution; otherwise, output the result.
4   Select the next generation.
5   Execute the MMA crossover on the population.
6   Execute the one-point mutation on the population and go to Step 2.

It first randomly generates an initial population and then calculates the fitness function of each chromosome. The fitness calculation includes many details, such as many independent calculations like fuzzy region conversion table, fuzzy support, and chromosome suitability. We all perform those operations on the GPU in parallel to reduce time costs. Since there is no fixed amount of data after L2, we allocate threads to each chromosome dynamically through a nested loop. After the LAll calculation is completed, the fitness value of each chromosome could be calculated and compared. The algorithm then outputs the best chromosome and mines the association rules through the best chromosome.

5 Experimental Evaluations

The main purpose of the extended method is to observe whether more fuzzy association rules can be obtained for the chromosomes generated when all frequent itemsets are considered. Therefore, we compare the original and the extended methods and observe the numbers of fuzzy association rules generated under different minimum supports. The experiment shows that the extended method GFM-GPU-LAll performs better in Table 5.

Table 5. Number of extracted association rules by the two methods.

	GPU-GFM-L1	GPU-GFM-LAll
Minimum Support	Association rules #	Association rules #
0.00015	34	48
0.00020	20	27
0.00025	3	15

However, scanning the database to calculate the fuzzy support requires more calculations, although GPU acceleration dramatically reduces the time consumption of the mining process. The time for a chromosome to find all frequent itemsets is about 0.5 s. However, multiple chromosomes need to be calculated in the extension method, which is not as good as the original method in terms of time performance. For example, if there are 120 chromosomes in a generation that need to be calculated, after 500 generations, the entire process will take nearly 30,000 s, which is about 8 h. The comparison is shown in Table 6.

Table 6. The execution time of the two methods.

	GPU-GFM-L1	GPU-GFM-LAll
Minimum Support	Execution time (sec.)	Execution time (sec.)
0.00015	41.62	28666.13
0.00020	40.54	28282.87
0.00025	40.76	27987.94

6 Conclusions and Future Work

Association rules mining has always been an interesting research topic because it can explore valuable relationships between items. Based on association rules mining, fuzzy theory can be used to explore transaction data with quantitative value. This kind of research is called fuzzy data mining. Determining the appropriate membership function in fuzzy data mining is an important issue. Researches such as genetic-fuzzy mining have been produced for finding suitable membership functions. However, genetic-fuzzy mining has a fatal time-consuming problem due to the high repeatability of the mining process. The research of our work is to use the property of parallel execution on GPUs to accelerate the processes with high repetitiveness, thereby significantly reducing the time cost. It is mainly divided into two methods: GFM-GPU-L1 and GFMGPU-LAll. GFM-GPU-L1 has a small-time cost, while GFM-GPU-LAll can explore more rules. Therefore, if users want to have better rules, they can use GFM-GPU-LAll. If users are more time-conscious, then GFM-GPU-L1 is preferred. We can learn the possibility of GPUs acceleration from the above experiments. A large number of highly repetitive processes can significantly reduce the number of executions to achieve acceleration. Especially the repetitive processes such as data mining and genetic algorithm are especially suitable for the GPUs. Since the current execution time of GFM-GPU-LAll still has a certain degree of time-consuming problem, in the future, we hope to make better usage of the GPUs architecture or consider the possibility of CPU multi-threading with GPU parallelism. In addition, we will try to design different structures to access data so that the item and chromosome data can have a larger upper limit.

References

1. Agrawal, R., Srikant, R.: Mining sequential patterns. In: Proceedings of the Eleventh International Conference on Data Engineering, pp. 3–14 (1995)
2. Chen, C.H., Hong, T.P., Tseng, V.S.: Genetic fuzzy mining with multiple minimum supports based on fuzzy clustering. Soft Comput. 15(12), 2319–2333 (2011)
3. Chen, C.H., He, J.S., Hong, T.P.: MOGA-based fuzzy data mining with taxonomy. Knowl.-Based Syst. 54, 53–65 (2013)
4. Chen, C.H., Huang, Y.Q., Hong, T.P.: An effective approach for genetic-fuzzy mining using the graphics processing unit. In: International Conference on Advances in Information Mining and Management, pp. 7–11 (2021)
5. Chen, C.H., Li, Y., Hong, T.P.: Type-2 genetic fuzzy mining with tuning mechanism. In: Conference on Technologies and Applications of Artificial Intelligence, pp. 296–299 (2015)
6. Hong, T.P., Kuo, C.S., Chi, S.C.: Trade-off between computation time and number of rules for fuzzy mining from quantitative data. Int. J. Uncertain. Fuzziness Knowl.-Based Syst. 9(5), 587–604 (2001)
7. Hong, T.P., Kuo, C.S., Chi, S.C.: Mining association rules from quantitative data. Intell. Data Anal. 3(5), 363–376 (1999)
8. Hong, T.P., Chen, C.H., Lee, Y.C., Wu, Y.L.: Genetic fuzzy data mining with divide-and-conquer strategy. IEEE Trans. Evol. Comput. 12(2), 252–265 (2008)
9. Hong, T.P., Liu, Y.Y., Wu, M.T., Chen, C.H., Wang, L.S.: Genetic fuzzy mining with MapReduce. In: IEEE International Conference on Systems, Man and Cybernetics, pp. 003294–003298 (2016)

10. Hong, T.P., Lee, Y.C., Wu, M.T.: Using the master-slave parallel architecture for genetic fuzzy data mining. IEEE Int. Conf. Syst. Man Cybern. **4**, 3232–3237 (2005)
11. Herrera, F., Lozano, M., Verdegay, J.L.: Fuzzy connectives based crossover operators to model genetic algorithms population diversity. Fuzzy Sets Syst. **92**, 21–30 (1997)
12. Kuok, C., Fu, A., Wong, M.: Mining fuzzy association rules in databases. SIGMOD Record **27**(1), 41–46 (1998)
13. Li, Z., et al.: A GPU based parallel genetic algorithm for the orientation optimization problem in 3D printing. In: International Conference on Robotics and Automation, pp. 2786–2792 (2019)
14. Ohira, R., Islam, M.S.: GPU accelerated genetic algorithm with sequencebased clustering for ordered problems. In: IEEE Congress on Evolutionary Computation, pp. 1–8 (2020)
15. Parodi, A., Bonelli, P.: A new approach of fuzzy classifier systems. In: Proceedings of Fifth International Conference on Genetic Algorithms, pp. 223–230 (1993)
16. Ting, C.K., Wang, T.C., Liaw, R.T., Hong, T.P.: Genetic algorithm with a structure-based representation for genetic fuzzy data mining. Soft. Comput. **21**(11), 2871–2882 (2016)
17. Wu, J.M., Lin, J.C., Fournier-Viger, P., Wiktorski, T., Hong, T., Pirouz, M.: A GA-based framework for mining high fuzzy utility itemsets. In: IEEE International Conference on Big Data, pp. 2708–2715 (2019)
18. Wang, T.C., Liaw, R.T.: Multifactorial genetic fuzzy data mining for building membership functions. In: IEEE Congress on Evolutionary Computation, pp. 1–8 (2020)
19. Wang, C.H., Hong, T.P., Tseng, S.S.: Integrating fuzzy knowledge by genetic algorithms. IEEE Trans. Evol. Comput. **2**, 138–149 (1993)
20. Yousef, A.H., Salama, C., Jad, M.Y., El-Gafy, T., Matar, M., Habashi, S.S.: A GPU based genetic algorithm solution for the timetabling problem. In: International Conference on Computer Engineering & Systems, pp. 103–109 (2016)

Efficient Classification with Counterfactual Reasoning and Active Learning

Azhar Mohammed$^{(\boxtimes)}$, Dang Nguyen, Bao Duong, and Thin Nguyen

Applied Artificial Intelligence Institute (A2I2), Deakin University, Geelong, Australia
{mohammedaz,d.nguyen,duongng,thin.nguyen}@deakin.edu.au

Abstract. Data augmentation is one of the most successful techniques to improve the classification accuracy of machine learning models in computer vision. However, applying data augmentation to *tabular data* is a challenging problem since it is hard to generate synthetic samples with labels. In this paper, we propose an efficient classifier with a novel data augmentation technique for tabular data. Our method called **CCRAL** combines causal reasoning to learn counterfactual samples for the original training samples and active learning to select useful counterfactual samples based on a *region of uncertainty*. By doing this, our method can maximize our model's generalization on the unseen testing data. We validate our method analytically, and compare with the standard baselines. Our experimental results highlight that **CCRAL** achieves significantly better performance than those of the baselines across several real-world tabular datasets in terms of accuracy and AUC. Data and source code are available at: https://github.com/nphdang/CCRAL.

Keywords: Data augmentation · Classification · Counterfactual reasoning · Active learning · Tabular data

1 Introduction

Recently, machine learning has become one of the most successful tools for supporting decisions, and it has been applied widely to many real-world applications including face recognition [36], security systems [3], disease detection [22], or recommended systems [33]. Two core components of a machine learning tool are the algorithm and the data. The algorithm can be classified into two mainstreams, namely classification and clustering while the data can be in different formats, e.g. tabular or image.

When dealing with images in computer vision applications, machine learning models (or classifiers) often leverage *data augmentation* techniques to improve the classification accuracy [14]. The main idea is that given an image of 'dog', if we rotate or flip the image, then we still recognize the object in the image as a 'dog'. By doing this geometric transformation, the label of an image is unchanged but we can obtain different variants of the image, helping the machine learning classifier to observe more data and improve its generalization. In addition to

N. T. Nguyen et al. (Eds.): ACIIDS 2022, LNAI 13757, pp. 27–38, 2022.
https://doi.org/10.1007/978-3-031-21743-2_3

geometric transformation, other data augmentation techniques are mix-up [43] and cut-mix [38].

In spite of a great success in computer vision, applying data augmentation to tabular data is challenging. There are three main reasons. First, an image is typically invariant to a small modification, e.g. flip, zoom, or rotation whereas a small change for a record in tabular data can result in a totally different outcome. All features (i.e. pixels) in images are i.i.d (independent and identical distributed) whereas each feature in tabular data (e.g. Sex or Age) has different ranges of values. Finally, one transformation operator can be applied to all features in images whereas each feature in tabular data often requires a relevant transformation operator depending on the type of the feature (continuous, discrete or categorical).

Our Method. We propose an efficient classification method with a new data augmentation technique for tabular data. Our method has two main steps. First, we use causal reasoning to learn counterfactual samples for the original training samples. Each counterfactual sample is a variant of an original sample whose all feature values are the same except the intervened feature. Since the counterfactual samples may have different outcomes from the original ones, we obtain their labels via a matching method. Second, we augment counterfactual samples to real samples to create a new training set to train the classifier. Since not all counterfactual samples are useful, we select the meaningful ones that potentially improve the classification performance using an active learning based method. Our active learning is an uncertain-based approach. It determines samples that are difficult to predict, then obtains their counterfactual version to enrich the training data. Using both real and counterfactual samples, our classifier improves its generalization, resulting in a better accuracy on unseen testing samples.

Our Contribution. To summarize, we make the following contributions.

1. We propose **CCRAL** (*Classifier with Causal Reasoning and Active Learning*), a novel method for classification with data augmentation in tabular data. To the best of our knowledge, **CCRAL** is the first method that combines both causal reasoning and active learning to train a classifier with synthetic samples in tabular data.
2. We develop an efficient framework to generate synthetic data. It consists of two steps: (1) it creates counterfactual samples via sample matching and (2) it selects useful counterfactual samples via active learning.
3. We demonstrate the benefits of our method on five real-world tabular datasets, where our method is significantly better than the standard classifier in both accuracy and AUC measures.

The rest of the paper is organized as follows. In Sect. 2, we briefly outline the fundamentals of data augmentation methods, the generation of counterfactual data, and active learning in the literature. We describe our proposed framework **CCRAL** with an algorithm and illustrate the *region of uncertainty* in Sect. 3. Our experimental settings, datasets, results are presented in Sect. 4, where we evaluate the performance of **CCRAL** and compare it with two existing methods. Finally, we conclude our work in Sect. 5.

2   Related Works

Data Augmentation: It is a process of augmenting newly generated data to the existing training set for improving the model's robustness. It can be performed by a minor alteration to the existing data. For example, in computer vision data augmentation is used to enhance deep learning models by *flipping, color spacing, injecting noise, random erasing* to reduce the bias in the classifier to favor more frequently presented training examples [11,18]. It can also be performed by generating synthetic data to act as a regularizer and reduce over-fitting while training machine learning models [37]. Some algorithms such as *data wrapping,* SMOTE and MaxUp modify real-world examples to create augmented datasets [4,8,17]. However, these methods are exclusively useful for either specific kinds of data. For example, image recognition dataset or to improve the performance of a particular algorithm like AGCN [38].

Counterfactual Augmented Data (CAD): Another popular method is to augment data is by using counterfactual reasoning to improve the generalization of the model. CAD can be generated by using existing machine learning algorithms by matching closely related samples within the training set, for example, POLYJUICE to generate text and *counterfactual image generation* for generating images by using generative adversarial networks [27,40]. Generating diverse sets of realistic counterfactuals has proven to improve the model's training efficiency and overall results [26]. For example, in classification problems, the models trained on CAD were not sensitive to spurious features unlike modified data [7,21]. While, in discrimination and fairness literature counterfactual data substitution and CAD helped to mitigate gender bias by replacing duplicate text and handling conditional discrimination respectively [25,44]. However, counterfactually augmented data does not always generalize better than unaugmented datasets of the same size and may also hurt the model's robustness [19]. There is a significant gap to explore on the quantity and quality of counterfactual data needed to be augmented on original dataset by an effective learning process such that, the model generalizes better and is robust across various environments.

Active Learning: It is a process that learns by an interaction between oracle and learner agent, it resolves the problem of costly data labeling in the learning process to improve the obtained model by making it efficient [9,32]. It can also be implemented on existing classification and predictive algorithms to optimize a model's performance when compared with state-of-the-art methods [10]. For example, in classification problems, logistic regression yielded remarkably better results by implementing the simplest suggested active learning method [23,34,41]. There are lots of effective approaches such as margin-based methods [13] and uncertainty sampling-based methods to optimize this process [16,35]. By using the uncertainty sampling-based learning process we can measure how certain a probabilistic classifier's prediction is and, obtain counterfactual versions of uncertain samples from the *region of uncertainty* to improve the model's transportability and robustness.

3 Framework

3.1 Problem Definition

Let $f(x)$ be a classifier and $\mathcal{D} = \{x_i, y_i\}_{i=1}^N$ be a dataset. Each $y_i \in \{0, 1\}$ is a binary *true label*. Given a sample $x_i \in \mathcal{D}$, $f(x_i)$ provides a probability (called *predicted score*) that x_i belongs to label 1 (i.e. $f(x_i) = P(y_i = 1 \mid x_i)$ and $f(x_i) \in [0, 1]$). We denote the *predicted label* of x_i as $\hat{y}_i \in \{0, 1\}$, where $\hat{y}_i$ is the rounding of $f(x_i)$ (i.e. $\hat{y}_i = 1$ if $f(x_i) \geq 0.5$, otherwise $\hat{y}_i = 0$).

Definition 1. *(Accuracy). We define accuracy as $P(\hat{y} = y)$, which means the percentage of samples in $\mathcal{D}$ predicted correctly by $f(x)$.*

Problem Statement. Given a training set $\mathcal{D}_{tr} = \{x_i, y_i\}_{i=1}^N$ and a *hold-out* test set $\mathcal{D}_{te} = \{x_i, y_i\}_{i=1}^M$, our goal is to learn a classifier $f(x)$ using $\mathcal{D}_{tr}$ such that $f(x)$ maximizes its accuracy on $\mathcal{D}_{te}$. This is the traditional classification problem in machine learning [5].

3.2 Proposed Method CCRAL

A typical way to solve the above problem is to train the classifier $f(x)$ using the available samples in the training set $\mathcal{D}_{tr}$, which tries to minimize a loss function measuring the difference between the true labels y and the predicted labels $\hat{y}$. Although this approach is straightforward, it often does not achieve good results.

Our method to solve the classification problem described in Sect. 3.1 is novel. Our main idea is that instead of using only training samples in $\mathcal{D}_{tr}$, we try to obtain more training samples, which is very helpful in improving the generalization of the classifier. When the classifier observes more training samples, it is more robust and its classification accuracy is often improved on *unseen* test samples. This process is often called *data augmentation*, which has become the state-of-the-art method to improve the performance of deep learning models in computer vision [37].

Our approach, called <u>C</u>lassifier with <u>C</u>ausal <u>R</u>easoning and <u>A</u>ctive <u>L</u>earning (**CCRAL**), has two main steps: (1) learning counterfactual samples using causal reasoning and (2) training a classifier with both real and counterfactual samples using active learning.

Learning Counterfactual Samples. We are dealing with the classification task on *tabular data*. Typically, a tabular dataset includes a mix of different types of features. They can be continuous, binary, or categorical features. Following the standard approach in causal reasoning [39], given the training set $\mathcal{D}_{tr}$ we select one binary feature T as the *treatment feature*. For example, the treatment feature can be Sex = "male/female" or Marital_Status = "single/married".

After determining the treatment feature T, we can obtain the counterfactual of any sample $x \in \mathcal{D}_{tr}$. Given a sample x_i, assume that its treatment feature has value 0 (i.e. $T_i = 0$), we then change the value of the treatment feature to

1. By doing this way, we now have the counterfactual sample $\bar{x}_i$ of x_i, which is the same as x_i except that the treatment feature of $\bar{x}_i$ has value 1 instead of 0. Since $\bar{x}_i$ is not a real sample, we do not have its label. To find the label $\bar{y}_i$ of $\bar{x}_i$, we use the sample matching approach that computes the distance between $\bar{x}_i$ and other samples $x' \in \mathcal{D}_{tr}$, and uses the label of the nearest sample as the label of $\bar{x}_i$ [6]. The formulation is retrieved the label of $\bar{x}_i$ is as follows:

$$\bar{y}_i = y(\operatorname*{argmin}_{x' \in \mathcal{D}_{tr}} d(\bar{x}_i, x')), \tag{1}$$

where $d(\bar{x}_i, x')$ is the function computing the distance between the counterfactual sample $\bar{x}_i$ and a sample $x' \in \mathcal{D}_{tr}$. Any distance can be used, for example, Euclidean, cosine, or Manhattan distances. In our case, we use the Euclidean distance. The function $\operatorname{argmin}_{x' \in \mathcal{D}_{tr}} d(\bar{x}_i, x')$ returns the sample that is nearest to $\bar{x}_i$, and $y(x_i)$ is the function that returns the label of an sample $x_i \in \mathcal{D}_{tr}$.

Training Classifier with Real and Counterfactual Samples. Using Eq. (1), we can generate the counterfactual version of any sample $x \in \mathcal{D}_{tr}$. The next question is how to use these counterfactual samples to improve the classification. Should we create the counterfactual counterpart for each sample, and augment them to the original training data to train the classifier? Using all counterfactual samples might not be a good solution. First, these counterfactual samples are unreal samples, they might add noises to the training data. Second, the quality of the labels of the counterfactual samples depend on how we compute the distance in Eq. (1). Finally, in some cases, if there were not very similar samples with the counterfactual sample $\bar{x}_i$, then the label $\bar{y}_i$ would be random.

To overcome the three above challenges when using the counterfactual samples as training data, we propose an *active learning* based method. We first train a classifier $f(x)$ using samples x_i in the training data $\mathcal{D}_{tr}$. Once we have learned the classifier $f(x)$, we use it to predict the score for each sample $x_i \in \mathcal{D}_{tr}$.

Since the classifier $f(x)$ has been trained with $\mathcal{D}_{tr}$, $f(x)$ predicts confidently the labels for most of the samples in $\mathcal{D}_{tr}$, where their predicted scores are close to 0 or 1. However, some samples are difficult to predict their outcomes, where their scores are close to the decision boundary (i.e. their scores are close to 0.5). We call these samples are *uncertain samples*.

To determine which samples are uncertain, we define an *uncertain region* as follows:

$$0.5 - \alpha \le f(x) \le 0.5 + \alpha, \tag{2}$$

where α is the *region margin*, $0.5 - \alpha$ is the lower region margin, and $0.5 + \alpha$ is the upper region margin.

From Eq. (2), if any training sample x_i whose predicted score $f(x_i)$ is in the uncertain region, then it will be the uncertain sample. Figure 1 illustrates the uncertain region and the uncertain samples.

Since the classifier $f(x)$ is very confused about the label of uncertain samples. It could be useful if we used their counterfactual version for the training process. Let $\mathcal{U} = \{x_1, x_2, ..., x_n\}$ be the set of uncertain samples. Following the

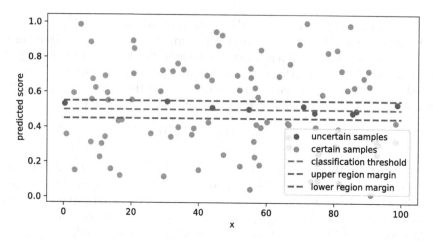

Fig. 1. Illustration of uncertain region and uncertain samples. Uncertain samples are indicated by green circles while the uncertain region is formed by two dashed green lines, the upper region margin and the lower region margin. (Color figure online)

process in Sect. 3.2, we learn counterfactual version for each sample $x_i \in \mathcal{U}$. We then augment these counterfactual samples $\bar{\mathcal{U}} = \{\bar{x}_1, \bar{x}_2, ..., \bar{x}_n\}$ to the original training set $\mathcal{D}_{tr}$ i.e. we have the new training set $\mathcal{D}'_{tr} = \mathcal{D}_{tr} \cup \bar{\mathcal{U}}$. Finally, we train the classifier $f(x)$ again with the new training set $\mathcal{D}'_{tr}$.

Since the region margin α has values being in the range of $[0, 0.5]$, we use a grid search (or Bayesian optimization [28]) to find the α that derives the best classifier $f(x)$ measured on a validation set $\mathcal{D}_{va}$. In particular, at each search iteration, we expand the uncertain region by increasing the value of α, and obtain more uncertain samples. We then find the counterfactual counterparts of these uncertain samples. Finally, we train the classifier $f(x)$ with real training samples along with the counterfactual samples and measure its accuracy on a validation set. The final classifier is the classifier whose accuracy is highest on the validation set, and this final classifier will be evaluated on the hold-out test set.

Algorithm 1 summarizes our method **CCRAL**.

4   Experiments and Discussions

We conduct extensive experiments on five real-world tabular datasets to evaluate the classification performance (accuracy and AUC) of our method **CCRAL**, comparing it with two strong baselines.

4.1   Datasets

To create an environment for comprehending counterfactual reasoning involved in our method **CCRAL**, we choose five real-world tabular datasets that have at least one binary feature that intrigues one's causal thinking. These datasets

---

Algorithm 1: The proposed **CCRAL** algorithm.

Input: $\mathcal{D} = \{x_i, y_i\}_{i=1}^N$: training set, K: # of iterations

1 split $\mathcal{D}$ into a (smaller) training set $\mathcal{D}_{tr}$ and a validate set $\mathcal{D}_{va}$;
2 define a grid of margins $[\alpha_1, \alpha_2, ..., \alpha_K]$;
3 train a classifier $f(x)$ on $\mathcal{D}_{tr}$;
4 select a binary feature T as the treatment feature;
5 **for** *each sample* $x_i \in \mathcal{D}_{tr}$ **do**
6   generate its counterfactual sample $\bar{x}_i$ by changing the value of the treatment feature of x_i;
7   compute its counterfactual label $\bar{y}_i = y(\text{argmin}_{x' \in \mathcal{D}_{tr}} d(\bar{x}_i, x'))$ (see Equation (1));
8 use $f(x)$ to predict a score $f(x_i)$ for each sample $x_i \in \mathcal{D}_{tr}$;
9 **for** $k = 1, 2, ..., K$ **do**
10   find $\mathcal{U}_k = \{x_1, x_2, ..., x_n\}$, where x_i is an uncertain sample i.e. $0.5 - \alpha_k \le f(x_i) \le 0.5 + \alpha_k$ (see Equation (2));
11   generate new training data $\mathcal{D}_{tr}^k = \mathcal{D}_{tr} \cup \bar{\mathcal{U}}^k$ where $\bar{\mathcal{U}}^k = \{\bar{x}_1, \bar{x}_2, ..., \bar{x}_n\}$ is the counterfactual of $\mathcal{U}^k$;
12   train $f_k(x)$ on $\mathcal{D}_{tr}^k$;
13   evaluate accuracy acc_k of $f_k(x)$ on $\mathcal{D}_{va}$;
14 return the best classifier $f_{k^*}(x)$, where $k^* = \text{argmax}_k acc_k$;

---

were often used to evaluate fairness-aware and causal inference machine learning algorithms [15, 29, 42].

Table 1 shows characteristics of each dataset along with the selected treatment feature and the respective outcome.

Table 1. Characteristics of five tabular datasets. We denote N: the number of samples, M: the number of features, T: the treatment feature, and y: the class feature.

Dataset	N	M	T	$T = 1$	$T = 0$	y	$y = 1$	$y = 0$
german	1,000	20	Sex	"male"	"female"	Credit	"good"	"bad"
bank	4,521	14	Marriage	"married"	"single"	Subscription	"yes"	"no"
twins	4,821	52	Weight	"heavier"	"lighter"	Mortality	"alive"	"death"
compas	4,010	10	Sex	"male"	"female"	Rearrested	"no"	"yes'
adult	30,162	13	Sex	"male"	"female"	Income	">50K"	"<50K"

german: this dataset describes each individual's credit score whether she/he has a good or bad credit score [12]. It has 1,000 samples and 20 features. We use *Sex* as the treatment feature.

bank: this dataset is about direct marketing campaigns of individuals for term deposit subscriptions. The outcome of this data is whether a person is subscribed or not depending upon the marketing and duration campaigned. *Marriage* is the treatment feature in this dataset.

twins: this dataset consists of around 5,000 records of twin's birth collected during the period of 1989–1991 in the U.S. [1]. It is a popular benchmark dataset in causality researches [24]. The outcome corresponds to the mortality of each twin's during the first year of birth. We choose twins of the same gender to replicate the counterfactual. The treatment feature is the twin's *weight*.

compas: this dataset includes a collection of data in Broward country, Florida about the use of the COMPAS risk assessment tool and has the data regarding felonies and charges on the degree of the arrest [2]. This dataset has the treatment feature *Sex* with an outcome of getting rearrested within two years.

adult: this dataset is the collection of individual data of their income recorded during the 1994 U.S census [20]. The outcome is a person's income. If the income is greater than $50K, then it is labeled as "1". Otherwise it is "0". This dataset has 30,162 samples and 13 features. We select *Sex* as the treatment feature.

4.2 Baselines and Evaluation

We compare our method **CCARL** with two strong baselines.

1. **Standard**: this method uses available training samples to train a classifier.
2. **Counterfactual**: this method uses the counterfactual samples of all original training samples in the training process. In other words, it fixes $\alpha = 0.5$ in Eq. (2).

For a fair comparison, we measure the accuracy and AUC of each method on the same hold-out test set. We also use the same classifier for all methods, namely the Support Vector Machine (SVM) with the linear kernel and $C = 1$ for the regularization. Note that other machine learning classifiers can be used with our method. We use the default search range $[0, 0.5]$ for α, and set the number of iterations $K = 10$. We evaluate methods on each dataset in five times with different train-test data splits, and report the averaged accuracy and AUC.

4.3 Results

Figure 2 shows the accuracy of each method on five datasets. It can be seen that our method **CCRAL** is much better than the standard classifier on all datasets. On *german* (a small dataset), Standard achieves only 61.0% whereas **CCRAL** achieves 70.0%, resulting in 9% better. On *adult* (a very large dataset), the accuracy of Standard is 79.28% compared to 82.82% of our **CCRAL**. On this dataset, our method achieves around 3% gains over the standard classifier.

Compared to the Counterfactual method, **CCRAL** is comparable on three datasets *bank*, *twins*, and *adult* while it is much better on two datasets *german* and *compas*. This shows that using all counterfactual samples in the training process was not a good solution since they might add noise to the training data, as we discussed in Sect. 3.2. Our method which uses active learning to select useful counterfactual samples is a more efficient approach to train the classifier.

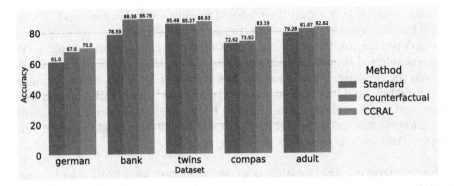

Fig. 2. The averaged classification accuracy of two baselines Standard, Counterfactual, and our method **CCRAL** on each dataset.

We also report the AUC of each method in Fig. 3. Our **CCRAL** is the best method, where it significantly outperforms two baselines Standard and Counterfactual. **CCRAL** always outperforms the standard classifier by a large margin across all datasets. Compared to Counterfactual, our method shows a great improvement, where it achieves 3–9% gains over Counterfactual. Again, this suggests that using active learning to select useful counterfactual samples is a much better strategy than using all counterfactual samples for training the classifier.

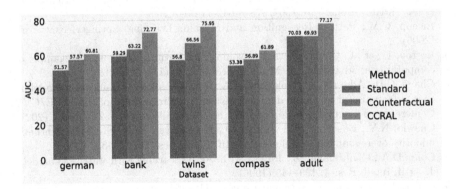

Fig. 3. The averaged AUC of two baselines Standard, Counterfactual, and our method **CCRAL** on each dataset.

5   Conclusion

In this paper, we have introduced an efficient classifier (named **CCRAL**) with a novel data augmentation technique for tabular datasets. We generate counterfactual data by flipping the binary value of the treatment feature of original training samples, and obtain their labels by using a matching method. We use active learning to select useful counterfactual samples based on a *region of uncertainty* depending on the predicted scores of the original training samples. We augment selected counterfactual samples to the set of original training samples to train the classifier. We demonstrate the efficacy of **CCRAL** on five standard real-world tabular datasets. The obtained results show that **CCRAL** generalizes better and is more robust towards unseen testing samples, where it significantly outperforms other methods. Our approach can be conceptually extended to other types of data such as sequences [31] and graphs [30].

Acknowledgment. This research is partly supported by NHMRC Ideas Grant GNT2002234.

References

1. Almond, D., Chay, K.Y., Lee, D.S.: The costs of low birth weight. Q. J. Econ. **120**(3), 1031–1083 (2005)
2. Angwin, J., Larson, J., Mattu, S., Kirchner, L.: Machine bias. ProPublica **23**(2016), 139–159 (2016)
3. Apruzzese, G., Colajanni, M., Ferretti, L., Marchetti, M.: Addressing adversarial attacks against security systems based on machine learning. In: Proceedings of the International Conference on Cyber Conflict, vol. 900, pp. 1–18 (2019)
4. Baird, H.S.: Document image defect models. In: Baird, H.S., Bunke, H., Yamamoto, K. (eds.) Structured Document Image Analysis, pp. 546–556. Springer, Heidelberg (1992). https://doi.org/10.1007/978-3-642-77281-8_26
5. Bishop, C.M.: Pattern Recognition and Machine Learning. Springer, New York (2006)
6. Bottou, L., et al.: Counterfactual reasoning and learning systems: the example of computational advertising. J. Mach. Learn. Res. **14**(11), 3207–3260 (2013)
7. Chang, C.H., Adam, G.A., Goldenberg, A.: Towards robust classification model by counterfactual and invariant data generation. In: Proceedings of the IEEE/CVF Conference on Computer Vision and Pattern Recognition, pp. 15212–15221 (2021)
8. Chawla, N.V., Bowyer, K.W., Hall, L.O., Kegelmeyer, W.P.: SMOTE: synthetic minority over-sampling technique. J. Artif. Intell. Res. **16**, 321–357 (2002)
9. Cohn, D.A., Ghahramani, Z., Jordan, M.I.: Active learning with statistical models. J. Artif. Intell. Res. **4**, 129–145 (1996)
10. Collet, T., Pietquin, O.: Active learning for classification: an optimistic approach. In: Proceedings of the Symposium on Adaptive Dynamic Programming and Reinforcement Learning, pp. 1–8 (2014)
11. DeVries, T., Taylor, G.W.: Improved regularization of convolutional neural networks with cutout. arXiv preprint arXiv:1708.04552 (2017)
12. Dua, D., Graff, C.: UCI machine learning repository (2019). https://archive.ics.uci.edu/ml
13. Ducoffe, M., Precioso, F.: Adversarial active learning for deep networks: a margin based approach. arXiv preprint arXiv:1802.09841 (2018)
14. Fawaz, H.I., Forestier, G., Weber, J., Idoumghar, L., Muller, P.A.: Data augmentation using synthetic data for time series classification with deep residual networks. arXiv preprint arXiv:1808.02455 (2018)

15. Friedler, S.A., Scheidegger, C., Venkatasubramanian, S., Choudhary, S., Hamilton, E.P., Roth, D.: A comparative study of fairness-enhancing interventions in machine learning. In: Proceedings of the Conference on Fairness, Accountability, and Transparency, pp. 329–338 (2019)
16. Gal, Y., Islam, R., Ghahramani, Z.: Deep Bayesian active learning with image data. In: Proceedings of the International Conference on Machine Learning. pp. 1183–1192 (2017)
17. Gong, C., Ren, T., Ye, M., Liu, Q.: Maxup: lightweight adversarial training with data augmentation improves neural network training. In: Proceedings of the IEEE Conference on Computer Vision and Pattern Recognition, pp. 2474–2483 (2021)
18. Hernández-García, A., König, P.: Data augmentation instead of explicit regularization. arXiv preprint arXiv:1806.03852 (2018)
19. Huang, W., Liu, H., Bowman, S.R.: Counterfactually-augmented SNLI training data does not yield better generalization than unaugmented data. arXiv preprint arXiv:2010.04762 (2020)
20. Kamiran, F., Karim, A., Zhang, X.: Decision theory for discrimination-aware classification. In: Proceedings of the IEEE International Conference on Data Mining, pp. 924–929 (2012)
21. Kaushik, D., Hovy, E., Lipton, Z.C.: Learning the difference that makes a difference with counterfactually-augmented data. arXiv preprint arXiv:1909.12434 (2019)
22. Kumar, V.B., Kumar, S.S., Saboo, V.: Dermatological disease detection using image processing and machine learning. In: Proceedings of the International Conference on Artificial Intelligence and Pattern Recognition, pp. 1–6 (2016)
23. Lewis, D.D., Gale, W.A.: A sequential algorithm for training text classifiers. In: Proceedings of the International ACM SIGIR Conference on Research and Development in Information Retrieval, pp. 3–12 (1994)
24. Louizos, C., Shalit, U., Mooij, J., Sontag, D., Zemel, R., Welling, M.: Causal effect inference with deep latent-variable models. arXiv preprint arXiv:1705.08821 (2017)
25. Maudslay, R.H., Gonen, H., Cotterell, R., Teufel, S.: It's all in the name: mitigating gender bias with name-based counterfactual data substitution. arXiv preprint arXiv:1909.00871 (2019)
26. Mothilal, R.K., Sharma, A., Tan, C.: Explaining machine learning classifiers through diverse counterfactual explanations. In: Proceedings of the Conference on Fairness, Accountability, and Transparency, pp. 607–617 (2020)
27. Neal, L., Olson, M., Fern, X., Wong, W.K., Li, F.: Open set learning with counterfactual images. In: Proceedings of the European Conference on Computer Vision, pp. 613–628 (2018)
28. Nguyen, D., Gupta, S., Rana, S., Shilton, A., Venkatesh, S.: Bayesian optimization for categorical and category-specific continuous inputs. In: AAAI. vol. 34, pp. 5256–5263 (2020)
29. Nguyen, D., Gupta, S., Rana, S., Shilton, A., Venkatesh, S.: Fairness improvement for black-box classifiers with gaussian process. Inf. Sci. **576**, 542–556 (2021)
30. Nguyen, D., Luo, W., Nguyen, T., Venkatesh, S., Phung, D.: Learning graph representation via frequent subgraphs. In: SDM, pp. 306–314. SIAM (2018)
31. Nguyen, D., Luo, W., Nguyen, T.D., Venkatesh, S., Phung, D.: Sqn2Vec: learning sequence representation via sequential patterns with a gap constraint. In: Berlingerio, M., Bonchi, F., Gärtner, T., Hurley, N., Ifrim, G. (eds.) ECML PKDD 2018. LNCS (LNAI), vol. 11052, pp. 569–584. Springer, Cham (2019). https://doi.org/10.1007/978-3-030-10928-8_34

32. Nissim, N., Boland, M.R., Tatonetti, N.P., Elovici, Y., Hripcsak, G., Shahar, Y., Moskovitch, R.: Improving condition severity classification with an efficient active learning based framework. J. Biomed. Inform. **61**, 44–54 (2016)
33. Portugal, I., Alencar, P., Cowan, D.: The use of machine learning algorithms in recommender systems: a systematic review. Expert Syst. Appl. **97**, 205–227 (2018)
34. Settles, B.: Active Learning. Morgan & Claypool Publishers, Cham (2012)
35. Settles, B., Craven, M., Ray, S.: Multiple-instance active learning. Adv. Neural. Inf. Process. Syst. **20**, 1289–1296 (2007)
36. Sharif, M., Bhagavatula, S., Bauer, L., Reiter, M.K.: Accessorize to a crime: Real and stealthy attacks on state-of-the-art face recognition. In: Proceedings of the ACM SIGSAC Conference on Computer and Communications Security, pp. 1528–1540 (2016)
37. Shorten, C., Khoshgoftaar, T.M.: A survey on image data augmentation for deep learning. J. Big Data **6**(1), 1–48 (2019)
38. Walawalkar, D., Shen, Z., Liu, Z., Savvides, M.: Attentive cutmix: an enhanced data augmentation approach for deep learning based image classification. arXiv preprint arXiv:2003.13048 (2020)
39. Wang, J., Mueller, K.: The visual causality analyst: an interactive interface for causal reasoning. Trans. Vis. Comput. Graphics **22**(1), 230–239 (2015)
40. Wu, T., Ribeiro, M.T., Heer, J., Weld, D.S.: POLYJUICE: generating counterfactuals for explaining, evaluating, and improving models. In: Proceedings of the Annual Meeting of the Association for Computational Linguistics (2021)
41. Yang, Y., Loog, M.: A benchmark and comparison of active learning for logistic regression. Pattern Recogn. **83**, 401–415 (2018)
42. Zafar, M.B., Valera, I., Rodriguez, M.G., Gummadi, K.P., Weller, A.: From parity to preference-based notions of fairness in classification. arXiv preprint arXiv:1707.00010 (2017)
43. Zhang, H., Cisse, M., Dauphin, Y.N., Lopez-Paz, D.: mixup: Beyond empirical risk minimization. arXiv preprint arXiv:1710.09412 (2017)
44. Žliobaite, I., Kamiran, F., Calders, T.: Handling conditional discrimination. In: Proceedings of the IEEE International Conference on Data Mining, pp. 992–1001 (2011)

Visual Localization Based on Deep Learning - Take Southern Branch of the National Palace Museum for Example

Chia-Hao Tu and Eric Hsueh-Chan Lu(⊠)

Department of Geomatics, National Cheng Kung University, No. 1, University Rd,
701 Tainan City, Taiwan (R.O.C.)
luhc@mail.ncku.edu.tw

Abstract. Visual localization uses images to regress camera position and orientation. It has many applications in computer vision such as autonomous driving, augmented reality (AR) and virtual reality (VR), and so on. The convolutional neural network simulates biological vision and has a good image feature extraction ability, so using it in visual localization can improve regression accuracy. Although our team has built an image indoor localization model for Southern Branch of the National Palace Museum, this paper tries to use new network and loss function to achieve better positioning accuracy. In this paper, we use ResNet-50 as backbone network, and change the output layer to 3-dimensional position and 4-dimensional orientation quaternion, and use learnable weights loss function. We compare different pretrained models and normalization methods, and the best result improves the positioning accuracy by about 60%.

Keywords: Visual localization · Deep learning · Convolutional neural network

1 Introduction

With the mature development of global navigation satellite system (GNSS), people can obtain accurate ground position information through GNSS and provide various applications. If it is indoor, satellite positioning will fail, so indoor positioning technology is to solve the positioning problem when there is no satellite signal. Common technologies for indoor positioning include inertial navigation and wireless positioning. Inertial navigation uses an inertial measurement unit (IMU) to assist in positioning. IMU includes an accelerometer and a gyroscope, which can measure the change in pose of an object. The pose at this moment can be obtained by adding the pose change obtained by the measurement to the pose at previous moment. This method requires a given initial value, and as time increases, it will accumulate errors and make later trajectory drift. Positioning accuracy depends on the price of IMU. If higher positioning accuracy is required, IMU is more expensive. Wireless positioning, such as infrared, Wi-Fi, Beacon, ZigBee, etc., is to arrange wireless transmitters in indoor field. User uses wireless receiver to measure the strength of wireless signal from each transmitter to calculate the user's location.

N. T. Nguyen et al. (Eds.): ACIIDS 2022, LNAI 13757, pp. 39–50, 2022.
https://doi.org/10.1007/978-3-031-21743-2_4

This method needs to arrange transmitters in advance, and wireless signal is susceptible to interference and positioning failure. Visual localization uses visual characteristics of image to compare the features with images of database or 3D model of the field, and then calculate camera pose of the image. This method needs to collect many images or build a 3D model in advance, but there will be no accumulation of errors and less susceptible to interference. At present, visual localization has been widely used in the field of computer vision, such as robot positioning and navigation, autonomous driving, augmented reality (AR) and virtual reality (VR), and so on.

Visual localization methods can be roughly divided into the following three categories: image retrieval, learning model and 3D structure [19]. Image retrieval collects many images and their poses, and stores in a database. When positioning, an input image is compared with images in the database for similarity, and the camera poses are interpolated from similar images. This method is simple, fast, and easy to implement, but the accuracy is poor. Learning model is divided into training and testing stage. On training stage, we also need to collect many images and their poses to train the model to learn to regress camera pose. On testing stage, we only need to input an image to the model to get the camera pose. This method is more accurate than image retrieval, but it takes time to train model first. 3D structure will first establish a 3D model of the scene. During positioning, it will establish the match between two-dimensional feature points of the image and three-dimensional feature points of the 3D model, and then calculate the camera pose. This method has the highest accuracy, but the calculation time is longer and with the larger 3D model is, the longer the time is, and it fails easily in positioning on images with textureless features. Due to the rapid development of deep learning in recent years, convolutional neural networks (CNN) have been successfully applied to various visual tasks, such as image classification, object detection, etc. CNN simulates biological vision and has good image feature extraction capability, so the use of CNN in visual localization can more effectively extract features and improve accuracy of regression. At present, more and more scientists have invested in research of visual localization based on deep learning [20], showing that CNN is more capable of processing textureless features, and showing the accuracy of CNN visual localization can approach or even exceed those methods based on 3D structure and have real-time prediction speed simultaneously [16, 17].

Although our team has established an image indoor localization model based on deep learning for Southern Branch of the National Palace Museum [18], a better convolutional neural network and visual localization loss function have been proposed later, so this paper tries to use a new network and loss function to achieve better positioning accuracy. We use ResNet-50 as backbone network [5], change the output layer to a 7-dimensional camera pose, including a 3-dimensional position and a 4-dimensional orientation quaternion, and use learnable weights to combine position and orientation in loss function [8]. We Compare with last year results of our team's experiment in Southern Branch of the National Palace Museum, the best result in this paper improves positioning accuracy by about 60%.

In the following papers, Chapter 2 summarizes related works on CNN and deep learning-based visual localization. Chapter 3 describes the visual localization model of this paper, including CNN architecture and loss function. Chapter 4 analyzes the experiments of pretrained model, normalization method and loss function.

2 Related Work

2.1 Convolutional Neural Network

LeCun et al. proposed the first CNN LeNet in 1998, which is composed of 2 convolutional layers and 3 fully connected layers [1]. Convolutional layers will downsample feature maps to reduce the size. The network uses convolution operations to extract image features, and automatically learns the parameters of convolution kernel through gradient calculation and backpropagation, so that the network learns to extract relevant features. With upgrading of computer hardware and emergence of big data, CNN has gained attention and caused a deep learning boom. Krizhevsky et al. proposed AlexNet in 2012, which is composed of 5 convolutional layers and 3 fully connected layers [2], and proposed ReLU activation function which accelerates network convergence and dropout method which reduces network overfitting. VGG proposed by Simonyan et al. in 2014 has a deeper network architecture [3]. For example, VGG16 is composed of 13 convolutional layers and 3 fully connected layers. In the same year, Google team Szegedy and others proposed GoogLeNet, not only to deepen the network, but also to increase the width by Inception architecture [4]. The same convolution layer extracts features of different scales through convolution kernels of different sizes, which helps to improve network performance. When the number of network layers increases, it is easy to cause the problem of back propagation gradient disappearance. In 2016, He et al. proposed residual learning to enable neural networks to have a deeper architecture such as the deepest ResNet has 152 layers [5]. Different from the network directly outputs features, residual network learns residual features and adds to the input as output features. Compared with the original direct learning features, residual learning is more stable. When residuals are zero, the network layer only performs identical mapping and will not degrade network performance. This paper will use ResNet as CNN model for visual localization, enabling CNN to have a deeper architecture to improve the performance of feature extraction.

2.2 Visual Localization Based on Deep Learning

In 2015, Kendall et al. first proposed PoseNet, a camera poses regression model based on GoogLeNet [6]. Pose includes 3-dimensional position and 4-dimensional orientation quaternion. The authors proved that transfer learning can reduce training time and achieve lower error. The following year, Kendall et al. used Bayesian CNN to estimate uncertainty of the model [7], detected whether there are scenes in the input image, and improved location accuracy of large-scale outdoor datasets. In 2017, Kendall et al. proposed learnable weights loss function to combine position and orientation to balance scale difference between position and quaternion values [8]. Wu et al. and Naseer et al. branched network into two parts to respectively output position and orientation [9, 10], and proposed data augmentation to increase the diversity of dataset by synthesizing images. Wu et al. further used trigonometric function to solve that orientation has periodic problems [9]. Wang et al. and Clark et al. proposed a model combining recurrent neural network (RNN) [11, 12]. The model connects two LSTM (Long Short-Term Memory) layers after CNN to extract temporal features. The model Clark et al. proposed specifically uses two-way LSTM (Bi-LSTM) and adopts mixture density network to output pose distribution to solve perceptual aliasing problem [12]. Walch et al. combined LSTM in different way [13]. The authors thought that high dimensionality of the output from fully connected layer at the end of PoseNet may cause overfitting, so they design a different form of LSTM to reduce the dimensionality and select useful features for pose regression. Melekhov et al. proposed an hourglass-shaped encoder-decoder network [14]. Similar to fully convolutional network, hourglass network deconvolves feature maps to recover size, to maintain detailed features. In 2018, Brahmbhatt et al. proposed MapNet [15], which adds a restriction on relative pose between two images in loss function and uses additional sensor data such as IMU and GNSS to train MapNet without pose groundtruth. On testing stage, MapNet combines with pose graph optimization (PGO) to smooth the predicted trajectory. Valada et al. proposed VLocNet, which combines visual localization and visual odometry with multi-task learning [16]. Both tasks base on ResNet and share weighs. VLocNet adds last pose prediction as feature into the model. The authors proposed a loss function to consider relative pose of continuous images. In the same year, Radwan et al. improved VLocNet and proposed VLocNet + +, which additionally considers semantic segmentation tasks [17]. Researches after 2020 apply the latest deep learning technology to visual localization. GrNet uses graph neural network (GNN) to aggregate information from nearby images [21]. AtLoc builds the network with self-attention [22]. MS-Transformer builds the network with transformer, and the network can locate in multi-scenes [23]. This paper will use learnable weights loss function to automatically learn the weights of positions and quaternions to balance scale difference.

3   Proposed Method

The model we proposed is composed of a CNN. The input is a 224 × 224 RGB image and the output is 7-dimensional camera pose. We use ImageNet pretrained model to speed up network convergence and achieve better performance. We perform z-score normalization on input image, so that training data can be scaled to the same scale, which is helpful for network to learn features.

3.1   Network Architecture

This paper uses ResNet as backbone network to extract image features [5], and its architecture is shown in Fig. 1. Convolution is mainly divided into 5 layers, and stride = 2 is used between each layer to reduce feature map. The Brackets in Fig. 1 are residual units. The second to fifth large layers are composed of several residual units. Global average pooling layer reduces feature maps to one-dimensional vector and then connects to output layer. ResNet was originally designed for image classification task. The architecture in Fig. 1 is used on ImageNet dataset with 1000 categories, so the dimension of output layer is 1000. After convolution layers extract image features, we need to regress camera pose, so we change the dimension of output layer to 7, including 3-dimensional position and 4-dimensional orientation quaternion, and removed softmax layer.

layer name	output size	18-layer	34-layer	50-layer	101-layer	152-layer
conv1	112×112			7×7, 64, stride 2		
				3×3 max pool, stride 2		
conv2_x	56×56	$\begin{bmatrix} 3\times3,64 \\ 3\times3,64 \end{bmatrix}\times2$	$\begin{bmatrix} 3\times3,64 \\ 3\times3,64 \end{bmatrix}\times3$	$\begin{bmatrix} 1\times1,64 \\ 3\times3,64 \\ 1\times1,256 \end{bmatrix}\times3$	$\begin{bmatrix} 1\times1,64 \\ 3\times3,64 \\ 1\times1,256 \end{bmatrix}\times3$	$\begin{bmatrix} 1\times1,64 \\ 3\times3,64 \\ 1\times1,256 \end{bmatrix}\times3$
conv3_x	28×28	$\begin{bmatrix} 3\times3,128 \\ 3\times3,128 \end{bmatrix}\times2$	$\begin{bmatrix} 3\times3,128 \\ 3\times3,128 \end{bmatrix}\times4$	$\begin{bmatrix} 1\times1,128 \\ 3\times3,128 \\ 1\times1,512 \end{bmatrix}\times4$	$\begin{bmatrix} 1\times1,128 \\ 3\times3,128 \\ 1\times1,512 \end{bmatrix}\times4$	$\begin{bmatrix} 1\times1,128 \\ 3\times3,128 \\ 1\times1,512 \end{bmatrix}\times8$
conv4_x	14×14	$\begin{bmatrix} 3\times3,256 \\ 3\times3,256 \end{bmatrix}\times2$	$\begin{bmatrix} 3\times3,256 \\ 3\times3,256 \end{bmatrix}\times6$	$\begin{bmatrix} 1\times1,256 \\ 3\times3,256 \\ 1\times1,1024 \end{bmatrix}\times6$	$\begin{bmatrix} 1\times1,256 \\ 3\times3,256 \\ 1\times1,1024 \end{bmatrix}\times23$	$\begin{bmatrix} 1\times1,256 \\ 3\times3,256 \\ 1\times1,1024 \end{bmatrix}\times36$
conv5_x	7×7	$\begin{bmatrix} 3\times3,512 \\ 3\times3,512 \end{bmatrix}\times2$	$\begin{bmatrix} 3\times3,512 \\ 3\times3,512 \end{bmatrix}\times3$	$\begin{bmatrix} 1\times1,512 \\ 3\times3,512 \\ 1\times1,2048 \end{bmatrix}\times3$	$\begin{bmatrix} 1\times1,512 \\ 3\times3,512 \\ 1\times1,2048 \end{bmatrix}\times3$	$\begin{bmatrix} 1\times1,512 \\ 3\times3,512 \\ 1\times1,2048 \end{bmatrix}\times3$
	1×1			average pool, 1000-d fc, softmax		
FLOPs		1.8×10^9	3.6×10^9	3.8×10^9	7.6×10^9	11.3×10^9

Fig. 1. ResNet architecture (excerpt from [5])

Residual unit treats learned features as residuals. When input is x and learned residual is F(x), add the input x and the residual F(x) so that final output feature is F(x) + x, When the residual is zero, performance of the network will not be degraded, to solve disappearance or explosion of gradient caused by too many network layers.

It can be seen from Fig. 1 that when the size of feature maps reduces a half, the number of feature maps doubles. This principle is to maintain the complexity of network, but it will cause the input and output between residual units cannot be added due to different dimensions. Therefore, He et al. use 1 × 1 convolution to transform the input dimension to make it the same as the output dimension.

3.2 Loss Function

We give the ground truth of position and orientation quaternion of each image in the training dataset. The network learns to regress camera pose by supervised learning, such as loss function (1). $\mathcal{L}_x$ is the loss function of position, $\mathcal{L}_q$ is the loss function of quaternion, and γ is norm. We divide predicted quaternion by its vector length to make it a unit quaternion.

$$\mathcal{L}_x = \| x - \hat{x}_\gamma \|, \quad \mathcal{L}_q = \left\| q - \frac{\hat{q}}{\|\hat{q}\|} \right\|_\gamma \tag{1}$$

Due to the different unit of position and quaternion, $\mathcal{L}_x$ and $\mathcal{L}_q$ have different scale. If we direct add $\mathcal{L}_x$ and $\mathcal{L}_q$ as loss function, training may be difficult to output the optimal position and orientation. In PoseNet, the authors use a scale factor β to balance $\mathcal{L}_x$ and $\mathcal{L}_q$ [6]:

$$\mathcal{L} = \mathcal{L}_x + \beta \mathcal{L}_q \tag{2}$$

However, finding the best β requires a lot of experiments, which is time-consuming, and there will has different optimal β depending on training scene. Kendall et al. later proposed a method which automatically learns the weights of loss function between multi-tasks and apply it to camera pose regression [8]. It treats position and orientation regression as two tasks and using homoscedastic uncertainty between tasks as the weight of $\mathcal{L}_x$ and $\mathcal{L}_q$. Homoscedastic uncertainty is a measure of uncertainty which is not dependent on input data. It stays constant for all input data and varies between different tasks. With this property, we can take homoscedastic uncertainty as the weight to combine losses of different tasks. The final loss function used in this paper is shown as Eq. (3). $\hat{s}_x$ and $\hat{s}_q$ are related to homoscedastic uncertainty, and they are automatic learning parameters. Only initial values need to be set, and then the best values will be found after network training converges.

$$\mathcal{L} = \mathcal{L}_x e^{-\hat{s}_x} + \hat{s}_x + \mathcal{L}_q e^{-\hat{s}_q} + \hat{s}_q \tag{3}$$

4 Experiments

This paper uses the dataset of Southern Branch of the National Palace Museum our team collected before [18]. The dataset has a total of 46,584 images, including 862 positions, and each position has 54 orientations. Our team collects scene images and their camera poses through indoor mobile mapping platform, which is divided into mapping system and positioning system. The mapping system uses the LadyBug5 panoramic camera, which can take images from 6 angles at the same time, stitch them into a panoramic image and then simulate it as mobile phone images; The positioning system includes IMU and GNSS instruments. It can obtain outdoor position through receiving GNSS signals and indoor position through IMU. This paper uses last year's experimental results as a benchmark for comparison. Last year, our team used GoogLeNet as backbone network and used Places dataset pretrained model which pretrained for 800 iterations. The normalization method of input data was subtracting mean image of training dataset. The loss function only considered position, and the positioning error of training 30,000 iterations is 0.38 m.

This paper uses ResNet-50 as backbone network for experiments, and compares the effects of different pretrained models, image normalization methods and loss functions on positioning error. The following experiments use Zenfone 2 simulation image of Southern Branch of the National Palace Museum. Image input size is 224 × 224. All images are randomly cut into 5 parts for 5-fold cross validation. The following results are median of 5 experiments. Each experiment is trained for 300 epochs (about 100,000 iterations). This paper uses GeForce RTX 3090 for network training, and a cross validation training takes about 40 h.

(a) Mobile mapping platform   (b) Examples of the dataset

Fig. 2. Southern branch of the National Palace Museum dataset

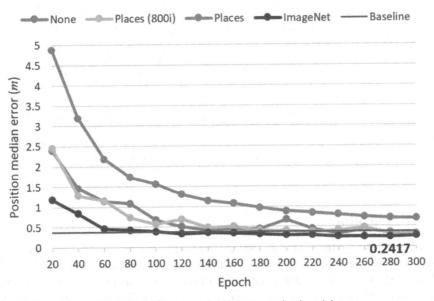

Fig. 3. Position error of different pretrained models

4.1 Pretrained Model

First, we want to find the best pretrained model. We first replace backbone network of the benchmark with ResNet-50, and then load different pretrained models. The positioning error comparison of different pretrained models is shown in Fig. 3. Baseline is the benchmark experiment. The result shows that the error of Places (800i) (Places dataset pretrained model pretrained for 800 iterations) and None (without pretrained model) have no significant difference, which means that only pretraining 800 iterations does not achieve the effect of transfer learning. Place (Places dataset pretrained model pretrained to convergence) has the highest error, and ImageNet pretraining effect is the best, the error is 0.24 m. We believe that the reason may be the difference between two pretrained model tasks. Places dataset is scene recognition task, and ImageNet dataset is image recognition task. The feature extracted by image recognition relative to scene recognition is more related to visual localization. Using ImageNet pretrained model can accelerate the network convergence and achieve better accuracy.

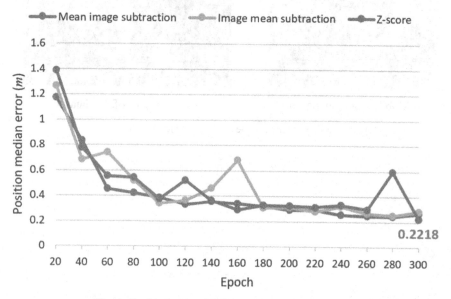

Fig. 4. Position error of different normalization methods

4.2   Normalization

We use ImageNet pretrained model and then want to find the best normalization method. The positioning error comparison of different normalization methods is shown in Fig. 4. Mean image subtraction is that input image subtracts mean image of training dataset, image mean subtraction is that input image subtracts mean RGB value of training dataset by channels, and z-score is that input image subtracts RGB mean and divides by RGB standard deviation. From the result, although mean image subtraction is relatively stable, it is found that three normalization methods have no obvious difference, which means that normalization has little effect on visual localization of Southern Branch of the National Palace Museum. Because z-score is common use and it has lowest error 0.22 m, we use z-score normalization for the next experiment.

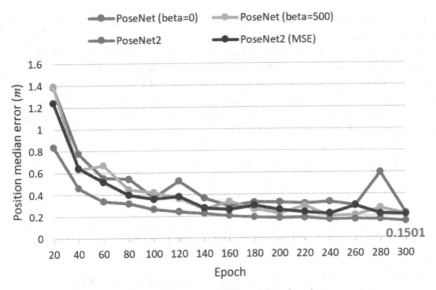

Fig. 5. Position error of different loss functions

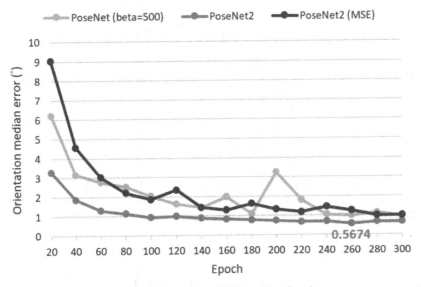

Fig. 6. Orientation error of different loss functions

4.3 Loss Function

In this section, we want to investigate the influence of different loss functions on positioning error. As shown in Figs. 5 and 6, we design four loss functions as shown in Eqs. 4, 5, 6, 7. The results show that the error of PoseNet (beta = 0) is relatively higher than the other three. It means that regressing position and orientation at the same time helps to reduce positioning error. We believe that the features extracted by regressing orientation are related to positioning. The position and orientation error of PoseNet2 are the lowest. The position error is 0.15 m, and the orientation error is 0.57°, which means that the loss function with learnable weights combined with position and orientation can effectively balance the scale between the two. The error of PoseNet2 (MSE) with square term is higher than that of PoseNet2. We speculate that the reason is the square term is more susceptible to errors, and the loss function cannot accurately reflect the difference between predicted value and ground truth.

1. PoseNet (beta = 0): Only consider position [6].

$$\mathcal{L} = \left(x - \hat{x}\right)^2 + \beta \left(q - \frac{\hat{q}}{\|\hat{q}\|}\right)^2, \beta = 0 \tag{4}$$

2. PoseNet (beta = 500): Use the same value as the original paper [6].

$$\mathcal{L} = \left(x - \hat{x}\right)^2 + \beta \left(q - \frac{\hat{q}}{\|\hat{q}\|}\right)^2, \beta = 500 \tag{5}$$

3. PoseNet2: Use the same initial values of parameters as the original paper [8].

$$\mathcal{L}_x = |x - \hat{x}|, \mathcal{L}_q = \left|q - \frac{\hat{q}}{\|\hat{q}\|}\right|,$$
$$\mathcal{L} = \mathcal{L}_x e^{-\hat{s}_x} + \hat{s}_x + \mathcal{L}_q e^{-\hat{s}_q} + \hat{s}_q, \hat{s}_x = 0, \hat{s}_q = -3 \tag{6}$$

4. PoseNet2 (MSE): Replace MAE with MSE [8].

$$\mathcal{L}_x = \left(x - \hat{x}\right)^2, \mathcal{L}_q = \left(q - \frac{\hat{q}}{\|\hat{q}\|}\right)^2,$$
$$\mathcal{L} = \mathcal{L}_x e^{-\hat{s}_x} + \hat{s}_x + \mathcal{L}_q e^{-\hat{s}_q} + \hat{s}_q, \hat{s}_x = 0, \hat{s}_q = -3 \tag{7}$$

We summarize the last epoch results of all experiments in Table 1. The lowest position error is 0.15 m, which uses ResNet-50, ImageNet pretrained model, z-score normalization and learnable weights loss function. Compared with the baseline, 0.38 m, which uses GoogLeNet, Places dataset pretrained model which pretrained for 800 iterations, mean image subtraction and position loss function, we reduce the error by about 60%.

Table 1. Comparison of position median error with all experiments

Pretrained model	None	Places (800i)	Places	ImageNet
	0.2942 m	0.3223 m	0.7003 m	0.2665 m
Normalization	Mean image subtraction	Image mean subtraction		Z-Score
	0.2665 m	0.2844 m		0.2218 m
Loss function	PoseNet ($\beta = 0$)	PoseNet ($\beta = 500$)	PoseNet 2	PoseNet 2 (MSE)
	0.2218 m	0.2139 m	**0.1501** m	0.2136 m

5 Conclusion and Future Work

This paper adopts new CNN model and loss function for visual localization of Southern Branch of the National Palace Museum. In terms of CNN, we use ResNet to replace the original GoogLeNet. Residual learning of ResNet enables CNN to have deeper architecture and improve performance. We also changed the output to regress position and orientation quaternion simultaneously, which the experiment result shows that can have better position accuracy than only regressing position. In terms of loss function, we use the loss function of learning weight which can automatically balance the scale difference between position and quaternion. Compared with the experiment result of South Branch of the Palace Museum last year, position accuracy has been improved by about 60%. In the future, we will combine visual localization with other related visual tasks such as visual odometry by multi-task learning. The features of related tasks will increase the performance of visual localization. We will also combine 3D information such as depth maps to learn 3D features of scene or add constraints to assist in localization.

Acknowledgement. This research was supported by Ministry of Science and Technology, Taiwan, R.O.C. under grant no. MOST 109–2121–M–006–013–MY2 and MOST 109–2121–M–006–005.

References

1. LeCun, Y., Bottou, L., Bengio, Y., Haffner, P.: Gradient-based learning applied to document recognition. Proc. IEEE **86**(11), 2278–2324 (1998)
2. Krizhevsky, A., Sutskever, I., Hinton, G.E.: Imagenet classification with deep convolutional neural networks. Adv. Neural. Inf. Process. Syst. **25**, 1097–1105 (2012)
3. Simonyan, K., Zisserman, A.: Very deep convolutional networks for large-scale image recognition. In: International Conference on Learning Representations (2015)
4. Szegedy, C., et al.: Going deeper with convolutions. In: IEEE Conference on Computer Vision and Pattern Recognition, pp. 1–9 (2015)
5. He, K., Zhang, X., Ren, S., Sun, J.: Deep residual learning for image recognition. In: IEEE Conference on Computer Vision and Pattern Recognition, pp. 770–778 (2016)

6. Kendall, A., Grimes, M., Cipolla, R.: Posenet: A convolutional network for real-time 6-dof camera relocalization. In: IEEE International Conference on Computer Vision, pp. 2938–2946 (2015)
7. Kendall, A., Cipolla, R.: Modelling uncertainty in deep learning for camera relocalization. In: IEEE International Conference on Robotics and Automation, pp. 4762–4769 (2016)
8. Kendall, A., Cipolla, R.: Geometric loss functions for camera pose regression with deep learning. In: IEEE Conference on Computer Vision and Pattern Recognition, pp. 5974–5983 (2017)
9. Wu, J., Ma, L., Hu, X.: Delving deeper into convolutional neural networks for camera relocalization. In: IEEE International Conference on Robotics and Automation, pp. 5644–5651 (2017)
10. Naseer, T., Burgard, W.: Deep regression for monocular camera-based 6-dof global localization in outdoor environments. In: IEEE/RSJ International Conference on Intelligent Robots and Systems, pp. 1525–1530 (2017)
11. Wang, S., Clark, R., Wen, H., Trigoni, N.: Deepvo: towards end-to-end visual odometry with deep recurrent convolutional neural networks. In: IEEE International Conference on Robotics and Automation, pp. 2043–2050 (2017)
12. Clark, R., Wang, S., Markham, A., Trigoni, N., Wen, H.: Vidloc: a deep spatio-temporal model for 6-dof video-clip relocalization. In: IEEE Conference on Computer Vision and Pattern Recognition, pp. 6856–6864 (2017)
13. Walch, F., Hazirbas, C., Leal-Taixe, L., Sattler, T., Hilsenbeck, S., Cremers, D.: Image-based localization using lstms for structured feature correlation. In: IEEE International Conference on Computer Vision, pp. 627–637 (2017)
14. Melekhov, I., Ylioinas, J., Kannala, J., Rahtu, E.: Image-based localization using hourglass networks. In: IEEE International Conference on Computer Vision Workshops, pp. 879–886 (2017)
15. Brahmbhatt, S., Gu, J., Kim, K., Hays, J., Kautz, J.: Geometry-aware learning of maps for camera localization. In: IEEE Conference on Computer Vision and Pattern Recognition, pp. 2616–2625 (2018)
16. Valada, A., Radwan, N., Burgard, W.: Deep auxiliary learning for visual localization and odometry. In:IEEE International Conference on Robotics and Automation, pp. 6939–6946 (2018)
17. Radwan, N., Valada, A., Burgard, W.: Vlocnet++: deep multitask learning for semantic visual localization and odometry. IEEE Robot. Autom. Lett. 3(4), 4407–4414 (2018)
18. Lu, E.H.C., Ciou, J.M.: Integration of convolutional neural network and error correction for indoor positioning. ISPRS Int. J. Geo. Inf. 9(2), 74 (2020)
19. Xin, X., Jiang, J., Zou, Y.: A review of visual-based localization. In: International Conference on Robotics, Intelligent Control and Artificial Intelligence, pp. 94–105 (2019)
20. Shavit, Y., Ferens, R.: Introduction to camera pose estimation with deep learning. arXiv preprint arXiv:1907.05272 (2019)
21. Xue, F., Wu, X., Cai, S., Wang, J.: Learning multi-view camera relocalization with graph neural networks. In: IEEE/CVF Conference on Computer Vision and Pattern Recognition, pp. 11372–11381 (2020)
22. Wang, B., Chen, C., Lu, C.X., Zhao, P., Trigoni, N., Markham, A.: Atloc: attention guided camera localization. AAAI Conf. Artif. Intell. 34(6), 10393–10401 (2020)
23. Shavit, Y., Ferens, R., Keller, Y.: Learning multi-scene absolute pose regression with transformers. In: IEEE/CVF International Conference on Computer Vision, pp. 2733–2742 (2021)

SimCPSR: Simple Contrastive Learning for Paper Submission Recommendation System

Duc H. Le[1,2], Tram T. Doan[1,2], Son T. Huynh[1,2,3], and Binh T. Nguyen[1,2,3(✉)]

[1] University of Science, Ho Chi Minh City, Vietnam
ngtbinh@hcmus.edu.vn
[2] Vietnam National University, Ho Chi Minh City, Vietnam
[3] AISIA Research Lab, Ho Chi Minh City, Vietnam

Abstract. The recommendation system plays a vital role in many areas, especially academic fields, to support researchers in submitting and increasing the acceptance of their work through the conference or journal selection process. This study proposes a transformer-based model using transfer learning as an efficient approach for the paper submission recommendation system. By combining essential information (such as the title, the abstract, and the list of keywords) with the aims and scopes of journals, the model can recommend the Top K journals that maximize the acceptance of the paper. Our model had developed through two states: (i) Fine-tuning the pre-trained language model (LM) with a simple contrastive learning framework. We utilized a simple supervised contrastive objective to fine-tune all parameters, encouraging the LM to learn the document representation effectively. (ii) The fine-tuned LM was then trained on different combinations of the features for the downstream task. This study suggests a more advanced method for enhancing the efficiency of the paper submission recommendation system compared to previous approaches when we respectively achieve 0.5173, 0.8097, 0.8862, 0.9496 for Top 1, 3, 5, 10 accuracies on the test set for combining the title, abstract, and keywords as input features. Incorporating the journals' aims and scopes, our model shows an exciting result by getting 0.5194, 0.8112, 0.8866, 0.9496 respective to Top 1, 3, 5, and 10. We provide the implementation and datasets for further reference at https://github.com/hduc-le/SimCPSR.

Keywords: Paper submission recommendation · Contrastive learning · Sentence embedding · Recommendation system

1 Introduction

Recommendation systems have become more and more popular in almost all fields, and people are using these systems in different industries such as retail, media, news, streaming service, and e-commerce. See the benefit of recommendation systems in the development of the economy; many companies built recommendation systems that utilize historical data of customers to give them some

© The Author(s), under exclusive license to Springer Nature Switzerland AG 2022
N. T. Nguyen et al. (Eds.): ACIIDS 2022, LNAI 13757, pp. 51–63, 2022.
https://doi.org/10.1007/978-3-031-21743-2_5

relevant suggestions to meet customer satisfaction and improve the company's product. Various well-known recommendation systems include the recommendation systems of Spotify or Netflix in streaming services, Amazon in e-commerce branches, Google, Facebook, and Youtube in media. Besides, the recommendation system in academics has gained importance in recent years, including the paper submission system [7,12,17,19], the knowledge-based recommendation system [16], and paper suggestion [1]. Selecting a suitable journal for submitting new work is not easy for almost researchers, including young and experienced people. To support the researchers in choosing a relevant journal to increase their work acceptance opportunities, Wang and his coworkers proposed the recommendation system for computer science publications [19] in the early stage and developed it by many other researchers later.

In this work, we aim to investigate the paper submission recommendation problem using general paper information and the aims and scopes of the journals as sufficient attributes in our method. Our target is to extract the semantic relationship between the paper submission and journal through those available features as well as possible. We expect the papers and journals' representations to be well-encoded, meaning close together for those that are semantically related and far away for contrast in the embedding space. We tackle the problem by applying the transformer architecture [18] as an encoder to extract the input feature effectively and utilize the contrastive learning framework [3,5] to enhance the model's robustness in the downstream task. The experimental results show that the proposed approach mostly outperforms the previous works. For example, in using the title, abstract, and keywords combined with journals' aims and scopes for training, our model gets 0.5194 for Top 1 accuracy and 0.9496 for Top 10 accuracy.

The contribution of our work can be described as follows:

(a) We propose a new framework for the paper submission recommendation problem using the transformer architecture, which shows a significant advance to tackle. The experimental results in Sect. 4 show that our approach has competitive performance compared to the previous works.
(b) Leveraging contrastive learning, the powerful method of sentence embedding, we enhance the framework's robustness in learning semantic relationships among documents or sentences.
(c) Our method provides a basic framework that can be extended further by applying other transformer models to achieve better performance in the paper submission recommendation problem.

The paper can be organized as follows. First, Sect. 2 provides a brief overview of the approaches that tackled this problem using well-known tools and algorithms. Then, in Sect. 3, we describe our main methods, and Sect. 4 shows the details of chosen experiments, including the training configurations, dataset, and achieved results. Finally, the paper ends with the conclusion and our discussion on future work.

2 Related Work

The idea of the paper recommendation system had developed by Wang, and his coworker in computer science publications [19]. Wang used the Chi-square statistic and the term frequency-inverse document frequency (TF-IDF) for feature selection from the abstract of each paper submission and utilized linear logistic regression to classify relevant journals or conferences. The accuracy in his study is 61.37% for the Top 3 recommended results. Son and colleagues later developed and proposed a new approach to improve the paper submission recommendation algorithm's performance using other additional features. The proposed method in [17] uses TF-IDF, the Chi-square statistic, and the one-hot encoding technique to extract parts from available information in each paper. They applied two machine learning models, namely Logistics Linear Regression (LLR) and Multi-layer Perceptrons (MLP), to the different combinations of features from the paper submission. Their proposed methods achieved outperformed results for the Top 3 accuracy, especially 88.60% for the LLR model and 89.07% for the MLP model when using a group of features, including the title, abstract, and keywords.

Regarding the problem, Dac et al. lately proposed a new approach [12] by applying two embedding methods, GloVe [14] and FastText [2], combining to Convolutional Neural Network (CNN) [9] and LSTM [6] for feature extraction. They considered seven features groups: title, abstract, keywords, title + keyword, title + abstract, keyword + abstract, and title + keyword + abstract for training progress. The experimental results show that the combination of S2RSCS [17] and CNN + FastText, namely the proposed S2CFT [12] model has the best performance with the Top 1 accuracy is 68.11% when using a mixture of attribute title + keyword + abstract, the accuracy at Top 3, 5, and 10 are 90.8%, 96.25%, and 99.21% respectively.

Son and his coworkers continued to propose a new approach to the paper recommendation system. In their study [7], besides valuable papers' attributes, they used additional information from the "Aims and Scopes" of journals for input data. They collected a dataset containing 414512 articles and the corresponding aims and scopes in the journal of these papers. They built a new architecture that still uses FastText [2] for embedding, and the input data is available information of paper submission; they created a new feature by measuring the similarity between paper submission and the journals' aims and scopes. Their proposed method is a practical approach to solving the paper submission recommendation problem.

In recent years, transformer architectures have succeeded in various fields of natural language processing (NLP) and computer vision (CV). One of the most significant works dedicated to that success is the Vanilla Transformer architecture [18]. It is active through adopting the attention mechanism and differentially weighting the significance of each part of the input data. The popularity and efficiency of the transformer models are unquestionable; it contributes to carrying many other models taken off, the most famous being BERT (Bidirectional Encoder Representations from Transformers) [4]. BERT is pre-trained with large amounts of textual data and fine-tuned to achieve new state-of-the-art results in many NLP tasks such as semantic textual similarity (STS) or sentence classification. In addition, BERT has been considered a powerful embedding method for documents or sentences by sending those to the BERT layers and taking the average of the output layer or the output of the first token (the [CLS] token) to get the fixed-size embeddings.

Although BERT shows its impression on many NLP tasks, it exists some limitations in sentence embedding by standard approaches. Recently, the Contrastive Learning [3] framework has become state-of-the-art in sentence embedding. Its idea conceptually describes a technique that aims to pull similar samples together and push dissimilar ones far away in the embedding space by the contrastive objective. One can apply contrastive learning to unlabeled and labeled data. As a result, fine-tuning transformer models by contrastive objective [3,5,15] become an efficient method to perform better input representation, not only textual data but also image data [3].

3   Methodology

In this section, we present our approach for the paper submission recommendation problem and the evaluation metrics.

3.1   Contrastive Learning

Contrastive learning recently has become one of the popular frameworks that can be applied for supervised and unsupervised learning machine learning tasks. This technique aims to learn similar/dissimilar representations from data constructed from a set of paired samples (x_i, x_i^+) semantically related effectively by pulling identical sample pairs together and pushing dissimilar ones apart in embedding space. The most significant promise of contrastive learning is utilizing a pre-trained language transformer model that was then fine-tuned with the contrastive objective to encode input into a good representation that can boost the performance of many downstream tasks.

We first leverage the supervised training dataset (as shown in Sect. 4.2) to construct a set of paired samples, $\mathcal{D} = \{(x_i, x_i^+)\}_{i=1}^m$, for the contrastive fine-tuning process, in which we denote x_i as the i'th sample that consists of the title, abstract, keywords and x_i^+ as semantically corresponding aims and scope. We follow the contrastive learning framework described in [5]. For a mini-batch of N pairs, let $\mathbf{h}_i$ and $\mathbf{h}_i^+$ denote the embedding or latent representation of x_i and x_i^+; the contrastive objective was defined as:

$$\ell_i = -\log \frac{e^{sim(\mathbf{h}_i, \mathbf{h}_i^+)/\tau}}{\sum_{i=1}^N e^{sim(\mathbf{h}_i, \mathbf{h}_j^+)/\tau}}, \tag{1}$$

where τ is a temperature hyper-parameter and $sim(\mathbf{h}_1, \mathbf{h}_2)$ is the cosine similarity, which is:

$$sim(\mathbf{h}_1, \mathbf{h}_2) = \frac{\mathbf{h}_1^T \mathbf{h}_2}{\|\mathbf{h}_1\| \cdot \|\mathbf{h}_2\|}.$$

3.2   Modeling

In this study, we build a two-state model containing two consecutive procedures, conceptually fine-tuning the pre-trained LM with a simple contrastive learning framework as an encoder to encode each document or sentence into sentence embedding efficiently. Then, the fine-tuned LM can be applied to different combinations of the features for the downstream task to classify for the Top K accuracy on groups of attributes.

Fine-Tuning: As mentioned in Sect. 3.1, we consider the aims and scope of the journal as a positive sample of the paper's title, abstract, and keywords, respectively. Finally, we perform fine-tuning the pre-trained Distil-RoBERTa (a distilled version of the pre-trained RoBERTa [10]) via the contrastive objective (Eq. 1) on the set of paired samples to fine-tune all parameters as depicted in Fig. 1.

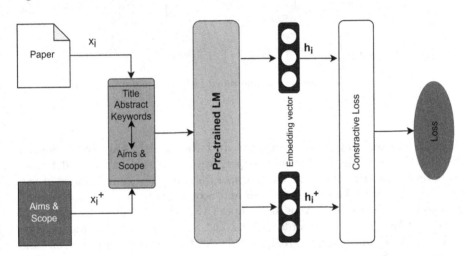

Fig. 1. Fine-tuning the pre-trained LM progress with a simple contrastive learning framework where x_i and x_i^+ are considered two semantically related samples and those corresponding representation $\mathbf{h}_i$, $\mathbf{h}_i^+$.

Downstream Task: We consider the fine-tuned LM as a backbone for the classification task. Therefore, we train it on different combinations of the features, either using available attributes of each paper submission or combining the paper information and the journals' aims. In this section, we describe two different use-cases in our experiments.

Models Using Paper Information. For this case, we extract helpful information from each paper submission, including Title (T), Abstract(A), and Keywords (K). This information is combined into seven different combinations of attributes: Title(T), Abstract(A), Keywords(K), Title + Abstract (TA), Title

+ Keywords (TK), Abstract + Keywords (AK), Title + Abstract + Keywords (TAK). The fine-tuned LM plays as the encoder to encode the batch of inputs into 768-dimensional embeddings, which are further forward propagated through an additional linear layer with ReLU activation and dropout to avoid overfitting. The last linear layer with Softmax activation acts as a classifier to output the N-probabilities that the given paper could belong to the respective journal. To identify the K-relevant journals that maximize the probability of acceptance, we choose the Top K of maximum values. The illustration for this case is in Fig. 2.

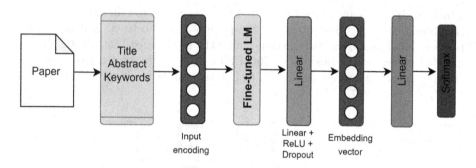

Fig. 2. The architecture of the fine-tuned LM using paper information.

Models Using Paper Information and Journals' Aims and Scope.
Besides the available attributes on paper, we use Journals' Aims and Scopes as potential external features. We end up with seven new combinations of features for the input data, Title + Aims and Scopes (TS), Abstract + Aims and Scopes (AS), Keywords + Aims and Scopes (KS), Title + Abstract + Aims and Scopes (TAS), Title +Keywords + Aims and Scopes (TKS), Abstract + Keywords + Aims and Scopes (AKS), Title + Abstract + Keywords + Aims and Scopes (TAKS).

Conceptually, we define two sub-branches and the main branch of this architecture. In the first sub-branch, we use the pre-trained Distil-RoBERTa model to encode input data into input encoding and reuse the fine-tuned LM to encode the external feature into the embeddings and pass it through a linear layer with ReLU activation for dimensional reduction. The second sub-branch is the same as the model using paper information described previously. Remarkably, the paper feature, output from the second sub-branch, will contribute to the following steps. First, we extract the similarity between it and external features produced from the first sub-branch using the cosine similarity and then concatenate it with these cosine features. Then, in the main branch, we feed that joined information to a linear layer and softmax activation to compute the probability of paper submission belonging to the journals and sort them in descending order to return the top list of recommended items. The illustration for the architecture can be found in Fig. 3.

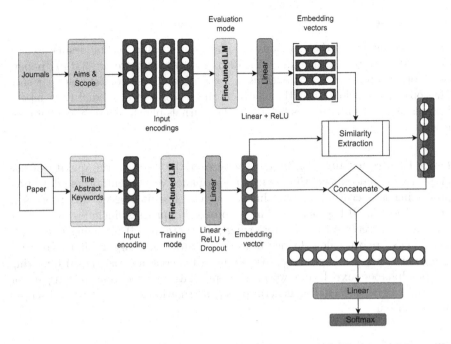

Fig. 3. The architecture of the fine-tuned LM using paper information and journal's aims.

3.3 Evaluation Metrics

As the previous approach [7], Son et al. used Accuracy@K to evaluate the performance of the proposed model, where K = 1, 3, 5, 10. The Accuracy@K is the ratio between the number of correct items at each K and the K number of recommended items, described in the formula follows:

$$P_{Top-K} = \frac{\text{The number of relevant items}}{\text{The number of viewed items}} \tag{2}$$

4 Experiments

This section will present how we conducted the experiments for the proposed methods and compare them with the previous approaches for the main problem.

4.1 Experimental Settings

Our experiments are run on Google Colab Pro, mostly with Tesla P100-PCIE 16 GB VRAM GPU accelerator, and implemented in the PyTorch framework [13]. In addition, we utilize the HuggingFace library [20], one of the most popular NLP frameworks that provide APIs to download and use pre-trained transformer models easily.

4.2   Datasets

Our experiments use the same dataset and preprocessing techniques as used in Son et al. 's work [7], including 414512 papers, where there are 331464 papers used for training and 83048 ones used for testing. These consist of paper submissions' information (the title, abstract, and list of keywords) and their labels. Besides, 351 aims and scopes belong to the journals play as external contributed features.

Data Preprocessing. Excluding the size of the dataset, the quality of the preprocessed data affects the model's performance very essentially, so data preprocessing is a crucial step in almost machine learning tasks; this process is more critical in NLP problems. The preprocessing progress includes some steps such as (1) lowercase text, (2) removing punctuations, (3) removing URLs, (4) removing stop words downloaded from the Natural Language Toolkit (NLTK6), and some other ones we define, (5) removing unnecessary spacing,(6) removing not-be-alphabet text. In our work, we apply data preprocessing techniques for two kinds of data, including general paper information and the aims and scopes of the journal.

4.3   Training Details

We start by using a pre-trained Distil-RoBERTa model and then fine-tune it using the set of paired examples, where each instance is constructed by pairing the title, abstract, and keywords with the relevant aims and scope of the journal. The pre-trained tokenizer of the model will encode each of those features into input representation. In order to extract the sentence embeddings, we ultimately feed the model the generated input encodings and select the first tokens of the last hidden states. Finally, using the contrastive objective (Eq. 1), we adjust the model's parameters to make it more reliable in extracting the semantic relationship between sentences. During the experiment, we found that the AdamW [11] optimizer, a variant of the optimizer Adam [8] uses weight decay to avoid overfitting, is the efficient choice for deep networks like our architecture.

According to Sect. 3.2, we solve the classification problem as the downstream task in which we train the fine-tuned Distil-RoBERTa model on different combinations of features and use the AdamW to optimize Cross-Entropy loss. Finally, we put the Softmax layer to achieve the Top K values representing the probability of given inputs belonging to corresponding labels and compute the accuracy at each K value as described in Sect. 3.3

4.4   Results

Our experiments are done on different combinations of attributes using two models defined previously: **Models using paper information** and **Models using paper information and Journals' Aims and Scope.** We compare the performance of our proposed model with the approach [7] (namely Approach A), the experimental results described in Table 1 and Fig. 4.

Firstly, for models only using paper information, the results show that the proposed method performs better than Approach A for combinations of attributes such as A, TK, TA, AK, and TAK. Especially, our approach has the best outcome for the group of features TAK in Accuracy@K as 0.5173, 0.8097, 0.8862, and 0.9496, where K = 1, 3, 5, and 10. Meanwhile, Approach A with the same input has 0.4852, 0.7856, 0.8624, and 0.9333, respectively. However, the proposed model's performance is lower than Approach A for the title (T) and keywords (K) inputs at the Top 3 and 5 in the accuracy; this difference is slight. Besides, the accuracy of the proposed model at the Top 1 surpasses the previous method for the title and keywords (TK) input; specifically, the Accuracy@K (with K = 1) of the proposed model is 0.3721 and 0.4022; meanwhile, the performance of Approach A is 0.3542, 0.3333, respectively. In addition, compared to the accuracy of Approach A in the Top 10, our approach gives better performance for the title (T) feature, and it has a little lower outcome for the keywords (K) attribute.

Secondly, for models using paper information and Journals' Aims and Scope, the proposed method's performance surpasses Approach A when using types of input data such as AS, TAS, AKS, and TAKS. For example, the best performance of the proposed approach in Accuracy@K (K=1, 3, 5, 10) is 0.5194, 0.8112, 0.8866, and 0.9496 when using all features (TAKS), while the performance of Approach A is lower than, which are 0.5002, 0.7889, 0.8627, and 0.9323 respectively. Although for the remaining input groups (namely TS, KS, and TKS), the Accuracy@K (K=1, 3, 5, 10) of the proposed model is lower than Approach A,

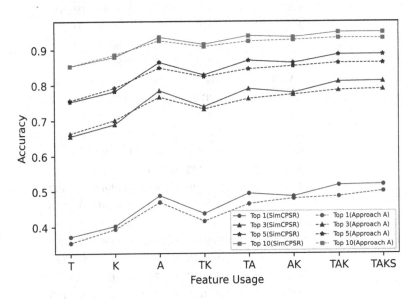

Fig. 4. The performance between the proposed SimCPSR approach and Approach A for different features.

excluding only one case Accuracy@10 with TKS as input data, the accuracy of the proposed approach is slightly greater than Approach A.

Finally, one can see that both the proposed method and approach A can help to improve performance when using additional information "Aims and Scopes" of the journals. Using Aims and Scopes' results in our proposed method is better than not using one. Except in the cases of using types of features such as title (T), the keyword (K), title, and keyword (TK), using the additional attribute "Aims and Scopes" is not helpful to increase performance. For example, the

Table 1. The performance of the proposed models compared to Son et al.'s results in [7]

Method	Feature Usage	Top 1	Top 3	Top 5	Top 10
Approach A	T	0.3542	**0.6634**	**0.7561**	0.8532
	TS	**0.4015**	**0.6991**	**0.7971**	0.8951
	K	0.3933	**0.7008**	**0.7919**	**0.8852**
	KS	**0.4284**	**0.7256**	**0.8189**	**0.9075**
	A	0.4691	0.7661	0.8482	0.9253
	AS	0.477	0.7662	0.8488	0.9258
	TK	0.4157	0.7315	0.8232	0.9084
	TKS	**0.4475**	**0.7490**	**0.8302**	0.9127
	TA	0.4644	0.7613	0.8448	0.9233
	TAS	0.4828	0.7754	0.8536	0.9276
	AK	0.4791	0.7730	0.8530	0.9273
	AKS	0.4951	0.7830	0.8602	0.9304
	TAK	0.4852	0.7856	0.8624	0.9333
	TAKS	0.5002	0.7889	0.8627	0.9323
SimCPSR	T	**0.3721**	0.6555	0.7526	**0.8533**
	TS	0.3737	0.6553	0.7513	0.8523
	K	**0.4022**	0.6892	0.7822	0.8792
	KS	0.4015	0.6921	0.7839	0.8784
	A	**0.4875**	**0.7842**	**0.8639**	**0.9351**
	AS	**0.4886**	**0.7849**	**0.8642**	**0.9353**
	TK	**0.4372**	**0.7382**	**0.8280**	**0.9145**
	TKS	0.4367	0.7354	0.8268	**0.9128**
	TA	**0.4935**	**0.7892**	**0.8689**	**0.9385**
	TAS	**0.5014**	**0.7920**	**0.8729**	**0.9428**
	AK	**0.4853**	**0.7782**	**0.8622**	**0.9350**
	AKS	**0.5030**	**0.7964**	**0.8765**	**0.9435**
	TAK	**0.5173**	**0.8097**	**0.8862**	**0.9496**
	TAKS	**0.5194**	**0.8112**	**0.8866**	**0.9496**

best performance of the proposed method in the Accuracy@K (K=1, 3, 5, 10) is 0.5194, 0.8112, 0.8866, and 0.9496 when using combinations of attributes TAKS; meanwhile, the outcome when not using the aims of the journals are 0.5173, 0.8097, 0.8862 and 0.9496 respectively.

In summary, our proposed approach, SimCPSR, has outperformed performance compared to the previous method. Especially the proposed method applied contrastive objectives with pre-trained LMs in learning sentence representation for embedding input data, which is an efficient idea and practical for the sentence embedding method because of its simplicity and ease of training. Therefore, our proposed method can be considered a new approach to the paper submission recommendation system when applying a new embedding method. Remarkably, using additional information "Aims and Scopes" can help improve models' efficiency. Besides, we can see the importance of the abstract feature; all models using input data containing this feature have better performance than the models using input without abstract information. Therefore, the abstract factor is essential for the paper submission recommendation problem.

5 Conclusion and Further Works

We presented a transformer-based approach to the paper submission recommendation system with simple contrastive learning. The proposed method utilized a simple supervised contrastive objective to fine-tune all parameters in the pre-trained LM for embedding input data and the fine-tuned LM to train different combinations of the features using available paper information and the journal's aims for downstream task. The experimental results show that the proposed approach has competitive performance and is an advanced method for enhancing the efficiency of the paper submission recommendation system. Our study especially gives a practical approach to efficient sentence embedding; based on this method's idea, one can apply it to other studies in learning effective sentence representation in recommendation systems separately and natural language processing in general. Furthermore, we will continue improving the proposed algorithm's performance and apply them to different datasets that belong to various areas in the future.

Acknowledgement. Son Huynh Thanh was funded by Vingroup JSC and supported by the Master, Ph.D. Scholarship Programme of Vingroup Innovation Foundation (VINIF), Institute of Big Data, code VINIF.2021.ThS.18.

References

1. Bai, X., Wang, M., Lee, I., Yang, Z., Kong, X., Xia, F.: Scientific paper recommendation: a survey. IEEE Access **7**, 9324–9339 (2019). https://doi.org/10.1109/ACCESS.2018.2890388
2. Bojanowski, P., Grave, E., Joulin, A., Mikolov, T.: Enriching word vectors with subword information. Trans. Assoc. Comput. Linguist. **5**, 135–146 (2017)

3. Chen, T., Kornblith, S., Norouzi, M., Hinton, G.E.: A simple framework for contrastive learning of visual representations. CoRR abs/2002.05709 (2020). https://arxiv.org/abs/2002.05709
4. Devlin, J., Chang, M.W., Lee, K., Toutanova, K.: BERT: pre-training of deep bidirectional transformers for language understanding. In: Proceedings of the 2019 Conference of the North American Chapter of the Association for Computational Linguistics: Human Language Technologies, Vol. 1 (Long and Short Papers), pp. 4171–4186. Association for Computational Linguistics, Minneapolis, Minnesota (2019). https://doi.org/10.18653/v1/N19-1423. https://aclanthology.org/N19-1423
5. Gao, T., Yao, X., Chen, D.: SimCSE: simple contrastive learning of sentence embeddings (2021). https://doi.org/10.48550/ARXIV.2104.08821
6. Hochreiter, S., Schmidhuber, J.: Long short-term memory. Neural Comput. 9(8), 1735–1780 (1997)
7. Huynh, S.T., Dang, N., Huynh, P.T., Nguyen, D.H., Nguyen, B.T.: A fusion approach for paper submission recommendation system. In: Fujita, H., Selamat, A., Lin, J.C.-W., Ali, M. (eds.) IEA/AIE 2021. LNCS (LNAI), vol. 12799, pp. 72–83. Springer, Cham (2021). https://doi.org/10.1007/978-3-030-79463-7_7
8. Kingma, D.P., Ba, J.: Adam: a method for stochastic optimization (2014). arxiv:1412.6980: Published as a conference paper at the 3rd International Conference for Learning Representations, San Diego (2015)
9. LeCun, Y., Bengio, Y., Hinton, G.: Deep learning. Nature 521, 436–444 (2015). https://doi.org/10.1038/nature14539
10. Liu, Y., et al.: Roberta: a robustly optimized BERT pretraining approach. CoRR abs/1907.11692 (2019). https://arxiv.org/abs/1907.11692
11. Loshchilov, I., Hutter, F.: Fixing weight decay regularization in Adam. CoRR abs/1711.05101 (2017). https://arxiv.org/abs/1711.05101
12. Nguyen, D.H., Huynh, S., Huynh, P., Cuong, D.V., Nguyen, B.T.: S2CFT: a new approach for paper submission recommendation. In: SOFSEM (2021)
13. Paszke, A., et al.: PyTorch: an imperative style, high-performance deep learning library. In: Wallach, H., Larochelle, H., Beygelzimer, A., d' Alché-Buc, F., Fox, E., Garnett, R. (eds.) Advances in Neural Information Processing Systems 32, pp. 8024–8035. Curran Associates, Inc. (2019). https://papers.neurips.cc/paper/9015-pytorch-an-imperative-style-high-performance-deep-learning-library.pdf
14. Pennington, J., Socher, R., Manning, C.D.: Glove: global vectors for word representation. In: Empirical Methods in Natural Language Processing (EMNLP), pp. 1532–1543 (2014). https://www.aclweb.org/anthology/D14-1162
15. Reimers, N., Gurevych, I.: Sentence-BERT: sentence embeddings using Siamese BERT-networks. CoRR abs/1908.10084 (2019). https://arxiv.org/abs/1908.10084
16. Samin, H., Azim, T.: Knowledge based recommender system for academia using machine learning: a case study on higher education landscape of Pakistan. IEEE Access 7, 67081–67093 (2019). https://doi.org/10.1109/ACCESS.2019.2912012
17. Son, H.T., Phong, H.T., Dac, N.H.: An efficient approach for paper submission recommendation. In: 2020 IEEE Region 10 Conference (TENCON), pp. 726–731 (2020)
18. Vaswani, A., et al.: Attention is all you need (2017). https://doi.org/10.48550/ARXIV.1706.03762
19. Wang, D., Liang, Y., Xu, D., Feng, X., Guan, R.: A content-based recommender system for computer science publications. Knowl.-Based Syst. 157, 1–9 (2018). https://doi.org/10.1016/j.knosys.2018.05.001

20. Wolf, T., et al.: Transformers: state-of-the-art natural language processing. In: Proceedings of the 2020 Conference on Empirical Methods in Natural Language Processing: System Demonstrations, pp. 38–45. Association for Computational Linguistics, Online (2020). https://www.aclweb.org/anthology/2020.emnlp-demos.6

Frequent Closed Subgraph Mining: A Multi-thread Approach

Lam B. Q. Nguyen[1], Ngoc-Thao Le[2], Hung Son Nguyen[3], Tri Pham[4], and Bay Vo[2(✉)]

[1] Faculty of Information and Communications, Kien Giang University, Kien Giang, Vietnam
[2] Faculty of Information Technology, HUTECH University, Ho Chi Minh City, Vietnam
vd.bay@hutech.edu.vn
[3] Faculty of Mathematics, Informatics and Mechanics, University of Warsaw, Warsaw, Poland
[4] Institute for Computational Science and Technology (ICST), Ho Chi Minh City, Vietnam

Abstract. Frequent subgraph mining (FSM) is an interesting research field and has attracted a lot of attention from many researchers in recent years, in which closed subgraph mining is a new topic with many practical applications. In the field of graph mining, GraMi (GRAph MIning) is considered state-of-the-art, and many algorithms have been developed based on the improvement of this approach. In 2021, we proposed the CloGraMi algorithm based on GraMi to mine closed frequent subgraphs from a large graph rapidly and efficiently. However, with NP time complexity and extremely high cost in terms of running time, graph mining is always a challenging problem for all researchers. In this paper, we propose a parallel processing strategy aiming to improve the execution speed of our Clo-GraMi algorithm. Our experiments on six datasets, including both undirected and directed graphs, with different sizes, including large, medium and small, show that the new algorithm significantly reduces running time and improves performance, and has better performance compared to the original algorithm.

Keywords: Subgraph mining · Closed subgraph mining · Multi-thread · Parallel

1 Introduction

Because graphs have a non-linear structure with NP complexity [1–3], graph mining has always been an interesting and challenging research area. Graphs are used to simulate, store, analysis and solve a wide variety of problems in both the scientific and commercial fields [4–6]. Graph mining [7–10] is the premise for many different studies and practical applications, in areas such as social networks, maps analysis, telephone networks, bioinformatics, chemical compounds, crime investigation, etc. Closed frequent subgraph mining [11–13] is a relatively new field in this context, with a few studies already having been published [11–16]. Extending this field is our motivation to study and research this topic.

For example, and as in [12], a commercial company needs to analyze data that has been collected from its customers, and the sales department needs to find the groups of loyal customers (who frequently buy the company's products) and the features of the

N. T. Nguyen et al. (Eds.): ACIIDS 2022, LNAI 13757, pp. 64–77, 2022.
https://doi.org/10.1007/978-3-031-21743-2_6

relationships among the customer groups in order to have a flexible business strategy. In Fig. 1, graph G is used to simulate all the customers in the dataset, and each node represents a customer of this company (labeled A, B, C or D), in which each edge in the graph represents the relationship of each pair of customers (labeled x, y, z, t or w). The two subgraphs S_1, S_2 are two samples of groups of loyal customers.

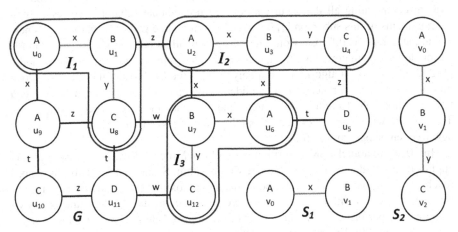

Fig. 1. G – a large graph – and its two subgraphs S_1 and S_2

The above example leads to very useful problems in practice, in which users need to find all the groups of loyal customers as well as the largest number of purchases for that group, as this can help the sales department to adjust the business strategies of the company. This problem is useful not only for business [3, 17], but also in many other fields [18–21] such as recommendation systems, crime investigation, information retrieval systems, graph clustering, decision support systems, traffic flow, map model analysis, etc.

In 2014, the GraMi algorithm [1, 2, 22, 23] was proposed with the aim to effectively mine all frequent subgraphs in a large graph. And then, in 2020, we introduced the SuGraMi algorithm [2] based on GraMi, which can find all frequent subgraphs and their support. In 2021, we proposed an algorithm to mine closed frequent subgraphs, the CloGraMi algorithm [12], based on SuGraMi. In this paper, our contribution is proposing an effective strategy to improve our CloGraMi algorithm by parallel processing, it aims to reduce the running time, and we call it PCGraMi (Parallel Closed Graph Mining).

Our paper consists of six sections: Sect. 1 introduces the problem of closed graph mining and our contribution; Sect. 2 surveys relevant algorithms; Sect. 3 consists of the main definitions in the field of graph mining; Sect. 4 presents the PCGraMi algorithm in detail; Sect. 5 describes the results of our evaluation studies; Finally, the conclusions and future work are discussed in Sect. 6.

2 Related Work

The GraMi algorithm [1–3] was proposed in 2014 as a fast frequent subgraph mining algorithm in a single large graph. This algorithm solves the FSM problem by introducing a new technique to store only templates of generated candidates. It searches isomorphisms and only marks the corresponding values in domains of those templates, without enumerating all the occurrences of each candidate subgraph (corresponding isomorphisms of this subgraph [1, 2, 24, 25]) from a large graph. Because of this new technique, GraMi is different from previous grow-and-store methods [24, 26]. GraMi has modelled the support of a subgraph as a CSP (constraint satisfaction problem) model [1, 2, 22]. The GraMi algorithm iteratively unfolds the CSP model until it discovers a minimum set of occurrences for a candidate subgraph to be correctly identified as a frequent subgraph. To reduce the search time, all remaining occurrences are ignored. This process will repeat by extending (adding new edges) a frequent subgraph until no new frequent subgraphs are found.

In 2016, a parallel FSM algorithm called ScaleMine [27] was proposed, and this applies distributed computing to the GraMi algorithm. ScaleMine is a two-phase approach [27, 28]: first, the algorithm performs an approximation to quickly identify the frequent subgraphs with high occurrence probability in the dataset; then, the exact calculations are performed using the results of the approximation phase to achieve better load balancing [29]. Experiments using this algorithm have extended to 8,192 cores of the Cray XC40 system, enabling it to solve a large graph with one billion edges and having faster mining performance than existing solutions. In 2020, we proposed a parallel approach PaGraMi [2] based on the original algorithm GraMi, using multi-thread processing in a computer with a multi-core CPU, combined with the SoGraMi algorithm [2] to reduce the search space of GraMi by pruning infrequent candidate subgraphs.

The process of generating candidate subgraphs needs lots of memory to store a huge number of candidates [1, 3, 30, 31]. Algorithms such as GraMi [1], SoGraMi [2], WeGraMi [25] and CCGraMi [23] have their own efficient strategies for pruning and thus reducing the number of infrequent candidates. The goals of all the mentioned algorithms are to reduce the costs of the generating phase [3], and improve the efficiency of memory usage and running time. Other parallel algorithms, such as ScaleMine [27] and Arabesque [32], have been proposed as new parallel algorithms using a message transfer interface on distributed systems, but they still lack a load balancing method for clusters or their machines in the system [3].

In 2018, the SSIGRAM algorithm (Spark-based Single Graph Mining) [33] was proposed as a parallel algorithm for FSM based on Spark. It parallelizes the extending and evaluating the support of subgraphs on all distributed "workers". In addition, a new heuristic-based search strategy was proposed with three optimizations: (1) load balancing, (2) top-down pruning and (3) pre-search pruning during support evaluation, helping to significantly improve performance. The support evaluation of candidates are simultaneously carried out at the "executors" in the Spark cluster. The evaluation results demonstrate that the SSIGRAM algorithm can perform significantly better than GraMi algorithm on all four different real datasets. Furthermore, it is capable of operating at lower support thresholds. While the SIGRAM [34] algorithm enumerates all intermediate steps using the maximum independent sets (MIS) measurements. However, these steps have NP complexity, thus the method is extremely expensive in practice [29].

In 2021, we introduced the CCGraMi algorithm [23], which is based on connected components in a large graph with two main contributions: finding isomorphisms of subgraphs in each connected component of the large graph instead of searching in the whole large graph to reduce searching time, and early pruning of the domains of candidate subgraphs based on the size of the connected components to reduce storage space and searching time on those domains. We also proposed the CloGraMi algorithm in 2021 [12] to mine closed subgraphs, moreover we have three effective strategies aiming to optimize this algorithm: a level order traversal strategy to reduce the time for the search, early determining of closed subgraphs, and setting a constraint for pruning of candidates that are non-closed subgraphs.

In 2022, because the GraMi algorithm [1] lacked an effective strategy to balance its search space, we introduced the BaGraMi algorithm [22] to address this issue. This decreases the size of both the search space for all generated subgraphs and domain for each candidate subgraph [35], and thus helps balance the search space of the original GraMi algorithm [2]. The main contributions of this algorithm are to reduce the invalid assignments in the domain of a candidate and the number of infrequent subgraphs, enhancing the mining performance. Our algorithm not only reduces the running time but also the memory requirements of the mining process.

According to our survey [3] in 2022, most related studies focus on FSM [7–10, 36], and there are few algorithms in the field of closed frequent subgraph mining [11, 13–16, 37, 38], and thus we propose to improve our CloGraMi algorithm by using a parallel processing strategy to improve the execution speed of this algorithm on multi-core personal computers. The goal of this algorithm is to boost the performance of the mining closed frequent subgraphs from a single large graph process.

3 Definitions

Definition 1. [1, 2, 12]: Let G be a large *graph*, $G = (V, E, L)$. In which, V denotes the set of all the nodes, E denotes the set of all the edges and L is a function to assign labels to all the nodes/edges, respectively.

Definition 2. [1, 12, 22]: A graph $S = (V_S, E_S, L_S)$ is a *subgraph* of $G = (V, E, L)$ iff $V_S \subseteq V$, $E_S \subseteq E$ and $L_S(v) = L(v)$ for $\forall v \in V_S$; $LS((u, v)) = L((u, v))$ for $\forall (u, v) \in ES$.

Definition 3. [1, 12, 23]: Let $S = (V_S, E_S, L_S)$ be a subgraph of $G = (V, E, L)$. A *subgraph isomorphism I* of S to G is an injective function $f : V_S \rightarrow V$ satisfying:
   (a) $L_S(v) = L(f(v)), \forall v \in V_S$
   (b) $(f(u), f(v)) \in E$ and $L_S(u, v) = L(f(u), f(v)), \forall (u, v) \in E_S$.

Example 1: In Fig. 1 [12], the subgraph S_2 has three distinct isomorphisms I_1, I_2 and I_3 as presented in Table 1.

Table 1. Three distinct isomorphisms of S_2.

Subgraph S_2	v_0	v_1
Isomorphism I_1	u_0	u_1
Isomorphism I_2	u_2	u_3
Isomorphism I_3	u_6	u_7

The notation u is used to indicate all the nodes in the large graph G and the notation v is used to indicate all the nodes in the subgraph S. Each node $v \in V_S$ has a domain D containing all the nodes u that have the same node label with v and they can be assigned to v. The assignments u to v are used to list, mark and count the number of isomorphisms of subgraph S in the large graph G.

Example 2: As shown in Fig. 2, the domain of $v_0 \in S$ is $D = \{u_0, u_2, u_6, u_9\}$.

Definition 4. [22]: An assignment of a node $u \in V_G$ in the domain of node $v \in V_S$ is *valid* if there exists an isomorphism I that assigns u to v, *invalid* otherwise.

Example 3: In Fig. 2, the nodes u_0, u_2, and u_6 in the domain are valid assignments because there exist three isomorphisms I_1, I_2 and I_3 assigning u_0, u_2, and u_6 to v_0, respectively; u_9 is an invalid assignment in the domain of S_2 because there is no isomorphism that assigns u_9 to v_0.

Definition 5. [2, 22]: *The support of S in G (denoted by $s_G(S)$) is the minimum number of all distinct valid assignments of $\forall v \in V_S$.* In other words:

$$s_G(S) = min\{t | t = |F(v)|, \forall v \in V_S\}.$$

Example 4 [2, 12]: The support of subgraph S_2 in large graph G:

$$s_G(S_2) = min(|F(v_0)|, |F(v_1)|, |F(v_2)|) = min(3, 3, 3) = 3.$$

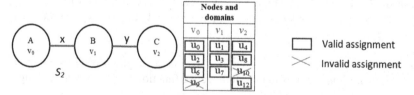

Fig. 2. Valid and invalid assignments for a subgraph

Definition 6. [3]: A subgraph S in a large graph G is called a *frequent subgraph* if $s_G(S) \geq \tau$, in which τ is a given frequency threshold.

Remark: We use $\tau = 2$ for all our examples in this paper.

Example 5 [12]: As $s_G(S_2) = 3 \geq \tau$, it is a frequent subgraph in G.

Definition 7. [3, 12]: A frequent subgraph S is a *closed frequent subgraph* if there does not exist any supergraph S' ($S \subset S'$) whose the support is equal to that of S (called as "closed subgraph" in short).

Example 6 [12]: S_1 is a non-closed subgraph because $S_1 \subset S_2$ and $s_G(S_2) = s_G(S_1) = 3$, and S_2 is a closed subgraph (see detail in [12]).

In the CloGraMi algorithm [12], we propose traversing the search tree by level to identify closed subgraphs early, in which in order to determine if the subgraph S at level n on the search tree is a closed subgraph the mining program only needs to compare S with frequent subgraphs at level $(n + 1)$ in the search tree, instead of comparing this frequent subgraph with all mined frequent subgraphs. In this work, a new parallel processing strategy for our CloGraMi algorithm is proposed. Whereas, each branch in the search tree corresponding to an edge in the *FrequentEdge* list [2, 12] will be assigned to a concurrent thread, thus many branches in the search tree will be executed simultaneously, each frequent subgraph at level n on the branches will be mined and compared with the frequent subgraph at level $(n + 1)$. In this PCGraMi algorithm, many frequent subgraphs are generated and compared at the same time to find out the closed subgraphs to decrease the running time in the comparison to CloGraMi, our original algorithm.

4   Proposed Method

In the SuGraMi algorithm [2], as $s_G(S) \leq s_G(S')$ with S' be a child of S on the search tree was proven, based on this Downward Closure Property (DCP) [25] we introduced a level-wise traversal strategy [12] on the search tree, which can quickly identify closed subgraphs. Each frequent subgraph S of size k compares its support only with subgraphs S' with size $(k + 1)$ being children of S [12]. We propose a parallel processing strategy to improve the speed of the CloGraMi algorithm in this paper, in which each branch of the search tree will be assigned to a thread, the threads will generate candidate subgraphs, check the support, and determine closed subgraphs on the same branch of the search tree.

Example 7: For the graph G in Fig. 1, after evaluating all edges' support and pruning the infrequent edges, we have a *FrequentEdges* list consisting of edges as in Fig. 3:

$$FrequentEdges = \{A \xrightarrow{x} B; B \xrightarrow{y} C; C \xrightarrow{z} D\}$$

When the mining process generates the ith thread to process the edge e_i in the *FrequentEdges* list, the edges $e_0, e_1, \cdots, e_{i-1}$ have already been processed in other threads, so they will be removed from the *FrequentEdges* list of this thread [2] as in Fig. 4, and thus main program can implement without generating duplicate subgraphs.

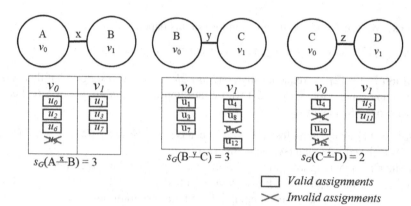

Fig. 3. The frequent edges list

Our algorithm consists of three steps.

(1) Each edge in the *FrequentEdges* list is assigned to a separate thread (Line 4 to Line 8).

(2) These threads are then executed simultaneously (Line 6 and Line 7) based on the number of available processor cores. The running time of this parallel phase will be the time of the largest thread in the list (usually the first thread because it needs to combine with all remaining edges).

(3) Collecting closed subgraphs (Line 9 and Line 10) returned from these threads (collection time is very small compared to mining time).

```
Algorithm: PCGraMi

Input: A graph G, a frequency threshold τ
Output: All closed subgraphs S of G

1   ClosedSubgraphsList ← ∅ //the list of all closed subgraph in graph G
2   Let FrequentEdges is a set of frequent edges in G
3   Let R be the result set for all simultaneously threads
4   foreach e ∈ FrequentEdges do
5      Generate a new thread and do
6         r ← ∅
7         r ← r ∪ extendSubgraph (e, G, τ, FrequentEdges)
8      Remove e from FrequentEdges
9   foreach r ∈ R do
10     ClosedSubgraphsList ← ClosedSubgraphsList ∪ r
11  return ClosedSubgraphsList
```

At Line 7 in our PCGraMi algorithm, the program needs a function *extendSubgraph*( ) to recursively extend frequent subgraphs, as follows.

Algorithm: extendSubgraph

Input: A frequent subgraph S, a frequency threshold τ, a set of frequent edges FrequentEdges of G
Output: All closed subgraphs in G extended from S

```
1  cS ← ∅ //closed subgraphs list
2  gS ← ∅ //generated subgraphs list
3  fS ← ∅ //frequent subgraphs list
4  foreach e ∈ FrequentEdges and v ∈ S do
5    if e can extend u then
6      Let Ext be the extension of S by adding e
7      if Ext is not already generated then
8        gS ← gS U Ext
9  clo ← true //a condition for being a closed subgraph
10  foreach g ∈ gS do
11    if sG(g)≥ f then
12      fS ← fS U g
13      if sG(g)=sG(S) then
14        clo ← false
15 if clo = true then
16    cS ← cS U S
17 foreach S' ∈ fS do
18    cS ← cS U extendSubgraph(e,f,FrequentEdges,G)
19 return cS
```

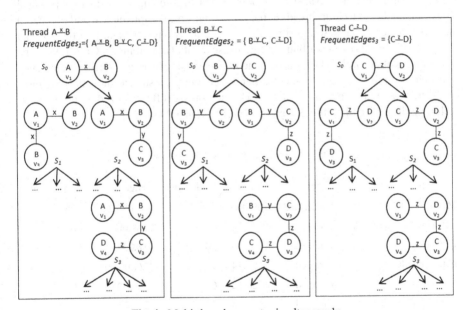

Fig. 4. Multi-threads execute simultaneously

Consider the example given in Fig. 4, instead of sequentially processing each edge in this *FrequentEdges* list as the CloGraMi algorithm, our parallel algorithm will generate three threads to process these three edges at the same time. The main program then collects the results from these three threads to get the total results. The three phases are generating all three threads, parallel mining and collecting the results of PCGraMi algorithm. The first and third phases, which are generating threads and collecting the results, require a very short time compared to the mining phase, only from 1% to 2% of the total time for the program. As such, the cost of these two extra phases of our PCGraMi algorithm is insignificant compared to CloGraMi, but they significantly reduce the time of the mining phase. The limitation of parallel approaches is the memory requirements. Because the main program needs to store and process multiple candidate subgraphs at the same time, more storage space is needed for parallel mining than sequential processing.

5 Experimental Results

To demonstrate the effectiveness of our parallel strategy, we record and compare the running times for new algorithm PCGraMi with the original CloGraMi algorithm. All of our experiments were carried out on a multi-core personal computer running the Windows 10 operating system, equipped with a Core i5 CPU having four physical cores 3.2 GHz, eight threads, JavaSE Development Kit 8. A computer with four physical cores can handle up to four threads at the same time, so we implemented and compared the performance of the new algorithm PCGraMi with two threads and four threads and compared the results to those obtained with the sequential algorithm CloGraMi. With the two versions of PCGraMi (PCGraMi_2 and PCGraMi_4), we try to reduce the frequency thresholds τ until they cannot execute, then we collect the results to show the efficiency of our method. Our experiments were performed on six different sized graph datasets, including directed and undirected graphs, as shown in Table 2.

Table 2. The features of six real datasets.

Datasets	Type	Nodes	Node labels	Edges	Edge labels
MiCo	Undirected	100,000	29	1,080,298	106
GitHub social network	Undirected	37,700	60	289,003	60
Facebook	Undirected	4,389	20	88,235	36
p2p-Gnutella09	Directed	8,114	25	26,013	40
CiteSeer	Directed	3,312	6	4,732	101
Email-Eu-core network	Directed	1,005	50	25,571	50

– With the MiCo dataset (Fig. 5.a) [1, 2]: This is an undirected data set describing Microsoft co-authoring information with over one million edges and 100,000 nodes. The nodes in this dataset represent all Microsoft authors and their labels are the authors' interests. A collaboration between a pair of authors is represented as an edge, the edge's label represents the number of co-authored articles. This is a large dataset, so the mining thresholds are also high. There are only two edges in the *FrequentEdges* list, so our parallel strategy generates only two processing threads corresponding to these two edges. The running time of the PCGraMi parallel algorithm with two threads is from 80% to 83% of that needed by the CloGraMi algorithm at the survey thresholds, and with four threads is roughly 71%. At the final threshold $\tau = 9{,}250$, CloGraMi needs 1,505.372 s for the mining process, while PCGraMi (with 4 threads) only needs 1,081.115 s.

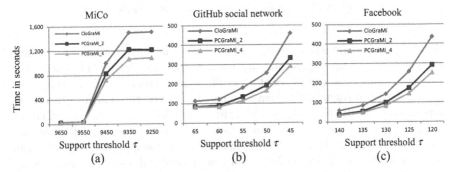

Fig. 5. The running time on the undirected datasets.

- With the GitHub social network dataset (Fig. 5.b) [12]: This is an undirected medium size dataset collected in June 2019 using the public API. This dataset represents the network of GitHub developers, the nodes in this graph represent developers, while the edges describe the mutual follower relationships between a pair of developers. This network dataset can be obtained from https://snap.stanford.edu/data/. It consists of 37,700 nodes and 289,003 edges. PCGraMi with two threads can reduce the running time to 72% of that needed by CloGraMi. The maximum running time (with four threads) can be decreased to 61% of that needed by the CloGraMi algorithm at $\tau = 55$, at the last threshold $\tau = 45$, CloGraMi needs 458.485 s, while PCGraMi only needs 300.211 s to complete the process.
- With the Facebook dataset (Fig. 5.c) [2, 25]: This directed dataset was obtained at: http://snap.stanford.edu/data/, and consists of 4,389 nodes, 88,235 edges, which are the 'circles' (or 'friends lists') of the Facebook social network. The data is collected from the social network user surveyed through the Facebook app. User data has been anonymized by Facebook, replacing each user's internal ids with a new value. Our new PCGraMi algorithm can reduce the running time to 66% that needed by the original CloGraMi algorithm with two processing threads, and to 56% with four threads, at the last threshold $\tau = 120$, CloGraMi needs 434.768 s to process while PCGraMi with four threads takes only 251.663 s.

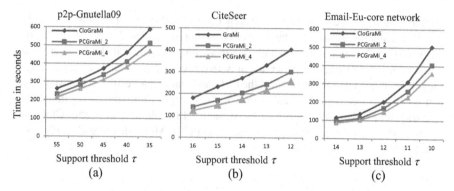

Fig. 6. The running time on three directed datasets.

- With p2p-Gnutella09 dataset (in Fig. 6.a) [2, 22]: This graph dataset was obtained from http://snap.stanford.edu/data/, and is a directed graph dataset of a sequence of snapshots from the Gnutella peer-to-peer file sharing network. All nine snapshots were collected in August 2002 from the Gnutella network. Hosts in this network are represented as nodes in the dataset, while the connections among the Gnutella hosts are the edges of pairs of nodes. This dataset has 8,114 nodes with 26,013 directed edges. Our improved algorithm has a running time equal to approximately 87% (with two simultaneous threads) and 79% (with four processing threads) of that needed by CloGraMi. At the threshold $\tau = 35$, the original algorithm CloGraMi needs 590.122 s, while PCGraMi_4 needs 473.134 s to complete the mining process.
- With the CiteSeer dataset (Fig. 6.b) [1, 23]: This directed graph dataset includes 3,312 nodes (each node in this graph corresponds to a publication) and 4,732 directed edges, they are citations between a pair of publications (citations are directed edges in this graph). In this graph, each node has a label representing a field of computer science. Each directed edge evaluates the similarity (labelling from 0 to 100) to between a pair of publications. Our proposed algorithm has the running time reduced by approximately 73% (with two threads) and 63% (with four threads) of CloGraMi's consumption, at the last threshold $\tau = 12$, the original algorithm CloGraMi needs 403.475 s but the PCGraMi parallel algorithm with four threads only needs 258.119 s.
- With the Email-Eu-core network dataset (Fig. 6.c) [12]: This dataset can be acquired from https://snap.stanford.edu/data/. It is a medium size directed graph consisting of 1,005 nodes with 25,571 directed edges. This network in a large European research institution was generated by the users using email data, but the users' information was anonymized for the sake of privacy. Each directed edge means that a user in this institution sent at least one email to another user. The parallel algorithm's running time can be reduced to 80% (with two parallel threads) and 71% (with four processing threads) of CloGraMi's, at the last threshold $\tau = 10$, CloGraMi needs 504.182 s while PCGraMi with 4 threads only needs 358,198 s.

6   Conclusion and Future Work

We have surveyed new algorithms and the literature on frequent subgraph mining and closed subgraph mining in recent years. There are few studies on closed subgraph mining, and such algorithms are very costly to implement. Therefore, we improved our closed subgraph mining algorithm CloGraMi by proposing an efficient parallel strategy, which is to concurrently explore multiple subgraphs at the same time using multi-threading. This parallel strategy can reduce the running time of the mining process. We conducted several experiments and compared PCGraMi with the original algorithm CloGraMi on six real datasets with different sizes (both directed and undirected), our new parallel algorithm has proved to overcome at all thresholds and on all selected datasets.

In the future, we have two promising research directions for closed subgraph mining: applying a parallel strategy to high-performance computing (HPC) to use the computing power to mine larger graphs with a smaller frequency threshold, and proposing an efficient pruning strategy for non-closed subgraphs that reduces the memory requirements of current algorithms.

Acknowledgement. This work was supported by Institute for Computational Science and Technology (ICST) – Ho Chi Minh City and the Department of Science and Technology (DOST) – Ho Chi Minh City under grant no. 23/2021/HĐ-QKHCN.

References

1. Elseidy, M., Abdelhamid, E., Skiadopoulos, S., Kalnis, P.: Grami: frequent subgraph and pattern mining in a single large graph. Proc. VLDB Endow. **7**(7), 517–528 (2014)
2. Nguyen, L.B.Q., Vo, B., Le, N.-T., Snasel, V., Zelinka, I.: Fast and scalable algorithms for mining subgraphs in a single large graph. Eng. Appl. Artif. Intell. **90**, 103539 (2020)
3. Nguyen, L.B.Q., Zelinka, I., Snasel, V., Nguyen, L.T.T., Vo, B.: Subgraph mining in a large graph: a review. Wiley Interdiscip. Rev. Data Min. Knowl. Discov. e1454 (2022)
4. Velampalli, S., Jonnalagedda, V.R.M.: Frequent subgraph mining algorithms: framework, classification, analysis, comparisons. In: Satapathy, S.C., Bhateja, V., Raju, K.S., Janakiramaiah, B. (eds.) Data Engineering and Intelligent Computing. AISC, vol. 542, pp. 327–336. Springer, Singapore (2018). https://doi.org/10.1007/978-981-10-3223-3_31
5. Borrego, A., Ayala, D., Hernández, I., Rivero, C.R., Ruiz, D.: CAFE: knowledge graph completion using neighborhood-aware features. Eng. Appl. Artif. Intell. **103**, 104302 (2021)
6. Fox, J., Roughgarden, T., Seshadhri, C., Wei, F., Wein, N.: Finding cliques in social networks: a new distribution-free model. SIAM J. Comput. **49**(2), 448–464 (2020)
7. Song, Q., Wu, Y., Lin, P., Dong, L.X., Sun, H.: Mining summaries for knowledge graph search. IEEE Trans. Knowl. Data Eng. **30**(10), 1887–1900 (2018)
8. Chehreghani, M.H., Abdessalem, T., Bifet, A., Bouzbila, M.: Sampling informative patterns from large single networks. Futur. Gener. Comput. Syst. **106**, 653–658 (2020)
9. Chen, Y., Zhao, X., Lin, X., Wang, Y., Guo, D.: Efficient mining of frequent patterns on uncertain graphs. IEEE Trans. Knowl. Data Eng. **31**(2), 287–300 (2018)
10. Iqbal, R., Doctor, F., More, B., Mahmud, S., Yousuf, U.: Big data analytics and computational intelligence for cyber-physical systems: recent trends and state of the art applications. Futur. Gener. Comput. Syst. **105**, 766–778 (2020)

11. Demetrovics, J., Quang, H.M., Anh, N.V., Thi, V.D.: An optimization of closed frequent subgraph mining algorithm. Cybern. Inf. Technol. **17**(1), 3–15 (2017)
12. Nguyen, L.B.Q., Nguyen, L.T.T., Zelinka, I., Snasel, V., Nguyen, H.S., Vo, B.: A method for closed frequent subgraph mining in a single large graph. IEEE Access (2021)
13. Karabadji, N.E.I., Aridhi, S., Seridi, H.: A closed frequent subgraph mining algorithm in unique edge label graphs. In: International Conference on Machine Learning and Data Mining in Pattern Recognition, pp. 43–57 (2016)
14. Yan, X., Han, J.: Closegraph: mining closed frequent graph patterns. In: Proceedings of the Ninth ACM SIGKDD International Conference on Knowledge Discovery and Data Mining, pp. 286–295 (2003)
15. Bendimerad, A., Plantevit, M., Robardet, C.: Mining exceptional closed patterns in attributed graphs. Knowl. Inf. Syst. **56**(1), 1–25 (2017). https://doi.org/10.1007/s10115-017-1109-2
16. Acosta-Mendoza, N., Gago-Alonso, A., Carrasco-Ochoa, J.A., Martínez-Trinidad, J.F., Medina-Pagola, J.E.: Mining generalized closed patterns from multi-graph collections. In: Iberoamerican Congress on Pattern Recognition, pp. 10–18 (2017)
17. Jia, Y., Zhang, J., Huan, J.: An efficient graph-mining method for complicated and noisy data with real-world applications. Knowl. Inf. Syst. **28**(2), 423–447 (2011)
18. Nejad, S.J., AhmadiAbkenari, F., Bayat, P.: A combination of frequent pattern mining and graph traversal approaches for aspect elicitation in customer reviews. IEEE Access **8**, 151908–151925 (2020)
19. Jie, F., Wang, C., Chen, F., Li, L., Wu, X.: A framework for subgraph detection in interdependent networks via graph block-structured optimization. IEEE Access **8**, 157800–157818 (2020)
20. Guan, H., Zhao, Q., Ren, Y., Nie, W.: View-based 3D model retrieval by joint subgraph learning and matching. IEEE Access **8**, 19830–19841 (2020)
21. Karwa, V., Raskhodnikova, S., Smith, A., Yaroslavtsev, G.: Private analysis of graph structure. ACM Trans. Database Syst. **39**(3), 1–33 (2014)
22. Nguyen, L., et al.: An efficient and scalable approach for mining subgraphs in a single large graph. Appl. Intell. 1–15 (2022)
23. Nguyen, L.B.Q., Zelinka, I., Diep, Q.B.: CCGraMi: an effective method for mining frequent subgraphs in a single large graph. MENDEL **27**(2), 90–99 (2021)
24. Ullmann, J.R.: An algorithm for subgraph isomorphism. J. ACM **23**(1), 31–42 (1976)
25. Le, N.-T., Vo, B., Nguyen, L.B.Q., Fujita, H., Le, B.: Mining weighted subgraphs in a single large graph. Inf. Sci. (Ny) **514**, 149–165 (2020)
26. Seeland, M., Girschick, T., Buchwald, F., Kramer, S.: Online structural graph clustering using frequent subgraph mining. In: Joint European Conference on Machine Learning and Knowledge Discovery in Databases, pp. 213–228 (2010)
27. Abdelhamid, E., Abdelaziz, I., Kalnis, P., Khayyat, Z., Jamour, F.: Scalemine: scalable parallel frequent subgraph mining in a single large graph. In: SC'16: Proceedings of the International Conference for High Performance Computing, Networking, Storage and Analysis, pp. 716–727 (2016)
28. Yan, X., Han, J.: gspan: Graph-based substructure pattern mining. In: 2002 IEEE International Conference on Data Mining, Proceedings, pp. 721–724 (2002)
29. Dhulipala, L., Blelloch, G.E., Shun, J.: Theoretically efficient parallel graph algorithms can be fast and scalable. ACM Trans. Parallel Comput. **8**(1), 1–70 (2021)
30. Thomas, L.T., Valluri, S.R., Karlapalem, K.: Margin: Maximal frequent subgraph mining. ACM Trans. Knowl. Discov. from Data **4**(3), 1–42 (2010)
31. Farag, A., Abdelkader, H., Salem, R.: Parallel graph-based anomaly detection technique for sequential data. J. King Saud Univ. Inf. Sci. **34**(1), 1446–1454 (2022)

32. Teixeira, C.H.C., Fonseca, A.J., Serafini, M., Siganos, G., Zaki, M.J., Aboulnaga, A.: Arabesque: a system for distributed graph mining. In: Proceedings of the 25th Symposium on Operating Systems Principles, pp. 425–440 (2015)
33. Qiao, F., Zhang, X., Li, P., Ding, Z., Jia, S., Wang, H.: A parallel approach for frequent subgraph mining in a single large graph using spark. Appl. Sci. **8**(2), 230 (2018)
34. Kuramochi, M., Karypis, G.: Finding frequent patterns in a large sparse graph. Data Min. Knowl. Discov. **11**(3), 243–271 (2005)
35. Kepner, J.: Keynote talk: large scale parallel sparse matrix streaming graph/network analysis. In: Proceedings of the 34th ACM Symposium on Parallelism in Algorithms and Architectures, p. 61 (2022)
36. Bouhenni, S., Yahiaoui, S., Nouali-Taboudjemat, N., Kheddouci, H.: A survey on distributed graph pattern matching in massive graphs. ACM Comput. Surv. **54**(2), 1–35 (2021)
37. Güvenoglu, B., Bostanoglu, B.E.: A qualitative survey on frequent subgraph mining. Open Comput. Sci. **8**(1), 194–209 (2018)
38. FournierViger, P., et al.: A survey of pattern mining in dynamic graphs. Wiley Interdiscip. Rev. Data Min. Knowl. Discov. **10**(6), e1372 (2020)

Decision Support and Control Systems

Complement Naive Bayes Classifier for Sentiment Analysis of Internet Movie Database

Christine Dewi[1,2] and Rung-Ching Chen[1(✉)]

[1] Department of Information Management, Chaoyang University of Technology, Taichung, Republic of China
crching@cyut.edu.tw
[2] Faculty of Information Technology, Satya Wacana Christian University, Salatiga, Indonesia
christine.dewi@uksw.edu

Abstract. Sentiment analysis (SA), often known as opinion mining, is the subjective examination of a written text. Moreover, SA is a critical technique in today's artificial intelligence (AI) field for extracting emotional information from huge amounts of data. The study is based on the Internet Movie Database (IMDB) dataset, which comprises movie reviews and the positive or negative labels that are connected with them. Our research experiment's objective is to identify the model with the best accuracy and the most generality. Text preprocessing is the first and most critical phase in a Natural Language Processing (NLP) system since it significantly impacts the overall accuracy of the classification algorithms. The experiment implements the Term Frequency-Inverse Document Frequency model (TFIDF) to feature selection and extractions. The following classifiers are used in this work: Linear Model and Naïve Bayes. Besides, we explore the possible options of loss functions such as *square_hinge, huber, modified_huber, log, epsilon_insensitive, perceptron,* and *modified_huber.* ComplementNB achieves the highest accuracy, 75.13%, for both classification reports based on our experiment result.

Keywords: IMDB · Sentiment analysis · Naïve Bayes · Complement NB · Sentiment classifications

1 Introduction

Due to the enormous rise of digital information, sentiment analysis is a fast-emerging topic of study. Sentiment analysis (SA) is a critical technique in today's field of artificial intelligence (AI) for extracting emotional information from huge amounts of data [1]. Sentiment analysis [2, 3] has improved over the years using several machine learning and dictionary-based algorithms to improve accuracy. Prior knowledge also plays an important role in conveying the polarity of views well with the advent of deep learning algorithms [4] in sentiment analysis. SA is often used in film reviews [5, 6], food reviews [7, 8], in addition to microblog data [9], stock markets [10], news articles [11], or political debates [12, 13], offers some valuable insights for companies to improve company plans through new customer acquisition.

© The Author(s), under exclusive license to Springer Nature Switzerland AG 2022
N. T. Nguyen et al. (Eds.): ACIIDS 2022, LNAI 13757, pp. 81–93, 2022.
https://doi.org/10.1007/978-3-031-21743-2_7

In the industrial context, businesses primarily use SA to collect and assess client feedback. The fields of natural language processing (NLP) and SA are inextricably linked [14]. The majority of the material on the internet comes in the form of natural language, which robots cannot understand owing to its complexity and inter-word semantics [15]. NLP is analyzing these natural texts to create things that the machine can understand for SA. The Internet Movie Database (IMDB) dataset comprises 50,000 reviews, equally split between 25,000 train and 25,000 test reviews. Positive or negative movie reviews are categorized, and the task is to guess the sentiment of an unseen review. The sentiment of a movie review is usually associated with a different rating, which can be used for classification dilemmas. It can be used as a reference instrument for movie preference [16, 17].

The following are the significant contributions of this work: (1). The study is based on the IMDB dataset, which comprises movie reviews and the positive or negative labels that are connected with them. (2). The goal of our study experiment is to find the model with the highest accuracy and the greatest generality. (3). The following classifiers are used in this work: Linear Model and Naïve Bayes. Different techniques, including Count Vectorizer, Term Frequency-Inverse Document Frequency model (TFIDF)Vectorizer, minimum-maximum number of words, and max features, are implemented. We can observe that the complement naive Bayes model performs well compared to other methods.

The rest of the paper is structured in the following manner. Section 2 introduces the related works. The third section describes the research process. Section 4 discusses the experiment's findings and conclusions. Section 5 summarizes the conclusion and future research endeavors.

2   Related Work

2.1   Sentiment Analysis (SA)

Researchers have been working on various recommendation algorithms based on text data supplied by internet users over the last couple of years. Zirn et al. [18] developed a completely automated system for fine-grained SA at the sub-sentence level, incorporating several sentiment lexicons, neighborhood links, and discourse linkages.

Appel et al. [19] established a hybrid strategy based on ambiguity management, semantic rules, and a sentiment lexicon using Twitter sentiment and movie review datasets. The authors evaluated the performance of their suggested hybrid system to that of conventional supervised algorithms such as Naive Bayes (NB) and Maximum Entropy (ME). The suggested approach outperforms supervised methods in terms of precision and accuracy. Similarly, Pang et al. [20] used unigram, bigram, and unibigram models to analyze IMDB movie review data SA. They classified the data into two classes using NB, ME, and SVM classifiers [21].

Sentiment analysis has improved over the years using several dictionary-based and machine learning algorithms to improve accuracy. Prior knowledge has also played a significant role in properly conveying the polarity of views with the emergence of deep learning algorithms in SA.

2.2 Complement Naïve Bayes Classifier

Naive Bayes (NB) is a well-known classification technique in data mining [22]. The NB classification model computes a class's posterior probability based on its posterior probability distribution using the word distribution in the text. The model is based on bag of words (BOWs) feature extraction, which does not consider the location of the word in the text. It makes use of the Bayes Theorem to forecast the likelihood that a given feature set belongs to a certain label in a given situation.

$$P(label|features) = \frac{P(label) * P(features|label)}{P(features)} \qquad (1)$$

where $P(label|features)$ quantifies the likelihood that a feature set is associated with a given label. $P(label)$ is the label's previous estimate. $P(label|features)$ denotes the likelihood that the provided feature set is associated with this label, and P(features) is the previous estimate for the occurrence of this particular feature collection. However, this categorization technique involves a basic assumption, namely that words in a review, category pair appear independently of other terms.

When dealing with discrete values, such as word counts, Complement Naïve Bayes classification is the best option. Further, we expect it to show the best accuracy and count the occurrences of a word in the complement to the class. The classification result is the class with the lowest sum of weights for each word in the message, which is represented by the class with the lowest value. The Bernoulli formula is very similar to the multinomial formula, with the exception that the input is a set of Boolean values (whether the word is in the message or not), rather than a set of frequencies. Consequently, the algorithm explicitly penalizes the absence of a feature (word in the message is absent from the vocabulary), whereas the multinomial approach makes use of the smoothing parameter for the values that are not present in the algorithm [23].

Lee et al. [24] developed a novel approach for determining feature weights in naive Bayesian learning by averaging the Kullback–Leibler measure across feature values. Additionally, they proposed a novel weight assignment paradigm for categorization learning, nicknamed the value weighting approach. Rather than weighting each attribute, they assign a different weight to each value.

In [25], the authors present a weighting attributes to Alleviate Naive Bayes' Independence Assumption (WANBIA), that optimizes the negative conditional log-likelihood or the mean squared error objective functions. Their experiment conduct rigorous analyses and demonstrate that WANBIA is a viable alternative to state-of-the-art classifiers such as Random Forest and Logistic Regression [26, 27].

2.3 Analysis Metrics

We used conventional performance indicators to evaluate our model's performance on IMDB datasets, notably the F1 and accuracy scores and their related class support divisions. Precision and recall are defined in Formula (2) and Formula (3) [28]. Moreover, accuracy and F1 are defined in Formula (4) and Formula (5) [29, 30].

$$Precision = \frac{TP}{TP + FP} \qquad (2)$$

$$Recall = \frac{TP}{TP + FN} \tag{3}$$

$$Accuracy score = \frac{TP + TN}{TP + TN + FP + FN} \tag{4}$$

$$F1 = \frac{2 x Precision x Recall}{Precision + Recall} \tag{5}$$

where True Positive (TP) is the number of reviews sorted properly into the appropriate sentiment classifications. Next, False Positive (FP) is the number of reviews assigned to an emotion class to which they do not belong. Hence, False Negative (FP) is the number of reviews labeled as not belonging to a sentiment category in which they really fit [31].

3 Methodology

3.1 Research Workflow

This section will describe our research workflow, as shown in Fig. 1. Furthermore, the IMDB dataset is gathered and imported. Pre-processing is applied to the data to eliminate noise and tidy it up before further processing. This research used the Beautiful Soup 4 library [32] for the scraping. Beautiful Soup is a Python library for pulling HTML and XML files data. Next, word embedding is a technique for expressing words in a low-dimensional space, most typically represented as real-valued vectors. It enables words with similar meanings and semantics to be expressed more closely together than words with less comparable meanings and semantics to be rendered farther apart. Therefore, the experiment implements the Term Frequency-Inverse Document Frequency model (TFIDF) to feature selection and extractions. Hence, the TFIDF vector converts text documents to a matrix of TFIDF features.

The following methods are used to resolve the classification problem: Logistic Regression (LR), Bernoulli Naïve Bayes (BernoulliNB), Complement Naïve Bayes (ComplementNB), and Linear Support Vector Machine (Linear SVM) with SGDClassifier and different loss type (*square_hinge, huber, modified_huber, log, epsilon_insensitive, perceptron, modified_huber*). Finally, the classification procedure is carried out, sentiment analysis will show the positive or negative result, and the methods used are analyzed.

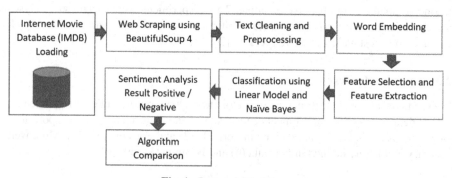

Fig. 1. Research Workflow.

3.2 Internet Movie Database (IMDB)

There are 25,000 training data, 25,000 test data, and 50,000 unlabeled data in the IMDB dataset [33]. The IMDB dataset is a binary sentiment classification dataset comprised of movie reviews extracted from the internet movie database [34]. Dataset training documents are highly polarized. There is a 1:1 ratio of negative to positive for labeled documents. The vectors for the documents are constructed using all of the documents in the collection, including train, test, and unlabeled data. Table 1 contains several samples drawn from the IMDB dataset.

Table 1. Examples of movie reviews for each class.

No	Review	Sentiment
1	*A wonderful little production. < br / > < br / > The filming technique is very unassuming- very old-time-BBC fashion and gives a comforting, and sometimes discomforting, sense of realism to the entire piece. < br / > < br / > The actors are extremely well-chosen- Michael Sheen not only "has got all the polari" but he has all the voices down pat too! You can genuinely see the seamless editing guided by the references to Williams' diary entries; not only is it well worth the watching, but it is a terrifically written and performed piece. A masterful production about one of the great masters of comedy and his life. < br / > < br / > The realism comes home with the little things: the fantasy of the guard, which, rather than using the traditional 'dream' techniques, remains solid then disappears. It plays on our knowledge and senses, particularly with the scenes concerning Orton and Halliwell, and the sets (particularly of their flat with Halliwell's murals decorating every surface) are well done*	Positive
2	*There's a family where a little boy (Jake) thinks there's a zombie in his closet & his parents are fighting all the time. < br / > < br / > This movie is slower than a soap opera... And suddenly, Jake decides to become Rambo and kill the zombie. < br / > < br / > OK, first of all when you're going to make a film you must Decide if it's a thriller or a drama! As a drama, the movie is watchable. Parents are divorcing & arguing like in real life. And then we have Jake with his closet, which ruins all the film! I expected to see a BOOGEYMAN similar movie, but instead, i watched a drama with some meaningless thriller spots. < br / > < br / > 3 out of 10 just for the well playing parents & descent dialogs. As for the shots with Jake: ignore them*	Negative

4 Experiment and Result

4.1 Experiment Results

The performance result of each classifier describes in Table 2. Our works analyze the classification report for the bag of words and the classification report for TFIDF features for all models. The following methods are used in our experiment: Logistic Regression (LR), Bernoulli Naïve Bayes (BernoulliNB), Complement Naïve Bayes (ComplementNB), and Linear Support Vector Machine (Linear SVM) with SGDClassifier and different loss type (*square_hinge, huber, modified_huber, log, epsilon_insensitive, perceptron, modified_huber*).

Linear classifiers, especially for SVM and logistic regression, were trained on SGD data using dense or sparse arrays of floating-point values as features. The loss parameter determines the appropriate model type, and by default, a linear support vector engine is used. A regularizer is a penalty applied to a loss function that lowers the model parameters to zero using either the squared Euclidean norm L2 or the absolute Euclidean norm L1 or a mixture of the two Elastic Nets. If the regularizer causes the parameter update to exceed 0.0, it is trimmed to 0.0 to facilitate sparse model learning and online feature selection. Our works explore the possible options of loss functions such as *square_hinge, huber, modified_huber, log, epsilon_insensitive, perceptron,* and *modified_huber.* When shuffle is set to True, the random state is used for shuffling the data. Pass an int for reproducible output across multiple function calls, and our works implement *random_state = 150*.

In the Linear classifiers SVM with SGD training group, SGDClassifiers with loss functions "log" achieves the best performance 75.07% accuracy in classification report for the bag of words. Furthermore, in the classification report for TFIDF features, SGDClassifiers with loss function "perceptron" exhibits the optimum accuracy of 74.99%. The minimum accuracy achieves when employing the loss function "huber." It gets a 51.04% accuracy classification report for the bag of words and a 50.07% classification report for TFIDF features.

Based on Table 2, ComplementNB achieves the highest accuracy, 75.13%, for both classification reports. ComplementNB was created to address the "severe assumptions" imposed by the ordinary Multinomial Naive Bayes classifier. It is especially well-suited for unbalanced data sets. Naive Bayes is often used as a baseline in text classification because it is fast and easy to implement [35]. Therefore, we apply the sklearn class as follows: *sklearn.naive_bayes.ComplementNB (*, alpha = 1.0, fit_prior = True, class_prior = None, norm = False).* On the other hand, BernoulliNB only achieves 54.71% accuracy for both classification reports. The statistical performance of Naïve Bayes classifiers shows in Table 3.

Table 2. Performance result of each classifiers.

Models	Classifiers	Loss	random_state	Classification report for the bag of words	Classification report for TFIDF features
SGD classifier	Linear Model	*squared_hinge*	150	0.7503	0.7496
SGD classifier	Linear Model	*Huber*	150	0.5104	0.5007

(*continued*)

Table 2. (*continued*)

Models	Classifiers	Loss	random_state	Classification report for the bag of words	Classification report for TFIDF features
SGD classifier	Linear Model	*modified_huber*	150	0.7503	0.7496
SGD classifier	Linear Model	*log*	150	0.7507	0.7456
SGD classifier	Linear Model	*epsilon_insensitive*	150	0.732	0.6038
SGD classifier	Linear Model	*squared_epsilon_insensitive*	150	0.7503	0.7496
SGD classifier	Linear Model	*perceptron*	150	0.7499	0.7499
SGD classifier	Linear Model	*hinge*	150	0.6116	0.5046
Logistic regression	Linear Model	-	-	0.7512	0.75
Bernoulli NB	Naïve Bayes	-	-	0.5471	0.5471
Complement NB	**Naïve Bayes**	-	-	**0.7513**	**0.7513**

Table 3. Statistic performance of Naïve Bayes classifiers.

Items	BernoulliNB svm_bow_score				ComplementNB svm_bow_score			
	Precision	Recall	F1-score	Support	Precision	Recall	F1-score	Support
Positive	0.52	1	0.69	4993	0.75	0.76	0.75	4993
Negative	0.97	0.1	0.18	5007	0.75	0.75	0.75	5007
Accuracy			0.55	10000			0.75	10000
Macro avg	0.75	0.55	0.43	10000	0.75	0.75	0.75	10000
Weighted avg	0.75	0.55	0.43	10000	0.75	0.75	0.75	10000
	svm_tfidf_score				svm_tfidf_score			
Positive	0.52	1	0.69	4993	0.75	0.76	0.75	4993
Negative	0.97	0.1	0.18	5007	0.75	0.75	0.75	5007
Accuracy			0.55	10000			0.75	10000

(*continued*)

Table 3. (*continued*)

Items	BernoulliNB svm_bow_score				ComplementNB svm_bow_score			
	Precision	Recall	F1-score	Support	Precision	Recall	F1-score	Support
Macro avg	0.75	0.55	0.43	10000	0.75	0.75	0.75	10000
Weighted avg	0.75	0.55	0.43	10000	0.75	0.75	0.75	10000

Figure 2 illustrates the confusion matrix for Linear Classifiers. In Fig. 2(a) our work implements SGDClassifiers with loss function = square_hinge. The TN value is 3851, FN value 1341, FP value 1156, and TP value 3652. Moreover, in Fig. 2(g), our works

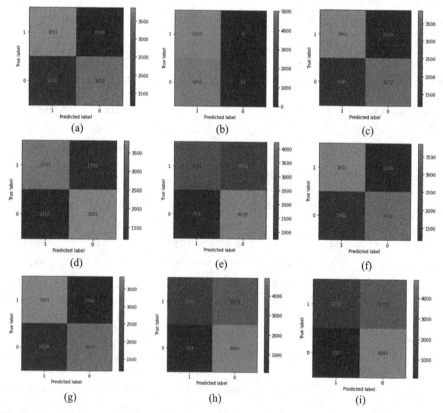

Fig. 2. Confusion matrix Linear Classifiers. (a) SGDClassifiers loss = square_hinge, (b) SGDClassifiers loss = huber, (c) SGDClassifiers loss = modified_huber, (d) SGDClassifiers loss = log, (e) SGDClassifiers loss = epsilon_insensitive, (f) SGDClassifiers loss = squared_epsilon_insensitive, (g) SGDClassifiers loss = perceptron, (h) SGDClassifiers loss = modified_huber = hinge, (i) Logistic Regression.

utilized SGDClassifiers with loss function = perceptron. The TN value is 3845, FN value 1339, FP value 1162, and TP value 3654.

The confusion matrix for the Naïve Bayes algorithm is described in Fig. 3. Figure 3(a) describe the confusion matrix Naïve Bayes classifiers for multivariate Bernoulli models. Therefore, the confusion matrix for complement naïve Bayes classifiers is shown in Fig. 3(b). The TN value is 1232, FN value 109, FP value 3775, and TP value 4884.

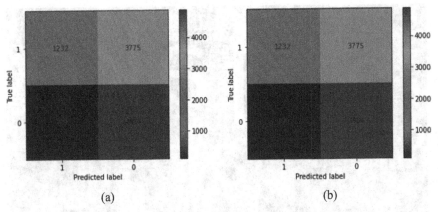

(a)   (b)

Fig. 3. Confusion matrix Naïve Bayes Algorithm (a) Naive Bayes classifier for multivariate Bernoulli models. (b) The Complement Naive Bayes classifier.

Figures 4 and 5 describe the positive and negative words using Word Cloud [36]. As we can see in Fig. 4, word cloud for the positive text includes fresh air, show, perform, origin, idea, drop, write, complete, etc. Our research utilized the function *WordCloud(width*

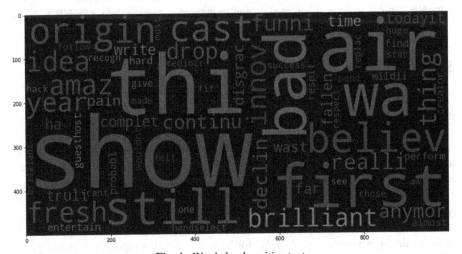

Fig. 4. Word cloud positive text.

= *1000, height = 500,max_words = 500,min_font_size = 5)*. The experiment shows the *positive_words* with *interpolation = 'bilinear'*.

Some examples of negative text generate by word cloud explain in Fig. 5. The negative text includes thriller, divorce, fight, parent, movie, descent, drama, kill, must, etc. Our experiment illustrates the *max_words = 500* and *negative_word* with *interpolation = 'bilinear'*.

Fig. 5. Word cloud negative text.

Some advantages of the Naive Bayes Algorithm are: (1). The NB method is efficient and can save a significant amount of time. (2). NB is well-suited for tackling challenges involving multi-class prediction. (3). If the premise of feature independence remains true, NB can outperform other models and require less training data. (4). NB is more appropriate to categories than to numerical input variables.

5 Conclusions

The study is based on the IMDB dataset, which comprises movie reviews and the positive or negative labels that are connected with them. We conduct a series of experiments to find the model with the highest accuracy and the greatest generality. The following classifiers are used in this work: Linear Model and Naïve Bayes. Different techniques, including Count Vectorizer, Term Frequency-Inverse Document Frequency model (TFIDF)Vectorizer, minimum-maximum number of words, and max features, are implemented. Besides, we explore the possible options of loss functions such as *square_hinge, huber, modified_huber, log, epsilon_insensitive, perceptron,* and *modified_huber*. Based on our experiment result, we can conclude: (1) ComplementNB achieves the highest accuracy, 75.13%, for both classification reports. (2) In the group of Linear classifiers SVM with SGD training, SGDClassifiers with loss functions "log" achieve the best performance 75.07% accuracy in classification report for the bag of words. (3) In the classification report for TFIDF features, SGDClassifiers with loss function "perceptron" exhibit the optimum accuracy of 74.99%.

We will explore the other loss function in future research to increase our performance results. We also want to combine sentiment analysis with Shapley Additive Explanations (SHAP) for explainable artificial intelligence (XAI).

Acknowledgment. This paper is supported by the Ministry of Science and Technology, Taiwan. The Nos are MOST-107-2221-E-324-018-MY2 and MOST-109-2622-E-324-004, Taiwan.

References

1. Kumar, S., Gahalawat, M., Roy, P.P., Dogra, D.P., Kim, B.G.: Exploring impact of age and gender on sentiment analysis using machine learning. Electronics. **9**(2), 374 (2020). https://doi.org/10.3390/electronics9020374
2. Kumar, S., Yadava, M., Roy, P.P.: Fusion of EEG response and sentiment analysis of products review to predict customer satisfaction. Inf. Fusion. **52**, 41–52 (2019). https://doi.org/10.1016/j.inffus.2018.11.001
3. Dewi, C., Chen, R.-C.: Human activity recognition based on evolution of features selection and random forest. In: 2019 IEEE International Conference System Man Cybernetics, pp. 2496–2501 (2019). https://doi.org/10.1109/SMC.2019.8913868
4. Kim, J.H., Kim, B.G., Roy, P.P., Jeong, D.M.: Efficient facial expression recognition algorithm based on hierarchical deep neural network structure. IEEE Access. **7**, 41273–41285 (2019). https://doi.org/10.1109/ACCESS.2019.2907327
5. Manek, A.S., Shenoy, P.D., Mohan, M.C., R, V.K.: Aspect term extraction for sentiment analysis in large movie reviews using Gini Index feature selection method and SVM classifier. World Wide Web **20**(2), 135–154 (2016). https://doi.org/10.1007/s11280-015-0381-x
6. Dos Santos, C.N., Gatti, M.: Deep convolutional neural networks for sentiment analysis of short texts (2014)
7. Pontiki, M., Galanis, D., Papageorgiou, H., Manandhar, S., Androutsopoulos, I.: SemEval-2015 Task 12: Aspect Based Sentiment Analysis (2015). https://doi.org/10.18653/v1/s15-2082
8. Dewi, C., Chen, R.-C., Yu, H., Jiang, X.: Robust detection method for improving small traffic sign recognition based on spatial pyramid pooling. J. Ambient. Intell. Humaniz. Comput. **12**, 1–18 (2021). https://doi.org/10.1007/s12652-021-03584-0
9. Cao, D., Ji, R., Lin, D., Li, S.: A cross-media public sentiment analysis system for microblog. Multimedia Syst. **22**(4), 479–486 (2014). https://doi.org/10.1007/s00530-014-0407-8
10. Ren, R., Wu, D.D., Wu, D.D.: Forecasting stock market movement direction using sentiment analysis and support vector machine. IEEE Syst. J. **13**(1), 760–770 (2019). https://doi.org/10.1109/JSYST.2018.2794462
11. Shapiro, A.H., Sudhof, M., Wilson, D.J.: Measuring news sentiment. J. Econom. **228**, 221–243 (2020). https://doi.org/10.1016/j.jeconom.2020.07.053
12. Abercrombie, G., Batista-Navarro, R.: ParlVote: a corpus for sentiment analysis of political debates (2020)
13. Dewi, C., Chen, R.C., Liu, Y.T., Jiang, X., Hartomo, K.D.: Yolo V4 for advanced traffic sign recognition with synthetic training data generated by various GAN. IEEE Access **9**, 97228–97242 (2021). https://doi.org/10.1109/ACCESS.2021.3094201
14. Chatterjee, S., Chakrabarti, K., Garain, A., Schwenker, F., Sarkar, R.: JUMRv1: a sentiment analysis dataset for movie recommendation. Appl. Sci. **11**(20), 9381 (2021). https://doi.org/10.3390/app11209381

15. Dewi, C., Chen, R.-C., Liu, Y.-T., Tai, S.-K.: Synthetic Data generation using DCGAN for improved traffic sign recognition. Neural Comput. Appl. 33(3), 1–15 (2021). https://doi.org/10.1007/s00521-021-05982-z
16. Chen, R.-C., Dewi, C., Zhang, W.-W., Liu, J.-M.: Integrating gesture control board and image recognition for gesture recognition based on deep learning. Int. J. Appl. Sci. Eng. 17(3), 237–248 (2020)
17. Dewi, C., Chen, R.-C., Jiang, X., Yu, H.: Deep convolutional neural network for enhancing traffic sign recognition developed on Yolo V4. Multimed. Tools Appl. 81, 37821–37845 (2022). https://doi.org/10.1007/s11042-022-12962-5
18. Zirn, C., Niepert, M., Strube, M., Stuckenschmidt, H.: Fine-grained sentiment analysis with structural features. In: Proceedings of 5th International Joint Conference National Language Process (2011)
19. Appel, O., Chiclana, F., Carter, J., Fujita, H.: Successes and challenges in developing a hybrid approach to sentiment analysis. Appl. Intell. 48(5), 1176–1188 (2017). https://doi.org/10.1007/s10489-017-0966-4
20. Pang, S., Lee, B., Vithyanathan, L.: Thumbs up? Sentiment classification using machine learning techniques. Proc. Inst. Civ. Eng. Transp. 172(2), 1–5 (2019)
21. Dewi, C., Chen, R.C.: Random forest and support vector machine on features selection for regression analysis. Int. J. Innov. Comput. Inf. Control 15(6), 2027–2037 (2019). https://doi.org/10.24507/ijicic.15.06.2027
22. Chen, S., Webb, G.I., Liu, L., Ma, X.: A novel selective naïve Bayes algorithm. Knowl. Based Syst. 192, 105361 (2020). https://doi.org/10.1016/j.knosys.2019.105361
23. Dewi, C., Chen, R., Liu, Y., Yu, H.: Various generative adversarial networks model for synthetic prohibitory sign image generation. Appl. Sci. 11, 2913 (2021)
24. Lee, C.H., Gutierrez, F., Dou, D.: Calculating feature weights in naive Bayes with Kullback-Leibler measure (2011). https://doi.org/10.1109/ICDM.2011.29
25. Zaidi, N.A., Cerquides, J., Carman, M.J., Webb, G.I.: Alleviating Naive bayes attribute independence assumption by attribute weighting. J. Mach. Learn. Res. 14, 1947–1988 (2013). https://doi.org/10.13039/501100000923
26. Dewi, C., Chen, R.-C., Hendry, Hung, H.-T.: Experiment improvement of restricted Boltzmann machine methods for image classification. Vietnam J. Comput. Sci., 8(3), 1–16 (2021). https://doi.org/10.1142/S2196888821500184
27. Dewi, C., Chen, R.-C., Tai, S.-K.: Evaluation of robust spatial pyramid pooling based on convolutional neural network for traffic sign recognition system. Electronics 9(6), 889 (2020). https://doi.org/10.3390/electronics9060889
28. Chen, R.-C., Dewi, C., Huang, S.-W., Caraka, R.E.: Selecting critical features for data classification based on machine learning methods. J. Big Data 7(1), 1–26 (2020). https://doi.org/10.1186/s40537-020-00327-4
29. Dewi, C., Chen, R.-C., Liu, Y.-T.: Wasserstein generative adversarial networks for realistic traffic sign image generation. In: Nguyen, N.T., Chittayasothorn, S., Niyato, D., Trawiński, B. (eds.) Intelligent Information and Database Systems. LNCS (LNAI), vol. 12672, pp. 479–493. Springer, Cham (2021). https://doi.org/10.1007/978-3-030-73280-6_38
30. Tai, S., Dewi, C., Chen, R., Liu, Y., Jiang, X., Yu, H.: Deep learning for traffic sign recognition based on spatial pyramid pooling with scale analysis. Appl. Sci. 10(19), 6997 (2020). https://doi.org/10.3390/app10196997
31. Dewi, C., Chen, R.-C., Yu, H.: Weight analysis for various prohibitory sign detection and recognition using deep learning. Multimed. Tools App. 79(43–44), 32897–32915 (2020). https://doi.org/10.1007/s11042-020-09509-x
32. Richardson, L.: Beautiful Soup Documentation Release 4.4.0 (2019)
33. Lakshmipathi, N.: IMDB Dataset of 50K Movie Reviews. Kaggle (2019)

34. Dew, C., Chen, R.C., Liu, Y.-T.: Taiwan stop sign recognition with customize anchor. In: ICCMS 2020, February 26–28, 2020 Brisbane QLD, pp. 51–55, Australia (2020)
35. Rennie, J.D.M., Shih, L., Teevan, J., Karger, D.: Tackling the poor assumptions of naive bayes text classifiers. In: Proceedings, Twentieth International Conference on Machine Learning, vol. 2 (2003)
36. Tessem, B., Bjørnestad, S., Chen, W., Nyre, L.: Word cloud visualisation of locative information. J. Locat. Based Serv. 9(4), 254–272 (2015). https://doi.org/10.1080/17489725.2015.1118566

Portfolio Investments in the Forex Market

Przemysław Juszczuk$^{(\boxtimes)}$ (iD) and Jan Kozak (iD)

Department of Machine Learning, University of Economics in Katowice,
1 Maja, 40-287 Katowice, Poland
{przemyslaw.juszczuk,jan.kozak}@ue.katowice.pl

Abstract. Investing in the forex market seems to be an especially challenging task due to the large variety of dependencies related to instruments. Among the crucial aspects that should be considered is the correlation between the currency pairs. In this article, we derive a general investing schema considering the signal generation based on the well-known classification methods and verify the quality of these signals with the idea of portfolio building. To do so, we derive a two-stage process, where the first stage is devoted to deriving the classifier capable of generating the trading signals on the forex market. We use the set of the most popular market indicators, and the decision about the potential buy (or sell) signal is dependent on the values of these indicators. Eventually, we derive the classifier in which quality is measured on the basis of accuracy, recall, and precision. Further, we use signals generated by the classifier to adjust the account balance of the decision-maker and estimate the relation between the quality of classification and the final account balance.

Experiments are performed using the trading system implemented by the authors on the real-world data covering several years.

Keywords: Financial data · Portfolio · Classification

1 Introduction

Investing in the forex market seems to be an especially challenging task – a decentralized global market with currency pairs as instruments open a wide range of opportunities for investors. Significant volatility, a variety of tools like technical analysis, fundamental analysis, and sentiment analysis with the possibility of trade 24 h/day, five days a week seems to be a competitive investment opportunity compared to the well-known stock market or bonds market. At the same time, it is still considered a relatively safe option compared to the cryptocurrency market, which is often related to a very high correlation between instruments.

The possibility of trading and earning profits on the market is based mainly on the tools emerging from technical and fundamental analyses. Significantly, technical analysis is still considered a critical trading tool among decision-makers [7]. Trading systems based on fuzzy sets [10], neural networks [22], text mining

© The Author(s), under exclusive license to Springer Nature Switzerland AG 2022
N. T. Nguyen et al. (Eds.): ACIIDS 2022, LNAI 13757, pp. 94–105, 2022.
https://doi.org/10.1007/978-3-031-21743-2_8

[17], tend to generate trading opportunities for a single currency pair. Even in the case where multiple instruments are taken into account simultaneously, it is still a relatively new field with high risk related to the strong correlation among the instruments. Examples of works extending the portfolio idea on the forex market are still difficult to find.

It is because classical financial models like Markowitz [15] or Merton [16] consider both expected mean and risk measured based on the correlation among instruments. Such correlation on the forex market is visibly higher than on the stock market, which can be a severe drawback. However, it is not clear how much impact such correlation could have on the overall results on the portfolio derived based on the set of currency pairs.

In this article, we try to fill the gap related to the portfolio problem on the forex market and move towards the investing process based on the set of instruments simultaneously, rather than consider a set of signals independently. To do so, we present the investing approach involving the classification methods used to generate signals. Further, these signals are considered simultaneously, leading to the situation that the decision-maker portfolio could include several currency pairs for a given time. Profit/loss from these currency pairs is adjusted to the decision-maker portfolio. However, we assume that no additional information about the correlation among instruments is considered.

Our main goal is to investigate whether the quality of signals classification performed on the data is directly related to the profits achieved by the decision-maker at the end of the investing period. Therefore, we implement a trading system including the signals classification module and investing module to verify that. Furthermore, we compare the classification quality performed on the well-known algorithms with the final account balance measured in dollars.

The article is organized as follows: in the next section, we briefly describe the works related to trading systems and the portfolio selection problem, mainly focusing on the forex market. The third section includes the proposed solution in which the overall flowchart of the proposed system and the investing process are described. The fourth section presents the experiments performed on the real-world data, while the last section concludes.

2   Related Works

Forex market (Foreign currency exchange) is the high-volatility market, where currencies are traded. There are numerous factors related to the present situation for the single currency, economic, political, and psychological factors that affect the current value of the currency pair [6]. Thus it is challenging to find an effective way to predict the future value and direction of the instruments on the forex market. There were numerous attempts related to the rule-based trading systems based on various market indicators like moving average [14] or Bollinger Bands [2]. However, due to the chaotic nature and high data volatility, many approaches combining the classical rules and technical indicators are combined with machine learning methods and optimization techniques. Examples of such works using the genetic algorithms can be found in [5]. Complex hybrid models combining the market indicators and machine learning techniques can be found in [18]. Comparative study for both: genetic algorithms and various methods from machine learning adapted for several different trading rules was presented in [3].

In general, a lot of articles are devoted to the use of neural networks as an element of trading systems on the forex. Examples including classical neural networks [4] and self-organizing maps [18]. A detailed survey of articles used for financial forecasting related to deep learning, in general, can be found in [20].

Relatively small number of articles is devoted to the fundamental analysis [11], text mining [17], and news analysis in general [12]. The last element is also connected with the sentiment analysis on the market. At the same time, a little place is devoted to the portfolio analysis on the forex market [1,19].

Numerous studies have described the profitability of technical analysis across several financial markets. There is no consensus related to the overall profitability of market indicators. An example of work pointing out the advantages of technical analysis and market indicators is [9]. Negative opinions about the efficiency of these tools were presented in papers like [21]. Despite diverse opinions, there is no doubt that the systems based on technical analysis and market indicators are very popular tools for practitioners.

3 Proposed Methodology

In this section, we describe the idea of our trading system based on the flowchart presented in Fig. 1.

Please note that the classification method and the investing algorithm described in this section can be freely modified and work independently from each other. Thus, the signal generation mechanism can be selected from the well-known methods from the literature or can be a simple approach based on a single technical indicator. Therefore, the overall system flowchart can be divided into four separate fragments:

- derive the data divided into learning and testing data;
- build a classifier based on the learning data;
- invest in the instruments from the testing data according to signals derived by the decision tree;
- update account balance according to the portfolio.

Our approach used the real-world data as a decision table with currency pair values and technical indicator values as conditional attributes. The decision class is represented by a discrete set including one of the following values: *STRONG BUY, BUY, WAIT, SELL, STRONG SELL*, where *BUY (STRONG BUY)* is the signal to open the position on the given currency pair, *SELL (STRONG SELL)* is the signal of closing the position (if the signal for this instrument was previously generated), while the *WAIT* is just a skip for the given instrument

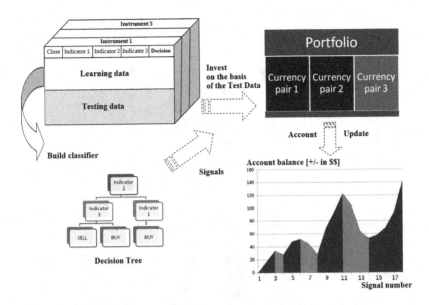

Fig. 1. The flowchart of the proposed system

at a time t. More details about the estimating signal value will be given further in this section.

In our system we use the following notations and concepts:

- t will be a discrete moment of time (reading) in which the instrument (currency pair) value and the market indicator values are calculated;
- T will be a time period, for which the whole investing process occurs. T consists of large number of t;
- CP will be a full set of currency pairs available in the system with cp_i as i-th currency pair;
- time period – will be a time, which has to pass between two successive readings;
- I – will be a set of indicators available in the system;
- PT – will be the portfolio of the decision-maker. This set is initially empty, however, the currency pairs cp are added to the portfolio, while the signals are observed on the market;
- c – will be the counter indicating the number of readings, for which the given cp is already in the portfolio;
- max – is the maximal number of readings, for which the cp_i can be present in the portfolio;
- p – is the number of readings which must pass to evaluate the decision for a given signal.

The general idea of the single indicator (example for a CCI indicator) is presented in Fig. 2. The trading rule for this particular indicator CCI, and currency pair cp_i in time t calculated on the past n readings is defined as follows:

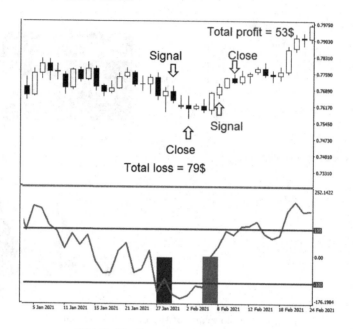

Fig. 2. Trading rule example – schema

$$f(CCI, cp_i, t) = true \ \ if \ \ CCI_n^l(t-1) < l_1 \ \ AND \ \ CCI_n^l(t) > l_1, \qquad (1)$$

where l_1 will be the predefined level for the particular indicator. In Fig. 2 this level can be observed as the blue bottom line in the lower part of the chart. The signal is generated when the indicator value (in two successive readings) crosses this predefined level. All indicators used in the research are based on the same idea where the signal is observed only if the value of the indicator crosses the predefined level. Whereas the decision for the data is calculated based on the formula:

$$decision = \begin{cases} STRONG\ BUY & \text{if } (price(t+p) - price(t)) > 3 \cdot \epsilon, \\ STRONG\ SELL & \text{if } (price(t+p) - price(t)) < 3 \cdot (-\epsilon), \\ BUY & \text{if } (price(t+p) - price(t)) > \epsilon, \\ SELL & \text{if } (price(t+p) - price(t)) < -\epsilon, \\ WAIT & \text{otherwise.} \end{cases} \qquad (2)$$

This considers the difference between the instrument's value for the time t, where the signal occurred, and the time after p readings. The difference between these two values indicates the decision class (BUY, if the difference was positive, $SELL$, is the difference was negative, and $WAIT$ if the price was within the ϵ range). Additionally, to initially clear the data for the classification process, the simple preprocessing schema was adapted. It covered the following steps:

- no empty values were observed in the data;
- all conditional attributes (except the Price) were discretized (classical interval discretization with the maximum of 20 intervals was performed);
- number of readings for all currency pairs was the same.

Especially the last condition will be crucial further in adapting the investing schema.

The preprocessed data acquired from the raw data is used as an input for the classification algorithm used to derive the decision for the system; we used the classical algorithm known from the literature – CART algorithm. One should know that the algorithm selection could visibly impact the results. Our initial assumption was to estimate the overall quality of the approach with the well-known approach from literature (without any additional modification). However, the selection of the algorithm should be investigated in further experiments.

Our goal in this trading system module is to initially estimate the quality of the classification based on the measures known from the literature. However, the classical confusion matrix used for the binary decision class is replaced here in such a way that a decision class including objects belonging to more than two classes is divided into subsets, and the following notation is introduced: TP – denotes all cases adequately classified to the selected class; TN – denoted all cases for which the proper assignment to the classes besides the selected class was made; FP – all cases incorrectly classified to the selected class; FN – all cases incorrectly classified to the classes besides the selected class. In addition, we used the following formulas:

$$acc(S) = \frac{TP + TN}{TP + TN + FP + FN},\tag{3}$$

where S is the selected dataset. For the precision measure we used the equation:

$$prec(S) = \frac{TP}{TP + FP},\tag{4}$$

while for the recall the following equation was used:

$$rec(S) = \frac{TP}{TP + FN}.\tag{5}$$

3.1 The Investing Process

Initially, acquired data is divided into two subsets. The first subset is used to derive the classifier, while the second subset measures classification quality. At the same time, the testing data will be used in the investing process (which will be performed simultaneously with the process of measuring the quality of classification). The schema for the investing process is presented in Algorithm 1.

The currency pairs can be added to the portfolio PT during the whole T. However, two different conditions should be satisfied – the decision for a given currency pair in reading t is BUY (or $STRONG\ BUY$), and the currency pair

---

Algorithm 1: The investing process

---

begin

1 | Create the empty set $PT = \emptyset$
2 | Select set of currency pairs CP and number of readings T
3 | Set additional parameters (number of decision classess and counter c)
4 | **for** *each reading t in T* **do**
5 | **for** *each cp_i in CP* **do**
6 | Set decision class for cp_i according to formula (2)
7 | **if** *decision for cp_i is (BUY OR STRONG BUY) AND $cp_i \notin PT$* **then**
   | add cp_i to PT
   | set counter $c = 0$ for cp_i
8 | **if** *decision for cp_i is (SELL OR STRONG SELL) AND $cp_i \in PT$* **then**
   | remove cp_i from PT
   | update the value of account in $
9 | **for** *each cp_i in PT* **do**
   | increase counter c
10 | **if** $c = max$ *for cp_i* **then**
   | remove cp_i from PT
   | update the value of account in $
11 | close all opened positions
12 | derive the final account balance to the decision-maker

---

is not in the portfolio at reading t. One should know that for any cp_i added to the PT, a counter c is set to 0. The counter c indicates the length (in a number of readings) for which the currency pair is already in the PT. We assumed that the instrument could be removed from the portfolio under the following circumstances:

- the decision for a currency pair in the portfolio is set to *STRONG SELL*;
- the maximal number of readings for the currency pair was achieved.

Eventually, after moving through all readings in T, all remaining transactions in the set PT are closed, the account balance is updated, and the final account balance is presented to the decision-maker.

In both cases, the current account balance measured in dollars is updated according to the following formula:

$$balance(t) = balance(t-1) + K \cdot (price(t+p) - price(t)), \qquad (6)$$

where $balance(t)$ is account value in dollars in reading t, $price(t+p)$ is the instrument value after p readings, while $price(t)$ is the instrument value in reading, where instrument was added to the portfolio PT. Finally, the K value is the constant value used to measure changes in dollars. For currency pairs with Japanese yen, K is set to 100, while for remaining currency pairs, it is set to 1000.

4   Numerical Experiments

In the experiments, we used 7 different currency pairs: AUDCAD, AUDCHF, AUDJPY, AUDNZD, AUDUSD, USDCAD, and USDCHF. For these instruments, we calculated values of six different market indicators: Bulls indicator, CCI (Commodity Channel Index), DM (DeMarker), OSMA (Oscillator of Moving Average), RSI (Relative Strength Index), and Stoch (Stochastic Oscillator). In Table 1 we can see summary including number of readings (size of T) for each dataset. One should know that there is no consensus about the length of the financial data, which should be considered in the literature. However, we selected the data covering a large period. Thus the obtained results could be (in some limited way) extrapolated to other periods. At the same time, we selected the D1 and H4 time windows, limiting the noise's impact on the obtained results. The time window selection is closely related to the presented approach, and it should be rather changed in the case with the small time windows (like 5 of 1 min per reading).

Table 1. Number of readings for datasets used in the experiments

Data set	All		Rising		Side	
Time window	D1	H4	D1	H4	D1	H4
No. of readings	2940	17666	827	4968	1499	8993

The whole experiment was divided into two separate phases. In the first phase, we perform the classification. However, we assumed that the training data would cover only 50% of all data; thus, it can be expected that overall results could be slightly worse than in the common division 70% training and 30% learning data. In the following section, we identify set 1 as the All data for the D1 time window, set 2 as the All data for the H4 time window. The number 3 and 4 are related to the data in the rising trend (Rising D1 and Rising H4, respectively), while the two last numbers 5 and 6 are included data for the consolidation (side trend) D1 and H4.

We performed the classification with the use of the CART algorithm, and the results for the aggregated data can be seen in Table 2. Thus both: *BUY* and *STRONG BUY* (the same occurs for the *SELL* and *STRONG SELL*) classes indicate the same price direction; we decided to aggregate this data. Please note that the results in the table represent the average classification efficiency for all instruments in the given dataset. One should remember that we used 50% of the data in the experiments as a testing set. The most important observation is the classification quality for the time window D1 (sets 1, 3, and 5), which are visibly higher than the remaining sets. It can be related to the fact that it is relatively easier to correctly classify the objects for a large time window (each new reading is derived every day), where the market noise is limited. However, since we aggregated the *BUY* and *STRONG BUY* (and the *SELL* and *STRONG SELL*) classes, presented results leave visible room for improvements. It is essential when we differentiate the results according to decision classes. In actual results, we focused instead on driving the decision about the general instrument value direction (BUY or $SELL$); however, from the point of view of the decision-maker, it could be essential to know the strength of the movement as well. Moreover, the number of decision classes was set arbitrarily. Still, it is possible to include more decision classes corresponding to the trend's strength (in such an example, the decision class depends directly on the range of the instrument value movement). In border cases, it is even possible to move towards the regression task, where instead of predicting the decision class, we rather expect the exact instrument value.

Table 2. Classification measures for all datasets ($-$ denotes the situation, for which no object was classified to this class). $S.\ BUY$ denotes the $STRONG\ BUY$ class, while $S.\ SELL$ denotes the $STRONG\ SELL$ class

Set	Precision			Recall			Accuracy
	BUY/S. BUY	SELL/S. SELL	WAIT	BUY/S. BUY	SELL/S. SELL	WAIT	
1	43.83%	46.50%	-	56.00%	50.67%	-	41.67%
2	25.75%	28.25%	24.75%	75.00%	10.25%	24.50%	25.50%
3	47.75%	47.75%	-	71.00%	40.25%	-	46.25%
4	28.67%	25.00%	20.67%	23.67%	44.50%	12.00%	27.00%
5	43.80%	41.00%	-	51.00%	56.50%	-	41.80%
6	27.00%	43.25%	19.25%	26.75%	35.33%	32.50%	25.00%

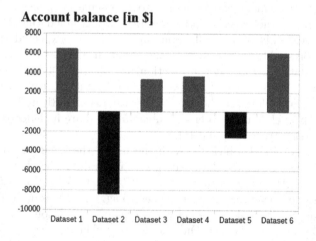

Account balance [in $]

Fig. 3. Preliminary results for the account balance calculation

The second part of the experiments is focused on the aspect related to the overall profit achieved in the decision-maker's account balance at the end of

Portfolio size

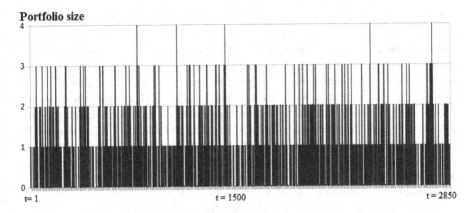

Fig. 4. Example portfolio composition variability over time

the investing period. These results are presented in Fig. 3. One should know that despite the overall good results (positive profit in four out of six) datasets, these results are ambiguous in connection to the classification results. First of all, the experiment design is somewhat limited, and some strict assumptions have been made. The most important observation is the fact that we do not have a clear answer about the impact of the classification quality on the account balance. It seems that the correlation between these two aspects is limited, and overall good classification quality does not positively impact the account balance. Moreover, presented classification results are here somehow averaged (average classification value overall used currency pairs). Thus there can be cases where a single currency pair draws the whole portfolio towards the positive account balance.

Additionally, we checked the variability of the portfolio (in terms of the size of the portfolio) for the example dataset. We selected four currency pairs. Thus the maximal portfolio size for this example is four. These results are presented in Fig. 4. It can be seen that the maximal portfolio size equal to 4 was observed only for several readings. Moreover, it lasted relatively short. Thus it can be assumed that the portfolio composition variability is very high.

5 Conclusions

In this article, we proposed the idea of the trading system, in which the trading signals were derived based on the classifier (decision tree). At the same time, instruments were added to the portfolio according to these signals. It leads to introducing the portfolio on the forex market. First, we presented the flowchart of the trading system, in which the data was divided into two sets. The first set derived the classifier, while the second test is used as an input for the portfolio module. Eventually, the account balance (measures in US dollars) was derived at the end of the investing process.

Our main goal was to investigate if there is a strict relation between the quality of classification introduced by the decision tree and the final account balance presented to the decision-maker. We used a large set of real-world data, including data in rising trends (bull market, where instruments values are rising) and the side trend (consolidation, where no dynamic moves are observed). In addition, two different time windows were used: H4 and D1. Despite the diversity of data used in the experiments, results were ambiguous, and no direct relation between the quality of classification and the account balance was observed. It concludes that these two problems (data classification and portfolio management on the forex market) should be considered independently. Thus the relatively good classification of the data does not imply the overall positive account balance at the end of the investing period.

As stated in the experimental section, the analyzed classification problem could be easily transformed into a regression case. Instead of predicting the decision classes, we rather focus on deriving the exact value of the instrument. The common assumption in the investing field is to know the general direction of the instrument value; however, for ongoing transactions learning the exact instrument value range could be vital.

References

1. Amiri, M., Zandieh, M., Vahdani, B., Soltani, R., Roshanaei, V.: An integrated eigenvector-DEA-TOPSIS methodology for portfolio risk evaluation in the FOREX spot market. Expert Syst. App. **37**, 509–516 (2010)
2. Bollinger, J.: Bollinger on Bollinger Bands, McGraw-Hill, (2002)
3. de Brito, R. F. B., Oliveira, A. L. I.: Comparative study of forex trading systems builtwith SVR+GHSOM and genetic algorithms optimization of technical indicators, in: Proceedings of the 2012 24th IEEE International Conference on Tools withArtificial Intelligence, IEEE, pp. 351–358, (2012)
4. Carapuco, J., Neves, R., Horta, N.: Reinforcement learning applied to Forex trading. Appl. Soft Comput. **73**, 783–794 (2018)
5. Deng, S., Yoshiyama, K., Mitsubuchi, T., Sakurai, A.: Hybrid method of multiple kernel learning and genetic algorithm for forecasting short-term foreign exchangerates. Comput. Econ. **45**, 49–89 (2015)
6. Edwards, D.: Risk Management in Trading: Techniques to Drive Profitability of Hedge Funds And Trading Desks, Wiley Finance (2014)
7. Gehrig, T., Menhhoff, L.: Extended evidence on the usage of technical analysis in foreign exchange. Int. J. Fin. Econ. **11**(4), 327–338 (2006)
8. Hryshko, A., Downs, T.: System for foreign exchange trading using genetic algorithms and reinforcement learning. Int. J. Syst. Sci. **35**(13–14), 763–774 (2004)
9. Hsu, P.-H., Taylor, M.P., Wang, Z.: Technical trading: is it still beating the foreign exchange market? J. Int. Econ. **102**, 188–208 (2016)
10. Juszczuk, P., Kruś, L.: Soft multicriteria computing supporting decisions on the Forex market. Appl. Soft Comput. **96**, 106654 (2020)
11. Kaltwasser, P.R.: Uncertainty about fundamentals and herding behavior in the FOREX market. Phys. A Statist. Mech. App. **389**(6), 1215–1222 (2010)
12. Kocenda, E., Moravcova, M.: Intraday effect of news on emerging European forex markets: an event study analysis. Econ. Syst. **42**(4), 597–615 (2018)

13. Korczak, J., Lipinski, P.: Evolutionary building of stock trading experts in a real-time system. In: Proceedings, 2004 Congress on Evolutionary Computation, pp. 940–947. IEEE (2004)
14. Larsen, F.: Automatic stock market trading based on technical analysis. Master's thesis, Norwegian University of Science and Technology (2007)
15. Markowitz, H.: Portfolio selection. J. Finan. **7**(1), 77–91 (1952)
16. Merton, R.: An analytic derivation of the efficient portfolio frontier. J. Finan. Quant. Anal. **7**(4), 1851–1872 (1972)
17. Nassirtoussi, A.K., Aghabozorgi, S., Wah, T.Y., Ngo, D.C.L.: Text mining of news-headlines for FOREX market prediction: a multi-layer dimension reduction algorithm with semantics and sentiment. Expert Syst. App. **42**(1), 306–324 (2015)
18. Ni, H., Yin, H.: Exchange rate prediction using hybrid neural networks and trading indicators. Neurocomputing **72**, 2815–2823 (2009)
19. Petropoulos, A., Chatzis, S.P., Siakoulis, V., Vlachogiannakis, N.: A stacked generalization system for automated FOREX portfolio trading. Expert Syst. App. **90**, 290–302 (2017)
20. Sezer, O.B., Gudelek, M.Y., Ozbayoglu, A.M.: Financial time series forecasting with deep learning?: a systematic literature review: 2005–2019. App. Soft Comput. **90**, 106181 (2020)
21. Shynkevich, A.: Predictability in bond returns using technical trading rules. J. Bank. Finan. **70**, 55–69 (2016)
22. Yao, J., Tan, C.L.: A case study on using neural networks to perform technical forecasting of forex. Neurocomputing **34**, 79–98 (2000)

Detecting True and Declarative Facial Emotions by Changes in Nonlinear Dynamics of Eye Movements

Albert Śledzianowski[1]([✉]), Jerzy P. Nowacki[1], Andrzej W. Przybyszewski[1], and Krzysztof Urbanowicz[2]

[1] The Faculty of Information Technology, Polish-Japanese Academy of Information Technology, Warsaw 02-008, Poland
albert.sledzianowski@gmail.com, przy@pjwstk.edu.pl
[2] Move2Mind, Boston, MA, USA

Abstract. In our previous work we have showed that we can improve classifications of facial emotions (FE) by extending a dataset with chaotic dynamics parameters of eye movements (EM). This time we wanted to confirm our results using public and independently created video sources and for this purpose we chose the Affectiva-MIT Facial Expression Dataset (AM-FED). Our purpose was to find out whether we can estimate Happiness through non-linear dynamics of EM also in independent video data. We have calculated EM chaotic dynamics in video recordings of the AM-FED database and performed estimations of Happiness calculated with parameters provided by the Open Face library (OF). We also calculated correlation between these parameters and parameters attached to the AM-FED database using our own method based on sliding windows and proposed a method of using its output parameter with a short algorithm. We have observed that true Happiness was connected to a moderate value of negative correlation with EM chaotic dynamics in the case when smile was not present, while for declarative "Smile" parameters we observed a moderate positive value. By using EM chaotic dynamics correlation we have estimated the difference between posed smiles and true Happiness with the XGBoost classifier, with accuracy results of 0.75 (ROC-AUC 0.9) and precision of 0.8 (tested with dataset of 0.33). We are proposing EM chaotic dynamics parameters as an extension for estimations of Happiness based only on facial muscles activity. We think that this approach can confirm the authenticity of Happiness in various cases and also introduce the distinction between real and declarative FE into Computer Vision. It also can bring solution in cases when lower part of the face is hidden, i.e. when is covered by a protective mask.

Keywords: Machine learning · Eye moves · Smile · Happiness · Facial emotions · Mimicry · Emotion authenticity · Noise · Chaos · Nonlinear dynamical systems · Eye move chaosity · Eye movement chaosity

N. T. Nguyen et al. (Eds.): ACIIDS 2022, LNAI 13757, pp. 106–116, 2022.
https://doi.org/10.1007/978-3-031-21743-2_9

1   Introduction

There is lack of publications related to the EM chaotic dynamics during different emotions, especially in the context of their use in estimation of the facial emotions. This is the continuation of our research on the complex dynamical system describing EM presented in our previous article [1], where we have proved a positive, statistically significant correlation between the value of EM chaotic dynamics and the intensity of the Happiness (as FE).

In our previous experiment we have simultaneously recorded FE and EM in 49 subjects and analyzed responses to the video stimuli with Happiness and Contempt [1]. We have described EM by nonlinear dynamical system and observed that Happiness was positively correlated with EM chaotic behavior, while both the linear and the noise components were mostly negatively correlated with this FE [1]. In the case of our data, we have observed that while Happiness intensity increases, the EM becomes more chaotic and behaves less noisy [1].

As a continuation of our research project, we wanted to test our findings on public domain videos with FE classified by Facial Action Coding System (FACS) [2]. For this purpose we have chosen the "AM-FED" database created by researchers of the Affectiva Inc, the MIT Media Lab and the Robotics Institute of Carnegie Mellon University, providing both human ("Smile") and algorithmic ("Smile V2") types of classification, showing the faces of people watching amusing video clips [3]. We wanted to compare the differences in EM chaotic dynamical properties between people who show the true (Happiness) or only the declarative FE (Smile). We have used the OpenFace (OF) library for Action Units (AU) estimations and for estimations of gaze vectors used to calculate the chaotic dynamics of the EM [4]. We used the FACS algorithm for Happiness estimations in the AM-FED videos. For Smile parameters, estimations have been provided with this database [3].

Seeing is an active process. Our eyes are continuously moving and searching the environment in order to capture interesting objects within the fovea which is the high resolution region of the retina. As these interactions are complex, EM must response with fast complex movements. Therefore, describing the EM behavior as a nonlinear dynamic system seems natural. The EM changes may have different attractors that determine their behavior in time. As we have previously demonstrated, in natural conditions EM behaviors are dominating by chaos or by noise. Chaotic behaviors were described in different parts of the human body: in the heart rate [5], in the brain activity recorded by EEG and described by fractal dimension or in the gait kinematics of healthy subjects that have chaotic properties [6] which are lost in Parkinson's diseases [7]. However, still not many researchers describe EM by nonlinear dynamics, even if processes in the retina show a complex, multi-attractor behaviors [8].

E. Paulson discussed the reading processes and concluded that it can be described as a self-similar, nonlinear dynamical system dependent on the reader and text characteristics that can be explained by the chaos theory principle. Chaotic dynamics can explain difficulty in predicting the nature of a reader's eye movement regressions [9]. K.M.Hampson and E.A.H. Mallen observed that the eye aberration dynamics are chaotic and they characterized it by using techniques from the chaos theory by measuring the monochromatic aberration [10]. W. Richards et al. have used the EM data collected from perceptual tasks: binocular rivalry, fixation patterns during search, simple multi-stable percepts, and perceptual segments in several movies suggested that the mechanisms underlying our percepts might be modeled as nonlinear, deterministic systems that exhibit chaotic behavior. The eye scanning strategy appears to be controlled by nonrandom, dynamical spatial representation, but not a temporal one [11]. Harezlak et al. classifies EM as a signal exposing features characteristic of the chaotic natures, simultaneously emphasized that obtaining confidence in differentiating chaotic and noise behaviors requires the application of various approaches [12]. S. S. M. Chanijani analyzed 3 types of students and concluded that information entropy of EM is higher for novice people. C. Astefanoaei et al. found properties of chaos in the saccadic EM temporal series collected from a healthy subjects by estimating the correlation dimensions [13, 14]

In our work, we have compared results of Smile and Happiness classification in context of the EM chaotic dynamics. We propose to use the EM chaotic parameters as an additional data in the method of FE classification. In particular, we were interested if EM chaotic dynamics can improve estimations of authenticity of expressed Happiness and if it can help to solve the common mistakes in the automated FE classifications, like mistaking a worry face of pain with a smile.

2   Methods

The AM-FED database consists of 242 facial videos (168,359 frames) recorded by web-application at the frequency of 14 FPS and with the resolution of 320×240 in a real user's conditions. The viewers watched one of three intentionally amusing videoclips [3]. All AM-FED's videos were estimated by certified FACS coders and by automated facial expression analysis [3]. Each video was labeled frame-by-frame for the presence of AUs, by at least 3 FACS trained specialists (chosen from 16) [3]. The AM-FED coders labeled activity of AU2, AU4, AU5, AU9, AU12, AU14, AU15, AU17, AU18 and AU26. This database also provides its own index for the "Smile" and according to the authors, smiles were labeled and in this dataset are distinct from the labels for AU12. The AU12 may occur also in an expression that would not necessarily be given the label of smile (e.g. a grimace) and this is why they labeled for presence of the "smile" rather than AU12 [3]. Researchers agreed that AU activity is present in case of consent of over 50% of the coders and assumed that a label is not present if 100% agreed (the mean percentage agreement across classifications was 0.98) [3]. This database also includes the results of the baseline estimation for smile ("smile V2") using an algorithm based on landmarks detection and tracking, Histogram of Orientated Gradients computed for the face regions and classification based on the Support Vector Machine with the RBF kernel [3].

As previously mentioned, we used the OF library to extract from AM-FED videos the information about eye directions expressed as the gaze vectors (x,y,z) and also to detect, set and track landmarks describing the shape of face elements in the video and evaluate the presence and the intensity of specific AU [1,4].

We used this data to estimate Happiness, first by calculating its presence (HP) as:

$$HP = P(AU06) \cdot P(AU12) \tag{1}$$

and next, for non-zero presence outputs calculating its intensity (HI) as:

$$HI = \frac{I(AU06) + I(AU12)}{2} \cdot HP \tag{2}$$

where P stands for a function which expresses the presence of emotion in AU, I - function which expresses intensity of given AU, n is the number of AU's defined for happiness.

Our methods of the EM chaotic dynamical analysis described in our previous article calculates 3 parameters: chaos, noise and linear [1]. The EM chaotic dynamics parameters were calculated in the window which size was determined experimentally and then we calculated Pearson's correlation coefficient for the AM-FED's "Smile", "smile V2" and for "Happiness. We used a simplified model to calculate the noise level, which does not require correlation entropy [15]. We basically counted the average line in the recurrence diagram, which must be greater than 1. In our case, the noise level was in the range of NTS = up to 50% i.e. the standard noise deviation is 50% of the standard deviation of the data. From the equation in previous noise estimation article [15], to estimate 50% of the NTS, we need an average line of length <n> and from this we calculate minimal number of data Min(no-data) as stated in Eq. 3.

$$\langle n \rangle (NTS = 0.5) = \frac{2 - 0.5^p}{1 - 0.5^p} = \frac{2 - 0.5^{0.3441717}}{1 - 0.5^{0.3441717}} = 5.71$$

$$Min(no_{data}) = e^{5.71} = 301 \tag{3}$$

From the Eq. 22 presented in the article on Noise-level estimation we can calculate the minimum value of parameter p = 0.3441717, so the minimum amount of data to estimate 50% of the noise level is 301 items [15]. The standard deviation and the mean change over time in our non-stationary data, thus we needed the window as small as possible but larger than 301, so that the stationarity within the window will be maintained to minimize the error.

We also wanted to keep the window data as stationary as possible, because the noise estimation error is also due to the non-stationarity. In order to reveal the optimal window size we calculated Noise, Linear and Chaos separately for different widths and the optimal window size of the stationary data was around 400 items. In our previous article we performed analysis to make sure that the number of points is not too small, thus we calculated cross-correlations to different reference window widths between 300 and 1000 items. After this analysis we concluded that the window of 300 points was too small, but width of 400 brought statistically significant results and showed the optimal reference point for the entire estimation in the center of the window [1].

In our dataset we also wanted to include different correlations and provided it by recreating correlation data through the dependency between EM chaotic dynamics and chaos means shifted to different positions in the window.

3   Results

As previously mentioned, we calculated the EM chaotic dynamics parameters basing on gaze vectors time series obtained from the AM-FED videos. Then we calculated the Pearson's correlation coefficient between EM chaotic dynamics parameters shifted to the center of the window and AM-FED's "Smile", "Smile V2" and "Happiness" calculated on the basis of the OF's AU classifications. Figure 1 presents results of this correlation calculations performed on averaged data from AM-FED videos, where the p-value of correlation calculated for a particular data was <0.05. For Smile it was 161 AM-FED videos, for "Smile V2" 240 and for Happiness 190. The total number of all video frames with calculated p-value <0.05 was 317783. As we can see on both Figs. 1a and 1b a smile is accompanied by a positive correlation with the chaotic EM, while in the case of Happiness we can see negative correlation (while "Smile" and "Smile

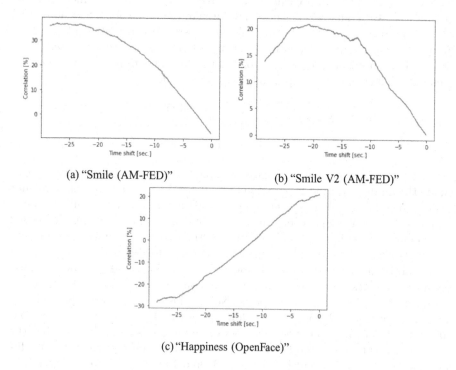

(a) "Smile (AM-FED)"   (b) "Smile V2 (AM-FED)"

(c) "Happiness (OpenFace)"

Fig. 1. Average correlations of EM chaotic dynamics

V2" are simultaneously not present). Because we assumed that both "Smile" parameters could simultaneously occur with Happiness, we calculated this factor and speculated that Happiness could occur in 65% of smile frames (73008 of AU06's non-zero intensity for 111708 non-zero intensity data frames of AU12).

Just to remind, according to the AMFED database authors, smiles are distinct from the labels of AU12, as AU12 may occur also in other expressions, like a grimace. For both Smile and Happiness the opposite direction is visible when analyzing the correlation results with the chaotic EM (while maintaining similar low-average levels). In our opinion, this parameter distinguishes between posed smile and true FE of Happiness. The chaotic EM is visible for all data positively correlated with Happiness, negative correlation is only visible for Happiness not confirmed by results of "Smile" estimations.

This observation was earlier confirmed with very similar positive statistically significant correlations levels [1]. Therefore, we propose to introduce the chaotic EM parameters into the FACS-based automated methods of Happiness estimation along with a decision mechanism that would determine whether the intensity of AU06 and AU12 can be classified as true or not. The resolving mechanism could be based on short algorithm realizing the following formula:

$$[AU12]Activity = TrueSmile = \\ [AU06]Intensity \cdot [Correlation(EyeChaos, [AU06]Intensity) > 0] \quad (4)$$

Figure 2 presents proposed model modifying to the standard FACS approach by adding parameters of chaotic EM to the estimation of the emotion. The final decision should be based on previously described window size estimation, method of windows centring and calculation of correlation coefficient between EM chaotic dynamics and intensity of Happiness.

We tested this new approach on a simplified dataset and for this purpose we used both Smile parameters and additionally: happiness intensity, centered EM chaos mean and EM chaos mean shifted by different thresholds (including different multiplications as described in the "Methods"). We normalized dataset with the min-max approach and as classification target we chose "Smile" or "Smile V2" binarized for zero and non-zero intesities. We tested estimations on our data with the XGBoost classifier, which might be described as a decision tree with gradient boosting optimization [16].

When we were building the dataset for estimations, we first checked if classification of smile parameters by Happiness data was random and after this confirmation, we started to add additional EM chaos parameters. Interestingly, when we started to do it, the accuracy started then to rise to a level of 0.78. Fig3 presents confusion matrixes for both types of smile estimation and for both types of dataset (with and without the chaotic dynamics data).

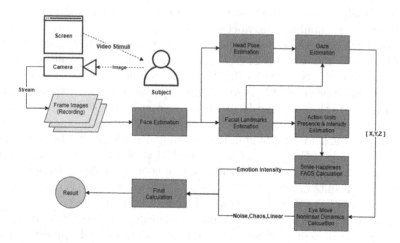

Fig. 2. Proposed improvement to Happiness estimation model basing on FACS and EM chaos calculations

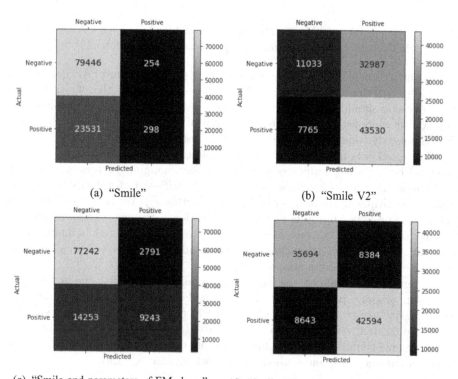

(a) "Smile"

(b) "Smile V2"

(c) "Smile and parameters of EM chaos"

(d) "Smile V2 and parameters of EM chaos"

Fig. 3. Confusion matrixes.

For a test dataset of 0.33 of the data and the XGBoost's classification we achieved "Smile" accuracy of 0.76 (ROC-AUC = 0.86) with precision of 0.44 and "Smile V2" accuracy of 0.74 (ROC-AUC = 0.93) with precision of 0.80.

The confusion matrixes show a significant difference between the results of classification of "Smile" and "Smile V2". We think that it is easier for an algorithm to estimate the results of another algorithm and this could be a reason for this differences. The results of human analysis could be more subjective and based on observations that are intuitive for humans, including complexity of negotiations carried out by human coders. This aspects could make classification more difficult for a machine learning methods, using this set of data.

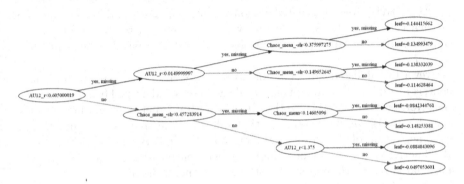

Fig. 4. The plot of decision tree for "Smile" attribute with Xgboost's regression and 2 levels of depth.

Figure 4 presents the decision tree generated for the "Smile" attribute with the Xgboost regression (XGBRegressor) with 2 levels of depth and the dataset used for classifications. We used different methods for visualisation, but we think that regression have a greater degree of freedom and thus are more reliable and conveys the best tree information. As we can see, two parameters are taking major part in the estimation process: the mean of EM chaos shifted to the window center ("Chaos_meansh") and the intensity of AU12 ("AU12_r"). However, at the first tree node the AU12 is taken into account only at very low intensity (<0.01) and later correlation between 'Chaos_meansh' and "Smile" is well visible. If 'Chaos_meansh' is small, then "Smile" is also low and if is greater than the threshold, the "Smile" value is greater too. From this decision decomposition we can see that EM chaos plays a more important role than the intensity of AU12.

4 Discussions

It should be noted, that both Smile and Happiness were estimated from the same video recordings, but partly from different time series when only AU12 (Smile) or both AU06 and AU12 are active (Happiness). Stewart, Bucy, and Mehu by using FACS Action Units definitions differentiate between Posed Smile (also called "non-felt" or false when only AU12 is active - lip corners pulled up and at an angle only) and true Happiness or Enjoyment (also called "felt" or "Duchenne") when both AU12 and AU06 are active (lip corners pulled up and at an angle, muscles surrounding the eyes contracted) [17]. We think that this approach has a reference to the differences between AM-FED smile parameters and Happiness calculated over the OF's estimations.

Happiness, besides the AU12 activity, is additionally accompanied by activity of the AU06. At the same time, this activity (AU06) seems to be connected to the change in correlation (from negative to positive) between the EM chaotic dynamics and the intensity of this emotion.

Researches so far differentiated between "Smile" and "Happiness" through the AU activity and postulated various emotional bases and social functions for each type of smile. As smiles are varying in relation to the social context, they have different consequences for observers and different diversity of expression [17].

Stewart et al. listed 5 different kinds of expressions: posed, enjoyment, amusement, controlled and contempt smiles. They might be identified by morphological characteristics of the face, like the direction of lip corner pulling, different muscular "controls" in the lower face that influence the shape of the mouth and by co-activation of muscles surrounding the eyes [17]. By distinguishing in the AU06 activity, authors found differences between posed smile and face expression related to the Happiness. According to authors, the most prominent signal of smiling is the pulling of lip corners, prototypically up and at an angle by the zygomatic muscle (AU12), but the activation of the orbicularis oculi muscle (AU6) is the most common facial component that could reflect the pleasantness of an emotional experience as well as its intensity and authenticity [17].

Posed smiles occur when an individual attempts to, either just signal positive emotion or to mask other emotions. In both cases the AU06 is generally not contracted [17]. In contrast, enjoyment smiles in addition to activation of AU12, activates also the upper face AUs and this way regulates the eye aperture and reinforces the impression in the recipient, that the smile was "felt". This is why enjoyment smiles have been strongly associated with feelings of amusement and happiness, as well as behaviors like cooperation. It might lead to "facial feedback" as the attribution of emotion by a viewers will be affected by specific facial display morphology [17].

We might say, that posed smiles might not be as strongly associated with the felt emotion of happiness as enjoyment smiles. In our study, we found that they are additionally accompanied by a change in the correlation between the chaotic EM and estimated emotion intensity. Perhaps this reaction has a communication dimension of social signaling. It might be related to the expectation of feedback on the observer's face, which is occurring by searching the view field and thus, in a change in the dynamics of EM. Here, facial expressions connected to Happiness should be interpreted as displaying different levels of emotional information, on the other hand, posed smiles seem to be a lighter signals, not triggering a feedback reaction in our social and emotional system.

We think that such observations could be also used in medical analysis. One example could be for Bradykinesia with Parkinson's disease. The Bradykinesia can be defined as slow motions and is often characterized by difficulties in initiating, maintaining or voluntarily synchronizing a motions [18]. One of the most difficult aspects of the Bradykinesia, is the fact that it ultimately affects all striated muscles [18]. Therefore, as the disease progresses, people with PD experience chewing difficulties, dysphagia, and difficulty speaking and expressing their faces [18]. Such a person can give the impression of being stiff and wooden-faced. In this regard, we think that the analysis of the EM chaotic dynamics could provide be a support in the interpretation of facial emotions.

5   Conclusions

Happiness allows individuals to build and strengthen social connections and facial expressions are very important features of communication for most of the primates [19,20]. The humans through their social and emotional abilities are excellent in distinguishing between posed and true Happiness, but existing methods in computer science often are wrong and rather unable to define the real FE. The method presented in this article uses EM chaotic dynamics correlation together with the FACS estimations, which can bring significant support and improvement in these problematic processes. We see the sense of further research in this area, as it can help in FE classification of people with mimicry problems, like Parkinson's disease patients with facial effects of the Bradykinesia ("poker face") or just people having problems with proper Happiness mimicry. This method can also bring solution in Happiness detection for partially covered face i.e. covered by mask worn during the epidemic).

Declaration of Competing Interest
OpenFace 2.0 Facial Behavior Analysis Toolkit contributed to the development of the software methods used in the analysis presented in this article. The authors declare no conflict of interests.

References

1. Sledzianowski, A., et al.: Face emotional responses correlate with chaotic dynamics of eye movements. In: Procedia Computer Science, vol. 192, pp. 2881–2892 (2021). Knowledge-Based and Intelligent Information And Engineering Systems: Proceedings of the 25th International Conference KES (2021)
2. P. Ekman and W.V. Friesen. Facial Action Coding System. Consulting Psychologists Press, Sunnyvale (1978)
3. McDuff, D., el Kaliouby, R., Senechal, T., Amr, M., Cohn, J.F., Picard, R.: Affectiva-MIT facial expression dataset (AM-FED): naturalistic and spontaneous facial expressions collected "in-the-wild". In 2013 IEEE Conference on Computer Vision and Pattern Recognition Workshops, pp. 881–888 (2013)
4. Baltrusaitis, T., Zadeh, A., Lim, Y.C., Morency, L.-P.: OpenFace 2.0: facial behavior analysis toolkit. In: 2018 13th IEEE International Conference on Automatic Face Gesture Recognition (FG 2018), pp. 59–66 (2018)

5. Glass, L.: Synchronization and rhythmic processes in physiology. Nature **410**, 277–284 (2001)
6. Piórek, M., Josiński, H., Michalczuk, A., Świtoński, A., Szczesna, A.: Quaternions and joint angles in an analysis of local stability of gait for different variants of walking speed and treadmill slope. Inf. Sci. **384**, 263–280 (2017)
7. Afsar, O., Tirnakli, U., Marwan, N.: Recurrence quantification analysis at work: Quasi-periodicity based interpretation of gait force profiles for patients with parkinson disease. Sci. Rep. **8**, 1–12 (2018)
8. Przybyszewski, A.W., Linsay, P.S., Gaudiano, P., Wilson, C.M.: Basic difference between brain and computer: integration of asynchronous processes implemented as hardware model of the retina. IEEE Trans. Neural Netw. **18**, 70–85 (2007)
9. Paulson, E.J.: Viewing eye movements during reading through the lens of chaos theory: how reading is like the weather. Read. Res. Q. **40**(3), 338–358 (2005)
10. Hampson, K.M., Mallen, E.A.H.: Chaos in ocular aberration dynamics of the human eye. Biomed. Opt. Express. **3**(5), 863–877 (2012)
11. Richards, W., Wilson, H.R., Sommer, M.A.: Chaos in percepts? Biol. Cybern. **70**, 345–349 (2004)
12. Harezlak, K., Kasprowski, P.: Searching for chaos evidence in eye movement signals. Entropy **20**(1), 1–11 (2018)
13. Chanijani, S.S.M., Klein, P., Bukhari, S.S., Kuhn, J., Dengel, A.: Entropy based transition analysis of eye movement on physics representational competence. In: UbiComp Adjunct, pp. 1027–1034 (2016)
14. Astefanoaei, C., Pretegiani, E., Optican, L.M., Creanga, D., Rufa, A.: Eye movement recording and nonlinear dynamics analysis - the case of saccades. Rom J. Biophys. **23**, 81–92 (2013)
15. Urbanowicz, Krzysztof, Hołyst, Janusz A.: Noise-level estimation of time series using coarse-grained entropy. Phys. Rev. E **67**, 1–14 (2003)
16. Chen, T., Guestrin, C.: XGBoost: a scalable tree boosting system. In: Proceedings of the 22nd ACM SIGKDD International Conference on Knowledge Discovery and Data Mining, KDD 2016, pp. 785–794. ACM, New York, (2016)
17. Stewart, P.A., Bucy, E.P., Mehu, M.: Strengthening bonds and connecting with followers: a biobehavioral inventory of political smiles. Polit. Life Sci. **34**(1), 73–92 (2015)
18. Galvan, A., Wichmann, T.: Pathophysiology of parkinsonism. Clin. Neurophysiol. Off. J. Int. Fed. Clin. Neurophysiol. **119**(7), 1459–1474 (2008)
19. Taylor, S.: Tend and befriend biobehavioral bases of affiliation under stress. Current Direct. Psychol. Sci. **15**, 273–277 (2006)
20. Taylor, S., Klein, L., Lewis, B., Gruenewald, R.G., Updegraff, J.: Biobehavioral responses to stress in females: tend-and-befriend, not fight-or-flight. Psychol. Rev. **107**, 411–429 (2000)

Impact of Radiomap Interpolation on Accuracy of Fingerprinting Algorithms

Juraj Machaj$^{(\boxtimes)}$ ⓘ and Peter Brida ⓘ

University of Zilina, Univerzitna 1, 010 26 Zilina, Slovakia
juraj.machaj@uniza.sk, peter.brida@feit.uniza.sk

Abstract. The positioning using Wi-Fi signals and fingerprinting algorithms achieved a lot of attention lately. The main drawback of the fingerprinting localization is the process of radiomap creation, which is labour intensive and time-consuming procedure. Therefore, some solutions for crowdsourcing and dynamic map creation were proposed. However, the problem of these is that users usually don't move regularly thru all the areas, which leads to undersampling of certain parts of the localization area. In this paper interpolation algorithms are used to increase radiomap density and thus reduce the problem with under-sampling. To evaluate the impact of interpolation on the performance of fingerprinting algorithms the NN, KNN and WKNN algorithms were tested on dynamic radiomap without interpolation as well as with radiomaps created using linear, inverse distance weight and Kriging interpolations. The paper shows the results achieved in the real-world scenario and shows the impact of the interpolation algorithms on the performance of the above-mentioned localization algorithms.

Keywords: Localization · Fingerprinting · Dynamic radiomap · Interpolation · PF-PDR · Wi-Fi

1 Introduction

Localization in indoor environments received a lot of attention from the research community in recent years. The reason for this is due to a large number of emerging location-based services that can be implemented by service providers [1], however, traditional GNSS based localization is not suitable for the indoor environment. This is due to the fact that GNSS signals are attenuated significantly by the building walls and GNSS receivers can receive mainly reflected signals, which may lead to higher localization errors [2].

Indoor localization systems introduced by the research community can be based on different measurements, including ultrasound signals, data from IMU sensors, data from cameras, visible light signals as well as wireless signals. Each of the enabling technology has its own advantages and disadvantages.

The most attention is attracted by localization utilizing measurements from wireless technologies [3] since most of the wireless signals can penetrate thru obstacles. Moreover, wireless signals are already available in indoor environments, since wireless networks are widely used for data transfers in indoor spaces.

N. T. Nguyen et al. (Eds.): ACIIDS 2022, LNAI 13757, pp. 117–128, 2022.
https://doi.org/10.1007/978-3-031-21743-2_10

The wireless technologies enabling indoor positioning include Wi-Fi [4], cellular networks [5], Bluetooth [6], UWB [7] as well as RFID [8]. While UWB and RFID can provide higher localization accuracies, their communication range is limited, and it is usually required to set up new infrastructure for these networks. Thus, the implementation cost of RFID and UWB positioning systems can be high. On the other hand, Wi-Fi signals are widely available in an indoor environment, therefore the implementation cost for the localization system can be reduced drastically.

That is probably the reason why Wi-Fi based positioning become extremely popular in the research community. There has been a vast number of solutions based on measurements of RSS, AoA, ToA as well as CSI data [4]. The location of mobile nodes in Wi-Fi systems can be estimated using either distance-based methods or fingerprinting algorithms.

The distance-based methods typically use lateration algorithms, which require estimation of the distance between transmitter and receiver [9]. Therefore, information about eradiated power of signal and placement of transmitters is required together with an accurate propagation model. However, this may be problematic due to multipath signal propagation and possible blockage of the direct path between transmitter and receiver in the indoor environment.

On the other hand, fingerprinting algorithms can benefit from attenuation caused by obstacles in the area, since these obstacles will attenuate radio signal and cause higher differences in signal parameters at different positions. Therefore, fingerprinting algorithms are more robust in the multipath environment [10] and thus more popular in the research community.

On the other hand, fingerprinting algorithms require a radiomap database to be able to perform localization. The radiomap is usually created using measurements at predefined spots, which is a time-consuming task. The other, approach is to model the radiomap using simulation tools. While this approach does not require a lot of time the accuracy of the radiomap may not be high enough to provide positioning service with low localization error. There are multiple reasons for this, the first being that the signal is affected by the obstacles between transmitter and receiver, but accurate modelling of all obstacles in the area is not quite possible. Moreover, propagation models used in the simulations may not be accurate enough. On top of that, information about the placement of some APs may not be accurate or available at all.

Recently a lot of effort was invested into crowdsourcing of radiomap measurements [11, 12], which may reduce the time required for radiomap creation, however, does introduce some new challenges. For example, it is required to estimate the position of the mobile user, this may be done by using IMU data, however, in such case it is required to estimate starting position of the mobile device. Moreover, creating a radiomap from crowdsourced data may introduce novel challenges due to differences in reported RSS from different devices. This may be caused by differences in antenna design, electronic circuits as well as software running of the mobile devices [13]. Moreover, the crowdsourced data may generate high-density radiomap in some indoor areas, like corridors, and low-density radiomap in other parts of the environment, i.e. offices, meeting rooms, etc. Since the accuracy of fingerprinting algorithms depends on the density of the radiomap, it is better to have an even distribution of reference points in the area, thus the performance of the localization system will be more stable.

In this paper, we focus on the performance evaluation of interpolation algorithms that may be used to increase the density of radiomap in order to improve localization accuracy in areas with low radiomap density.

The rest of the paper is organized as follows, in Sect. 2 related work in the area of fingerprinting and radiomap interpolation is presented, Sect. 3 introduces experimental scenario and discuss achieved results and Sect. 4 concludes the paper.

2 Related Work

The fingerprinting localization framework is based on a comparison of measured signal characteristics, typically RSS measurements, with the data stored in the radiomap database. The fingerprinting localization framework can be divided into two phases, i.e. offline and online phase [14]. During the offline phase, the radiomap is created. This can be done by performing either measurements at predefined positions, by simulations or by crowdsourcing. Each of these approaches has its pros and cons.

For example, performing measurements at the predefined positions can provide realistic data, which is beneficial for the performance of the localization system, however, it takes a large amount of time especially when the positioning system is deployed in large areas. On the other hand, simulations require a highly accurate model of the environment as well as information about the positions and settings of all APs. Though, simulated RSS values still may not be realistic and thus achievable accuracy of the system may be lower.

Crowdsourcing of RSS measurements for radiomap attracted a lot of attention since it may provide realistic information without demands on time-consuming labour, thus making deployment of the system more cost-effective. However, there may be issues caused by noisy information about the position where data were crowdsourced. Moreover, different users may use devices with different receiver designs, thus the localization algorithm needs to be able to compensate for these differences [13].

2.1 Fingerprinting Localization

In recent years a lot of attention was paid to the development of localization algorithms that will improve localization accuracy. Previously proposed algorithms can be divided into three main groups: deterministic algorithms, statistic algorithms and machine learning-based algorithms.

The basic assumption in the case of deterministic algorithms is that RSS samples, or other signal parameters, depend on the position of the mobile device [14]. On the other hand, statistic algorithms are based on assumption that the position of the mobile device is random and can be chosen based on the posterior probability calculated from the measured data and data stored in the radiomap [15].

In the machine learning approach, different solutions based on neural networks and deep learning were proposed, all these methods formulate the positioning process as a classification problem [16].

In this work, we have implemented deterministic algorithms since these seem to achieve similar accuracy as statistical algorithms and can generalize better than ML algorithms which need to be re-trained in a case when radiomap has changed, i.e. when new calibration measurements are included in the radiomap or when a new area is about to be covered by the localization system. The position estimate in the case of deterministic algorithms is given by the following estimator:

$$[P_x, P_y] = \frac{\sum_{j=1}^{n} \omega_j \gamma_j}{\sum_{j=1}^{n} \omega_j} \tag{1}$$

where $[P_x, P_y]$ represents the position estimate, ω_j stands for the weight given by Euclidean distance between RSS samples from measurements and data stored in the radiomap for j-th reference point (RP) and γ_j represents the position of the j-th reference point in the radiomap database and n is the total number of RPs stored in the radiomap database.

In case when only the point with the highest weight is selected the algorithm is called NN (Nearest Neighbour), when RPs with K highest weights are considered the algorithm is called WKNN (Weighted K-Nearest Neighbour) and KNN (K-Nearest Neighbour) algorithm select RPs with K highest weights and sets all of them to 1, so the final position is given by the centre of gravity of selected RPs [14, 15].

2.2 Dynamic Radiomap

The solution which utilizes data from the Wi-Fi receiver and IMU sensor built-in smart device for dynamic radiomap creation was proposed in [17]. The track of the user can

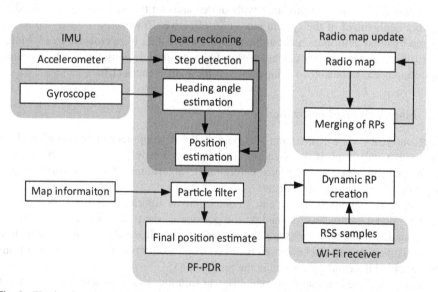

Fig. 1. The implemented solution for dynamic radiomap creation using PF-PDR and radiomap update by merging of RPs with high similarity.

be estimated using the particle filter pedestrian dead reckoning (PF-PDR) algorithm which utilizes information about obstacles in the indoor area to improve the accuracy of the PDR position estimates [18]. The block diagram of the implemented solution is presented in Fig. 1.

The implemented solution estimates the position of the mobile node from IMU measurements collected by the mobile device with the sampling rate of 100 Hz, the data is first used to estimate the heading and steps of the user. Then these data are used to estimate the position of the user, which is used in the update step of the particle filter. The implemented particle filter takes into account information from the PDR algorithm as well as information about obstacles, i.e. walls and other structures, to estimate the final position of the user in the given update step.

When RSS data are sampled from the Wi-Fi receiver, which is done once is 5 s to allow the user to travel some distance between measurements, a dynamic RP can be created using information from PF-PDR as position and RSS samples as a fingerprint.

In the radiomap update process, the generated dynamic RPs are compared with RPs that are stored in the radiomap. When RPs have a similar position as well as RSS samples RPs are merged, and the resulting position is given as the average position of merged RPs and assigned RSS samples are the average values of RSS samples from original RPs.

2.3 Interpolation Algorithms

In this work, we focus on the implementation of interpolation algorithms to improve the density of the dynamic radiomap. Since the performance of the fingerprinting positioning depends mainly on the number of APs in the area and radiomap density, the implementation of the interpolation should improve achievable accuracy.

The interpolation can be defined as a process of estimation of missing values from a set of known values. In the implementation of interpolation algorithms positions of APs and obstacles in the area were not taken into account. In the experiment, we have implemented three interpolation methods, i.e. linear interpolation, inverse distance weight interpolation and kriging interpolation.

The interpolation points were generated in the grid with a 2 m distance, covering the whole localization area. However, not all the generated points were used in the interpolation process. Points that were outside the area covered by existing RP from the dynamic map as well as points that were less than 1.5 m from existing RPs were removed and were not considered in the interpolation process.

In the first step of both linear and inverse distance weight interpolations, the data from existing RPs were processed by the Delaunay triangulation algorithm [19]. The Delaunay triangulation maximizes the observed minimum angle of the tessellation, therefore, resulting in smoother surfaces than other data-independent triangulation methods.

In the second step, each generated RP is assigned a tessellation triangle from Delaunay trilateration based on its coordinates. Thus, it will be possible to perform RSS interpolation using measured data from existing reference points placed on vertices of the tessellation triangle.

In the **linear interpolation** the interpolated RSS values RSS_{int} at the generated reference point with the position $[x_{int}, y_{int}]$ can be estimated as follows:

$$RSS_{int} = \alpha_1 RSS_1 + \alpha_2 RSS_2 + \alpha_3 RSS_3, \tag{2}$$

where RSS_i stands for RSS values measured at i-th reference points at i-th vertex of the tessellation triangle and α_i represents weights used for estimation of the RSS values the weights inside the tessellation triangle given by:

$$\alpha_1 = \frac{(y_2 - y_3)(x_{int} - x_3) + (x_3 - x_2)(y_{int} - y_3)}{(y_2 - y_3)(x_1 - x_3) + (x_3 - x_2)(y_1 - y_3)}, \tag{3}$$

$$\alpha_2 = \frac{(y_3 - y_1)(x_{int} - x_3) + (x_1 - x_3)(y_{int} - y_3)}{(y_2 - y_3)(x_1 - x_3) + (x_3 - x_2)(y_1 - y_3)}, \tag{4}$$

$$\alpha_3 = 1 - \alpha_1 - \alpha_2, \tag{5}$$

where x_i and y_i are coordinates of the i-th reference point placed at the i-th vertex of the tessellation triangle. From Eqs. (3)–(5) it is clear that $\alpha_1 + \alpha_2 + \alpha_3 = 1$ [20].

On the other hand, **inverse distance weighting interpolation** (IDW) takes into account that signal attenuation is not linear and thus RSS samples collected at RPs closer to the new generated RP should have a greater influence on interpolation [21]. This type of interpolation method is sometimes referred to as kernel smoothers. In the case of IDW the RSS value at the generated reference points can be calculated as follows:

$$RSS_{int} = \frac{\sum_i \beta_i RSS_i}{\sum_i \beta_i}, \tag{6}$$

where β_i stands for weight defined as an inverse power function of distance:

$$\beta(d) = d^{-n}. \tag{7}$$

where d is the Euclidean distance between points and n is the constant which defines the steepness of the function, i.e., the higher the value n the higher the impact of the closest RP.

The third implemented interpolation method is the **Kriging interpolation**, which is widely utilized mainly in GIS applications [22]. However, its application in radiomap interpolation for Wi-Fi fingerprinting was described as well [23].

The Kinging interpolation is based on finding mathematical functions which can be used to create a spatial structure that can be used to estimate values in new locations [23]. The basic assumption is that the spatial random variable, i.e. RSS, is spatially dependent, which means that closer points are more similar than those further away. Moreover, it is assumed that the spatial random variable is stationary, i.e. has a correlation between relative location and there is no correlation between absolute locations. This the theoretical semi-variance ϑ can be defined as follows:

$$\vartheta(d) = \frac{1}{2} V[RSS_{RPi} - RSS_{RPj}], \tag{8}$$

where d is the distance between points RPi an RPj, V is the variance and RSS_{RPi} represents the RSS measurements at the point RPi. Then, the spatial covariance function C can be used to estimate the Kriging predictor. The relationship between ϑ and C is given by:

$$\vartheta(d) = C(0) - C(d). \qquad (9)$$

For experimental data, the semi-variance $\hat{\vartheta}(d)$ can be mathematically estimated using:

$$\hat{\vartheta}(d) = \frac{1}{2N(d)} \sum_{N(d)} \left[RSS_{RPi} - RSS_{RPj} \right]^2, \qquad (10)$$

where $N(d)$ is the number of existing reference points that are within the distance d. Thus, it is possible to estimate experimental as well as theoretical semi-variogram models defined by:

$$\hat{\vartheta}(d) = S \left[1 - \exp\left(-3 \left(\frac{d}{R} \right)^E \right) \right], \qquad (11)$$

where S represents the sill parameter and R represents the range parameter and the exponent $0 \leq E \leq 2$ was set to 1.5 in the implementation.

3 Experimental Scenario and Achieved Results

The experiments were performed at the Department of Multimedia and Information Communication Technology at the University of Zilina. The dynamic radiomap was created in the localization area of 912 m², which consists of lecture rooms laboratories offices and corridors. The localization area is shown in Fig. 2, in the figure the red squares represent the position of RPs from the dynamic radiomap, black dots represent positions of RPs for interpolation and blue circles represent spots used for position estimation.

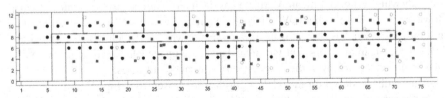

Fig. 2. The localization area with positions RPs from the dynamic map (red squares), RPs from interpolation (black dots) and spots for position estimation (blue circles) (Color figure online).

The dynamic radiomap was created during a single walk of the user in the area. It is important to note here that the median localization accuracy of the implemented PFPDR algorithm was approximately 0.5 m. Therefore, we can assume that dynamic reference points have reasonably good accuracy and can provide reasonably good initial data for interpolation algorithms.

The position estimation was performed using a dynamic radiomap as well as radiomaps created by three interpolation algorithms described in Subsect. 2.3. Moreover, the position was estimated using NN, KNN and WKNN algorithms. Both KNN and WKNN algorithms used $K = 3$ RPs to estimate the position of the mobile device. From Fig. 2 it can be seen that some estimated positions were placed on a border or outside of the area covered by a radiomap. Therefore, higher positioning errors can be expected in some cases. However, this scenario can still represent realistic conditions in the case when a dynamic radiomap is used for fingerprinting localization.

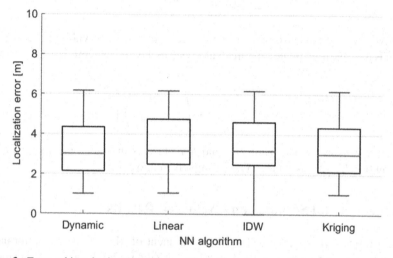

Fig. 3. Error achieved using the NN algorithm with dynamic and interpolated radiomaps.

Results achieved with dynamic and interpolated radiomaps using NN algorithms are shown in Fig. 3. From the figure, it can be seen that interpolation of the radiomap does not have a significant impact on the median (depicted as a red line) localization accuracy which is not significantly affected by the interpolation of the dynamic radiomap. However, it can be seen that IDW interpolation helped to improve localization accuracy in some cases, which is shown by the lower value of whisker, which represents the lower 25% of localization errors. On the other hand, the upper whisker represents 75% of localization error, however, this value is very similar for all cases.

The results achieved for the KNN algorithm are presented in Fig. 4, from the results it can be seen that in this case, a higher density of radiomap due to interpolation had some impact on the localization accuracy. This result is expected since the KNN algorithm combines the positions of 3 RPs therefore with a higher density of radiomap selected RPs can be closer to the real position of the mobile device. It can be seen that the lowest median localization error was achieved with linear interpolation, however, localization using IDW had the lowest deviation, thus it may provide the most reliable position estimates. It is important to note that Kriging interpolation achieved results very similar to the dynamic radiomap, therefore, it can be concluded that Kriging interpolation does not help to improve the localization accuracy of the KNN algorithm.

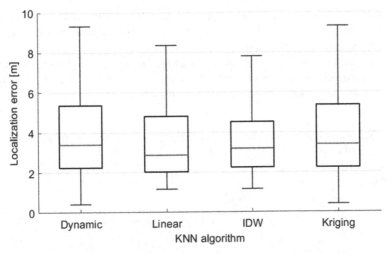

Fig. 4. Error achieved using the KNN algorithm with dynamic and interpolated radiomaps.

Furthermore, Fig. 5 shows the results achieved by the WKNN algorithm. From the results, it is obvious that in this case the lowest localization error was achieved when the radiomap was interpolated using linear interpolation. Moreover, in the case of linear interpolation, the deviation of localization error was the lowest in the case of linear interpolation. Thus, it seems that linear interpolation, although being the simplest interpolation method, can provide reasonably good results when the WKNN positioning algorithm is used.

The difference between the results achieved by KNN and WKNN algorithms is caused because of the fact that in WKNN algorithms accuracy of the RSS samples at

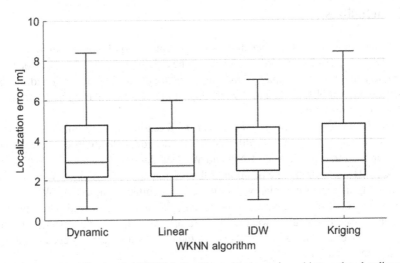

Fig. 5. Error achieved using the WKNN algorithm with dynamic and interpolated radiomaps.

the interpolated points has a higher impact. This is due to the fact that weights used to calculate the position estimate depend on the comparison of measured RSS samples with the RSS samples in the radiomap.

The results achieved for all the algorithms and radiomaps are summarized in Table 1 for easier comparison. In the table mean localization errors and standard deviations achieved in all the algorithms and scenarios are presented.

Table 1. Mean and standard deviation of localization error [m] achieved by algorithms with dynamic radiomap and different interpolations.

Algorithm	Radiomap			
	Dynamic	Linear	IDW	Kriging
NN	3.51 ± 1.97	3.78 ± 2.19	3,49 ± 1.8	3.5 ± 1.97
KNN	4.05 ± 2.65	3.78 ± 2.26	3,79 ± 2.1	4.05 ± 2.6
WKNN	3.77 ± 2.4	3.59 ± 1.99	3.63 ± 2.11	3.77 ± 2.4

From the table, it is clear that the lowest mean localization error was achieved in the case of the NN algorithm and radiomap with IDW interpolation. However, when compared to the performance of NN and dynamic radiomap without interpolation is not significant since both mean value and standard deviation are very similar.

Results achieved by the WKNN algorithm in the case of radiomap with linear interpolation are the best amongst WKNN and KNN results in all scenarios. Therefore, it seems linear interpolation can provide the best results when algorithms that take multiple reference points into account during the localization.

4 Conclusions

In the paper, we have implemented and tested multiple interpolation techniques, namely linear, IDW and Kriging, in order to increase the density of radiomap used for finger-printing localization. The impact of the interpolated and original dynamic radiomap on the performance of NN, KNN and WKNN algorithms was tested using data collected in a real-world environment.

From the achieved results it is clear that interpolation had minimum effect on the simplistic NN algorithm since the difference in achieved localization accuracy was negligible. On the other hand, when KNN and WKNN algorithms, which use multiple (in our case 3) RPs to estimate the position of the mobile node, the impact of interpolation was more significant. Interpolation had the most significant impact on WKNN algorithms where the difference between measured RSS samples and RSS data stored in a radiomap is used to calculate weights that are used to calculate the resulting position estimate. In the case of the WKNN algorithm, the best results were achieved when linear interpolation was used.

Acknowledgement. This work has been partially supported by the Slovak VEGA grant agency, Project No. 1/0588/22 "Research of a location-aware system for achievement of QoE in 5G and B5G networks", and Operational Program Integrated Infrastructure 2014–2020 of the project: Innovative Solutions for Propulsion, Power and Safety Components of Transport Vehicles, code ITMS 313011V334, co-financed by the European Regional Development Fund.

EUROPEAN UNION
European Regional Development Fund
OP Integrated Infrastructure 2014 – 2020

MINISTRY
OF TRANSPORT
AND CONSTRUCTION
OF THE SLOVAK REPUBLIC

References

1. Basiri, A., et al.: Indoor location based services challenges, requirements and usability of current solutions. Comput. Sci. Rev. **24**, 1–12 (2017). https://doi.org/10.1016/j.cosrev.2017.03.002

2. Diggelen, F.V.: SYSTEM DESIGN & TEST-GNSS accuracy-lies, damn lies, and statistics-this update to a seminal article first published here in 1998 explains how statistical methods can create many different. GPS World **18**, 26–33 (2007)

3. Xiao, J., Zhou, Z., Yi, Y., Ni, L.M.: A survey on wireless indoor localization from the device perspective. ACM Comput. Surv. **49** 25:1–25:31 (2016). https://doi.org/10.1145/2933232

4. Liu, F., et al.: Survey on WiFi-based indoor positioning techniques. IET Commun. **14**, 1372–1383 (2020). https://doi.org/10.1049/iet-com.2019.1059

5. Gorak, R., Luckner, M., Okulewicz, M., Porter-Sobieraj, J., Wawrzyniak, P.: Indoor localisation based on GSM signals: multistorey building study. Mob. Inf. Syst. **2016**, 2719576 (2016). https://doi.org/10.1155/2016/2719576

6. Pelant, J., et al.: BLE device indoor localization based on RSS fingerprinting mapped by propagation modes. In: 2017 27th International Conference Radioelektronika (RADIOELEK-TRONIKA), pp. 1–5 (2017). https://doi.org/10.1109/RADIOELEK.2017.7937584

7. Pala, S., Jayan, S., Kurup, D.G.: An accurate UWB based localization system using modified leading edge detection algorithm. Ad Hoc Netw. **97**, 102017 (2020). https://doi.org/10.1016/j.adhoc.2019.102017

8. Wang, J., Dhanapal, R.K., Ramakrishnan, P., Balasingam, B., Souza, T., Maev, R.: Active RFID Based Indoor Localization. IEEE, New York (2019)

9. Liu, H., Darabi, H., Banerjee, P., Liu, J.: Survey of wireless indoor positioning techniques and systems. IEEE Trans. Syst. Man Cybern. Part C (Appl. Rev.) **37**, 1067–1080 (2007). https://doi.org/10.1109/TSMCC.2007.905750

10. He, S., Chan, S.-G.: Wi-Fi fingerprint-based indoor positioning: recent advances and comparisons. IEEE Commun. Surv. Tutorials **18**, 466–490 (2016). https://doi.org/10.1109/COMST.2015.2464084

11. Wilk, P., Karciarz, J., Swiatek, J.: Indoor radio map maintenance by automatic annotation of crowdsourced Wi-Fi fingerprints. In: 2015 International Conference on Indoor Positioning and Indoor Navigation (IPIN), pp. 1–8 (2015). https://doi.org/10.1109/IPIN.2015.7346933

12. Ma, L., Fan, Y., Xu, Y., Cui, Y.: Pedestrian dead reckoning trajectory matching method for radio map crowdsourcing building in WiFi indoor positioning system. In: 2017 IEEE International Conference on Communications (ICC), pp. 1–6 (2017). https://doi.org/10.1109/ICC.2017.7996457

13. Machaj, J., Brida, P., Majer, N.: Challenges introduced by heterogeneous devices for Wi-Fi–based indoor localization. Concurrency Comput. Pract. Exper. **32** (2020). https://doi.org/10.1002/cpe.5198
14. Bahl, P., Padmanabhan, V.N.: RADAR: an in-building RF-based user location and tracking system. In: Proceedings IEEE INFOCOM 2000. Conference on Computer Communications. Nineteenth Annual Joint Conference of the IEEE Computer and Communications Societies (Cat. No. 00CH37064), vol. 2, pp. 775–784 (2000). https://doi.org/10.1109/INFCOM.2000.832252
15. Honkavirta, V., Perala, T., Ali-Loytty, S., Piche, R.: A comparative survey of WLAN location fingerprinting methods. In: Navigation and Communication 2009 6th Workshop on Positioning, pp. 243–251 (2009). https://doi.org/10.1109/WPNC.2009.4907834
16. Ahmadi, H., Bouallegue, R.: Exploiting machine learning strategies and RSSI for localization in wireless sensor networks: a survey. In: 2017 13th International Wireless Communications and Mobile Computing Conference (IWCMC), pp. 1150–1154 (2017). https://doi.org/10.1109/IWCMC.2017.7986447
17. Brida, P., Machaj, J., Racko, J., Krejcar, O.: Algorithm for dynamic fingerprinting radio map creation using IMU measurements. Sensors. **21**, 2283 (2021). https://doi.org/10.3390/s21072283
18. Racko, J., Brida, P., Perttula, A., Parviainen, J., Collin, J.: Pedestrian dead reckoning with particle filter for handheld smartphone. In: 2016 International Conference on Indoor Positioning and Indoor Navigation (IPIN), pp. 1–7 (2016). https://doi.org/10.1109/IPIN.2016.7743608
19. de Berg, M., Cheong, O., van Kreveld, M., Overmars, M.: Computational Geometry: Algorithms and Applications. Springer, Heidelberg (2008). https://doi.org/10.1007/978-3-540-77974-2
20. Talvitie, J., Renfors, M., Lohan, E.S.: Distance-based interpolation and extrapolation methods for RSS-based localization with indoor wireless signals. IEEE Trans. Veh. Technol. **64**, 1340–1353 (2015). https://doi.org/10.1109/TVT.2015.2397598
21. Ezpeleta, S., Claver, J.M., Pérez-Solano, J.J., Martí, J.V.: RF-based location using interpolation functions to reduce fingerprint mapping. Sensors. **15**, 27322–27340 (2015). https://doi.org/10.3390/s151027322
22. Stein, M.L.: Interpolation of Spatial Data: Some Theory for Kriging. Springer, New York (1999). https://doi.org/10.1007/978-1-4612-1494-6
23. Jan, S.-S., Yeh, S.-J., Liu, Y.-W.: Received signal strength database interpolation by Kriging for a Wi-Fi indoor positioning system. Sensors **15**, 21377–21393 (2015). https://doi.org/10.3390/s150921377

Rough Set Rules (RSR) Predominantly Based on Cognitive Tests Can Predict Alzheimer's Related Dementia

Andrzej W. Przybyszewski[1,2]([✉]) [iD], Kamila Bojakowska[3], Jerzy P. Nowacki[1], Aldona Drabik[1], and BIOCARD Study Team

[1] Polish-Japanese Academy of Information Technology, 02-008 Warszawa, Poland
{przy,jerzy.nowacki,drabik}@pjwstk.edu.pl
[2] Department of Neurology, UMass Medical School, Worcester, MA 01655, USA
[3] Department of Neurology, Central Clinical Hospital MSWiA, 02-507 Warszawa, Poland
kama.435@wp.pl

Abstract. Technology progress helped us to live better and longer, but aging is the major factor related to ND (neurodegenerative diseases) such as Alzheimer's or Parkinson's disease. Alzheimer's disease (AD) correlated neurodegenerative processes begin over 30 years, whereas cognitive changes begin about 15–11 years, before the first AD symptoms.

The purpose of our study was to predict if cognitive 'healthy' subjects might get AD dementia soon. We have analyzed Biocard data from the project that started with 349 normal subjects were followed over 20 years with over 150 different attributes. Subjects were evaluated every year by neurologists with the global score CDR (Clinical Dementia Rating) parameters to determine if a particular individual is normal, has Mild Cognitive Impairment, or has dementia. We have used classification based on CDRSUM (sum of boxes) as a more precise and quantitative general index than the global score to provide more information on patients with mild dementia. CDRSUM values for prodromal patients are: 0.0 normal; (0.5–4.0): questionable cognitive impairment; (0.5–2.5): questionable impairment; (3.0–4.0) very mild dementia; (4.5–9.0) – mild dementia. We have obtained rough set rules (RSR) from Model1: 149 patients classified as AD, MCI, and normal; and Model2: 40 patients with AD. By using Model1 classified by neurologists as 21 normal (CDR = 0) subjects, with our classification based on RSR, we have obtained 8 subjects with CDRSUM > 0: all 8 subjects were above 0.75, one subject between 0.75 and 1.25, and 5 subjects between 0.75 and 2.25, and two subjects were above 2.25. These subjects might have questionable cognitive impairment. Using Model2 we found with RSR that two subjects had CDRSUM between 4.5 and 6.5, which means they might have mild dementia (4.5–9.0). RSR consist of algorithms that might predict future cognitive AD-related impairments in individual, normal, healthy subjects.

Keywords: Neurodegeneration · Rough set theory · Intelligent predictions

© The Author(s), under exclusive license to Springer Nature Switzerland AG 2022
N. T. Nguyen et al. (Eds.): ACIIDS 2022, LNAI 13757, pp. 129–141, 2022.
https://doi.org/10.1007/978-3-031-21743-2_11

1 Introduction

Cognitive changes are dominating in the most common neurodegenerative disease (ND) - Alzheimer's disease (AD). In most cases of AD, neurodegeneration starts in the hippocampus and frontal cortex, and it is related to memory and orientation problems. With the disease progression, other brain regions become also affected.

As each patient has dissimilar neurodegeneration development and compensation in consequence symptoms might be various and finding optimal treatment is an art for an experienced neurologist.

We have estimated disease progression with sets of psychophysical attributes found as the most meaningful in patients from the BIOCARD study [1, 2] and combined them with the results of the APOE.

The risk of AD increases and the age-at-onset decreases with the number of APOE4 alleles [3, 4]. A single APOE4 allele increases risk 2–4 fold and having two APOE4 alleles increases the risk about 8–12 fold [4]. The APOE4 allele also drives the age-at-onset down, APOE4 carriers are, on average, about 12 years younger than non-carriers [3, 4] However having a single APOE allele ε2 reduces the risk AD by about 40%, and being homozygous for APOE ε2 reduces the risk even more APOE2 homozygotes have a 66% reduction in AD risk compared to +2/+3 carriers, an 87% reduction in AD risk compared to APOE3 homozygotes, and a 99.6% reduction in AD risk when compared to APOE4 homozygotes [5].

This study is the continuation of the rough set theory application to follow predominantly the cognitive changes in neurodegenerative diseases (ND) such as Parkinson's [6] and now Alzheimer's diseases.

2 Methods

We have analyzed data from normal subjects (N), Mild Cognitive Impairment (MCI), and Alzheimer Disease (AD) patients divided into three main groups:

- **Model1** consists of 149 subjects with 40 normal (N), 40 AD, and 69 MCI
- **Model2** consists of 40 AD patients.
- **TestGr** consists of 21 normal (N) subjects

All subjects had the following neuropsychological tests performed every year: Logical Memory Immediate (LOGMEM1A), Logical Memory Delayed (LOGMEM2A), Trail Making, Part A (TrailA - connecting time in sec of randomly placed numbers), Trail Making Part B (TrailB - connecting time in a sec of randomly placed numbers and letters), Digit Symbol Substitution Test (DSST), Verbal Fluency Letter F (FCORR), Rey Figure Recall (REYRECAL), Paired Associate Immediate (PAIRED1), Paired Associate Delayed (PAIRED2), Boston Naming Test (BOSTON). In addition, we have subjects' age (years), APOE genotype; individuals who are *ApoE-4* carriers vs. non-carriers (digitized as 1 vs. 0), and CDRSUM (sum of boxes) as a precise and quantitative general index of the Clinical Dementia Rating [7].

*Data used in preparation of this article were derived from BIOCARD study, supported by grant U19 – AG033655 from the National Institute on Aging. The BIOCARD study team did not participate in the analysis or writing of this report, however, they contributed to the design and implementation of the study. A listing of BIOCARD investigators can be found on the BIOCARD website (on the 'BIOCARD Data Access Procedures' page, 'Acknowledgement Agreement' document).

2.1 Theoretical Basis

Our analysis was performed with help of rough set theory (RST) invented by Zdzislaw Pawlak [8]).

In the standard RST procedure, our data were inserted in the decision table with rows that stand out the actual attributes' values for the different or for the same subject, and columns were linked to diverse attributes. Following [8] an information system is a pair $S = (U, A)$, where U, and A are nonempty finite sets. The set U is the universe of objects, and A is the set of attributes. If $a \in A$ and $u \in U$, the value $a(u)$ is a unique element of V (where V is a value set). The *indiscernibility relation IND(B)* of any subset B of A is defined after [8]: $(x, y) \in IND(B)$ *iff* $a(x) = a(y)$ for every $a \in B$ where the value of $a(x) \in V$. This relation divides A into *elementary granules* and it is the basis of RST. In the information system S set $B \subset A$ of is a reduct if $IND (B) = IND (A)$ and it cannot be further reduced. Other important RST properties such as *lower approximation* and *upper approximation* were defined and discussed in [8, 9].

In this work, we have used different intelligent algorithms implemented in RSES 2.2 software such as: exhaustive algorithm, genetic algorithm [10], covering algorithm, or LEM2 algorithm [11].

In addition to the classical [8] information system is its extension of the decision table to a triplet: $S = (U, C, D)$ where the set of attributes A is divided into C *as* condition and D as decision attributes [12]. In a single row, there are many conditions and only one decision attribute, all related to specific tests of the individual subject or patient. Names and values of classification attributes related to the value of the decision attribute give a unique rule. One difficult problem in the medical field is related to contradictory measurements (results). In this case, doctors often are using averaging techniques. RSR are considering and solving problems with contradictory rules, but the main feature of RSR is to generalize individual measurements (rules) and universal principles (knowledge) but rules have different confidence. There are always true rules related to the *lower approximation set* and rules that are only partly true associated with the *upper approximation set*. The difference between upper and lower approximation sets is the called border set. If this set is non-empty, it is related to uncertain rules.

We have based our approach on the mechanisms in the visual brain related to advanced processes of complex objects' recognition [13]. The processes in the higher visual brain areas that are related to different objects classification are using RSR to find upper and lower approximations of the retinal image [13]. These approximations are compared with the different objects' models (images) saved in the visual cortex. In the next steps of the object recognition (classification) lower visual areas are tunned to extract the properties of the chosen Model (the difference between upper and lower approximations becomes smaller). If the border set becomes empty, we recognize the object. We use this approach by proposing different Models to approximate the actual (future) cognitive state of tested normal subjects.

We have used RSR determined by RSES 2.2 [14] which generalizes rules from the decision table to process different patterns related to an individual patient sets of measurements. In our previous publication, related to Parkinson's disease patients, we demonstrated that the rough set theory application provides better results than other ML methods [6].

3 Results

As described above in the Methods section we had three groups of subjects: Model1 (149 subjects), Model2 (40 AD patients) and TestGr - test group (21 subjects).

3.1 Statistical Results

The subjects from group Model1 n = 149, Model2 n = 40, TestGr n = 21.

Table 1. Statistics of our data

P#	Age	Lgm1A	Lgm2A	TrailA	TrailB	DSST	Fcorr	CDRSUM
Model1	76.4 ± 8.6	16.1 ± 4.6	15.2 ± 5.3	39.6 ± 20	99.9 ± 59	46.5 ± 14	15.4 ± 5	1.3 ± 1.8
Model2	78.5 ± 12	12.2 ± 4.4	10.1 ± 5.2	5.03 ± 30	151 ± 78	37.3 ± 13	12.6 ± 6	3.5 ± 2.5
TestGr	76.6 ± 8.4	18.0 ± 3.0	17.5 ± 3.9	30.7 ± .12	64.2 ± 21	57 ± 13	8.2 ± 5	0 ± 0

Table 1 presents the statistical calculations for the Model1 and Model2 test normal subjects TestGr as mean ± SD. The age of subjects in different groups is similar, but other parameters show differences: Lgm1A (LOGMEM1A) is smallest for AD patients and largest for N, Lgm2A (LOGMEM2A) has similar changes as Lgm1A, execution functions: TrailA and TrailB are growing from N to AD, DSST is decreasing from N to AD in a similar way as Fcorr (FCORR). We did not show other parameters because of a lack of space. The changes in the CDRSUM are obvious as in normal subjects its values are 0 and significantly larger for AD patients. There are large differences between the values of individual subjects, so the mean values were not statistically significant.

3.2 RSR for Reference of Model1 Group

We have placed Model1 data in the following information table (Table 2):

The complete Table 2 has 149 rows, and 14 columns, there are shown the following condition attributes: P# - the number given to each patient, age –age of the subject, Lgm1A -Logical Memory Immediate, Lgm2A - Logical Memory Delayed, TrailA - Trail Making Part A, TrailB -Trail Making Part B, DSST - Digit Symbol Substitution

Table 2. Part of the decision table for Model1 subjects

P#	Age	Lgm1A	Lgm2A	Trail A	Trail B	DSST	Gcorr	Reycrl	APOE	...	CDRSUM
67643	74	9	8	40	208	35	14	18	1	...	0.5
70407	88	8	5	66	150	21	21	10	0	...	4.6
102541	71	15	25	25	202	52	52	23.5	0	...	1
119156	92	7	34	34	386	40	40	10.5	0	...	3.5
139134	81	6	51	51	60	49	49	6	1	...	2.5
142376	76	18	54	54	50	19	19	12	0	...	0

Table 3. Discretized table extract for above (Table 1) Model1 subjects

P#	Age	Lgm1A	Lgm2B	TrailA	TrailB	...APOE...	CDRSUM
67643	"(73.5,86.5)"	"(−5.5,16.5)"	"(5.5,16.5)"	"(31.5,48.0)"	"(137.5,Inf)"	...1...	"(−Inf,0.75)"
70407	"(86.5,Inf)"	"(−Inf,10.0)"	"(1.0,5.5)"	"(48.0,143.0)"	"(137.5,Inf)"	...0...	"(1.25,Inf)"
102541	"(−Inf,73.5)"	"(12.5,16.0)"	"(5.5,16.5)"	"(23.5,28.5)"	"(78.0,114.0)"	...0...	"(0.75,1.25)"
119156	"(86.5,Inf)"	"(−Inf,10.0)"	"(5.5,16.5)"	"(31.5,48.0)"	"(78.0,114.0)"	...0...	"(1.25,Inf)"
139134	"(73.5,86.5)"	"(−Inf,10.0)"	"(1.0,5.5)"	"(48.0,143.0)"	"(78.0,114.0)"	...1...	"(1.25,Inf)"
142376	"(73.5,86.5)"	"(16.0,20.5)"	"(5.5,16.5)"	"(48.0,143.0)"	"(137.5,Inf)"	...0...	"(−Inf,0.75)"

Test, Fcorr - Verbal Fluency Letter F, Reyrcl - Rey Figure Recall, APOE - *ApoE* genotype, ... CDRSUM – the sum of boxes - index of the Clinical Dementia Rating.

Table 3 is a discretized table for six patients: 67643 to 142376. Significant condition attributes were age, Lgm1A (LOGMEM1A), Lgm2A (LOGMEM2A), TrailA, TrailB, APOE, and others not shown in Table 2 like DSST, Fcorr (FCORR), Reyrcl (REYRECA), PAIRED1, BOSTON. Not significant was: PAIRED2.

We have used RSES 2.2 for Model1 group discretization with the local cuts [RSES]. There were the following 3 ranges of the decision attribute CDRSUM: "(−Inf, 0.75)", "(0.75, 1.25)", "(1.25, Inf)".

We obtained 2581 rules using the exhaustive algorithm for Model1 subjects, and as an example, we present below 10 rules filled by the most cases:

$$(DSST = {}^{"}(50.5, Inf)^{"})\&(REYRECAL = {}^{"}(15.75, 25.75)^{"})\&(age = {}^{"}(73.5, 86.5)^{"})$$
$$=> (CDRSUM = {}^{"}(-Inf, 0.75)^{"}[16])16 \qquad (1)$$

$$(LOGMEM\,1A = {}^{"}(21.5, Inf)^{"})\&(BOSTON = {}^{"}(29.5, Inf)^{"}) => (CDRSUM$$
$$= {}^{"}(-Inf, 0.75)^{"}[16])16 \qquad (2)$$

$$(LOGMEMA = {}^{"}(1.0, 5.5)^{"}) => (CDRSUM = {}^{"}(1.25, Inf)^{"}[8])8 \qquad (3)$$

$$(LOGMEM\,2A = {}^{"}(5.5, 16.5)^{"})\&(TRAILA = {}^{"}(31.5, 48.0)^{"})\&(REYRECAL$$

$$= \text{"} (Inf, 15.75)\text{"})\&(BOSTON = \text{"} (-Inf, 26.5)\text{"}) => \left(CDRSUM = \text{"} (1.25, Inf)\text{"} [7] \right)7$$
(4)

$$(TRAILA = \text{"}(31.5, 48.0)\text{"})\&(DSST = \text{"}(-Inf, 47.5)\text{"})\&$$
$$(REYRECAL = \text{"}(-Inf, 15.75)\text{"})\&(BOSTON = \text{"}(-Inf, 26.5)\text{"})$$
$$=> (CDRSUM = \text{"}(1.25, Inf)\text{"}[7])7$$
(5)

$$(FCORR = \text{"}(-Inf, 10.5)\text{"})\&(REYRECAL = \text{"}(-Inf, 15.75)\text{"})\&(APOE = 1)$$
$$=> (CDRSUM = \text{"}(1.25, Inf)\text{"}[7])7$$
(6)

$$(TRAILB = \text{"}(114.0, 137.5)\text{"})\&(APOE = 1) => (CDRSUM = \text{"}(1.25, Inf)\text{"}[6])6$$
(7)

$$(LOGMEMIA = \text{"}(16.0, 20.5)\text{"})\&(BOSTON = \text{"}(-Inf, 26.5)\text{"})\&$$
$$(age = \text{"}(73.5, 86.5)\text{"}) => (CDRSUM = \text{"}(0.75, 1.25)\text{"}[5])5$$
(8)

$$(TRAILB = \text{"}(51.0, 78.0)\text{"})\&(BOSTON = \text{"}(-Inf, 26.5)\text{"})\&$$
$$(age = \text{"}(73.5, 86.5)\text{"}) => (CDRSUM = \text{"}(0.75, 1.25)\text{"}[5])5$$
(9)

There is the following interpretation of the above equations: Eq. 1 claims for 16 cases that if DSST is above 50.5 and REYRECAL is between 15.75 and 25.75 and the patient's age is between 73.5 and 86.5 years then CDRSUM is below 0.75 that means questionable impairment. The Eq. 3 if LOGMEM2A (Logical Memory Delayed) is between 1.0 and 5.5 then CDRSUM is larger than 1.25 which means that patient has the least questionable impairment (8 cases). Equation 10 states that if TrailB is between 51 and 78 s, and Boston naming test result is below 25.5, and the patient's age is between 73.5 and 86.5 years then CDRSUM is between 0.75 and 1.25, which means that patient has a questionable impairment, and it is fulfilled in 5 cases.

We have used the above general rules from Model1 to predict CDRSUM of the TestGr.

Table 4. Confusion matrix for CDRSUM of TestGr group by rules obtained from Model1 by *local cuts* [13]

Predicted				
Actual	"(−Inf, 0.75)"	"(1.25, Inf)"	"(0.75, 1.25)"	ACC
"(−Inf, 0.75)"	17. 0	2.0	2.0	0.81
"(1.25, Inf)"	0.0	0.0	0.0	0.0
"(0.75, 1.25)"	0.0	0.0	0.0	0.0
TPR	1.0	0.0	0.0	0.0

TPR: True positive rates for decision classes; ACC: Accuracy for decision classes: the global coverage was 1.0 and the global accuracy was 0.81, the coverage for decision classes was 1.0, 0.0, 0.0.

Table 4 results are that for the 17 patients' prediction and tests results agreed. In two patients' predictions gave CDRSUM values above 1.25 and in two other patients, they were between 0.57 and 1.25 where tests results were 0 (below 0.75).

We were interested in those normal subjects who had predicted values of the CDR-SUM > 0. From Table 4 there were two subjects with predicted values of CDRSUM = (0.75, 1.25)

$$(Pat = 284424)\&(LOGMEM\,1A = \text{"}(16.0, 20.5)\text{"}))\&$$
$$(LOGMEM\,2A = \text{"}(5.5, 16.5)\text{"})\&(TRAILA = \text{"}(48.0, 143.0)\text{"})\&$$
$$(TRAILB = \text{"}(78.0, 114.0)\text{"})\&(DSST = \text{"}(-Inf\,.47.5)\text{"})\&$$
$$(FCORR = \text{"}(21.5, Inf\,)\text{"})\&(REYRECAL = \text{"}(15.75, 25.75)\text{"})\&(PAIRDI =$$
$$\text{"}(12.0, 21.5)\text{"})\&(BOSTON = \text{"}(26.5, 29.5)\text{"})\&(age = \text{"}(86.5, Inf\,)\text{"})$$
$$\&(APOE = 0) => (CDRSUM = \text{"}(0.75, 1.25)\text{"}\tag{10}$$

$$(Pat = 558865)\&(LOGMEM\,1A = \text{"}(16.0, 20.5)\text{"}))\&(LOGMEM\,2A$$
$$= \text{"}(5.5, 16.5)\text{"})\&(TRAILA = \text{"}(31.5, 48.0)\text{"})\&(TRAILB = \text{"}(78.0, 114.0)\text{"})\&$$
$$(DSST = \text{"}(-Inf, 47.5)\text{"})\&(FCORR = \text{"}(21.5, Inf\,)\text{"})\&(REYRECAL = \text{"}(15.75, 25.75)\text{"})\&$$
$$(PAIRDI = \text{"}(12.0, 21.5)\text{"})(BOSTON = \text{"}(26.5, 29.5)\text{"})\&(age = \text{"}(73.5, 86.5)\text{"})\&$$
$$(APOE = 1) => (CDSUM = \text{"}(0.75, 1.25)\text{"}\tag{11}$$

In both patients, bad executive functions (TrialA and TrialB) play a significant role in the possible questionable impairment [7].

From Table 4 there were two subjects with predicted values of CDRSUM = (1.25, Inf)):

$$(Pat = 164087)\&(LOGMEM\,1A = \text{"}(12.5, 16.5)\text{"}))\&(LOGMEM\,2A$$
$$= \text{"}(5.5, 16.5)\text{"}\&(TRAILA = \text{"}(34.5, 42.0)\text{"})\&(TRAILB = \text{"}(83.0, 114.0)\text{"})\&$$
$$(DSST = \text{"}(47.5, 50.5)\text{"})\&FCORR = \text{"}(10.5, Inf\,)\text{"})\&(REYRECAL$$
$$= \text{"}(15.75, 26.5)\text{"})\&(PAIRDI = \text{"}(12.0, 21.5)\text{"})\&(BOSTON = \text{"}(-Inf, 26.5)\text{"})\&$$
$$(age = \text{"}(73.0, 76.5)\text{"})\&$$
$$(APOE = 1) => (CDRSUM = \text{"}(1.25, Inf\,))\text{"}\tag{12}$$

$$(Pat = 401297)\&(LOGMEM\,1A = \text{"}(12.5, 16.5)\text{"})\&(LOGMEM\,2A$$
$$= \text{"}(5.5, 16.5)\text{"}\&(TRAILA = \text{"}(-Inf, 21.0)\text{"})\&(TRAILB = \text{"}(-Inf, 50.0)\text{"})\&$$
$$(DSST = \text{"}(69.0, Inf\,)\text{"}\&(FCORR = \text{"}(10.5, Inf\,)\text{"})\&$$
$$(REYRECAL = \text{"}(-Inf, 15.75)\text{"})\&(PAIRDI = \text{"}(12.0, 21.5)\text{"})\&$$
$$(BOSTON = \text{"}(26.5, Inf\,)\text{"})\text{"})\&(age = \text{"}(-Inf, 73.0)\text{"})\&$$
$$(APOE = 1) => (CDRSUM = \text{"}(1.25, Inf\,)\text{"}\tag{13}$$

136   A. W. Przybyszewski et al.

The first patient Eq. 13 has affected executive functions (TrialA and B), and the second patient (Eq. 14) has good executive functions, but bad the Rey Figure Recall REYRECAL.

Table 5. Confusion matrix for CDRSUM of TestGr group by rules obtained from Model1 by the *global cuts* [14]

Predicted				
Actual	"(−Inf, 0.75)"	"(2.25, Inf)"	"(0.75, 1.25)"	ACC
"(−Inf, 0.75)"	15. 0	2.0	4.0	0.71
"(2.25, Inf)"	0.0	0.0	0.0	0.0
"(0.75, 2.25)"	0.0	0.0	0.0	0.0
TPR	1.0	0.0	0.0	0.0

TPR: True positive rates for decision classes; ACC: Accuracy for decision classes: the global coverage was 1.0 and the global accuracy was 0.714, the coverage for decision classes was 1.0, 0.0, 0.0.

In Table 5 we have also used RSES 2.2 for Model1 group discretization with the *global cuts* [14]. There were the following 3 ranges of the decision attribute CDRSUM: "(−Inf, 0.75)", "(0.75, 2.25)", "(2.25, Inf)". We have obtained 776 rules with the genetic algorithm for Model1 subjects. After removing rules related to single support cases, we have got 324 rules that in the confusion matrix gave only three subjects with CDRSUM = (0.75, 2.25) – we found the excluded subject and marked it below, but the same two subjects with CDRSUM = (2.25, Inf).

As above, we were interested in those normal subjects who had predicted values of the CDRSUM > 0. From Table 3 there were four subjects with CDRSUM = (0.75, 2.25) that with values between (0.5–2.5) might have a questionable impairment [7]:

$(Pat = 204670)\&(LOGMEM\,1A = "(15.5, 20.5)")\&(LOGMEM\,2A$
$= "(16.5, Inf)")\&(TRAILA = "(-Inf, 23.5)")\&(TRAILB$
$= "(-Inf, 74.5)")\&(FCORR = "(16.5, Inf)")\&(REYRECAL = "(25.25, Inf)")\&$
$(PAIRD2 = "(6.5, Inf)")\&(age = "(-Inf, 76.5)")\&(APOE = 1)$
$=> (CDRSUM = "(0.75, 2.25)"$
$\hspace{10cm}(14)$

$(Pat = 463437)\&(LOGMEM\,1A = "(15.5, 20.5)")))\&(LOGMEM\,2A$
$= "(16.5, Inf)")\&TRAILA = "(23.5, 35.5)")\&(TRAILB = "(74.5, 153.0)")\&$
$(FCORR = "(-Inf, 16.5)")\&(REYRECAL = "(15.75, 25.25)")\&(PAIRD2$
$= "(-Inf, 6.5)")\&(age = "(-Inf, 76.5)")\&(APOE = 0)$
$=> (CDRSUM = "(0.75, 2.25)"$
$\hspace{10cm}(15)$

$(Pat = 558865)\&(LOGMEM\,1A = "(15.5, 20.5)"))\&(LOGMEMA$

$= "(-Inf, 16.5)")\&(TRAILA = "(23.5, 35.5)")\&(TRAILB = "(74.5, 153.0)")\&$
$(FCORR = "(16.5, Inf)")\&(REYRECAL = "(15.75, 25.25)")\&$
$(PAIRD2 = "(-Inf, 6.5)")\&(age = "(76.5, Inf)")\&(APOE = 1)$
$=> (CDRSUM = "(0.75, 2.25)"$ (16)

$(Pat = 808698)\&(LOGMEM1A = "(15.5, 20.5)"))\&(LOGMEM2A$
$= "(16.5, Inf)"\&(TRAILA = "(-Inf, 23.5)")\&(TRAILB = "(-Inf, 74.5)"\&$
$(FCORR = "(-Inf, 16.5)")\&(REYRECAL = "(15.75, 25.25)")\&$
$(PAIRD2 = "(6.5, Inf)")\&(age = "(-Inf, 76.5)")\&(APOE = 1)$
$=> (CDRSUM = "(0.75, 2.25)"$ (17)

As you may notice, rules for all 4 patients are similar, but with some significant differences, e.g., in Eqs. 16 and 17 both patients are slow in the executive actions; TrailA – connecting randomly place numbers, and TrialB - connecting numbers and letters – cognitive task. In the other two patients, it seems that the combination of different factors plays a major role in questionable impairment [7].

By using rules from Model1 group discretization with the *global cuts* [14] we got the following predictions:

$(Pat = 284424)\&(LOGMEM1A = "(15.5, 20.5)"))\&(LOGMEM2A$
$= "(-Inf, 16.5)")\&(TRAILA = "(35.5, Inf)")\&(TRAILB = "(74.5, 153.0)")\&$
$(FCORR = "(16.5, Inf)")\&(REYRECAL = "(15.75, 25.25)")\&$
$(PAIRD2 = "(6.5, Inf)")\&(age = "(76.5, Inf)")\&(APOE = 0)$
$=> (CDRSUM = "(0.75, 2.25)"$ (18)

In our new Model1 with global cuts there were two patients were classified as above *Pat* = 204670 and *Pat* = 808698, but we got a new patient *Pat* = 284424 that we have not seen above (Eq. 19). He/she has problems with the executive actions (TrailA and B).
From Table 5 there were two subjects with CDRSUM = (2.25, Inf)):

$(Pat = 164087)\&(LOGMEM1A = "(-Inf, 15.5)"))\&(LOGMEM2A$
$= "(-Inf, 16.5)")\&(TRAILA = "(35.5, Inf)")\&(TRAILB = "(74.5, 153.0)")\&$
$(FCORR = "(-Inf, 16.5)")\&(REYRECAL = "(15.75, 25.25)")\&(PAIRD2$
$= "(-Inf, 6.5)")\&(age = "(-Inf, 76.5)")\&(APOE = 1)$
$=> (CDRSUM = "(2.25, Inf)"$ (19)

$(Pat = 401297)\&(LOGMEM1A = "(-Inf, 15.5)")\&$
$(LOGMEM2A = "(-Inf, 16.5)")\&(TRAILA = "(35.5, Inf)")\&$
$(TRAILB = "(74.5, 153.0)")\&(FCORR = "(-Inf, 16.5)")\&$
$(REYRECAL = "(15.75, 25.25)")\&(PAIRD = "(-Inf, 6.5)")\&$
$(age = "(-Inf, 76.5)")\&$

$$(APOE = 1) => (CDRSUM = "(2.25, \text{ Inf})" \tag{20}$$

Notice that there are the same patients that were classified above (Eqs. 13, 14) with CDRSUM = (1.25, Inf). Parameters in the above equations (Eqs. 20, 21) have different values than in Eqs. 13, 14 but the interpretation is similar. The first patient *(Pat = 164087)* has affected executive functions (TrialA and B), especially TrialB related to the cognitive impairments. The second patient *(Pat = 401297)* has good (short times executions) executive functions, but bad the Rey Figure Recall REYRECAL.

3.3 RSR for Reference of Model2 Group

We have placed Model2 data in the following information table (Table 4):

Table 6. Part of the decision table for Model2 patients

P#	age	Lgm1A	Lgm2A	TrailA	TrailB	DSST	Fcorr	Reyrcl	APOE	...	CDRSUM
70407	88	8	5	66	150	21	21	10	0	...	4.5
119156	92	7	34	34	386	40	20	10.5	0	...	3.5
155699	94	9	3	76	119	14	9	7	1	...	4
265499	91	13	8	54	239	27	6	7	0	...	5
268713	79	10	5	29	189	32	12	11	1	...	3.5
299967	69	17	17	42	82	48	18	13	1	...	2

The complete Table 6 has 40 rows, and 14 columns, there are shown the following condition attributes: P# - the number given to each patient, age –age of the subject, Lgm1A -Logical Memory Immediate, Lgm2A - Logical Memory Delayed, TrailA - Trail Making Part A, TrailB -Trail Making Part B, DSST - Digit Symbol Substitution Test, Fcorr -Verbal Fluency Letter F, Reyrcl - Rey Figure Recall, APOE - *ApoE* genotype, … CDRSUM – the sum of boxes- index of the Clinical Dementia Rating.

Table 7. Discretized table extract for above (Table 6) Model2 AD patients

P#	Age	Lgm1A	TrailA	TrailB	DSST	Fcorr	Recyrcl	APOE	...	CDRSUM
70407	*	"(−Inf,11.5)"	*	*	"(−Inf,39.5)"	"(20.5,Inf)"	*	*	...	"(4.5,6.5)"
119156	*	"(−Inf,11.5)"	*	*	"(39.5,Inf)"	"(16.5,20.5)"	*	*	...	"(−Inf,4.5)"
155699	*	"(−Inf,11.5)"	*	*	"(−Inf,39.5)"	"(−Inf,12.5)"	*	*	...	"(−Inf,4.5)"
265499	*	"(11.5,13.5)"	*	*	"(−Inf,39.5)"	"(−Inf,12.5)"	*	*	...	"(4.5,6.5)"
268713	*	"(−Inf,11.5)"	*	*	"(−Inf,39.5)"	"(Inf,12.5)"	*	*	...	"(−Inf,4.5)"
299967	*	"(17.0,Inf)"	*	*	"(39.5,Inf)"	"(16.5,20.5)"	*	*	...	"(−Inf,4.5)"

Table 7 is a part of the discretized table for six (from all 40) AD patients: 70407 to 299967. Significant condition attributes were: Lgm1A *(LOGMEM1A)*, DSST, Fcorr, and others not shown in Table 7 like the BOSTON test. Not significant condition attributes were age, Lgm2A (LOGMEM2A), TrailA, TrailB, Reyrcl (REYRECAL), PAIRED1, PAIRED2, and APOE.

We had obtained 58 rules for Model2 subjects, and as an example, we present below 6 rules filled by the most cases:

$$(LOGMEM\,1A = \,"(17.0, Inf\,)") => (CDRSUM = \,"(-Inf\,, 4.5)"[7])\,7 \qquad (21)$$

$$(FCORR = \,"(12.5, 16.5)")\&(BOSTON = \,"(24.5, Inf\,)")$$
$$=> (CDRSUM = \,"(-Inf\,, 4.5)"[5])\,5 \qquad (22)$$

$$(LOGMEM\,1A = \,"(-Inf\,, 11.5)")\&(DSST = \,"(39.5, Inf\,)")$$
$$\&(BOSTON = \,"(24.5, Inf\,)") => (CDRSUM = \,"(-Inf\,, 4.5)"[4])\,4 \qquad (23)$$

$$(DSST = \,"(39.5, Inf\,)")\&(FCORR = \,"(20.5, Inf\,)")$$
$$=> (CDRSUM = \,"(-Inf\,, 4.5)"[2])\,2 \qquad (24)$$

$$(LOGMEM\,1A = \,"(13.5, 15.5)")\&(FCORR = \,"(12.5, 16.5)")$$
$$=> (CDRSUM = \,"(-Inf\,, 4.5)"[2])\,2 \qquad (25)$$

$$(LOGMEM\,1A = \,"(13.5, 15.5)")\&(FCORR = \,"(-Inf\,, 12.5)")$$
$$=> (CDRSUM = \,"(4.5, 6.5)"[2])\,2 \qquad (26)$$

Interpretation: Eq. 22, if Logical Memory Immediate (LOGMEM1A) is above 17 then CDRSUM is below 4.5; Eq. 27 if Logical Memory Immediate (LOGMEM1A) is above between 13.5 and 15.1 and the Verbal Fluency (Letter F - FCORR) is poor (below 12.5) then CDRSUM is between 4.5 and 6.5, which means the mild dementia [7]. However, notice that this rule was fulfilled in only two cases in our 40 AD patients, whereas the rule Eq. 24 was fulfilled in 7 cases.

Table 8. Confusion matrix for CDRSUM of TestGr group by rules obtained from Model2-group

Predicted					
Actual	"(4.5,6.5)"	"(4.5,6.5)"	"(−Inf,4.5)"	"(6.5,Inf)"	ACC
"(−Inf,4.5)"	0.0	0.0	0.0	0.0	
"(6.5,Inf)"	2.0	19.0	0.0	0.9	
"(6.5,Inf)"	0.0	0.0	0.0	0.0	
TPR	0.0	1.0	0.0		

TPR: True positive rates for decision classes; ACC: Accuracy for decision classes: the global coverage was 1.0 and the global accuracy was 0.90, the coverage for decision classes was 0.0, 1.0, 0.0.

$$(Pat = 164087)\&(LOGMEM\,1A = \,"(13.5, 15.5)")\&$$
$$(DSST = \,"(39.5, Inf\,)")\&(FCOR = \,"(-Inf\,, 12.5)")\&$$
$$(BOSTON = \,"(24.5, Inf\,)")\&(APOE = 1) \qquad (27)$$
$$=> (CDRSUM = \,"("(4.5, 6.5)"$$

$$(Pat = 776254)\& (LOGMEM\ 1A = "(13.5, 15.5)")\&$$
$$(DSST = "(39.5, Inf)")\&(FCOR = "(16.5, 20.5)")\&$$
$$(BOSTON = "(24.5, Inf)")\&(APOE = 0)$$
$$=> (CDRSUM = "("(4.5, 6.5)" \tag{28}$$

Results from Table 8 can be interpreted that from 21 'normal' subjects, 19 subjects have CDRSUM values below 4.5, which means that some of them may have very mild dementia or questionable impairment in the worst case [7]. However, two subjects had CDRSUM between 4.5 and 6, which means that they might have mild dementia [7]. The first patients, with his/her results described in Eq. 28 seems to be easier to interpret at least partly based on Eq. 27 with poor logical memory intermate and poor verbal fluency also with sensitive genetics with APOE = 1. It is more difficult to interpret Eq. 29.

4 Discussion

We have applied rough set theory and its rules (RSR) as the granular computing to estimate a possible disease progression in normal subjects from the BIOCARD study. We used intelligent granular computing with RSR to investigate test results set as granules for individual patients. To estimate their properties, we need to have a Model that has meaning and tells us what the importance of the pattern (granule) is. In fact, our granules are complex (c-granules) as they are changing their properties with the time of the neurodegeneration development till become like granules of patients with dementia or PD [15]. In this work, we have limited our test to the static granules (in one time moment) and we have tried to estimate what is the meaning of a particular, individual granule. We have used two models: Model1 has granules related to normal subjects, MCI, and AD patients. On this basis, we have obtained a large set of rules that have represented subjects' different stages of the disease from normal to dementia. We have tested several of such models mostly changing normal subjects and getting different rules, which we have applied to other normal subjects and estimated what 'normal' means. Also, rules can be created with different granularity and algorithms that might give different classifications. Therefore, we were looking for classifications that are universal e.g., they give similar results with different sets of rules. Model1 has given us rules that are subtle and determine the beginning of possible symptoms. In the next step, we used a more advanced model – Model2 that gave rules based on AD patients. We got higher values of the CDRSUM that gave us only classifications of the possible subjects with mild dementia. Looking into different rules, some of them are easy to interpret, but other patients' granules look relatively normal. As it is the first, to our knowledge, work that estimate distinctive complex pattern of individual patients' symptoms. These results open possibilities for early (preclinical) AD diagnosis based on neuropsychological testing that can be performed remotely.

References

1. Albert, M., et al.: the BIOCARD research team, cognitive changes preceding clinical symptom onset of mild cognitive impairment and relationship to *ApoE* genotype. Curr. Alzheimer Res. **11**(8), 773–784 (2014)

2. Albert, M., et al.: Predicting progression from normal cognition to mild cognitive impairment for individuals at 5 years Brain. **141**(3), 877–887 (2018)

3. Corder, E.H., et al.: Gene dose of apolipoprotein E type 4 allele and the risk of Alzheimer's disease in late onset families. Science **261**, 921–923 (1993)

4. Farrer, L.A., et al.: Effects of age, sex, and ethnicity on the association between apolipoprotein E genotype and Alzheimer disease. A meta-analysis. JAMA **278**, 1349–1356 (1997)

5. Husain, M.A., Laurent, B., Plourde, M.: APOE and Alzheimer's Disease: From Lipid Transport to Physiopathology and Therapeutics REVIEW published, 17 February 2021. https://doi.org/10.3389/fnins.2021.630502

6. Przybyszewski, A.W., Kon, M., Szlufik, S., Szymanski, A., Koziorowski, D.M.: Multimodal learning and intelligent prediction of symptom development in individual Parkinson's patients. Sensors **16**(9), 1498 (2016). https://doi.org/10.3390/s16091498

7. O'Bryant, S.E., Waring, S.C., Cullum, C.M., et al.: Staging dementia using clinical dementia rating scale sum of boxes scores: a Texas Alzheimer's research consortium study. Arch Neurol. **65**(8), 1091–1095 (2008)

8. Pawlak, Z.: Rough Sets: Theoretical Aspects of Reasoning About Data. Kluwer, Dordrecht (1991)

9. Bazan, J., Nguyen, H.S., et al.: Desion rules synthesis for object classification. In: Orłowska, E. (ed.) Incomplete Information: Rough Set Analysis, pp. 23–57. Physica, Heidelberg (1998)

10. Bazan, J., Nguyen, H.S., et al.: Rough set algorithms in classification problem. In: Polkowski, L., Tsumoto, S., Lin, T. (eds.) Rough Set Methods and Applications, pp. 49–88. Physica, Heidelberg (2000)

11. Grzymała-Busse, J.: A new version of the rule induction system LERS. Fundamenta Informaticae **31**(1), 27–39 (1997)

12. Bazan, J.G., Szczuka, M.: The rough set exploration system. In: Peters, J.F., Skowron, A. (eds.) Transactions on Rough Sets III. LNCS, vol. 3400, pp. 37–56. Springer, Heidelberg (2005). https://doi.org/10.1007/11427834_2

13. Przybyszewski, A.W.: The neurophysiological bases of cognitive computation using rough set theory. In: Peters, J.F., Skowron, A., Rybiński, H. (eds.) Transactions on Rough Sets IX. LNCS, vol. 5390, pp. 287–317. Springer, Heidelberg (2008). https://doi.org/10.1007/978-3-540-89876-4_16

14. Bazan, J.G., Szczuka, M.: RSES and RSESlib - a collection of tools for rough set computations. In: Ziarko, W., Yao, Y. (eds.) RSCTC 2000. LNCS (LNAI), vol. 2005, pp. 106–113. Springer, Heidelberg (2001). https://doi.org/10.1007/3-540-45554-X_12

15. Przybyszewski, A.W.: Parkinson's disease development prediction by C-granule computing. In: Nguyen, N.T., Chbeir, R., Exposito, E., Aniorté, P., Trawiński, B. (eds.) ICCCI 2019. LNCS (LNAI), vol. 11683, pp. 296–306. Springer, Cham (2019). https://doi.org/10.1007/978-3-030-28377-3_24

Experiments with Solving Mountain Car Problem Using State Discretization and Q-Learning

Amelia Bădică[1], Costin Bădică[1(✉)], Mirjana Ivanović[2],
and Doina Logofătu[3]

[1] University of Craiova, Craiova, Romania
costin.badica@edu.ucv.ro
[2] University of Novi Sad, Novi Sad, Serbia
[3] Frankfurt University of Applied Sciences, Frankfurt am Main, Germany

Abstract. The aim of this paper is to explore the model of the Mountain Car Problem. We provide insight into the physics behind the model. We present some experimental results obtained by numerically simulating the model. We also propose a reinforcement learning approach for deriving an optimal control policy combining model discretization and Q-learning.

Keywords: Dynamic system · Mountain car problem · Q-learning · SARSA

1 Introduction

The Mountain Car Problem is a standard benchmark for experimenting with reinforcement learning algorithms. Our aim is to provide new theoretical and experimental insights into the Mountain Car Problem. Although this problem has a tradition of more than three decades in the literature of reinforcement learning, we could not find in the literature a complete and physically accurate explanation of its dynamic model.

The main contributions of our work are outlined as follows: i) accurate presentation of the physics details of developing the dynamic model of the Mountain Car Problem; ii) new insights into random walk experimental results obtained for the Mountain Car Problem; iii) detailed investigation and comparison of standard Q-learning and SARSA algorithms for solving the digitized version of the Mountain Car Problem.

We start in Sect. 2 with a brief overview of related works on developments of the Mountain Car model in the reinforcement learning literature. In the first half of Sect. 3 we present the detailed development of the Mountain Car model starting from the first principles from physics. In the second half of Sect. 3 we present new insights into experimental results that we obtained by performing a sequence of random walks on the Mountain Car Model. Section 4 is dedicated to

N. T. Nguyen et al. (Eds.): ACIIDS 2022, LNAI 13757, pp. 142–155, 2022.
https://doi.org/10.1007/978-3-031-21743-2_12

introduce the Q-learning and SARSA algorithms, our proposed implementations as well as a discussion of the experimental results obtained. For the experiments we have used a digitized (i.e. finite state) version of the Mountain Car Problem. The last section concludes and points to future works.

2 Related Works

Although the Mountain Car Problem has been proposed more than 30 years ago as a benchmark problem for optimal control and reinforcement learning algorithms, we could not find in the scientific literature a complete and physically accurate explanation of its dynamic model.

The Mountain Car Problem was firstly proposed in Moore's PhD thesis [3, Chapter 4.3] and was referenced in one of his early papers [4]. In these works, the hill-shaped driving track of the car is described by a polynomial function that defines the height of the hill depending on the horizontal vehicle location. Moreover, the model is using a continuous model for the speed of the car depending on the pedal height.

The Moore's proposed model is referenced in the experimental part of [5] as benchmark for the reinforcement learning with replacing eligibility traces approach. However, the details of the Mountain Car problem considered in this work are not given in [5].

Another dynamic model of the Mountain Car Problem is presented in [6, Chapter 4.5.2]. This model is closer to the model that we have used in our paper. The shape of the hill is the same, but the values of the other parameters differ. Moreover, although some physics details are included in the model presentation, its complete physics explanation is missing.

OpenAI Gym is a software framework for bench-marking reinforcement learning algorithms [1]. On one hand it includes a number of dynamic models, and on the other hand it allows definition of new models. Those models are defined as environment model. The idea is that reinforcement learning algorithms are encapsulated into agents that interact with the environment through a standard programming interface. OpenAI Gym includes two Mountain Car models: "Mountain Car v0" that is similar to the model that we used in our paper and "Mountain Car Continuous v0" that differs from our model by the fact that the action space is continuous, rather than discrete.

Mountain Car Problem was used in many online tutorials that introduce the practical programming of reinforcement learning algorithms. One such a video tutorial is [8]. Here the author presents an approach for solving the Mountain Car Problem included in OpenAI Gym using Q-learning and state discretization. We have used it as a base implementation for our experiments. Nevertheless, we extended this implementation in many directions (see Sect. 4).

Standard reference [7, Chapter 10.1] also includes a model of the Mountain Car Problem that is similar to the one used in our paper. However, the details of the physics of the model are missing. The model is introduced as a mathematical object, suitable for experimenting with various reinforcement learning algorithms.

3   Modeling the Mountain Car Problem

3.1   Physics of the Mountain Car Problem

The Mountain Car problem assumes an autonomous car driving on a one-dimensional track that follows a mountain range described by the equation:

$$y = \sin \omega x \quad \text{for} \quad x \in [-\frac{3\pi}{2\omega}, \frac{\pi}{2\omega}] \tag{1}$$

Note that Eq. (1) describes a function that spans a full cycle of "sinus" function thus modeling a valley between a left and right hill. The car is supposed to start its driving episode at initial position x_0 somewhere in the valley in the vicinity of $-\frac{\pi}{2\omega}$ and its goal is to climb the rightmost hill. The goal can be described as the car reaching a location $x \geq x_g$, where the "goal position" x_g is a real value in the left vicinity of $\frac{\pi}{2\omega}$.

The car engine is controlled by a thruster that can either thrust left or right with equal force or no thrust at all. The force of the thruster is not strong enough to defeat gravity and accelerate up the slope to the top of the right hill. The solution assumes the movement of the car in the opposite direction to the goal (i.e. to the left) to accumulate enough inertia to defeat the slope, even if slowing down up to the top of the hill.

The standard physical environment of the Mountain Car is shown in Fig. 1. In this case $\omega = 3.0$ and the curve describing the mountain range $y = \sin 3x$ is defined on interval $[-\frac{\pi}{2}, \frac{\pi}{6}] \approx [-1.57, 0.52]$. Note that in the standard Mountain Car model, as defined in [7, Chapter 10.1] and [1], this interval is set to $[-1.2, 0.6]$. If the position of the car reaches a value below the lower bound then the location is clipped to the lower bound and the car velocity is reset to 0. The goal of the car is to reach the top of the hill at a location $x \geq 0.5$. The initial position x_0 is defined in the vicinity of $-\frac{\pi}{6} \approx -0.52$, while the initial velocity of the car is set to 0. The current implementation from OpenAI Gym assumes that x_0 is a uniformly distributed random number in interval $[-0.6, -0.4]$.

The dynamics of the car is described with the following equation in vector format:

$$m \cdot \overrightarrow{a_t} = \overrightarrow{F_t} + \overrightarrow{G} + \overrightarrow{F_f} \tag{2}$$

Projecting Eq. (2) on the movement direction we obtain:

$$m \cdot a_t \cos \theta = F_t \cdot \cos \theta - m \cdot g \cdot \sin \theta - k_f \cdot v_t \cdot \cos \theta \tag{3}$$

Note that g denotes the gravitational constant, m is the mass of the car and k_f represents the friction coefficient. F_t is the force of the thruster and it represents the input that controls the dynamics of the model. Simplifying Eq. (3) by $m \cdot \cos \theta$ we obtain:

$$a_t = \frac{F_t}{m} - g \cdot \tan \theta - \frac{k_f}{m} \cdot v_t \tag{4}$$

Note that $\tan \theta$ is the slope of the tangent to the curve representing the mountain range. If $y = f(x)$ is the equation of this curve, it follows that $\tan \theta =$

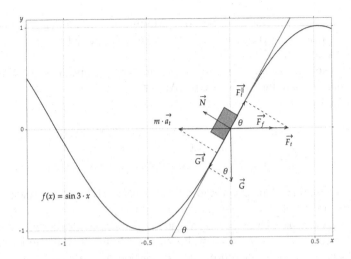

Fig. 1. Mountain car physical environment.

$\frac{df(x)}{dx}(x_t) = f'(x_t) = \omega \cdot \cos \omega x_t$. Substituting $v_t = \dot{x}_t$ and $a_t = \ddot{x}_t$, Eq. (4) becomes:

$$\ddot{x}_t = \frac{F_t}{m} - g \cdot \omega \cdot \cos \omega x_t - \frac{k_f}{m} \cdot \dot{x}_t \tag{5}$$

Differential Eq. (5) gives the dynamics of the car under the control of the thrust force. The system state is captured by the pair $(x_t, v_t) = (x_t, \dot{x}_t)$ containing the location and the velocity of the car in horizontal direction. The initial state is $(x_0, 0)$ where x_0 is the initial location of the car in the vicinity of $-\frac{\pi}{2\omega}$. This equation is usually presented in discretized form (6) with a time step $h > 0$ using Euler's numerical integration method.

$$\dot{x}_{t+h} = \dot{x}_t + h \cdot (\frac{F_t}{m} - g \cdot \omega \cdot \cos \omega x_t - \frac{k_f}{m} \cdot \dot{x}_t)$$
$$x_{t+h} = x_t + h \cdot \dot{x}_t \tag{6}$$

Note that Eqs. (6) still represent a discrete time continuous model, as the time is discrete but the states are continuous.

The aim of the Mountain Car problem is to reach the goal location on the top of the right hill. Therefore each action of the car is rewarded accordingly. In the weaker version of the problem it is assumed that the car does not know the goal. So the only way to perceive the goal is when it was reached. Therefore the car is rewarded a small negative value (usually -1) for each reached state that does not achieve the goal.

3.2 Model Exploration Using Random Walk and Numerical Simulation

In this section we provide an experimental investigation of the Mountain Car model as defined by Eq. (6) using numerical simulation. We have performed experiments using our own hand-crafted Mountain Car model. We chose this approach after noticing that the simulation using our model was faster than the one provided by OpenAI Gym.

We set the Mountain Car model parameters to obtain a mathematical model similar to that included in OpenAI Gym [1] and standard reference ([7, Chapter 10.1]):

$$\omega = 3.0 \qquad m = 1000.0 \qquad g = 0.0025/3.0$$
$$k_f = 0 \qquad h = 1.0 \qquad x_g = 0.5 \tag{7}$$
$$F_t \in \{-1, 0, 1\} \quad x_t \in [-1.2, 0.6] \quad v_t \in [-0.07, 0.07]$$

Note the bounds of position and velocity in Eqs. (6). While the bounds of location are pretty obvious, taking into account Eq. (1), the bounds of velocity are far from obvious. Therefore we decided to evaluate them experimentally by performing a series of random walks in the Mountain Car model given by Eqs. (6). Each random walk represents an episode of the simulation, in which the car starts from the initial state and drives until reaching the goal state.

The simulation is described by Algorithm 1. This algorithms takes the Mountain Car environment Env and the total number $ITMAX$ of episodes and returns the bounds of position and velocity $(pmin, pmax, vmin, vmax)$, as well as the vector $Episodes$ containing the length of each episode. We assume that the API of the Mountain Car environment supports the following methods:

- *reset()* that resets the environment to its initial state as a random value in the vicinity of $-\frac{\pi}{2\omega}$. This method returns the state and the value of *done* that is reset to *False*.
- *step(action)* that takes an action from the action space and determines the next state of the environment. If the goal is reached, flag *done* is set to *True*.
- *random_action()* that returns an action of the car by sampling the action space.
- Note that methods *reset()* and *step(action)* return also the reward, but this is not used in Algorithm 1.

We have implemented Algorithm 1 in Python 3.7.3 on an x64-based PC with a 2 cores/4 threads Intel i7-5500U CPU at 2.40 GHz and running Windows 10.

The total simulation time for 2000 episodes was 1057.793 s, i.e. approximately 18 min. Episode lengths were distributed in the interval $[865, 324776]$ with an average value of 42209.101 and standard deviation of 37989.213.

Figure 2 presents the 80 bins histogram of episode lengths. It is interesting to observe that the shape of this histogram suggests a log-normal distribution of the episode length [2]. This observation is also supported by the high value obtained for the standard deviation that is typical for log-normal distribution. However, observe that the shape is clipped and highly skewed to the left side, as there is a strictly positive minimum length of an episode.

The minimum and maximum bounds of velocity that were recorded during this simulation were $vel\_min = -0.06587993$ and $vel\_max = 0.05977063$, thus

Algorithm 1. $MountainCarRandomWalk(Env, ITMAX)$ algorithm for determining the position and velocity bounds in Mountain Car problem using a series of random walks. The algorithms returns also the vector of lengths of each driving episode from initial state to the goal state.

Require: Env. The environment model of the Mountain Car problem.
  $ITMAX$ The number of episodes.
Ensure: $Episodes$. Vector of lengths of each episode.
  $pmin, pmax, vmin, vmax$. Position and velocity bounds.
1: $Episodes \leftarrow []$
2: $(pmin, pmax, vmin, vmax) \leftarrow (0,0,0,0)$
3: **for** $k = 1, ITMAX$ **do**
4:    $(p, v, done) \leftarrow Env.reset()$
5:    $count \leftarrow 0$
6:    **while** $\neg done$ **do**
7:       $action \leftarrow Env.random\_action()$
8:       $(p, v, done) \leftarrow Env.step(action)$
9:       $(pmin, pmax) \leftarrow (\min(pmin, p), \max(pmax, p))$
10:      $(vmin, vmax) \leftarrow (\min(vmin, p), \max(vmax, p))$
11:      $count \leftarrow count + 1$
12:      $Episodes.append(count)$
13:   **end while**
14: **end for**
15: **return** $Episodes, pmin, pmax, vmin, vmax$

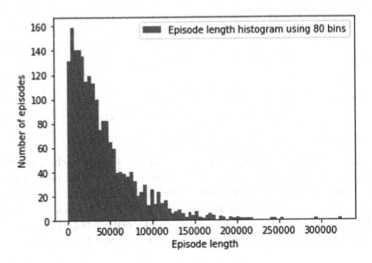

Fig. 2. The 80 bins histogram of episode length using data from 2000 episodes.

confirming the bounding box $[-0.07, 0.07]$ set for the velocity in the Mountain Car model. The bounds recorded for the position were $pos\_min = -1.2$ and $pos\_max = 0.52564615$, again confirming expectations.

4   Optimal Control Using State Discretization and Q-Learning

4.1   Q-Learning and SARSA Algorithms

We model our autonomous car as an intelligent agent that is able to perceive its environment through sensors and act upon that environment through actuators, as shown in Fig. 3. In particular we are interested to employ the technique of reinforcement learning to let the car "learn" an optimal acting policy through self-driving episodes. Reinforcement learning aims to optimize agent action based on action feedback as punishments and/or rewards. An agent percept in this model will consist of a pair *state, reward*. The *state* represents the agent observation of the environment state, while *reward* is a real value that locally estimates the "goodness" of that state from the agent perspective.

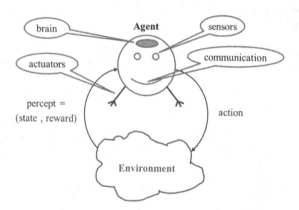

Fig. 3. Elements of an intelligent agent.

The agent strategy specifies what the agent must do in each observed environment state. If agent takes action a in state s then environment will transit in state s'. We assume that the environment is Markovian, i.e. its next state depends only on its current state and agent action. If E denotes the set of environment states and A denotes the set of agent actions, the strategy is called *policy* and it is defined as function π mapping each state to an action:

$$\pi : E \to A \tag{8}$$

Let us assume that the reward of the agent by taking action a in state s is denoted as $R(s, a)$ and that the initial state of the environment is s_0. An agent following a given policy π will generate the following agent run:

$$r = s_0 \xrightarrow{a_0 = \pi(s_0)} s_1 \xrightarrow{a_1 = \pi(s_1)} \ldots \xrightarrow{a_n = \pi(s_n)} s_{n+1} \cdots \tag{9}$$

The utility perceived by the agent for the run $[s_0, a_0, s_1, a_1, s_2, a_2 \ldots]$ accumulates all the rewards that were perceived by the agent in each visited state. We assume a potentially infinite horizon of the agent, i.e. the length of the agent run is unbounded. As there are goal states, the agent run is terminated when a goal state is reached. In order to guarantee the convergence of the agent utility in the presence of unbounded runs, we assume a discounted additive utility of the agent, defined as follows:

$$U^\pi(s = s_0) = U([s_0, a_0, s_1, a_1, \ldots, s_n, a_n, \ldots]) = \sum_{n \geq 0} \gamma^n \cdot R(s_n, a_n) \tag{10}$$

Equation (10) suggests the recursive definition of $U^\pi(s)$ for each transition $s \xrightarrow{a = \pi(s)} s'$ given by Eq. (11).

$$U^\pi(s) = R(s, a) + \gamma \cdot U^\pi(s') \tag{11}$$

Assuming that the discount factor is $0 < \gamma < 1$ and the set of rewards $\{R(s, a)\}_{s \in E, a \in A}$ is bounded, the utility of the agent is well defined by Eq. (10). The aim of reinforcement learning is to determine the optimal policy π that maximizes the value of $U^\pi(s)$ for all states s. The corresponding maximum value $U(s)$ represents the utility of the agent in state s by running its optimal policy.

$$U(s) = \max_{\pi(s) = a \in A} U^\pi(s) \tag{12}$$

Maximizing for π in Eq. (11) we obtain recurrence (13) for agent utility. This is a model-based equation, as it explicitly uses the transition model $s \xrightarrow{a} s'$ of the environment. In a realistic setting, this model is not available to the agent.

$$U(s) = \max_{a \in A, s \xrightarrow{a} s'} (R(s, a) + \gamma \cdot U(s')) \tag{13}$$

Q-learning aims to determine the values $Q(s, a)$ that estimate the utility of the agent when taking action a in environment state s. Then the optimal policy is defined as:

$$\pi(s) = \underset{a \in A}{\operatorname{argmax}} \, Q(s, a) \tag{14}$$

We can derive a model-based recursive equation defining Q values, following Eq. (13). However, as we assume that the transition model of the environment is not available, for the computation of Q values we shall use temporal difference learning. If the currently observed state is s then the agent will choose a certain action a using a given learning strategy. This produces reward $R(s, a)$ and next state s'. Then the value $Q(s,a)$ is updated using Eq. (15). Hyper-parameter α (learning rate) can be set to a constant small positive value or it can be computed as the harmonic sequence $\alpha_n = A/(B + C \cdot n)$ (A, B, C are positive constants) depending on the number n of times action a was taken in state s. Note that for $C = 0$ the learning factor is constant $\alpha = A/B$.

$$Q(s,a) = (1 - \alpha) \cdot Q_{OLD}(s,a) + \alpha \cdot Q_{NEW}(s,a)$$
$$= Q_{OLD}(s,a) + \alpha \cdot (Q_{NEW}(s,a) - Q_{OLD}(s,a)) \tag{15}$$

We consider two related algorithms that follow Eq. (15) for updating Q values for each observed transition:

- Q-learning that computes Q_{NEW} using Eq. (16).

$$Q_{NEW}(s,a) = R(s,a) + \gamma \cdot \max_{a' \in A} Q(s',a') \tag{16}$$

- SARSA that computes Q_{NEW} using Eq. (17). Here a' is the agent action in state s' determined using the learning strategy.

$$Q_{NEW}(s,a) = R(s,a) + \gamma \cdot Q(s',a') \tag{17}$$

The learning algorithm follows an epsilon-Greedy strategy that mixes exploration and exploitation. Exploration favors random actions, while exploitation favors the locally best actions based on current Q value. The choice between the two possibilities is determined by a Bernoulli experiment with a probability $0 \leq \varepsilon \leq 1$. As the learning proceeds, ε is decreased so exploitation (i.e. Greedy choice) is favored in the limit.

4.2   State Discretization

State discretization strategy follows the idea from [8]. Its effect is to transform our discrete time continuous model (6) into a finite state model. The size of the resulting model is controlled by the discretization step.

Let us consider a state variable with domain $[a,b)$ and a natural number $n \in \mathbb{N}$ representing the number of states of the finite state representation. The discretization step is computed as $h = (b-a)/n$. Each value $x \in [a,b)$ is mapped to a natural number $digi(x) \in \{0,1,\dots,n-1\}$ according to Eq. (18).

$$digi(x) = k \in \{0,1,\dots,n-1\} \iff x \in I_k = [a + k \cdot h, a + (k+1) \cdot h) \tag{18}$$

Discretization in Q-learning is useful to control the size of the table $Q[s,a]$. In the Mountain Car problem there are two state variables: position and velocity. If the finite state representation uses n_p positions and n_v velocities then the size of Q table is $3 \cdot n_p \cdot n_v$.

State discretization is used as follows. Before performing the update of the Q values, the observed state is digitized. Then the action determination according to the learning strategy and the update itself will use the digitized value of the state. This basically means that for all states $s \in I_k$ for which $digi(s) = k$ the algorithm will determine the same value in the Q table.

4.3   Experimental Results

We have implemented the Q-learning and SARSA algorithms presented in this section. The starting point of our implementation is [8]. We have extended the implementation in many directions, as follows:

- We have added the SARSA algorithm, not present in [8].
- We have added our own implementation of the Mountain Car model, following the complete physical model developed in Sect. 3. Using our model, the learning runs considerably faster than using the OpenAI Gym provided model.
- We have added two additional visualizations of the optimal policy and of the model exploration during learning process.
- We have added a procedure for evaluating the policy computed by the algorithms.

A training session using either Q-learning or SARSA involves performing a sequence of N_E episodes. Each episode is a sequence of agent steps, following the learning algorithm. An episode can finish either by reaching a maximum preset number of steps N_{LE} or by reaching the goal state. Differently from [8], we imposed a significantly larger upper bound of each episode of the learning session. This decision was taken based on the observation that more than 50% of the episodes of a series of random walks actually reached the goal state after a reasonable number of steps (below 30000).

The probability ε of selecting exploration versus exploitation starts from an initial value ε_s and it is decreased at a constant rate after each episode until it reaches the minimum value $\varepsilon_f < \varepsilon_s$. The rate of decrease is $\Delta\varepsilon = (\varepsilon_s - \varepsilon_f)/N_E$.

The policy determined by the learning algorithm was evaluated by running a fixed number of episodes N_{TE}, each with a maximum number of steps N_{TS}. Then we determined the percent of episodes that reached the goal state acc, as well as their average length len. For each episode we recorded the value of the accumulated rewards. Then we plotted the moving average of each N_m consecutive accumulated rewards.

The description and the values of the parameters that we have used in our experiments are summarized below:

- $N_E = 4000$. Number of learning episodes.
- $N_{LE} = 30000$. Maximum number of steps of a learning episode.
- $\varepsilon_s = 0.6$. $\varepsilon_f = 0.01$. Maximum (initial) and minimum (final) value of the probability of selecting exploration versus exploitation during learning.
- $N_{TE} = 1000$. Maximum number of episodes for testing the policy computed by the learning algorithm.
- $N_{TS} = 500$. Maximum number of steps of running each episode for testing the policy computed by the learning algorithm.
- $n_p, n_v \in \{10, 20, 30, 40\}$. Number of discretization steps for position and velocity.
- $\gamma = 0.99$. Discount factor.

- $A = 1.0$, $B = 10.0$, $C = 0.01$. Parameters of the harmonic learning factor α_n.
- $N_m = 50$. Number of rewards used to compute and plot the moving average of the accumulated rewards per each learning episode.

Our experimental results are presented in Tables 1 and 2. Each table entry is defined for a specific value of the discretization parameters n_l and n_v. Topmost halves of each table refer to Q-learning, while bottom half of each table refers to SARSA.

Table 1 presents the accuracy (percentage of successful test episodes) and the average length of the car trajectory (from successful episodes) generated by the policy computed by the algorithm. First of all observe that all the 32 algorithm runs produced an accuracy higher than 73%, only 3 runs produced an accuracy below 80%, 6 runs produced an accuracy below 90% and 24 runs produced an excellent accuracy higher than 95%. In 9 runs the accuracy was actually 100%.

The average of the length of the shortest trajectory was 140.99 and it was obtained by SARSA algorithm for $n_p = 20$ and $n_v = 30$ (shown in bold underlined in Table 1). Combined with the excellent accuracy 99.90%, clearly this case can be considered as providing the best result among all the cases. Note also that close and consistent results for $n_p = 20$ and $n_v = 30$ were also obtained by the Q-learning algorithm.

Table 1. Accuracy (acc) and average solution length (len) obtained with Q-learning (top) and SARSA (bottom) algorithms.

n_p/n_v	10	20	30	40
10	80.10%, 169.75	96.20%, 150.54	99.20%, 154.68	100%, 243.46
20	90.10%, 172.05	100%, 157.97	99.60%, 147.26	100%, 160.73
30	94.40%, 158.01	98.40%, 179.66	100%, 171.51	100%, 145.14
40	77.30%, 157.77	97.80%, 150.23	99.30%, 147.64	99.30%, 195.60
10	81.00%, 174.16	96.70%, 154.68	99.30%, 155.96	100%, 151.28
20	86.00%, 168.38	100%, 156.47	99.90%, **140.99**	73.50%, 154.38
30	95.50%, 165.31	98.50%, 157.80	100%, 168.21	100%, 149.57
40	76.30%, 208.01	97.90%, 147.78	99.70%, 176.92	95.10%, 184.15

Table 2 presents the total number of iterations performed by each experimental case during learning. Note that the values were different as we did each training episode until the goal state was reached, by setting a high value for the maximum number of iterations per episode (to force closing really long-running training episodes). The shortest number of iterations was done for training the Q-learning algorithm for $n_p = 20$ and $n_v = 30$. Moreover, in this case the accuracy of the solution was 100%. However, as regarding its optimality, The average of the length of the shortest trajectory was 157.97 (see cell in bold underline in Table 2), i.e. higher than the best case 140.99.

In our opinion, the differences between the results obtained by Q-learning and SARSA were not significant. Therefore, we also checked if there is some statistical similarity in the results produced by the algorithms. We found a positive correlation (0.983) between the number of iterations produced by each algorithm, as well as a positive correlation 0.686 between the accuracy of the algorithms. However, the average lengths of the trajectories generated by the solution policies were uncorrelated (0.038).

We also checked if there are any correlations between the results within each algorithm. Among all the situation, we found interesting the negative correlation between the number of iterations and the accuracy of the solution in both algorithms (-0.46 for Q-learning and -0.33 for SARSA). One possible explanation could maybe that the higher number of iterations indicates some difficulties in the learning process, difficulties that are also explained by the lower accuracy of the solution.

As part of our numerical simulation program for learning the optimal policy and testing the solution, we have also developed three visualizations for presenting: i) the convergence of the accumulated rewards during learning; ii) the optimal policy; iii) the exploration done by the car agent during the learning process.

Table 2. Number of iterations performed by Q-learning (top) and SARSA (bottom) algorithms.

n_p/n_v	10	20	30	40
10	1332618	1161777	1245702	1285570
20	1306056	**1156701**	1178533	1244548
30	1474829	1236062	1272904	1373627
40	1563608	1347271	1394002	1510739
10	1346977	1182048	1227957	1323528
20	1347295	1176168	1215110	1291964
30	1442344	1258570	1312203	1411848
40	1585404	1358674	1437671	1548015

Figure 4 presents the accumulated reward convergence process for the case that we considered the best: SARSA with $n_p = 20$ and $n_v = 30$.

Figure 5 presents the policy computed by SARSA algorithm with $n_p = 20$ and $n_v = 30$. The color codes represents actions performed by the policy in each state, while small rectangle patches represent the digitized states of the Mountain Car model.

Figure 6 is an example plot showing how the car agent is exploring the environment by taking push right moves. Similar visualizations were produced for no push, push left as well as total number of actions taken by the agent car in each state of the environment.

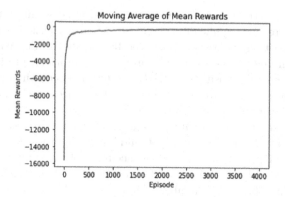

Fig. 4. Convergence of accumulated rewards during learning in SARSA with $n_p = 20$ and $n_v = 30$, plotted as the moving average of results obtained for $N_m = 50$ consecutive episodes.

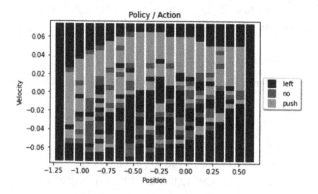

Fig. 5. Policy computed by SARSA with $n_p = 20$ and $n_v = 30$.

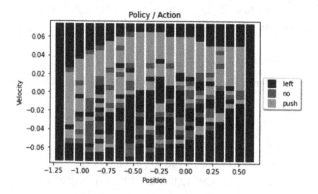

Fig. 6. Number of push right actions per state during SARSA learning for $n_p = 20$ and $n_v = 30$.

5   Conclusions and Future Work

In this paper we provided a theoretical and experimental analysis of the Mountain Car Problem. We have explained the physics of the dynamic model of the Mountain Car. We have used the model to perform a sequence of random walks in order to better understand the problem. Then we have shown how the problem can be solved by combining state discretization with Q-learning and SARSA algorithms. We provide detailed experimental results regarding the computational effort incurred by these algorithms, as well as the accuracy and optimality of the solutions produced by the algorithms. As future work, this research can be extended by: i) incorporating different reinforcement learning algorithms for the Mountain Car problem; ii) providing similar analysis for other standard benchmark problems in reinforcement learning.

References

1. Brockman, G., et al.: Openai gym. arXiv preprint arXiv:1606.01540 (2016)
2. Limpert, E., Stahel, W.A., Abbt, M.: Log-normal Distributions across the sciences: keys and clues: on the charms of statistics, and how mechanical models resembling gambling machines offer a link to a handy way to characterize log-normal distributions, which can provide deeper insight into variability and probability-normal or log-normal: that is the question. BioScience **51**(5), 341–352 (2001). https://doi.org/10.1641/0006-3568(2001)051[0341:LNDATS]2.0.CO;2
3. Moore, A.W.: Efficient memory-based learning for robot control. Ph.D. thesis, Computer Laboratory, University of Cambridge, Cambridge CB3 0FD, United Kingdom (October 1990)
4. Moore, A.W.: Variable resolution dynamic programming: efficiently learning action maps in multivariate real-valued state-spaces. In: Birnbaum, L.A., Collins, G.C. (eds.) Machine Learning Proceedings 1991, pp. 333–337. Morgan Kaufmann, San Francisco (1991). https://doi.org/10.1016/B978-1-55860-200-7.50069-6
5. Singh, S.P., Sutton, R.S.: Reinforcement learning with replacing eligibility traces. Mach. Learn. **22**(1), 123–158 (1996). https://doi.org/10.1023/A:1018012322525
6. Sugiyama, M.: Statistical Reinforcement Learning. Modern Machine Learning Approaches. Chapman and Hall/CRC, Boca Raton (2015)
7. Sutton, R.S., Barto, A.G.: Reinforcement Learning: An Introduction, 2nd edn. The MIT Press, Cambridge (2020)
8. Tabor, P.: Q learning with just NumPy. Solving the mountain car. Tutorial (2019). https://www.youtube.com/watch?v=rBzOyjywtPw&t=3s. Accessed 7 Jan 2022

A Stable Method for Detecting Driver Maneuvers Using a Rule Classifier

Piotr Porwik$^{(\boxtimes)}$, Tomasz Orczyk, and Rafal Doroz

Faculty of Science and Technology, Institute of Computer Science,
University of Silesia, Bedzinska 39, 41-200 Sosnowiec, Poland
{piotr.porwik,tomasz.orczyk,rafal.doroz}@us.edu.pl

Abstract. Traffic accidents and vehicle mishandling are significant problems in road transportation, affecting human lives. Various studies suggest that driver behavior is a key factor in the most road accidents and contributes significantly to fuel consumption and emissions. Improvements in driver behavior can be achieved by providing feedback to drivers on their driving behavior. The identification of risky and wasteful maneuvers allows the evaluation of driver behavior. This allows the elimination of irresponsible drivers who pose a danger in traffic, and at the same time, it allows the reduction of maintenance and repair costs of the vehicle fleet. This paper presents the first stage of a driver profiling method based on the analysis of signals coming from the vehicle CAN bus and auxiliary device containing a GPS receiver and an IMU unit. No additional equipment is needed, what is an advantage of the proposed method.

1 Introduction

Around the world, road traffic is growing at a tremendous rate, year after year. This is due to economic needs, supply chains to hard-to-reach places or the convenience of individual car including car long- or short-term leasing and rental services. Unfortunately, the increase in traffic is also associated with an increase in the number of accidents involving drivers and other road users. Every year, thousands of people lose their lives or are seriously injured in road accidents. In Europe in 2019, according to European Union statistics, accidents occur on highways (9%), in urban areas (38%) and on rural roads (53%) [3]. Road accidents generate costs related to treatment and rehabilitation of people, repair of cars or repair of road infrastructure. This is always a big burden on the budget of any country. Any erroneous decision by a driver can lead to a dangerous traffic incident that must be effectively addressed. This problem becomes even more complex when the traffic is heterogeneous and involves different types of vehicles.

The issue of vehicle damage and insurance is very important from a business point of view because it always involves losses for the company. After an accident, the car must be repaired and tangible and intangible damages must

N. T. Nguyen et al. (Eds.): ACIIDS 2022, LNAI 13757, pp. 156–165, 2022.
https://doi.org/10.1007/978-3-031-21743-2_13

be covered, which is often the cause of legal disputes. This is also important for car rental companies, where car recalls generate economic losses. In these types of companies, customers do not take care of the rental cars and often damage them through irresponsible driving or mishandling. Some of these behaviors are difficult to detect and only manifest themselves after some time in the poor condition of the vehicle. This user behavior is increasingly prompting companies to install devices in vehicles that monitor the driver's driving [6]. This makes it possible, in particularly drastic cases, to refuse to rent the vehicle again or, in the case of other companies, to send the driver to retraining courses

Most of the studies have used machine learning methods - artificial neural networks (ANN) and attention-based deep neural network (ADNet) [8], adaptive neuro-fuzzy inference systems (ANFIS) [2], convolutional networks (CNN) [5,9], functional principal component analysis (PCA) [4] with projection into a low dimensional space, as well as support vector machines (SVM) [6]. Most studies devoted to analyzing driver behavior focus on analyzing the age and gender of the driver, look at the impact of using additional devices while driving, check for the presence of vehicles on side streets, the intensity of pedestrian, bicycle or vehicle traffic in front of, from behind or from the opposite direction. A comprehensive review of these parameters, the reader can find in works [2,6,7]. The proposed solutions can mainly be used in laboratory simulators, where prediction of driver behavior can be measured, and then some recommendations can be formed. As a result, this leads to increased road safety by imparting this knowledge during driver training.

In real conditions, observation of driver's behavior (e.g. whether he uses cell phone) and observation of traffic where the vehicle is moving are not possible. This happens, for example, at car rental companies, where once a car is rented, the driver is not observed and their driving style is not recorded. This often leads to improper use of the vehicle, and sometimes to deliberate violations of road safety rules. As a result, this leads to frequent vehicle breakdowns and even road accidents. In general, driving can be divided into two basic categories: safe or aggressive driving. Safe (eco-friendly) driving results in lower CO_2 emissions, fewer fatalities, and safer roads. Aggressive driving results in more air pollution around the roads, more fuel consumption, more accidents, and less safety for road users.

Some driving simulator studies have analyzed driver behavior under various road conditions [6]. In addition, the study focused on driver behavior at intersections with traffic lights. Longitudinal acceleration and longitudinal speed were measured, as well as throttle pressure, deceleration during braking, and brake pedal force. However, these measurements were carried on simulators and have not been conducted under actual road conditions. Due to the above limitations, in our study, we analyze the data coming from the CAN (Controller Area Network) bus, an accelerometer, and a gyroscope. In our research, additional devices to detect eye, head and body movements have been eliminated. This is expensive equipment, not always installed, and not always effective. The use of road edge detection systems was also eliminated because shadows or road lighting limit the effectiveness of such a solution.

2   Data Logging

The driver's driving behavior is evaluated on the basis of driving parameters read from the CAN bus. The CAN (Controller Area Network) bus is an automotive bus standard designed to provide communication between microcontrollers and devices of a car. CAN is a serial communication bus using a two-wire twisted pair cable. Messages can be transmitted at a maximum rate of 1 Mbps. The CAN protocol gives high noise immunity and reliability [1]. In addition to CAN bus messages, data from additional sensors installed in the vehicle were used. The data from the sensors were synchronized with the CAN. The research used a precise three-axis digital accelerometer and gyroscope operating in three orthogonal directions: X, Y and Z. It is a so-called IMU (Inertial Measurement Unit) sensor. The idea of how such a sensor works on the car board is shown in Fig. 1.

Table 1. List of monitored driving parameters

Variables	Units	Source
Longitudinal velocity	m/s	CAN
Acceleration in X,Y,Z direction	m/s²	IMU
Angular speed around X,Y,Z axis	°/s	IMU
Brake pedal force	N	CAN
Throttle opening angle	°	CAN
Clutch pedal force*	N	CAN
ABS, ESP, ASR Intervention	Yes/No	CAN
Turn signals	On/Off	CAN
Steering wheel angle	°	CAN
Geographical coordinates	°	GPS
Heading	°	GPS

*) for cars with a manual transmission

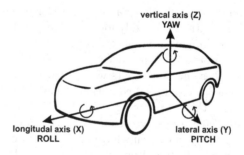

Fig. 1. Measured accelerations and rotations.

The parameters listed in Table 1 are measured using car's on-board equipment. Only the 3-axis accelerometer/gyroscope (MPU6050) and GPS sensor are installed as additional external devices. This reduces the cost of vehicle equipment.

2.1 Data Stream Forming

As mentioned before, our system is using at least three different data sources: the CAN bus (which actually transfers data from various in-car control units, like Engine Control Unit, ABS/ESP/ASR Unit, Chassis/Body Control Module and others), the 6 DoF IMU (3-axis gyroscope + accelerometer), and the GPS receiver. Some of these units send data in regular time periods, others send it on event based regime, and other need to be polled. Data send over CAN bus is often in a raw form, e.g. Analog-to-Digital converter (ADC) data. This really isn't a problem for classification algorithms, but different car manufacturers use different data scaling, so it is a good idea to scale these raw data to the physical units, so the resulting ML algorithm would be universal across different cars, and only data decoding routines need to be changed. The same concern is in case of IMU, global integrated circuits shortages may force choosing another IMU in the future, so it is safer to use unified (physical) units. The problem with GPS receiver is a little different, it returns its data in a well standardized NMEA 0183 format, yet if we want to approximate some values in the re-sampling process, it is required to decode these frames.

For the security reasons, on-board CAN bus devices can't be actively queried, as this might affect their functions, so we may only passively listen to the inter-node communication over a CAN bus, and capture frames containing relevant information, whenever they occur. The IMU device can be actively polled or be free running, as well as GPS, but GPS receivers are known to feed their data with a significant delay. This brings a problem of forming a data stream for a further analysis. Such a data stream preferably should have a constant time step, and contain no missing values. So, low or irregular sample rate parameters must be stored and updated once a new value occurs, and all streams should also be time adjusted to the GPS data stream (so delayed by a few seconds). Delay of GPS stream is constant and may be determined, by analyzing time offsets between wave forms of GPS reported speed, and ABS (CAN) reported speed.

2.2 Data Collection

Data for identifying and scoring various maneuvers was collected on a closed test track with two types of surface (concrete slabs, and skid plate), by several drivers assisted by a pilot, using a proprietary, multichannel data logging device, with a sampling rate up 100 Hz. Drivers were following the route prepared by road safety specialists. Additionally, during the experiment, the driver assistant (pilot) marked the beginning and end of each maneuver, using a digital tablet with specialized software. The result of one of such drives is shown on Fig. 2, where track marks are depicted as small triangles, and the bright area shows the skid plate. Tagged maneuvers are marked by colors (red - moving off, brown - sequence of turns, violet - left turn, blue - bypassing obstacle, orange - braking in left turn). The test track was driven in different configurations, and directions, 300 times, by randomly selected drivers (of different genders).

It should be noted that maneuvers manually selected by the assistant-pilot do not match perfectly with either the track markings or the maneuvers detected by GPS. It clearly follows from observation of trajectories in Fig. 2. Thus, their direct usefulness for machine learning is limited. They may, however, be used for verifying the correctness of maneuvers detected by our algorithm.

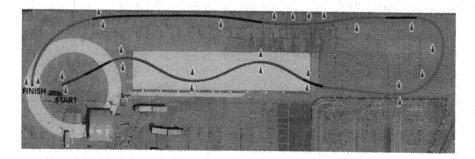

Fig. 2. GPS track tagged by a pilot with a maneuver markers. (Color figure online)

During the drive, data from the CAN bus, IMU and GPS were recorded. Based on these signals, it is possible to detect the type of maneuver the car driver is performing at any given time. The result of such detection can be seen on Fig. 3 (colors illustrate maneuvers detected by our software, red - moving off, violet - turning right, brown - turning left, green - braking).

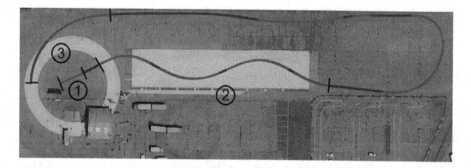

Fig. 3. GPS track tagged by software with detected maneuvers. (Color figure online)

Based on these measurements, a rule classifier model was built. The operations of the classifier present Algorithms 1, 2 and 3. Automatic detection of maneuvers has been done using experimentally established rules, which are based on observing the telemetry data. Effects of detecting 4 different maneuvers on 3 sections marked on Fig. 3 as ①, ②, and ③, have been shown, on the background of actual telemetric data from this drive, on Fig. 4, 5 and 6. It can be seen that the maneuver recognition based on the rule classifier is more accurate compared to GPS detection and assistant-pilot indications.

Algorithm 1. An algorithm for detecting car moving off

Input: $Speed, STaccelerationX, LTaccelerationX, ThrottleIncrease$
Output: $IsMovingOff$
if $(!IsMovingOff \land ThrottleIncrease \land STaccelerationX >= 0.2 \land Speed < 5)$
then
  $IsMovingOff \leftarrow TRUE$
else if $(IsMovingOff \land LTaccelerationX < 0.2)$ **then**
  $IsMovingOff \leftarrow FALSE$
end if

Algorithm 2. An algorithm for detecting car making a turn

Input: $SteeringToLeft, SteeringToRight, GyroscopeZ$
Output: $IsTurningLeft, IsTurningRight$
if $((!IsTurningLeft \lor !IsTurningRight) \land SteeringToLeft \land GyroscopeZ >= 0.5)$
then
  $IsTurningLeft \leftarrow TRUE$
else if $(IsTurningLeft \land GyroscopeZ < 0.5)$ **then**
  $IsTurningLeft \leftarrow FALSE$
end if
if $((!IsTurningLeft \lor !IsTurningRight) \land SteeringToRight \land GyroscopeZ <= -0.5)$ **then**
  $IsTurningRight \leftarrow TRUE$
else if $(IsTurningRight \land GyroscopeZ > -0.5)$ **then**
  $IsTurningRight \leftarrow FALSE$
end if

Algorithm 3. An algorithm for detecting car braking

Input: $Speed, STaccelerationX, LTaccelerationX, ThrottleDecrease$
Output: $IsBraking$
if $(!IsBraking \land BrakeIncrease \land STaccelerationX <= -0.4)$ **then**
  $IsBraking \leftarrow TRUE$
else if $(IsBraking \land LTaccelerationX > -0.4)$ **then**
  $IsBraking \leftarrow FALSE$
end if

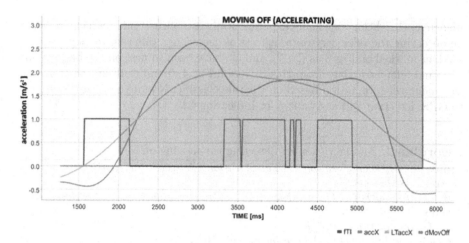

Fig. 4. Telemetry data for section ① (moving off/accelerating) from Fig. 3

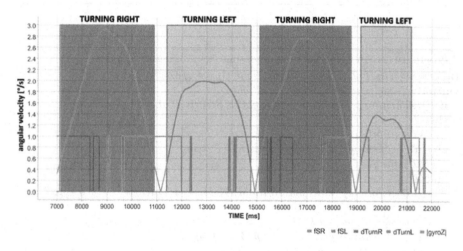

Fig. 5. Telemetry data for section ② (sequence of turns) from Fig. 3

The above graphs show the smoothed signals collected during driving, from the CAN bus and, the previously mentioned, IMU sensors. Names of these signals are collected in Table 2 which should be read together with Table 1.

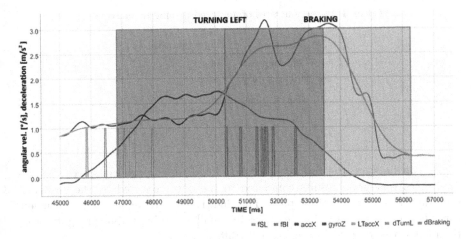

Fig. 6. Telemetry data for section ③ (braking while turning left) from Fig. 3

Table 2. Symbols used on graphs and in pseudocode.

Symbol on graph	Variable name	Description
accX	STaccelerationX	Momentary (short term) acceleration in X axis
LTaccX	LTaccelerationX	Averaged (long term) acceleration in X axis
gyroZ	GyroscopeZ	Angular speed around Z axis
fTI	ThrottleIncrease	Throttle is increasing
fBI	BrakeIncrease	Brake pressure is increasing
fSL	SteeringToRight	Steering wheel is turning right
fSR	SteeringToLeft	Steering wheel is turning left
dMovOff	IsMovingOff	Moving Off maneuver detection
dTurnL	IsTruningLeft	Turning left maneuver detection
dTurnR	IsTruningRight	Turning right maneuver detection
dBrake	IsBraking	Braking maneuver detection

3   Evaluation of the Model

The correctness of the maneuver detection rules was verified on 7 loops of a 3 km long public road under normal urban traffic conditions. The road was passed in different directions. The obtained maneuver detection rates in this environment are shown in Table 3.

Table 3. Detection rates of analyzed driving maneuvers

No	Maneuver	Detection rate
1	Moving off	100.0%
2	Turning left	85.5%
3	Turning right	95.0%
4	Braking	100.0%

As can be seen, the proposed method for detecting car maneuvers on the road gives very good results compared to the data collected by the on-board GPS device and the assistant-pilot. The slightly lower detection ratio of the left turns is probably related to the generally smaller turning radius of right turns, than left turns, due to right hand side traffic.

4 Conclusions and Further Work

Presented algorithm - maneuvers detection is only the first stage for assessing safety of maneuvers, and evaluating style of driving. It allows marking fragments of telemetry data stream, for secondary features extraction and further evaluation. In the future, the research presented in the article will be supplemented with the assessment of the driver's driving style. In future research, we plan to find an optimal set of parameters for detecting maneuvers and determining driver driving style. The driving style will be evaluated based on additional parameters such as driving speed, steering wheel jerking or acceleration. Such solutions are especially expected by companies with large fleets of vehicles, where repair and service costs are high.

Acknowledgments. This research was done in cooperation with the Lincor Software Sp. z o.o. sp. k., and was supported by the Polish National Center for Research and Development under the grant no. POIR.01.01.01-00-0057/21 (Opracowanie i weryfikacja w warunkach rzeczywistych narzdzia opartego o algorytmy uczenia nadzorowanego umoliwiajacego ocen jazdy kierowcy i podwyszanie jej bezpieczestwa) for Lincor Software, Warszawa, Poland. Project co-financed by EU Smart Growth Operational Programme in the years 2014–2020.

References

1. CAN Specification v2.0. Robert Bosch GmbH (1991)
2. Eftekhari, H.R., Ghatee, M.: Hybrid of discrete wavelet transform and adaptive neuro fuzzy inference system for overall driving behavior recognition. Transport. Res. F: Traffic Psychol. Behav. **58**, 782–796 (2018)
3. European Road Safety Observatory, European Commission: Annual statistical report on road safety in the EU 2020. https://ec.europa.eu/transport/road_safety/system/files/2021-07/asr2020.pdf. Accessed May 2022

4. Guardiola, I., Leon, T., Mallor, F.: A functional approach to monitor and recognize patterns of daily traffic profiles. Transport. Res. Part B: Methodol. **65**, 119–136 (2014)
5. Huang, X., Sun, J., Sun, J.: A car-following model considering asymmetric driving behavior based on long short-term memory neural networks. Transport. Res. Part C: Emerg. Technol. **95**, 346–362 (2018)
6. Karri, S.L., De Silva, L., Lai, D.T.C., Yong, S.Y.: Identification and classification of driving behaviour at signalized intersections using support vector machine. Int. J. Autom. Comput. **18**, 480–491 (2021)
7. Singh, H., Kathuria, A.: Profiling drivers to assess safe and eco-driving behavior - a systematic review of naturalistic driving studies. Accid. Anal. Prevent. **161**, 106349 (2021)
8. Xiao, W., Liu, H., Ma, Z., Chen, W.: Attention-based deep neural network for driver behavior recognition. Futur. Gener. Comput. Syst. **132**, 152–161 (2022)
9. Zheng, J., Suzuki, K., Fujita, M.: Car-following behavior with instantaneous driver-vehicle reaction delay: a neural-network-based methodology. Transport. Res. Part C: Emerg. Technol. **36**, 339–351 (2013)

Deep Learning Models

Using Deep Transformer Based Models to Predict Ozone Levels

Manuel Méndez, Carlos Montero, and Manuel Núñez[✉][iD]

Universidad Complutense de Madrid, Madrid, Spain
{manumend,cmonte09,manuelnu}@ucm.es

Abstract. Ozone (O3) is an air pollutant that has harmful effects in human health when its concentration exceeds a certain level. Therefore, it is important to advance in methods that can appropriately predict O3 levels. In this paper we present a new model to estimate 4 h, 12 h, 24 h, 48 h and 72 h ahead O3 concentration levels. We rely on Deep Transformer Networks. Interestingly enough, these models were originally developed to be used in Natural Language Processing applications but we show that they can be successfully used in classification problems. In order to evaluate the usefulness of our model, we applied it to predict O3 levels in the centre of Madrid. We compare the results of our model with four baseline models: two LSTMs and two MLPs. Accuracy (*Acc*) and Balanced Accuracy (*BAC*) are the metrics employed to evaluate the goodness of all the models. The results clearly show that our Deep Transformer based Network obtains the best results.

Keywords: Air quality prediction · Deep learning · Transformer networks

1 Introduction

Air pollution is one of the major problems currently faced by humanity. It causes seven million deaths every year according to international organisms [5]. Pollutants like ozone (O3), nitrogen dioxide (NO2), sulphur dioxide (SO2), carbon monoxide (CO) and particulates matter (PM2.5, PM10) are some of the most common air pollutants [4]. They are also the pollutant included in the air quality index measurement. In this paper we focus on O3 (ozone). Ozone is a colourless gas located in the atmosphere. It is one of the most common existing pollutants and, as such, it is one of the pollutants taken into account to determine the air quality index. The exposure to this pollutant has severe effects in human health such as eyes and nose irritation and inflammation, lung function reduction, exacerbation of respiratory diseases and increased susceptibility to diseases infection,

This work has been supported by the Spanish MINECO/FEDER projects FAME (RTI2018-093608-B-C31) and AwESOMe (PID2021-122215NB-C31) and the Region of Madrid project FORTE-CM (S2018/TCS-4314), co-funded by EIE Funds of the European Union.

N. T. Nguyen et al. (Eds.): ACIIDS 2022, LNAI 13757, pp. 169–182, 2022.
https://doi.org/10.1007/978-3-031-21743-2_14

among others [4]. Ozone concentration levels depend on complex processes that happen in the atmosphere. Precursor gases, such as nitrogen oxides, are chemically transformed into ozone when they are exposed to solar light. Since sunlight is needed for ozone formation, its concentration highly depends on the time of the day and on the current meteorological conditions. Moreover, human emission of these precursor pollutants by fabrics and traffic have affected the increase of ozone concentration in the last decades. Most major cities around the world have specific protocols to deal with high concentrations of pollutants. In the case of Madrid, used in this paper as case study, when the $O3$ concentration exceeds a certain level, the authorities take measures to reduce the $O3$ effects in population. These measures range from recommendations to reduce physical exercise outdoors for vulnerable people to prohibitions of outdoor activities, specially sport activities [2]. Therefore, it is very important to be able to estimate future $O3$ levels in order to alert the population of possible future recommendations or restrictions.

We present a new model to predict the $O3$ concentration level. Our approach applies a novel Transformers-based model to predict the $O3$ concentration level. Unlike classical time series forecasting models, Transformers do not process the data ordered. On the contrary, they process the entire sequence and use self-attention techniques to find dependencies between variables. We considered an optimised Transformers-based model and our preliminary experiments revealed that it was a good candidate to overcome the accuracy and the balanced accuracy of usual time series classification networks such as MLP or LSTM.

In order to evaluate the usefulness of our proposal, we compare its results with the ones produced by four neural network baselines models. We also make an analysis of the variation of the model accuracy and balanced accuracy depending on the modification of three hyperparameters in short-term and in long-term cases. For this, we use as case of study the task of predicting $O3$ levels in the centre of Madrid for the next 4, 12, 24, 48 and 72 h. We consider a total of fourteen predictors variables. Our results show that our proposal is better than the baseline models based on the evaluation metrics. In average, the balanced accuracy of our proposal is 4.3% better than the one corresponding to the best baseline model.

The rest of the paper is organised as follows. Section 2 reviews related work. In Sect. 3 we present background concepts such as the baseline models and different evaluation metrics that we use. In Sect. 4 we present the main characteristics of the problem that we want to solve and of the model that we construct to confront the problem. In Sect. 5 we present our experiments and discuss the obtained results. Finally, in Sect. 6, we give conclusions and outline some directions for future work.

2   Related Work

Several studies have used either statistical machine learning techniques or deep learning models to forecast pollutant concentrations. Paoli *et al.* [19] develop an optimised MLP network to forecast $O3$ concentration in Corsica. This model is more accurate than other baseline models and properly detects ozone peaks. Li *et al.* [15] propose a Random Forest model, leveraging its capacity to work with numerical and non-numerical variables, to predict the concentration of three air pollutants. Other models have been developed to forecast pollutant concentrations such as an improved ARIMA (Liu *et al.*) [16], LSTM networks (Seng *et al.* [21]) and multi-linear regression (Jato-Espino *et al.*) [13]. In addition to machine learning techniques, collective information can be compiled to forecast air quality. Palak *et al.* [18] present a collective framework to predict air pollution in places where no meters are available. The combination of monitoring and CEP is also a good approach to forecast air quality. In this line, Díaz *et al.* [8] considered Petri Nets while Corral-Plaza *et al.* [7] and Semlali *et al.* [20] used an IoT approach.

Our approach is based on Transformer Networks. Transformers were proposed by Vaswani *et al.* [22]. Originally, they were developed as a Natural Language Processing tool that improves classical LSTM networks and Recurrent Neural Networks (RNN) in tasks such as text classification and translation. Its potential is based on self-attention layers, which estimate the attention weights between input variables. In recent years, Transformers have been used to solve tasks in other fields such as image recognition, multi-class classification and time series prediction. Wu *et al.* [23] present a deep model based on Transformers to influenza-like illness forecasting. Results show that the proposed method is more precise than baseline methods such as LSTM or ARIMA. Dosovitsky *et al.* [9] show that a pure Transformer application can overcome classical CNNs in image classification tasks.

To the best of our knowledge, Deep Transformer based Models have not been used to analyse air pollutants. Although machine learning techniques have been used to forecast ozone concentration values [6, 19] (that is, as part of a regression model), we are not aware of their use to predict ozone levels as such (that is, as part of a classification model).

3   Preliminaries

In this section we will review some concepts that we will use along the paper. Specifically, we will discuss the baseline models that will be used to compare our approach with and the evaluation metrics that will be used to measure different models quality.

3.1   Baseline Models

In order to assess the usefulness of our approach, we will compare it with two classical algorithms that are very suitable to solve the same problem: *LSTM networks* and *MLP networks*. Although these models were defined some time ago, they are currently and frequently used both to predict the behaviour of complex systems and as baseline models [10, 14, 17].

MLP Networks [12]. An MLP network is a computational model inspired by a human brain whose objective is to find relationships between data. It is composed by three types of layers: input layer, hidden layer and output layer. The input layer receives the input data to be processed. A number of hidden layers are placed between the input and output layers. Data flows from input to output in forward direction. Each layer is composed by a number of simple processing elements called *neurons*. Neurons are trained using the back-propagation learning algorithm to minimise a loss function.

The mathematical operations that occur in every neuron in hidden and output layers are, respectively, given by the following expressions:

$$H_x = f(b_1 + W_1 \cdot x) \qquad O_x = f'(b_2 + H_x \cdot W_2)$$

being x an input vector, b_1 and b_2 bias vectors, W_1 and W_2 weight matrices and f and f' activation functions. Usual activation functions are the RELU and sigmoid functions.

$$RELU(a) = \max(0, a) \qquad Sig(a) = \frac{1}{1 + e^{-a}},$$

where a is the input data.

Hyperparameters such as the *number of hidden layers* (h), *number of neurons in each hidden layer* (h_n), *number of epochs* (ep), *dropout rate* (dr), *learning rate* (lr), and *batch size* (bs) are optimised to get the best accuracy in MLP networks.

LSTM Networks [24]. An RNN network does not have a defined layers structure. Actually, it allows random connections between neurons, developing temporality and providing memory to the network. This makes RNNs well suited algorithms in fields such as Natural Language Processing and sequence data processing. However, if a long context is needed, we have the long-term dependencies problem, that is, the gradual forgetfulness of previous information of the network. In order to solve it, LSTM networks were developed.

The key to LSTMs is the cell state (C_t) that runs straight down the entire chain. The information flows along it almost without modifications. The LSTM either updates or discards information in the cell state by using structures called *gates*. These gates are composed by a sigmoid layer, which outputs a number between 0 and 1 that indicates how much information must be let through, and a multiplicative element. LSTM cells have three gates: input gate (i_t), forget gate (f_t) and output gate (o_t). The mathematical operations that occur in each gate are as follows:

$$f_t = \sigma(W_f \cdot [h_{t-1}, x_t] + b_f)$$
$$i_t = \sigma(W_i \cdot [h_{t-1}, x_t] + b_i)$$
$$\tilde{C}_t = tanh(W_c \cdot [h_{t-1}, x_t] + b_C)$$
$$C_t = f_t \cdot C_{t-1} + i_t \cdot \tilde{C}_t$$
$$o_t = \sigma(W_o[h_{t-1}, x_t] + b_o)$$
$$h_t = o_t \cdot tanh(C_t)$$

where x_t is the input data in time t, h_t is the hidden state in time t, each W_x is a weight matrix and each b_x is a bias vector.

Hyperparameters such as the *number of neurons in LSTM layer* (n), *number of epochs* (ep), *dropout rate* (dr), *learning rate* (lr) and *batch size* (bs) are optimised to get the best accuracy in the LSTM network.

3.2 Performance Evaluation Metrics

We will consider accuracy (Acc) because it is the most used evaluation measure in multi label classification models [11]. In addition, as in all cases our data is imbalanced, we will also use Balanced Accuracy (BAC), which gives the same weight to all the categories. Given $m \in \mathbb{N}$, let M be an $m \times m$ confusion matrix. The formal definitions of these measures are:

$$Acc = \frac{\sum_{k=1}^{m} M_{k,k}}{\sum_{k=1}^{m}\sum_{i=1}^{m} M_{k,i}} \qquad BAC = \frac{\sum_{k=1}^{m} \frac{M_{k,k}}{\sum_{i=1}^{m} M_{k,i}}}{m}$$

Intuitively, Acc is the ratio between the number of correctly predicted observations and total number of observations, while BAC computes these ratios for each category.

4 Problem Description and Our Model

In this section we present the main concepts underlying our case of study. We also present a definition and description of our Deep Transformer based Network. Finally, we set the hyperparameters of each baseline model that will be used to compare with our model.

4.1 Problem Description

In order to evaluate the proposal model, we decided to use pollutant data because its concentration usually depends on a great amount of factors such as meteorology, industrial emissions, transportation and use of chemical products. Among the different pollutants, we choose ozone because it has a behaviour along the year that diverges from the rest of pollutants influencing the Air Quality Index. We wish to evaluate the goodness of our proposal in a classification problem and using a previous *standard* that we cannot influence. Therefore, we categorise ozone values but take into account its hazard as determined by official standards of Madrid Council [2].

The ozone concentration level predicting problem is formulated as a supervised machine learning task. Specifically, our goal is to predict the category of the target variable y in k hours time, that is, $y(h + k)$. We will use a vector of 56 predictor variables, 14 by hour, and using the last 4 available values. We have 12 continuous hourly predictor variables and 2 (both calendar variables) categorical ones. The target variable is a categorical one. Table 1 shows all the predictor variables (* denotes a categorical variable).

Table 1. Predictor variables by type

Type	Variables
Pollutant	$NO, CO, NO2, NOX, SO2, O3$
Meteorological	$rain, maxtemp, mintemp, medtemp, maxpress, minpress$
Calendar	$week\ day^{*}, type\ of\ day^{*}$

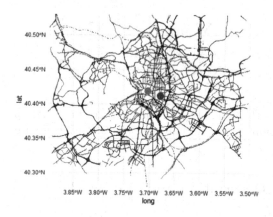

Fig. 1. City of Madrid. The red mark points at 'Plaza del Carmen Station' while the blue one points at 'El Retiro Station'. (Color figure online)

Our proposal, as well as the baseline models, will be evaluated in a real case study. Specifically, we consider data collected in Madrid from 01/01/2017 to 31/07/2021. Pollutant data is hourly collected by dedicated sensors, placed in stations distributed in the city, and it is publicly available at the Madrid City Hall website [2] database. We also use this database to extract calendar data. In addition, meteorological data can be extracted from AEMET [1], the State Meteorology Agency of Spain. We transform daily variables in hourly variables to standardise all of them. We consider two stations located in the centre of Madrid (see Fig. 1): 'Plaza del Carmen Station' and 'El Retiro'.

Data has been pre-processed by inputting previous observation variable values in not available and outliers values. All the variables have also been scaled. Ozone continuous data runs from $0\,\mu g/m^3$ to $373\,\mu g/m^3$. Following the guideline of Madrid City Hall, we classify it in a three class target variable. In Table 2 we show the categories chosen and the number of existing observations in each of them. We obtain a final data-set with 38.064 observations.

Table 2. Ozone levels

Category (values)	Ozone state	Number of observations
0 (0–60)	Good	26.985
1 (60–120)	Medium	10.509
2 (>120)	Harmful	570

4.2   Deep Transformer Based Models

In this section we describe the main characteristics of our model, which is based on Deep Transformer Networks [22]. The model architecture is composed by a number of Transformer encoder networks (*num_transformers*) and a final MLP network to make the classification. Each transformer encoder is composed by a *normalisation* layer, which applies a transformation to maintain the mean of the previous layer activation close to 0 and the standard deviation close to 1. Then, we add the essence of the Transformer model: an *attention* layer. It uses an attention mechanism to learn the contextual relation between variables. This mechanism avoids the requirement of the recurrent connections in the neural network. This layer has a set of hyperparameters, which must be controlled, such as the number of attention heads (*num_head*). A number of heads greater than 1 (Multi-Head Attention) allows the model to jointly attend to information from different sub-spaces of representation at different positions [22]. The size of each one (*head_size*) and the dropout probability (*dropout*) are other hyper-parameters to consider. The next step is to add the feed forward part of the transformer encoder, which is composed by a *normalisation layer* and by two one dimension *convolutional layers* with a kernel size equal to one. A dropout probability is applied in the first convolutional layer, which works with a ReLU activation function. The convolutional layers create a convolutional kernel with the normalisation layer, in this case, in a temporal dimension $t = 1$. In the first convolutional layer, the number of output filters is chosen by the *filters* hyperparameter. In the second one, the number of output filters is equal to the dimension of the input shape. In order to reduce the output tensor of the set of Transformer Encoders, we add a one dimension *Average Pooling Layer*. Finally, we add an MLP network to make the classification. This MLP uses ReLU as activation function in the hidden layer and softmax as activation function in the output layer. We also apply here a dropout probability (*mlp_dropout*). MLP has just one hidden layer. The number of neurons in the hidden layer of the MLP (*hid_layer*) is a hyperparameter to consider.

In order to train the network we use the gradient descent learning method. We apply also an *Early Stopping*, which is a regulation method to stop the training when the error diminution between two consecutive iterations is less than a previously set threshold (usually, 0.0001).

The chosen loss function has been the *sparse categorical cross-entropy* and we consider *Adam* as weights optimiser. The number of epochs and the batch size will be two of the hyperparameters to fit.

We use python, specifically the *keras* library, to implement our model.

In Fig. 2 we use *Netron* [3] to show the structure of our Deep Transformer based Network with the following hyperparameters: $num\_transformers = 1$, $num\_head = 2$, $head\_size = 2$, $dropout = 0.2$, $filters = 1$, $mlp\_dropout = 0.15$ and $hid\_layer = 500$. In order to reduce the size of the graphical representation, we use low hyperparameters values. The interested reader can visit https://github.com/MMH1997/TransformerNetworks where it is possible to see the architecture with other hyperparameters and deeply analyse each component of the model.

4.3   MLP and LSTM Networks

In this section we describe the baseline models with which we will compare our proposal. In order to make an unbiased comparison, we compare our proposal with two LSTMs and two MLPs, with respectively more/less trainable parameters than our proposal. We use the *keras* library in python to implement these models.

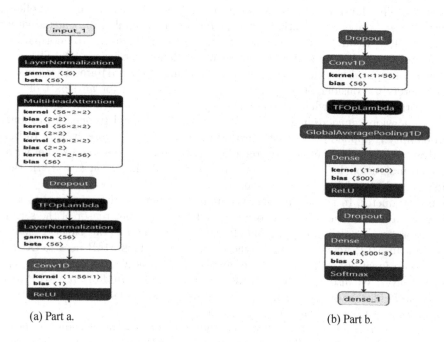

(a) Part a.   (b) Part b.

Fig. 2. Deep transformer based model structure

– *MLP networks.* The first MLP model (*MLP1*) has $h = 5$ hidden layers. The number of neurons in each hidden layer are 512, 256, 128, 32 and 32. It results in a total of 198.691 trainable parameters. The second MLP model (*MLP2*) also has $h = 5$ hidden layers. The number of neurons in each hidden layer are 64, 64, 16, 8 and 8. It results in a total of 9.083 trainable parameters. Both models have the following hyperparameters: $ep = 25$, $dr = 0.2$, $lr = 0.0001$ and $bs = 16$. Each hidden layer is activated by a ReLU function and the output layer by a softmax function. Both output layers have three neurons (each of them returns the probability of each category). In both models, the loss function has been *sparse categorical cross-entropy*.

- *LSTM networks.* Our first LSTM model (*LSTM1*) has $n = 150$ neurons in the LSTM layer. This layer is connected to three hidden layers having, respectively 15, 5 and 5 neurons. Hidden layers are activated by a ReLU function. Again, the output layer has three neurons (each of them returns the probability of each category) and is activated by a softmax function. This model has a total of 123.593 trainable parameters. Our second LSTM model (*LSTM2*) has $n = 12$ neurons in the LSTM layer. This layer is connected to an output layer that has three neurons and is activated by a softmax function. This model only has 3.251 trainable parameters.

Both models have the following hyperparameters: $ep = 25$, $dr = 0.2$, $lr = 0.0001$ and $bs = 16$. In these two models, the loss function has also been the *sparse categorical cross-entropy.*

Table 3. *Acc* by model and time in advance

Model	4 h	12 h	24 h	48 h	72 h	Average
Proposal	**0.891**	**0.730**	**0.772**	0.753	**0.753**	**0.776**
LSTM1	0.872	0.642	0.753	0.716	0.700	0.729
LSTM2	0.881	0.651	0.762	0.741	0.722	0.744
MLP1	**0.891**	0.707	0.768	0.747	0.736	0.765
MLP2	0.882	0.701	0.771	**0.757**	0.730	0.764

Table 4. *BAC* by model and time in advance

Model	4 h	12 h	24 h	48 h	72 h	Average
Proposal	0.816	**0.471**	**0.570**	**0.484**	**0.480**	**0.540**
LSTM1	0.730	0.346	0.467	0.428	0.411	0.448
LSTM2	0.705	0.351	0.491	0.467	0.446	0.467
MLP1	**0.819**	0.462	0.520	0.471	0.451	0.518
MLP2	0.686	0.429	0.517	0.482	0.437	0.496

5   Experiments

In this section we present the experiments that we performed to validate our model. In all the experiments we used 75% of the observations as training set and 25% of the observations as testing set. Since the difference between the results is not very large, in order to reduce the error we compute the average of 20 repetitions of the experiment. The programming code and data used in these experiments are freely available at https://github.com/MMH1997/TransformerNetworks.

5.1   Comparison Between Models

We apply the proposed model and the baseline models to predict the categories in the next 4, 12, 24, 48 and 72 h. In this experiment, we use the following hyperparameters in the Deep Transformer based Model: $num\_transformers = 2$, $num\_head = 15$, $head\_size = 5$, $dropout = 0.2$, $filters = 25$, $mlp\_dropout = 0.15$, $hid\_layer = 5000$, $bs = 8$ and $epochs = 25$. This choice results in a total of 65.375 trainable parameters.

Comparing the model with the previously defined MLP and LSTM networks, we appreciate that, in general, our developed model obtains better results, particularly in long-term cases. In terms of accuracy (see Table 3), our model obtains higher accuracy than all the baselines models, except in the 4 h case, where it obtains the same accuracy as $MLP1$, and in the 12 h case, where our model is slightly worse than $MLP2$. In terms of Balanced Accuracy (see Table 4), the proposed model obtains the best accuracy in all cases except in the 4 h case, where our model is worse than $MLP1$. It is worth to mention the 24 h case, where the difference between the BAC of the proposed model and the one of the baseline models is greater than 5 points. We think that this is due, in part, to the capability of our model to detect minority class cases, while the baseline models are not able to detect them when predicting 24 h in advance.

By the resolve of the task, the presence of a minority class in our data which implies the called *imbalanced class problem* must taken into account. We have tried to solve this problem by applying typical techniques to diminish it such as over-sampling, under-sampling or re-sampling. However, none of them were effective, probably due to the high number of variables and the dynamism in data. This ineffectiveness was reflected on the inability of all models (proposed and baseline) to get a BAC higher than 0.33, that is, models just classify all the observations in the same category.

The obtained values show the typical increase of Acc and BAC as time in advance is reduced, that is, we expect better results in predictions 4 h in advance than in predictions in a longer time. However, this pattern does not hold if we consider the 12 h case. In fact, there is a clear reduction of Acc and BAC in all models with respect to the 24 h or 48 h cases. After a careful review of the factors that influence ozone concentration, we realised that meteorological data, in particular and as we advanced in the introduction of the paper the presence/absence of solar light, are very relevant. In more technical terms, the correlation between the values of the predictors at time t and the ones at times $t - 24$ and $t - 48$ is higher than the corresponding to time $t - 12$. In fact, the difference of several predictor variables in the 12 h case is usually high, not only concerning meteorological variables but also other pollutant concentrations $(NO, CO, SO2, NO2, NOX)$. For example, in most cases, the average value of these pollutants at noon is the double than the average value at midnight.

In the next experiments, we will analyse the optimisation of the proposal model hyperparameters.

5.2   Hyperparameters Optimisation

In this section we will analyse the evolution of BAC and Acc depending of the values of three hyperparameters: num_head, $head\_size$ and $filters$. We performed two sets of experiments: prediction 4 h in advance and prediction 24 h in advance. We chose these two cases because they are the shortest term and the most *typical* prediction (what will happen the next day at the same time). It is worth to mention that our preliminary experiments showed a similar behaviour for, on the one hand, all short term cases and, on the other hand, all long term cases.

The following values were chosen to evaluate each hyperparameter:

- num_head: 3, 6, 9, 12, 15, 18;
- $head\_size$: 3, 6, 9, 12, 15, 18;
- $filters$: 5, 10, 15, 20, 25.

The rest of hyperparameters used in this experiment are set to: $num\_transformers = 1$, $dropout = 0.2$, $mlp\_dropout = 0.15$, $hid\_layer = 5000$, $bs = 8$ and $epochs = 25$. Finally, the number of trainable parameters in the combinations range from 27.947 to 101.712.

We have evaluated BAC and Acc in all the possible combinations of the hyperparameters values previously mentioned. In order to provide a unique result for each experiment, we compute the average value of the metrics evaluated for each hyperparameter value. For example, if we fix $head\_size = 6$, we compute the average of all the values returned from the 30 experiments corresponding to all the combinations of num_head and $filters$ such that $head\_size = 6$. In the two cases studies, Acc values remain almost constant with the hyperparameters modifications. Therefore, we focus our analysis on the variations of BAC values.

The experiments corresponding to the 4 h in advance case clearly show that low values of $head\_size$ and num_head return higher values of BAC. If we consider $filters$, we observe that the maximum BAC values are achieved when $filters = 10$. In higher $filters$ values, BAC values remain constant. Interestingly enough, the absolute maximum BAC value (0.8228) is achieved in a combination of hyperparameters values that does not correspond to the general conclusions: $head\_size = 18$, $num\_head = 15$ and $filters = 25$.

If we consider the 24 h in advance case, we observe that an increase of $head\_size$ values suggest a small decrease of BAC. Unlike the previous case, maximum BAC values are achieved when $num\_head = 15$. For higher values, BAC seems remain constant. Maximum BAC values are achieved when $filters = 15$. However, experiments show that changes in $filters$ values do not imply significant modifications in BAC. Unlike the previous case, the absolute maximum BAC value seems correspond to the general conclusions. It is achieved when $head\_size = 9$, $num\_head = 15$ and $filters = 5$.

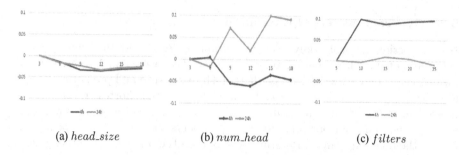

(a) $head\_size$ (b) num_head (c) $filters$

Fig. 3. BAC percentage change by hyperparameter value in short-term and long-term cases.

In Fig. 3 we show how the obtained values vary. From these results we can extract the following conclusions:

- The variation of BAC values depending on the performance of $head\_size$ is similar in both the short term and long term cases.
- In short term cases, an increase of num_head suggests an increase of BAC. However, in long term cases, the increase in the value of the hyperparameter suggests a decrease of BAC values.
- In long term cases, the modification of $filters$ values seems to have no effect in BAC. In contrast, in short term cases, its increase implies an increase of BAC.

6   Conclusions and Future Work

In this paper, we have proposed a novel Deep Transformer based Network to classify the ozone levels in the centre of Madrid using fourteen predictor variables. Our experiments show that this model overcomes, in general terms, the accuracy and the balanced accuracy of four baselines neural network models. The capability of the proposed model to detect minority class observations, particularly, in 24 h in advance case is one of the main advantages of proposed model respect to the baseline methods.

We have also made an analysis of three hyperparameters of the proposal model in short term and in long term cases. This analysis suggests us that the behaviour of balanced accuracy depending on hyperparameters modifications differs between long term and short term cases.

We consider some lines for future work. First, in order to increase accuracy, we would like to perform a deeper analysis on the optimisation of the hyperparameters, in particular, concerning whether we need similar adjustments for short-term and long-term prediction. For example, in Fig. 3 we can see that $head_size$ has a similar behaviour in both cases, while it is very different for num_head and $filters$. Second, taking into account the high quality of the proposed model, we would like to produce a similar model for other data. In particular, we would like to apply this model to the rest of pollutants included in the air quality index measurement so that it is possible to classify them by levels. Combining this idea with Complex Event Processing technologies, air quality index could be accurately predicted. We would also like to adapt our model to other air quality indexes that, in particular, might have data in a format that it is not compatible with the one that we have used. Finally, we would like to compare our proposal with other classification models such as Random Forest, ARIMA and Support Vector Machine.

References

1. AEMET Open Data. https://opendata.aemet.es/centrodedescargas/productosAEMET?. Accessed 25 Oct 2021
2. City of Madrid. https://www.mambiente.madrid.es/opencms/export/sites/default/calaire/Anexos/Procedimiento_ozono.pdf. Accessed 20 Oct 2021
3. Netron open source tool. https://netron.app/. Accessed 15 Nov 2021
4. New South Wales Government. https://www.health.nsw.gov.au/environment/air/Pages/common-air-pollutants.aspx. Accessed 20 Oct 2021
5. World Health Organisation. https://www.who.int/health-topics/air-pollution. Accessed 20 Oct 2021
6. Castelli, M., Gonçalves, I., Trujillo, L., Popoviǎ, A.: An evolutionary system for ozone concentration forecasting. Inf. Syst. Front. **19**(5), 1123–1132 (2017)
7. Corral-Plaza, D., Boubeta-Puig, J., Ortiz, G., García de Prado, A.: An Internet of things platform for air station remote sensing and smart monitoring. Comput. Syst. Sci. Eng. **35**(1), 5–12 (2020)
8. Díaz, G., Macià, H., Valero, V., Boubeta-Puig, J., Cuartero, F.: An intelligent transportation system to control air pollution and road traffic in cities integrating CEP and colored Petri nets. Neural Comput. Appl. **32**(2), 405–426 (2020)
9. Dosovitskiy, A., et al.: An image is worth 16x16 words: transformers for image recognition at scale. In: 9th International Conference on Learning Representations, ICLR 2021, pp. 1–21 (2021)
10. Fakir, M.H., Kim, J.K.: Prediction of individual thermal sensation from exhaled breath temperature using a smart face mask. Build. Environ. **207**, 108507 (2022)
11. Grandini, M., Bagli, E., Visani, G.: Metrics for multi-class classification: an overview (2020). https://arxiv.org/abs/2008.05756
12. Hastie, T., Tibshirani, R., Friedman, J.: The Elements of Statistical Learning: Data Mining, Inference, and Prediction. Springer, New York (2009). https://doi.org/10.1007/978-0-387-84858-7
13. Jato-Espino, D., Castillo-Lopez, E., Rodriguez-Hernandez, J., Ballester-Muñoz, F.: Air quality modelling in Catalonia from a combination of solar radiation, surface reflectance and elevation. Sci. Total Environ. **624**, 189–200 (2018)
14. Laubscher, R., Rousseau, P.: An integrated approach to predict scalar fields of a simulated turbulent jet diffusion flame using multiple fully connected variational autoencoders and MLP networks. Appl. Soft Comput. **101**, 107074 (2021)
15. Li, J., Shao, X., Zhao, H.: An online method based on random forest for air pollutant concentration forecasting. In: 37th Chinese Control Conference, CCC 2018, pp. 9641–9648. IEEE (2018)
16. Liu, T., Lau, A.K.H., Sandbrink, K., Fung, J.C.H.: Time series forecasting of air quality based on regional numerical modeling in Hong Kong. J. Geophys. Res. Atmos. **123**(8), 4175–4196 (2018)

17. Middya, A.I., Roy, S., Chattopadhyay, D.: CityLightSense: a participatory sensing-based system for monitoring and mapping of illumination levels. ACM Trans. Spat. Algorithms Syst. **8**(1), Article 5 (2021)
18. Palak, R., Wojtkiewicz, K., Merayo, M.G.: An implementation of formal framework for collective systems in air pollution prediction system. In: Nguyen, N.T., Iliadis, L., Maglogiannis, I., Trawiński, B. (eds.) ICCCI 2021. LNCS (LNAI), vol. 12876, pp. 508–520. Springer, Cham (2021). https://doi.org/10.1007/978-3-030-88081-1_38
19. Paoli, C., Notton, G., Nivet, M.-L., Padovani, M., Savelli, J.-L.: A neural network model forecasting for prediction of hourly ozone concentration in Corsica. In: 10th International Conference on Environment and Electrical Engineering, EEEIC 2011, pp. 1–4. IEEE (2011)
20. Semlali, B.B., El Amrani, C., Ortiz, G., Boubeta-Puig, J., García de Prado, A.: SAT-CEP-monitor: an air quality monitoring software architecture combining complex event processing with satellite remote sensing. Comput. Electr. Eng. **93**, 107257 (2021)
21. Seng, D., Zhang, Q., Zhang, X., Chen, G., Chen, X.: Spatiotemporal prediction of air quality based on LSTM neural network. Alex. Eng. J. **60**(2), 2021–2032 (2021)
22. Vaswani, A., et al.: Attention is all you need. In: 31st Conference on Neural Information Processing Systems, NIPS 2017, pp. 1–11 (2017)
23. Wu, N., Green, B., Ben, X., O'Banion, S.: Deep transformer models for time series forecasting: the influenza prevalence case (2020). https://arxiv.org/abs/2001.08317
24. Yu, Y., Si, X., Hu, C., Zhang, J.: A review of recurrent neural networks: LSTM cells and network architectures. Neural Comput. **31**(7), 1235–1270 (2019)

An Ensemble Based Deep Learning Framework to Detect and Deceive XSS and SQL Injection Attacks

Waleed Bin Shahid$^{(\boxtimes)}$ (ID), Baber Aslam, Haider Abbas(ID), Hammad Afzal(ID), and Imran Rashid(ID)

National University of Sciences and Technology, Islamabad 44000, Pakistan
{waleed.shahid,ababer,haider,hammad.afzal,irashid}@mcs.edu.pk
https://www.mcs.nust.edu.pk

Abstract. Safeguarding websites is of utmost importance nowadays because of a wide variety of attacks being launched against them. Moreover, lack of security awareness and widespread use of traditional security solutions like simple Web Application Firewalls (WAFs) has further aggravated the problem. Researchers have moved towards employing sophisticated machine learning and deep learning based techniques to counter common web attacks like the SQL injection (SQLi) and Cross Site Scripting (XSS). Lately, keen interest has been taken in tackling these attacks through cyber deception. In this paper, we propose an ensemble based deep learning approach by combining Convolutional Neural Network (CNN) and Long Short Term Memory (LSTM) models. This detection framework also contains a Session Maintenance Module (SMM) which maintains user state in an otherwise stateless protocol by analyzing cookies thereby providing further optimization. The proposed framework detects SQLi and XSS attacks with an accuracy of 99.83% and 99.47% respectively. Moreover, in order to engage attackers, a deception module based on dockers has been proposed which contains deceptive lures to engage the attacker. The deceptive module has the capability to detect zero-days and is more efficient when compared to other similar solutions.

Keywords: SQL injection attacks · XSS · Deep learning · Deception

1 Introduction

Internet has grown very rapidly in recent years providing outstanding benefits to its users all across the globe. This has enabled businesses, corporations and technically sound people across the world to share data and information in a simple and easy way. This massive data sharing and information dissemination

Sponsored by the Higher Education Commission (HEC), Pakistan through its initiative of National Center for Cyber Security for the affiliated lab National Cyber Security Auditing and Evaluation Lab (NCSAEL), Grant No: 2(1078)/HEC/ME/2018/707.

N. T. Nguyen et al. (Eds.): ACIIDS 2022, LNAI 13757, pp. 183–195, 2022.
https://doi.org/10.1007/978-3-031-21743-2_15

is mainly carried out through hundreds of millions of websites which are growing day by day [1]. These web applications are connected with back-end databases which contain useful and at times critical organizational data reserved and available only for authorized and legitimate users. Unfortunately, this overwhelming use and resulting ease in data/information sharing and dissemination has come with a price as people with malicious intent do their best to compromise these web applications to serve their illicit purpose. They either subvert web application's security to gain illegitimate access to critical data stored in back-end database or want to disrupt the availability of the website by turning it down or damage it's integrity by defacing it. Another important aim of the attackers is to find vulnerabilities in the target web application to place a malicious script or payload there with an aim to infect everyone visiting that website.

OWASP [2] lists common web attacks based on the ease of exploitation, difficulty in patching and their frequency of occurrence. SQL injection (SQLi) and Cross Site Scripting (XSS) attacks are a very common opportunity for attackers to compromise the website, its back-end database and the visiting benign users. For instance, attackers exploit XSS bugs in order to inject the vulnerable target website with malicious JavaScript so that all normal users visiting the website also get infected and start sending the desired information/data to attacker. Figure 1 explains how an XSS attack takes place.

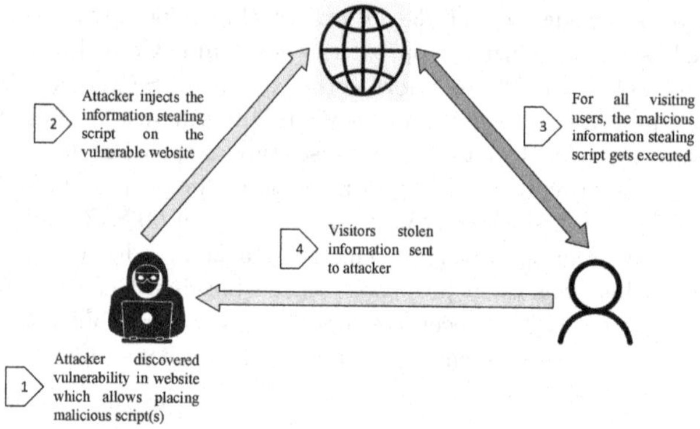

Fig. 1. A cross site scripting attack

SQL injection attacks take place when attackers supply untrusted data or commands to a website especially in he authentication fields. These SQL payloads and statements are executed by the vulnerable website resulting in providing the attacker illegitimate access to the back-end database. Security researchers have used various techniques like Web Application Firewall (WAF) to counter these web attacks [3]. WAFs have many drawbacks as they are unable to deal with sophisticated attackers, custom payloads and scripts, fail against zero-days and are mostly rule-based lacking contextual or behavioral analysis of the incoming HTTP request. This persuaded researchers to come up with machine learning and later deep learning based approaches to counter XSS and SQL injection attacks. Apart from detecting and stopping malicious HTTP requests, researchers are taking ever increasing interest in absorbing these attacks by deceiving the attackers with the help of deceptive lures. This helps in studying and analysing the attack methodologies and attacker behaviour in a lot more detail. The proposed framework achieves considerable success in firstly detecting these two common web attacks and later absorbing them by engaging the attacker with the help of deceptive lures. The main research contribution of this paper is as follows:

1. An ensemble based deep learning framework using CNN and LSTM classifiers has been proposed which detects SQL injection and Cross Site Scripting attacks with very high accuracy.
2. The attack detection module is supported by a State Maintenance Module that helps maintaining the state of attacker resulting in successful attacker profiling and categorization, which strengthens both attack detection and deception.
3. A light-weight high interaction docker based deception module which comprises of a docker daemon that controls and manages SQL injection and XSS attack dockers where deceptive lures have intentionally been placed to engage attackers in real time.

1.1   Background Study

Many research works have used deep learning based approaches in order to counter web application attacks. Luo et al. [4] proposed an ensemble classification model based with three classifiers. The research gives high accuracy and low false positive values. Niu et al. [5] proposed framework to counter web attacks by joining Convolutional Neural Network (CNN) with Gated Recurrent Unit (GRU) in order to yield high accuracy. Both research works used the CSIC data [6] for model training and testing. Tae-Young Kim and Sung-Bae Cho [7] combined CNN and LSTM models to deeply analyze the incoming web traffic with high accuracy. Adem Tekerek [8] used CNN to detect web attacks with high accuracy. Yao Pan et al. [9] used end to end deep learning for detecting XSS and SQLi attacks. Fawaz Mahiuob et al. [10] used an artificial neural network approach for detecting XSS attacks by dynamically extracting traffic features.

All these approaches lack the key feature of analyzing state of the incoming traffic thereby lacking the ability to maintain attacker's profile overtime. Few research works have focused on profiling attackers in order to enhance the attack detection capability [11]. Moreover, these techniques only counter attacks and do not have any deception module to engage the attacker. Nevertheless, many stand alone deception systems have been proposed which are based on docker containers. SMartin Valicek et al. [12] used dockers to create high interaction

honeypots for Linux and Windows. Gaspari et al. [13] also used dockers to propose an architecture which empowers the production system for active defense.
Andronikos Kyriakou et al. [14] proposed a docker based honeypot approach by
deploying different standalone honeypots.

Most of the deception solutions do not offer customized deception and are
static in nature. Moreover, they are not coupled with attack detection modules.
Since the proposed technique is based on an ensemble approach using CNN and
LSTM models along with a Session Maintenance Module which helps in maintaining the attacker's state thereby augmenting the performance and efficacy of
the deception module.

2 Proposed Detection and Deception Technique

The proposed framework for detecting and deceiving SQLi and XSS attacks is
shown in Fig. 2. Here, the attacker's request first passes through the detection
module where the SMM does the profiling and the ensemble based classifier
detects and forwards it to the attack specific dockers.

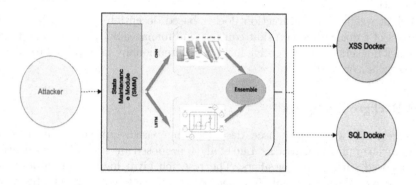

Fig. 2. The proposed framework

For the purpose of training and testing the deep learning classifiers (CNN
and LSTM), we employed publicly available benchmark datasets [15,16] which
contain both benign and anomalous XSS and SQL injection request packets.

2.1 Data Preparation and Feature Selection

The dataset containing XSS and SQL injection attack packets is in the form
of network flows, therefore, it is important to understand the significance of
these flows which would facilitate in picking only the necessary and relevant
flows for model training. These bi-directional flows reflect quite a lot about the
nature of traffic as they carry detailed information about the traffic being sent

to and received from the target website. For that matter, it is not necessary that all network flows will play a decisive role in determining the nature of HTTP request. If a network flow does not relate with the type of attack, there is no point in using it as a feature for model training. For that matter, while selecting/choosing the network flows for training our deep learning classifiers, it was made sure that:

- All flows which had the value *Zero/NULL* were eliminated as they do not participate at all in any sort of classification.
- All network flows which had static information like source and destination socket information were eliminated as it does not play any part in determining either XSS or SQL attacks.
- Network flows which actually carry information about XSS and SQL injection payload and scripts etc. are preferred. For instance, total bytes in the HTTP request, HTTP request body, destination packets and length of the HTTP response are some key network flows which carry immense value while dealing specifically with XSS and SQL injection attacks.

As a result only *29* bi-directional network flows were chosen out of whom *14* carried numerical data. Since LSTM and CNN classifiers deal with numerical data, therefore, the remaining *15* network flows needed to be converted from categorical to numerical form. For that purpose the pre-processing label encoder in *sklearn* [17] was used which labels in a way such that the label values only range from *0* to *total_classes-1*.

2.2 Using the Ensemble Based Deep Learning Classifiers

Ensemble based deep learning approaches help reduce variance of neural networks by combining predictions from various models because the bias added by combining these predictions helps reduce variance of a single employed neural network model [18].

CNN Classifier. For training the convolutional neural network based classifier, the feature set is defined as S which comprises of features $(s_1, s_2, \ldots, s_{29})$. We then apply the embedding, dropout, convolution layers to give the output as shown in Eq. 1 where A_{ReLU} denotes the activation function, win denotes the window size, W represents the kernel weight and λ represents bias for a particular feature s_i.

$$C_l^1 = A_{ReLU} \left(\sum_{i=1}^{win} W_l^c . s_x^c + \lambda_{sl}^c \right) \tag{1}$$

After convolution, the pooling, dense and activation layers are applied after which the model is projected onto an output layer that has two neurons for two classes by using the *Softmax* activation function. Table. 1 shows the CNN model with all its layers and parameters when applied on the SQL dataset.

Table 1. CNN model output for SQL injection attacks

Layer (type)	Output shape	Parameters
Embedding	(None, 33, 256)	4968192
Dropout	(None, 33, 256)	0
Conv1D	(None, 31, 250)	192250
Pooling	(None, 250)	0
Dense	(None, 250)	62750
Dropout	(None, 250)	0
Activation	(None, 250)	0
Dense	(None, 10)	2510
Activation	(None, 10)	0

LSTM Classifier. The same dataset containing both SQL injection and XSS attack requests was also used to train the LSTM classifier. A standard neural network algorithm typically comprises of an input activation along with the output activation. Both these layers relate with each other through an activation function. In case of LSTM classifier, the input activation after its application gets multiplied by some factor and later added due to the recurrent self-connection. Table 2 shows the LSTM model with all its layers when applied on the XSS dataset.

Table 2. LSTM model output for XSS attacks

Layer (type)	Output shape	Parameters
Input	(None, 10, 28)	0
LSTM	(None, 10, 128)	80384
Flatten	(None, 1280)	0
Output (Dense)	(None, 10)	1280

Both deep learning models have been combined using the ensemble approach in order to come up with a final and optimal classification.

2.3   State Maintenance Module

The state maintenance module is responsible for maintaining the attacker's state by analyzing cookies as they are provided by the website to the visiting users in order to provide a better user experience. Therefore, they contain some data specific to the visiting users. Every time they visit the website, they are being identified through the cookies they bring. If the ensemble based detection module labels an incoming request as malicious, the SMM would log this activity so that in future the user is directly diverted to the deception module by just analyzing the cookie field. This helps in optimizing the proposed framework of attack deception and deception. Moreover, the SMM is also capable of finding out any mutation, presence of any script or unwanted content in the cookie field.

2.4   Deception Module to Lure/Engage Attackers

After successfully detecting SQL and XSS attacks, attack packets are forwarded to respective dockers via the docker daemon. These dockers contain deceptive lures to engage the attackers. The prime purpose of using the docker approach is the operational flexibility, design simplicity, easy runtime spawning and memory efficiency it provides, unlike virtual machines [19]. Moreover, dockers share the common Operating System (OS) which is not a problem because both SQL and XSS attacks reside at the application layer and do not interfere with the underlying OS.

Docker Daemon. The *docker daemon* which is securely controlled by the administrator and is responsible for integrating the XSS and SQL dockers with the ensemble based deep learning module.

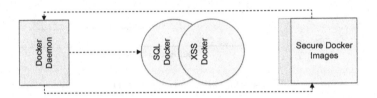

Fig. 3. The Docker Daemon controlling attack specific dockers

The daemon also fetches securely placed docker images when required and performs centralized log management as shown in Fig. 3.

Securing Docker Containers. In order to base the deception module on docker containers, it was important to prioritize the security of all dockers so that attackers fail to compromise security of the docker daemon or the attack specific dockers. For that purpose, all dockers were regularly updated and patched against known vulnerabilities (other than the ones intentionally placed), were run with limited permissions and were enabled with Docker Trust Management.

SQL Injection Docker. This docker is aimed to engage attackers launching SQL injection attacks by providing them the requisite lures. Moreover, any extension in the proposed deception module would connect this docker with other attack and pre-attack dockers further enhancing the deceptive capability. This docker contains few authentication pages which belong to the actual website by removing the SQL related validation and sanitization checks so that whenever an attacker generates an HTTP request containing an SQL payload, he/she will be routed towards this docker by the docker daemon, lured in and engaged to extract useful attack related information. The SQL injection docker is further explained in Fig. 4.

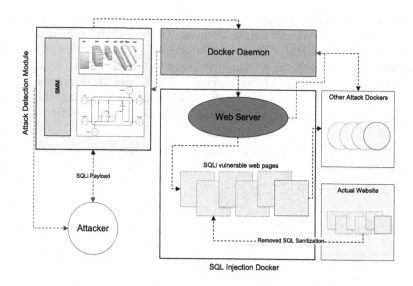

Fig. 4. SQL injection attack docker

XSS Docker. XSS attacks are applicable only on those web pages which basically require user input in some form because malicious JavaScript after execution results in a compromise thereby providing the attacker with victim's information in an unlawful way. In order to build the XSS docker, the original website, where the entire deception module was deployed, was patched against all cross site loopholes. The difference in this docker is that it contains both XSS vulnerable web pages and replicas of some secure/patched web pages from the actual website. These were added to thwart the possibility of finding the deceptive system by an attacker who could possible perform a comparative analysis of received responses by coming from an altogether different identity. On the other hand, if attackers share links of the vulnerable XSS pages with anyone, he/she will also be diverted to this docker by the docker daemon (Fig. 5).

3   Discussion, Performance Analysis and Testing

In order to test the proposed deep learning models we made use of the testing dataset which was separated earlier from the training dataset. Both the CNN

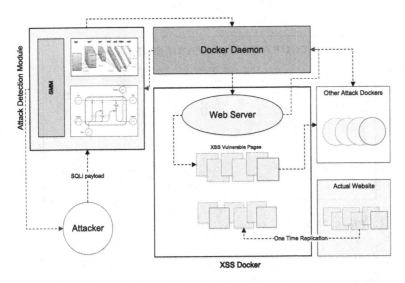

Fig. 5. Cross site scripting docker

and LSTM models were executed on this testing dataset comprising of SQL and XSS traffic (attack and benign requests) for *10* epochs in order to maximize accuracy and minimize validation loss. Later the ensemble classifier made the final classification decision by combining the CNN and LSTM classifications. Figures 6 and 7 show the validation accuracy and validation loss in detecting XSS attacks by both classifiers. It is evident that the LSTM based deep learning classifier outperforms the CNN classifier in accurately detecting cross site scripting attacks. Similarly, both deep learning models were tested on the SQL injection dataset comprising of both benign and malicious requests in order to yield a very high accuracy and minimized validation loss as shown in Figs. 8 and 9. Here, the CNN classifier yields better performance as compared to LSTM in accurately predicting SQL injection attacks, thereby contributing more in the ensemble based decision.

As far as the performance of the deception module is concerned, it was observed that by using the docker based approach, no major delay was observed in response time as HTTP responses by both the attack dockers were very minutely greater than response time of the actual website as shown in Eq. 2, thereby not letting the attacker notice about presence of any deception module.

$$\mathbb{T}_d \leqq \mathbb{T}_w \tag{2}$$

It was observed during testing, that for *141* malicious SQL requests, the SQL attack docker successfully engaged the attacker for *89* requests by reaching a maximum engagement level of *5*.

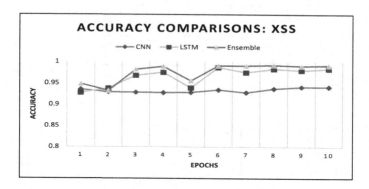

Fig. 6. Validation Accuracy (XSS): CNN, LSTM, Ensemble

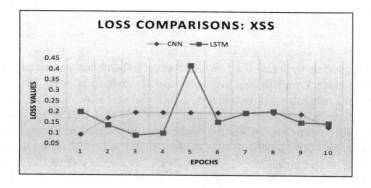

Fig. 7. Validation Loss (XSS): CNN vs. LSTM

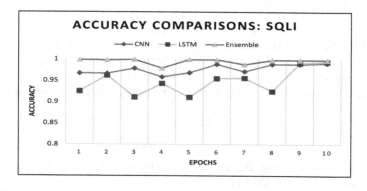

Fig. 8. Validation Accuracy (SQLi): CNN, LSTM, Ensemble

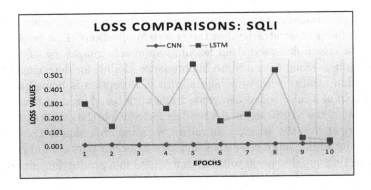

Fig. 9. Validation Loss (SQLi): CNN vs. LSTM

3.1 Comparative Analysis

The proposed deception framework was compared with Glastopf [20], another honeypot solution for countering web application attacks, in Table 3. It was observed that our technique is more optimized because of the State Maintenance Module which helps in recording the attacker profile and then dealing with all future requests from the attacker without much of processing. This feature is the hallmark of the proposed framework and helps it to detect zero-day attacks because attackers send a lot of reconnaissance and scanning packets to the target website before injecting it with zero-day attack payloads. Therefore, once detected in the pre-attack phase, the SMM would help detect the unknown zero-day attacks without much problem.

Table 3. Proposed deception module vs. Glastopf

Functionality	Glastopf [20]	Proposed framework
Maintaining attacker' state	No	Yes
Effectiveness of deceptive lures	Low	High
Runtime Docker spawning capability	No	Yes
Zero day attacker engagement	No	Possible
Centralized control	No	Yes
Customization	Difficult	Easy

4 Conclusion and Future Work

The rapidly evolving threat landscape has motivated security researchers to protect websites which carry useful data along with the back-end systems holding critical organizational data. Researchers have focused on using advanced techniques to counter these attacks that target web applications. This research work proposes an ensemble based deep learning approach comprising of CNN and LSTM classifiers along with a State Maintenance Module for detecting XSS and SQL injection attacks with very high accuracy. Deception is being carried out with the help of a full-fledged deception framework based on docker containers. In this module, the highly secure docker daemon controls the attack dockers that have lures to engage the attacker. In future, we intend to enhance the deception framework by adding more pre-attack and attack specific dockers and expose the entire detection and deception solution to live data in a production environment.

References

1. Lindsay Liedke.: 100+ Internet Statistics and Facts for 2020. http://www.websitehostingrating.com/internet-statistics-facts/. Accessed 29 Mar 2021
2. The Open Web Application Security Project.: OWASP Top Ten. http://owasp.org/www-project-top-ten/. Accessed 25 Mar 2021
3. Clincy, V., Shahriar, H.: Web application firewall: network security models and configuration. In: 2018 IEEE 42nd Annual Computer Software and Applications Conference (COMPSAC), vol. 01, pp. 835–836 (2018)
4. Luo, C., Tan, Z., Min, G., Gan, J., Shi, W., Tian, Z.: A novel web attack detection system for internet of things via ensemble classification. IEEE Trans. Ind. Inform. **01**, 1 (2020). https://doi.org/10.1109/TII.2020.3038761(2018)
5. Niu, Q., Li, X.: A high-performance web attack detection method based on CNN-GRU model. In: 2020 IEEE 4th Information Technology, Networking, Electronic and Automation Control Conference (ITNEC), vol. 01, pp. 804–808 (2020). https://doi.org/10.1109/ITNEC48623.2020.9085028
6. Giménez, C.T., Villegas, A.P., Marañón, G.Á.: HTTP DATASET CSIC 2010. http://www.isi.csic.es/dataset/. Accessed 8 Nov 2021
7. Kim, T.-Y., Cho, S.: Web traffic anomaly detection using C-LSTM neural networks. Expert Syst. Appl. **106** (2018). https://doi.org/10.1016/j.eswa.2018.04.004
8. Tekerek, A.: A novel architecture for web-based attack detection using convolutional neural network. Comput. Secur. **100**, 102096 (2021). https://doi.org/10.1016/j.cose.2020.102096
9. Pan, Y., et al.: Detecting web attacks with end-to-end deep learning. J. Internet Serv. Appl. **10** (2019). https://doi.org/10.1186/s13174-019-0115-x
10. Mokbal, F.M.M., Dan, W., Imran, A., Jiuchuan, L., Akhtar, F., Xiaoxi, W.: MLPXSS: an integrated XSS-based attack detection scheme in web applications using multilayer perceptron technique. IEEE Access 7, 100567–100580 (2019). https://doi.org/10.1186/s13174-019-0115-x
11. Shahid, W.B., Aslam, B., Abbas, H., Khalid, S.B., Afzal, H.: An enhanced deep learning based framework for web attacks detection, mitigation and attacker profiling. J. Netw. Comput. Appl. **198**, 103270 (2022)
12. Valicek, M., Schramm, G., Pirker, M., Schrittwieser, S.: Creation and integration of remote high interaction honeypots. In: 2017 International Conference on Software Security and Assurance (ICSSA), pp. 50–55 (2017). https://doi.org/10.1186/s13174-019-0115-x

13. De Gaspari, F., Jajodia, S., Mancini, L.V., Panico, A.: AHEAD: A New Architecture for Active Defense, pp. 11–16. Association for Computing Machinery (2016). https://doi.org/10.1145/2994475.2994481
14. Kyriakou, A., Sklavos, N.: Container-based honeypot deployment for the analysis of malicious activity. In: 2018 Global Information Infrastructure and Networking Symposium (GIIS), pp. 1–4 (2017). https://doi.org/10.1109/GIIS.2018.8635778
15. The TON IoT Datasets. http://research.unsw.edu.au/projects/toniot-datasets. Accessed 7 Oct 2021
16. Stratosphere Lab: A labeled dataset with malicious and benign IoT network traffic. http://www.stratosphereips.org/datasets-iot23. Accessed 4 Oct 2021
17. sklearn.preprocessing.LabelEncoder.   http://scikit-learn.org/stable/modules/generated/sklearn.preprocessing.LabelEncoder.html. Accessed 8 Nov 2021
18. Polikar, R.: Ensemble Machine Learning, pp. 1–34. Springer, New York (2012). https://doi.org/10.1007/978-1-4419-9326-7
19. Shahid, W.B., Aslam, B., Abbas, H., Afzal, H., Khalid, S.B.: A deep learning assisted personalized deception system for countering web application attacks. J. Inf. Secur. Appl. **67**, 103169 (2022)
20. Mphago, B., Mpoeleng, D., Masupe, S.: Deception in web application honeypots: case of Glastopf. In: International Journal of Cyber-Security and Digital Forensics, vol. 6, pp. 179–185. The Society of Digital Information and Wireless Communications (2017)

An Image Pixel Interval Power (IPIP) Method Using Deep Learning Classification Models

Abdulaziz Anorboev[1] ⓘ, Javokhir Musaev[1] ⓘ, Jeongkyu Hong[1](✉) ⓘ,
Ngoc Thanh Nguyen[2] ⓘ, and Dosam Hwang[1](✉) ⓘ

[1] Yeungnam University, Daegu, Republic of Korea
jhong@yu.ac.kr, dos-amhwang@gmail.com
[2] Wroclaw University of Science and Technology, Wroclaw, Poland
Ngoc-Thanh.Nguyen@pwr.edu.pl

Abstract. The implementation of deep learning (DL) in various fields is becoming common. In addition, demand for higher accuracy models is increasing continuously at the same rates as the growth of other fields of science. Using all DL tools in the development of computer vision (CV) is a fundamental aspect of its future development. Considering all such tools, we conducted research on the effect of data representation on the final classification accuracy and proposed image pixels' double representation (IPDR) and image pixels' multiple representations (IPMR) for skipping certain pixels in the images in a dataset. Because the image pixel values range from 0 to 255, we proposed including all knowledge from different intervals of pixels. With IPDR, we trained the model using a dataset and obtained the prediction probabilities for the classification task. Next, we created two different datasets from an existing dataset. The first dataset took only image pixels of higher than 127, with all other image pixels in the dataset changed to zeros. The second dataset took only image pixels equal to or lower than 127. These two created datasets were trained on the same model architecture and their prediction accuracies for classification were ensembled with the prediction accuracies of the main model. With the IPMR method, we applied the same method as previously described, although instead of two intervals, from 0 to 127, and 127 to 255, we used, multiple intervals of 50 (i.e., [0:50], (50:100], (100:150], (150:200], and (200:255]) for the Cifar10 dataset. The number of intervals depends on the dataset, and applying our method, we achieved 89.46%, 98.90%, and 73, 38% accuracies on the Fashion MNIST, MNIST, and Cifar10 datasets, respectively, whereas their original classification accuracies under classic training were 89.27%, 98.65%, and 71.29%, respectively. As the advantage of using this method, it can be applied to any classification task and gives only extra knowledge on the trained data. As another simplicity of this method, it can be used with other DL ensemble models simultaneously.

Keywords: Image pixel double representation · Image pixel multiple representation · Model ensemble · Prediction scope

© The Author(s), under exclusive license to Springer Nature Switzerland AG 2022
N. T. Nguyen et al. (Eds.): ACIIDS 2022, LNAI 13757, pp. 196–208, 2022.
https://doi.org/10.1007/978-3-031-21743-2_16

1 Introduction

Computer Vision (CV) has recently begun a new era with the development of DL and the introduction of neural networks (NNs). CV has moved from basic experiments in science and has started to implement its knowledge on real-life projects. Sectors such as the automobile industry, IoT, medicine, and finance, among many others, are becoming primary consumers of CV products. At the same time, the need for high accuracy models is growing dramatically. To meet this need, many studies are being conducted on DL and CV. Even a small increase in the model accuracy will affect future development of this field.

As the history of DL shows, various algorithms and models in each subfield of DL have been developed to solve various problems. Many algorithms and tools have been proposed to solve such problems and fill in the gaps in this field. One is ensemble learning, which allows DL models to share their knowledge through the use of ensemble methods.

The power of a DL ensemble has been presented by various researchers [1–7]. Different approaches have been applied to the models, data, and algorithms that allow utilizing the knowledge of ensemble units. The results of ensemble models were presented in [8–20], including medicine, financial risk assessment, and oil price prediction, among other fields.

Any positive change in DL models can help with future development. Hence, we proposed the Image Pixel Interval Power (IPIP) Ensemble method for DL classification tasks. We proposed two sub methods (IPDR and IPMR), which describes IPIP to make other datasets out of original dataset that is used for a DL classification task. Using IPDR, we copied the original dataset and replaced all pixels of lower than or equal to 127 with zeros. As a result, we obtained a second dataset for training. To create the third dataset, we copied the original dataset again, and this time, we replaced all pixel values that are higher than 127 with zeros. At the second step, we trained three datasets with three models and their ensemble prediction results. The result of this method was extremely positive. We increased the accuracy of the model, which was 98.84%, to 99.11%. IPMR uses more than two intervals to create datasets out of the original dataset and applies them in training, utilizing the same process as with IPDR. The key point to mention here is the increase in accuracy to close to 100%. In many DL ensemble models, it is difficult to achieve better results while the model has already achieved an accuracy of near 100%. The prediction scope of an ensemble is usually located in the prediction scope of the main model and does not allow increasing the accuracy. However, with our method, we partially solved this problem and obtained better results for our models.

Another attractive feature of this model is that it does not affect the training of the main model because it is trained separately and includes nearly all knowledge of the main model.

The structure of the work consists of the following. Section 1 introduces the paper, giving general information regarding ensemble learning and the content of the study. Section 2 provides information regarding the related studies and problems that have been found in ensemble learning. Section 3 provides solutions to the problems described in Sect. 2 and gives a detailed explanation of proposed methodology. Section 4 presents the experiments and results of our research, including information regarding the dataset, base method, training setup, evaluation metrics, experiment results, and discussions. Finally, Sect. 5 provides some concluding remarks regarding this research and areas of future study.

2   Related Works

An image pixel interval was one of the remaining topics in DL classification and ensemble models. The focus using an image pixel representation is applied in [21], where the authors recommend applied transformers for a computer vision task dividing images into patches. In [21], the authors achieved a state-of-art result in computer vision using one of the last methods in natural language processing. Another study [22] represented images as semantic visual tokens. The low-level features of the images were extracted using a convolutional neural network and used for further training in Visual Transformers [22]. The use of Visual Transformers outperformed the accuracy of ResNet on ImageNet. In addition, in [23], the authors introduced an MLP-mixer that proved there is no need for convolutional neural networks or an attention-based model. The MLP-mixer also used image patches as an input but included two types of layers: one MLP applied independently to each patch and another MLP applied across image patches. The other approach to our method was ensemble learning. We ensembled the prediction probabilities of the models. Reviewing ensemble learning, we found a gap in which most of the models in a DL ensemble used model ensembles to develop their methods. The majority of studies that varied the training data focused on changing the training data samples, batches, or methods of feeding the data into the model, rather than applying insight regarding the image pixels into the models. Surveys [1, 6, 7, 10, 23, 24] on ensemble learning have shown that the main studies have evaluated the effect of model ensembles and the combinations of whole datasets. Reviewing all of the abovementioned research, we found a gap in that none of the studies considered the effect of pixel variance on the model or its insight, which can add more power to the final data classification. This motivated us to study the power of image pixel intervals in DL classification tasks.

3   Proposed Methodology

To address the gap found in the literature review, we proposed the IPIP method, which studies image pixel variance and includes two sub methods: IPDR and IPMR. IPIP is described through the following steps:

- Creating datasets from the original dataset using IPDR or IPMR
- Training the main dataset with the model architecture
- Training the created datasets with a model chosen for these datasets
- Ensembling the prediction probabilities from each model
- Accepting the maximum predictions as the predicted class

IPIP is described using IPDR and IPMR. The main contribution of IPIP is to use datasets copied from the original dataset, leaving certain interval pixel values. The difference in the number of intervals encouraged us to make an initially double representation of the main dataset and multiple representation of the main dataset. IPDR is a simple double representation of the main dataset. With IPDR, we create two zero arrays (dataset_1 and dataset_2) with the same size as the main dataset. In our experiments, we used the MNIST dataset. For this dataset we created two arrays with a size of $60000 \times 32 \times 32 \times 3$ all filled with zeros. For dataset_1, we took only pixel values from the main dataset that belongs to the $[0:127]$ interval and copied and pasted them at the same position in dataset_1. Dataset_2 was also built using the same method as dataset_1, except the pixel value interval for dataset_2 was $(127:255]$. All values higher than 127 were copied and pasted at the appropriate position in dataset_2. The rest of the training process is presented in Fig. 1.

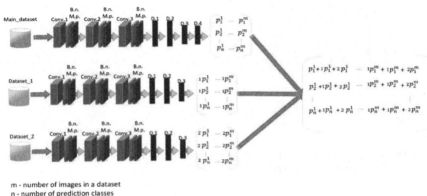

Fig. 1. IPIP: IPDR method architecture. p – prediction probability. The pattern from the right side is ensemble of prediction outcome.

As illustrated in Fig. 1, we started creating the datasets from main dataset. Dataset_1 and dataset_2 are the results of IPDR, which was used for further training of the model architecture shown in Figs. 3 and 4. The main dataset was trained in the architecture proposed in Fig. 3, whereas the other datasets created (dataset_1 and dataset_2) were trained using the model presented in Fig. 4. After training the models, we ensemble the prediction probabilities of each class with the corresponding probabilities from the two other models. Depending on the datasets applied, we use IPDRor IPMR to create datasets from the main dataset and train them using IPIP.IPMR shows the opportunity of creating more datasets and applying its knowledge on the classification task. We present the working scheme of IPIP with IPMR, which can use multiple intervals and n datasets for training, in Fig. 2. Then, the main dataset is trained in the model architecture shown in Fig. 3, and the other datasets are trained using the model architecture shown in Fig. 4. Their classification prediction probabilities were ensembled in the final step to achieve the final classification result. Finding the best number of intervals can be explained through the true prediction scope of each dataset, which adds extra knowledge to final model. We started with IPDR saving resources and decreasing the cost of training.

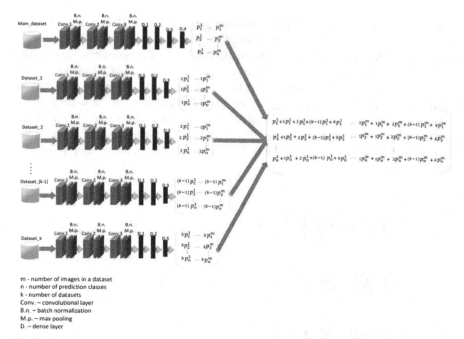

m - number of images in a dataset
n - number of prediction classes
k - number of datasets
Conv. – convolutional layer
B.n. – batch normalization
M.p. – max pooling
D. – dense layer

Fig. 2. IPIP: IPMR method architecture. p – prediction probability. The pattern from the right side is ensemble of prediction outcome.

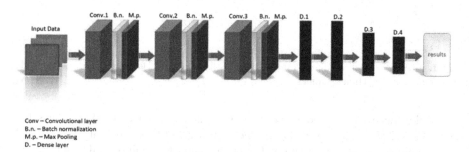

Conv – Convolutional layer
B.n. – Batch normalization
M.p. – Max Pooling
D. – Dense layer

Fig. 3. The model architecture for main dataset.

If IPIP with IPDR does not return the expected result, training can be continued using IPIP with IPMR, where the models number more than two and can represent a difference in the prediction scopes. The goal of our proposed method was to find the optimal way to obtain better insight from the image pixels. This led us to start our training from IPIP with IPDR, whereas IPIP with IPMR requires training the model by spending more time and power.

As the next important point, we used two different model architectures with different numbers of parameters. The main model has 226,122 trainable parameters, as presented in Fig. 3, whereas the model for the other created datasets has 160,330 trainable parameters. We chose a smaller model for the created datasets to avoid an overuse of time

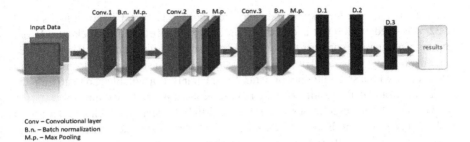

Conv – Convolutional layer
B.n. – Batch normalization
M.p. – Max Pooling
D. – Dense layer

Fig. 4. The model architecture for created datasets.

and power during IPIP implementation. In general, the larger the model is, the higher the results achieved. The architecture for the main model, illustrated in Fig. 3, consists of three convolutional layers with 32, 64, and 128 filters, and four dense layers with 256, 256, 128, and the number of class nodes for each dataset, respectively. In addition, we used max pooling with a 2×2 filter and batch normalization layers in both model architectures. The filter used for the convolutional layers in both models had a size of 3×3. The architecture of the model, presented in Fig. 4, includes three convolutional layers with 32, 64, and 128 filters, and three dense layers, which have following numbers of nodes: 256, 128, and the number of classes in each dataset. The proof of the effect of our method is shown in the results described in the Unique True Prediction (UTP) metrics introduced in this study. Each model uses different datasets, and as shown in Figs. 1 and 2, classification probabilities from trained models are ensembled to achieve the final classification probabilities. Each model can predict a certain number of images that differ from the prediction of the other models. When we ensemble these predictions, the prediction area of the ensemble model includes some parts of the UTP of the models. In our study, to train the models and ensemble their prediction probabilities, we used different datasets that were created by applying IPIP with IPDR or IPMR.

4   Experiments and Results

In this section, we provide comprehensive information about the experiments and their results. This section also includes test results of the used model and dataset. Its comparison to our base model and the evaluation is the main part of this section.

4.1   Datasets

In our research, we used the Cifar10[1], MNIST, and Fashion MNIST datasets. The first dataset was Cifar10. This dataset is considered to be one of the most popular datasets available and is used in image classification tasks. Furthermore, Cifar10 includes a sufficient number of images to train and obtain reasonable results for classification models. The dataset is a subset of a dataset of 80 million tiny images, which were

---

[1] https://www.cs.toronto.edu/~kriz/cifar.html.

collected by Alex Krizhevsky, Vinod Nair, and Geoffrey Hinton. Cifar10 consists of 32 × 32 × 3 sized 10 000 test and 50000 train images that belongs to 10 classes. In our training we did not preprocess the images. For our method we used Cifar10 and created other datasets out of Cifar10 by applying our method. For a simple model similar to our architecture with a small number of parameters, Cifar10 is a suitable choice to train the model and evaluate the results with the requirements of the proposed method.

The next dataset used in our research is the MNIST[2] dataset, which was developed by LeCun, Cortes, and Burges. The dataset consists of handwritten digits that belongs to 10 classes. It has 60,000 training sets and 10,000 test sets 28 × 28 greyscale images which represents digits from 0 to 9. The MNIST dataset was originally selected and experimented with by Burges and Cortes, who used bounding-box normalization and centering. LeCun's version uses centering based on center of mass within in a larger window. Extremely light-task DL models were used to train this dataset.

In addition, this dataset easily achieves a high accuracy with convolutional neural networks. This feature helped us during our experiments to check the effect of the proposed method against high accuracy models.

The next dataset is Fashion MNIST[3], which includes 60,000 training images and 10,000 test images. The dataset consists of Zalando's article images and has a size of 28 × 28. The number of images in the dataset is the same as that of the MNIST dataset, although this dataset includes 10 different classes of fashion.

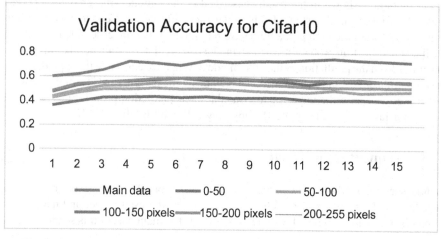

Fig. 5. Validation accuracy of Cifar10 and created datasets with certain pixel intervals.

4.2 Baseline Method

In our study, we used classic training as a baseline method for the experiment evaluations. The reason to choose classic training underlies the difficulty of finding an alternative

---

[2] http://yann.lecun.com/exdb/mnist/.

[3] https://www.kaggle.com/zalando-research/fashionmnist.

similar method that can be used to compare the results. Most of the DL ensemble models focus on model architectures and data representations through image level preprocessing in not researching the pixel level variances. The main objective of this method is to apply it in various combinations and with many other DL ensembles simultaneously. Another motivation for conducting this research was to utilize the knowledge from the pixel-level variance by changing the data representation and to achieve a better accuracy at the final classification step. As a result, we achieved better results on three different datasets like the MNIST, Fashion MNIST and Cifar10 datasets.

4.3 Training Setup

Python 3.6.12 and TensorFlow 2.1.0 were used to build the model architecture for the proposed method. For our experiments, we used a 12-GB Nvidia Titan-XP with CUDA 10.2 on a computer with an Intel core-i9 11900F CPU and 64 GB of RAM. For our training, we randomly initialized the weight for the model and trained for certain epochs. To train the model, we used an Adam optimizer with a default learning rate of 0.001 and sparse categorical loss function. The model was prepared for training three different datasets. Each model was set to train for 15 epochs. The results from the 15th epoch were used to build a new ensemble and were evaluated using the metrics focused on in this study.

4.4 Evaluation Metrics

We proposed a method that mainly focuses on the accuracy of the model. Many other studies include different metrics such as the F1 score, Recall, IoU, and ROC. For our research, we chose two metrics that meaningfully explains the achievements of the method on the different datasets. The accuracy is the ratio of true predictions to the total number of cases that are used to evaluate the model. Equation 1 shows the calculation of the accuracy.

$$\text{Accuracy} = \frac{TP + TN}{TP + TN + FP + FN} \tag{1}$$

TP - true predicted positive results
TN - true predicted negative results
FP - false predicted positive results
FN - false predicted negative results

$$UTP(X, Y) = X - X \cap Y \tag{2}$$

UTP - Unique True Prediction
X - Prediction Scope of a model
X Y - Prediction Scope of a model Y

The next evaluation metric is UTP, which identifies the percentage of unique predictions for each model with respect to another model. In Eq. (2), UTP(X, Y) finds unique true predictions of model X with respect to model Y. This metric explains why our proposed model achieved better results than the main model where we trained only main dataset. The index of true predicted images differs for each model, even when they have the same accuracy. This leads the ensemble to achieve better results.

4.5 Experimental Results and Discussions

In this section of our study, we introduced experiment results and their detailed explanations. In addition, this study includes the validation accuracy and UTP as the evaluation metrics. The reason for choosing only these two metrics is because we only focused on an explanation of our study and for clarifying future studies in this field. We calculated the prediction scope and found the unique true prediction of each member of the ensemble with respect to the main model. Figure 5 shows the accuracy of the validation data for the Cifar10 dataset. Figure 6 shows the validation accuracy of the MNIST dataset, and Fig. 7 represents the validation accuracy for the Fashion MNIST dataset. Table 1 shows the test set accuracy of the main and ensemble models for each dataset. Table 2 shows the UTP of the model trained using the dataset created by the authors with respect to the model, which was trained with the original dataset. Because our ensembles have different numbers of ensemble members, our method achieved better results on each of the datasets.

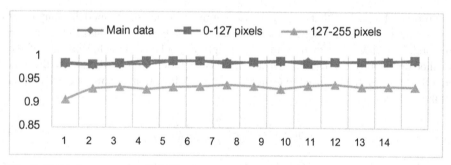

Fig. 6. Validation accuracy of MNIST dataset and created datasets with certain pixel intervals.

As mentioned above, the experiments started from copying the original dataset and ignoring some pixels that do not lie within a certain interval. For the Cifar10 dataset, we applied the same processes described in the methodology part of our study. As the main dataset, we chose the Cifar10 dataset and made five copies of it. Consequently, for the 0–50 interval dataset, we give a value of zero to all pixels that do not belong to the 0–50 pixel value interval. For the second copy of the dataset, we repeated the same processes applied for the first one except the interval was from 50 to 100 pixels. For the third copy, the same process was repeated using a 100–150 pixel value intervals. For the fourth copy of the original dataset, we take only 150–200 pixel values, making all other pixels zero. For the last copy of the dataset, we repeated the same process using 200–255 pixel value intervals. All six datasets were trained using the same model architecture, and their final prediction probabilities were ensembled. Figure 5 shows the validation accuracy of Cifar10 and the five newly created datasets from Cifar10. As the results show, there are no dramatic changes in each of the epochs. Increasing the accuracy of the model becomes a more difficult task when there is no high variance in the accuracies of the different epochs. In our model, Cifar10 achieved an accuracy of approximately 70% during each epoch. For the last epoch, 71.29% accuracy was achieved on the test set, and after applying our method, we obtained an accuracy of 73.38%. The next dataset is the MNIST dataset, which consisted of 28 × 28 greyscale images. From this dataset, we created two other datasets. The first one ignored all pixel values that are higher than 127. For this dataset, we replaced all pixel values that are higher than 127 to zeros. The second dataset was created using higher pixel values than 127.

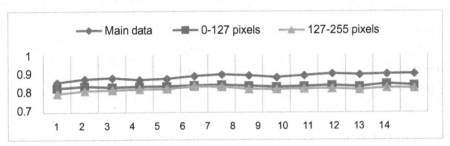

Fig. 7. Validation accuracy of Fashion MNIST dataset and two created datasets that has 0–127 and 127–255 pixel values.

Pixel values of less than 127 were replaced with zeros. The same model architecture was used to train the MNIST and two other created datasets. The validation accuracy of the MNIST dataset and two other datasets, which included pixel values of 0–127 and 127–255, respectively, are presented in Fig. 5. MNIST and 0–127 pixel-valued datasets have a very similar trend on the validation accuracy. The figure shows that the 0–127 interval includes more knowledge of the original dataset than any other part. By contrast, the 127–255 pixel-valued dataset has a lower validation accuracy. However, all three datasets reach an extremely high accuracy. Hence, increasing it was a challenging problem for us. When we ensemble the prediction accuracy of these three models trained on three datasets (MNIST, 0–127 and 127–255 pixel valued), we attained better accuracy than simply training the MNIST dataset on the utilized model architecture. Therefore, we achieved 98.90% accuracy using our method when the original model achieved an accuracy of 98.65%.

As Fig. 7 shows, the last dataset trained using the proposed method was Fashion MNIST. For this dataset, we used the same process as MNIST and created two other datasets out of Fashion MNIST. Table 1 shows the final results of our proposed method where we obtained better results on the three datasets. The results of classic training, using only the original dataset, and training using the model, were lower than the results of our method, which we note as an ensemble model. After applying our method, we

Table 1. Test set accuracy for each of the datasets and models

Datasets	Main model	Ensemble model
Cifar10	0.7129	0.7338
MNIST	0.9865	0.9890
Fashion MNIST	0.8927	0.8946

achieved 89.46% accuracy. The reason for achieving better results can be explained using Table 2.

Table 2. UTP (Secondary Models, Main Model)

Datasets	UTP
Cifar10	0.1097
MNIST	0.0069
Fashion MNIST	0.0353

The UTP of one model with respect to another returns a unique true prediction of first model, which shows a difference in the indexes of the true predicted image in the models. Although one model can achieve a lower accuracy, the other model that achieved a better accuracy cannot truly predict some of the true predicted images of the first model. This explains the increase in the ensemble of our method. Analyzing Table 2, we can see that for each of the datasets created, some extra prediction space can be added to the main model. In a perfect implementation of the method, final predictions can return better results by percent, as presented in Table 2. Considering the error after an ensemble that occurs when the classification probabilities are added, not all of the possible prediction space are added to the final model. This part of our work shows future studies that could be successful for this topic.

5 Conclusion

Utilizing the knowledge of image pixel variance gave our proposed model better results than simply using the whole image. We trained three different datasets and in each of them achieved better results after using the proposed approach. The reason for these increases was the difference in prediction scope of each model. We tried partially to solve the generalization problem of DL by using the variance ensemble of the image pixels. It is important to note that using any knowledge to increase the result of the model will help with the development of the DL field in all studies. In this research, we used the difference from the true prediction as a key metric that explains the growth of our results. In addition, choosing the best interval for the pixel values also adds more knowledge to the model. As presented in the UTP results, some of the models have a

10% different UTP, which provided the chance to increase the model by up to 10%. In our study, we used only a small part of this opportunity. Using all of this knowledge remains as a future area of our study.

The future works should concern in using ontology structures for storing and processing meta-knowledge about images. Such a knowledge base can be very useful in achieving higher accuracy and effectiveness [25–27].

Acknowledgements. This work was supported by the National Research Foundation of Korea (NRF) grant funded by the Korea government (MSIT) (No. 2022R1F1A1074641).

References

1. Vega-Pons, S., Ruiz-Shulcloper, J.: A survey of clustering ensemble algorithms. Int. J. Pattern Recogn. Artif. Intell. **25**(3), 337–372 (2011)
2. Zhou, Z.H.: Ensemble learning. In: Li, S.Z., Jain, A. (eds.) Encyclopedia of Biometrics, pp. 270–273. Springer, Boston (2009). https://doi.org/10.1007/978-0-387-73003-5_293
3. Lappalainen, H., Miskin, J.W.: Ensemble learning. In: Girolami, M. (ed.) Advances in Independent Component Analysis, pp. 75–92. Springer, London (2000). https://doi.org/10.1007/978-1-4471-0443-8_5
4. Alqurashi, T., Wang, W.: Clustering ensemble method. Int. J. Mach. Learn. Cybern. **10**(6), 1227–1246 (2018). https://doi.org/10.1007/s13042-017-0756-7
5. Krogh, A.: Neural Network Ensembles, Cross Validation, and Active Learning
6. Sagi, O., Rokach, L.: Ensemble learning: a survey. Wiley Interdisc. Rev.: Data Min. Knowl. Discov. **8**(4) (2018)
7. Dong, X., Yu, Z., Cao, W., Shi, Y., Ma, Q.: A survey on ensemble learning. Front. Comput. Sci. **14**(2), 241–258 (2019). https://doi.org/10.1007/s11704-019-8208-z
8. Webb, G.I., Zheng, Z.: Multistrategy ensemble learning: reducing error by combining ensemble learning techniques. IEEE Trans. Knowl. Data Eng. **16**(8), 980–991 (2004)
9. Faceli, K., de Carvalho, A., Carlos, M., de Souto, P.: Multi-objective clustering ensemble. Classical Weightless Neural Systems View project Feature Extraction and Selection Analysis in Biological Sequences View project See Profile (2007). https://www.researchgate.net/publication/220515994
10. Gomes, H.M., Barddal, J.P., Enembreck, A.F., Bifet, A.: A survey on ensemble learning for data stream classification. ACM Comput. Surv. **50**(2), 1–36 (2017)
11. Qummar, S., et al.: A deep learning ensemble approach for diabetic retinopathy detection. IEEE Access **7**, 150530–150539 (2019)
12. Gaikwad, D.P., Thool, R.C.: Intrusion detection system using bagging ensemble method of machine learning. In: Proceedings - 1st International Conference on Computing, Communication, Control and Automation, ICCUBEA 2015, pp. 291–295 (2015)
13. Hamori, S., Kawai, M., Kume, T., Murakami, Y., Watanabe, C.: Ensemble learning or deep learning? application to default risk analysis. J. Risk Financ. Manag. **11**(1), 12 (2018)
14. Zhao, Y., Li, J., Yu, L.: A deep learning ensemble approach for crude oil price forecasting. Energy Econ. **66**, 9–16 (2017)
15. Yu, Z., et al.: Incremental semi-supervised clustering ensemble for high dimensional data clustering. IEEE Trans. Knowl. Data Eng. **28**(3), 701–714 (2016)
16. Sarmadi, H., Entezami, A., Saeedi Razavi, B., Yuen, K.V.: Ensemble learning-based structural health monitoring by Mahalanobis distance metrics. Struct. Control Health Monit. **28**(2) (2021)

17. Yu, L., Wang, S., Lai, K.K.: Credit risk assessment with a multistage neural network ensemble learning approach. Expert Syst. Appl. **34**(2), 1434–1444 (2008)
18. Xiao, Y., Wu, J., Lin, Z., Zhao, X.: A deep learning-based multi-model ensemble method for cancer prediction. Comput. Methods Programs Biomed. **153**, 1–9 (2018)
19. Zhou, T., Lu, H., Yang, Z., Qiu, S., Huo, B., Dong, Y.: The ensemble deep learning model for novel COVID-19 on CT images. Appl. Soft Comput. **98** (2021)
20. Galicia, A., Talavera-Llames, R., Troncoso, A., Koprinska, I., Martínez-Álvarez, F.: Multi-step forecasting for big data time series based on ensemble learning. Knowl.-Based Syst. **163**, 830–841 (2019)
21. Dosovitskiy, A., et al.: An Image is Worth 16 × 16 Words: Transformers for Image Recognition at Scale (2020). http://arxiv.org/abs/2010.11929
22. Wu, B., et al.: Visual Transformers: Token-based Image Representation and Processing for Computer Vision (2020). http://arxiv.org/abs/2006.03677
23. Tolstikhin, I., et al.: MLP-Mixer: An all-MLP Architecture for Vision (2021). http://arxiv.org/abs/2105.01601
24. Krawczyk, B., Minku, L.L., Gama, J., Stefanowski, J., Woźniak, M.: Ensemble learning for data stream analysis: a survey. Inf. Fusion **37**, 132–156 (2017)
25. Duong, T.H., Nguyen, N.T., Jo, G.S.: A hybrid method for integrating multiple ontologies. Cybern. Syst. **40**(2), 123–145 (2009)
26. Duong, T.H., Nguyen, N.T., Jo, G.S.: A method for integration of WordNet-based ontologies using distance measures. In: Lovrek, I., Howlett, R.J., Jain, L.C. (eds.) KES 2008. LNCS (LNAI), vol. 5177, pp. 210–219. Springer, Heidelberg (2008). https://doi.org/10.1007/978-3-540-85563-7_31
27. Pietranik, M., Nguyen, N.T.: A multi-atrribute based framework for ontology aligning. Neurocomputing **146**, 276–290 (2014)

Meta-learning and Personalization Layer in Federated Learning

Bao-Long Nguyen[1,2], Tat Cuong Cao[1,2], and Bac Le[1,2(✉)]

[1] Faculty of Information Technology, University of Science,
Ho Chi Minh City, Vietnam
[2] Vietnam National University, Ho Chi Minh City, Vietnam
lhbac@fit.hcmus.edu.vn

Abstract. Federated learning systems are confronted with two challenges: systemic and statistical. Non-IID data is acknowledged to be a primary component in causing statistical challenges. To address the federated learning system's substantial performance loss on non-IID data, we offer the `FedMeta-Per` algorithm (which combines meta-learning methods and personalization layer approaches into a federated learning system). In terms of performance and personalization, `FedMeta-Per` has been shown in experiments to outperform typical federated learning algorithm, algorithms using personalization layer techniques and algorithms using meta-learning in system optimization.

Keywords: Federated learning · Non-IID · Meta-learning · Personalization layer

1 Introduction

In the era of the Internet of Things, facing a huge amount of data at edge devices generated every second, the centralized training process presents many disadvantages. First, data at edge devices often contains sensitive information about users. When it has to be transmitted to a server for training a machine learning model, this information can be exposed, seriously affecting data privacy. Second, the cost of transmitting information between the server and the clients is very expensive because a large amount of data must be transferred to the server. Third, the server must have large computing power and storage to conduct the training process.

Edge computing [12] was born to push the computation and storage of data from the server to clients or edge devices, reducing the load on the server. Furthermore, the computation is moved to or near the data generation site, which saves huge amounts of communication costs and reduces computational latency. Not only that, the process of exchanging data between the server and the clients no longer takes place, greatly increasing the user's data privacy.

Based on the idea of edge computing, *federated learning* (FL) [18] was born to protect user data privacy by allowing machine learning models to be trained

N. T. Nguyen et al. (Eds.): ACIIDS 2022, LNAI 13757, pp. 209–221, 2022.
https://doi.org/10.1007/978-3-031-21743-2_17

on separate datasets that are distributed across edge devices. Specifically, the goal of the FL system is to find a global parameter w_G^* such that the system error is minimized:

$$w_G^* = \arg\min_{w_G} f_{global}(w_G)$$

$$= \arg\min_{w_G} \frac{1}{n} \sum_{i=1}^{n} f_{local}(w_i) \tag{1}$$

where, n is the number of clients participating in the system, f_{global} is the error of the whole system, $f_{local}(w_i)$ is the error of the client i when it uses the parameter w_i.

This training method not only inherits the benefits of edge computing (lowering hardware costs and latency, increasing data security), but also has the potential to increase personalization for each user. However, the clients participating in the system are often heterogeneous in terms of storage and computing capacity. The data distributed across these clients is also not uniformly distributed (non-IID). This creates two main challenges of an FL system: the Systemic Challenge and the Statistical Challenge [15]. As for the statistical challenge, [21] research indicates that FL's traditional algorithm - Federated Averaging (FedAvg) suffers from severe performance loss on non-IID data.

Task. This study aims to partially solve the statistical problem. Specifically, we improve the performance as well as the personalization per user of the FL system on non-IID data in the supervised learning problem by meta-learning (ML) algorithms [9] and personalization layer (PL) technique.

Contribution. The study proposes the FedMeta-Per algorithm, a combination of ML algorithms (MAML [7], Meta-SGD [16]) and PL techniques (FedPer [1], LG-FedAvg [17]), which helps to increase accuracy and personalize for each user. Experimentally, the research shows the effectiveness in increasing the accuracy as well as the personalization of the proposed algorithm compared with the algorithm FedAvg; algorithms that incorporate ML into FL [3] (FedMeta(MAML), FedMeta(Meta-SGD)) and algorithms that use PL (FedPer) on a FL system of 50 users, containing the data of two datasets MNIST [5] and CIFAR-10 [13] respectively.

2   Related Work

Since the FedAvg algorithm was first introduced in 2017, a lot of work has been done to improve the accuracy of the system on non-IID data. Here, the study reviews a few typical works in the realization of this goal.

Meta-learning - based approach. ML is a new learning method that allows the learning model to gain experience by performing many different tasks in the same task distribution. This results in ML models being able to adapt quickly to new tasks after only a few training steps with limited training data. This is an important discovery, playing a role in bringing machine learning closer to human learning [8].

In the context of non-IID data, each user, with different needs, generates datasets with very different distributions. If we consider each user as a task, training the global model on a set of users (FL system) is equivalent to training the machine learning model on a task distribution (ML approach). Research [3] proposes framework FedMeta and experimentally illustrates by two algorithms FedMeta(MAML), FedMeta(Meta-SGD). The results show that the global model has fast adaptability and faster convergence than FedAvg on the LEAF dataset [2]. Study [6] proposes the combination of MAML on the CIFAR-10 dataset, and study [11] proposes the combination of Reptile [19] on the EMNIST-62 dataset [4] to the FL system, to increase personalization for each client.

An FL system, on the other hand, will have new clients over time. One downside for systems that use ML in training is that they treat clients that have been around for a long time in the system and those that have just joined the system equally. Therefore, although global models are able to adapt quickly to new data, their performance can be further improved.

Personalization layer-based approach. To capture the data properties, which are highly personalized for different users, dividing the deep neural network into personalization layers and base layers and maintaining the personalization layers at each client (base layers will be co-trained by the clients) is suggested. Accordingly, study [1] proposes the FedPer algorithm with the idea of maintaining fully-connected layers of the deep neural network at the clients with the goal of increasing personalization. Experimentally, FedPer achieved much higher results than FedAvg when tested on the data sets CIFAR-10 and CIFAR-100. In addition, study [17] with the algorithm LG-FedAvg proposes to maintain the feature extraction layers of the deep neural network at the clients with the goal of capturing the features of each different client. The results obtained are even better than FedPer in most cases when testing on a system consisting of 10 to 100 users containing data from the data sets CIFAR-10, CIFAR- 100, and Omniglot [14].

The similarity between these two methods lies in the training base layers. These layers are trained and synthesized using the FedAvg algorithm. Therefore, they are not able to adapt quickly to the new data set. In addition, during the inference phase, the FedPer algorithm takes the average of the parameters of the personalization layers and then combines them with the parameters of the base layers to conduct the test. The parameters of the personalization layers, which are highly personalized, are now averaged, which can cause the classification efficiency to be significantly reduced when the data is strong non-IID. The algorithm LG-FedAvg proposes performing an ensemble test between personalization layers. However, only personalization layers trained on the same data as the test data will really perform well. Therefore, in some cases, ensemble test results are not high because there are too many personalization layers that have never worked with the same data as the data during the test.

3   Proposed Method

The proposed method of this study is a combination of maintaining personalization layers at the edge device of the algorithm FedPer and using meta-learning in training base layers of the model. With this combination, the research aims at two goals: (1) - Fast adaptability on new dataset of ML algorithms, (2) - High personalization ability of PL techniques.

Fast adaption. As mentioned above, the base layers of algorithms using PL techniques are not really strong because they are trained on the FedAvg algorithm, which leads to them being slow to adapt to the new data sets. The training of base layers by ML algorithms is expected to provide these base layers with the ability to quickly adapt to the data set of clients, overcoming the disadvantages of algorithms using PL techniques.

High personalization. The personalization of the FL system combined with ML comes from the fact that the system allows the global model to perform a fine-tune on part of the client data before going into testing on the rest of the client data. Allowing the FL system to maintain a portion of the deep neural network (personalization layers) on each client also greatly increases the ability to personalize each user. Here, our algorithm proposes a combination of the two methods of personalization enhancement mentioned above, which is expected to further increase the personalization of the FL system.

Specifically, the deep neural network is divided into two parts: the base layers include feature extraction layers, and the personalization layers contain fully-connected layers. The base layers are trained against two ML algorithms (MAML and Meta-SGD). The personalization layers are maintained by separate clients. With this combination, we call our proposed algorithm FedMeta-Per.

Algorithm 1. FedMeta-Per (Server)

1: Initialize w_B^0 for MAML or (w_B^0, α_B^0) for Meta-SGD
2: **for** $t = 0, 1, 2, \ldots$ **do**
3:     Sample a subset C_t of m clients
4:     **for** $c_i \in C_t$ **do**
5:         $w_{B(i)}^{t+1} \leftarrow \text{ModelTrainingMAML}(c_i, w_B^t)$ for MAML
6:         $(w_{B(i)}^{t+1}, \alpha_{B(i)}^{t+1}) \leftarrow \text{ModelTrainingMetaSGD}(c_i, w_B^t, \alpha_B^t)$ for Meta-SGD
7:
8:     $n_i = \left| \mathcal{D}_{train(i)}^{query} \right|, N_m = \sum_{i=0}^{m} n_i$
9:     Aggregate $w_B^{t+1} = \sum_{i=0}^{m} \frac{n_i}{N_m} w_{B(i)}^{t+1}$ for MAML
10:     Aggregate $(w_B^{t+1}, \alpha_B^{t+1}) = \sum_{i=0}^{m} \frac{n_i}{N_m} (w_i^{t+1}, \alpha_{B(i)}^{t+1})$ for Meta-SGD

Training phase. On the server, the Algorithm 1 is implemented to coordinate the training activities. These operations include: parameter initialization of base layers; distributing this parameter to clients; synthesizing the parameters of the new global base layers. This process is summarized in Table 1. For the MAML algorithm, the base layers parameters include only part of the global model parameters (w_B^t at round t). Meanwhile, the parameter of the Meta-SGD algorithm's base layers contains an extra part of the learning rate at each client (α_B^t at round t), which is the same size as w_B^t. This learning rate value will be well explained in the training process of the clients. These values will be distributed to a subset C_t of m clients to conduct distributed training. The clients then train on their own dataset and return the parameters of the base layers ($w_{B(i)}^{t+1}$ for MAML and $(w_{B(i)}^{t+1}, \alpha_{B(i)}^{t+1})$ for Meta-SGD at round t) to the server. The parameters of the new base layers are summarized as follows:

$$\text{MAML: } w_B^{t+1} = \sum_{i=0}^{m} \frac{n_i}{N_m} w_{B(i)}^{t+1} \tag{2}$$

$$\text{Meta-SGD: } (w_B^{t+1}, \alpha_B^{t+1}) = \sum_{i=0}^{m} \frac{n_i}{N_m} (w_i^{t+1}, \alpha_{B(i)}^{t+1}) \tag{3}$$

where, n_i is the number of data samples of the client's query set $c_i \in C_t$, N_m is the total number of data samples on the query set of m of the client participating in global training.

Table 1. Server activity summary at round t^{th}

	Algorithm	
	MAML	Meta-SGD
Send	w_B^t	(w_B^t, α_B^t)
Receive	Weights: $w_{B(i)}^{t+1}, i \in [1,m]$	Parameters: $(w_{B(i)}^{t+1}, \alpha_{B(i)}^{t+1}), i \in [1,m]$
Aggregate	w_B^{t+1}	$(w_B^{t+1}, \alpha_B^{t+1})$

Either Algorithm 2 or Algorithm 3, is implemented at the clients, corresponding to the training methods of the type MAML and Meta-SGD. To train the model on bilevel optimization-based ML algorithms , the client data is divided into support set ($\mathcal{D}_{train(i)}^{support}$) and query set ($\mathcal{D}_{train(i)}^{query}$). The training steps at the client include: merging parameters of base layers and personalization layers to form complete model parameters; performing training according to ML algorithms; new model parameter resolution.

Specifically, for clients that train on the MAML algorithm, at the tth step of the global training process, client c_i will receive an initialization weight w_B^t of the base layers sent by the server. Next, c_i merges the weight of the base layers w_B^t with the weight of the personalization layers $w_{P(i)}^t$ to get the weight of the complete model w_i^t . w_i^t is then trained as follows:

$$\text{Train: } \hat{w}_i^{t+1} \leftarrow w_i^t - \alpha \nabla_{w_i^t} f_{local}\left(w_i^t, \mathcal{D}_{train(i)}^{support}\right) \tag{4}$$

$$\text{Meta-train: } w_i^{t+1} \leftarrow w_i^t - \beta \nabla_{w_i^t} f_{local}\left(\hat{w}_i^{t+1}, \mathcal{D}_{train(i)}^{query}\right) \tag{5}$$

Algorithm 2. FedMeta-Per (MAML Client)

1: **ModelTrainingMAML**(c_i, w_B^t):
2: Sample support set $\mathcal{D}_{train(i)}^{support}$ and query set $\mathcal{D}_{train(i)}^{query}$
3: **if** $t = 0$ **then**
4:     Initialize $w_{P(i)}^t$
5: **else**
6:     Load $w_{P(i)}^t$ from the external memory
7: $w_i^t \leftarrow w_B^t \oplus w_{P(i)}^t$ ▷ Merge w_B^t and $w_{P(i)}^t$ to form w_i^t
8: Training:

$$\hat{w}_i^{t+1} \leftarrow w_i^t - \alpha \nabla_{w_i^t} f_{local}\left(w_i^t, \mathcal{D}_{train(i)}^{support}\right)$$

$$w_i^{t+1} \leftarrow w_i^t - \beta \nabla_{w_i^t} f_{local}\left(\hat{w}_i^{t+1}, \mathcal{D}_{train(i)}^{query}\right)$$

9: $w_{B(i)}^{t+1}, w_{P(i)}^{t+1} \leftarrow w_i^{t+1}$ ▷ Resolve w_i^{t+1} to form $w_{B(i)}^{t+1}$ and $w_{P(i)}^{t+1}$
10: Send $w_{B(i)}^{t+1}$ to server and store $w_{P(i)}^{t+1}$

---

where, $\alpha, \beta \in [0, 1]$ are the learning rates.

For clients that use the `Meta-SGD` algorithm in training, the base layers of the deep neural network include two parameters: the weight of the base layers w_B^t and the learning rate of the base layers α_B^t; personalization layers of the network include two parameters $w_{P(i)}^t$ and $\alpha_{P(i)}^t$ (same size as $w_{P(i)}^t$). Parameter merging also takes place between the parameters of base layers and personalization layers to form the complete parameter set w_i^t and α_i^t. It should be noted that `Meta-SGD` configures α as a learning rate array of the size of the model weight and treats α as a trainable parameter. Thus, the algorithm performs element-wise multiplication between α and the derivative of the error function in Equation 6 and performs the derivative with respect to α in Equation 7.

$$\text{Train: } \hat{w}_i^{t+1} \leftarrow w_i^t - \alpha_i^t \circ \nabla_{w_i^t} f_{local}\left(w_i^t, \mathcal{D}_{train(i)}^{support}\right) \tag{6}$$

$$\text{Meta-train: } (w_i^{t+1}, \alpha_i^{t+1}) \leftarrow (w_i^t, \alpha_i^t) - \beta \nabla_{(w_i^t, \alpha_i^t)} f_{local}\left(\hat{w}_i^{t+1}, \mathcal{D}_{train(i)}^{query}\right) \tag{7}$$

Inference phase. Based on the testing process of study [17], we divide users into two types: local clients and new clients. For local clients, the personalization layers are reused to achieve a high degree of personalization. For new clients, we do not use ensemble testing techniques as suggested by study [17], but rather use each of the previously built personalization layers combined with base layers to operate on the new dataset. During the fine-tuning process, the client selects the parameter set with the smallest error. Then use this set of parameters to process test data.

Algorithm 3. FedMeta-Per (Meta-SGD Client)

1: **ModelTrainingMetaSGD**(c_i, w_B^t, α_B^t):
2: Sample support set $\mathcal{D}_{train(i)}^{support}$ and query set $\mathcal{D}_{train(i)}^{query}$
3: **if** $t = 0$ **then**
4:     Initialize $(w_{P(i)}^t, \alpha_{P(i)}^t)$
5: **else**
6:     Load $(w_{P(i)}^t, \alpha_{P(i)}^t)$ from the external memory
7: $w_i^t \leftarrow w_B^t \bigoplus w_{P(i)}^t$ ▷ Merge w_B^t and $w_{P(i)}^t$ to form w_i^t
8: $\alpha_i^t \leftarrow \alpha_B^t \bigoplus \alpha_{P(i)}^t$ ▷ Merge α_B^t and $\alpha_{P(i)}^t$ to form α_i^t
9: Training:

$$\hat{w}_i^{t+1} \leftarrow w_i^t - \alpha_i^t \circ \nabla_{w_i^t} f_{local}\left(w_i^t, \mathcal{D}_{train(i)}^{support}\right)$$

$$(w_i^{t+1}, \alpha_i^{t+1}) \leftarrow (w_i^t, \alpha_i^t) - \beta \nabla_{(w_i^t, \alpha_i^t)} f_{local}\left(\hat{w}_i^{t+1}, \mathcal{D}_{train(i)}^{query}\right)$$

10: $w_{B(i)}^{t+1}, w_{P(i)}^{t+1} \leftarrow w_i^{t+1}$ ▷ Resolve w_i^{t+1} to form $w_{B(i)}^{t+1}$ vá $w_{P(i)}^{t+1}$
11: $\alpha_{B(i)}^{t+1}, \alpha_{P(i)}^{t+1} \leftarrow \alpha_i^{t+1}$ ▷ Resolve α_i^{t+1} to form $\alpha_{B(i)}^{t+1}$ vá $\alpha_{P(i)}^{t+1}$
12: Send $(w_{B(i)}^{t+1}, \alpha_B^{t+1})$ to server and store $(w_{P(i)}^{t+1}, \alpha_{P(i)}^{t+1})$

4 Numerical Experiments

Dataset. This study uses 50 clients for training (containing 75% of the data) and 50 clients for testing (containing 25% of the data). The clients contain the data from the two datasets, MNIST [5] and CIFAR-10 [13], respectively. To simulate Non-IID data, each client is configured to contain only 2/10 labels. The amount of data between clients is uneven (see Table 2). During testing, each local user has the same data distribution as that of a training client; each new user has a data distribution that is completely different from the data distribution of all the clients participating in the training. At each client, the data is divided into a support set (containing 20% of the data) and a query set (containing 80% of the data).

Table 2. Statistics on MNIST and CIFAR-10 (non-IID data)

Dataset	#clients	#samples	#classes	#samples/client				#classes/client
				min	mean	std	max	
MNIST	50	69,909	10	135	1,398	1,424	5,201	2
CIFAR-10	50	52,497	10	506	1,049	250	1,986	2

Metrics. The study evaluates the model through the following metrics: accuracy in correlation with all data points (acc_{micro}), accuracy in correlation with all clients (acc_{macro}) and F1-score ($F1_{macro}$).

Experiments. This study performs centralized and distributed experiments on the same model architecture. For distributed training, the study experimented on FedAvg, FedPer, FedMeta(MAML), FedMeta(Meta-SGD). The results are compared with the proposed algorithm FedMeta-Per. For a fair comparison, the study allows the global model obtained by the FedAvg and FedPer algorithms to perform a fine-tune one or several times on the client's support set during the inference phase. This is the idea of FedAvgMeta and FedPerMeta. Details of the experimental settings can be found in Appendix A.

5   Results and Discussion

Centralized and Decentralized. The results of centralized training and distributed training (FedAvg on IID and non-IID data) are presented in Table 3. It is easy to see that the centralized training model achieves higher results on all metrics. Results obtained on FedAvg (IID data) are 7% to 8% lower because the global model captures the features indirectly (through parametric averaging at the server). However, when the input data is non-IID, the metrics decrease from 7% to 13% on MNIST and do not reach convergence on CIFAR-10. Metrics on non-IID data also differ quite a lot (3%–6% on MNIST and 3%–4% on CIFAR-10) compared to differences on IID data (less than 1%). This proves that the uneven data distribution across the clients has adversely affected the classification quality.

Table 3. Classification results (%) of centralized and decentralized (FedAvg on IID and non-IID data) using MNIST and CIFAR-10 datasets. Best results per metrics are boldfaced.

		acc_{micro}	acc_{macro}	$F1_{macro}$
MNIST	Centralized	**97.07**	–	**97.04**
	FedAvg (IID data)	90.36	90.34±2.24	90.12 ± 2.29
	FedAvg (local client)	85.03	82.14 ± 14.76	79.43 ± 16.83
	FedAvg (new client)	83.92	81.69 ± 19.71	77.66 ± 22.54
CIFAR-10	Centralized	**61.91**	–	**61.72**
	FedAvg (IID data)	53.83	53.83 ± 3.14	53 ± 3.21
	FedAvg (local client)	19.02	19.29 ± 25.11	16.85 ± 23.92
	FedAvg (new client)	24.63	24.83 ± 22.57	20.52 ± 20.45

Convergence ability. The results on metrics of FedMeta-Per and the algorithms used in the comparison are presented in Table 4. To solve the problem of non-IID data, the technique of fine-tuning localizes the FL system in a simple form (algorithm FedAvgMeta), in order to improve the adaptability and personalization of the model on the new data sets. However, the results obtained in Table 4 as well as in Fig. 2 are not very satisfactory (low convergence on the

MNIST dataset and not on the CIFAR-10 dataset). The reason for `FedAvgMeta` not converging is that the adaptability of the global model to the new dataset is very low. The PL technique is also used in the form of two algorithms, `FedPer` and `FedPerMeta`. However, the results obtained are even worse than `FedAvg` and `FedAvgMeta`. This is understandable given that the model architecture used by study [1] is a pre-trained network `MobileNet-v1` [10], a more complex deep neural network than this study (see Appendix A.1). This also proves that using the PL technique is not enough for handling non-IID data.

Table 4. Classification results (%) of FedAvg, FedPer and FedMeta-Per using MNIST and CIFAR-10 datasets. Best results per metrics are boldfaced.

		CIFAR-10			MNIST		
		acc_{micro}	acc_{macro}	$F1_{macro}$	acc_{micro}	acc_{macro}	$F1_{macro}$
Local client	FedAvg	19.02	19.29 ± 25.11	16.85 ± 23.92	85.03	82.14 ± 14.76	79.43 ± 16.83
	FedPer	13.22	12.99 ± 19.39	10.52 ± 14.91	77.29	75.48 ± 14.84	72.32 ± 15.99
	FedAvgMeta	40.3	38.47 ± 31.52	33.81 ± 30.61	84.84	81.56 ± 16.68	78.31 ± 19.8
	FedPerMeta	18.57	17.48 ± 22.55	14.54 ± 18.67	75.91	74.11 ± 16.2	71.22 ± 16.77
	FedMeta(MAML)	69.02	68.76 ± 14.86	61.14 ± 20	92.99	91.14 ± 5.99	90.16 ± 6.28
	FedMeta(Meta-SGD)	78.63	78.73 ± 11.59	72.87 ± 18.31	98.02	96.35 ± 4.62	95.80 ± 5.51
	FedMeta-Per(MAML)	**86.6**	**86.52 ± 6.31**	**85.33 ± 6.77**	**99.37**	**99.12 ± 1.29**	**98.94 ± 1.6**
	FedMeta-Per(Meta-SGD)	85.61	85.68 ± 7.22	85.08 ± 7.32	98.92	98.15 ± 3.32	98.20 ± 2.94
New client	FedAvg	24.63	24.83 ± 22.57	20.52 ± 20.45	83.92	81.69 ± 19.71	77.66 ± 22.54
	FedPer	14.4	14.52 ± 20.15	10.66 ± 13.79	78.3	76.19 ± 18.79	72.72 ± 19.3
	FedAvgMeta	43.39	43.54 ± 18	35.14 ± 17.22	84.34	82.37 ± 17.42	78.78 ± 19.31
	FedPerMeta	13.33	13.57 ± 19.62	10.05 ± 13.17	77.47	75.56 ± 20.33	72.60 ± 21.37
	FedMeta(MAML)	61.69	61.64 ± 12.49	50.76 ± 19.2	92.96	91.88 ± 5.88	90.02 ± 7.34
	FedMeta(Meta-SGD)	68.36	67.89 ± 15.11	60.24 ± 21.52	96.39	93.53 ± 8.39	89.31 ± 14.56
	FedMeta-Per(MAML)	64.22	63.70 ± 12.29	53.68 ± 19.06	93.6	93.57 ± 5.58	91.83 ± 6.43
	FedMeta-Per(Meta-SGD)	**69.97**	**69.13 ± 14.63**	**62.42 ± 20.94**	**96.62**	95.88 ± 3.58	**94.85 ± 4.61**

When comparing the convergence ability of `FedMeta` and `FedMeta-Per`, the fact that both algorithms have the participation of ML and have similarities in the convergence process (Fig. 1) shows that ML has a certain impact on the FL system.

The ability of ML in the proposed algorithm in comparison with algorithms `FedAvg`, `FedAvgMeta`, `FedPer`, and `FedPerMeta` is shown in Fig. 2. Accordingly, ML provides the system with fast adaptability on new clients, leading to `FedMeta-Per` convergence much faster than the old algorithms, reaching convergence thresholds of about 60%-85% on CIFAR-10 and about a 20% improvement in accuracy on MNIST. More specifically, this convergence comes from performing a single fine-tune step on 20% of the data at the client, once again demonstrating the very fast adaptability of the FL-ML system.

In addition, thanks to the construction of good personalization layers, the testing process on local clients also achieves a higher degree of convergence when comparing the proposed algorithm with `FedMeta` (Fig. 1). However, in the face of a strange data distribution of new clients, these two algorithms have almost equivalent convergence ability.

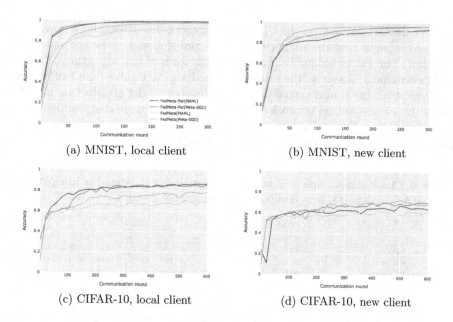

(a) MNIST, local client   (b) MNIST, new client

(c) CIFAR-10, local client   (d) CIFAR-10, new client

Fig. 1. acc_{micro} of FedMeta-Per and FedMeta. The end result and convergence of FedMeta-Per is significantly higher than that of FedMeta on local clients. On new clients, even though FedMeta-Per reaches a higher value, the difference in convergence is not so great.

Personalization ability. Observing the standard deviation of the algorithms FedMeta, FedAvg and FedPer in Table 4, we can see the standard deviation in the results of FedMeta is much smaller than the rest of the algorithms. It shows that the personalization of the FL system has been improved many times thanks to the training of ML algorithms as well as the fine-tuning execution on the support set to capture the characteristics of a new data set. However, as mentioned above, this ability can be completely improved. With new suggestions during the inference phase, FedMeta-Per further reduces the standard deviations on the scales. Specifically, after combining the fine-tuning of the model with the support set of local data and maintaining the personalization layer for each client, the standard deviation of the local clients decreases from 4% to 14% on CIFAR-10 and from 1% to 5% on MNIST while the averages are still superior. On new clients, the mean values are higher but the standard deviation, in some cases, is not actually smaller than the previous algorithms. This is because the personalization layers in the deep neural network of new clients have not really matched that user yet. However, over time, when new clients participate in one or a few steps of global training, they will build up good personalization layers for themselves, resulting in the metrics reaching the same value as local clients.

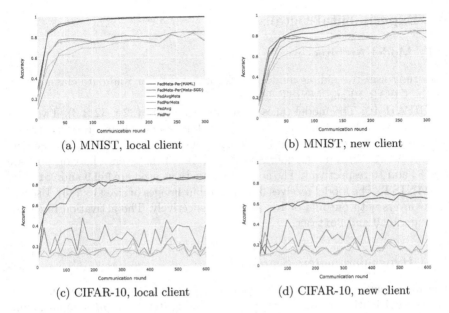

(a) MNIST, local client

(b) MNIST, new client

(c) CIFAR-10, local client

(d) CIFAR-10, new client

Fig. 2. acc_{micro} of FedMeta-Per, FedAvg, FedAvgMeta, FedPer and FedPerMeta. FedMeta-Per's convergence and accuracy outperform the rest of the algorithms.

6 Conclusion

Through the research process, we combine ML algorithms and PL techniques into the FL system, to improve the accuracy as well as the ability to personalize the machine learning model for each user of the system on non-IID data. Experimentally, we prove the effectiveness of the proposed algorithm `FedMeta-Per` compared with algorithms `FedAvg`, `FedPer`, and `FedMeta` over 50 users with two datasets, CIFAR-10 and MNIST. In which, it can be explained that achieving high results is based on two inherited factors: (1) - Fast adaptability to new data set of the proposed algorithm inherited from ML algorithms, (2) - High personalization ability for each user inherited from personalization layers of PL technique and model fine-tuning on the support set of ML algorithms.

In the future, bilevel optimization-based algorithms (`FO-MAML` [7], `iMAML` [20], `Reptile`) can be integrated into the system. The search and clustering of users so that each user finds the best set of partial parameters should also be further studied to increase the performance of the system. In addition, the issue of information security between the server and the client should also be considered to make the system more practical.

Acknowledgements. This research is funded by University of Science, VNU-HCM under grant number CNTT 2022–09.

A   Experimental Details

A.1   Model Architecture

The study uses two simple models for feature extraction and data classification for the CIFAR-10 and MNIST datasets.

CIFAR-10. The model takes input images of size $(32 \times 32 \times 3)$. Two convolution layers (kernel of size (5×5), channel numbers of 6 and 16 respectively) are used for feature extraction. Following each convolution layer is a MaxPooing layer of size (2×2). The classifier consists of three linear layers whose outputs are 120, 84 and 10 respectively. The activation functions used are ReLU and Softmax.

MNIST. The model receives flattened input images of size (1×784). Use two linear layers with outputs of 100 and 10, respectively. The activation functions used are ReLU and Softmax.

A.2   Hyper-parameters Searching

The hyper-parameters of the FL system of the study include: the number of clients participating in training in a global training step ($\#clients/round$), number of local training steps ($\#epochs$), number of global training steps ($\#rounds$), amount of data in a data batch ($batch\_size$), number of personalization layers for algorithms using PL techniques ($\#per\_layers$) and the learning rates used in the optimization of deep neural networks by mini batch gradient descent.

To accommodate the limited configuration client hardware in the Horizontal FL scenario, we limits $\#epochs = 1$ and $batch\_size = 32$. From the survey of FL experiments in recent studies, the number of clients participating in global training was selected as 2, 5, and 10, respectively. In which, $\#clients/round = 5$ gives a higher result and consumes an acceptable computational cost. $\#per\_layers$ is searched in $\{1, 2, 3\}$ (calculated from the last linear layer). Experimental results show that using the last linear layer as the personalization layer gives the best results. The learning rates are presented in Table 5.

Table 5. Tuning learning rate for algorithms. Empty cells indicate that hyper-parameters cannot be found for the model to converge.

	CIFAR-10	MNIST
FedAvg, FedAvgMeta	–	10^{-5}
FedPer, FedPerMeta	–	10^{-5}
FedMeta(MAML) (α, β)	$(0.01, 0.001)$	$(0.001, 0.001)$
FedMeta(Meta-SGD)(α, β)	$(0.001, 0.001)$	$(0.001, 5 \times 10^{-4})$
FedMeta-Per(MAML)(α, β)	$(0.001, 0.005)$	$(0.001, 0.001)$
FedMeta-Per(Meta-SGD)(α, β)	$(0.01, 0.01)$	$(0.001, 5 \times 10^{-4})$

References

1. Arivazhagan, M.G., Aggarwal, V., Singh, A.K., Choudhary, S.: Federated learning with personalization layers. arXiv preprint arXiv:1912.00818 (2019)
2. Caldas, S., et al.: LEAF: a benchmark for federated settings. arXiv preprint arXiv:1812.01097 (2018)
3. Chen, F., Luo, M., Dong, Z., Li, Z., He, X.: Federated meta-learning with fast convergence and efficient communication. arXiv preprint arXiv:1802.07876 (2018)
4. Cohen, G., Afshar, S., Tapson, J., van Schaik, A.: EMNIST: an extension of MNIST to handwritten letters. arXiv preprint arXiv:1702.05373 (2017)
5. Deng, L.: The MNIST database of handwritten digit images for machine learning research. IEEE Signal Process. Mag. **29**(6), 141–142 (2012)
6. Fallah, A., Mokhtari, A., Ozdaglar, A.: Personalized federated learning: a meta-learning approach. arXiv preprint arXiv:2002.07948 (2020)
7. Finn, C., Abbeel, P., Levine, S.: Model-agnostic meta-learning for fast adaptation of deep networks. In: International Conference on Machine Learning, pp. 1126–1135. PMLR (2017)
8. Harlow, H.F.: The formation of learning sets. Psychol. Rev. **56**(1), 51 (1949)
9. Hospedales, T., Antoniou, A., Micaelli, P., Storkey, A.: Meta-learning in neural networks: a survey. arXiv preprint arXiv:2004.05439 (2020)
10. Howard, A.G., et al.: MobileNets: efficient convolutional neural networks for mobile vision applications. arXiv preprint arXiv:1704.04861 (2017)
11. Jiang, Y., Konečný, J., Rush, K., Kannan, S.: Improving federated learning personalization via model agnostic meta learning. arXiv preprint arXiv:1909.12488 (2019)
12. Khan, W.Z., Ahmed, E., Hakak, S., Yaqoob, I., Ahmed, A.: Edge computing: a survey. Future Gener. Comput. Syst. **97**, 219–235 (2019)
13. Krizhevsky, A., Hinton, G., et al.: Learning multiple layers of features from tiny images (2009)
14. Lake, B.M., Salakhutdinov, R., Tenenbaum, J.B.: Human-level concept learning through probabilistic program induction. Science **350**(6266), 1332–1338 (2015)
15. Li, T., Sahu, A.K., Talwalkar, A., Smith, V.: Federated learning: challenges, methods, and future directions. IEEE Signal Process. Mag. **37**(3), 50–60 (2020)
16. Li, Z., Zhou, F., Chen, F., Li, H.: Meta-SGD: learning to learn quickly for few-shot learning. arXiv preprint arXiv:1707.09835 (2017)
17. Liang, P.P., et al.: Think locally, act globally: federated learning with local and global representations. arXiv preprint arXiv:2001.01523 (2020)
18. McMahan, B., Moore, E., Ramage, D., Hampson, S., y Arcas, B.A.: Communication-efficient learning of deep networks from decentralized data. In: Artificial intelligence and statistics, pp. 1273–1282. PMLR (2017)
19. Nichol, A., Achiam, J., Schulman, J.: On first-order meta-learning algorithms. arXiv preprint arXiv:1803.02999 (2018)
20. Rajeswaran, A., Finn, C., Kakade, S.M., Levine, S.: Meta-learning with implicit gradients. In: Advances in Neural Information Processing Systems, vol. 32 (2019)
21. Zhao, Y., Li, M., Lai, L., Suda, N., Civin, D., Chandra, V.: Federated learning with Non-IID data. arXiv preprint arXiv:1806.00582 (2018)

ETop3PPE: EPOCh's Top-Three Prediction Probability Ensemble Method for Deep Learning Classification Models

Javokhir Musaev⬤, Abdulaziz Anorboev⬤, Huyen Trang Phan⬤,
and Dosam Hwang$^{(\boxtimes)}$⬤

Yeungnam University, Daegu, Republic of Korea
dosamhwang@gmail.com

Abstract. The rapid growth in the field of deep learning (DL) has increased research interests in this area, including computer vision (CV), for which high-quality products have been produced. Many fields, including medicine, the automobile industry, finance, education, and the military have used state-of-art CV results. Any change in CV affects human life through the abovementioned sector. Considering the importance of the development in CV, we proposed Epoch's Top-Three Prediction Probability Ensemble (Etop3PPE) method for DL classification problems. Our method focuses on the use of lost knowledge during training and an optimal way to increase the accuracy of the model. Each epoch during training represents a different prediction space, which is the key to the development of the proposed method. We used the top-three prediction probabilities of each image for the classification task from different epochs of training and ensembled them into the best model prediction probabilities. Applying our method, we partially solved the ensemble error problem in the models and increased the accuracy of our model from 89.32% to 90.91%. We added 32.8% lost knowledge from the prediction spaces of the training epochs that were used during the ensembling. We used the Cifar10dataset to evaluate our method. In addition, we compared the results of our method with those of the classic ensemble (ensembling all prediction probabilities into the best model). The result was surprising, our method overcome by 16% in case of adding lost knowledge of different epochs.

Keywords: Prediction space · Top3 maximum prediction probabilities · Ensemble

1 Introduction

Deep learning (DL) has recently started a new era with the introduction of neural networks (NNs), and next-generation DL models and methods have been developed. As the quality of DL products continues to increase, numerous new topics are being studied. The implementation of DL in various fields such as medicine, IoT, the military and automobile industries, finance, and many other areas has increased the amount of funding spent on research in this field. In addition, the number of tasks in DL-based CV is growing

© The Author(s), under exclusive license to Springer Nature Switzerland AG 2022
N. T. Nguyen et al. (Eds.): ACIIDS 2022, LNAI 13757, pp. 222–233, 2022.
https://doi.org/10.1007/978-3-031-21743-2_18

continuously. Particularly during the spread of Covid 19, detection has become one of the main tools used to check the temperature and mask compliance of individuals. Other implementations, from medical diagnosis to heavy industry, are increasing significantly in number. Any change in DL adds an extra opportunity for the future development of the field.

To fill in the gap of lost knowledge occurring during training, we conducted research on DL model ensemble methods. Ensemble learning is an extremely powerful DL method. Its effect on the results has been studied by numerous researchers [1]–[4]. Various ensemble models and methods contributing to the development of ensemble learning in DL have been proposed. Despite a large number of studies, gaps requiring a solution exist, and improvements in CV have been proposed. Our initial studies showed that the true prediction scope of the worse models differs from that of the best models. In our study, we attempted to manipulate the prediction scope of the training epochs. We chose the DL classification task to study and improve the use of ensemble methods. Our experiment results show that each epoch can truly predict different images better than the other one, even when their accuracy is quite small. We attempted to find an optimal method to use this knowledge in our research. In our previous study, we used the maximum epoch prediction probabilities and achieved a higher accuracy than that achieved through classic training. In this study, we developed our previous study by applying the top-three maximum prediction probabilities to the ensemble model. We used the VGG50 model pretrained with ImageNet to train the Cifar10 dataset. We replaced the last layer of VGG50 with a dense layer, including 10 nodes that are equal to the number of classes in Cifar10. Next, we resized the Cifar10 dataset into $224 \times 224 \times 3$ sized images and trained the model for 20 epochs using callbacks to save the best epoch from among these 20 epochs. After achieving an accuracy of 86.08% for the first 20 epochs, we trained 20 more epochs, with an accuracy of 89.32%. The second model was chosen as the base model. The following step was to check the prediction spaces of the first model that saved after the first 20 epochs. Its 4.85% true prediction rate differed from the true predictions of the second model. We were motivated to use this insight to increase the accuracy by applying the true prediction space of this second model. We used the prediction probabilities of the first model with an accuracy of 86.08% and selected the top-three or three maximum prediction probabilities for each image in the test set and ensemble them into the corresponding position within the prediction probabilities of the second model with an accuracy of 89.32%. As a result, we increased the accuracy of the model from 89.32% to 90.91%.

Our study consists of the following sections. Section I introduces the proposed method and provides general information regarding ensemble learning and the content of this study. Section II provides information regarding related studies and problems found in ensemble learning. Section III provides solutions to the problems described in Section II and provides a detailed explanation of the proposed methodology. Section IV describes the experiments and results of our research, including information regarding the dataset, base method, training setup, evaluation metrics, experiment results, and discussions. Finally, Section V provides some concluding remarks regarding this research and areas for future study.

2 Related Works

Many researchers have studied ensemble learning and their applications in various fields. The combination of data representation and model ensembles [4] has been the focus of numerous researchers. The bias-variance tradeoff and cross-validation ensembles were among the forms of DL ensembles that were studied in [5]. Other forms of DL ensembles include varying the training data, models, and combinations. If we look at applications throughout various sectors, we can find numerous applications of ensemble learning in credit risk assessment [6], oil price forecasting [7], multi-step forecasting for big data time series, [8] and an ensemble learning model for Covid-19 based on CT images [9]. In [6], the authors studied a credit risk assessment and proposed a multistage neural network ensemble for risk assessment using a bagging sampling method for generating training data. In addition, different models were used in [6] to train the dataset, and the results were scaled into unit intervals followed by fusion. For oil price prediction [7], the authors used stacked denoising autoencoders to model the nonlinear and complex relationships of oil prices with its features. Another implementation [8] of ensemble learning conducted on the forecasting of big data time series used decision trees, gradient boosted trees, and a random forest to develop an ensemble model. Weights for ensemble members were computed using a weighted least squares method. The next successful study on ensemble learning focused on Covid-19 detection using 2933 lung CT images obtained from different sources. The authors initialized the model parameters using transfer learning and three pretrained DL models: AlexNet, GoogleNet, and ResNet. These models were used to extract features from all images and the final dense layer with the softmax function used for classification. The final accuracy is higher than that of the component classifiers of the ensemble. Effective approach to prevent asthma is to control it using data from asthma patients. [10] studied 90 asthma patients during 9 months and collected data of the patients from specialized hospital for pulmonary diseases in Tehran. Authors [10] proposed new ensemble learning algorithm with combining physicians' knowledge in the form of a rule-based classifier and supervised learning algorithms to detect asthma control level in a multivariate dataset with multiclass response variable. The model outcome resulting from the balancing operations and feature selection on data yielded the accuracy of 91.66%. The next implementation [11] of ensemble learning was dedicated to improve medical image segmentation. [11] proposed new methods to improve the segmentation probability estimation without losing performance in a real-world scenario that has only one ambiguous annotation per image. Authors marginalize the estimated segmentation probability maps of networks that are encouraged to under-segment or over-segment with the varying Tversky loss without penalizing balanced segmentation. In addition, study proposed a unified hypernetwork ensemble method to alleviate the computational burden of training multiple networks. Proposed approaches successfully estimated the segmentation probability maps that reflected the underlying structures and provided the intuitive control on segmentation for the challenging 3D medical image segmentation. Following research [12] studied an analysis on ensemble learning optimized medical image classification with deep convolutional neural networks. In this work, a reproducible medical image classification pipeline was proposed to analyze the performance impact of augmenting, stacking, and bagging methods. The pipeline includes state-of-the-art pre-processing and image augmentation methods and nine deep convolution neural network architectures. Four different popular medical image datasets were used to evaluate the research. It was applied on four popular medical imaging datasets with varying complexity. The results were evaluated with an F-1 score which showed that stacking, augmentation, and bagging increased the results up to 13%, 4 and 10%, respectively. Another research [13] studied performance analysis of hyperparameter optimization methods for ensemble learning with small and medium-sized medical datasets. For this task, the study analyzed Grid search, Random search, and Bayesian optimizations for an ensemble classifier. One more problem that studied in ensemble learning is medical diagnosis [14] with imbalanced data. The proposed method includes data pre-processing, training base classifier and final ensemble. In the data pre-processing step, they introduced the extension of Synthetic Minority Oversampling Technique (SMOTE) by integrating it with cross-validated committees filter (CVCF) technique. It allowed them to synthesize the minority sample and thereby balance the input instances by filtering the noisy examples. In the classification phase, they introduced ensemble support vector machine (ESVM) classification technique followed by the weighted majority voting strategy. Also, they proposed simulated annealing genetic algorithm (SAGA) to optimize the weight vector and thereby enhance the overall classification performance. The proposed ensemble learning method was tested on nine imbalanced medical datasets and achieved better results than other state-of-the-art classification models. The next ensemble learning study [15] proposes a disease prediction model (DPM) to provide an early prediction for type 2 diabetes and hypertension based on individual's risk factors data. DPM consists of isolation forest (iForest) based outlier detection method to remove outlier data, synthetic minority oversampling technique tomek link (SMOTETomek) to balance data distribution, and ensemble approach to predict the diseases. Four datasets were used to build the model and extract the most significant risks factors. The authors claims that the proposed DPM achieved highest accuracy when compared to other models and previous studies. Following research [16] studied magnetic resonance (MR) images and engines that effect the quality of the images. To solve one of the challenging problems in medical images [16] proposed an ensemble learning and deep learning framework that improves MR image resolution. Authors utilized five commonly used super-resolution algorithms and achieved enlarged image datasets with complementary priors. Subsequently, GAN is trained to generate super-resolution MR images. At the final step, another GAN is used for ensemble learning that synergizes the outputs of GANs into the final MR super-resolution images. Results of the study showed that achievements of the ensemble outperformed any single GAN's results. [17]–[23] proved that ensemble learning adds a significant improvement to the models and extends prediction spaces of the models.

After our literature review, we found a gap in which the epoch prediction probabilities and the effect of their prediction probabilities were not thoroughly studied. In this research, we learned the effect of the number of top prediction probabilities on the ensemble model.

3 Proposed Method

To address the gap found during the literature review and learn more insight from images using a DL ensemble, we proposed Epoch's Top Three Prediction Probability Ensemble (ETop3PPE) method for DL classification problems. When we studied ensemble learning and its forms, we found that not all insights of the models were optimally applied to the ensemble model. Varying the data, model, or their combinations yields better results than the main component of the ensemble. Despite this, there is still an opportunity to add extra knowledge to the final ensemble model using the knowledge of different epochs. The motivation for developing the proposed method was the true prediction spaces of different models. In our previous research, we studied the maximum prediction probabilities of the epoch and their effects on the ensemble models. In this research, we studied the effect of the number of top prediction probabilities on the ensemble models. When a model is trained with a dataset for the classification task, we achieve the classification probabilities for each image in the dataset. In case of 50000x32x32x3 sized dataset that has 10 classes and 50000 images with 32x32x3 sizes, prediction probability size equals to 50000x10, 10 classification probability for each images. We studied the true predictions of the epochs during training and determined that each epoch found different images better than the other epochs. For instance, if epoch 10 can truly predict images from the 1000th position to the 45000th position in the dataset, another epoch can be found from the 500th position to the 43000th position of the images. This shows that when we ensemble these two epochs, a certain number of images from the 500th position to the 1000th position can be accurately predicted in addition to the positions from the 1000th to 45000th positions.

Figure 1 illustrates ETop3PPE for deep learning classification models and includes the following steps:

1. In the initial step, we uploaded the data and resized it to $224 \times 224 \times 3$ and rescaled each pixel by dividing it by 255.
2. The pre-processed dataset was fed to the pretrained ResNet50 model with ImageNet dataset.
3. The model was trained for 20 epochs and best epoch was saved when considering the validation accuracy.
4. The model was trained for 20 more epochs, and for this interval of training, we saved the best model to evaluate its validation accuracy.
5. A high accuracy epoch was chosen as the main model, and a lower accuracy epoch was chosen as the secondary model.
6. The top-three prediction probabilities of the secondary model of each image are added to the corresponding prediction probabilities of the main model.
7. The maximum prediction probabilities of the main model were selected for each image for classification.

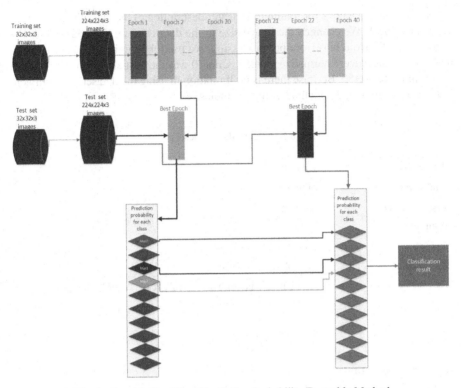

Fig. 1. Epoch's Top-Three Prediction Probability Ensemble Method

In this research, we used the Cifar10 dataset; hence, for the final layers of the ResNet50 model, we applied a dense layer with 10 nodes. The advantage of this model is that it can be used with other DL ensemble models and only adds more knowledge to the ensemble. We present the detailed practical effects of the method in the next section.

4 Experiments and Results

In this section, we provide comprehensive information regarding the experiments and their results. In addition, this section includes the test results of the model and dataset used. Here, we provide a clearer view of our method through experiments. We used the Cifar10 dataset to evaluate the proposed method.

4.1 Dataset

We used one of the popular datasets from the image classification field with a sufficient number of images and reliable labeled data, i.e., the Cifar10[1] collected by Krizhevsky, Nair, and Hinton, which is popular in classification tasks. In addition, the size of the

---

[1] https://www.cs.toronto.edu/~kriz/cifar.html.

dataset was advantageous for training. A clearer description of the classification is provided in Table 1. We changed the image size in the dataset from 32 × 32 × 3 to 224 × 224 × 3. To avoid bias from pre-processing, we used the minimum pre-processing tools. The images were normalized to 255. Cifar10 was the best choice in our research for training ResNet50. Because there is no limitation in using the dataset, this will help in further developing the method in future studies.

Table 1. Cifar10 dataset description.

Dataset name	Cifar10
Total number of images	60000
Size of images	32x32x3
Train set	50000
Test set	10000
Size of dataset	163 mb (python version)
Class names	"airplane", "automobile", "bird", "cat", "deer", "dog", "frog", "horse", "ship", "truck"

4.2 Training Setup

We used Python 3.9 and Python 2.1.0 of the TensorFlow framework in our training. The experiments were conducted using a 12 GB NVidia Titan-XP GPU with CUDA 10.2 on a computer with an Intel Core-i9 11th generation CPU and 64 GB of RAM. In our training, we initialized the weights with pretrained weights from ImageNet. In addition, we used a sparse categorical loss function for our training and chose 20 epochs for the first step, followed by an additional 20 epochs. We trained the model during different intervals and chose the best models for representing more knowledge.

4.3 Evaluation Metrics

In our training, we focused on accuracy as a main metric and used a unique true prediction (UTP) to explain the success of the method on an ensemble, which is the ratio of true predictions to the total number of cases used to evaluate the model. Equation (1) shows the calculation of the accuracy achieved.

$$Accuracy = \frac{TP + TN}{TP + TN + FP + FN} \tag{1}$$

TP- true predicted positive results

TN-true predicted negative results

FP-false predicted positive results

FN- false predicted negative results

$$UTP(X, Y) = X - X \subset Y \tag{2}$$

UTP - Unique True Prediction

X - Prediction Scope of a model X

Y - Prediction Scope of a model Y

The next evaluation metric is UTP, which identifies the percentage of unique predictions for each model with respect to another prediction. In Eq. (2), UTP(X, Y) finds the UTP of model X with respect to model Y. These metrics explain why our proposed model achieved better results than the main model, where we trained only the main dataset. The indices of the true predicted images are different in each model, despite having the same accuracy. This leads the ensemble to achieve better results.

4.4   Experiment Results and Discussions

In this part of our study, we introduced a detailed explanation of the results and the reasons for achieving these results. Moreover, we evaluated our method using the accuracy and UTP. We used accuracy metrics because our main focus was on the effect of the epochs on the final prediction. We used the UTP metric to explain why better results were achieved.

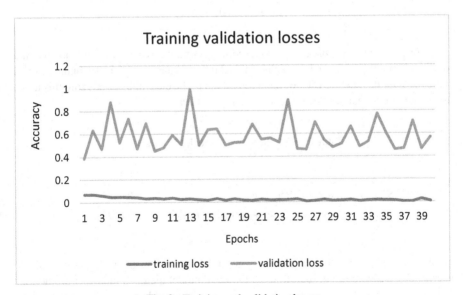

Fig. 2. Training and validation losses

Figure 2 shows the training and validation losses in our training and illustrates losses that include high bias as it reached a training loss extremely close to zero; however, the accuracies of the validation were still higher than 0.4. The same trend is presented in Fig. 3, where the training and validation accuracies of the models are presented. Although the training accuracy reached 100%, the validation accuracy still did not reach 90%. As in many models, there is a generalization problem in that, although the training data are learned extremely well, not all features of the validation data can be extracted as training data.

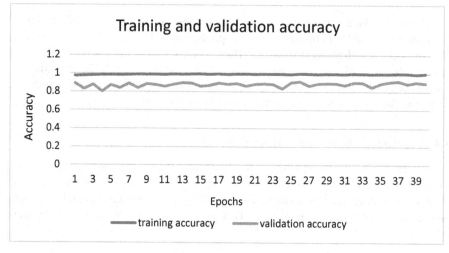

Fig. 3. Training and validation accuracies

To overcome this problem, we proposed increasing the accuracy of the model using additional knowledge from the epochs. Table 2 presents the UTP of the secondary model to the main model and the UTP of the main model to the ensemble. Analyzing this table, we can conclude that 4.85% of the knowledge was available for use.

Table 2. UTP (Secondary model, Main model)

Models	UTP
Secondary model to Main model	0.0485
Main model to (Main model + secondary model)	0.0175

We used only 32% of existing knowledge, and the rest of the knowledge was available as long as the prediction probabilities of the main model included incorrect predictions that affected the ensemble. Each component of the ensemble has both positive effects and unwanted side effects on the prediction space of the ensemble. After ensembling the models, we lost 1.75% of the true predictions from the main model, but still achieved better results than using only the prediction probabilities of the main model. Hence, when we analyzed Table 3, the accuracy of the main model was lower than that of the ensemble model, which was created by ensembling prediction probabilities of the main model and the top-three prediction probabilities of the secondary model.

Table 3. Accuracies

Models	Accuracy
Secondary model	86.08
Main model	89.32
Secondary model + Main model	90.83
Proposed model	**90.91**

After applying our method, we increased the accuracy of the model to 90.91%. When we ensembled all prediction probabilities of the secondary model into the prediction probabilities of the main model, we obtained an accuracy of 90.83%. Our method achieved the expected results after training with Cifar10. Moreover, this method can be used simultaneously with other ensemble models.

5   Conclusions

In this study, we used a unique true prediction space as the main tool to find a gap and tried to fill it by applying the ETop3PPE method. When we used our method with the ResNet50 pretrained model and Cifar10 dataset, we achieved better results than when only using the ensemble component and adding all of the probabilities of the secondary model into the appropriate prediction probabilities of the main model. We explained our method results with an enlarged prediction space for the ensemble model. As a result, we were able to increase the accuracy of the model on the Cifar10 dataset from 89.32% to 90.91%. In addition, we used 32.8% of the extra knowledge of the secondary model. There is still a huge area of research remaining to be conducted on the epoch knowledge.

In the future work we plan to use ontology for building a knowledge base for meta information about the CV objects [24–26]. This base should be useful for processing the images.

References

1. Alqurashi, T., Wang, W.: Clustering ensemble method. Int. J. Mach. Learn. Cybern. **10**, 1–18 (2018). https://doi.org/10.1007/s13042-017-0756-7
2. Abbasi, S.-O., Nejatian, S., Parvin, H., Rezaie, V., Bagherifard, K.: Clustering ensemble selection considering quality and diversity. Artif. Intell. Rev. **52**(2), 1311–1340 (2018). https://doi.org/10.1007/s10462-018-9642-2
3. Vega-Pons, S., Ruiz-Shulcloper, J.: A survey of clustering ensemble algorithms. Int. J. Pattern Recogn. Artif. Intell. **25**(3), 337–372 (2011). https://doi.org/10.1142/S0218001411008683
4. Dong, X., Yu, Z., Cao, W., Shi, Y., Ma, Q.: A survey on ensemble learning. Front. Comp. Sci. **14**(2), 241–258 (2019). https://doi.org/10.1007/s11704-019-8208-z

5. Krogh, A.: Neural network ensembles, cross validation, and active learning
6. Yu, L., Wang, S., Lai, K.K.: Credit risk assessment with a multistage neural network ensemble learning approach. Expert Syst. Appl. **34**(2), 1434–1444 (2008). https://doi.org/10.1016/j. eswa.2007.01.009
7. Zhao, Y., Li, J., Yu, L.: A deep learning ensemble approach for crude oil price forecasting. Energ. Econ. **66**, 9–16 (2017). https://doi.org/10.1016/j.eneco.2017.05.023
8. Galicia, A., Talavera-Llames, R., Troncoso, A., Koprinska, I., Martínez-Álvarez, F.: Multistep forecasting for big data time series based on ensemble learning. Knowl.-Based Syst. **163**, 830–841 (2019). https://doi.org/10.1016/j.knosys.2018.10.009
9. Zhou, T., Lu, H., Yang, Z., Qiu, S., Huo, B., Dong, Y.: The ensemble deep learning model for novel COVID-19 on CT images. Appl. Soft Comput. **98** (2021). https://doi.org/10.1016/j.asoc.2020.106885
10. Khasha, R., Sepehri, M.M., Mahdaviani, S.A.: An ensemble learning method for asthma control level detection with leveraging medical knowledge-based classifier and supervised learning. J. Med. Syst. **43**(6), 1–15 (2019). https://doi.org/10.1007/s10916-019-1259-8
11. Hong, S., et al.: Hypernet-ensemble learning of segmentation probability for medical image segmentation with ambiguous labels (2021). Available: http://arxiv.org/abs/2112.06693
12. Müller, D., Soto-Rey, I., Kramer, F.: An analysis on ensemble learning optimized medical image classification with deep convolutional neural networks
13. Kadam, V.J., Jadhav, S.M.: Performance analysis of hyperparameter optimization methods for ensemble learning with small and medium sized medical datasets. J. Discrete Math. Sci. Crypt. **23**(1), 115–123 (2020). https://doi.org/10.1080/09720529.2020.1721871
14. Liu, N., Li, X., Qi, E., Xu, M., Li, L., Gao, B.: A novel ensemble learning paradigm for medical diagnosis with imbalanced data. IEEE Access **8**, 171263–171280 (2020). https://doi.org/10.1109/ACCESS.2020.3014362
15. Fitriyani, N.L., Syafrudin, M., Alfian, G., Rhee, J.: Development of disease prediction model based on ensemble learning approach for diabetes and hypertension. IEEE Access **7**, 144777–144789 (2019). https://doi.org/10.1109/ACCESS.2019.2945129
16. Lyu, Q., Shan, H., Wang, G.: MRI super-resolution with ensemble learning and complementary priors. IEEE Trans. Comput. Imaging **6**, 615–624 (2020). https://doi.org/10.1109/tci.2020.2964201
17. Lahmiri, S., Bekiros, S., Giakoumelou, A., Bezzina, F.: Performance assessment of ensemble learning systems in financial data classification. Intell. Syst. Acc. Financ. Manag. **27**(1), 3–9 (2020). https://doi.org/10.1002/isaf.1460
18. Ni, J., Zhang, L., Tao, J., Yang, X.: Prediction of stocks with high transfer based on ensemble learning. J. Phys.: Conf. Ser., 1651, 1 (2020). https://doi.org/10.1088/1742-2596/1651/1/012 124ssss
19. Gaikwad, D.P., Thool, R.C.: Intrusion detection system using bagging ensemble method of machine learning. In: Proceedings - 1st International Conference on Computing, Communication, Control and Automation, ICCUBEA 2015, pp. 291–295 (2015). https://doi.org/10.1109/ICCUBEA.2015.61
20. Rehman Javed, A., Jalil, Z., Atif Moqurrab, S., Abbas, S., Liu, X.: Ensemble adaboost classifier for accurate and fast detection of botnet attacks in connected vehicles. Trans. Emerg. Telecommun. Technol. (2020). https://doi.org/10.1002/ett.4088
21. Bynagari, N.B.: Anti-Money Laundering Recognition Through The Gradient Boosting Classifier
22. Pham, B.T., et al.: Ensemble modeling of landslide susceptibility using random subspace learner and different decision tree classifiers. Geocarto Int. (2020). https://doi.org/10.1080/10106049.2020.1737972

23. Soares, R.G.F., Chen, H., Yao, X.: A cluster-based semisupervised ensemble for multiclass classification. IEEE Trans. Emerging Top. Comput. Intell. **1**(6), 408–420 (2017). https://doi.org/10.1109/TETCI.2017.2743219
24. Pietranik, M., Nguyen, N.T.: A multi-atrribute based framework for ontology aligning. Neurocomputing **146**, 276–290 (2014)
25. Duong, T.H., Nguyen, N.T., Jo, G.S.: A method for integration of wordnet-based ontologies using distance measures. In: Lovrek, I., Howlett, R.J., Jain, L.C. (eds.) KES 2008. LNCS (LNAI), vol. 5177, pp. 210–219. Springer, Heidelberg (2008). https://doi.org/10.1007/978-3-540-85563-7_31
26. Nguyen, N.T.: Conflicts of ontologies – classification and consensus-based methods for resolving. In: Gabrys, B., Howlett, R.J., Jain, L.C. (eds.) KES 2006. LNCS (LNAI), vol. 4252, pp. 267–274. Springer, Heidelberg (2006). https://doi.org/10.1007/11893004_34

Embedding Model with Attention over Convolution Kernels and Dynamic Mapping Matrix for Link Prediction

Thanh Le[1,2(✉)] [iD], Nam Le[1,2] [iD], and Bac Le[1,2] [iD]

[1] Faculty of Information Technology, University of Science,
Ho Chi Minh City, Vietnam
{lnthanh,lhbac}@fit.hcmus.edu.vn, 18120061@student.hcmus.edu.vn
[2] Vietnam National University, Ho Chi Minh City, Vietnam

Abstract. Knowledge Graph Completion, especially its sub-task link prediction attracts the attention of the research community and industry because of its applicability as a premise for developing several potential applications. Knowledge graph embedding (KGE) shows promising results to solve this problem. This paper focuses on the neural networks-based approach for KGE, which can extract features from the graphs better than other groups of embedding methods. The ConvE model is the first work using 2D convolution over embeddings and stacking multiple nonlinear feature layers to model knowledge graphs. However, its computation is inefficient and does not preserve translation between entity and relation embedding. Therefore, dynamic convolution was designed to solve limited representation capability issues and show the promised performance. This work introduces a mixture model that incorporates attention into performing the convolutional operation on projection embeddings. The TransD idea is used to project entity embedding from entity space to relation space. Then, it is stacked with relation embedding to perform dynamic convolution over stacked embedding without reshaping, following the idea that comes from Conv-TransE. So the translational property between the entity and the relation is preserved, and their diversity is considered. We experimented on benchmark datasets and showed how our proposed model is better than baseline models in terms of MR, MRR, and Hits@K.

Keywords: Knowledge graph embedding · Link prediction · Dynamic convolution · Attention over convolution kernel

1 Introduction

In 2012, Google introduced "knowledge graphs" and released Google Knowledge Graphs, leading to many research directions in mining and application. This kind of knowledge base aims to describe the entire natural world. KGs organizes data from multiple sources and associates entities through relations. It has brought

N. T. Nguyen et al. (Eds.): ACIIDS 2022, LNAI 13757, pp. 234–246, 2022.
https://doi.org/10.1007/978-3-031-21743-2_19

many vital applications, such as recommendation systems, automatic question answering, semantic search systems. In terms of structure, it has a graph structure consisting of entities (nodes) associated with specific relations (edges). An edge is usually of the form (head entity, relation, tail entity) indicates an existing relation between head and tail entity.

In the set problems about the mining knowledge bases, Knowledge Graph Completions is often the first concern because the more data, the better the data mining results. Therefore, it is also known as the link prediction problem in a particular aspect. The reason for missing associate information is that KGs are often constructed manually or semi-automatically, leading to missing the information between entities. FreeBase is one of the most extensive knowledge bases, and according to a recent statistic, much personal information is missing. More specifically, about 99% have no ethnicity, date of birth information or relations between humans [23]. The more complete the data, the more positive influential the knowledge base mining is.

Most of the current research focuses on knowledge graph embedding, which encodes entities and relations into a continuous low-dimensional vector space and uses these embedding to predict new relations (new entities). Translation distance-based models like TransE [1] and its extensions such as TransH [22], TransR [13], TransD [4], RotaHS [8] or semantic matching-based models like DistMult [24], ComplEx [20] are simple and effective approaches, but learn less expressive features. Graph network based-models such as R-GCN [16], SACN [17] are based on the encoder-decoder framework. The encoder encodes neighbour information for nodes. The decoder is a model of knowledge graph embedding to compute scores for triples. However, its complexity is very high due to graph operations and graph computations. Neural network-based models like ConvE [3], ConvKB [14], HyperConvKB [9] and Conv-TransE [17] use convolution operation over embeddings and multiple layers of non-linear features and achieve state-of-the-art performance on standard benchmark datasets for knowledge graph completion. Some modes like NAGAN-ConvKB [11], fusions GAN and CNN to generate negative samples and producing a better graph embedding.

In this paper, we propose a model to solve the incompleteness of KGs. Our model takes advantage of dynamic convolutional, which has attention over convolution kernels, which are fast and lightweight. Its computation is efficient and has more representation power since these kernels are aggregated in a nonlinear way via attention [2]. Using the TransD idea, we project entity embedding from entity space to relation space, using projection matrix and vector multiplication, which improves the calculation efficiency and solves the problem of too many parameters. We take projected entity embedding, stack with relation embedding. Then dynamic convolution is performed over stacked embedding without reshaping following the idea from Conv-TransE [17] to preserve the translational characteristic and consider the diversity of the entity and relation, which is very useful in our task.

Our contributions are summarized as the list below:

- We present the ConvAP model, fusion TransD idea, and Conv-TransE idea, improving the calculation efficiency and considering the diversity of objects in the knowledge base while preserving the translation characteristic between entities and relations. Besides that, the representation power has been enhanced by using dynamic convolution.
- We perform our model on the standard FB15k-237, WN18RR, kinship, and UMLS datasets and show improvement over some recent models in terms MR (Mean Rank), MRR (Mean Reciprocal Rank), Hits@K (K=1, 3, 10).

2   Related Work

Knowledge graph embedding has recently become a potential approach with promising performance in link prediction tasks, especially entity prediction and relational prediction tasks. Several models have been proposed, and we can categorize them into: (i) Translation distance-based models; (ii) Semantic matching-based models; (iii) Graph network-based models; (iv) Neural network-based models.

In the group of translation distance-based models, TransE is the first model using the idea of Word Embedding, considering relation as a translation from head entity to tail entity. Because of fast and straightforward calculation, TransE can be used in large knowledge graphs. However, the TransE [1] model has some drawbacks, one of which is too simple to model complex relations such as reflexive, one-to-many, many-to-one, and many-to-many. After that, there are many proposed models, in which the improved models such as TransH [22], CTransR/TransR [13], TransD [4]. TransH introduces hyperplanes for relation-specific, where we can map entities. One of the disadvantages of the TransH [22] model is that the entity and relation are in the same semantic space, and it is not valid in the real world. TransR proposed to embed entities and relations in different entity embedding and relation embedding spaces. Although TransR has significantly improved performance in the link prediction and triple classification task, it still has many parameters because it uses the projection matrix and matrix multiplication. TransD was proposed to solve that problem by introducing two projection vectors. TransD replaces projection matrix and matrix multiplication in TransR by using projection vectors and vector multiplication, which reduces the model parameters, improving the calculation performance and model predictions.

The next category is semantic matching based-models, which use the semantic similarity association between entities and relations. Some popular models are RESCAL [15], DistMult [24], ComplEx [20]. RESCAL is the first model learning based on three-way tensor factorization, using the relations matrix to model the interactions between entities and relations. However, RESCAL has many parameters, so the computational cost increases. The DistMult model was proposed. It improves based on RESCAL and decomposes the relation matrix into a diagonal matrix, thereby reducing the number of parameters for the model. Instead of using real-valued space as in previous works, ComplEx [20], Thanh Le et al. [7] use the complex vector space to model complex relations like symmetric and anti-symmetric relations.

Recently, applying neural networks and graph networks for knowledge bases has received much attention from many researchers. R-GCN [16] is the first model to use graph convolution operators for multi-relational data. However, it suffers from parameter overloading problems. Then, there are many works based on the encoder-decoder framework to overcome R-GCN's problem, such as SACN [17]. SACN proposed a new end-to-end architecture, including two components, encoder and decoder. The encoder is WGCN, which has weight learning capability that adapts information from dc neighbours using local aggregation, and the decoder is the Conv-TransE model. CompGCN proposed a novel framework for Graph Convolutional, incorporating entity (node) embedding and relation embedding in a multi-relational graph and using entity-relation composition operators from KGE techniques to generalize current GCN methods.

Models based on neural networks take advantage of several network architectures like the convolutional neural network to extract features from graphs. ConvE [3] is the first model using 2D convolutions over embeddings to extract more feature interactions. Due to using the convolution operation, ConvE can focus on the local feature on the stacking embedding, which means the local relations between different dimensional entities embedding or relation embedding. Thus ConvE does not use global relations, ignores the translational characteristic, and affects the model's performance. ConvKB [14] is another model using a convolution operator to define a score function to get a score for each triplet. Conv-TransE [17] was introduced as a decoder of SACN [17], based on the ConvE model. In order to keep the translation property, the model does not reshape stacked embedding of entity embedding and relation embedding and do 2D convolution like ConvE. To the best of our knowledge, InteractE [21] is another model that aims to maximize the interactions between entity embeddings and relation embeddings to the best of our knowledge. The model uses three essential methods: Feature Permutation, Checkered Reshaping and Circular Convolution. InteractE improves the link prediction performance on multiple benchmark datasets. The circular convolution proposed in the model allows extracting as many features as possible, but its computational complexity is still high.

The neural network approach shows potential because of using convolution operation to capture local relations between different dimensional embedding. We use the idea that no reshaping embedding comes from Conv-TransE to preserve translational characteristics, which is helpful in link prediction task. Furthermore, incorporating translational lightweight models like TransD can enhance performance.

3   Background

This section presents the preliminaries, including the basic definition and link prediction formalization. We also present baseline models from which we propose our improvements.

Following previous works and recent surveys [10], we define a knowledge graph as $\mathcal{G} = \{(h, r, t)\} \subseteq \mathcal{E} \times \mathcal{R} \times \mathcal{E}$ mean a set of triples (facts), each triple consisting of a relation $r \in \mathcal{R}$ and head entity h, tail entity $t \in \mathcal{E}$. Each triple (h, r, t) denotes a relations r between the head entity h and the tail entity t in KGs. Link prediction problem has the objective is learning a scoring function $\psi : \mathcal{E} \times \mathcal{R} \times \mathcal{E} \rightarrow \mathbb{R}$ giving an implausibility score for an input triple (h, r, t).

3.1   Dynamic Convolution

To perform better convolution in our model, we use dynamic convolution, which incorporates the attention mechanism over convolution kernels. *Dynamic convolution* is a specific dynamic perceptron proposed by Chen et al. [2] to improve performance while reducing computational cost inside the neural network architecture. Figure 1 illustrates the static convolution, which has weight Wand the bias b. The Conv g is not dependent on the input. If we need to boost the performance, we can make the neural networks "deeper" or "wider". However, the computational cost and complexity increase significantly, so the network model works inefficiently. The convolution kernel is a function of the input in dynamic convolution, with different inputs that their convolutional kernels can process.

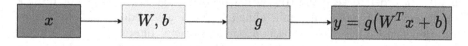

Fig. 1. Static convolution flow [2]

To compute kernel attentions $\{\pi_k(x)\}$, a pipeline for processing the information is proposed. Figure 2 illustrates this pipeline in detail and shows an overview of a dynamic convolution layer. Firstly, global average pooling squeezes the input information x. And then, passing two fully-connected layers (with a ReLU nonlinear function between them) and finally through the softmax function to generate normalized attention weights for K convolution kernels. Due to the small kernel size and the low computation cost for attention, dynamic convolution layers compute the convolution operation more efficiently and have more representation power than static convolution.

3.2   TransD Model

To help in considering the diversity of the entity and relation, we use TransD idea to project entity embeddings from entity space to relation space with the

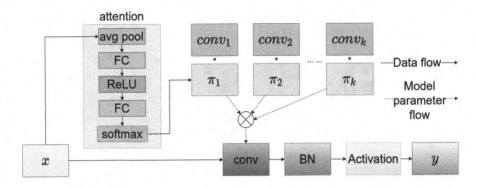

Fig. 2. A dynamic convolution layer [2]

least computation cost. Ji et al. proposed the TransD [4] model by using the mapping of head entities and tail entities to relation space. The projection vector of TransD [4] model is defined as

$$\mathbf{M}_{rh} = \mathbf{r}_p \mathbf{h}_p^\top + \mathbf{I}^{m \times n}, \quad \mathbf{M}_{rt} = \mathbf{r}_p \mathbf{t}_p^\top + \mathbf{I}^{m \times n}. \tag{1}$$

where $\mathbf{M}_{rh}, \mathbf{M}_{rt} \in \mathbb{R}^{m \times n}$ and are two mapping matrices, $\mathbf{h}, \mathbf{h}_p, \mathbf{r}, \mathbf{r}_p, \mathbf{t}, \mathbf{t}_p$ are vector of each element in a triplet (h, r, t), $\mathbf{h}, \mathbf{h}_p, \mathbf{t}, \mathbf{t}_p \in \mathbb{R}^n$, $\mathbf{r}, \mathbf{r}_p \in \mathbb{R}^m$ and $\mathbf{I}^{m \times n}$ is an identity matrix, p marks the projection vectors. The projected vector embedding can be written as $\mathbf{h}_\perp = \mathbf{M}_{rh}\mathbf{h}, \quad \mathbf{t}_\perp = \mathbf{M}_{rt}\mathbf{t}$. Then the scoring function for TransD is defined like TransH, TransR:

$$\psi_r(\mathbf{h}, \mathbf{t}) = ||\mathbf{h}_\perp + \mathbf{r} - \mathbf{t}_\perp||_2^2 \tag{2}$$

4 The Proposed Model

This section presents the proposed model - ConvAP (Convolutional with attention over convolution kernels on projection embeddings).

The main idea of our model is to define two vectors for each entity and relation. As in the TransD [4] model, the first vector represents the meaning of an entity or relation, and the other vector is the projection vector. The projection vector represents how to project a real embedding into a relation vector space. After that, the score is determined by a dynamic convolution over 2D embedding, which has no reshape after stacking $\mathbf{e}_{sp}$ and $\mathbf{e}_r$. The architecture is summarised in the Fig. 3.

Given an input triple (s, r, o), the encoding component maps entities $s, o \in \mathcal{E}$ to their distributed embedding representations $\mathbf{e}_s, \mathbf{e}_o \in \mathbb{R}^d$. For each entity embedding, mapping matrix $\mathbf{M}_r$ is defined to project entity from entity space to relation space:

$$\mathbf{M}_r = \mathbf{r}_p \mathbf{e}_p^\top + \mathbf{I}^{dt \times d} \tag{3}$$

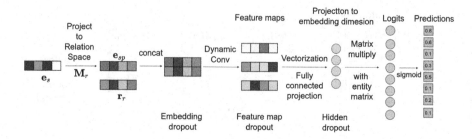

Fig. 3. The proposed ConvAP model

where $\mathbf{r}_p$, $\mathbf{e}_p$ are relation and entity projection vector, respectively and $\mathbf{I}$ is identity matrix.

With the mapping matrices, the projected vector is defined as $\mathbf{e}_{sp} = \mathbf{M}_r \mathbf{e}_s$. In this way, the advantage of the TransD model is taken, replacing the transfer matrix by using the product of two projection vectors of an entity-relation. After that, we stack $\mathbf{e}_{sp}$ and $\mathbf{r}_r$ without reshaping stacked embedding and apply dynamic convolution over this embedding. Intuitively, the translation characteristic is preserved. Furthermore, the performance is boosted by using attention over the convolution kernels.

The stacked of $\mathbf{e}_{sp}$ and $\mathbf{r}_r \in \mathbb{R}^d$ is used as the input for a dynamic convolution layer. After that, the dynamic convolution layer returns feature maps $\mathcal{T} \in \mathbb{R}^{c \times d}$, where c is the number of feature maps. We apply vectorization on this feature map tensor to obtain a vector. Then, it is projected into d-dimensional space using linear transformation donated by the matrix $\mathbf{W} \in \mathbb{R}^{c \times d \times d}$. Before applying the activation function, it is multiplied with $\mathbf{e}_o$ via the inner product. This way is equivalent to negative sampling, where the negative sample includes entities not in the dataset and the other tail entities linked with head and relation are positive. Formally, the scoring function is defined as:

$$\psi_r(\mathbf{e}_s, \mathbf{e}_o) = f(vec(f(DynamicConv([\mathbf{e}_{sp}, \mathbf{r}_r]))\mathbf{W}))\mathbf{e}_o \tag{4}$$

where $\mathbf{r}_r \in \mathbb{R}^d$ is a relation parameter depending on r and f is nonlinear function.

For training our model parameters, the logistic sigmoid function $\sigma(.)$ is used like ConvE to the scoring function: $p(\mathbf{e}_s, \mathbf{e}_r, \mathbf{e}_o) = \sigma(\psi_r(\mathbf{e}_s, \mathbf{e}_o))$, minimizing the binary cross-entropy loss function:

$$\mathcal{L} = -\frac{1}{N} \sum_i (t_i.log(p_i) + (1 - t_i)log(1 - p_i)) \tag{5}$$

where t is the label vector with dimension $\mathbb{R}^{1 \times 1}$ for one-to-one scoring or $\mathbb{R}^{1 \times d}$ for one-to-many scoring. If the relations exists, the element of vector t are ones and zero otherwise.

We also use dropout on the embedding like ConvE model, feature maps after the convolution operation, and the hidden units after the fully connected layer.

Due to saturation of output non-linearities at the labels, following ConvE, Adam optimizer [5] and label smoothing is used to avoid overfitting problem [18].

5   Experiments and Result Analysis

This section presents our experiments, including the description of the benchmark datasets, hyperparameter setup, and analyses of the results.

5.1   Benchmark Datasets

To evaluate the performance of link prediction we use some standard benchmark datasets: **Kinship** (Alyawarra Kinship) [12], **UMLS** (Unified Medical Language Systems) [6], **WN18RR** [3], **FB15k-237** [19]. Their statistic information is detailed in the Table 1. Each dataset is split into three sets for training, validation and testing.

Table 1. Statistical Information on KG Benchmarks

Dataset	Entities	Relations	Training	Validation	Test	Mean d_{in}	Median d_{in}
FB15k-237	14,541	237	272,115	17,535	20,466	18.71	8
WN18RR	40,943	11	86,835	3034	3134	2.12	1
Kinship	104	25	8544	1068	1074	82.15	82.5
UMLS	135	46	5216	652	661	38.63	20

5.2   Experimental Setup

In our Link Prediction experiments, the proportion of correct entities ranked in top k, where k is 1, 3 and 10 (Hits@1, Hits@3, Hits@10), and the mean reciprocal rank (MRR) are reported. Let r_h is rank in the set that generated by replacing head entity and r_t is rank in the set that generated by replacing tail entity and number of test set is n_t. Hits@K and Mean Reciprocal Rank (MRR) is computed as follow:

$$Hits@K = \frac{1}{2 \times n_t} \sum_i p\left[r_h \leq K\right] + p\left[r_t \leq K\right]$$

$$\text{where } p(x \leq K) = \begin{cases} 1, & \text{if } x \leq K \\ 0, & \text{otherwise} \end{cases} \tag{6}$$

$$MRR = \frac{1}{2 \times n_t} \sum_i \left(\frac{1}{r_h} + \frac{1}{r_t}\right) \tag{7}$$

To avoid some corrupted triples, we use the filtered setting [1] to filter out all valid triples before ranking: the lower the MR score, the higher MRR or, the higher Hits@K, the better performance.

Following the SACN experimental setup, we also used grid search to determine our hyperparameters during the training. We manually specify the hyperparameters for each dataset. For input dropout is fixed 0.1, learning rate in range {0.001, 0.002, 0.003, 0.004}, dropout rate is selected in the range {0.1, 0.2, 0.3, 0.4}, embedding size is in range {100, 200, 300}, number of channels is in range {50, 100, 200, 300} and the kernel size from {1, 3, 5}. Based on the results obtained from the experiments, the best hyperparameters for our proposed model are shown in Table 2.

Table 2. Best hyperparamters for ConvAP where d is embedding dimension, c is number of out channels, k is kernel size and p_1, p_2, p_3 are input dropout, feature map dropout and hidden dropout, respectively.

Dataset	d	c	k	p_1	p_2	p_3
FB15k-237	100	100	5	0.1	0.2	0.3
WN18RR	200	200	3	0.1	0.2	0.4
Kinship	100	100	5	0.1	0.2	0.3
UMLS	200	200	3	0.1	0.2	0.3

We compare our model - ConvAP with state-of-the-art baselines, including: **Semantic matching based-models**: DistMult [24]; ComplEx [1,20]; **Neural networks based-models**: ConvE [3], ConvKB [14]; **Graph Neural Networks based-models**: R-GCN [16].

5.3   Results

Main Results. On FB15k-237, ConvAP has relatively comparable performance to the ConvE. In particular, our model improved 2.041% performance on Hits@10 to compare with ConvE and improved significantly compared with DistMult or ComplEx. Unfortunately, when compared with Conv-TransE, our model does not improve much; the results are pretty equal due to our model's lack of ability to use context information outside the local receptacle. On WN18RR, we also obtain similar results to the experiment on the FB15k-237. It continues to have more performance improvements than the ConvE, ComplEx, R-GCN for all evaluation metrics. Table 3 presents the results in detail. On the Kinship and UMLS dataset, the results are detailed in Table 4.

At all, ConvAP got better results for all evaluation metrics compared with ConvE and others baseline models. More specifically, ConvAP promotes interaction between entity and relation embeddings by applying dynamic projection to relation space for entity embeddings, even if ConvE and ConvAP have identical experimental results.

Convergence Analysis. The convergence of the model is one of the critical factors for a model. An experiment is conducted on the FB15K-237, WN18RR, and

Table 3. Results of the link prediction tasks by Mean Reciprocal Rank (MRR) and HITS@K on two benchmark datasets WN18RR and FB15k-237. Best results are in bold, and the second-best results are underlined.

Model	FB15k-237				WN18RR			
	MRR	Hits@1	Hits@3	Hits@10	MRR	Hits@1	Hits@3	Hits@10
TransE	0.25	0.17	0.28	0.42	0.18	0.02	0.29	0.44
DistMult	0.24	0.16	0.26	0.42	0.43	0.39	0.44	0.49
ComplEx	<u>0.25</u>	0.16	<u>0.28</u>	0.43	0.44	<u>0.41</u>	**0.46**	**0.51**
R-GCN	<u>0.25</u>	0.15	0.26	0.42	–	–	–	–
ConvE	**0.32**	**0.24**	**0.35**	<u>0.49</u>	**0.46**	0.39	0.43	0.48
ConvAP (ours)	**0.32**	<u>0.23</u>	**0.35**	**0.50**	<u>0.45</u>	**0.42**	**0.46**	**0.51**

Kinship for our model. Figure 4 shows the experimental process on FB15k237 and WN18RR. The chart shows four lines with different colours, including MRR and Hits@K. ConvAP has achieved the best performance after around 100 epochs. On WN18RR, our model also achieves high performance and balance for a short period, about 100 epochs. A similar result when experimenting on FB15K237 also shows this.

Table 4. Results of the link prediction tasks by Mean Reciprocal Rank (MRR) and HITS@K on two benchmark datasets Kinship and UMLS. Best results are in bold, and the second-best results are underlined.

Model	Kinship				UMLS			
	MRR	Hits@1	Hits@3	Hits@10	MRR	Hits@1	Hits@3	Hits@10
DistMult	0.516	0.367	0.581	0.867	0.868	0.821	–	0.967
ComplEx	0.823	0.733	0.899	0.971	0.894	0.823	–	0.995
ConvE	<u>0.833</u>	<u>0.738</u>	<u>9.17</u>	**0.981**	<u>0.902</u>	**0.928**	-	**0.997**
ConvKB	0.614	0.4362	0.755	0.953	–	–	–	–
R-GCN	0.109	0.003	0.088	0.239	–	–	–	–
ConvAP (ours)	**0.834**	**0.744**	**0.918**	<u>0.978</u>	**0.903**	0.84	**0.95**	<u>0.98</u>

Kernel Size Analysis. Kernel size is essential in convolution operations. To evaluate its effects on our model, we conducted an experiment on the Kinship dataset. The init embedding size is 100; the drop rate is 0.2; and the number of kernels is 100. With different kernel sizes, the results of the evaluation measures are affected.

As a result shown in Fig. 5, with the kernel size $2 \times ks$, which $ks = \{1, 3, 5\}$, the reception fields proportional to K, thus evaluation metrics also affected. This inference can be generalized to larger datasets like FB15k-237, WN18RR,

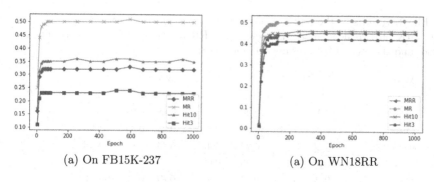

(a) On FB15K-237   (a) On WN18RR

Fig. 4. The convergence study on FB15K-237 and WN18RR

or YAGO3-10. The optimal kernel size depends heavily on experimental data and specific tasks.

Embedding Dimension Analysis The initialization embedding dimension is essential for link prediction tasks, so it is essential to analyze this hyperparameter. We continue to conduct experiments on kinship, which are dense enough to evaluate model performance in a short time. The default init kernel size is set to 3, kernels are 100, and the drop rate is 0.2. In Fig. 6, all evaluation metrics are affected by the number of dimensions initialized, and the optimal initial embedding size depends on the experimental data and our task.

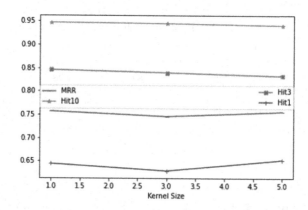

Fig. 5. Kernel analysis for ConvAP on Kinship dataset

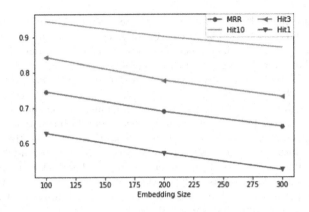

Fig. 6. Initial embedding dimension analysis for ConvAP on Kinship dataset

6 Conclusion and Future Research Directions

This work introduced a mixture model that uses dynamic convolution, improving the performance of computation operations. Besides that, taking advantage of the dynamic mapping matrix coming to project entity embedding to relation space, the model considers multiple types of entities and keeps the model complexity balanced. The ConvAP model's performance has improved performance better than the state-of-the-art model like ConvE. ConvAP has done quite well in convergence time when it has achieved relatively high performance in a few epochs. We plan to integrate the ConvAP model into the Graph Neural Networks framework in the future. Our model plays a role as a decoder module and combines local structure information to improve our performance.

Acknowledgements. This research is funded by the University of Science, VNU-HCM, Vietnam under grant number CNTT 2022-02 and Advanced Program in Computer Science.

References

1. Bordes, A., Usunier, N., García-Durán, A., Weston, J., Yakhnenko, O.: Translating embeddings for modeling multi-relational data. In: NIPS (2013)
2. Chen, Y., Dai, X., Liu, M., Chen, D., Yuan, L., Liu, Z.: Dynamic convolution: attention over convolution kernels. In: 2020 IEEE/CVF Conference on Computer Vision and Pattern Recognition (CVPR), pp. 11027–11036 (2020)
3. Dettmers, T., Minervini, P., Stenetorp, P., Riedel, S.: Convolutional 2D knowledge graph embeddings. In: AAAI (2018)
4. Ji, G., He, S., Xu, L., Liu, K., Zhao, J.: Knowledge graph embedding via dynamic mapping matrix. In: ACL (2015)
5. Kingma, D.P., Ba, J.: Adam: a method for stochastic optimization. CoRR abs/1412.6980 (2015)

6. Kok, S., Domingos, P.M.: Statistical predicate invention. In: ICML 2007 (2007)
7. Le, T., Huynh, N., Le, B.: Link prediction on knowledge graph by rotation embedding on the hyperplane in the complex vector space. In: Farkaš, I., Masulli, P., Otte, S., Wermter, S. (eds.) Artificial Neural Networks and Machine Learning - ICANN 2021, pp. 164–175. Springer, Cham (2021). https://doi.org/10.1007/978-3-030-86365-4_14
8. Le, T., Huynh, N., Le, B.: RotatHS: rotation embedding on the hyperplane with soft constraints for link prediction on knowledge graph. In: Nguyen, N.T., Iliadis, L., Maglogiannis, I., Trawiński, B. (eds.) Computational Collective Intelligence, pp. 29–41. Springer, Cham (2021). https://doi.org/10.1007/978-3-030-88081-1_3
9. Le, T., Nguyen, D., Le, B.: Learning embedding for knowledge graph completion with hypernetwork. In: Nguyen, N.T., Iliadis, L., Maglogiannis, I., Trawiński, B. (eds.) Computational Collective Intelligence, pp. 16–28. Springer, Cham (2021). https://doi.org/10.1007/978-3-030-88081-1_2
10. Le, T., Nguyen, H., Le, B.: A survey of the link prediction on static and temporal knowledge graph. J. Res. Devel. Inf. Commun. Technol. (2021)
11. Le, T., Pham, T., Le, B.: Negative sampling for knowledge graph completion based on generative adversarial network. In: Nguyen, N.T., Iliadis, L., Maglogiannis, I., Trawiński, B. (eds.) Computational Collective Intelligence, pp. 3–15. Springer, Cham (2021). https://doi.org/10.1007/978-3-030-88081-1_1
12. Lin, X.V., Socher, R., Xiong, C.: Multi-hop knowledge graph reasoning with reward shaping. In: EMNLP (2018)
13. Lin, Y., Liu, Z., Sun, M., Liu, Y., Zhu, X.: Learning entity and relation embeddings for knowledge graph completion. In: AAAI (2015)
14. Nguyen, D.Q., Nguyen, T.D., Nguyen, D.Q., Phung, D.Q.: A novel embedding model for knowledge base completion based on convolutional neural network. In: NAACL (2018)
15. Nickel, M., Tresp, V., Kriegel, H.P.: A three-way model for collective learning on multi-relational data. In: ICML (2011)
16. Schlichtkrull, M., Kipf, T., Bloem, P., van den Berg, R., Titov, I., Welling, M.: Modeling relational data with graph convolutional networks. In: ESWC (2018)
17. Shang, C., Tang, Y., Huang, J., Bi, J., He, X., Zhou, B.: End-to-end structure-aware convolutional networks for knowledge base completion. In: AAAI (2019)
18. Szegedy, C., Vanhoucke, V., Ioffe, S., Shlens, J., Wojna, Z.: Rethinking the inception architecture for computer vision. In: 2016 IEEE Conference on Computer Vision and Pattern Recognition (CVPR), pp. 2818–2826 (2016)
19. Toutanova, K., Chen, D., Pantel, P., Poon, H., Choudhury, P., Gamon, M.: Representing text for joint embedding of text and knowledge bases. In: EMNLP (2015)
20. Trouillon, T., Welbl, J., Riedel, S., Gaussier, É., Bouchard, G.: Complex embeddings for simple link prediction. In: ICML (2016)
21. Vashishth, S., Sanyal, S., Nitin, V., Agrawal, N., Talukdar, P.P.: InteractE: improving convolution-based knowledge graph embeddings by increasing feature interactions. In: AAAI (2020)
22. Wang, Z., Zhang, J., Feng, J., Chen, Z.: Knowledge graph embedding by translating on hyperplanes. In: AAAI (2014)
23. West, R., Gabrilovich, E., Murphy, K.P., Sun, S., Gupta, R., Lin, D.: Knowledge base completion via search-based question answering. In: Proceedings of the 23rd International Conference on World Wide Web (2014)
24. Yang, B., Tau Yih, W., He, X., Gao, J., Deng, L.: Embedding entities and relations for learning and inference in knowledge bases. CoRR abs/1412.6575 (2015)

Employing Generative Adversarial Network in COVID-19 Diagnosis

Jakub Dereń and Michał Woźniak$^{(\boxtimes)}$ (iD)

Wrocław University of Science and Technology,
Wybrzeże Wyspiańskiego 27, 50 -370 Wrocław, Poland
michal.wozniak@pwr.edu.pl

Abstract. In recent years, many papers and models have been developed to study the classification of X-ray images of lung diseases. The use of transfer learning, which allows using already trained network models for new problems, could allow for better results in the COVID-19 disease classification problem. However, at the beginning of the pandemic, there were not very large databases of SARS-CoV-2 positive patient images on which a network could perform learning. A solution to this problem could be a Generative Adversarial Network (GAN) algorithm to create new synthetic data indistinguishable from the real data using the available data set. It would allow training a network capable of performing classification with greater accuracy on a larger and more diverse number of training data. Obtaining such a tool could allow for more efficient research on how to solve the global COVID-19 pandemic problem. The research presented in this paper aims to investigate the impact of using a Generative Adversarial Network for COVID-19-related imaging diagnostics in the classification problem using transfer learning.

Keywords: Pattern classification · Data augmentation · GAN · COVID-19 · Transfer learning

1 Introduction

Since the first infections were reported in November 2019 in Wuhan, China, SARS-CoV-2 has spread worldwide, including reaching Europe in early 2020. The World Health Organization (WHO) has declared ca the center of the pandemic. From the very beginning, a critical element in the fight against the pandemic has become its correct diagnosis and the choice of the correct therapeutic approach for the individual course of the disease [19]. Chest X-ray has been selected as one of the diagnostic tools, although it should be noted here that it is relatively insensitive in detecting pulmonary abnormalities in the early stages of the disease [3]. However, it can be very useful to monitor the progression of pulmonary involvement in COVID-19, especially in patients with advanced disease.

With a new pathogen and little history of disease in patients, the problem was how to rapidly acquire a large volume of imaging information that would

© The Author(s), under exclusive license to Springer Nature Switzerland AG 2022
N. T. Nguyen et al. (Eds.): ACIIDS 2022, LNAI 13757, pp. 247–258, 2022.
https://doi.org/10.1007/978-3-031-21743-2_20

allow the development of an automated tool to aid in the rapid diagnosis of COVID-19 [4]. It is also worth noting here that the need to develop such a tool was also related to the unification of diagnostics because, especially in the early stages of the pandemic, including the lack of guidelines for describing x-rays - radiological findings could nevertheless vary among radiologists.

To obtain a good quality diagnostic system, it was necessary to collect a sufficiently large database containing images of COVID-19 sufferers and healthy patients. At the beginning of the pandemic, the main problem was obtaining images of infected individuals, so the so-called problem of building a diagnostic system based on imbalanced data was encountered., i.e., the disproportion between the number of observations from different classes [9]. One of the techniques used to balance the distributions is data augmentation, which involves adding synthetic minority class objects to the data set. Among the various techniques that can be chosen to implement such a process, it was decided to use a generative adversarial network (GAN) [1]. It allows two competing generator and discriminator modules to generate synthetic chest x-ray images of individuals infected with Sars-Cov-2. With such a technique, a sufficient volume of images can be obtained to adequately train a deep network allowing diagnosis in the indicated range.

The main goal of this work is to evaluate whether the use of Generative Adversarial Network for generating synthetic images and transfer learning techniques is helpful in the problem of learning deep models dedicated to the COVID-19 classification problem.

2   Proposed Framework

This section shows how the data was prepared, how synthetic images were generated, and how transfer learning techniques were used in the task of classifying lung images of COVID-19 patients.The pipeline of the proposed framework is presented in Fig. 1.

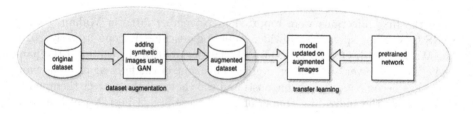

Fig. 1. The pipeline of the proposed framework.

2.1   Data Augmentation

The Generative Adversarial Network (GAN) proposed by Goodfellow [6] was used to generate the synthetic images. GANs are now popular tools for learning complex distribution characteristics. At the same time, GANs show outstanding performance in various areas, including image synthesis [8], image-to-image or text-to-image translation, photo blending, etc. The idea of GAN [14] is to train two models simultaneously, a so-called generator that learns to generate false data using feedback from a so-called discriminator. The task of the generator is to estimate the probability distribution of real samples to generate samples with a distribution close to the real distribution. The discriminator estimates the probability that a given sample comes from the real data distribution rather than being generated. The idea of the GAN is presented in Fig. 2

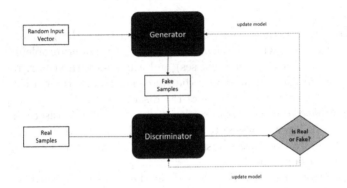

Fig. 2. Block diagram of the Generative Adversarial Network.

An extension to the GAN model, CGAN (Conditional Generative Adversarial Network) [2,11] was implemented to generate separate examples for the COVID and Normal classes. It enables an additional variable to be added to each generated data, specifying the class of the problem.

Preprocessing. For the GAN model learning process, all images were preprocessed, i.e., they were rescaled to 256×256 pixels to standardize all data to a size that would allow correct working with the available images. All images were also loaded in RGB format, and the pixel value was scaled from $[0, 255]$ to $[-1, 1]$.

Discriminator. We built a network model having multiple inputs and outputs, allowing additional information about the class of the created image. The method configuring the model takes as input the number of classes from which the data was generated (in the case of the problem presented above, these are the COVID and Normal classes) and the shape of the target image $(256, 256, 3)$. Initially, an *Input* layer is defined, taking the class of the data to be handled. The class is then processed by the *Embedding* layer. The output from that layer is passed to the *Dense* layer having a linear activation. Those activations are transformed into a single activation map (256, 256) and are combined with the input image [2]. This process is followed by downsampling.

The model of this network is then based on convolutional neural networks combined with the use of the activation function *LeakyReLU*, recommended for the construction of the GAN model [2, 10]. That function takes a slope factor of 0.2. The first normal layer contains eight filters, the next values of the number of filters are multiplied times two by the number of filters in the previous layer. In the output layer, we employed *sigmoid* as the activation function to obtain the probability that the generated data with the class label is true. The model was trained using *binary crossentropy* as the objective function and *Stochastic Gradient Descent* as the optimization method [7].

Generator. The method configuring the model takes the number of classes from which the data will be generated (in the case of the problem presented above, the COVID and Normal classes) and a fixed-length vector, randomized from a Gaussian distribution. First, the *Input* layer takes the class of the data to be handled. Then, as in the case of the discriminator, this class is processed by the layer *Embedding*. The process also looks similar to the discriminator. The output from this layer is passed to the layer with linear activation. However, in this case, the activations are transformed into a single map (4,4). This process is followed by upsampling.

The model of this network is based on the transposed convolution layers *Conv2DTranspose* combined with the use of the activation function *LeakyReLU* recommended for the construction of the GAN generator model [2, 10]. The function takes a slope factor of 0.2. individual hidden transposed convolution layer of the generator model has 256 filters. In the output layer, the activation function *tanh* was used. It is differentiable and monotonic. It is used for the classification of binary problems. Both this function and the *sigmoid* function are used in feed-forward networks [13].

2.2 Transfer Learning

Canonical machine learning methods attempt to build models dedicated to solving specific problems while being trained on data from scratch, i.e., they do not use the knowledge that could have been acquired while learning other tasks. The idea of transfer learning is to employ an already existing model as a starting point to learn how to solve a new task. Thus attempt to transfer the previously learned knowledge to the new task [20]. Transfer learning techniques are often used for deep models, which have high capacity and are often trained on large data sets. Then there is an attempt to learn the original model on samples from the new task [21].

Having created an environment to allow training using the GAN architecture, an environment was implemented to perform experiments using generated data and extracted real data from the repository. We chose the transfer learning method to complete the needed training. We decided to use **VGG16** and **VGG19** (available in the Keras library) as the pre-trained models.

The first implementation step was to perform data preprocessing. The images needed to be scaled to 224×224 pixels with values from 0 to 1. Moreover, the input images were converted from the RGB model to the BGR model. The color channel was adjusted by centering it and adapting it to the ImageNetcite dataset [5], which we used to train the VGG16 VGG19 networks [15].

We used *categorical crossentropy* [18] as the objective function and *Stochastic Gradient Descent* as the optimization method.

3 Experimental Evaluation

Goal. The experimental study aimed to answer whether the augmentation of minority class images using GAN and transfer learning can improve the quality of x-ray image classification toward COVID-19 diagnosis.

Used datasets. In order to do the learning process of the GAN model and classification, it was necessary to select a corresponding dataset for the given problem. The collection *COVID-19 Radiography Database* from the kaggle repository [12] was used to investigate the posed thesis. It consists of several thousand X-ray images of 299×299 pixels, divided into four classes:

- COVID (3616 images)
- Lung Opacity - (6012 images)
- Normal - (10192 images)
- Viral Pneumonia (1345 images)

As the aim of this study was to propose an effective tool for diagnosis of covid-19 patients, hence only images of healthy (Normal) subjects and those diagnosed with covid-19 (COVID) were selected from the above database. This choice is also justified by the need for extensive computational resources when performing the GAN model learning process on high-resolution images. For this reason, the number of images was limited to 1000 COVID class images and 2819 Normal class images (the proportions of the number of images present in the selected data set were preserved). Examples of the images used are shown below in Fig. 3[1].

---

[1] https://www.kaggle.com/tawsifurrahman/covid19-radiography-database, June 2021.

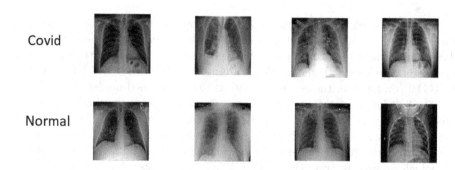

Fig. 3. Examples of images provided by the selected dataset

Implementation and Reproducibility. Complete source code, sufficient to repeat the experiments, was made available at public avalialble repository[2]. The proposed algorithm, as well as the experiments described in this work, were implemented in the Python programming language. We also used *Keras* library [17].

3.1 Using GAN to Generate Synthetic Images

The resulting dataset was imbalanced, i.e., the imbalance ratio was ca. 1:3. Hence it was decided to choose a method how to balance it. One of the popular data augmentation methods is oversampling, i.e., adding the number of minority class objects (COVID) to the original set equals the number of majority class objects (Normal). We may distinguish two main approaches, i.e., random oversampling or generation of synthetic objects. We decided to generate synthetic minority class X-ray images from the used dataset to provide more diversity in the learning set.

One of the problems that had to be solved was the appropriate choice of generation parameters. Since it was impossible to determine in advance what number of epochs was needed to generate realistic images of covid-19 patients, the subjective expert evaluation was used every ten epochs to determine whether the generated images were already of satisfactory quality.

Let us show exemplars of generated images after a fixed number of epochs. The following images present the generated images after 10 (Fig. 4), 50 (Fig. 5), 100 (Fig. 6), and 150 (Fig. 7) epochs. As can be seen, especially after a few epochs, they do not resemble the real images. According to expert judgment, it was considered that satisfactory quality was obtained after 180 epochs, and these images were used in the next step to learn the classifier.

A dozen examples of generated images labelled with class *COVID* and class *Normal* are presented in Fig. 8.

---

[2] https://github.com/jderen/Covid-19-GAN-Results.

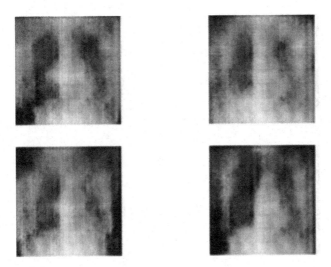

Fig. 4. Result of the generation of four images by the model generator after 10 epochs

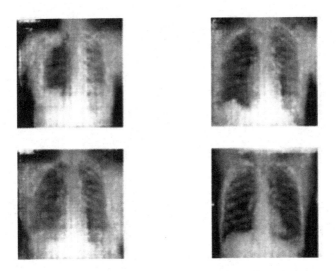

Fig. 5. Result of the generation of four images by the model generator after 50 epochs

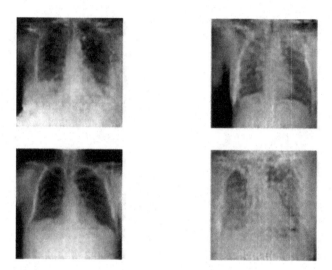

Fig. 6. Result of the generation of four images by the model generator after 100 epochs

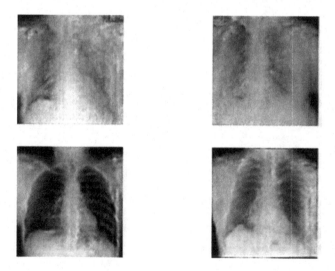

Fig. 7. Result of the generation of four images by the model generator after 150 epochs

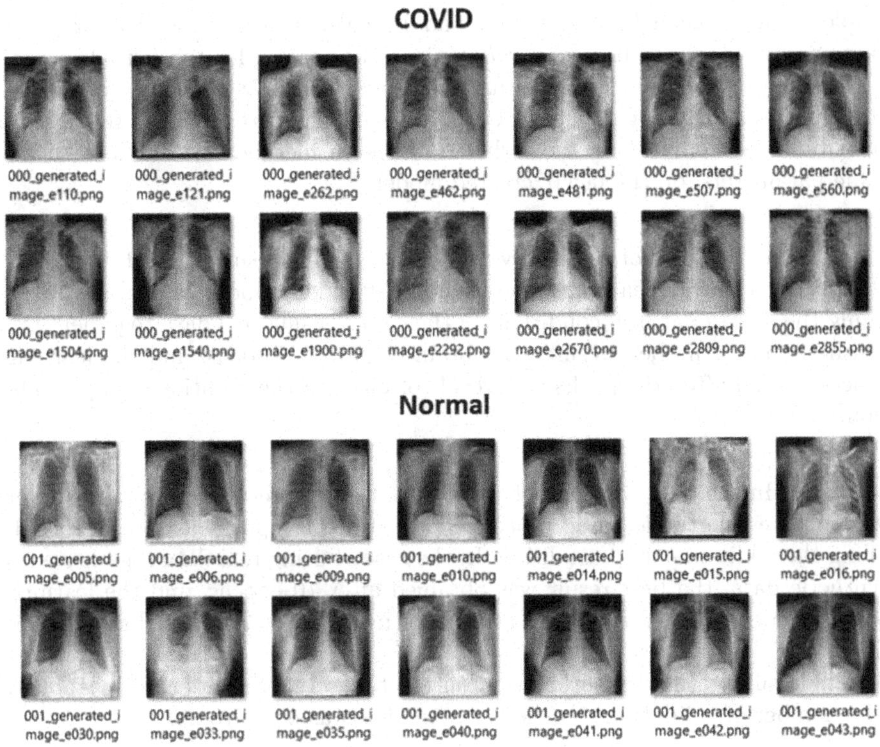

COVID

Normal

Fig. 8. Result of generating images by the model generator after 180 epochs

3.2 Transfer Learning

The final step of the research was to perform classification with transfer learning, using real and synthetic data. For this purpose, the following research method was proposed:

1. Performing a training of the transfer learning model with real data only to proceed with the classification
2. Performing a training of the transfer learning model with real and generated data, to proceed with the classification
3. Comparison of the results obtained from the two experiments

Section 2.2 mentioned that two pre-trained models were selected to perform transfer learning: VGG16 and VGG19.

Data Preparation and Used Methods. During the experimental evaluation, we divided the real data of 2819 Normal class images and 1000 COVID class images into training, validation and test data, in a ratio of 70% : 20% : 10%. The maximum number of epochs in the post-training process was 250. Still, we employed also *ModelCheckpoint* and an additionally stop criterion *EarlyStopping* that stops the learning process if *val_loss* stops decreasing for another 50 epochs.

After post-training the model using only real data, 2819 generated Normal class images and 1000 generated COVID class images were added to the data in the training set only. As in the earlier case, the *fit* method was called, with the same parameters, on the newly created model.

VGG16. In the case of using only real data, the best result was obtained after 69 epochs, while the learning process ended after 111 epochs as a result of fulfilling the additional stop condition. In the case of using real data together with synthetic data, the best result was obtained after 71 epochs, and the learning process ended after 79 epochs as a result of meeting the additional stop condition.

VGG19. In the case of using only real data, the best result was obtained after 91 epochs, while the learning process ended after 117 epochs as a result of fulfilling the additional stop condition. In the case of using real data together with synthetic data, the best result was obtained after 70 epochs, and the learning process ended after 76 epochs as a result of meeting the additional stop condition.

The results of models' evaluation on the test set were presented in Table 1. The number of validation steps for both models was equal to 20.

Table 1. Classification accuracy [%] of the selected models trained on real and augmented dataset

	real	real+synthetic
VGG16	0.944	0.956
VGG19	0.919	0.944

Lessons Learned. The purpose of the study was to answer the question of whether the use of transfer learning and data augmentation based on the generation of synthetic images using GAN can improve the quality of classification. Based on the results obtained, the answer to both questions is positive. It was possible to obtain synthetic images that resembled real images in expert opinion, and adding them to the dataset resulted in an increase in classification quality for both trained networks. However, we have to emphasize that it required more epochs to obtain the final model.

4   Conclusion

This paper discussed how to improve the classification quality of lung x-ray images using the data augmentation technique employing GAN and transfer learning. Although the proposed approach showed good quality compared to the approach without augmentation, it is important to note the problems encountered during the research. Performing research using deep network architectures requires a lot of computational power, which is generally impossible with local resources, hence an attractive approach is to use cloud resources of the type of platform used in this work. The second problem was to define a stopping criterion for GAN learning. It was determined during the augmentation process based on an expert assessment of the quality of images after a given number of epochs. However, the authors are aware that this way of evaluation may have negatively affected the quality of synthetic objects and, consequently, the quality of final predictive models. It seems appropriate to use methods that can automatically understand image information [16] to evaluate the usefulness of images for learning. It may be that even distorted images (e.g., obtained after a few epochs and judged by the expert as not very realistic) can help build a proper representation for the classification system.

The study also formulated potential directions for further research.

- The selection of different real data sets allows for a greater diversity of generated images, allowing for better classification results.
- Increasing the number of lung disease classes would also allow classifying of other lung diseases besides COVID-19.
- Work on selecting an adequate stop criterion in the augmentation process, which would allow obtaining synthetic images of higher quality.

Acknowledgements. This work is supported in part by the Research Fund of Department of Systems and Computer Networks, Faculty of ICT, Wroclaw University of Science and Technology and by the CEUS-UNISONO programme, with funding from the National Science Centre, Poland under grant agreement No. 2020/02/Y/ST6/00037.

References

1. Aggarwal, A., Mittal, M., Battineni, G.: Generative adversarial network: an overview of theory and applications. Int. J. Inf. Manag. Data Insights **1**(1), 100004 (2021). https://doi.org/10.1016/j.jjimei.2020.100004. ISSN 2667-0968
2. Brownlee, J.: How to develop a conditional GAN (CGAN) from scratch. https://www.machinelearningmastery.com/how-to-develop-a-conditional-generative-adversarial-network-from-scratch/ (2021)
3. Cleverley, J., Piper, J., Jones, M.M.: The role of chest radiography in confirming covid-19 pneumonia. BMJ **370** (2020). https://www.bmj.com/content/370/bmj.m2426https://doi.org/10.1136/bmj.m2426
4. Cyganek, B., et al.: A survey of big data issues in electronic health record analysis. Appl. Artif. Intell. **30**(6), 497–520 (2016). https://doi.org/10.1080/08839514.2016.1193714
5. Deng, J., Dong, W., Socher, R., Li, L.J., Li, K., Fei-Fei, L.: ImageNet: a large-scale hierarchical image database. In: 2009 IEEE Conference on Computer Vision and Pattern Recognition, pp. 248–255 (2009). https://doi.org/10.1109/CVPR.2009.5206848

258   J. Dereń and M. Woźniak

6. Goodfellow, I.J., et al.: Generative adversarial nets. In: Proceedings of the 27th International Conference on Neural Information Processing Systems, Vol. 2, pp. 2672–2680 NIPS 2014. MIT Press, Cambridge, MA, USA (2014)
7. Hong, Y., Hwang, U., Yoo, J., Yoon, S.: How generative adversarial networks and their variants work. ACM Comput. Surv. **52**(1), 1–43 (2020). https://doi.org/10.1145/3301282
8. Karras, T., Aila, T., Laine, S., Lehtinen, J.: Progressive growing of GANs for improved quality, stability, and variation. CoRR arXiv: abs/1710.10196 (2017)
9. Krawczyk, B.: Learning from imbalanced data: open challenges and future directions. Prog. Artif. Intell. **5**(4), 221–232 (2016)
10. Langr, J., Bok, V.: GANs in Action: deep learning with generative adversarial networks. Manning (2019). https://www.books.google.pl/books?id=HojvugEACAAJ
11. Mirza, M., Osindero, S.: Conditional generative adversarial nets (2014). http://arxiv.org/abs/1411.1784
12. Rahman, T.: COVID-19 radiography database. https://www.kaggle.com/tawsifurrahman/covid19-radiography-database (2021)
13. Sharma, S.: Activation functions in neural networks. https://www.towardsdatascience.com/activation-functions-neural-networks-1cbd9f8d91d6 (2021)
14. Shorten, C., Khoshgoftaar, T.M.: A survey on image data augmentation for deep learning. J. Big Data **6**, 60 (2019). https://doi.org/10.1186/s40537-019-0197-0
15. Simonyan, K., Zisserman, A.: Very deep convolutional networks for large-scale image recognition (2014). http://arxiv.org/abs/1409.1556
16. Szczepaniak, P.S., Tadeusiewicz, R.: The role of artificial intelligence, knowledge and wisdom in automatic image understanding. J. Appl. Comput. Sci. **18**(1), 75–85 (2010). https://www.it.p.lodz.pl/file.php/12/2010-1/jacs-1-2010-Szczepaniak-Tadeusiewicz.pdf
17. Team, K.: Keras documentation: the functional API. https://www.keras.io/guides/functional_api/ (2021)
18. Wang, X., Kodirov, E., Hua, Y., Robertson, N.: Instance cross entropy for deep metric learning (2019). http://arxiv.org/abs/1911.09976
19. Wasilewski, P., Mruk, B., Mazur, S., Półtorak-Szymczak, G., Sklinda, K., Walecki, J.: COVID-19 severity scoring systems in radiological imaging a review. Pol. J. Radiol. **85**(1), 361–368 (2020). https://doi.org/10.5114/pjr.2020.98009
20. Yang, Q., Zhang, Y., Dai, W., Pan, S.J.: Transfer learning. Cambridge University Press, Cambridge (2020). https://doi.org/10.1017/9781139061773
21. Zhuang, F., et al.: A comprehensive survey on transfer learning. In: Proceedings of the IEEE, pp. 1–34 (2020). https://doi.org/10.1109/JPROC.2020.3004555

SDG-Meter: A Deep Learning Based Tool for Automatic Text Classification of the Sustainable Development Goals

Jade Eva Guisiano[1,3(✉)], Raja Chiky[2], and Jonathas De Mello[3]

[1] LISITE - ISEP, Paris, France
Jade.guisiano@etu.sorbonne-universite.fr
[2] 3DS OUTSCALE, Paris, France
[3] United Nations Environment Program, Paris, France

Abstract. The 17 Sustainable Development Goals (SDGs) are a "shared blueprint for peace and prosperity for people and the planet, now and into the future". Since 2015, they help pointing out pathways to solve interlinked challenges being faced globally. The monitoring of SDGs is essential to assess progress and obstacles to realise such shared goals. Streams of SDG-related documents produced by governments, academia, private and public entities are assessed by United Nations teams to measure such progress according to each SDG, requiring labelling to proceed to more in-depth analyses. Such laborious task is usually done by the experts, and rely on personal knowledge of the links between the documents contents and the SDGs. While UNEP has experts in many fields, links to the SDGs that are outside their expertise may be overlooked. In this context, we propose to solve this problem with a multi-label classification of texts using Bidirectional Encoder Representations from Transformers (BERT). Based on this method, we designed the SDG-Meter, an online tool able to indicate to the user in a fully automatic way the SDGs linked to their input text but also to quantify the degree of membership of these SDGs.

Keywords: NLP · Deep learning · Sustainable development goals

1 Introduction

The 2030 Agenda for Sustainable Development, adopted by all United Nations Member States in 2015, provides a "shared blueprint for peace and prosperity for people and the planet, now and into the future". At its heart are the 17 Sustainable Development Goals (SDGs), which are an urgent call for action by all countries - developed and developing. They recognize that ending poverty and other deprivations must go hand-in-hand with strategies that improve health and education, reduce inequality, and spur economic growth —all while tackling climate change and working to preserve our oceans and forests. The UN Secretary

© The Author(s), under exclusive license to Springer Nature Switzerland AG 2022
N. T. Nguyen et al. (Eds.): ACIIDS 2022, LNAI 13757, pp. 259–271, 2022.
https://doi.org/10.1007/978-3-031-21743-2_21

General presents every year an Annual SDG progress report, aiming to identify challenges and achievements from different countries and regions.

As part of the work of the United Nations Environment Programme (UNEP), multiple documents (policy recommendations, toolkits, reports, project submissions, progress reports, etc.) are analysed by SDG experts for properly indexing them according to SDGs. Experts have to read the document and identify which SDGs are mentioned, or are related to it. These mapping exercises are time-consuming and subjective to personal knowledge of the links between the document contents and the SDGs. While UNEP has experts in many fields, links to the SDGs that are outside their expertise may be overlooked. This work, based on the subjectivity of each expert, introduces a bias in the overall mapping work, which therefore does not allow for a comparison of the associations made for each text. This association work (labeling) based on text analysis therefore requires increased knowledge of the terminology of each SDG to achieve an optimal analysis. The improvement of the whole text labeling process consists firstly in adopting automation of this process (time saving for the experts), and secondly, in designing a single intelligence with a solid knowledge of the terminologies of each SDG (bias reduction).

Aiming to address these issues, we propose the online tool SDG-Meter which allows to automatically classify a text—without expert intervention—according to the 17 SDGs but also to quantify each link between the SDGs and a text. The SDG-Meter is based on the use of a powerful deep learning algorithm which is currently one of the most powerful and efficient algorithms in the field of Natural Language Processing. Following a series of tests, it was confirmed that our tool is able to perform a classification identical to that of an expert.

This paper starts by presenting the state of the art of multi-label classification methods applied to SDGs, and detailing the functioning of the BERT (Bidirectional Encoder Representations from Transformers) algorithm which is at the heart of our method. The functioning of the SDG-Meter via the implementation of the BERT algorithm in a web solution will also be presented. The paper will then explain the method used to build our training database. Finally, we present the results of our SDG-Meter in comparison with an expert manual classification on 5 different texts.

2   State-of-the-Art

Various methods have recently been developed with the objective of classifying text according to the 17 SDGs. (i.e. multi-label classification task).

The Organisation for Economic Co-operation and Development (OECD) has developed several methods for classifying texts according to the SDGs. For example, in order to assess the volume and alignment of financial flows with the SDGs, OECD has developed the online tool SDG tracker [14][1]. This tool links the project description of financial contributions from bilateral providers or multilateral organizations to the SDGs they address. Their approach is based on a

---

[1] https://sdg-tracker.org.

combination of the XGBoost (Boosted Gradiant) [3] and ULMFiT [7] (Universal Language Model Fine-Tuning for Text Classification) algorithms followed by a manual verification by experts for each output. This tool by tracking global progress towards the SDGs, permits to the users to explore progress on all of the SDG indicators for which data is available at the global, regional and country-level. OECD has also recently proposed the Sustainable Development Goals Business Actions Identifier (SDG-BAI) [12] allowing to identify, now automatically, the contributions of companies to the SDGs. The latter is composed of 17 XGBoost sub-algorithms, each of which provides a binary classification of whether the text is related to a specific SDG. Finally, the SDG Pathfinder[2] online tool still from OECD combine automatic and manual approaches for associating OECD and its partners texts with SDGs which rely on the use of UNSILO Classify[3] commercial algorithms which permits to extract automatically all the concepts from text and then compare these concept to each SDG terminology. However, in this case, the goal is not to predict a specific label, but rather to define clusters in the data. The features of these clusters are then manually mapped to labels (form of ontology). This method thus provides a comprehensive list of topics related to the SDGs and a model to detect topics in new texts and thus link new documents to the SDGs. This tool explores the key themes underlying the SDGs and the links between them, using analysis and data from six international organizations.

On the side of the United Nations agencies, various and similar methods have also been developed. The online tool Open-Source Approach to Classify Text Data by UN Sustainable Development Goals (OSDG)[4] [15], developed by the UNDP SDG AI Lab, aims to integrate various existing attempts to classify research according to Sustainable Development Goals based on the principles of ontology and domain recognition. The input texts of the OSDG tool are labeled with fields of study defined by the Microsoft Academic Graph tool and are cross-referenced with an evolving ontology to determine the degree of correlation between the two. And finally, via the determination of a specific threshold the numerical value of the relationship between the SDG and the text is interpreted as strong or moderate. This tool allows users to detect SDGs mentioned in the texts of their choice either by entering the DOI of an article or by manual entry in a dedicated insert. There is also the online tool. LinkedSDG[5] [8], developed by United Nations Department of Economic and Social Affairs (UN DESA), provides a tool to link a text to the SDGs it addresses and to establish the degree to which the text is linked to the target of each SDG. This method is based on the use of an ontology and the extraction of concepts from the input texts to classify them according to the distribution of concepts. This tool permits to users to detect the SDGs present in an input text (by entering a text link or by manual entry) but also to detect the precise targets of each SDG. Finally,

---

[2] https://sdg-pathfinder.org.
[3] https://unsilo.ai.
[4] https://osdg.ai.
[5] https://linkedsdg.officialstatistics.org/.

based on the same will as ours to be able to provide a method [10] capable of classifying document flows received by the United Nations, the UN Department Of Economic And Social Affairs (UNDESA) proposes an unsupervised classification method based on the Latent Dirichlet Allocation (LDA) [1] algorithm. The LDA automatically labels texts by statistically identifying logical subgroups in the data, i.e. classes/SDGs, each related to specific word lists. These results in a statistical model that can be used to classify another data set. For a given text the method is also able to assign a similarity score between the words of a text and the words related to each SDGs previously defined by the LDA.

There are also similar others methods developed independently of official organizations like Unitez [9] which has been designed to evaluate how SDGs are represented in Austrian research. The method is based on a manual selection of keywords list for each SDG, then a search algorithm is applied to scientific texts to retrieve one or more keywords linked to the SDGs. More related to the use of deep learning algorithms, a study [11] proposes SDG-based text classification by training the BERT classification algorithm on a Japanese corpus to identify SDGs related to input sentences and vectorize the semantics of the input sentences. Another method [16], has the goal to detect UN General Assembly (UNGA) resolutions implicitly related to SDGs. The authors propose a method to apply deep learning methods for text classification with the constraint of a small labelled dataset. They built an ensemble method that effectively combines generic (non domain-specific) deep learning based document obtained from Global Vectors for Word Representation (GloVe) [13] and Universal Sentence Encoder (USE) [2] with domain-specific Term Frequency–Inverse Document Frequency (TF-IDF) document similarities.

Although these methods are efficient in term of classification, they have some limitations such as the unsupervised methods always require time-consuming intervention of experts to map detected features to each SDG. Moreover, these methods are generally based on keyword analysis and do not take into account the context of the texts. Also in most of the case, tool and methods for SDG classification no not provide quantification of the link between a text and an SDG, which is useful to establish a more precise classification.

For these reasons, we decided to carry out additional research in order to build a method that would allow us to link a text to one or more SDGs that it deals with and also to quantify its degree of belonging to these different SDGs. Our tool, called SDG-Meter, is also based on the will to exclude any form of human intervention during the process of associating SDGs to texts (even for the training step). In addition to quantifying the text-SDGs relationships and saving time for our experts, our method will also allow external users to link their project/thoughts to the large fields of action by learning the SDGs that their project deals with and thus give them the possibility to better fit their project into the framework of the SDGs in question.

3   Multi-labeled Text Classification with BERT

As our objective is to link and also quantify the degree to which a text belongs to all 17 SDGs, a multi-labeled Text classification algorithm (supervised learning) was selected, which makes it possible to link a text not only to one label (SDG) but to several labels. More specifically, the recent and powerful algorithm Bidirectional Encoder Representations from Transformers (BERT) [4], which is a neural network-based technique for natural language processing was chosen.

3.1   BERT: Bidirectional Encoder Representations from Transformers

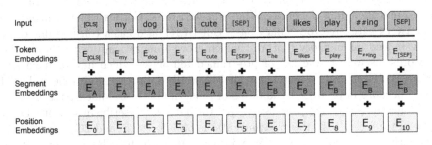

Fig. 1. BERT input representation [4]

BERT for "Bidirectional Encoder Representations from Transformers" is a neural network architecture developed by Google research teams. It can perform various NLP tasks such as text classification, translation, summarization, and question answering. BERT is a transformer-based architecture is a deep learning model based on the mechanism of self-attention, deferentially weighting the significance of each part of the input data which thus provides context for any position in the input sequence. It works by performing a small and constant number of steps. In each of them, it applies the attention mechanism to understand relationships between all words in a sentence. BERT architecture (BERT Base model in our case) consists of 12 Transformer encoders stacked together, 12 attention heads and 110 million parameters. Each Transformer encoder contains two sub-layers: a self-attention layer and a feed-forward layer.

As shown in the Fig. 1 The input to the encoder for BERT is a sequence of tokens, which are first converted into vectors and then processed in the neural network. But before processing, BERT needs as input a sequence of tokens, for that 3 elements have to be added to the initial input:

– Token embeddings: A [CLS] token is added to the input word tokens at the beginning of the first sentence and a [SEP] token is inserted at the end of each sentence.

- Segment embeddings: A marker indicating Sentence A or Sentence B is added to each token. This allows the encoder to distinguish between sentences.
- Positional embeddings: A positional embedding is added to each token to indicate its position in the sentence.

To solve NLP tasks, BERT perform a 2-steps process, starting by training a language model on a large unlabelled text corpus (unsupervised or semi-supervised). Then it Fine-tune this large model to specific NLP tasks (multi-label classification in our case) to utilize the large repository of knowledge this model has gained (supervised). BERT has two principal advantages compared to others NLP models. First, it is a "deeply bidirectional" model which means that BERT learns information from both the left and the right side of a token's context during the training phase. This bidirectionality is important for truly understanding the meaning of a text. Secondly, BERT is pre-trained on a huge corpus of unlabelled text including the entire Wikipedia and Book Corpus which confers with him a very extensive vocabulary base. But BERT also has a considerable drawback, indeed his memory footprint which grows quadratically with the number of words, as well as the use of pre-trained templates which have a fixed size. Then, BERT accepts a maximum of 512 words (approximately one full PDF page) because beyond that the limits of modern GPUs are quickly reached.

Concerning the data processing part, the dataset (Describe in section4.1) was split in train and test, the splitting ratio is 80% for train and 20% for the test. Then, the data need to be transform into a format which BERT can understand. For that the class "InputExample's" was create using the constructor provided in the BERT library with variable "text_a" containing the text to classify and "label" is the labels corresponding to our texts. In a second step, the text data was pre-process so that it matches the BERT's requirements (text in lower case, tokenization, etc.) Then, the BERT module was loaded and a single new layer was created that will be trained to adapt BERT to our multi-label classification task. The training phase has the goal to train the model to learn the main characteristics (words and their context) of the texts contained in our 17 classes (17 SDGs). Then the model keeps its training parameters to be applied to texts where their class is not given to the algorithm (But they was keep apart.), the algorithm with its parameters must be able to classify them correctly and once the classification is established, the class to which the algorithm classifies the text was compare to the one to which it is really assigned. Finally, to quantify the belonging of a text to a SDG, the prediction output was set in probability thanks to the use of a Logit function. Dropout was also included to prevent overfitting[6]. And finally the loss was computed between predicted and real label which permits to assess the accuracy of our model which has an accuracy of 98% (based on 400 tests).

---

[6] Overfitting occurs when the algorithm over-learns (overfit.)in other words, when it learns from data but also from patterns (diagrams, structures) which are not related to the problem, such as noise, thus degrading the performance of the algorithm.

3.2 SDG-Meter Tool

Once the BERT was trained on our database (using a 32GB GPU), the trained model and its parameters was saved so that it could be reapply for the classification of new texts. Then it was implemented in a web based tool that allow end users to achieve interaction through a website as shown in the Fig. 2. The library PyTorch[7] has been used for implementing the model. Pytorch is suitable for saving the model in python and deploying it on places such as websites and mobile applications.

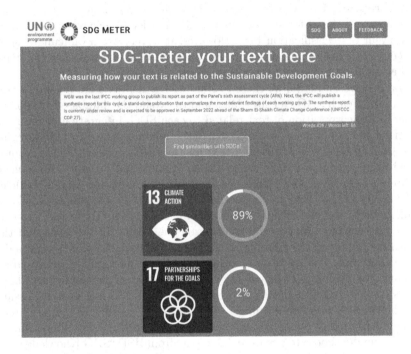

Fig. 2. SDG-Meter web interface

The used model is Fast-Bert[8] wich is based on Pytorch and consists in an encapsulated version of original BERT algorithm[9]. This library supports BERT and XLNet for both multi-Class and multi-Label text classification which is the main purpose of our approach. The front-end is written in HTML/CSS and for the back-end website, Flask[10] was used as a lightweight webpage to connect with our application based on python. The code can be found on the UNEP's

---

[7] https://pytorch.org/.

[8] https://pypi.org/project/fast-bert/.

[9] The original version of BERT is no longer available for the moment because its improved version "SMITH" is under development.

[10] https://flask.palletsprojects.com/.

Git repository[11]. As shown in the figure, users can enter the text of their choice in the dedicated box (with a strict maximum limit of 512 words) and in less than a second the list of detected SDGs appears in a decreasing order according to their probability of belonging to the text (expressed in %).

4   Experimentation

4.1   Dataset

Web-Scrapping Process. Deep learning methods often require a large dataset for training (of the order of a few thousand) in order to achieve a good level of accuracy in the classification task. However, it is very often difficult to obtain a large set of labeled data and difficult to mobilize an expertise for the manual annotation of so much data (very time-consuming task).

In a previous study [6] we chose to start with a 169 texts database describing the targets of each SDG (ref site/data) and to extend the database to 6000 texts via manual incorporation of text related to each SDG, integration of the definition of synonyms of the keywords of each SDG and finally the use of Markov chain (NLG) for automatic generation of text for the themes of each SDG. Despite a classification accuracy of 94% we found the quality and diversity of the generated texts to be very poor. Moreover, by generating texts for each SDG theme, the majority of the texts in our database were linked to a single SDG, which is problematic in the context of a multi-label classification. To do this, we chose to re-construct our database by focusing on the quality of the texts and their labels rather than on the quantity of the latter. We have therefore sought a data source containing quality texts associated to one or more SDGs (multi-labeled dataset). We also impose the condition that each text of the data source has to be associated to SDG manually by experts but not any other form of classification algorithm in order to not introduce additional bias that we cannot directly manage. After many researches and collaborations with the One Planet linguist team, we finally choose to exploit data coming from IISD "SDG Knowledge Hub" website[12]. Indeed, this website has all the required criteria; it provides users with labelled articles such as "News", "Guest Article" and "Policy Brief" so three different writing styles (important for the diversification of vocabulary and writing style). In order to exploit these data, we had to extract the texts and the labels of each article, for that we had to resort to the techniques known as "Webscrapping". This process allows, via the HTML encoding of the web site, to identify the contents of the texts and labels in order to extract them. We included this process in an algorithm written in Python allowing us to carry out this operation for each category of text, for each web page and for each article. We also add the condition that each article must have less than 512 words to be extracted (explanation of this limitation in the section below).

---

[11] https://github.com/UNEP-Economy-Division/SDG-Meter.
[12] https://sdg.iisd.org/.

Data Set Description thanks to the web-scrapping process, we could extract 724 labeled texts from "News", "Guest Article" and "Policy Brief" format. The texts in our database have an average length of 374 words and have an average of 4 labels(SDGs) with a minimum of one label and a maximum of 17 labels.

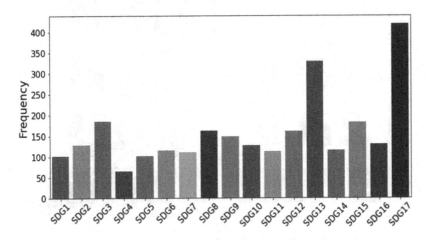

Fig. 3. Frequency of SDGs in the database

It can be noticed on the Fig. 3 that the distribution of the SDGs, thus of our classes, is unbalanced. Indeed, the texts extracted from IISD were written and labeled during the years 2020–2021, the main period of the COVID epidemic. The distribution of SDGs clearly shows a peak for SDG 3 "Good Health and Well Being". During these period various themes emerged in the run-up to Cop26, all of which are mainly related to SDG 13 "Climate Action", which is reflected in the distribution of classes. The largest peak is attributed to SDG 17 "Partnership for the Goals", which is explained by the fact that SDG groups a large number of sub-themes that are almost automatically associated with all texts. In contrast, it can be note that for this period and the selected texts, SDG 3 "Quality Education" is under-represented.

4.2   Test and Results

In order to test our SDG-Meter, new texts labeled by experts from IISD-SDG was selected. These have 1 to 4 labels including also under-represented labels in the training data (e.g. SDG 4 and 7) so that it can be test if despite their under-representation the SDG-Meter manages to identify them. It should be specify that the labels coming from IISD do not have an order according to their relation with the text, the order is numerical according to the SDG numbers. As for the SDG-Meter results, these indicate the probability of an SDG belonging to the text expressed as a percentage. Each class (SDG) is independent, so the sum of the probabilities for different SDGs can be different from 1.

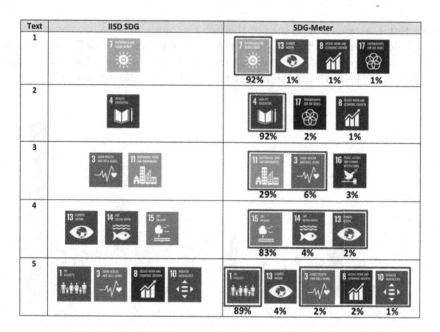

Fig. 4. SDG-Meter & IISD-SDG true labels comparisons

Text 1 : "Brief Suggests Five Key Actions to Reach Africa's Wind Energy Potential" This first text only related to SDG 7 "Affordable and Clean Energy" is mentioned as being based on the following keywords: Energy, Wind. **SDG-Meter:** it can be see that the SDG-Meter identifies SDG 1 in the first position SDG 7 with a probability of belonging to the text of 92% which therefore establishes a strong relationship with the text. It can also be see appearing in weak relation with the text, 1% of probability of belonging of SDG 13, 8 and 17. The SDG-Meter recognizes and assigns a high probability to SDG 7 Fig. 4.

Text 2 : "UN Embarks on Preparations for Education Summit" The second text, which is uniquely related to SDG 4 "Quality Education", addresses the following areas: Means of Implementation, Capacity Building & Education. **SDG-Meter:** SDG 4 has a probability of belonging to the 92% text, so a strong relationship is established with the former. SDG 17 "Partnerships for the Goals" is also detected with a probability of 2% and finally SDG 8 with a probability of 1%. According to the SDG-Meter, the text deals mostly with SDG 4.

Text 3 : "UNGA Event Highlights Potential Sources of Financing for Road Safety" The third text is related to SDG 3 "Good Health and Well-Being" and SDG 11 "Sustainable Cities and Communities" deals with the following areas: Health, Transport.

SDG-Meter: SDGs 11 and 3 are detected with respective membership percentages of 29% and 6%. A relationship is also found with SDG 16 "Peace, Justice and Strong Institutions" with a probability of 3.

Text 4 : "Environmental Leaders Selected for Contributions to Ecosystem Restoration" The third text is related to SDG 13 "Climate Action", SDG 14 "Life Below Water" and SDG 15 "Life on Land" and covers the following areas: Atmosphere, Biodiversity, Energy, Health, Conservation, Ecosystem Approach, Oceans & Coasts, Stakeholder Participation, Climate Change. **SDG-Meter:** In this case, the 3 SDGs mentioned by the IISD are all found by the SDG-Meter without additional SDGs. SDG 15 appears in first position with a probability of belonging to 83%, then SDG 14 with a probability of 4% and finally SDG 13 with a probability of 2%.

Text 5: "Environmental Leaders Selected for Contributions to Ecosystem Restoration" The third script is related to SDG 1 "No Poverty", SDG 3 "Good Health and Well-Being", SDG 8 "Decent Work and Economic Growth" and SDG 10 "Reduce Inequalities" and covers the following areas: Health, Poverty Eradication, Economics & Investment **SDG-Meter:** SDG 1 is detected with a high 89% probability of belonging to the text, SDGs 3 and 8 are detected with a 2% probability, and SDG 10 in last place with a 1% probability. The SDG 13 is also detected with a probability of 4%.

These results show us that the SDG-meter is able to provide a text analysis very similar to that of an expert. Indeed, we have seen that the SDG-Meter manages to find all the SDGs selected by the experts for each text. In the case of text 4, the results are identical. For the other texts, the SDG-Meter in addition to the SDGs indicated by the IISD finds by itself other SDGs related to the text that the IISD does not mention. In this case, it is difficult to know on which side the right answer lies, it is also possible that the SDG-meter via its deep learning is able to detect SDGs that the experts would not have noticed. As for example in text 4, the SDG 13 is detected with a probability of 4% which represents a significant percentage. On the other hand, it is possible to question the significance threshold for texts 1 and 2, which mention SDGs with probability percentages of around 1%. Text 5 also presents a threshold of 1% but for SDG 10 which is common between the IISD and SDG-Meter analyses and which therefore remains significant.

5 Conclusion

Within the various UN entities, experts regularly spend significant time classifying documents according to the 17 SDGs. In addition to being repetitive and time consuming, this approach introduces a large bias as the view from one expert to another varies, meaning that for the same text two experts may assign very different SDGs. Aiming to unify this classification system but also to automate this task, we propose the online tool SDG-meter. The latter is based on the use of the BERT deep learning algorithm which is able to provide excellent results (accuracy of 98%) with a limited amount of training data. Indeed, we have seen that for various texts our method is able to find the same SDGs as an expert, which means that this type of solution could in the future completely replace the current manual work of experts.

However, it is possible to contribute to its improvement on various points such as the improvement of training data sets and BERT itself. Indeed, as shown in Fig. 3 our initial dataset can be qualified as imbalanced because some SDGs are less represented than others in our dataset. This can lead to low predictive accuracy for the infrequent class tending to decrease the overall accuracy of the model. Increasing our dataset could therefore contribute to smooth the differences between the representations of each SDG and thus obtain a better accuracy. It would also be preferable, in the context of increasing our dataset, to obtain labeled texts from various sources in order to enrich the vocabulary related to the SDGs, but also to train our model on the various semantic structures. (research articles, official reports, news, etc.) Finally, via our database, the BERT model has been able to learn to recognize SDGs through the specificity of each text, but it does not know what an SDG is not. Indeed, if a text not related to the SDGs is introduce, the SDG Meter will still try to make the link. This leads to a noise during classification which is manifested by the presence of a very small percentage of SDGs not linked to a text. To remedy this problem, it would be necessary to include in the training database various texts that are certainly not related to the SDGs and to stipulate their non-membership in the SDGs by assigning a value of 0 to each class (SDG). Another important aspect concerns the word limit of the input texts imposed by BERT. Indeed, this limit allows for the moment to analyze only the equivalent of a full PDF page. The advanced version of BERT named "SMITH" is currently being designed and should allow to extend this limit to 2048 words. However, it would be better to avoid any word limit. Some methods [17] permit to avoid this limit as for example those based on the principle of sliding window but these last ones are known as being less powerful because they have insufficient long-range attentions. Another method, COgnize Long TeXts (CogLTX) [5], allows to overcome the BERT limit by identifying key sentences by training a judge model, concatenates them for reasoning, and enables multi-step reasoning. This last solution presents excellent results and seems to be a way to improve BERT.

References

1. Blei, D.M., Ng, A.Y., Jordan, M.I.: Latent Dirichlet allocation. J. Mach. Learn. Res. **3**, 993–1022 (2003)
2. Cer, D., et al.: Universal sentence encoder for english, pp. 169–174 (2018). https://aclanthology.org/D18-2029, https://doi.org/10.18653/v1/D18-2029
3. Chen, T., Guestrin, C.: XGBoost. In: Proceedings of the 22nd ACM SIGKDD International Conference on Knowledge Discovery and Data Mining (2016). https://doi.org/10.1145/2939672.2939785

4. Devlin, J., Chang, M.W., Lee, K., Toutanova, K.: BERT: pre-training of deep bidirectional transformers for language understanding. In: NAACL (2019)
5. Ding, M., Zhou, C., Yang, H., Tang, J.: CogLTX: Applying BERT to long texts. In: NeurIPS (2020)
6. Guisiano, J., Chiky, R.: Automatic classification of multilabel texts related to sustainable development goals (SDGs). In: TECHENV EGC2021. Montpellier, France (2021). https://hal.archives-ouvertes.fr/hal-03154261
7. Howard, J., Ruder, S.: Universal language model fine-tuning for text classification (2018)
8. Joshi, A.: A knowledge organization system for the united nations sustainable development goals. In: Verborgh, R., et al. (eds.) ESWC 2021. LNCS, vol. 12731, pp. 548–564. Springer, Cham (2021). https://doi.org/10.1007/978-3-030-77385-4_33
9. Körfgen, A.: It's a hit! mapping Austrian research contributions to the sustainable development goals. Sustainability 10, 3295 (2018)
10. LaFleur, M.: Art is long, life is short: an SDG classification system for DESA publications (2019). https://doi.org/10.2139/ssrn.3400135
11. Matsui, T., et al.: A natural language processing model for supporting sustainable development goals: translating semantics, visualizing nexus, and connecting stakeholders. Sustain. Sci (2022). https://doi.org/10.1007/s11625-022-01093-3
12. OCDE: Industrial policy for the sustainable development goals (2021). https://www.oecd-ilibrary.org/content/publication/2cad899f-en, https://doi.org/10.1787/2cad899f-en
13. Pennington, J., Socher, R., Manning, C.: GloVe: global vectors for word representation. In: EMNLP vol. 14, pp. 1532–1543 (2014). https://doi.org/10.3115/v1/D14-1162
14. Pincet, A., Okabe, S., Pawelczyk, M.: Linking aid to the sustainable development goals –a machine learning approach. In: OECD Development Co-operation Working Papers, vol. 52 (2019)
15. Pukelis, L., Puig, N., Srynik, M., Stanciauskas, V.: OSDG –open-source approach to classify text data by un sustainable development goals (SDGs) (2020)
16. Sovrano, F., Palmirani, M., Vitali, F.: Deep learning based multi-label text classification of UNGA resolutions. CoRR abs/2004.03455 (2020). https://arxiv.org/abs/2004.03455
17. Wang, Z., Ng, P., Ma, X., Nallapati, R., Xiang, B.: Multi-passage BERT: a globally normalized BERT model for open-domain question answering. In: Proceedings of the 2019 Conference on Empirical Methods in Natural Language Processing and the 9th International Joint Conference on Natural Language Processing (EMNLP-IJCNLP), pp. 5878–5882. Association for Computational Linguistics, Hong Kong, China (2019). https://aclanthology.org/D19-1599, https://doi.org/10.18653/v1/D19-1599

The Combination of Background Subtraction and Convolutional Neural Network for Product Recognition

Tin-Trung Thai[1,2], Synh Viet-Uyen Ha[1,2(✉)], Thong Duy-Minh Nguyen[1,2], Huy Dinh-Anh Le[1,2], Nhat Minh Chung[1,2], Quang Qui-Vinh Nguyen[1,2], and Vuong Ai-Nguyen[1,2]

[1] School of Computer Science and Engineering, International University, Ho Chi Minh City, Vietnam
[2] Vietnam National University, Ho Chi Minh City, Vietnam
hvusynh@hcmiu.edu.vn

Abstract. Multi-class retail product recognition is an important Computer Vision application for the retail industry. Track 4 of the AICITY challenge is introduced for the retail industry. This track focuses on the accuracy and efficiency of the automatic checkout process. However, due to the lack of training data for retail items in the real world, a synthetic data set is usually generated based on the 3d scanned items to produce training data for an automated checkout system. To overcome the difference informative representative appearance between training data and the real-world scenario in the test set provided by the AICITY organizer, our research focuses on analyzing and recognizing retail items by combining the traditional method and state-of-the-art Convolutional Neural Network (CNN) approach. This paper presents our proposed system for product counting and recognition for automated retail checkout. Our proposed method is ranked top 8 in the experimental evaluation in the 2022 AI City challenge Track-4 with an F1-score 0.4082.

Keywords: Computer vision · Background subtraction · Convolutional neural network · Product recognition

1 Introduction

Retail product recognition intends to advance product management. This recognition problem is vital in a lot of Computer Vision applications in real-world supermarkets, Convenience stores, retail execution management, and customer behavior observation. At the present, Barcode [1] is now universally applied technology in industries where automatic commodity identification is used. By scanning the barcode which is printed on each product, the management of every item can be easily simplified. However, it often takes time to manually locate the barcode position and assist the machine to identify the barcode on each item at the checkout counter. As retailing is enlarged at a fast pace, enterprises are

N. T. Nguyen et al. (Eds.): ACIIDS 2022, LNAI 13757, pp. 272–284, 2022.
https://doi.org/10.1007/978-3-031-21743-2_22

now intensively focusing on artificial intelligence (AI) applications to reshape the retail industry with online and offline experiences. With the support of many kinds of electronic devices for video recording is rapidly growing day by day. How to analyze and identify these items, which classify them, has become a key issue in the product recognition field. Product recognition, which is mainly based on computer vision, is used to replace the manual process of identifying and classifying products. The automated retail checkout is applied to localize and classify the items when a customer performs a checkout action by scanning items in front of the counter. Therefore, this system requires the correct identification of retail items with reliable accuracy. The AICITY Challenge is a competition that focuses on efficient systems in many tasks and pushes these technologies closer to practical integration. Our research work deals with the Track-4 introduced by the 6th AICITY Workshop: Multi-Class Product Counting & Recognition for Automated Retail Checkout. In our work, we research and analysis a framework that can both efficient performance and accuracy. We come up with combining traditional method for speed processing and deep neural net for accuracy. Our contributions to this paper are described as follows:

Firstly, detect the proposal region using the Background Subtraction model and Skin removal.

Secondly, classify each proposal region by the Inception-ResNet-v2 model.

Finally, track and count objects by simple algorithms.

The rest of the paper is structured as follows. Section 2 describes the related work. In Sect. 3, our proposed system for Multi-class product counting and Recognition for Automated Retail Checkout. Section 4 is our experiment and evaluation. Finally, the Conclusions are in Sect. 5.

2   Related Work

Object Detection : Object detection is an important key in object behavior analysis, where the detection module can support many tasks: object tracking, object counting, object recognition, etc. In the case of retail, there are many existing methods. The traditional method features like SIFT [2] and Harris corner [3] are used for detecting and recognizing retail commodities. In recent years, the research community has shifted attention to the deep learning [4] approach for object detection tasks. Convolutional Neural Network (CNN) has been proven to generate the best result for many image classification datasets. Two-stage detection methods [5–7] have demonstrated the effectiveness of CNN for accurately recognizing image objects, however, trade-off time execution for accuracy on high-end machines. Compared with the two-stage detection method, the one-stage models like MobileNet-SSD [8,9], YOLOv4-Tiny [10], and EfficientDet [11] require less computational complexity while preserving accuracy. Recently, [12] proposed a novel framework "Swin Transformer" for object detection. Even though the method achieves state-of-the-art result, it requires super hardware devices for training and testing.

Multi-object Tracking : Multi-Object Tracking (MOT) is one of the most widely investigated tasks as it is important in many computer vision applications. The typical motif of MOT problems is Tracking-by-detection (TBD) [13,14]. In TBD, object detector [6,7] is first applied for box detections , then a re-ID model [15,16] is applied to extract features. For tracklets, a motion model like the Kalman filter is used to compute appearance and motion affinities. However, TBD methods have 2 separated models detection and tracking, therefore they are usually computationally expensive.

Object Counting : Object counting is now more important because of its wide range of real world applications such as crowd surveillance, traffic monitoring, retail product management, etc. Two presentative approaches for object counting are Detection base method [17] and regression based methods [18,19]. The detection based methods count the number of objects by an object detector, while the regression based methods count objects by a density map in which represents the density of target objects at each corresponding location. However, most of these methods rely on a large training data set to train a counting model.

3   Methodology

Fig. 1. Some of the cases in the original training data that the images are missing data, dark, or unreal.

In track 4 of the AICITY challenge, the organizer provides a data set of 116,500 synthetic images which are created from a 3D scanned model and does not include any hand-labeled data. Each training image contains only one item under various environments. Even though the dataset contains more than 110,000 images, most of them are fake images that are generated from a synthetic system based on the 3d scan of items. Most of these images contain noise, are dark and small size, or even have different informative representative, this lead to bias and overfitting in the training model. Figure 1 shows some of the cases in the original training data where the images are missing data, dark, or unreal. The organizer also provides a test set video for testing. However, due to the

Fig. 2. The gap domain between the synthetic training data set and test set video. A-B are the synthetic images of item class 103 in training set and B-C is the class 103 in test set video.

gap domain between the test set and training data, it is hard to train an object detector based on the training data set. Figure 2 shows the different appearance of item 103 in the training data set and test set. To handle the gap domain between the training set and test set, we try to simulate items in the background like in the test set scenario. We put different items in the simulated background that is cropped from the test video and reproduce the training data set. We first train the object segmentation model based on the mmdetection framework [27]. Although achieving an accuracy of more than 0.9 in training, it detects and classifies items incorrectly in the test set scenarios. Figure 3 demonstrates the reproduced training set and the result of the object segmentation model with a test set video. We then observe that we need to narrow down the different appearances between the objects in the test set video and synthetic images in training data to achieve a better result. To scope this problem, it is necessary to crop the item regions in each frame and classify the category of each region. Firstly, we segment potential regions that may contain the item in each frame by using the background subtraction model. After that, a CNN-based classification is applied to each potential region to check whether there is an item in it and which category the item belongs to.

Fig. 3. The reproduce training data set and result of segmentation model with test set. A: The reproduce data set, B the result of segmentation model.

Our proposed method combines a traditional computer vision algorithm: background subtraction and a CNN-based classification for a Multi-Class product for Automated Retail Checkout. The system consists of 3 modules (As shown in Fig. 4): Background Subtraction model and Skin removal, Product Classification, Product tracking, and Counting. Firstly, the Background Subtraction model is applied to identify moving objects in each frame, followed by a skin removal which aims to detect the hand region and segment Proposal Regions for object classification. A CNN model is assigned to classify each segmented region and which class it belongs to. The final step is tracking and counting objects by calculating the distance between the bounding box of each identified object in each consecutive frame.

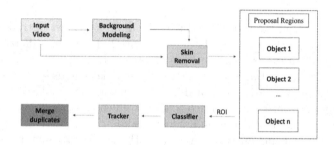

Fig. 4. The proposed framework.

3.1   Background Subtraction and Skin Removal

Background Subtraction. Background subtraction (BGS) is one of the most common ways to analyze representative information in video from static cameras. The background subtraction approach aims to isolate the moving object from a scene by generating a foreground mask. Processing a video stream to segment moving objects is an important step in many applications. To dynamically the background model, we adopted the K-nearest neighbors (KNN) background subtraction [21]. Hence, the KNN background model achieves well results for the videos that the background changes rapidly like in the test A video set. Moreover, the KNN algorithm is also efficient in computation and real-time processing - more than 40 fps. Applying KNN algorithms, our proposed system can work in every low-cost computer without requiring high computation. The KNN algorithm stores all available cases and classify the new samples by calculating the similarity. In each video, the training samples are represented as the n-dimensional feature vector. When an unknown sample needs to be classified whether it is a background or foreground, KNN search is applied to find the K nearest samples based on the Euclidean distance that is closest to the unknown sample. The density estimation formula is:

$$[t]p\left(x \mid x_i\right) \approx \frac{1}{TV} \sum_{m=1}^{N} b^m K\left(\frac{\|x_i - x\|}{D}\right) \tag{1}$$

where T is a reasonable time adaptation. The density function [21] is estimated by counting the number of sample k from the data set X that lies with the volume V of the kernel. The volume V is a hypersphere with diameter D. The kernel function $K(u) = 1$ if $u < 1/2$ and 0 otherwise. If the video sequence is assigned as foreground, the value of b_m is 0. If the value of $P(x|x_i)$ is greater than a certain threshold of T, the pixel is assigned background, otherwise, it is assigned as foreground.

After achieving a foreground mask (FG) from the input frame, we perform morphology operations: closing and opening to fill holes, and removing noises for constructing Refined Foreground Image (RFG) as follows:

$$RFG = (((FG \oplus E) \otimes E) \otimes E) \oplus E$$

where E is a 5×5 structuring kernel, $\otimes$ is erosion, $\oplus$ is dilation.

Skin Removal - Proposal Region Detection. In track 4 of the 2022 AIC-ITY challenge, it is necessary to determine the highest potential regions (called region proposals) that may contain the items. The region proposals with high probability scores are the location of objects in an image. Basically, in three state-of-the-art two-stage detection models (R-CNN, Fast R-CNN, Faster R-CNN). Selective search (R-CNN and Faster R-CNN) and the region proposal network (Faster R-CNN) are used in detecting proposal regions (highly poten-tial regions) that may contain candidate objects. However, the selective search takes plenty of time for detecting, and these proposal regions can be noisy, over-lapping, or may not contain the object. While the region proposal network works perfectly, it requires a large dataset for training and testing. In the testing sce-nario of track 4 in the 2022 AICITY challenge, a customer is hand-holding items and pretending to perform a check-out action in front of the counter in a natural manner. Because of the hand holding action and we have already segmented the moving objects, we propose a Skin Removal algorithm for detecting the pro-posal region in our system. Skin color extraction is experimentally considered one of the most common traditional methods for detecting face candidate regions [22]. Skin-based models have many advantages: not affected by posture and size, are so robust to occlusion and rotation, also require less computation, and are extremely fast. In our system, the skin removal module helps to erase the hand region when the customer is holding the item. It helps us to extract correctly the region that highly contains the object and increases the accuracy for object classification.

In our skin removal process, we combine HSV and YCbCr color space using the assumption of Gaussian distribution of skin tone color. Based on the Gaus-sian Distribution and the skin color in the test set videos, we research and figure out the suitable value for white and yellow skin color. Given a segmented mov-ing object frame F, F is converted into HSV and YCbCr color space, FHSV and FYCbCr respectively. All pixels in two converted images are classified as skin color as the following equations:

$$\text{FHSV } \{H : 0 \leq p(x,y) \leq 17\, S : 15 \leq p(x,y) \leq 170\, V : 0 \leq p(x,y) \leq 255$$
$$\text{FYCbCr } \{Y : 0 \leq p(x,y) \leq 255\, Cb : 85 \leq p(x,y) \leq 135\, Cr : 135 \leq p(x,y) \leq 180$$

where $p(x, y)$ is the pixel value at the coordinate point (x, y) in each color space element. By combining 2 results, we produce the final moving object segmentation without hand regions. We focus on the central region or the region of interest where a shopping tray is placed. Therefore, candidate regions are cropped in this central region in each frame and remove the region which has a size smaller than 50×50 pixels. Figure 5 shows the process in the first step module in our system.

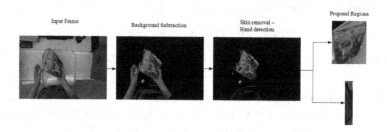

Fig. 5. The processes of proposal region detection.

3.2   Product Classification

After extracting the proposal region in the previous step, the problem is now Object Classification instead of Object Detection. Very deep convolutional neural networks have achieved prosperous results in the field of computer vision, especially in image recognition performance in recent years [23–25]. Our solution takes advantage of one of the state-of-the-art object classification models, the Inception-ResNet-v2 [24] which combines ResNet [25] and Inception architecture [26]. [24] has achieved tremendous results in the ImageNet dataset which has 1000 classes. In the other hand, [27] was proposed later and achieved better result in ImageNet dataset. The model take plenty of time for training (more than 3 times longer than [26]). when combining [26] with BGS and our proposed tracking and counting algorithms, the accuracy is improved. Therefore, it is reasonable to apply [24] for object classification in track 4 which contains 116 classes.

Training Data: To train the classifier, we select only 22,584 images out of 116,500 images from the 116 class dataset provided by the AICITY organizer and does not include any hand-labeled or external data . In the original 3D scanned dataset, there are a lot of images that represent the information on 3 sides of the object. When training all images from the original, the classifier does not work correctly due to the difference in informative representation in the training data set and testing video A set. To cope with these problems, we divide the original 116 class dataset into 3 categories based on the segmentation label images: Box shape, rectangle shape, and other shapes. The box shape is an image that has 5 or 6 corners detected in segmentation label images, while the rectangle shape is the image that has 4 corners detected in segmentation label images, the other images are classified as the other category. Then we select only the rectangle shape and other shapes to prepare the training data for the classifier. We randomly slip the date set 80% for training (18022 images belonging to 116 classes) and 20% for validation (4562 images belonging to 116 classes). For the augmentation, each image in the training dataset will be zoomed in and vertical flip.

Model: For the training model, we unfree a few top layers of a frozen model base and then add 3 new dense layers for the classification. This allows us to "fine-tune" the best feature representations in the base model in order to produce the best result for the classifier with the input image size is 150×150 pixels.

Loss Optimization: In the training process, the classifier is trained with the Adam optimizer and the cross-entropy loss function. We train our classifier from the pre-trained Imagenet models within 10 epochs to avoid overfitting.

3.3 Product Tracking and Counting

The tracking and counting algorithm is one of our major contributions to the proposed system. In our system, each tracklet consists of a bounding box of objects each that appears from frame t to $t + n$. Each bounding box contains 4 values $[x, y, w, h]$ - the position of the top left corner (x, y) and the width (w) and high (h). The value of classification-class, and probability of each object is also assigned to each box of each tracklet from frame t1 to tn. In each tracklet, we select the most occurrence class and assign the recognition value to each tracklet. Each tracklet is described by 4 characteristics :

$$
\begin{aligned}
\text{boxes} &= \left[\text{rec}^{t^2}, \text{rec}^{t^2}, \ldots, \text{rec}^{t^n} \right] \\
\text{classes} &= \left[\text{class}^{t^2}, \text{class}^{t^2}, \ldots, \text{class}^{t^n} \right] \\
\text{probabilites} &= \left[\text{prob}^{t^1}, \text{prob}^{t^2}, \ldots, \text{prob}^{t^n} \right] \\
\text{frames} &= \left[t^1, t^2, \ldots, t^n \right]
\end{aligned}
\tag{2}
$$

where boxes list stores the bounding box value of tracklet A, classes list stores the classified class value of tracklet A, probabilities list stores the classified class value of tracklet A. Finally, frames list stores the frame index that object A appears from frame t^1 to t^n.

Product Tracking. Inspired by the runner-up of track 1 in the 2021 AIC-ITY challenge, the team combines the IoU metrics, Mahalanobis distance, and histogram analysis. In our approach, we adopt a simple tracking algorithm for product tracking: distance between the center point of bounding boxes and IoU matching. Given set of available Tracklets $T = [T_1, T_2, \ldots, T_n]$ and a new classified object $O_m = [class, prob, rec]$ which contains class index, probability and bounding box at frame m, we need to find the best matched tracklet T_i that O_m belong to. For all available tracklets T from frame m-60 to frame m, calculate the Euclidean distance (Eq. 3) between the centroids of the last bounding box and bounding box of O_m and check if 2 boxes are intersection, and select the tracklet T_i that has the minimum distance. The detail tracking algorithm for matching object O_m and all tracklets is described in Algorithm 1.

$$d(p1, p2) = \|p1 - p2\|_2 \tag{3}$$

where $p1$, $p2$ are the center point of 2 bounding boxes

Product Counting. After the process of classifying and tracking, we analyze the tracklet and conduct the final result in each video in the test set. At this stage, we merge the tracklets together. Because of the inconsistency of candidate region detection is produced by the background subject in Sect. 3.1. The tracklet of the same object is not continuous, we have to check whether 2 tracklet belong to the same item or they are different items. As we have mentioned before, the class of each tracklet is the most occurring class and assigns the class value to the tracklet T_i. Given 2 tracklets A and B with class a and b respectively, if a and b are the same class and the 2 tracklets are less than *thresh* frame apart, then join 2 tracklets into 1 tracklet. Figure 6 shows the example of joining 2 tracklets together in the final step.

4 Experiment

4.1 Experimental Setup

Our experiments are tested using the test set A provided by the AI City Challenge 2022 which contains 5 videos in different scenarios. We tested our solution on the Google Colab Pro environment for both training the classifier and running the inference on the test set A video provided by the AICITY organizer.

4.2 Training Classifier

All the training images for the Classifier are resized to 150×150 and normalized. In the training stage, as we have mentioned in Sect. 3.2, we use the Adam optimizer with the initial learning rate $1e^{-4}$. We trained our model in 10 epochs with a batch size of 64, the training process is less than 1 h. After 10 epochs, our model achieves an accuracy of 0.9876 and an f1 score of 0.9671.

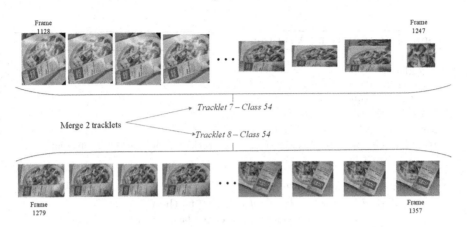

Fig. 6. Merge 2 tracklets together

4.3 Result and Discussion

The evaluation for track 4 is based on the model identification performance, measured by the F1 score. In our first trial, we didn't remove the result of classes 71 and 116, the evaluation result is very low, F1-Score 0.3265, Precision 0.2857, and Recall: 0.3810. Then, we remove the object or tracklet that belongs to class 71 and 116, since class 116 and 71 (Fig. 1 first and second item in second row from the left) are glare and white color which both have the same color with background. This problem causes noise and make it imbalance when training and testing the model. The final result of our proposed system is improved and the F1 - Score: 0.4082 with a precision: 0.3571 and recall: 0.4762. Table 1 shows our final score on the evaluation system.

Our team achieved a final score of F1 = 0.4082 and ranked 8 on the final Leaderboard. The final ranking results of the challenge track 4 are shown in Table 2. Although our proposed method is ranked 8th, it is still a good result. The first time, the organizer released the evaluation system, the constraints on how to evaluate and rank are complicated: items must be on the hand and appear in-tray. Our proposed system achieve the rank 8th on the final scoreboard, however, our solution is promising and does not require a super-powerful hardware device for execution. Our solution does not require extreme computation and a hand-labeled dataset for training a model. We only use the synthetic data provided by the organizer to train the classifier. Furthermore, our system combines the traditional method and the CNN approach, the system also has a competitively fast execution speed (about 18–20 fps). Although the top 1 achieves the result 1.0 score in the final result, it is invalid because the result is hand-labeled in the test set videos. Also, teams from rank 2 to rank 7 have a score from 0.4 to 0.48. All of the team are from prestigious universities, so their results are reliable and integrity. The reason why most of the team have a score lower than 0.5 is that the organizer provides synthetic data for training, external data for training is not allowed. Therefore, most of the teams have low scores.

Algorithm 1: Tracking Algorithm

Input:
Tracklets T;
New Object 0_m;
$minDis$, default = 1e9;
$matchedID$, default = None;
$checkMatched$, default = **False**;
Result: The identity of the best mached tracklet T_i in tracklet T or create a new tracklet

begin
  foreach T_i in Tracklets T **do**
    if the last frames value t^n of $T_i - m \leq 60$ **then**
      $CenterT_n$ = the center point of the last box rec^{t^n}
      $CenterO_m$ = the center point of bounding box of 0_m
      if $IOU(rec^{t^n}, 0_m)$ **then**
        Dis = distance($CenterT_n$, $CenterO_m$)
        if $Dis \leq minDis$ **then**
          $minDis = Dis$
          $matchedID$ = i
          $checkMatched$ = **True**
  if $checkMatched$ and $minDis \leq thresh$ **then**
    | match object 0_m to tracklet $T_m atchedID$
  else
    | create new tracklet for 0_m

Table 1. The evaluation results of our proposed system on the AI City evaluation

Methods	F1	Precision	Recal
116 classes	0.3265	0.2857	0.3810
114 classes (remove class 71 and 116)	0.4082	0.3571	0.4762

Table 2. The final ranking results of the track 4 on the test set A

Rank	Team Name	Score
1	BUPT-MCPRL2	1.0000
2	SKKU Automation Lab	0.4783
3	The Nabeelians	0.4545
...	...	...
8	**HCMIU-CVIP**	**0.4082**
9	CyberCore-Track4	0.4000
10	UTE-AI	0.4000
11	KiteMetric	0.3929
12	AICLUB@UIT	0.3922

5 Conclusions

In this paper, we propose a system for Multi-Class Product Counting and Recognition for Automated Retail Checkout. Our framework combines the traditional method: background subtraction model - Skin removal and Deep learning approach: Inception-Resnet-V2 for object recognition. Our proposed system achieves top-8 on the final leaderboard.

Acknowledgment. We would like to express a big thank to Ho Chi Minh City International University-Vietnam National University (HCMIU-VNU) for supporting our work. Additionally, we would like to express our appreciation to all of our colleagues for their contributions, which considerably aided in the revision of the manuscript.

References

1. Sriram, T., et al.: Applications of barcode technology in automated storage and retrieval systems. In: Proceedings of the 1996 IEEE IECON. 22nd International Conference on Industrial Electronics, Control, and Instrumentation, vol. 1, pp. 641–646 (1996)
2. LoweDavid, G.: Distinctive image features from scale-invariant keypoints. Int. J. Comput. Vis. (2004)
3. Christopher, G., Harris, M.J., Stephens, A.: Combined corner and edge detector. In: Alvey Vision Conference (1988)
4. LeCun, Y., Bengio, Y., Hinton, G.: Deep learn. nat. **521**(7553), 436–444 (2015)
5. Girshick, R.: Fast R-CNN. In: Proceedings of the IEEE International Conference on Computer Vision, pp. 1440–1448 (2015)
6. Girshick, R., Donahue, J., Darrell, T., Malik, J.: Rich feature hierarchies for accurate object detection and semantic segmentation. In: Proceedings of the IEEE Conference on Computer Vision and Pattern Recognition, pp. 580–587 (2014)
7. Girshick, R., Donahue, J., Darrell, T., Malik, J.: Rich feature hierarchies for accurate object detection and semantic segmentation. In: Proceedings of the IEEE Conference on Computer Vision and Pattern Recognition (CVPR) (2014)
8. Howard, A.G., et al.: MobileNets: Efficient convolutional neural networks for mobile vision applications. arXiv preprint arXiv:1704.04861 (2017)
9. Liu, W., et al.: SSD: single shot multibox detector. In: Leibe, B., Matas, J., Sebe, N., Welling, M. (eds.) ECCV 2016. LNCS, vol. 9905, pp. 21–37. Springer, Cham (2016). https://doi.org/10.1007/978-3-319-46448-0_2
10. Wang, C-Y., Bochkovskiy, A., Mark Liao, H.Y.: Scaled-YOLOV4: Scaling cross stage partial network. In: Proceedings of the IEEE/CVF Conference on Computer Vision and Pattern Recognition, pp. (13029–13038) (2021)
11. Tan, M., Pang, R., Le, Q.V.: EfficientDet: scalable and efficient object detection. In: Proceedings of the IEEE/CVF Conference on Computer Vision and Pattern Recognition, pp. (10781–10790) (2020)
12. Ze, L., et al.: Swin transformer: hierarchical vision transformer using shifted windows. In: Proceedings of the IEEE/CVF International Conference on Computer Vision (2021)
13. Yin, J., Wang, W., Meng, Q., Yang, R., Shen, J.: A unified object motion and affinity model for online multi-object tracking. In: Proceedings of the IEEE/CVF Conference on Computer Vision and Pattern Recognition, pp. 6768–6777 (2020)

14. Braso, G., Leal-Taixe, L.: Learning a neural solver for multiple object tracking. In: Proceedings of the IEEE/CVF Conference on Computer Vision and Pattern Recognition, pp. 6247–6257 (2020)
15. Bergmann, P., Meinhardt, T., Leal-Taixe, L.: Tracking without bells and whistles. In: Proceedings of the IEEE/CVF International Conference on Computer Vision, pp. 941–951 (2019)
16. Wojke, N., Bewley, A., Paulus, D.: Simple online and realtime tracking with a deep association metric. In: 2017 IEEE International Conference on Image Processing (ICIP), pp. 3645–3649. IEEE (2017)
17. Chattopadhyay, P., Vedantam, R., Selvaraju, R.R., Batra, D., Parikh, D.: Counting everyday objects in everyday scenes. In: Proceedings of the IEEE Conference on Computer Vision and Pattern Recognition, pp. 1135–1144 (2017)
18. Zhang, Y., Zhou, D., Chen, S., Gao, S., Ma, Y.: Single-image crowd counting via multi-column convolutional neural network. In: Proceedings of the IEEE Conference on Computer Vision and Pattern Recognition, pp. 589–597 (2016)
19. Ma, Z., Wei, X., Hong, X., Gong, Y.: Bayesian loss for crowd count estimation with point supervision. In: Proceedings of the IEEE/CVF International Conference on Computer Vision, pp. 6142–6151 (2019)
20. Kai, C., et al.: MMDetection: Open MMLab Detection Toolbox and Benchmark (2019)
21. Zivkovic, Z., et al.: Efficient adaptive density estimation per image pixel for the task of background subtraction. Pattern Recogn. Lett. **27**(7), 773–780 (2006)
22. Lucena, O., et al.: Improving face detection performance by skin detection post-processing. In: 2017 30th SIBGRAPI Conference on Graphics, Patterns and Images (SIBGRAPI), pp. 300–307 IEEE, (2017)
23. Simonyan, K., Zisserman, A.: Very deep convolutional networks for large-scale image recognition. arXiv preprint arXiv:1409.1556 (2014)
24. Szegedy, C., Ioffe, S., Vanhoucke, V., Alemi, A.A.: Inception-v4, inception-ResNet and the impact of residual connections on learning. In: Thirty-first AAAI Conference on Artificial Intelligence (2017)
25. He, K., Zhang, X., Ren, S., Sun, J.: Deep residual learning for image recognition. In: Proceedings of the IEEE Conference on Computer Vision and Pattern Recognition pp. 770–778 (2016)
26. Szegedy, C., Vanhoucke, V., Ioffe, S., Shlens, J., Wojna, Z.: Rethinking the inception architecture for computer vision. In: Proceedings of the IEEE Conference on Computer Vision and Pattern Recognition, pp. 2818–2826 (2016)
27. Tan, M., Le, Q.: EfficientNet: Rethinking Model Scaling for Convolutional Neural Networks. International conference on machine learning, PMLR (2019)

Strategy and Feasibility Study for the Construction of High Resolution Images Adversarial Against Convolutional Neural Networks

Franck Leprévost(ID), Ali Osman Topal(✉)(ID), Elmir Avdusinovic(ID),
and Raluca Chitic(ID)

University of Luxembourg, House of Numbers, 6, Avenue de la Fonte,
4364 Esch-sur-Alzette, Luxembourg
{Franck.Leprevost,Aliosman.Topal,Raluca.Chitic}@uni.lu,
Elmir.Avdusinovic.001@student.uni.lu

Abstract. Convolutional Neural Networks, that perform image recognition, assess images by first resizing them to their fitting input size. In particular, high resolution images are scaled down, say to 224×224 for CNNs trained on ImageNet. So far, existing attacks, that aim at creating an adversarial image that a CNN would misclassify while a human would not notice any difference between the modified and the unmodified image, actually work in the 224×224 resized domain and not in the high resolution domain. Indeed, attacking high resolution images directly leads to complex challenges in terms of speed, adversity and visual quality, that make these attacks infeasible in practice. We design an indirect strategy that addresses effectively this issue. It lifts to the high resolution domain any existing attack that works in the CNN's input size domain. The adversarial noise is of the same size as the original image. We apply this strategy to construct efficiently high resolution adversarial images of good visual quality that fool VGG-16 trained on ImageNet.

Keywords: High resolution adversarial images · Convolutional neural networks · Black-box attack · Evolutionary algorithm

1 Introduction

The profusion of images in our modern-day society and the need to analyze quickly the information they contain for a large series of applications (self-driving cars, face recognition and security controls, etc.) has led to the emergence of tools to automatically process and sort this type of data. Trained convolutional neural networks (CNNs) are among the dominant and most accurate tools for automatic object recognition and classification. Nevertheless, CNNs can be led to erroneous classifications by specifically designed adversarial images, that depend on the attack scenario considered. For instance, starting with an original image classified

© The Author(s), under exclusive license to Springer Nature Switzerland AG 2022
N. T. Nguyen et al. (Eds.): ACIIDS 2022, LNAI 13757, pp. 285–298, 2022.
https://doi.org/10.1007/978-3-031-21743-2_23

by a CNN in a given category, the target scenario essentially consists in choosing a target category, different from the original one, and in creating a variant of the original image that the CNN will classify in the target category, although a human would classify this adversarial image still in the original category, or would be unable to notice any difference between the original and the adversarial image.

Attacks, that intend to construct adversarial images, are classified according to the level of knowledge about the CNN at the disposal of the attacker. In this hierarchy, black-box attacks are the most challenging ones, since no knowledge about the architecture of the CNN (number and type of layers, weights, etc.) is assumed. Such attacks already exist. For instance (see also [13,14,20] for other black-box attacks, and [4,7,25] for gradient-based attacks) the paper [26] shows how an evolutionary-based algorithm successfully fooled 10 CNNs trained on ImageNet [11] to sort images of size 224×224 into 1000 categories ([8,9] provided a first version of this algorithm that fooled VGG-16 [5] trained on CIFAR-10 [17] to sort images of size 32×32 into 10 categories).

1.1   Attacks in the $\mathcal{R}$ Domain

So far, all such attacks — black-box or not — addressed images of moderate size, ranging from 32×32 (typically for CNNs trained on CIFAR-10) up to 224×224 (typically for CNNs trained on ImageNet), or resized to these values that the CNNs handle natively, what is called here the $\mathcal{R}$ domain. The construction of images, adversarial for the target scenario in this "traditional" context, is achieved by adding some carefully designed adversarial noise to the potentially resized original image in a process illustrated in Fig. 1.

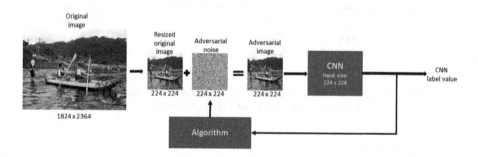

Fig. 1. Generating an adversarial image of size 224×224.

In particular, the adversarial noise created by all these attacks is in the $\mathcal{R}$ domain handled natively by the CNNs, so that the obtained adversarial images are as large as the CNN's input size. This means that these attacks explore a search space of size that does not depend on the size of the original image, but that coincides with the size of the CNN input (note *en passant* that the smaller the input size of the CNN, the easier the creation of adversarial noise).

1.2   Three Challenges Faced by Attacks in the $\mathcal{H}$ Domain

However, if the adversarial image should preserve almost all the details of an original image of large size, namely an image in what we call here the $\mathcal{H}$ domain, in particular of a high resolution (HR) image, the adversarial noise should have the same size as the original image, and consequently the adversarial image should as well have the same size as the original one. A key point is that the adversity character of a modified image is measured only when it is exposed to the CNN, hence when it is resized to fit into the $\mathcal{R}$ domain. The adversarial character should show up when the CNN proceeds to the classification of its resized version, as illustrated in the process given in Fig. 2.

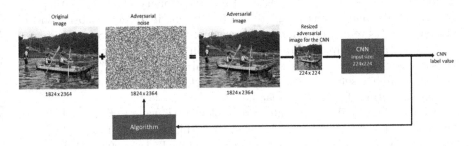

Fig. 2. Generating adversarial images in the HR domain.

Creating adversarial images of large size leads to three challenges in terms of speed, adversity and visual quality. Firstly, the complexity of the problem increases drastically with the size of the images, as the search space for the adversarial noise grows quadratically. For instance, the noise search space provided by the original image represented in Fig. 2 is 86 times larger than it is in the 224×224 domain. Secondly, the noise introduced in the $\mathcal{H}$ domain should be assessed as adversarial in the $\mathcal{R}$ domain: it should "survive" the resizing process to fit the CNN. In the example of Fig. 2, it would essentially mean that it survives a 86-fold squeezing process. Thirdly, the noise introduced in the $\mathcal{H}$ domain should be imperceptible to a human eye looking at the images at their native size, and not merely once they are reduced to fit the $\mathcal{R}$ domain.

Already the first challenge of simply getting a HR adversarial image is a very serious one. This is probably one reason for which, to the best of our knowledge, so far, no attack — black-box or not — has attempted to address large size images, in particular high-resolution (HR) images, by creating convenient adversarial noise in the $\mathcal{H}$ domain, so that the modified image, resized to the size handled natively by the CNN, becomes adversarial. Applying existing methods does not work, at least in reasonable time. Although efficient in the $\mathcal{R}$ domain, their extension to the $\mathcal{H}$ domain is not.

1.3   Our Contribution: A Strategy and a Feasibility Study

This paper is a first step towards the creation of adversarial noise of size of the original image, whatever this size may be. Our contribution is essentially twofold.

Firstly, we describe a strategy that leads to the construction of HR images in the $\mathcal{H}$ domain that are adversarial for the target scenario performed on a trained CNN (Sect. 3). The conceptual design of the strategy, that furthermore lists indicators relevant to the problem, is flexible enough to lift to the $\mathcal{H}$ domain attacks considered as efficient in the $\mathcal{H}$ domain.

Secondly, we perform a feasibility study of this strategy with 10 explicit HR images and on one CNN, namely VGG-16 trained on ImageNet. We lift to the $\mathcal{H}$ domain a black-box attack based on an evolutionary algorithm. We prove experimentally that our strategy is highly efficient in terms of speed and of adversity, and is reasonably efficient in terms of visual quality (Sect. 4).

Two sections complete this article. Section 2 fixes some notations about CNNs, formalizes the target scenario in general, and its "lifted" version in the context of high resolution images. Finally, Sect. 5 wraps up our findings, and provides directions for future research.

All algorithms and experiments were implemented using Python 3.8 [27] with NumPy 1.17 [19], TensorFlow 2.4 [1], Keras 2.2 [10], and Scikit 0.24 [28] libraries. Computations were performed on nodes with Nvidia Tesla V100 GPGPUs of the IRIS HPC Cluster at the University of Luxembourg.

2   CNNs and the Target Scenario

Convolutional neural networks (CNNs) performing image classification are train-ed on some large dataset $\mathcal{S}$ to sort images into predefined categories $c_1, \cdots, c_\ell$. The categories, and their number ℓ, are associated to $\mathcal{S}$, and are common to all CNNs trained on $\mathcal{S}$. The training phase of a CNN is essentially made in two steps. During the first step, the CNN is given both a series of training images, and, for each training image, a vector of length ℓ, where each real-value component assesses the probability that the training image represents an object in the corresponding category. During the second step, the CNN is challenged against a validation set of images that assess its ability to sort images accurately.

Once trained, a CNN can be exposed to images (typically) of the same size as those on which it was trained. In practice, given an input image $\mathcal{I}$, the trained CNN produces a classification output vector

$$\mathbf{o}_{\mathcal{I}} = (\mathbf{o}_{\mathcal{I}}[1], \cdots, \mathbf{o}_{\mathcal{I}}[\ell]),$$

where $0 \leq \mathbf{o}_{\mathcal{I}}[i] \leq 1$ for $1 \leq i \leq \ell$, and $\sum_{i=1}^{\ell} \mathbf{o}_{\mathcal{I}}[i] = 1$. Each component $\mathbf{o}_{\mathcal{I}}[i]$ of the output vector defines the c_i-label value measuring the probability that the image $\mathcal{I}$ belongs to the category c_i. Consequently, the CNN classifies the image $\mathcal{I}$ as belonging to the category c_k if $k = \arg\max_{1 \leq i \leq \ell}(\mathbf{o}_{\mathcal{I}}[i])$, and one denotes $(c_k, \mathbf{o}_{\mathcal{I}}[k])$ this outcome. The higher the label value $\mathbf{o}_{\mathcal{I}}[k]$, the higher the confidence that $\mathcal{I}$ represents an object of the category c_k.

2.1   The Target Scenario

Let $\mathcal{C}$ be a trained CNN as above, c_a be a category among the ℓ possible categories, and $\mathcal{A}$ an image classified by $\mathcal{C}$ as belonging to c_a. One denotes by τ_a its c_a-label value. The *target scenario* (c_a, c_t) performed on $\mathcal{A}$ requires first to select a category $c_t \neq c_a$, and then to construct an image $\mathcal{D}$ that is either a *good enough adversarial image* or a τ-*strong adversarial image* in the following sense.

In any case, one requires that $\mathcal{D}$ remains so close to $\mathcal{A}$, that a human can not notice any difference between $\mathcal{A}$ and $\mathcal{D}$. The quantities $L_2(\mathcal{A}, \mathcal{D})$ and $\epsilon(\mathcal{A}, \mathcal{D})$ assess numerically this human perception. The L_2-distance essentially evaluates the difference between the pixel values of $\mathcal{A}$ and $\mathcal{D}$, and ϵ controls (or restricts) the global maximum amplitude allowed for the value modifications of each individual pixel of $\mathcal{A}$ to obtain $\mathcal{D}$. Since the L_2-distance used in many works [3,6,15] as a numerical proof of similarity between original and adversarial images, we also use it in this paper.

A *good enough adversarial image* is an adversarial image that $\mathcal{C}$ classifies as belonging to the target category c_t, without any requirement on the c_t-label value beyond being strictly dominant among all label values. A τ-*strong adversarial image* is an adversarial image that $\mathcal{C}$ not only classifies as belonging to the target category c_t, but for which its c_t-label value $\geq \tau_t$ for some threshold value $\tau_t \in]0,1]$ fixed *a priori*. We write (c_t, τ_t) the outcome of the CNN's classification of $\mathcal{D}$ in this later case.

2.2   The Target Scenario Lifted to HR Images

In the experiments of Sect. 4, we shall consider a CNN $\mathcal{C}$ that handles images of size 224×224, and that is trained on ImageNet to classify images into 1000 categories. In our context, we ask $\mathcal{C}$ to give the dominating category, and the corresponding label value. Henceforth, $\mathcal{C}$'s classifications take values in

$$\mathcal{V} = \{(c_i, v_i), \text{ where } v_i \in]0,1] \text{ for } 1 \leq i \leq 1000\}.$$

To express the target scenario in the context of high resolution (HR) images, let $\mathcal{H}$ denote the set of HR images (of various sizes $h \times w$, whatever they are), and $\mathcal{R}$ denote the set of images of size natively adapted to $\mathcal{C}$. One assumes given a fixed *degradation function* ρ, that transforms any image $\mathcal{I}$ of $\mathcal{H}$ into an image $\rho(\mathcal{I})$ of $\mathcal{R}$. The well-defined composition of maps $\mathcal{C} \circ \rho$ from $\mathcal{H}$ to $\mathcal{V}$ allows $\mathcal{C}$ to classify, in particular, the reduced image $\mathcal{A}_a = \rho(\mathcal{A}_a^{\mathrm{hr}}) \in \mathcal{R}$ in some class c_a, with τ_a being the c_a-label value outputted by $\mathcal{C}$ for $\mathcal{A}_a$, so that $\mathcal{C}(\mathcal{A}_a) = (c_a, \tau_a)$.

In this context, an adversarial HR image for the (c_a, c_t) target scenario performed on $\mathcal{A}_a^{\mathrm{hr}} \in \mathcal{H}$ is an image $\mathcal{D}_t^{\mathrm{hr}}(\mathcal{A}_a^{\mathrm{hr}}) \in \mathcal{H}$ satisfying the two following conditions. On the one hand, a human should not be able to notice any visual difference between the original $\mathcal{A}_a^{\mathrm{hr}}$ and the adversarial $\mathcal{D}_t^{\mathrm{hr}}(\mathcal{A}_a^{\mathrm{hr}})$ HR images. On the other hand, $\mathcal{C}$ should classify the reduced adversarial image $\mathcal{D}_t(\mathcal{A}_a^{\mathrm{hr}}) = \rho(\mathcal{D}_t^{\mathrm{hr}}(\mathcal{A}_a^{\mathrm{hr}}))$ in the category c_t for a sufficiently convincing c_t-label value. The image $\mathcal{D}_t^{\mathrm{hr}}(\mathcal{A}_a^{\mathrm{hr}}) \in \mathcal{H}$ is then a *good enough adversarial image* or a τ-*strong adversarial image* if its reduced version $\mathcal{D}_t(\mathcal{A}_a^{\mathrm{hr}}) = \rho(\mathcal{D}_t^{\mathrm{hr}}(\mathcal{A}_a^{\mathrm{hr}}))$ is.

3 Attack Strategy for the Target Scenario on HR Images

We present here a strategy that attempts to circumvent the three challenges cited in the Introduction. The first step consists in getting an image in $\mathcal{R}$, adversarial against the image $\mathcal{A}_a \in \mathcal{R}$ reduced from $\mathcal{A}_a^{\mathrm{hr}} \in \mathcal{H}$. Although getting such adversarial images in the $\mathcal{R}$ domain is crucial for obvious reasons, the strategy does not depend on how they are obtained. The second step *lifts* this low-resolution adversarial image up to a high resolution image, called here the *HR tentative adversarial image*. The last step checks whether this HR tentative adversarial image becomes adversarial once reduced.

3.1 Construction of Adversarial Images in $\mathcal{H}$

The starting point is a large size image $\mathcal{A}_a^{\mathrm{hr}} \in \mathcal{H}$, and its reduced image $\mathcal{A}_a = \rho(\mathcal{A}_a^{\mathrm{hr}}) \in \mathcal{R}$, classified by $\mathcal{C}$ as belonging to a category c_a.

Step 1 assumes given an image $\widetilde{\mathcal{D}}_{t,\tilde{\tau}_t}(\mathcal{A}_a^{\mathrm{hr}}) \in \mathcal{R}$, that is adversarial for the (c_a, c_t) target scenario performed on $\mathcal{A}_a = \rho(\mathcal{A}_a^{\mathrm{hr}})$, for a c_t-label value exceeding a threshold $\tilde{\tau}_t$, whatever the method leading to such an adversarial image.

For Step 2, one needs a fixed *enlarging function* λ, that transforms any image of $\mathcal{R}$ into an image in $\mathcal{H}$. One applies λ to the low-resolution adversarial $\widetilde{\mathcal{D}}_{t,\tilde{\tau}_t}(\mathcal{A}_a^{\mathrm{hr}}) \in \mathcal{R}$ to obtain the HR tentative adversarial image $\mathcal{D}_{t,\tau_t}^{\mathrm{hr}}(\mathcal{A}_a^{\mathrm{hr}}) = \lambda(\widetilde{\mathcal{D}}_{t,\tilde{\tau}_t}(\mathcal{A}_a^{\mathrm{hr}})) \in \mathcal{H}$. Note that, although the *reduction function* ρ and the *enlarging function* λ have opposite purposes, these functions are not necessarily inverse one from the other. In other words, usually $\rho \circ \lambda \neq id_{\mathcal{R}}$ and $\lambda \circ \rho \neq id_{\mathcal{H}}$.

For Step 3, applying the reduction function ρ on this HD tentative adversarial image creates an image $\mathcal{D}_{t,\tau_t}(\mathcal{A}_a^{\mathrm{hr}}) = \rho(\mathcal{D}_{t,\tau_t}^{\mathrm{hr}}(\mathcal{A}_a^{\mathrm{hr}}))$ in the $\mathcal{R}$ domain. One runs $\mathcal{C}$ on $\mathcal{D}_{t,\tau_t}(\mathcal{A}_a^{\mathrm{hr}})$ to get its classification to obtain a classification in c_t.

The attack succeeds if $\mathcal{C}$ classifies this image in c_t, potentially for a c_t-label value exceeding the threshold value τ_t fixed in advance, and if a human is unable to notice any difference between the images $\mathcal{A}_a^{\mathrm{hr}}$ and $\mathcal{D}_{t,\tau_t}^{\mathrm{hr}}(\mathcal{A}_a^{\mathrm{hr}})$ in the $\mathcal{H}$ domain. Scheme 1 essentially summarizes the different steps:

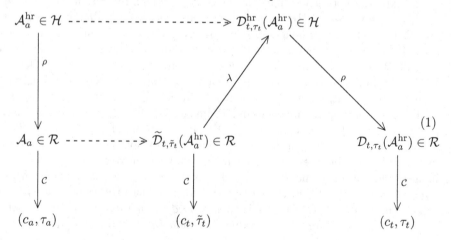

3.2 Indicators: The Loss Function $\mathcal{L}$ and L_2-distances

Although both $\widetilde{\mathcal{D}}_{t,\tilde{\tau}_t}(\mathcal{A}_a^{\mathrm{hr}})$ and $\mathcal{D}_{t,\tau_t}(\mathcal{A}_a^{\mathrm{hr}})$ stem from $\mathcal{A}_a^{\mathrm{hr}}$, and belong to the same set $\mathcal{R}$ of low resolution images, these images nevertheless differ in general, since $\rho \circ \lambda \neq id_{\mathcal{R}}$, what leads to the following consequences.

On the one hand, it justifies the necessity of the verification process performed in Step 3 on the HR tentative adversarial image, that checks whether its reduction indeed belongs to c_t. On the other hand, should it be the case, it implies as well that $\tilde{\tau}_t$ and τ_t differ. One defines the real-valued *loss function* $\mathcal{L}$ for a given $\mathcal{A}_a^{\mathrm{hr}} \in \mathcal{H}$ as

$$\mathcal{L}(\mathcal{A}_a^{\mathrm{hr}}) = \tilde{\tau}_t - \tau_t. \tag{2}$$

Our attack is effective if one can set accurately the value of $\tilde{\tau}_t$ either to match the threshold value τ_t, or to make sure that $\mathcal{D}_{t,\tau_t}(\mathcal{A}_a^{\mathrm{hr}})$ is a good enough adversarial image in the $\mathcal{R}$ domain, while controlling the distance variations between $\mathcal{A}_a^{\mathrm{hr}}$ and the adversarial $\mathcal{D}_{t,\tau_t}^{\mathrm{hr}}(\mathcal{A}_a^{\mathrm{hr}})$. For this, one needs to assess the statistical behavior of the loss function $\mathcal{L}$, and of the L_2 distance $L_2(\mathcal{A}_a^{\mathrm{hr}}, \mathcal{D}_{t,\tau_t}^{\mathrm{hr}}(\mathcal{A}_a^{\mathrm{hr}}))$. Note that the values of these quantities, and therefore the performances and adequacy of the resized adversarials to the addressed problem, clearly depend on the reducing and enlarging functions ρ and λ selected in the scheme.

4 Feasibility Study

The feasibility study is performed with the specific CNN VGG-16 trained on ImageNet, and with the 10 HR images $\mathcal{A}_1^{\mathrm{hr}}, \cdots, \mathcal{A}_{10}^{\mathrm{hr}}$ shown in Table 1. Out of them, 8 are taken from the Internet (under Creative Commons Licenses), and 2 are images from the French artist Speedy Graphito (pictured in [24], the corresponding files were graciously provided by the artist). Table 1 gives the size of each original HR image, the category c_a and the c_a-label value outputted by VGG-16 for $\mathcal{A}_a^{\mathrm{hr}}$. It also provides the target category c_t, chosen at random among the categories $\neq c_a$ of ImageNet, that is used for the target scenario (c_a, c_t) to perform on each $\mathcal{A}_a^{\mathrm{hr}}$. One interest of adding the two specific artistic images is that, while a human may have difficulties in classifying them in any category, VGG-16 does it, although with relatively small label values.

The selection of the functions ρ and λ relied on an evaluation of four interpolation methods that convert an image from one scale to another. The Nearest Neighbor [22], the Bilinear method [2], the Bicubic method [16] and the Lanczos method [12,21] are non-adaptive methods among the most common interpolation algorithms, and are available in python librairies. Our experiments led us to the combination $(\rho, \lambda) = $ (Lanczos, Lanczos), used in all further experiments.

We briefly describe the evolutionary algorithm $\mathrm{EA}^{\mathrm{target},\mathcal{C}}$ used as a blackbox attack against $\mathcal{C} = $ VGG-16 (Subsect. 4.1). We apply the strategy with the evolutionary algorithm and get the HR adversarial images that fool VGG-16 for the target scenario with the threshold value set to $\tau_t = 0.55$ (Subsect. 4.2). Finally, we discuss the visual quality of the obtained HR adversarial images, especially from a human point of view (Subsect. 4.3).

Table 1. For $1 \leq a \leq 10$, the image $\mathcal{A}_a^{\mathrm{hr}}$ classified by VGG-16 in the category c_a.

a	1	2	3	4	5	6	7	8	9	10
c_a	Cheetah	Eskimo Dog	Koala	Lamp Shade	Toucan	Screen	Comic Book	Sports Car	Binder	Coffee Mug
$w \times h$	910 × 604	960 × 640	910 × 607	2462 × 2913	910 × 607	641 × 600	1280 × 800	1280 × 800	1954 × 2011	1710 × 1740
$\mathcal{A}_a^{\mathrm{hr}}$										
	0.9527	0.3434	0.9976	0.5359	0.4581	0.7063	0.4916	0.4831	0.2825	0.0844
c_t	poncho	goblet	Weimaraner	weevil	wombat	swing	altar	beagle	triceratops	hamper

The HR ancestor images and one sample of an HR adversarial image per (c_a, c_t) combination can be retrieved from https://github.com/aliotopal/HRad-versImgs/blob/main/original-advers.md.

4.1 The Evolutionary Algorithm $\mathrm{EA}^{\mathrm{target},\mathcal{C}}$

Although there are different techniques for generating adversarial images, we obtained the adversarial images with an evolutionary algorithm in continuation of our previous works [8,9,26]. The pseudo code of the evolutionary algorithm $\mathrm{EA}^{\mathrm{target},\mathcal{C}}$ is given in Algorithm 1. The size of the population is set to 40 as a result of a series of experiments. The algorithm starts with 40 identical copies of the ancestor image. The objective of evolving an individual ind towards an image classified as c_t is encoded in the fitness function

$$fit(ind) = o_{ind}^{\mathcal{C}}[t] \tag{3}$$

Throughout the evolution, the individuals are continuously mutated and recombined to create population members with larger fitness.

Algorithm 1. EA attack pseudocode

1: **Input:** CNN $\mathcal{C}$, ancestor $\mathcal{A}$, perturbation magnitude α, maximum perturbation ϵ, ancestor class c_a, ordinal t of target class c_t, g current and X maximum generation;
2: Initialize population as 40 copies of $\mathcal{A}$, with I_0 as first individual;
3: Compute fitness for each individual;
4: **while** $(o_{I_0}[t] < \tau)$ & $x < X$ **do**
5:   Rank individuals in descending fitness order and segregate: elite 10, middle class 20, lower class 10;
6:   Select random number of pixels to mutate and perturb them with $\pm\alpha$. Clip all mutations to $(-\epsilon, \epsilon)$. The elite is not mutated. The lower class is replaced with mutated individuals from the elite and middle class;
7:   Cross-over individuals to form new population;
8:   Evaluate fitness of each individual;

To maintain the proximity of the evolved images with the ancestor image, the maximum pixel modification on individuals is limited to a fixed range $\epsilon =$

$[-16, 16]$ throughout the search process. The step size per selected pixel is set to $\alpha = \pm 1$. The individuals compete with each other until one of the EA's stop conditions is satisfied, namely until one individual satisfies $o_{ind}^{C}[t] \geq \tau$, or the maximum number of generations $X = 20,000$ is reached.

4.2 Running the Strategy to Get Adversarial Images with the EA

With the rescaling functions $(\rho, \lambda) = $ (Lanczos, Lanczos), we deploy the strategy detailed in Subsect. 3.1 with the evolutionary algorithm $EA^{target,C}$, the CNN VGG-16, and the ancestor images $\mathcal{A}_a^{hr}$. The goal is to create 0.55-strong HR adversarial images as well as good enough HR adversarial images for the target scenario (c_a, c_t) specified in Table 1.

Since different seed values for the EA lead to different results, we increased the robustness of the outcomes by performing 10 independent runs with random seeds for each (c_a, c_t) pair and ancestor $\mathcal{A}_a^{hr}$ given in Table 1, leading to altogether 100 trials. All runs terminated successfully in less than 20,000 generations. For each ancestor image $\mathcal{A}_a^{hr}$, Table 2 gives the average over the 10 independent runs of three indicators. $avgGens^{0.55}$ is the average number of generations required to obtain the 0.55-strong HR adversarial images $\mathcal{D}_{t,\tau_t}^{hr}(\mathcal{A}_a^{hr})$, $avgGens^{ge}$ is the average number of generations required to obtain good enough adversarial HR images $\mathcal{D}_{t,\tau_t}^{hr,ge}(\mathcal{A}^{hr})$, and $avg_{\tau_t}^{ge}$ is their average c_t-label values.

Table 2. Average performance in creating *good enough* and 0.55-*strong* HR adversarial images for the target scenario (c_a, c_t) performed on $\mathcal{A}_a^{hr}$

	$\mathcal{A}_1^{hr}$	$\mathcal{A}_2^{hr}$	$\mathcal{A}_3^{hr}$	$\mathcal{A}_4^{hr}$	$\mathcal{A}_5^{hr}$	$\mathcal{A}_6^{hr}$	$\mathcal{A}_7^{hr}$	$\mathcal{A}_8^{hr}$	$\mathcal{A}_9^{hr}$	$\mathcal{A}_{10}^{hr}$	Avg
$avgGens^{ge}$	7531	1868	2921	2736	2200	4288	2089	2223	9728	5104	**4069**
$avg_{\tau_t}^{ge}$	0.106	0.032	0.131	0.178	0.088	0.146	0.127	0.115	0.093	0.101	**0.112**
$avgGens^{0.55}$	9994	3985	3529	3212	2845	5188	3000	3377	15603	11770	**6250**

Table 2 shows that, on average, our algorithm creates *good enough* HR adversarial images in 4069 generations, and 0.55-*strong* HR adversarial images in 6250 generations. Measured by the number of additional generations required, the effort necessary to move up from a *good enough* HR adversarial image, that has a c_t-label value of 0.112 in average, to a 0.55-*strong* HR adversarial image is 53.6%. In terms of the average computational time, roughly 35 minutes were necessary to create a *good enough adversarial image*, and 55 minutes for a 0.55-*strong adversarial image*.

For each ancestor image $\mathcal{A}_a^{hr}$, one draws the convergence characteristics of the algorithm $EA^{target,C}$ for $\tilde{\tau}_t$ and for τ_t on the way to the HR 0.55-*strong adversarial image* $\mathcal{D}_{t,\tau_t}^{hr}(\mathcal{A}_a^{hr})$. The cases of $\mathcal{A}_7^{hr}$ and $\mathcal{A}_{10}^{hr}$, representative of the overall behavior, are given in Fig. 3 (the graphs are capped on the horizontal axis at their respective $avgGens^{0.55}$ values). Computation shows that the maximum values of $\mathcal{L}(\mathcal{A}_a^{hr}) = \tilde{\tau}_t - \tau_t$ for all ancestors are in the range $[7.9e - 03, 6.4e - 02]$.

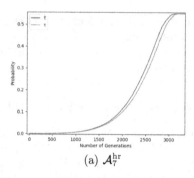
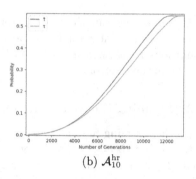

(a) $\mathcal{A}_7^{\mathrm{hr}}$ (b) $\mathcal{A}_{10}^{\mathrm{hr}}$

Fig. 3. Convergence characteristics of the EA for τ_t and $\tilde{\tau}_t$ for $\mathcal{A}_7^{\mathrm{hr}}$ and $\mathcal{A}_{10}^{\mathrm{hr}}$.

A consequence of the convergence graphs given in Fig. 3 and of the numerical maximum values of $\mathcal{L}(\mathcal{A}_a^{\mathrm{hr}})$ is that setting a threshold c_t-label value $\tilde{\tau}_t = \tau_t + 0.065$ (the largest $\mathcal{L}_{max}$ value is 0.064) seems a safe choice, at least if one aims at getting 0.55-*strong* HR adversarial images by our method.

4.3   Visual Quality

We first assess numerically the quality of the obtained HR adversarial images as compared to their HR ancestors by computing their L_2 differences. Experiments show that the average value of $L_2(\mathcal{A}_a^{\mathrm{hr}}, \mathcal{D}_{t,\tau_t}^{\mathrm{hr}}(\mathcal{A}_a^{\mathrm{hr}}))$ and of $L_2(\mathcal{A}_a^{\mathrm{hr}}, \lambda \circ \rho(\mathcal{A}_a^{\mathrm{hr}})$ remains comparable (close to 35400), so that, at least in average, our attack does not arm the numerical performance of the resizing functions.

Still, The "true" visual quality for a human eye is assessed by looking at the images either from some distance, or by zooming on some areas. For instance, the HR ancestor image $\mathcal{A}_7^{\mathrm{hr}}$, and the HR 0.55-*strong* adversarial image created by EA$^{\mathrm{target},\mathcal{C}}$ are pictured in Fig. 4. *Mutatis mutandis* for $\mathcal{A}_{10}^{\mathrm{hr}}$ in Fig. 5. At some distance, the HR adversarial seems to have a good visual quality as compared to its HR ancestor. Zooming shows that details from the HR ancestor image become blurry in the HR adversarial image.

This blurriness is not due to an inefficiency of our strategy, nor of the algorithm EA$^{\mathrm{target},\mathcal{C}}$, at least to a large extent, but comes from the fact that the used resizing functions loose high-frequency features on the HR adversarial images.

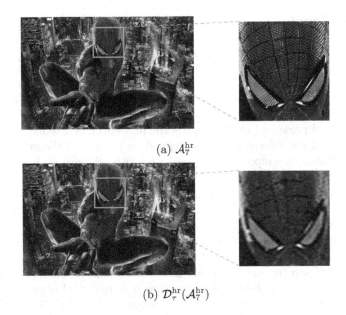

(a) $\mathcal{A}_7^{\mathrm{hr}}$

(b) $\mathcal{D}_\tau^{\mathrm{hr}}(\mathcal{A}_7^{\mathrm{hr}})$

Fig. 4. Visual comparison of the ancestor image $\mathcal{A}_7^{\mathrm{hr}}$ (a) and the obtained HR adversarial image (b) in the $\mathcal{H}$ domain.

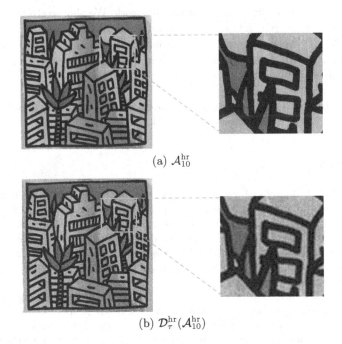

(a) $\mathcal{A}_{10}^{\mathrm{hr}}$

(b) $\mathcal{D}_\tau^{\mathrm{hr}}(\mathcal{A}_{10}^{\mathrm{hr}})$

Fig. 5. Visual comparison of the ancestor image $\mathcal{A}_{10}^{\mathrm{hr}}$ (a) and the obtained HR adversarial image (b) in the $\mathcal{H}$ domain.

5 Conclusion

Trained CNNs, performing image recognition, convert input images to some fixed and moderate size, say 224×224 for those trained on ImageNet typically. This process transforms the input image into a low resolution image that the CNN is able to analyze. Attacks, aiming at adversarial images fooling these CNNs, create some adversarial noise of size equal to the input size of the CNN.

The method presented in this work is the first effective attempt to make the search space for the adversarial noise depend on the size of the original image, and not on the CNN's input size. In particular, it is effective for high resolution images in terms of speed, adversity and visual quality. The designed strategy lifts any existing attack, efficient in the low resolution domain, to an attack that applies in the high resolution domain. We performed an experimental study for VGG-16 trained on ImageNet, that creates, thanks to the evolutionary algorithm $EA^{\text{target},\mathcal{C}}$, high resolution images adversarial for the target scenario, and that VGG-16 classifies in the target category with confidence ≥ 0.55. While the attack directly in the high resolution domain is not feasible in practice (we tried, and the c_t-label value of the best tentative adversarial remained of order $1.8e - 05$ after more than 40 h), our indirect method succeeds within 55'. Our method also gives the precise assessment of the value of the loss function the attack should aim at in order to obtain adversarial images in the $\mathcal{H}$ domain (0.065 if one aims at 0.55-strong adversarial images).

There are clearly many ways to extend this work. We plan to study alternative scaling functions, such as adaptive interpolation methods [14,18,30] or ML-base methods [23,29]. We also intend to apply our strategy to more CNNs, more high resolution images (of different nature, e.g. satellite images, medical images, etc.), and more attacks, black-box or not.

Acknowledgments. The authors express their gratitude to Speedy Graphito and to Bernard Utudjian for their interest in this work.

References

1. Abadi, M., et al.: TensorFlow: Large-scale machine learning on heterogeneous systems (2015). https://www.tensorflow.org/softwareavailablefromtensorflow.org
2. Agrafiotis, D.: Chapter 9 - video error concealment. In: Theodoridis, S., Chellappa, R. (eds.) Academic Press Library in signal processing, academic press library in signal processing, vol. 5, pp. 295–321. Elsevier (2014). https://doi.org/10.1016/B978-0-12-420149-1.00009-0, https://www.sciencedirect.com/science/article/pii/B9780124201491000090
3. Baluja, S., Fischer, I.: Adversarial transformation networks: learning to generate adversarial examples. arXiv preprint arXiv:1703.09387 (2017)
4. Mukherjee, I., Canini, K., Frongillo, R., Singer, Y.: Parallel boosting with momentum. In: Blockeel, H., Kersting, K., Nijssen, S., Železný, F. (eds.) ECML PKDD 2013. LNCS (LNAI), vol. 8190, pp. 17–32. Springer, Heidelberg (2013). https://doi.org/10.1007/978-3-642-40994-3_2

5. Blier, L.: A brief report of the heuritech deep learning meetup#5 (2016). https://heuritech.wordpress.com/2016/02/29/a-brief-report-of-the-heuritech-deep-learning-meetup-5/

6. Brendel, W., Rauber, J., Bethge, M.: Decision-based adversarial attacks: reliable attacks against black-box machine learning models. arXiv preprint arXiv:1712.04248 (2017)

7. Carlini, N., Wagner, D.: Towards evaluating the robustness of neural networks. In: 2017 IEEE Symposium on Security and Privacy (SP), pp. 39–57 IEEE (2017)

8. Chitic, R., Bernard, N., Leprévost, F.: A proof of concept to deceive humans and machines at image classification with evolutionary algorithms. In: Nguyen, N.T., Jearanaitanakij, K., Selamat, A., Trawiński, B., Chittayasothorn, S. (eds.) ACIIDS 2020. LNCS (LNAI), vol. 12034, pp. 467–480. Springer, Cham (2020). https://doi.org/10.1007/978-3-030-42058-1_39

9. Chitic, R., Leprévost, F., Bernard, N.: Evolutionary algorithms deceive humans and machines at image classification: an extended proof of concept on two scenarios. J. Inf. Telecommun. **5**, 1–23 (2020)

10. Chollet, F., et al.: Keras. https://keras.io (2015)

11. Deng, J., Dong, W., Socher, R., Li, L.J., Li, K., Fei-Fei, L.: The ImageNet Image Database (2009). http://image-net.org

12. Duchon, C.E.: Lanczos filtering in one and two dimensions. J. Appl. Meteorol. Climatol. **18**(8), 1016–1022 (1979)

13. Guo, C., Gardner, J., You, Y., Wilson, A.G., Weinberger, K.: Simple black-box adversarial attacks. In: International Conference on Machine Learning, pp. 2484–2493 PMLR (2019)

14. Hu, W., Tan, Y.: Generating adversarial malware examples for black-box attacks based on GAN. arXiv preprint arXiv:1702.05983 (2017)

15. Jere, M., Rossi, L., Hitaj, B., Ciocarlie, G., Boracchi, G., Koushanfar, F.: Scratch that! an evolution-based adversarial attack against neural networks. arXiv preprint arXiv:1912.02316 (2019)

16. Keys, R.: Cubic convolution interpolation for digital image processing. IEEE Trans. Acoust. Speech Sign. Process. **29**(6), 1153–1160 (1981)

17. Krizhevsky, A., Nair, V., Hinton, G.: CIFAR-10 (canadian institute for advanced research). http://www.cs.toronto.edu/kriz/cifar.html

18. Li, X., Orchard, M.T.: New edge-directed interpolation. IEEE Trans. Image Process. **10**(10), 1521–1527 (2001)

19. Oliphant, T.E.: A guide to NumPy. Trelgol Publishing USA (2006)

20. Papernot, N., McDaniel, P., Goodfellow, I., Jha, S., Celik, Z.B., Swami, A.: Practical black-box attacks against machine learning. In: Proceedings of the 2017 ACM on Asia Conference on Computer and Communications Security, pp. 506–519 (2017)

21. Parsania, P.S., Virparia, P.V.: A comparative analysis of image interpolation algorithms. Int. J. Adv. Res. Comput. Commun. Eng. **5**(1), 29–34 (2016)

22. Patel, V., Mistree, K.: A review on different image interpolation techniques for image enhancement. Int. J. Emerg. Technol. Adv. Eng. **3**(12), 129–133 (2013)

23. Schulter, S., Leistner, C., Bischof, H.: Fast and accurate image upscaling with super-resolution forests. In: Proceedings of the IEEE Conference on Computer Vision and Pattern Recognition (CVPR) (2015)

24. SpeedyGraphito: Mes 400 Coups. Panoramart (2020)

25. Szegedy, C., et al.: Intriguing properties of neural networks. arXiv preprint arXiv:1312.6199 (2013)

298   F. Leprévost et al.

26. Topal, A.O., Chitic, R., Leprévost, F.: One evolutionary algorithm deceives humans and ten convolutional neural networks trained on imagenet at image recognition. (Under review), pp. 67–480 (2022)
27. Van Rossum, G., Drake, F.L.: Python 3 Reference Manual. CreateSpace, Scotts Valley, CA (2009)
28. Van der Walt, S., et al.: The scikit-image contributors: scikit-image: image processing in Python. PeerJ **2**, (2014). https://doi.org/10.7717/peerj.453
29. Ye, M., Lyu, D., Chen, G.: Scale-iterative upscaling network for image deblurring. IEEE Access **8**, 18316–18325 (2020). https://doi.org/10.1109/ACCESS.2020.2967823
30. Zhang, X., Wu, X.: Image interpolation by adaptive 2-D autoregressive modeling and soft-decision estimation. IEEE Trans. Image Process. **17**(6), 887–896 (2008)

Using Deep Learning to Detect Anomalies in Traffic Flow

Manuel Méndez[1], Alfredo Ibias[2], and Manuel Núñez[1]($\boxtimes$) (iD)

[1] Universidad Complutense de Madrid, Madrid, Spain
{manumend,manuelnu}@ucm.es
[2] SANO-Centre for Computational Medicine, Krakow, Poland
a.ibias@sanoscience.org

Abstract. Uncertainty is an ever present challenge in data analysis. In particular, it is important to detect, as precisely as possible, unforeseen phenomena. In this paper we study the usefulness of two deep learning based methods (CNN auto-encoder and BiLSTM auto-encoder) to detect anomalies in situations that can be defined in terms of time series. In order to evaluate our approaches, we consider traffic flow data and perform experiments in two orthogonal scenarios: a guided scenario (training only with data considered as 'normal' after a naïve labelling) and a basic scenario. Our results show that if we train the models using only the considered 'normal' data, the obtained models do not achieve good results because none of them are able to detect all type of abnormal data correctly. In contrast, both models can detect all type of time series anomalies when we consider the basic scenario.

Keywords: Deep learning · Auto-encoders · Anomalies detection · Unsupervised learning

1 Introduction

Anomaly detection [15] is a process that consists in detecting different abnormal behaviours, called *outliers*, in a dataset and provide useful information about their causes. Applications of anomaly detection include fraud detection [13], heart abnormalities [20], fault detection [10] and spam classification [16], among many others. In addition, anomaly detection can be applied to the analysis of time series. For example, it can be applied to streaming data related to the Internet of Things [24] and the stock market [14]. In addition, it can be applied to traffic flow time series, the application field that we consider in this paper. This field presents a main issue that does not appear in the previously mentioned cases: the lack of knowledge about what an anomaly is. Therefore, this technique requires the application of unsupervised models. At a first glance, we might be tempted to think that

---

This work has been supported by the Spanish MINECO/FEDER projects FAME (RTI2018-093608-B-C31) and AwESOMe (PID2021-122215NB-C31) and the Region of Madrid project FORTE-CM (S2018/TCS-4314) co-funded by EIE Funds of the European Union.

anomalies are the same as extreme values. Nevertheless, ideally, a well-performed model should be able to detect not just the extreme values but the values that imply, for example, the failure of a given pattern or a sudden increase or decrease from the previous value. Summarising, the definition of either a "normal" or an "abnormal" data should be a time-dependent quantity [24].

Several deep learning methods have been proposed to identify abnormal data in time series. Recurrent neural networks (RNN), mainly long short-term memory (LSTM) networks based models, are the primary option. RNN are applied in two different scopes. On the one hand, RNN aims to predict the future values and, subsequently, compare them with the actual values or with predefined thresholds by determining whether the corresponding data is or not an abnormal value [4,12,21]. On the other hand, auto-encoders (AE) or variational auto-encoders (VAE) based on RNNs (specifically, on LSTMs) are developed. Their aim is to restore the original values and compare the actual and the reconstructed values. Then, a threshold is established. If the discrepancy between the actual and the reconstructed value is greater than the threshold, then the corresponding data will be considered as abnormal [8,11,22].

In this work, we present two models to detect anomalies in time series data by using previous data as input. In order to assess the usefulness of our models, we will apply them to traffic flow data obtained in the city of Madrid. First, we develop a CNN auto-encoder. In addition to the dependent variable, this model will use another eight independent variables related to traffic flow, that is, we will be working within a multivariate time series. The second model is a novel Bidirectional LSTM auto-encoder model. This model will only use as input the traffic flow data (dependent variable), that is, we will work within a univariate time series. These models will be applied in both a basic scenario and a guided scenario (training only with data considered as'normal' after a naïve labelling).

Anomaly detection in traffic flow by employing auto-encoders is an unexplored research line. There are some recent approaches dealing with this task by using other techniques. A method known as *local outlier factor* was employed in the city of Odense [7]. By leveraging the availability of video surveillance, an algorithm based on fuzzy theory was developed to detect anomalies in road traffic flow [18]. An original and novel method, which does not take into account the historical data but uses expert feedback to deal with the fluid definition of anomaly, was developed in Vienna and was able to provide high accuracy [1]. To detect anomalies in Zagreb traffic flow, a tensor based method was proposed [23]. An adapted k-nearest neighbours method was used with the same purpose in Beijing [5]. If the reader is interested in further discussion on this topic, there is a survey about urban traffic flow anomalies detection algorithms [6]. To the best of our knowledge, there are no scientific contributions that develop a Bidirectional auto-encoder to detect anomalies in a time series, despite the fact of its capability to obtain good results in unsupervised scenarios.

The rest of the paper is organised as follows. In Sect. 2 we present the main characteristics of the anomaly detection problem in traffic flow, emphasising the application to our specific case study and used data. In Sect. 3 we present the main characteristics of our two models. In Sect. 4 we present our experiments and analyse the obtained results. Finally, in Sect. 5 we give our conclusions and outline some directions for future work.

2   Problem Description

In this section, we will describe the data used in this work and, subsequently, we will present the two scenarios that we consider: a guided scenario and a basic one.

2.1   Data

We use hourly traffic flow data of a station located in Madrid. We will use the last available 30 months of data: from January 2018 to July 2020. We also have eight additional variables related to traffic flow data that will be used to allow the model to better understand the context. These eight variables are:

- (1, 2) *Hourly traffic flow* in two near *additional stations*.
- (3, 4, 5) *Daily average, maximum* and *minimum temperature*.
- (6) *Daily rainfall*.
- (7) *Type of day* (working day, weekend or public holiday).
- (8) *Time* stamp of the data.

The values of the weather variables (3–6) where measured in the nearest weather station, while the traffic flow concerning additional stations (1–2) was measured in the two nearest traffic stations to the target traffic station.

2.2   Scenarios

Before we describe the considered scenarios, it is important to know the types of abnormal values that we can expect in time series [19]. These are:

- A value is considered a *global anomaly* if it is far outside of the dataset range.
- A value is considered a *contextual anomaly* if it significantly deviates from other data in a similar context.
- A subset of the dataset is considered a *collective anomaly* if those values, as a subset, significantly deviate from the dataset, but individually the values are not considered anomalous by themselves.

The goal of this paper is to provide a method to automatically detect abnormal behaviour of traffic flow data in Madrid using two different scenarios.

In the basic scenario we use all the data for training, and the aim of the models is to analyse and detect values corresponding to abnormal data. Obviously, the efficacy of the model cannot be assessed with traditional methods. Thus, we will evaluate the models by graphically analysing whether the values considered as anomalies correspond to any type of abnormal value in time series and by checking whether any remarkable event happened in Madrid in the date in which the abnormal data is detected.

In the guided scenario the aim is the same but data is divided into two sets. The first set includes the values greater than a high percentile of the dataset, which we know for sure they are abnormal values. The second set includes the rest of the data values. Obviously, not all abnormal values of the time series are included in the first set. However, the goal of this division is to check whether training the algorithm without some abnormal values improves or not the overall performance of the algorithm.

Formally, if X is the input data, then the models in both scenarios aim to find a function f such that:

$$f(X, \delta) = y,$$

being $y = 0$ if the value is not considered an anomaly and $y = 1$ if the value is considered an anomaly. The δ parameter is a pre-set value whose role depends on the model and will be described in the next section.

3   Auto-encoder Models

In this section we will present the basics of the models that will be applied in this work: *CNN* and *BiLSTM* auto-encoder models. The interested reader can find additional information about these models in specialised work [9,17] that introduce the principles of LSTM and CNN neural networks. First, let us describe how an auto-encoder works.

A variational auto-encoder (VAE) is an algorithm composed by an encoder and a decoder. The idea is that the encoder will translate the input, X, from its high-dimensional space to a lower-dimensional space. Then, the decoder will take that lower-dimensional representation of the input, what we usually called the *latent vector*, and will try to reconstruct the original input in the high-dimensional space, producing $\tilde{X}$. Due to its nature, the decoder usually has a structure that mimics the one of the encoder, being its goal to obtain a value $\tilde{X}$ as similar as possible to the original input, X. The dissimilarity between the original and reconstructed data is defined by a loss function. Therefore, the goal of the model is to reduce, as much as possible, this value during the training process.

This model is trained in a unsupervised way due to the output trying to be the input and, therefore, there is no need for labels. We will use two versions of a VAE to create the encoder and decoder: one using CNNs and another using BiLSTMs. We will use these models over two scenarios: a *basic* scenario and a *guided* one.

In both scenarios we train our model and then compute the dissimilarity score of all the trained data. This dissimilarity score is computed by comparing the input data with the reconstructed output data using a dissimilarity measure. We will use these dissimilarity scores to compute a threshold δ. This threshold is usually set to have the $90 - 99$th percentile of dissimilarity scores as abnormal data. Then, for new data, if the model can reconstruct input data with a dissimilarity value lower than δ, then it will be considered *normal data*; otherwise, it will be classified as *abnormal data*.

Regarding our scenarios, the difference is in training. In the basic scenario, we train the model with all the data (normal and abnormal), while in the guided scenario we train the model with only the data that we consider normal (the ones in the second set explained before). The idea is that, in this second case, the algorithm will learn to represent better the normal data, as the abnormal data has been taken away from training. Consequently, we expect the reconstruction of abnormal data to be more deficient, what will allow us to detect them more easily.

3.1 CNN Auto-encoder Model

Let y be the traffic flow in the target station, $x^1, \ldots, x^8$ be the additional variables defined in Sect. 2.1 and t be a time step. In order to decide whether the traffic flow data in time step t, that is y_t, is an abnormal data, we will use the variables $y, x^1, \ldots, x^8$ from time step $t - 23$ to t. We reshape these values into a 24×9 matrix, where 24 is the number of values that we use in order to detect the anomalies. Therefore, we have

$$X = \begin{bmatrix} y_{t-23} & x^0_{t-23} & \cdots & x^8_{t-23} \\ y_{t-22} & x^0_{t-22} & \cdots & x^8_{t-22} \\ \cdots & \cdots & \cdots & \cdots \\ y_{t-1} & x^0_{t-1} & \cdots & x^8_{t-1} \\ y_t & x^0_t & \cdots & x^8_t \end{bmatrix}$$

In addition, we will use as dissimilarity loss function the *mean absolute error*:

$$Loss(X, \tilde{X}) = MAE = \frac{1}{216} \sum_{i=1}^{24} \sum_{j=1}^{9} |X_{i,j} - \tilde{X}_{i,j}|,$$

where i and j represent, respectively, the rows and columns of the matrix.

Concerning the specific structure of our CNN model, as we previously said, this model is composed by two parts: encoder and decoder. Our encoder has two convolutional layers of 32 and 64 filters, respectively, each of them with a kernel size of 3 and a stride of 2. Then, a layer vector is added in order to transform the data into a flatten vector. Finally, we add a dense layer with 8 units to generate the 8-dimensional latent vector. Our decoder is composed by a set of layers whose operation is exactly the opposite to that of the encoder. First, we add a dense layer, which increases the dimension to the dimension of the flatten layer, and then we add a reshape layer, which modifies the data shape into a matrix form. Finally, we add three transpose convolutional layers of 64, 32 and 1 filters, respectively, a kernel size of 3 and a stride of 2. Summarising, the inverse encoder process is the one used to reconstruct input data from the latent vector.

3.2 BiLSTM Auto-encoder Model

This model is also composed by an encoder and a decoder. In this case, BiLSTM layers are used because they leverage the advantages of BiLSTM and encoder-decoder architectures, respectively. The general behaviour of this model is similar to the previously described CNN model.

In order to determine whether traffic flow data in time step t, that is the y_t value, is an abnormal data, we will use the univariate time series of traffic flow, from time step $t-23$ to t. We use as input data a 24-sized vector, where 24 is the number of previous values that we use in order to detect anomalies. Therefore, we have the following:

$$X = [y_{t-23}, y_{t-22}, \ldots, y_{t-1}, y_t]$$

In this case, the dissimilarity is also measured between the input sequence and the output, that is, the reconstructed sequence. We use again the *mean absolute error* as dissimilarity loss function:

$$Loss(X, \tilde{X}) = MAE = \frac{1}{24} \sum_{t-23}^{t} |y_i - \tilde{y}_i|.$$

Concerning the specific structure of our model, the encoder is composed of a BiLSTM layer with 128 units followed by a dropout layer, with a rate of 0.2, which reduces the over-fitting of the model. Then, we add a repeat vector layer which repeats the input k times (one for time step). Our decoder also includes a BiLSTM layer with 128 units, which in this case returns the entire sequence, followed by a dropout layer with a rate of 0.2. Finally, a dense layer with one unit is added to the network in order to output each time step in the final output sequence. We would like to emphasise that in this model we do not use predictor variables, as in the previous case, because BiLSTM networks do not work optimally by using matrices as input. Therefore, using as input the matrix considered in previous case, even as a flatten vector, would imply a significant loss of the quality of the model.

4 Experiments

In this section we report on the experiments that we performed to evaluate both models in the considered scenarios.

4.1 Basic Scenario

We will analyse the results of applying the CNN and BiLSTM auto-encoder models to the basic scenario for Paseo de Santa María de la Cabeza traffic station (N E). In both cases, we will use as training set data 16.391 observations and as testing set data 5.448 observations. We train all the models by using 15 epochs and 16 as batch size.

For both models, we will plot three different graphs. The first graphs (see Fig. 1) are histograms with 50 bins that represent the distribution of the dissimilarity loss in training data. A higher concentration of dissimilarity loss with low representation in the extreme right of the graphs gives the possibility to choose higher thresholds that virtually will not return any false positive at time to detect anomalies. The second graphs (see Fig. 2) represent the dissimilarity score for the observations of the testing set and which of them are considered

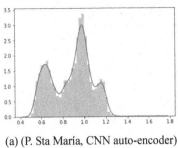

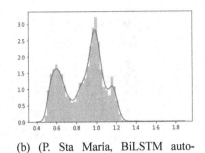

(a) (P. Sta María, CNN auto-encoder)

(b) (P. Sta María, BiLSTM auto-encoder)

Fig. 1. Histogram of the dissimilarity loss distribution in training data.

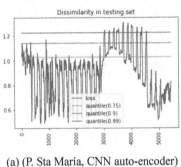

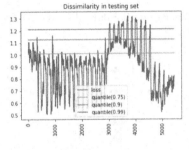

(a) (P. Sta María, CNN auto-encoder)

(b) (P. Sta María, BiLSTM auto-encoder)

Fig. 2. Specific data detected as anomalous by threshold.

anomalies depending on the selected threshold. The y-axis represents the dissimilarity loss and the x-axis represents each observation of the testing set. The chosen thresholds (represented by horizontal lines) are the 75th, 90th and 99th quantiles of dissimilarity loss in training data. The values above each horizontal line are the anomalies considered according to its respective threshold. The third pair of graphs (see Figs. 3 and 4) show the detected anomalies (plotted in the time series). The x-axis represents the date while the y-axis represents the hourly traffic flow. The different coloured points show the values considered abnormal data depending on the selected threshold. Obviously, the data considered abnormal by using a given quantile as threshold will be also considered abnormal by higher quantiles. In these graphs, we can analyse, graphically and by date, the reasons why different observations are considered anomalies.

Comparison Between Both Models. A simple glance shows a great similarity between the results in the two models. However, there are some remarkable differences that we would like to discuss. On average, the dissimilarity measure is smaller in the BiLSTM auto-encoder model and the thresholds are higher

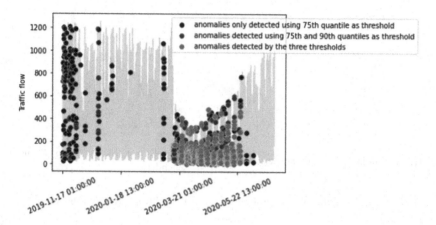

Fig. 3. Specific data detected as anomalous by threshold (P. Sta María, CNN auto-encoder).

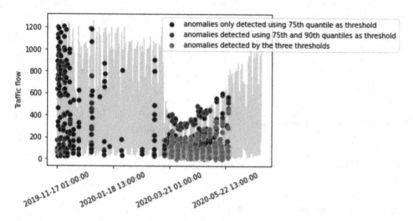

Fig. 4. Specific data detected as anomalous by threshold (P.Sta María, BiLSTM auto-encoder).

in the CNN auto-encoder model. Consequently, the BiLSTM auto-encoder will detect more possible outliers than the CNN auto-encoder. By using the 99th quantile as threshold (see Figs. 3 and 4), the anomalies are detected in a same range of dates that, interestingly enough, corresponded to the beginning of the restrictions caused by the COVID-19 pandemic. By using the 90th quantile as threshold, both models can also detect as anomalies data which corresponds to the Christmas festivities and to the last weekend of November. However, the BiL-STM auto-encoder detects more outliers in the range of dates that correspond to the restrictions (with 90th quantile as threshold, BiLSTM auto-encoder obtains 788 anomalies in total versus the 638 obtained by CNN auto-encoder). By using the 75th quantile as threshold, there are some detected anomalies which do not correspond to any important date or to any graphical deviation. Thus, we may conclude that they are not actual anomalies.

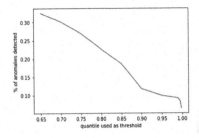 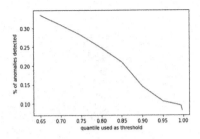

Fig. 5. % of anomaly detection by quantile used as threshold (CNN auto-encoder).

Fig. 6. % of anomaly detection by quantile used as threshold (BiLSTM auto-encoder).

As explained before, if we use smaller quantiles as threshold, then the number of detected anomalies increases. Figures 5 and 6 show this effect in both models. Interestingly enough, the percentage of anomalies remains practically constant in the CNN auto-encoder model from the 90th quantile up to the 99th quantile while in the BiLSTM auto-encoder model, this *stability* happens only from the 95th quantile up to the 99th quantile. The presence of this stability in a wider range implies a greater certainty on the anomalies detected by using as threshold the lower end of the range under consideration.

In conclusion, both models obtain similar results in the basic scenario. However, the BiLSTM auto-encoder model shows a better reconstruction of the input and, therefore, it can detect more anomalies with the same quantile set as threshold. Moreover, we should take into account that the CNN auto-encoder needs eight independent variables while the BiLSTM auto-encoder only needs the time series of the target variable.

4.2 Guided Scenario

We will analyse the results by using the CNN and BiLSTM auto-encoder models in the guided scenario for Paseo de Santa María de la Cabeza station (the one that we used in the basic scenario). We propose to consider for the abnormal set every value higher than the 90% and 99% of the range of traffic flow time series data. The idea is to analyse how modifying the number of abnormal data used for training affects the performance of the algorithm. We use 75th, 90th and 99th quantiles for the dissimilarity loss threshold.

In contrast with the basic scenario, we will use as training set only the data that is considered "normal", that is, those data points that are not in the abnormal set. Thus, ideally, the models will learn the representation of normal data and will obtain a small dissimilarity loss. Therefore, it is expected that when we consider the testing set (including, in particular, abnormal data), the dissimilarity loss will be greater.

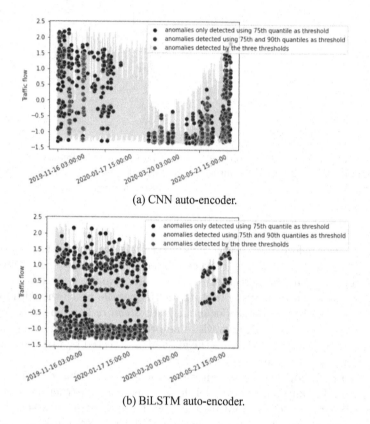

(a) CNN auto-encoder.

(b) BiLSTM auto-encoder.

Fig. 7. Specific data detected as anomalous by threshold (considering as lower limit of abnormal data the 90% of the data range).

Comparison Between Models Considering as Lower Limit of Abnormal Data the 90% of the Data Range. If we compare the models by considering as lower limit of abnormal data the 90% of the data range (see Figs. 7a and 7b) we appreciate a different behaviour between them. On the one hand, regarding the CNN auto-encoder model, the model with 99th quantile as threshold can detect only the beginning of specific and remarkable periods such as December public holidays, Christmas holidays or the most restrictive period caused by the COVID19 pandemic. However, by using either the 90th or 75th quantiles, the abnormal data detected are not apparently related to any remarkable date or event. On the other hand, regarding the BiLSTM auto-encoder model, the model with 99th quantile as threshold only detects three abnormal timesteps, which seems a very poor result. By using either the 90th or 75th quantiles, the model detects data of the extremes. That is, very high or very low values. But, for example, the model does not appreciate an abnormal behaviour in the COVID19 period. That is clearly a direct cause of the training method. In general, both models obtain worst results in this scenario than in the basic scenario. Never-

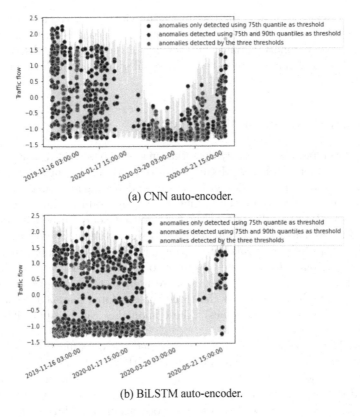

(a) CNN auto-encoder.

(b) BiLSTM auto-encoder.

Fig. 8. Specific data detected as anomalous by threshold (considering as lower limit of abnormal data the 99% of the data range).

theless, the CNN auto-encoder can detect, although in a poorly way, all type of abnormal data while the BiLSTM auto-encoder can detect only the abnormal data known as *global anomaly*.

Comparison Between Models Considering as Lower Limit of Abnormal Data the 99% of the Data Range. If we consider as lower limit of abnormal data the 99% of the data range (see Figs. 8a and 8b) we appreciate a similar behaviour to the previous case. As a first conclusion, we have that an increase of the lower limit of abnormal data implies an increase of the number of detected abnormal data. Regarding the CNN auto-encoder model, the model with 99th quantile as threshold detects exactly the same abnormal data as in the previous case. By using 90th or 75th quantiles, the detected abnormal data increases significantly but still there is no relation to any remarkable date or event. Regarding the BiLSTM auto-encoder model, the model with 99th quantile as threshold can detect more abnormal data than in previous case, but they are only *global anomalies*. By using the 90th and 75th quantiles, the model detects practically the same data as in the previous case. Again, we conclude that the CNN auto-encoder, especially using 99th quantile as threshold, can detect exclusively abnormal data (but not all) of the three type defined above, while the BiLSTM model can detect almost all global anomalies, but none of other type.

5   Conclusions and Future Work

We have developed two anomaly detection methods based on neural networks: CNN auto-encoder and BiLSTM auto-encoder. Both models have been applied in a guided and a basic scenarios with the goal of detecting abnormal data. In the guided scenario, we break up values higher than an established quantile into an abnormal set. Results indicate that the CNN auto-encoder, although can detect anomalies of the three types, cannot detect a significant number of them while BiLSTM auto-encoder can only detect anomalies of one type (global anomalies). Therefore, we conclude that the usefulness of these models in this scenario is limited. However, in the basic scenario, we appreciate that both models can virtually detect all anomalies of the three types (global anomalies, contextual anomalies and collective anomalies). These anomalies can be appreciated both graphically and by date.

Generally, we conclude that auto-encoder methods work fine to detect anomalies in time series in a basic scenario. Moreover, they work well with all types of abnormal data because they base their operation on the difference between the input and the reconstructed output. However, standard deep learning methods do not work as well because they base their operation on finding similar characteristics between data, but different types of anomalies are not similar to each other.

As future work, we would like to improve our proposal with better encoder/decoder architectures. We would like to combine our models with current work on Complex Event Processing [2,3] so that we are able to more accurately detect potentially anomalous values. Also, we would like to further analyse the performance of our proposals, looking for different scenarios with which to generalise our conclusions. Finally, other interesting idea is to use our auto-encoders in other type of time series urban data such as pollutant concentrations.

References

1. Alam, Md R., Gerostathopoulos, I., Amini, S., Prehofer, C., Attanasi, A.: Adaptable anomaly detection in traffic flow time series. In: 2019 6th International Conference on Models and Technologies for Intelligent Transportation Systems (MT-ITS), pp. 1–9 (2019)
2. Corral-Plaza, D., Medina-Bulo, I., Ortiz, G., Boubeta-Puig, J.: A stream processing architecture for heterogeneous data sources in the internet of things. Comput. Standard. Interfaces **70**, 103426 (2020)

3. Corral-Plaza, D., Ortiz, G., Medina-Bulo, I., Boubeta-Puig, J.: MEdit4CEP-SP: a model-driven solution to improve decision-making through user-friendly management and real-time processing of heterogeneous data streams. Knowl. Based Syst. **213**, 106682 (2021)
4. Ding, N., Ma, H., Gao, H., Ma, H., Tan, G.: Real-time anomaly detection based on long short-term memory and gaussian mixture model. Comput. Electr. Eng. **79**, 106458, (2019)
5. Djenouri, Y., Belhadi, A., Lin, J. C.-W., Cano, A.: Adapted k-nearest neighbors for detecting anomalies on spatio-temporal traffic flow. IEEE Access **7**, 10015–10027 (2019)
6. Djenouri, Y., Belhadi, A., Lin, J. C.-W., Djenouri, D., Cano, A.: A survey on urban traffic anomalies detection algorithms. IEEE Access **7**, 12192–12205 (2019)
7. Djenouri, Y., Zimek, A., Chiarandini, M.: Outlier detection in urban traffic flow distributions. In: 2018 IEEE International Conference on Data Mining (ICDM), pp. 935–940 (2018)
8. Gugulothu, N., Malhotra, P., Vig, L., Shroff, G.: Sparse neural networks for anomaly detection in high-dimensional time series (2018)
9. Hochreiter, S., Schmidhuber, J.: Long short-term memory. Neural Comput. **9**(8), 1735–1780 (1997)
10. Hong, Y., Wang, L., Kang, J., Wang, H., Gao, Z.: A novel application approach for anomaly detection and fault determination process based on machine learning. In: 2020 6th International Symposium on System and Software Reliability (ISSSR), pp. 1–5 (2020)
11. Hsieh, R.-J., Chou, J., Ho, C.-H.: Unsupervised online anomaly detection on multivariate sensing time series data for smart manufacturing, pp. 90–97 (2019)
12. Hundman, K., Constantinou, V., Laporte, C., Colwell, I., Söderström, T.: Detecting spacecraft anomalies using LSTMs and nonparametric dynamic thresholding. CoRR, abs/1802.04431 (2018)
13. Jiang, J., Chen, J., Gu, T., Raymond Choo, K.-K., Liu, C., Yu, M., Huang, W., Mohapatra, P.: Anomaly detection with graph convolutional networks for insider threat and fraud detection. In: MILCOM 2019–2019 IEEE Military Communications Conference (MILCOM), pp. 109–114 (2019)
14. Karthik, S., Supreetha, H.V., Sandhya, S.: Detection of anomalies in time series data. In: 2021 IEEE International Conference on Computation System and Information Technology for Sustainable Solutions (CSITSS), pp. 1–5 (2021)
15. Atzmueller, M., Arnu, D., Schmidt, A.: Anomaly detection and structural analysis in industrial production environments. In: Data Science – Analytics and Applications, pp. 91–95. Springer, Wiesbaden (2017). https://doi.org/10.1007/978-3-658-19287-7_13
16. Laorden, C., Ugarte-Pedrero, X., Santos, I., Sanz, B., Nieves, J., Bringas, P.G.: Study on the effectiveness of anomaly detection for spam filtering. Inf. Sci. **277**, 421–444 (2014)
17. LeCun, Y., et al.: Backpropagation applied to handwritten zip code recognition. Neural Comput. **1**(4), 541–551 (1989)
18. Li, Y., Guo, T., Xia, R., Xie, W.: Road traffic anomaly detection based on fuzzy theory. IEEE Access **6**, 40281–40288 (2018)
19. Rajeswari, A.M., Yalini, S.K., Janani, R., Rajeswari, N., Rajeswari, N., Deisy, C.: A comparative evaluation of supervised and unsupervised methods for detecting outliers. In: 2018 Second International Conference on Inventive Communication and Computational Technologies (ICICCT), pp. 1068–1073 (2018)

20. Ripan, R.C., Sarker, I.H., Furhad, M.H., Anwar,M.M., Hoque, M.M: An effective heart disease prediction model based on machine learning techniques. https://www.preprints.org/manuscript/202011.0744/ (2020)
21. Shen, L., Li, Z., Kwok, J.: Timeseries anomaly detection using temporal hierarchical one-class network. In: H. Larochelle, M. Ranzato, R. Hadsell, M. F. Balcan, and H. Lin, editors, In: Advances in Neural Information Processing Systems vol. 33, pp. 13016–13026 Curran Associates Inc (2020)
22. Su, Y., Zhao, Y., Niu, C., Liu, R., Sun, W., Pei, D.: Robust anomaly detection for multivariate time series through stochastic recurrent neural network. In: Proceedings of the 25th ACM SIGKDD International Conference on Knowledge Discovery & Data Mining, KDD '19, pp. 2828–2837, New York, NY, USA, (2019). Association for Computing Machinery
23. Tišljarić, L., Fernandes, S., Carić, T., Gama, J.: Spatiotemporal road traffic anomaly detection: a tensor-based approach. Appl. Sci. 11(24), 12017 (2021)
24. Werra, L.V., Tunstall, L., Hofer, S.: Unsupervised anomaly detection for seasonal time series. In: 2019 6th Swiss Conference on Data Science (SDS), pp. 136–137 (2019)

A Deep Convolution Generative Adversarial Network for the Production of Images of Human Faces

Noha Nekamiche[1(✉)], Chahnez Zakaria[1(✉)], Sarra Bouchareb[2(✉)], and Kamel Smaïli[3(✉)]

[1] École Nationale Supérieure d'Informatique, Algiers, Algeria
{hn_nekamiche,c_zakaria}@esi.dz
[2] Laboratoire LIMPAF Bouira University, Bouira, Algeria
sa.bouchareb@univ-bouira.dz
[3] Loria, Campus Scientifique, Vandoeuvre Lès-Nancy, France
smaili@loria.fr

Abstract. Generative models get huge attention by researchers in different topics of artificial intelligence applications, especially generative adversarial networks (GANs) which have demonstrated good performance in data generation. In this paper, we would like to explore the potential of this class of models in producing human faces images. For that, we will use Deep Convolutional Generative Adversarial Network (DCGAN). Since that, the evaluation of GANs is still difficult even with the existing metrics like Inception Scores (IS), Mode Score (MS), Kernel Inception Distance (KID), Fréchet Inception Distance (FID), Multi-Scale Structural Similarity (MS-SSIM), etc. Thus, the best possible evaluation remains that carried out by human evaluators. This is why we propose a new hybrid measure combining qualitative and quantitative evaluation, we called this measure: Measuring the Quality of the Features of an Image (MEQFI). The images produced with the DCGAN method were trained on three well known datasets from the literature and the results were evaluated with MEQFI.

Keywords: Image generation · Generative adversarial network · Deep convolutional generative adversarial network · Batch normalization

1 Introduction

Generative adversarial networks (GANs) are a class of generative models proposed by Ian J. Goodfellow et al. [6], which consist of two adversarial players, a Generator that produces new data from random input and a Discriminator which classifies the produced and the real data. Both the Generator and the Discriminator try to maximize their success and minimize the success of the other one. This approach is inspired from a learning paradigm that was introduced by Schmidhuber in 1990 [23]. In this paradigm, two different neural networks compete each other in minmax game. However, the simultaneous training of the two adversarial networks makes

N. T. Nguyen et al. (Eds.): ACIIDS 2022, LNAI 13757, pp. 313–326, 2022.
https://doi.org/10.1007/978-3-031-21743-2_25

the learning process noisy, unstable and hard to optimize. This leads to problems such as vanishing gradients [6] because of the adversarial training, mode collapse [5] which happens when the Generator produces a limited set of examples and the Discriminator is blocked in the local minimum and consequently it can't make any progress, especially when the model's parameters are destabilized. Several works have been developed to overcome these problems and improve the quality of the data generated, we find those which improve the architecture of the GAN by using additional classifiers or by adopting semi-supervised learning to guide the learning process and those which focus on modifying the objective function depending on the application domain. GANs have made great achievements since their introduction in image synthesis [2] [21], in style transfer [12], in image super-resolution [4] [15, 24] and in image-to-image translation [11]. Recently, applications have been extended to a wide range of fields, not only image generation, but also language and speech processing, security analysis, malware detection and chess program. One of the main challenges in machine learning is the collecting data and sizing datasets appropriately for relevant training. Thus, GANs can help to manage this problem by generating synthetic data to achieve data augmentation and dataset balancing.

In this paper, we focus on face generation tasks by using Deep Convolutional Generative Adversarial Networks (DCGAN) [20] and we introduce a human evaluation metric that calculates the score of each generated image based on human feedbacks. This paper is structured as follows: in Sects. 2 and 3 we summarize the previous related works done on GANs. Then we present the characteristics of the datasets used for our study, the experiment configurations as well as the experimental results with analysis in Sect. 4. After that, we review the used evaluation metrics and illustrate our evaluation metric in Sect. 5. Finally, we draw the conclusion from the model that we have studied in Sect. 6.

2   A Recall of the Genarative Adversial Networks (GAN)

The GAN model consists of two different Deep Neural Networks. The Generator Network: takes random inputs Z to learn how to produce fake data and try to make them very similar to the real data X (training data). Thus, the objective of the Generator is to produce undistinguished data from the real one. The Discriminator Network: takes as input fake and real data and tries to learn how to distinguish between them using a loss function. Thus, the objective of the Discriminator is to recognize which data are fake and which ones are real and this is what makes the Discriminator works like a classifier.

GANs implement two neural networks, a Generator G and a Discriminator D. They both play an adversarial game where the Generator functions as a forger who at all times tries to fool the expert (the Discriminator) by generating data that is increasingly similar to that of the training set. The Discriminator is on guard not to be fooled into identifying fake data from real data. The Discriminator and the Generator work simultaneously as in a min-max game with two players D and G in which each one tries to maximize its success and minimize the success of the other one, in other words they try to reach the Nash equilibrium. In this model, the training process is split into 2 parts:

- Discriminator update. The D Discriminator is a binary classifier. Its inputs are data without a priori knowledge about the quality of true or false data. As output, it produces a probability that the data is true or false (those produced by the Generator). The weights of the Discriminator network are updated in order to reduce its classification error. After several iterations of updating the weights, D is better able to distinguish between true and false examples.
- Generator update. For the Generator, it first generates some fake examples using the noise z as input, and then these are passed to the Discriminator. Because the Generator does not know the real examples, at first it does not know how to produce real data, and the Discriminator can easily distinguish between real and fake data without much error. The Generator aims to maximize its probability of producing an example given the class label corresponding to real data. Then the Discriminator acts as what we have already explained above and after updating the parameters of G and calculating the cost, the gradient is then propagated backwards and the parameters of the Generator are updated (Fig. 1).

The cost function is given by the Formula 1.

$$V(D,G) = \mathop{\mathbb{E}}_{x \sim p_{data}(x)} [log D(x)] + \mathop{\mathbb{E}}_{z \sim p_z(z)} [log(1 - D(G(z)))] \qquad (1)$$

The first item is the expectation of the log of the Discriminator output when the data is from the real data distribution. The second parameter is the expected value of the log of the quantity of one minus the Discriminator prediction on the fake samples. The method consists in maximising $D(x)$ and minimising $D(G(z))$. So that is why, $1 - D(G(z))$ is used in order to make the Generator and the Discriminator moving in consistent directions between the two terms in the equation.

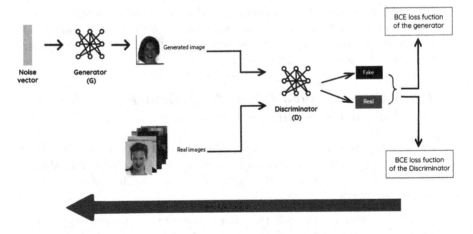

Fig. 1. GAN architecture.

3   Related Works Concerning the Variants of GANs

Several works have been done to improve the quality of the GANs, we can summarise them into two main variants:

3.1   Architecture-Variant

This category regroups works that improve the quality of the generated images by changing the architecture of the Generators and Discriminators, like: DCGAN which we will study in this paper, LAPGAN [3] which combines the model of conditional GAN (cGAN) with Laplacian pyramid (LP) framework which is constructed from the Gaussian pyramid (GP). Or by changing the latent space like cGAN [18] that uses label information into the Generator and the Discriminator in order to control the type of the generated data. The semi-supervised GAN (SSGAN) [19] has been inspired from the principle of semi-supervised learning and the ability of GANs to produce unlabelled data, so its Discriminator becomes a multi-class semi-supervised classifier with Softmax and Sigmoid activation functions for the classification of the fake and real examples. In [26], they used a modified architecture of GAN according to the application domain. CycleGAN [26] is used to achieve image-to-image translation when unpaired training data is not available. In [15], the method based on GAN generates high-resolution images from low-resolution images by upsampling, also it extends the loss by adding content loss.

3.2   Loss-Variant

This category regroups works that improve the performance of GANs by changing the type of the loss function, like: WGAN [1] which uses Earth-Mover (EM) or Wasserstein distance instead of using BCE loss function and here the Discriminator is used to fit the Wasserstein distance. In [1], they proved that WGAN overcome the vanishing gradient problem and remove the mode collapse partially. Others propose to add additional penalisation to the loss function or any type of normalisation applied to the network, like: WGAN-GP [7] which is a variant of WGAN with gradient penalty.

4   Deep Convolutional GAN: A Method Adopted for Human Faces Images Producing

Since fully connected layers reduce the quality of generated images in the GAN model, in [20], the authors introduced the concept of Deep Convolutional GAN (DCGAN) that permits to generate high-quality images. The authors replaced the pooling layers with:

- Strided Convolutional layers in the Discriminator to down-sample the images.
- Fractionally–strided convolutional layers in the Generator to up-sample the images.

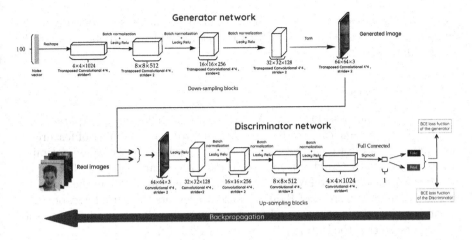

Fig. 2. DCGAN architecture.

Table 1. List of the 40 binary attributes extracted from the images of CelebA database.

Index	Definition	Index	Definition	Index	Definition	Index	Definition
1	5o'ClockShadow	11	Blurry	21	Male	31	Sideburns
2	ArchedEyebrows	12	BrownHair	22	MouthSlightlyOpen	32	Smiling
3	Attractive	13	BushyEyebrows	23	Mustache	33	StraightHair
4	BagsUnderEyes	14	Chubby	24	NarrowEyes	34	WavyHair
5	Bald	15	DoubleChin	25	NoBeard	35	WearingEarrings
6	Bangs	16	Eyeglasses	26	OvalFace	36	WearingHat
7	BigLips	17	Goatee	27	PaleSkin	37	WearingLipstick
8	BigNose	18	GrayHair	28	PointyNose	38	WearingHat
9	BlackHair	19	HeavyMakeup	29	RecedingHairline	39	WearingNecktie
10	BlondHair	20	HighCheekbones	30	RosyCheeks	40	Young

They removed the fully connected hidden layers, they also used batch normalization [10] in order to help the training process to become more stable by normalizing the inputs of each layer. The general architecture of the DCGAN is illustrated in the Fig. 2:

4.1 Datasets

To evaluate our DCGAN model, we used three databases of famous faces, the first is *CelebA* , which was initially collected by researchers from the Chinese University of Hong Kong MMLAB. This dataset is a large-scale face attribute database, it contains more than 200,000 images of more than 10,000 celebrities, it covers many images of different races, ages, genders, hairstyles, eye colors, etc. We summarise them in 40 binary attributes (see Table 1). The second is *CelebA-HQ* [13], which is an improved version of CelebA in terms of image quality. The third is the Labeled Faces in the Wild *LFW* dataset [9], which was created and

N. Nekamiche et al.

maintained by researchers at the University of Massachusetts. It contains 13,233 images of 5,749 people annotated with the same 40 binary attributes as in the CelebA database.

4.2 Configuration

For the implementation of DCGAN Model, we use the following hyperparameters: batch size = 128, image size = 64 ± 64, noise vector =100, number of features in both of the Generator and the Discriminator 64, and we use the Adam optimizer [14] to update both G and D neural networks with a learning rate of 0.0002 and β_1 =0.5. We also used the LeakyRelu activation function [17,25] in the generator and discriminator, instead of Relu, because it prevents the neural network from getting stuck in the death state, where the neural network only produces zeros for all outputs, so that the learning process no longer learns.

4.3 Results

For the test we use 12800 images from each dataset: CelebA, CelebA-HQ (only female faces) and LFW by testing several values of epochs: 100, 150 and 200.

In Figs. 3a, 3b and 3c, we give the illustration of the images produced by training the DCGAN on CelebA. While we analyse the loss function of Figs. 3d, 3e and 3f, we can remark that after the first epochs that the Generator learns

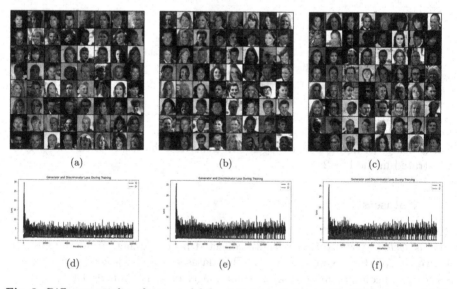

Fig. 3. Different results of our model from CelebA dataset with 100 batches of 128 images. (a), (b) and (c) represent the generated images after 100, 150 and 200 epochs respectively. (d), (e) and (f) the G (blue color) and D (orange color) losses after 100, 150 and 200 epochs respectively. (Color figure online)

to fool the Discriminator by creating more or less real images. The overall shape of the Generator's curve is decreasing, where the values are ranging from a loss of about 29,91 to 3,34. But unfortunately, the loss value gets stuck in the range [1.2 ... 6.11]. As for the discriminator from the beginning, it manages to distinguish the fake data, nevertheless for certain batches, it experiences some difficulties and is fooled by the Generator. And finally, regarding to these experiments, it is preferable to use only 100 epochs, increasing the number does not help to improve the training.

Concerning the experiment of Fig. 4 with a high quality images of women faces, the aspect of the images are more accurate, and we can remark also as for the experiment of Fig. 5 that for 100 epochs the results are better than the other two other experiments with 150 and 200 epochs. During the training, the initial generated images are not very good, the Discriminator was able to classify them easily, but as the training progresses the generation process becomes better and the quality of the synthesised images is more accurate. This makes the classification task difficult for the Discriminator because the generated images become more similar to the real ones.

For a general comment concerning the previous experiments, based on the loss functions of the Generator and Discriminator of each dataset ((d), (e) and (f) of Fig. 3, 4 and 5), the process oscillates in a narrow range, this is due to the adversarial training and the fact that the system reaches the Nash equilibrium, which is a state where the Discriminator can not distinguish between real

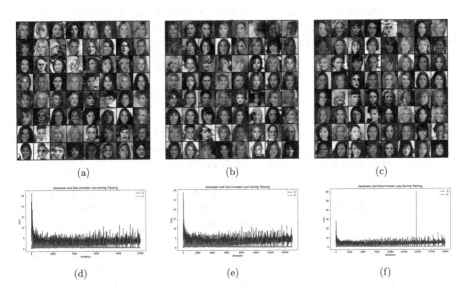

Fig. 4. Different results of our model from CelebA-HQ-females only dataset with 100 batches. (a), (b) and (c) represent the generated images after 100, 150 and 200 epochs respectively. (d), (e) and (f) the G (blue color) and D (orange color) losses after 100, 150 and 200 epochs respectively. (Color figure online)

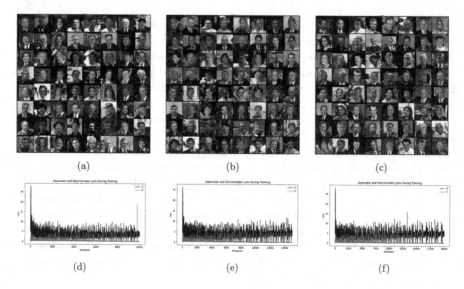

(a)   (b)   (c)

(d)   (e)   (f)

Fig. 5. Different results of our model from Labelled Faces in the Wild dataset with 100 batches. (a), (b) and (c) represent the generated images after 100, 150 and 200 epochs respectively. (d), (e) and (f) the G (blue color) and D (orange color) losses after 100, 150 and 200 epochs respectively. (Color figure online)

and fake images. But it is difficult to achieve this equilibrium because of the simultaneous training of the two adversarial networks. We also found that the Discriminator's loss value decays towards zero faster than the Generator's loss value.

5   Evaluation

One of the hardest tasks in this kind of process is how to find a good metric that allows to evaluate the images achieved by a process based on GANs algorithm. In the literature we identified two types of evaluation:

- Qualitative evaluation, where we will evaluate the model based on the human subjective evaluation [3] [16] and according to the feedback, one can get an idea about the quality of the produced images. This is what is done in several research areas in artificial intelligence.
- Quantitative evaluation, where a specific numerical score is calculated indicating the quality of the produced images. In this topic, we identified 24 quantitative metrics to evaluate GANs and the most used one is the Inception score (IS) which was proposed by Salimans et al. [22]. The idea is to

apply an inception model on each generated image to get a conditional label distribution $P(y|x)$, the ideal images are those which have a distribution obeying to a low entropy. The inception score has the lowest value of 1 and the highest value of the number of classes supported by the classification model, and with this measure, we expect that the model will produce divers images. The inception score is calculated by estimating the class conditional probabilities for each generated image. The images that have a higher ranking in one class compared to all other classes correspond to the best images. And therefore the conditional probability of all generated images must have low entropy. As it is a question here of obtaining a diversity of images, it is therefore necessary to use the marginal probability $\int_z P(y|x = G(z))dz$, that is to say the probability distribution of all the images generated. To verify this condition of obtaining a maximum of diversity of images, it is therefore necessary to maximize the entropy. The general formula of this score is a combination of these two distributions by using the Kullback-Leibler divergence (KL) between the conditional and the marginal distributions knowing that here, we are interested in the average of the KL divergence for all generated images. The inception score for a generator is given by:

$$IS(G) = exp(\mathbb{E}_{z \sim P(z)} KL(P(y|x) || P(y))) \qquad (2)$$

There is another measure named Fréchet Inception Distance (FID) which was proposed by Heusel et al. [8]. It measures the similarity between two distributions (real data x and generated data g) by applying embeddings on real and fake images to extract their features. The embedding is computed using a pre-trained convolutional network called Inception-v3 which was proposed by a team at Google. The formula of FID is given by:

$$FID(x, g) = ||\mu_x - \mu_g||_2^2 + Tr(\sum_x + \sum_g -2(\sum_x \sum_g)^{1/2}) \qquad (3)$$

where: μ_x and μ_g are the feature-wise mean of real and generated images, respectively. $\sum_x$ and $\sum_g$ are the covariance matrix of real and generated images, respectively.

5.1   Our Contribution

In this section, we propose a hybrid measure based on a qualitative evaluation for each present feature and which is summarized by a quantitative score. In the following, this score will be named Measuring the Quality of the Features of an Image (MEQFI). For that, we define a list of F features that will be used to calculate the score of each generated image. For each feature i, we attribute a mark based on human feedback to calculate its probability P_i. We multiply this probability by an activation coefficient α_i. This score is given by:

$$Score(m_k) = \frac{1}{F - I} \sum_{i=1}^{F} \alpha_i * P_i \tag{4}$$

where :

- m_k: is the kth produced image to evaluate.
- F: represents the number of features that we use to evaluate the quality of the generated image.
- I : represents the number of features ignored for some images from the generated images, for example: if the generated image is not smiling, we will eliminate the feature teeth.
- α_i: represents the activation coefficient of each feature. It is equal to zero if the corresponding feature is ignored otherwise, it is set to 1.
- P_i: represents the probability that the model succeeded in creating the attribute i. This probability is calculated as follows: $P_i = \frac{mark}{scale}$.

After calculating the score of each generated image, we calculate the average score over all the generated images.

5.2   Evaluation of the Images Produced by DCGAN with MEQFI

Fig. 6. Eight generated images from CelebA-HQ only female.

We fix a group of features {Nose, Eyes, Eyebrows, Mouth, Ears, Skin tone} to calculate the score of each image by using MEQFI. For that, we ask a group of evaluators to attribute a mark from one to five for each feature of each image. To illustrate the result on a concrete example, we give for the images of Fig. 6, the results in the Table 2 in terms of the MEQFI measure for each of these images and for each attribute.

Table 2. MEQFI Score on eight generated images from CelebA-HQ-females.

Image	Nose	Eyes	Eyebrowns	Mouth	Ears	Teeth	Skin Tone	Image Score (%)
	0,6	0,6	1	0,4	0	0,5	1	68,33
	0,5	0,6	1	0,4	0,5	0,5	0,9	62,86
	0,2	0,5	0,8	0,4	0,5	0,2	1	51,43
	0,4	0,3	0,6	0,4	0,5	0,2	0,8	45,71
	0,4	0,5	1	0,5	0,5	0	0,9	63,33
	0,4	0,6	1	0,5	0,4	0,6	1	64,23
	0,4	0,4	1	0,4	0	0,7	1	65,0
	0,4	0,5	0,7	0,6	0,4	0	0,9	58,33
Totale Score %								**59,91**

Each line of the Table 2 represents the application of MEQFI on the concerned image.

After having collected the human feedbacks of the 64 images for each of the three datasets, we calculate the score of each image with MEQFI, and we plot the results in the Fig. 7:

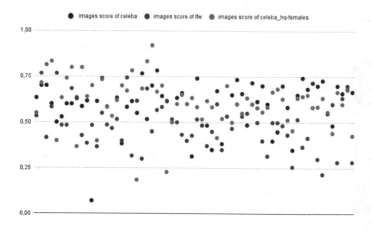

Fig. 7. MEQFI score for 64 images generated by DCGAN from the datasets (CelebA, CelebA-HQ-females, LFW). (Color figure online)

Where the blue, yellow and red dots correspond to the results on the CelebA, CelebA-HQ-females and LFW corpora respectively. According to the distribution of dots, we can remark that the yellow ones have higher scores (CelebA-HQ-females) than the blue dots (CelebA) and the red dots (LFW) at the end. This confirms what we discussed in Sect. 4.3. The total score on the 64 generated images of each dataset is given in the Table 3.

Table 3. MEQFI score of 64 DCGAN generated images from different datasets (CelebA, CelebA-HQ-females, LFW).

Dataset	Total Score (%)
CelebA	59,18
CelebA-HQ-female	60,74
LFW	49,12

6   Conclusion

In this work, we discuss the use of generative adversarial networks in human face imaging. We focus on using deep convolutional GANs to produce new data and to achieve data augmentation of human face images. Based on our experiments, the DCGAN architecture generates good images with similar properties to the training distribution, especially when the training dataset contains high quality images. The training process of DCGAN models is still unstable, even with the present techniques that help to improve the stability of the model. Indeed, we will address this issue in our further works. Observing the generated images, we remarked that the generated images from CelebA-HQ (only female faces) dataset are much better than the two other datasets because it contains high quality images, and also because we used only female faces only which reduces the noise in the generation process. In this article, we developed also a new measure to evaluate the quality of the produced images, this is the measure MEQFI that allows to take into account the evaluation of each annotator and combines them into a single score. The scores achieved by this measure are correlated to what we observed with the loss function during the learning step.

References

1. Arjovsky, M., Chintala, S., Bottou, L.: Wasserstein generative adversarial networks. In: International Conference on Machine Learning. In: PMLR, pp. 214–223 (2017)
2. Brock, A., Donahue, J., Simonyan, K.: Large scale GAN training for high fidelity natural image synthesis. In: arXiv preprint arXiv:1809.11096 (2018)
3. Denton, E.L., et al.: Deep generative image models using a Laplacian pyramid of adversarial networks. In: Advances in neural information processing systems 28 (2015)
4. Dong, H., et al.: Semantic image synthesis via adversarial learning. pp. 5706–5714 (2017)
5. Durall, R., et al.: Combating mode collapse in GAN training: an empirical analysis using hessian eigenvalues. p. 211218 (2021)
6. Goodfellow, I., et al.: Generative adversarial nets. In: Advances in neural information processing systems 27 (2014)
7. Gulrajani, I., et al.: Improved training of wasserstein gans. In: Advances in neural information processing systems 30 (2017)
8. Heusel, M., et al.: Gans trained by a two time-scale update rule converge to a local nash equilibrium. In: Advances in neural information processing systems 30 (2017)
9. Huang, G.B., Ramesh, M., Berg, T., Learned-Miller, E.: In: Labeled faces in the wild: a database for studying face recognition in unconstrained environments (2007)
10. Ioffe, S., Szegedy, C.: Batch normalization: accelerating deep network training by reducing internal covariate shift. pp. 448–456 (2015)
11. Isola, P., et al.: Image-to-image translation with conditional adversarial networks. pp. 1125–1134 (2017)
12. Karras, T., Laine, S., Aila, T.: A style-based generator architecture for generative adversarial networks. pp. 4401–4410 (2019)
13. Karras, T., et al. Progressive growing of gans for improved quality, stability, and variation. In: arXiv preprint arXiv:1710.10196 (2017)
14. Kingma, D.P., Adam, J.B.: A method for stochastic optimization. In: arXiv preprint arXiv:1412.6980 (2014)
15. Ledig, C., et al.: Photo-realistic single image super-resolution using a generative adversarial network. pp. 4681–4690 (2017)
16. Lopez, C., Miller, S.R., Tucker, C.S.: Human validation of computer vs human generated design sketches. vol. 51845, p. V007T06A015 (2018)
17. Maas, A.L., et al.: Rectifier nonlinearities improve neural network acoustic models. vol. 30(1), p. 3 (2013)
18. Mirza, M., Osindero, S.: Conditional generative adversarial nets. In: arXiv preprint arXiv:1411.1784 (2014)
19. Odena, A.: Semi-supervised learning with generative adversarial networks. In: arXiv preprint arXiv:1606.01583 (2016)

20. Radford, A., Metz, L., Chintala, S.: Unsupervised representation learning with deep convolutional generative adversarial networks. In: arXiv preprint arXiv:1511.06434 (2015)
21. Reed, S., et al.: Generative adversarial text to image synthesis. pp. 1060–1069(2016)
22. Salimans, T., et al. Improved techniques for training gans. In: Advances in neural information processing systems 29 (2016)
23. Schmidhuber,J.: Making the world differentiable: on using self-supervised fully recurrent n eu al networks for dynamic reinforcement learning and planning in non-stationary environm nts. (1990)
24. Wang, H., et al.: Image super-resolution using a improved generative adversarial network. pp. 312–315 (2019)
25. Xu, B., et al.: Empirical evaluation of rectified activations in convolutional network. In: arXiv preprint arXiv:1505.00853 (2015)
26. Zhu, J.Y., et al.: Unpaired image-to-image translation using cycle-consistent adversarial networks. pp. 2223–2232 (2017)

ECG Signal Classification Using Recurrence Plot-Based Approach and Deep Learning for Arrhythmia Prediction

Niken Prasasti Martono[(✉)], Toru Nishiguchi, and Hayato Ohwada

Department of Industrial Administration, Faculty of Science and Technology,
Tokyo University of Science, Noda, Japan
{niken,ohwada}@rs.tus.ac.jp, 7422537@ed.tus.ac.jp

Abstract. Automatic electrocardiogram (ECG) analysis is crucial in diagnosing heart arrhythmia but is limited by the performance of existing models owing to the high complexity of time series data analysis. Arrhythmia is a heart condition in which the rate or rhythm of the heartbeat is abnormal. The heartbeat may be excessively fast or slow or may have an irregular pattern. Research has shown that the use of deep Convolutional Neural Networks (CNNs) for time-series classification has several advantages over other methods.They are highly noise-resistant models and can very informatively extract deep features that are independent of time. Five classes of heartbeat types in the MIT-BIH arrhythmia database were classified using the resilient and efficient deep CNNs proposed in this study. The proposed method achieved the best score (95.8% accuracy) for arrhythmia detection using the deep learning classification method.

Keywords: Arrhythmia · Recurrence plot · Convolutional neural network · Electrocardiography · Deep learning

1 Introduction

Arrhythmia is a condition characterized by an abnormal heart rhythm. There are several types of arrhythmias, with each being associated with a pattern that allows it to be identified and classified [11]. An electrocardiogram (ECG) is a crucial instrument for diagnosing arrhythmias. Identifying and classifying arrhythmias from an ECG can be difficult because it is often essential to examine each heartbeat in the ECG readings. Owing to weariness, human error may occur during the examination of the ECG records. Computational approaches can be used for automatic classification. Automated computer analysis of typical electrocardiograms has become increasingly important for medical diagnosis as ECGs migrate from analog to digital [10]. However, the restricted efficacy of classical algorithms limits their use as a standalone diagnostic tool, and they can only offer supporting functions.

N. T. Nguyen et al. (Eds.): ACIIDS 2022, LNAI 13757, pp. 327–335, 2022.
https://doi.org/10.1007/978-3-031-21743-2_26

Over the years, a variety of automatic ECG categorization algorithms, based on signal processing techniques, have been presented. Commonly used methods include the wavelet transform, frequency analysis, support vector machines (SVMs), artificial neural networks (ANNs), decision trees, linear discriminant analysis, and Bayesian classifiers [1,5,9,13].

The implementation of deep-learning algorithms has recently emerged as the most popular strategy [7,14]. Deep convolutional neural networks (CNNs) have recently yielded impressive results in tasks such as image classification and speech recognition, and this technology can greatly improve healthcare and clinical practice. To date, the most effective systems have used supervised learning to automate examination diagnoses.

This research proposes a CNN-based ECG arrhythmia classification model, with recurrence plot-based data used in the training, to maximize the ability of the classifier to analyze time-series data. We identify and learn ECG data with the goal of improving the accuracy by constructing and optimizing neural networks. The ECG dataset was organized into five categories in our study to provide a general assessment of heart health, as well as an important and dependable reference for future diagnosis by the doctor. The obtained results indicate that deep CNNs with a recurrence plot can be used for arrhythmia classification. The results show that our classifier achieved an average accuracy of 95.8%, average precision of 75.8%, and average recall of 74.6%. The remainder of this paper is organized as follows. Sections 2 and 3 describe the materials and methods, including the analysis of using recurrence plots to produce the 2D segment representation of the ECG signal and the CNN architecture used in this study. Sections 4 and 5 present the results and discussion, as well as the scope for future work.

2   Data Set

In this study, we used the MIT-BIH arrhythmia dataset [8], which is well known for measuring arrhythmias and can be used in basic cardiac dynamic research. This database contains excerpts of 30 min double-channel recordings from 47 participants between 1975 and 1979. Each one had two ECG lead signals with an 11-bit resolution in the 10 mV range, digitized at 360 samples per second. These annotations were divided into five categories according to the AAMI EC57 standard [2]: normal beat, left bundle branch block (LBBB), right bundle branch block (RBBB), and atrial escape and nodal junction escape beats belong to the class N category; class V contains premature ventricular contraction (PVC) and ventricular escape beats; class S contains atrial premature (AP), aberrated premature (aAP), nodal junction premature (NP), and supraventricular premature (SP) beats; class F contains only fusion of ventricular and normal (fVN) beats, and class Q, which is known as an unknown beat, contains paced beats (P), fusion of paced and normal (fPN) beats, and unclassified beats. Figure 1 illustrates the differences between categories. The total number of training and test datasets used in this study is shown in Table 1. The total number of test and training data was decided by trial and error, and we chose the set of training and test datasets that gave the best results in classification performance.

In the next section, we propose a CNN-based ECG arrhythmia classification approach, which consists of three steps: ECG data preprocessing using recurrence plots, model training, and model evaluation. To train and test the CNN classifier, we preprocessed the MIT-BIH arrhythmia database and separated the preprocessed five categories of ECG data into mutually exclusive training and test sets. The CNN classifier was trained on the training set, and after the relevant training model was obtained, it was used to predict the categorization of ECG types in the test set.

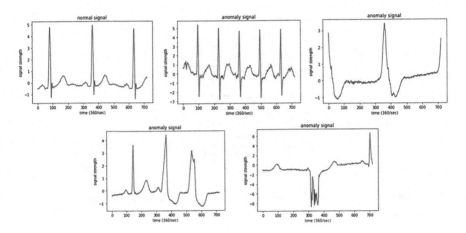

Fig. 1. Illustration of time series signal for each category (class N, class V, class S, class F, and class Q)

Table 1. Descriptions of arrhythmia signals and number of and samples

Label	Description	Train	Test
N	Non-ectopic beat	8963	10560
S	Supraventricular ectopic beat	1236	1057
V	Ventricular ectopic beat	993	2495
F	Fusion beat	495	296
Q	Unknown beat	7	8

3 Method

The overall process implemented in this study is illustrated in Fig. 2. First, data preprocessing is conducted, including signal labeling, splitting, normalization, and implementing the recurrence plot method on the ECG signal data.

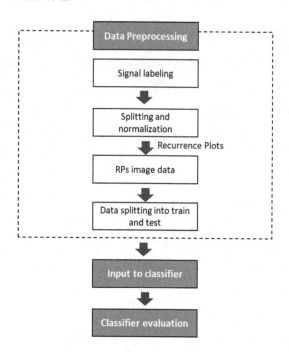

Fig. 2. The proposed overall process of arrhythmia classification

3.1 Time Series to Recurrent Plots

To better utilize the ECG data, we transform and preprocess the ECG data using recurrence plots. Recurrence plots (RPs) are advanced tools that can indicate how resemblances between orders vary over time [6]. These plots are generally used for qualitative evaluation of time series in dynamic systems. It is an advanced nonlinear data analysis technique for detecting latent dynamical patterns and nonlinearity in the data. It is a representation (or graph) of a square matrix, with the matrix elements corresponding to the times when the state of a dynamical system recurs (columns and rows correspond to a certain pair of times). The numerical expression for an RP is defined according to Eq. 1:

$$R_{ij}^{mn} = \theta(\epsilon_o - ||x_i - x_j||), x_i \in R^m, ij = 1 \cdots N. \tag{1}$$

where N is the number of considered states, ϵ_i is a threshold distance, θ is the Heaviside function, x_i and x_j are the observed subsequences at both points i and j; and $|| \cdot ||$ is the Euclidian norm. This study uses a modified version of an RP that incorporates color information. Color maps, rather than Equation (1), are utilized to create the image, allowing distances to be depicted in color. Figure 3 shows examples of typical RP modifications for various signal types. A matrix with values between 0.0 and 1.0 is used to make RP pictures. In the next step,

RPs were used as the input for building a classification model for arrhythmia detection.

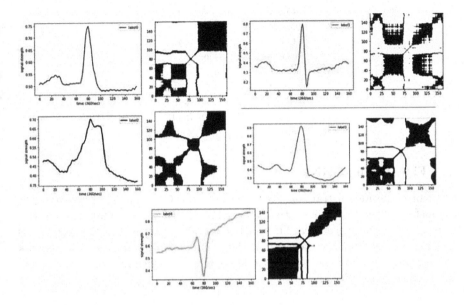

Fig. 3. Example of reccurence plot results for each class of ECG signals

3.2 Convolutional Neural Networks

Convolutional neural networks (CNNs), also known as ConvNet, are deep feedforward architectures that outperform networks with fully connected layers in terms of generalization [3]. CNNs were defined by Indolia et al. Al (2018) proposed a biologically inspired idea for hierarchical feature detectors. It can learn abstract traits and identify things accurately. The following are some reasons why CNNs are preferred over traditional models. First, the concept of weight sharing is deployed to reduce the number of parameters that need to be trained, resulting in enhanced generalization. CNNs can be trained smoothly and without overfitting because they have fewer parameters [4]. Table 2 presents the structure of the CNNs used in this study.

In this process, we also implemented an activation function to introduce nonlinear characteristics into the neural network model. A rectified linear unit (ReLU) is one of the most commonly used activation functions in CNNs. When the input is positive, there will be no gradient saturation problem. Furthermore, because the relationship is merely linear, the calculation speed is faster than that of the *sigmoid* and *tanh*.

Table 2. CNNs classifier structure used in this work

Layer	Type	Shape	Activation
0	Null	160,160	
1	Conv2D	158,158	ReLU
2	MaxPooling	79,70	
3	Conv2D	77,77	ReLU
4	MaxPooling	38,38	
5	Conv2D	36,36	ReLU
6	MaxPooling	18,18	
7	dense	64	ReLU
8	dense	32	ReLU
9	dense	5	Softmax

3.3   Fine Tuning CNNs

To finalize the model training, we fine-tuned the CNN model. Fine-tuning is a transfer learning technique that stores knowledge gained while solving one problem and applying it to a different but related problem [12]. Because CNNs are composed of numerous layers and a large number of parameters, the network training phase would benefit from the use of databases rich in examples to avoid the problem of overfitting.

3.4   Classification Performance Evaluation

Correct classification or misclassification is quantified using four metrics: true positive (TP), true negative (TN), false positive (FP), and false negative (FN). The classification results were analyzed using evaluation metrics, including precision, recall, F1-score, and accuracy based on the formula shown below.

$$Accuracy = \frac{TP+TN}{TP+TN+FP+FN}$$

$$Precision = \frac{TP}{TP+FP}$$

$$Recall = \frac{TP}{TP+FN}$$

$$F1 = \frac{2*TP}{2*TP+FP+FN}$$

4   Results and Discussion

Using all the sample data listed in Table 1, we can use the proposed method to classify and evaluate the ECG as an arrhythmia classifier. Table 3 presents the confusion matrix of the classifier for all samples based on class, and Table 4 lists the evaluation performance of the CNN classifier after fine-tuning the classifier.

Table 3. Prediction results of ECG signals

		Predicted label				
		N	S	V	F	Q
Actual label	N	10386	102	36	34	2
	S	145	899	12	1	0
	V	152	34	2270	37	2
	F	30	2	9	255	0
	Q	3	0	3	1	1

Table 4. Overall performance score

Evaluation	Score
Accuracy	0.958
Precision	0.758
Recall	0.746
F1-score	0.749

Table 5. Classification performance of ECG signals using proposed method

Label	Precision	Recall	F1-score	Support
N	0.969	0.984	0.976	10560
S	0.867	0.851	0.859	1057
V	0.974	0.91	0.941	2495
F	0.777	0.861	0.817	296
Q	0.2	0.125	0.154	8

Table 6. Classification performance using 5-fold cross validation

Evaluation	1st fold	2nd fold	3rd fold	4th fold	5th fold	Average
Accuracy	0.929	0.929	0.936	0.935	0.934	0.9326
Precision	0.694	0.692	0.706	0.706	0.7	0.6996
Recall	0.662	0.663	0.664	0.66	0.669	0.6636
F1-score	0.676	0.676	0.684	0.681	0.88	0.7194

The results indicate that our classifier could classify most of the classes well, with classes N and V exhibiting the best performance. For classes, N, S, V, and Q, the precision was above 70.0%, with 97.4% being the highest score. However, for class Q, owing to the very limited number of samples compared with the other classes, the performance result seems to be very low. The overall performance scores for all the five classes of signals are listed in Table 4.

In addition to the performance evaluation, as shown in Table 6 we implemented a 5-fold cross-validation method and calculated the accuracy, precision, recall, and F-1 score. Table 6 reveals that using 5-fold cross validation, the classifier has 93.3% accuracy, 70.0% precision, 66.4% recall, and 72.0% F-1 score. The low results for the last three evaluations could be attributable to the low prediction score of class Q.

5   Conclusion and Future Works

In this study, we proposed a CNN-based ECG arrhythmia classification algorithm. This study used a data reconstruction method; in this case, recurrence plots were used to prepare the training data. Using this method reduced the need for complex feature processing and calculation. The ECG records in the MIT-BIH arrhythmia database were processed to obtain 2-D outputs and used as model input data. Finally, the trained model classified the ECG signal into five classes: normal beat, supraventricular ectopic, ventricular ectopic, fusion, and unknown beats. The optimized CNNs model uses an RELU activation function, dropout, and other technologies to create a network architecture. Our proposed method performs well in the classification, with an overall average accuracy rate of 93.3% and best accuracy of 95.8% lead ECG data, which is superior to previous works on arrythmia prediction using deep CNNs and recurrence plot. It could accurately categorize ECG signals according to the results. In future, 12-lead ECG data, rather than 2-lead ECG data, should be used to create an automated prediction model for arrhythmia classification.

References

1. Celin, S., Vasanth, K..: ECG signal classification using various machine learning techniques. J. Med. Syst. **42**(12), 1–11 (2018). https://doi.org/10.1007/s10916-018-1083-6
2. Das, M.K., Ari, S.: ECG beats classification using mixture of features. Int. Schol. Res. Not. **2014**, 1–12 (2014). https://doi.org/10.1155/2014/178436
3. Indolia, S., Goswami, A.K., Mishra, S.P., Asopa, P.: Conceptual understanding of convolutional neural network- a deep learning approach. textbf132, 679–688. Elsevier B.V. (2018). https://doi.org/10.1016/j.procs.2018.05.069
4. Jeong, D.U., Lim, K.M.: Convolutional neural network for classification of eight types of arrhythmia using 2d time-frequency feature map from standard 12-lead electrocardiogram. Sci. Rep. **11**, 679–688 (2021). https://doi.org/10.1038/s41598-021-99975-6
5. Martis, R.J., Acharya, U.R., Min, L.C.: ECG beat classification using PCA, LDA, ICA and discrete wavelet transform. Biomed. Signal Process. Control **8**, 437–448 (2013). https://doi.org/10.1016/j.bspc.2013.01.005
6. Mathunjwa, B.M., Lin, Y.T., Lin, C.H., Abbod, M.F., Sadrawi, M., Shieh, J.S.: ECG recurrence plot-based arrhythmia classification using two-dimensional deep residual CNN features. Sensors **22**, 1660 (2022). https://doi.org/10.3390/s22041660

7. Mathunjwa, B.M., Lin, Y.T., Lin, C.H., Abbod, M.F., Shieh, J.S.: ECG arrhythmia classification by using a recurrence plot and convolutional neural network. Biomed. Signal Process. Control **64**, 102262 (2021). https://doi.org/10.1016/j.bspc.2020.102262

8. Moody, G., Mark, R.: MIT-BIH arrhythmia database (2005). https://physionet.org/content/mitdb/1.0.0/

9. Raj, S., Ray, K.C.: ECG signal analysis using dct-based dost and PSO optimized SVM. IEEE Trans. Instrum. Meas. **66**, 470–478 (2017). https://doi.org/10.1109/TIM.2016.2642758

10. Ribeiro, A.H., et al.: Automatic diagnosis of the 12-lead ECG using a deep neural network. Nat. Commun. textbf11, 1760 (12 2020). https://doi.org/10.1038/s41467-020-15432-4

11. da S. Luz, E.J., Schwartz, W.R., Cámara-Chávez, G., Menotti, D.: ECG-based heartbeat classification for arrhythmia detection: a survey. Comput. Meth. Prog. Biomed. **127**, 144–164 (2016). https://doi.org/10.1016/j.cmpb.2015.12.008

12. Taormina, V., Cascio, D., Abbene, L., Raso, G.: Performance of fine-tuning convolutional neural networks for hep-2 image classification. Appl. Sci. (Switzerland) **10**, 1–20 (10 2020). https://doi.org/10.3390/app10196940

13. Varatharajan, R., Manogaran, G., Priyan, M.K.: A big data classification approach using LDA with an enhanced SVM method for ECG signals in cloud computing. Multimedia Tools Appl. **77**, 10195–10215 (2018). https://doi.org/10.1007/s11042-017-5318-1

14. Zhang, H., et al.: Recurrence plot-based approach for cardiac arrhythmia classification using inception-ResNet-V2. Front. Phys.**12**, 648950 (2021). https://doi.org/10.3389/fphys.2021.648950

Internet of Things and Sensor Networks

Collaborative Intrusion Detection System for Internet of Things Using Distributed Ledger Technology: A Survey on Challenges and Opportunities

Aulia Arif Wardana[1]([✉]) [iD], Grzegorz Kołaczek[1] [iD], and Parman Sukarno[2] [iD]

[1] Wrocław University of Science and Technology, wybrzeże Stanisława Wyspiańskiego 27, 50-370 Wrocław, Poland
aulia.wardana@pwr.edu.pl, auliawardan@telkomuniversity.ac.id
[2] Telkom University, Terusan Buahbatu - Bojongsoang, Jl. Telekomunikasi No. 1, Kec. Dayeuhkolot, Jawa Barat 40257 Sukapura, Bandung, Indonesia

Abstract. This review presents the current state-of-the-art of the Distributed Ledger Technology (DLT) model used in the Collaborative Intrusion Detection System (CIDS) for anomaly detection in Internet of Things (IoT) network. The distributed IoT ecosystem has many cybersecurity problems related to anomalous activities on the network. CIDS technology is usually applied to detect anomalous activities on the IoT network. CIDS is suitable for IoT network because they have the same distributed characteristic. The use of DLT technology is expected to be able to help the IDS system accelerate detection and increase the accuracy of detection through a collaborative detection mechanism. This review will look more deeply at the placement strategies, detection method, security threat, and validation & testing method from CIDS with DLT-based for IoT network. This review also discusses the open issue and the lesson learned at the end of the review. The result is expected to produce the next research topic and help professionals design effective CIDS based on DLT for the IoT network.

Keywords: Intrusion detection · Internet of Things · Distributed ledger · Cybersecurity · Blockchain

1 Introduction

The IoT is made up of a large-scale network with heterogeneous characteristics. A huge amount of heterogeneous data is transferred across the IoT network in real time [1, 1]. Among the devices connected to the IoT network, there are fake devices that can be a threat to the IoT network. The huge amount of data that is sent over the IoT network is vulnerable to forgery and data manipulation, causing the integrity and confidentiality of the data to be disrupted. With this condition, IoT networks are potential targets for cybersecurity attacks. There are some types of attack in IoT network like

© The Author(s), under exclusive license to Springer Nature Switzerland AG 2022
N. T. Nguyen et al. (Eds.): ACIIDS 2022, LNAI 13757, pp. 339–350, 2022.
https://doi.org/10.1007/978-3-031-21743-2_27

Distributed Denial of Service (DDoS), routing attack, data transit attack, and botnet attack or malware. All these types of attacks are anomalous activities in the IoT network [3].

One of the technologies used to detect anomalies in the IoT network is Intrusion Detection System (IDS) [4]. IDS on a distributed network is not enough to work independently for detecting coordinated attacks, but must be also collaborative to accelerate detection in distributed systems. Therefore, CIDS is a suitable system to detect anomalies on the distributed system [5]. The IoT network is one of the distributed system type with heterogeneous network. This means that IDS should adapt the distributed characteristic of the IoT network using collaboration mechanism [6]. There are many types and approaches of CIDS in the network; one of them is using DLT for built CIDS system for detect anomalous activities on the network. The use of DLT in CIDS is for data synchronization, trust management, and alert exchange [7, 7].

This review aims to discuss open issues and future research related to the use of DLT for CIDS in the IoT network. This review gives insight about the usage of DLT system for CIDS in IoT network. This work also helps professionals and industries design and implement CIDS based on DLT for IoT networks.

1.1 Related Surveys on Collaborative Intrusion Detection Systems for IoT

In recent years, there has been a lot of research related to CIDS for IoT that has been published. Research [9] is reviewing the usage of blockchain to integrate with CIDS in IoT network. This research aims to support the researcher by giving insight on the overview of the integration of blockchain and CIDS to detect anomalies in IoT network. The review [4] is a survey of IDS in IoT. One of the IDS models discussed in this survey is CIDS for IoT network. The CIDS on the survey is applied multi-IDS agent in distributed architecture. The other review from [10] explains about security on IPv6 Routing Protocol for Low Power and Lossy Network (RPL). One strategy to secure the IPv6 RPL-based IoT network is using CIDS.

Based on all the previous reviews, this research motivates us to review about CIDS on IoT network. The usage of DLT also enables integration with CIDS for anomaly detection in IoT network. The contribution from this literature survey is summarized as follow:

- This research will review the integration DLT technology with CIDS in IoT networks for storage sharing, trust & audit management, and information exchange.
- This research is also review about the usage of DLT for collaborative learning in IDS to accelerate detection on IoT network.
- This research will review all two points before using the point of view from the placement strategies, detection method, security threat, and validation & testing method.

1.2 Overview of the Paper

This review of the literature is organized as follows: Sect. 2 consists of the DLT paradigm that should be clearly defined in this paper. Section 3 consists of all the research of CIDS with DLT-based for IoT network. The research will look into placement strategies, detection method, validation & testing method, and DLT platform in more depth. Section 4 consists of an open issue and related research topic that can generate for future research. In addition, the challenge of implementation and research in CIDS with DLT-based in IoT network will be discussed in Sect. 4. Finally, Sect. 5 consists of the conclusion from this review.

2 Distributed Ledger Technology

DLT is a term related to a system with multiple nodes or parties [11]. Multiple nodes or parties that connected on the networks called the ledger [12]. At each ledger, the data are stored in a protected block. Data protection in each block is due to a consensus algorithm used to verify every transaction [13]. The network in this technology has the Peer-to-Peer (P2P) characteristic [14]. Figure 1 is the architecture of the DLT system. The architecture will describe the preprocessing data flow in DLT. The first component in DLT called *transaction*, this component related to any planned change in every ledger. This process starts with an *unconfirmed transaction* that is processed by the end user and send to every ledger to confirm the transaction. The second component is *log* or *mempool*, this component related to storage in every validating node for save unconfirmed transaction and every node may have different log.

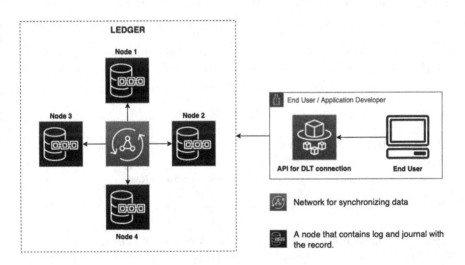

Fig. 1. DLT system architecture [11].

The third component is the *record*, this component related to selection of an unconfirmed transaction to be a candidate record. The unconfirmed transaction will be processed using the specified protocol to make it valid and send to every valid connected node. The fourth component is the *journal*; this component related to each valid node that received a candidate record for the verification process. The last component is the *ledger*, this component related to all connected nodes agree with the transaction of new record. The agreement process is produced by verifying processes that synchronize in every valid node [11].

3 Collaborative Intrusion Detection Systems Based on Distributed Ledger Technology in Internet of Things

This section will review up-to-date research related to CIDS based on DLT for the IoT network. This section will describe CIDS placement strategies, the CIDS detection method, the DLT platform, and method to validate and test the CIDS. After describing some of the previous main topics, a taxonomy diagram of the main discussion will be made.

Table 1. Research related to CIDS DLT-based for IoT network.

Reference	Placement strategies	Detection method	Security threat	Validation and testing	DLT platform
Putra et al., 2020 [15]	Decentralized	Hybrid	DoS, MITM, reconnaissance, and replay attacks	Theoretical	Ethereum
Li et al., 2019 [16]	Distributed	Signature Based	DoS, Malware, and Insider exploration	Real Approach	-
Mirsky et al., 2020 [17]	Decentralized	Anomaly Based	Buffer Overflow, Code-Reuse, and Replay Attack (Key Reinstallation Attack)	Real Approach	Ethereum
Kumar et al., 2020 [18]	Decentralized	Anomaly Based	DDoS	Simulation	Ethereum
Golomb et al., 2018 [19]	Distributed	Anomaly Based	Attacks Against the Agent and Attacks Against the CIoTA Chain	Real Approach	-
Hu et al., 2019 [20]	Distributed	Anomaly Based	DDoS, MITM, Tempeting Attack, Replay Attack	Simulation	-
Alkadi et al., 2021 [21]	Distributed	Other	DDoS, Malware, Sniffing, etc	Real Approach	Ethereum

Figure 2 illustrates the taxonomy of CIDS in IoT. Based on the taxonomy diagram, this review creates a table to summarize all systems related to CIDS DLT-based for IoT

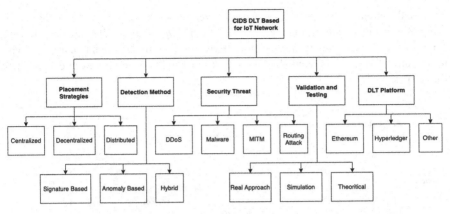

Fig. 2. Taxonomy from CIDS in IoT.

network. Table 1 is summarizes some research in the field CIDS DLT-based for IoT network. All existing research and solution in Table 1 will compare and explain more detail in every sub-section of this section.

3.1 CIDS Placement Strategies

CIDS have some architecture for deployment in the system. Figure 3 is the type of CIDS architecture. There are three types of CIDS architecture. It is centralized, decentralized, and distributed [5, 5, 5].

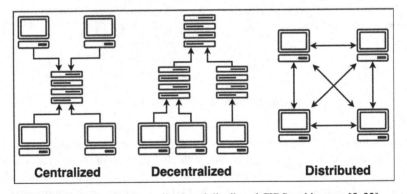

Fig. 3. Centralized, decentralized, and distributed CIDS architecture [5, 22].

The centralized CIDS architecture has a central node as *analysis unit*. All nodes that spread over the network will become a *detection unit*. If there are anomalies on the network, the detection unit will send to the analyze unit for analyze the anomaly. The detection unit will also be the *alert unit* when the analyze unit gives command to alert the anomalies on the network. The *decentralized CIDS* architecture used a hierarchical or multilayer IDS structure. This architecture is better than a centralized architecture because this architecture has a coordinator in every layer of hierarchy. This architecture also has an analysis unit in the top of the hierarchy. *Distributed CIDS* does not have a central node like two architectures before. Every node will become an analysis, monitoring, and alert unit. It means that every task is shared between nodes in every CIDS [5, 5, 5].

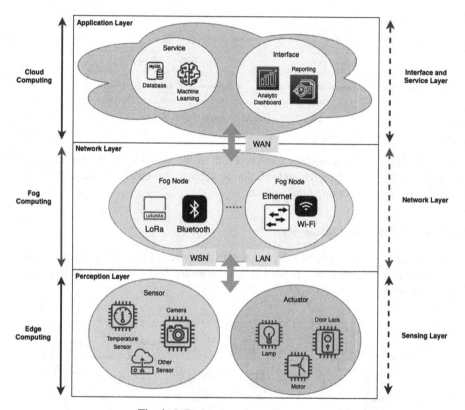

Fig. 4. IoT ecosystem layer [10, 10, 10].

In addition to the CIDS architecture, this study also reviews the existing architecture in the IoT ecosystem. Figure 4 is the layer in the IoT ecosystem. This layer is used to map data flow and connection between service & devices. The layer can also provide an overview for deploying IoT systems. IoT has two big category architectures, there are three-layer architecture and Software Oriented Architecture (SOA). This research will try to combine these two architectures for explanation in Fig. 2. *Application Layer* is top layer from IoT. This layer consists of many services or middleware for processing data. This layer is identical to cloud computing. *The network layer* is used for the transmission and deliver of all data from the perception layer to application layer. Protocol, security, and architecture are important in this layer. The *perception layer* is a source of information and data in IoT. This layer consists of a sensor, actuator, and other smart devices to collect data from the physical world [3, 3].

The choice of CIDS architecture and IoT architecture will affect the placement of the CIDS agent and component. Good placement of the CIDS agent and component will enhance collaboration performance in CIDS. Most of the CIDS in Table 1 used decentralized and distributed in their architecture. This architecture is suitable for IoT and DLT that have distributed characteristic architecture.

3.2 Detection Method

There are three types of detection methods in IDS, signature, anomaly, and hybrid method. A list of signatures is available that can be used to assess whether the sent packet sent is malicious or not. A data packet will be compared with an existing list. This method will protect the system from known types of attack. The anomaly method can detect an unknown attack type in the network. This method uses machine learning to detect anomalous behavior in the network [27, 27]. The hybrid method combines the signature and anomaly method in the IDS system [28]. The detection method in CIDS has the same name and category as ordinary IDS. But there are special characteristics in CIDS, and the detection method in CIDS is done collaboratively [5].

Most of the CIDS in Table 1 used anomaly-based detection. The anomaly method is chosen because it can detect unknown attacks and the learning process in CIDS with anomaly-based detection done collaboratively, so it is more effective.

3.3 Security Threat

CIDS is an excellent IDS system for dealing with coordinated attack types. Coordinated attacks type often attack distributed networks. There are several types of coordinated attacks that often attack such as DDoS, worm outbreak, and large-scale stealthy scans [5]. IoT is a distributed network that is possible to be exposed to such attacks. In addition to that, the IoT is also exposed to other attacks such as the Man-in-the-middle attack (MITM) and the routing attack. All these attacks must be detected quickly by the IDS [29]. The security threat is not just from outside attack, but also from an insider attack. Some CIDS system also detects insider attacks that attack his own system. CIDS DLT-based must consider the threat that will appear on DLT. According to research [30] there are some vulnerabilities in the DLT blockchain platform. The vulnerabilities such as incorrect cryptographic operation, identity vulnerability, manipulation-based attack, quantum vulnerability, reputation-based attack, vulnerability in service, network vulnerability, and privacy issue must be considered when design the CIDS with DLT-based.

Most of the attacks that are commonly detected by CIDS in **Table 1** are coordinated attacks such as DDoS, malware, and MITM. Therefore, CIDS is suitable to detect attacks in IoT network with distributed characteristic. Some research also has additional protection against insider attacks.

3.4 Validation and Testing Method

The validation method in this review consists of three types: real approach, simulation, and theoretical. The real approach method is to implement the CIDS with DLT based on real environment IoT (e.g., using a device). The simulation method is only to use virtual machines or simulation tools to implement CIDS with DLT-based. The theoretical method used a mathematical model, concept, or just ideas on the research. Most of the research in Table 1 used real approach for validation. This method is good to see the real performance for validation from the proposed system.

The system testing method in Table 1 used a different method. For the real approach, performance testing or overhead analysis is used mainly. For detection performance, some parameters are used, such as accuracy, precision, or time detection.

3.5 DLT Platform

DLT using the concept of distributed system to develop an application with no central authority [31]. There are several platforms that can be programmed to implement DLT. Ethereum and Hyperledger are examples of open source DLT platforms that are easy to use [32]. There are three types of DLT platform (i) private, (ii) permissioned, and (iii) permissionless. Private DLT means that the platform is installed on a private server and managed by an organization for internal use. Permissioned DLT means that the platform is maintained and used by multiple known organizations. Permissionless DLT means that the platform is public and everyone can use it for transactions on the platform [33]. The DLT system needs a lot of energy and resource to operate, so the choice of DLT platform model is very important [34].

Most of the CIDS in Table 1 use Ethereum as DLT platform. The advantage of Ethereum is developed by many developers. This condition makes Ethereum one of the most trusted platforms. The ease of developing a system and integrating it into a system is the widely used key to the DLT platform widely used for the research or development of other systems.

4 Future and Opportunities

After looking into other research on DLT-based CIDS for IoT network, this section will explain the lesson learned from the review. This section will explain the open issues, future research, and challenge of DLT-based CIDS for IoT network.

4.1 Challenge in Research

The implementation challenge is important to consider by researchers or industries who want to combine CIDS and DLT for anomaly detection in IoT networks. Therefore, this review will describe implementation challenges from some recently published research that has been reviewed before. There are some challenges to implement CIDS with DLT-based in the IoT network. Here are the details:

1. *Resource and Cost*: IoT has limited resource and energy in its devices, but DLT concept need more computational power and energy to validate the data. This condition requires the implementation of CIDS DLT-based model in IoT need to consider resource and cost.

2. *Insider Attack*: CIDS is used to detect outsider attack in IoT networks (e.g., DDoS, Malware, MITM, etc.), but the insider attack needs to be considered too. Many CIDS systems focus only on the detection model of outsider attack, but the CIDS model also needs to focus on insider attack like CIDS agent attack and DLT vulnerabilities. Researchers and industries must validate these security issues on their work (model or implementation).

3. *Complexity and Performance*: IoT is one of the distributed system with complex characteristic of the system. This happens because IoT integrates heterogeneous technology (software and hardware) into their ecosystem. The implementation of CIDS DLT-based in IoT will make the system more complex, so the design from CIDS DLT-based must be simple and effective with good performance.

4.2 Future Research and Opportunities

The emerge of collaboration between DLT and Artificial Intelligence (AI) will drive collaborative learning to be hot topic for research in the future. Decentralized learning is one of the future research agenda for collaboration between AI and DLT. The development of decentralized learning will create three research opportunities. Firstly, the good design of consensus algorithm in DLT will affect machine learning model training. Secondly, smart contract is programable, so simple AI computation in smart contract is possible. Lastly, computation in smart contract is heavy task while doing in on-chain. One of the solutions for this condition is using off-chain computation. [35, 35, 35]. CIDS with anomalies detection is used in machine learning, deep learning, or artificial intelligence to detect anomalies activities on the network. The learning process on CIDS using the collaborative concept. Therefore, collaborative learning and decentralized learning methods will be rising in the future for research in CIDS based on DLT.

There are many DLT platforms that exist until now. The integration model between the DLT platform and IoT system must be considered. There is some consideration about the adoption of DLT in the IoT, including scalability, privacy, processing power, storage, and consensus challenge [34]. Also, the integration model between the DLT platform and CIDS must be considered. There is some consideration about adoption of DLT in CIDS like signature & data sharing, collaboration alert & reporting, adaptive architecture for large system, collaborative real-time monitoring, and deployment simplification [38, 38, 38]. All parameters must be considered when choosing or designing the DLT platform for integrate with CIDS and IoT. It is also important to look at the research aspect before integrating the DLT platform with CIDS to detect anomalous activities in the IoT network.

5 Conclusion

The combination of DLT with CIDS to detect anomalous activities in the IoT network will improve the CIDS detection process from the CIDS. This paper presented an overview from some research in that field. This paper also reviews recently published research in that field for a relevant literature review. This literature review paper can help researchers and industries to see the challenge and future direction for adoption of CIDS with DLT system to detect anomalies in the IoT network.

This review finds three major challenges about implementation of CIDS DLT-based in IoT systems, namely: resource & cost, insider attack, and complexity & performance. This review also sees some opportunities for future research and implementation of collaborative learning and decentralization learning method for the CIDS DTL-based in IoT network. Also, this review gives some highlights of the aspects of choosing or designing DLT platform for the combination with CIDS for IoT network.

Acknowledgements. The first author would like to thank Wroclaw University of Science and Technology (WUST) and Narodowa Agencja Wymiany Akademickiej (NAWA) for funding this research through PhD scholarship and NAWA STER scholarship. Also, thanks to Telkom University for all the support during this research and PhD studies.

References

1. Al-Fuqaha, A., Guizani, M., Mohammadi, M., Aledhari, M., Ayyash, M.: Internet of Things: A Survey on Enabling Technologies, Protocols, and Applications. IEEE Commun. Surv. Tutorials 17(4), 2347–2376 (2015). https://doi.org/10.1109/COMST.2015.2444095
2. Wardana, A.A., Rakhmatsyah, A., Minarno, A.E., Anbiya, D.R.: Internet of Things Platform for Manage Multiple Message Queuing Telemetry Transport Broker Server. Kinet. Game Technol. Inf. Syst. Comput. Network, Comput. Electron. Contr. 4(3), 197–206 (2019). https://doi.org/10.22219/kinetik.v4i3.841
3. Lin, J., Yu, W., Zhang, N., Yang, X., Zhang, H., Zhao, W.: A survey on internet of things: architecture, enabling technologies, security and privacy, and applications. IEEE Internet Things J. 4(5), 1125–1142 (2017). https://doi.org/10.1109/JIOT.2017.2683200
4. Zarpelão, B.B., Miani, R.S., Kawakani, C.T., de Alvarenga, S.C.: A survey of intrusion detection in Internet of Things. J. Netw. Comput. Appl. 84(February), 25–37 (2017). https://doi.org/10.1016/j.jnca.2017.02.009
5. Zhou, C.V., Leckie, C., Karunasekera, S.: A survey of coordinated attacks and collaborative intrusion detection. Comput. Secur. 29(1), 124–140 (2010). https://doi.org/10.1016/j.cose.2009.06.008
6. Arshad, J., Azad, M.A., Amad, R., Salah, K., Alazab, M., Iqbal, R.: A review of performance, energy and privacy of intrusion detection systems for IoT. Electron. 9(4), 1–24 (2020). https://doi.org/10.3390/electronics9040629
7. Meng, W., Tischhauser, E.W., Wang, Q., Wang, Y., Han, J.: When intrusion detection meets blockchain technology: a review. IEEE Access, 6(c), 10179–10188 (2018). https://doi.org/10.1109/ACCESS.2018.2799854
8. Al'Aziz, B.A.A., Sukarno, P., Wardana, A.A.: Blacklisted IP distribution system to handle DDoS attacks on IPS Snort based on Blockchain. Proceeding - 6th Inf. Technol. Int. Semin. ITIS 2020 41–45 (2020). https://doi.org/10.1109/ITIS50118.2020.9320996
9. Benaddi, H., Ibrahimi, K. : A Review: Collaborative Intrusion Detection for IoT integrating the Blockchain technologies. In: Proceedings of the 2020 International Conference Wireless Networks Mobile Communication. WINCOM 2020,(2020). https://doi.org/10.1109/WINCOM50532.2020.9272464
10. Verma, A., Ranga, V.: Security of RPL based 6LoWPAN networks in the internet of things: a review. IEEE Sens. J. 20(11), 5666–5690 (2020). https://doi.org/10.1109/JSEN.2020.2973677

11. Rauchs, M., et al.: Distributed Ledger Technology Systems: A Conceptual Framework. SSRN Electron. J. (2018) https://doi.org/10.2139/ssrn.3230013
12. Kannengießer, N., Lins, S., Dehling, T., Sunyaev, A.: Trade-offs between Distributed Ledger Technology Characteristics. ACM Comput. Surv. 53(2), 42:1–42:37 (2020) https://doi.org/10.1145/3379463
13. El Ioini, N., Pahl, C.: A review of distributed ledger technologies. In: Panetto, H., Debruyne, C., Proper, H.A., Ardagna, C.A., Roman, D., Meersman, R. (eds.) On the Move to Meaningful Internet Systems. OTM 2018 Conferences. Lecture Notes in Computer Science, vol. 11230, pp. 277–288. Springer, Cham (2018). https://doi.org/10.1007/978-3-030-02671-4_16
14. Siano, P., De Marco, G., Rolan, A., Loia, V.: A survey and evaluation of the potentials of distributed ledger technology for peer-to-peer transactive energy exchanges in local energy markets. IEEE Syst. J. 13(3), 3454–3466 (2019). https://doi.org/10.1109/JSYST.2019.2903172
15. Putra, G.D., Dedeoglu, V., Kanhere, S.S., Jurdak, R.: Poster abstract: Towards scalable and trustworthy decentralized collaborative intrusion detection system for IoT. In: Proc. - 5th ACM/IEEE Conf. Internet Things Des. Implementation, IoTDI 2020, pp. 256–257 (2020). https://doi.org/10.1109/IoTDI49375.2020.00035
16. Li, W., Tug, S., Meng, W., Wang, Y.: Designing collaborative blockchained signature-based intrusion detection in IoT environments. Futur. Gener. Comput. Syst. 96, 481–489 (2019). https://doi.org/10.1016/j.future.2019.02.064
17. Mirsky, Y., Golomb, T., Elovici, Y.: Lightweight collaborative anomaly detection for the IoT using blockchain. J. Parallel Distrib. Comput. 145, 75–97 (2020). https://doi.org/10.1016/j.jpdc.2020.06.008
18. Kumar, P., Kumar, R., Gupta, G.P., Tripathi, R.: A distributed framework for detecting DDoS attacks in smart contract-based blockchain-IoT systems by leveraging fog computing. Trans. Emerg. Telecommun. Technol. 32(6), 1–31 (2021). https://doi.org/10.1002/ett.4112
19. Golomb, T., Mirsky, Y., Elovici, Y.: CIoTA: Collaborative Anomaly Detection via Blockchain (2018). https://doi.org/10.14722/diss.2018.23003
20. Hu, B., Zhou, C., Tian, Y.C., Qin, Y., Junping, X.: A Collaborative Intrusion Detection Approach Using Blockchain for Multimicrogrid Systems. IEEE Trans. Syst. Man, Cybern. Syst. 49(8), 1720–1730 (2019). https://doi.org/10.1109/TSMC.2019.2911548
21. Alkadi, O., Moustafa, N., Turnbull, B., Choo, K.K.R.: A deep blockchain framework-enabled collaborative intrusion detection for protecting IoT and cloud networks. IEEE Internet Things J. 8(12), 9463–9472 (2021). https://doi.org/10.1109/JIOT.2020.2996590
22. Vasilomanolakis, E., Karuppayah, S., Muhlhauser, M., Fischer, M.: Taxonomy and survey of collaborative intrusion detection. ACM Comput. Surv. 47(4), 1–33 (2015). https://doi.org/10.1145/2716260
23. Liao, H.J., Richard Lin, C.H., Lin, Y.C., Tung, K.Y.: Intrusion detection system: a comprehensive review. J. Netw. Comput. Appl. 36(1) 16–24 (2013). https://doi.org/10.1016/j.jnca.2012.09.004
24. Alaba, F.A., Othman, M., Hashem, I.A.T., Alotaibi, F.: Internet of things security: a survey. J. Netw. Comput. Appl. 88(April), 10–28 (2017). https://doi.org/10.1016/j.jnca.2017.04.002
25. Jiang, J., Li, Z., Tian, Y., Al-Nabhan, N.: A Review of Techniques and Methods for IoT Applications in Collaborative Cloud-Fog Environment. Secur. Commun. Netw. 2020, (2020). https://doi.org/10.1155/2020/8849181
26. Da Xu, L., He, W., Li, S.: Internet of things in industries: a survey. IEEE Trans. Ind. Informatics 10(4), 2233–2243 (2014). https://doi.org/10.1109/TII.2014.2300753
27. Warzynski, A., Kolaczek, G.: Intrusion detection systems vulnerability on adversarial examples. In: 2018 IEEE International Confernce Innovation Intelligence System Application INISTA 2018,pp. 31–34 (2018). https://doi.org/10.1109/INISTA.2018.8466271

28. Khraisat, A., Gondal, I., Vamplew, P., Kamruzzaman, J., Alazab, A.: A novel ensemble of hybrid intrusion detection system for detecting internet of things attacks. Electron. **8**(11), (2019). https://doi.org/10.3390/electronics8111210

29. Asharf, J., Moustafa, N., Khurshid, H., Debie, E., Haider, W., Wahab, A.: A review of intrusion detection systems using machine and deep learning in internet of things: Challenges, solutions and future directions. Electron. **9**(7), 1–4 (2020). https://doi.org/10.3390/electroni cs9071177

30. Dasgupta, D., Shrein, J.M., Gupta, K.D.: A survey of blockchain from security perspective. J. Banking Financial Technol. **3**(1), 1–17 (2018). https://doi.org/10.1007/s42786-018-00002-6

31. Antal, C., Cioara, T., Anghel, I., Antal, M., Salomie, I.: Distributed ledger technology review and decentralized applications development guidelines. Futur. Internet **13**(3), 1–32 (2021). https://doi.org/10.3390/fi13030062

32. Natarajan, H., Krause, S.K., Gradstein, H.L.: Distributed Ledger Technology (DLT) and Blockchain. FinTech Note, 1, pp. 1–60, (2017). http://hdl.handle.net/10986/29053%0Ahttp://documents.worldbank.org/curated/en/177911513714062215/pdf/122140-WP-PUBLIC-Distributed-Ledger-Technology-and-Blockchain-Fintech-Notes.pdf

33. Chowdhury, M.J.M., et al.: A comparative analysis of distributed ledger technology platforms. IEEE Access **7**, 167930–167943 (2019). https://doi.org/10.1109/ACCESS.2019.2953729

34. Farahani, B., Firouzi, F., Luecking, M.: The convergence of IoT and distributed ledger technologies (DLT): Opportunities, challenges, and solutions. J. Netw. Comput. Appl. **177** 102936 (2021). https://doi.org/10.1016/j.jnca.2020.102936

35. Pandl, K.D., Thiebes, S., Schmidt-Kraepelin, M., Sunyaev, A.: On the convergence of artificial intelligence and distributed ledger technology: a scoping Review and future research agenda. IEEE Access **8**, 57075–57095 (2020). https://doi.org/10.1109/ACCESS.2020.2981447

36. Harris, J.D., Waggoner, B.: Decentralized and collaborative AI on blockchain. In: Proceedings of the 2019 2nd IEEE International Conference Blockchain,Blockchain **2019**(2) 368–375 (2019). https://doi.org/10.1109/Blockchain.2019.00057

37. Montes, G.A., Goertzel, B.: Distributed, decentralized, and democratized artificial intelligence. Technol. Forecast. Soc. Change **141** 2018 354–358 (2019). https://doi.org/10.1016/j.techfore.2018.11.010

38. Li, W., Wang, Y., Li, J., Au, M.H.: Toward a blockchain-based framework for challenge-based collaborative intrusion detection. Int. J. Inf. Secur. **20**(2), 127–139 (2020). https://doi.org/10.1007/s10207-020-00488-6

39. Meng, W., Tischhauser, E.W., Wang, Q., Wang, Y., Han, J.: When intrusion detection meets blockchain technology: a review. IEEE Access **6**, 10179–10188 (2018). https://doi.org/10.1109/ACCESS.2018.2799854

40. Alkadi, O., Moustafa, N., Turnbull, B.: A review of intrusion detection and blockchain applications in the cloud: approaches, challenges and solutions. IEEE Access **8**, 104893–104917 (2020). https://doi.org/10.1109/ACCESS.2020.2999715

An Implementation of Depth-First and Breadth-First Search Algorithms for Tip Selection in IOTA Distributed Ledger

Andras Ferenczi$^{(\boxtimes)}$ and Costin Bădică📵

Faculty of Automatics, Computers and Electronics, University of Craiova, Craiova, Romania
ferenczi.andras.h5f@student.ucv.ro, costin.badica@edu.ucv.ro

Abstract. IOTA is one of the most promising distributed ledgers due to its performance, scalability, and fee-less transaction capability. The interaction with the IOTA ledger requires client software to perform complex logic. As this software is often running on low-power Internet of Things (IoT) devices, it is important for it to be highly optimized. Our work highlights the benefits of DAG-based distributed ledgers. We start with an overview of IOTA tip selection algorithms, including the Markov Chain Monte Carlo (MCMC) algorithm and time-based tip selection algorithm. The original IOTA Reference Implementation (IRI) of Cumulative Weight Calculation (CWC) is based on Breadth-First Search (BFS) algorithm. The main contribution of our paper is to show how the Depth-First Search (DFS) algorithm is a better alternative solution to the CWC problem. We describe a sample software implementation, a method for collecting a snapshot from the Tangle, and finally we discuss the experimental outcomes.

Keywords: IOTA · Tangle · Graph · Depth-first search · Breadth-first search · Cumulative weight

1 Introduction

Blockchain is a revolutionary technology that has the potential of changing the way our economic and social systems work. It all started when Satoshi Nakamoto published the bitcoin white paper [1] in 2008, and created and deployed a reference implementation of the software and the cryptocurrency ledger. The next phase followed in 2014 when Vitalik Butterin published his white paper [2] suggesting that blockchain can be used for other purposes besides cryptocurrencies, giving birth to the next evolution of the technology, the Ethereum Blockchain. Generally, blockchains are records of transactions that are replicated to all the nodes in the network. The transactions are grouped in blocks and are chained together in an immutable, chronological order. The integrity of ledger is maintained via various consensus methods, such as Proof of Work (a.k.a. mining),

N. T. Nguyen et al. (Eds.): ACIIDS 2022, LNAI 13757, pp. 351–363, 2022.
https://doi.org/10.1007/978-3-031-21743-2_28

Practical Byzantine Fault tolerance. With all the benefits blockchains have to offer, their downsides are multiple, including limited scalability, high latency, and high transaction costs to paid to miners that maintain the ledger integrity. Traditional blockchains are also highly inefficient, as strict ordering in a distributed environment means that only one of multiple blocks mined simultaneously will make it into the blockchain while the rest are discarded.

DAG (Directed Acyclic Graph) based distributed ledgers address these concerns. By allowing forks, DAG-Blockchains prevent the waste of blocks while considerably speeding up transaction confirmation. Wang et al. [12] in their systematization of knowledge (SoK) present a comprehensive overview of DAG-based blockchains, including the protocol design, mathematical modeling, and consensus mechanisms used by the current state of art. One such DAG-Blockchain system is IOTA [10]. For a transaction to be validated, the issuing node must validate two previous transactions, thus contributing to the ledger's integrity. This approach cuts out the need for miners, meaning fee-less transactions, making it ideal for industrial and IoT integration. The method used to select the transactions for verification is called the "Tip Selection Algorithm" and is implemented in IRI using Markov chain Monte Carlo (MCMC) algorithm. This random walk picks transactions based on their Cumulative Weight (CW), thus making its calculation an essential component of the node software. Cumulative Weight is the most critical parameter IOTA Tangle transactions. We will turn our attention to the specifics of the Cumulative Weight Calculation in the IRI implementation and will propose an alternative solution that significantly reduces its time complexity. This is especially important, since IOTA was built to run IoT devices with limited computing capacity.

Our experimental results showed a significant reduction of execution times as the number of transactions in the snapshot increases.

Our paper is structured as follows. In Sect. 1 we present an overview of blockchain technologies and highlight the benefits of DAG-based distributed ledgers. Section 2 presents an overview of IOTA tip selection algorithms, including the Markov Chain Monte Carlo (MCMC), based on Popov's paper [6] and a time-based tip selection, per Haglamuge's work [3]. In Sect. 3 we explain the problem and present the Breadth-First Search (BFS) algorithms, then explain how the original IOTA Reference Implementation (IRI) [5] Cumulative Weight Calculation (CWC) uses it. The main contribution of this paper is in Sect. 4, in which we explain how the Depth-First Search (DFS) algorithm works and show how it can be used as an alternative solution to the CWC problem. In Sect. 5 we describe a sample software implementation, a method for collecting a snapshot from the Tangle, and discuss the experimental outcomes. We summarize our conclusions in Sect. 6, where we also set the stage for our future research of IOTA's ultimate Chrysalis 2.0 release.

2   Related Works

Popov [7] presents a family of Markov Chain Monte Carlo (MCMC) algorithms that can be used – but not mandated – for selecting transactions for verification.

The proposed algorithm entails a number of random walkers, start from older transactions of the tangle and let them walk randomly towards the newer, unapproved transactions. In case of a weighted walk, the random walkers will favor choosing on transactions with higher cumulative weight. Whenever 2 conflicting transactions are found, the node runs the tip selection algorithm multiple times and decides on the transaction that is chosen indirectly more often. The transactions chosen by the random walk will be the ones verified by the new transaction. This approach discourages "lazy" nodes that favor older transactions to new ones and it also discourages "parasite chains", that can be built by an attacker to artificially inflate the weights of own transactions.

Halgamuge [3] in her paper presents a time-based tip selection mechanism and best approver selection. The proposed tip selection algorithm divides the tangle into equal timeframes that are configurable at the node level, and selects the two most verified transactions in the nearest timeframe. The proposed method results in reduction of malicious attacks, increased scalability, application-specific a priority levels, and adds cumulative importance to transaction request times.

3   Problem Description

The Tangle, which is the data structure behind IOTA, is a particular kind of directed acyclic graph(DAG), which holds transactions. Each transaction is represented as a vertex in the graph. When a new transaction joins the tangle, it chooses two previous transactions to approve, adding two new edges to the graph. Transactions that do not yet have approvers are called tips.The very first transaction in the tangle is called the genesis. The genesis does not approve any other transactions.

Transactions on the Tangle have their own weight, which is a numeric value. This value represents the sum of all transactions approved by the node that submitted this transaction. The Cumulative Weight is calculated by summing up the own weight plus the weights of all transactions that approved this transaction, either directly or indirectly (i.e. approvers of approvers). The higher the Cumulative Weight, the more likely is for a transaction to be confirmed. As new transactions are added to the tangle, the Cumulative Weight of our transaction increases over time.

Popov in The Tangle [7] suggests the use of the MCMC algorithm to select the 2 tips to approve. Given an interval within the tangle, a number of "walkers" are placed randomly on the graph and are performing random discrete-time walks toward the tips. The walks are performed via the edges in opposite direction to the approvals. The path of the first 2 to reach the tip will be used to approve the adjacent transactions. In a weighted walk, the walkers will favor transactions with higher cumulative weight. The formula that provides the weight of transactions uses a parameter α that regulates the importance of the cumulative weight.

Next, we will briefly explain the Breadth-First Search (BFS) and Depth-First Search (DFS) graph algorithms, as described by Cormen et al [4]. These algorithms are extensively used in both the existing, as well as, the proposed solutions.

3.1 Breadth-First Search

Given the graph $G = (V, E)$ and a source vertex s, BFS incrementally discovers the adjacent vertices of s and then does the same for the latter, until it exhausts all connected elements. BFS keeps a color attribute for each vertex, to keep track the stages of their exploration: prior to being discovered, these are *white*, at time of discovery are colored *gray* and finally *black* when all their connections were identified. In the process, the algorithm also assigns the parents of each vertex, thus building a Breath-first tree, the predecessor subgraph G_π. This is an important aspect we used in our implementation of IOTA tip weight calculation.

Algorithm 1. Breadth-first search algorithm

```
 1: procedure BFS(G, s)
 2:     for each vertex u ∈ G.V do
 3:         u.color = WHITE
 4:         u.d = ∞
 5:         u.π = NIL
 6:     end for
 7:     s.color = GRAY
 8:     s.d = 0
 9:     s.π = NIL
10:     Q = ∅
11:     ENQUEUE(Q, s)
12:     while Q ≠ ∅ do
13:         u = DEQUEUE(Q)
14:         for each v ∈ G.Adj[u] do
15:             if v.color == WHITE then
16:                 v.color = GRAY
17:                 v.d = u.d + 1
18:                 v.π = u
19:                 ENQUEUE(Q, v)
20:             end if
21:         end for
22:         u.color = BLACK
23:     end while
24: end procedure
```

In Algorithm 1 lines 2..6 prepare all vertices by coloring them *white* (undiscovered), setting their parents to Nil and setting the discovery time to infinity. It does the same for the source vertex, except sets the discovery time to zero, as *s* is "discovered" at this point. Line 10 initiates an empty queue and *s* is enqueued in line 11. Lines 12..23 will process every gray verticle, as follows. Line

13 extracts the next vertex, then the for loop iterates over all its connected vertices. If any of them is found to be undiscovered (*white*), it is colored **gray**, its discovery time is set to its parent's time plus one (line 17), then its parent is set and it goes into the queue (line 19). In order to calculate the time complexity of the BFS algorithm, we must note that lines 2 to 6 are executed for every vertex, hence it takes $\Theta(V)$. All queue operations take $O(1)$. The For loop is performed once for every vertex (V times). Since each vertex's adjacency list is scanned only once, the total number of times the For loop is invoked equals the number of edges of the graph. As a result, the total running time of the BFS is $O(V + E)$.

3.2 IRI Implementation of Cumulative Weight Calculation

The open-source IOTA Reference Implementation (IRI) in Java [5] The tip selection procedure (method) takes as a parameter the hash of an entry point (edge) transaction. This transaction (vertex) is a candidate to be approved by our IOTA client.

Algorithm 2 takes as input the graph and the entry point. Lines 2..4 initialize the weight map to be returned, a cache for storing fetched vertices and a stack used for the usual BFS processing. The original implementation pre-allocates sufficient memory to hold all vertices and their associated ratings (weights) for the "cone" that has the entry point at its tip. As the APPROVERS call is a direct call to a database that hosts a snapshot of the tangle, to avoid frequent disk access, the cache declared in line 3 is passed to this function for every call. Lines 4..22 are similar to lines 10..23 of Algorithm 1(BFS), with small exceptions: color, discovery time and parent information is not tracked. There is also additional functionality. As mentioned in the INTRODUCTION, the cumulative weight is calculated by adding up all the approvers and their approvers and so on, plus 1. Hence, lines 10, 11 set the rating to 1 for transactions that have zero approvers. Lines 13..17 are not different from the way BFS does the processing in Algorithm 1 lines 14..21. The interesting part of Algorithm 2 starts at line 18, where the current vertex is added to the adjacency list and then its rating is returned in variable w. The **GET-WEIGHT** function is shown in Algorithm 3. It is, basically, another search algorithm that builds up the ancestry graph for vertex instance from Algorithm 2, line 8 and returns its weight, which is equal to the size of the stack used to hold them. Back to Algorithm 2, line 20 adds the vertex and its weight to the weight map and the program proceeds to the next vertex in the stack. Lines 23..26 make sure that *entryPoint* is also present in the map and, finally, line 28 returns the map for further processing.

We will need to make an assumption, in order to perform a simplified complexity calculation. First, we will treat the APPROVERS call to be simply just an index lookup in an array. Just as in Algorithm 1: APPROVERS(G, vertex, cache) $\equiv G.Adj[vertex]$. Similarly to Algorithm 1, time complexity of stack operations (i.e., PUSH and PULL) in Algorithm 2 is $O(1)$ and in total is O(V), where V represents the number of vertices. Algorithm GET-WEIGHT has a time complexity of $O(E^2)$, so the total running time is $O(V + E^2)$. This may not be an issue if the tangle is sparely connected, meaning the number of approvals per transaction is low. Nevertheless, activity on the tangle is unpredictable and we couldn't find concrete statistics that could confirm one way or another.

Algorithm 2. IRI algorithm for returning cumulative weight map

---

```
 1: procedure GET-WEIGHT-MAP-IRI(G, entryPoint)
 2:     weightMap = ∅
 3:     cache = ∅
 4:     stack = ∅
 5:     adj = APPROVERS(G, entryPoint, cache)
 6:     PUSH(stack, adj)
 7:     while stack ≠ ∅ do
 8:         vertex = POP(stack)
 9:         adj = APPROVERS(G, vertex, cache)
10:         if adj = ∅ then
11:             PUT(weightMap, (vertex, 1))
12:         else
13:             for each a ∈ adj do
14:                 if a ∉ stack then
15:                     ADD(stack, a)
16:                 end if
17:             end for
18:             ADD(adj, vertex)
19:             w = GET-WEIGHT(G, adj, cache)
20:             PUT(weightMap, (vertex, w))
21:         end if
22:     end while
23:     if entryPoint ∉ weightMap then
24:         size = SIZE(weightMap)
25:         PUT(weightMap, (entryPoint, size + 1))
26:     end if
27:     return weightMap
28: end procedure
```

---

4   Proposed Solutions

Our proposed solution uses a variant of the Depth-First Search algorithm.

4.1   Depth-First Search (DFS)

Similarly to BFS, DFS [4] uses the **adjacency-list** representation of a graph. The algorithm starts at a root vertex on the graph and goes as far as possible along the branch before backtracking. The process continues until all branches – and hence vertices – are discovered. Whenever DFS discovers a vertex v while scanning the adjacency list of a vertex u, it sets its predecessor attribute $v * \pi$ to

Algorithm 3. IRI algorithm for returning transaction weight

```
 1: procedure GET-WEIGHT(G, adj, cache)
 2:     vertices = CLONE(adj)
 3:     while vertices ≠ ∅ do
 4:         vertex = POP(vertices)
 5:         vertexAdj = APPROVERS(G, vertex, cache)
 6:         for each a ∈ vertexAdj do
 7:             if a ∉ adj then
 8:                 PUSH(adj, a)
 9:                 PUSH(vertices, a)
10:             end if
11:         end for
12:     end while
13:     return SIZE(adj)
14: end procedure
```

u. As the algorithm handles vertices, it sets their status (i.e. *color*) accordingly. Vertices are *white* prior to being discovered, *gray* when first encountered, and finally *black* when their adjacency list has been exhausted. This technique is required to ensure that each vertex is handled exactly once in the process. The algorithm also maintains 2 timestamps on vertices. The first, $v.d$ is recorded when the vertex is discovered, and $v.f$ recorded at the time when the vertex is colored *black*.

Algorithm 4. Depth-first search algorithm

```
 1: procedure DFS(G)
 2:     for each vertex u ∈ G.V do
 3:         u.color = WHITE
 4:         u.π = NIL
 5:     end for
 6:     time = 0
 7:     for each vertex u ∈ G.V do
 8:         if u.color == WHITE then
 9:             DFS-VISIT(G, u)
10:         end if
11:     end for
12: end procedure
```

Algorithm 4 first iterates over each vertices to set their status to **WHITE**, i.e. not visited. The second loop iterates again over all vertices in the graph and calls **DFS-VISIT** on them.

Algorithm 5 increments time, sets the color of vertex v to **GRAY**, sets the discovery time for v. Then, it iterates over all adjacent vertices and calls iteratively **DFS-VISIT(G, v)**. Once done, it marks u's color to **BLACK** (completed).

Algorithm 5. Helper procedure for visiting in depth branches

```
1: procedure DFS-VISIT(G, u)
2:     time = time + 1
3:     u.d = time
4:     u.color = GRAY
5:     for each vertex v ∈ Adj[u] do
6:         if v.color == WHITE then
7:             v.π = u
8:             DFS-VISIT(G, v)
9:         end if
10:    end for
11:    u.color = BLACK
12:    time = time + 1
13:    u.f = time
14: end procedure
```

The running time of **DFS** is $\Theta(V + E)$, as in Algorithm 5 lines 2..4 and 7..11 respectively take $\Theta(V)$ each; **DFS-VISIT** is called once for each vertex and since:

$$\sum_{v \in G} |Adj[v]| = \Theta(E)$$

4.2 Our Solution

We take a slightly different approach to the problem. We will be using DFS and ensure that every vertex and edge is visited exactly once. The core procedure is Algorithm 6 it is derived from Algorithm 4. It initializes the **weightMap** to be returned, the **stack** and the same **cache** that was described in the previous section for the IRI Implementation. Line 6 returns the vertices adjacent to *entryPoint*. Lines 7..11 prepares the vertices for the Depth-first search. Lines 12..16 have identical functionality to Algorithm 4 lines 7..11. Line 14 calls DFS-VISIT – again, same functionality as Algorithm 5. The latter is represented in Algorithm 7 . It is performing the recursion, and when it completes it for each node, it increments its weight and also updates its parent's accordingly in one go. Making the same assumptions as in prior section, i.e. APPROVERS(G, vertex, cache) ≡ G.Adj[vertex], this algorithm's time complexity is identical to that of the Depth-first search algorithm: O(V + E). The experimental results below confirm this.

Algorithm 6. Helper procedure for visiting in depth branches

1: **procedure** GET-WEIGHT-MAP-DFS($G, entryPoint$)
2:    $time = 0$
3:    $weightMap = \emptyset$
4:    $stack = \emptyset$
5:    $cache = \emptyset$
6:    $vertices = $ APPROVERS($G, entryPoint$, cache)
7:    **for each vertex** $u \in vertices$ **do**
8:        $u.color = $ WHITE
9:        $u.w = 0$
10:       $u.\pi = Nul$
11:   **end for**
12:   **for each** $vertex \in vertices$ **do**
13:       **if** $vertex.color == WHITE$ **then**
14:           DFS-VISIT($G, vertex, weightMap, cache$)
15:       **end if**
16:   **end for**
17:   $size = $ SIZE($weightMap$)
18:   PUT($weightMap, (entryPoint, size + 1)$)
19:   **return** weightMap
20: **end procedure**

Algorithm 7. Recursive procedure to explore in-depth branches

1: **procedure** DFS-VISIT($G, vertex, weightMap, cache$)
2:    $vertex.color = GRAY$
3:    $adjVerticles = $ APPROVERS($G, vertex, cache$)
4:    **for each** $v \in adjVerticles$ **do**
5:        **if** $v.color = WHITE$ **then**
6:            $v.color = GRAY$
7:            $v.\pi = vertex$
8:            DGS-VISIT($G, vertex, weightMap, cache$)
9:        **end if**
10:   **end for**
11:   $vertex.color = BLACK$
12:   $vertex.w = vertex.w + 1$
13:   **if** $vertex.\pi \neq Null$ **then**
14:       $vertex.\pi.w = vertex.\pi.w + vertex.w$
15:   **end if**
16: **end procedure**

5 Experimental Results

As expected, the DFS-based implementation of the Cumulative Weight Calculation performed far better than the BFS-based one. The code of the sample implementation is available at [11].

5.1   Sample Implementation

For experimentation we needed a Tangle snapshot. For this, we used the IOTA client software to listen to Tangle events from an existing public live node on the mainnet. As IOTA 1.0 mainnet is no longer active, we collected the snapshot from the current Chrysalis 1.5 network. We made the assumption that the 2 versions on the Tangle have similar distributions. The pseudo code for our program is presented in Algorithm 8.

The program returns a map, $referenceMap$, having as keys the $transactionId$ and as value an array of $approverId$, that represents all approvals of $transactionId$ at any given time. In line 3 we subscribe to IOTA message submissions and approval events using the client libraries. Lines 6..14 handle messages notifications in the form of a $transactionId$ and an array of $approverId$-s. We then make sure that each has an entry in the map and the approval is added to the respective $transactionId$.

Algorithm 8. Tangle snapshot recording

```
 1: procedure COLLECT-SNAPSHOT(iotaNode, duration)
 2:     referenceMap = ∅
 3:     SUBSCRIBE-EVENT(iotaNode, SUBMIT, APPROVE)
 4:     SUBSCRIBE-TIMER(duration)
 5:     upon event message do
 6:         if message.transactionId ∉ referenceMap then
 7:             PUT(referenceMap, (transactionId, ∅))
 8:         end if
 9:         for all approverId ∈ message.approvers do
10:             if approverId ∉ referenceMap then
11:                 PUT(referenceMap, (approverId, ∅))
12:             end if
13:             INSERT(referenceMap, (transactionId, approverId))
14:         end for
15:     upon event timer do
16:         UNSUBSCRIBE-EVENT(SUBMIT, APPROVE)
            wait for all processes to end
17:         return referenceMap
18: end procedure
```

Algorithm 9 is a simple test harness for comparing executions times. Just as the random walks require the execution of the CWC algorithms, we are invoking it using both approaches (Algorithm 2 and Algorithm 6.)

Algorithm 9. Test harness

1: **procedure** GENERATE-STATS(*referenceMap*)
2:    *size* = SIZE(*referenceMap*)
3:    *G* = CREATE-GRAPH(*referenceMap*)
4:    **for** *entryId* ∈ (0..*size*) **do**
5:       TIMER-START(*systemTime*)
6:       GET-WEIGHT-MAP-IRI(*G*, *entryId*)
7:       *duration* = TIMER-STOP(*systemTime*)
8:       LOG(G, entryId, duration)
9:       TIMER-START(*systemTime*)
10:       GET-WEIGHT-MAP-DFS(*G*, *entryId*)
11:       *duration* = TIMER-STOP(*systemTime*)
12:       LOG(G, entryId, duration)
13:    **end for**
14: **end procedure**

5.2   Experiments and Discussion

The experiments were performed on a desktop computer equipped with Intel Corporation Xeon E3-1200 v6/7th Gen Core Processor with 32 GB RAM and 240 GB SSD, running Ubuntu 18.04. After executing Algorithm 9, we generated the reports below using python and packages **pyplot** and **pandas**. The BFS algorithm uses significant amount of resources, as it loads the whole graph in memory. We averaged the data by number of vertices (transactions) to reduce the noise. To make more sense of these graphs, we also interpolated them by using the functions from the **scipy.interpolate** python package. The initial line graph is colored blue and the interpolation in red.

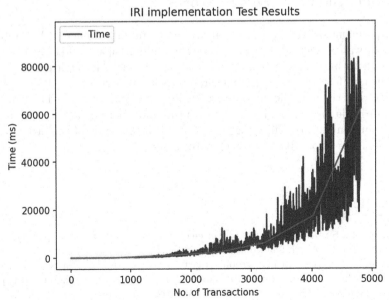

The graph of the testing results for the IRI implementation response times increase significantly with the number of edges.

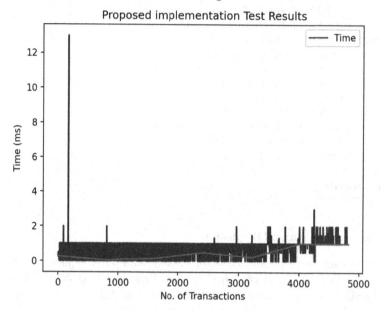

The Depth-first search-based implementation, although has a noisy graph, performs better than the IRI implementations. Even with thousands of edges the typical response time is around 1–2 ms.

6   Conclusion and Future Works

Our experiment shows that there is always room for improvements. As the IOTA client software typically runs on low-power limited capacity IoT devices, it is important to implement the most optimal algorithms. Although IRI is decommissioned we felt this was a useful exercise. As of this writing, the IOTA project is undergoing the most significant rewrite in the project's history. The Chrysalis 1.5 [8] upgrade brought higher speed, security and efficiency and the 2.0 release [9], the final milestone, will bring complete decentralization of the protocol. We are passionate about IOTA and plan to analyze in detail its new releases in upcoming paper(s).

References

1. Nakamoto, S.: Bitcoin: a peer-to-peer electronic cash system (white paper) (2008). https://bitcoin.org/bitcoin.pdf. Accessed 5 Feb 2022
2. Buterin, V.: A next-generation smart contract and decentralized application platform - Ethereum white paper (2014) https://www.ethereum.org/pdfs/EthereumWhitePaper.pdf. Accessed 5 Feb 2022

3. Halgamuge, M.N.: optimization framework for best approver selection method (BASM) and best tip selection method (BTSM) for IOTA tangle network: blockchain-enabled next generation Industrial IoT. Comput. Netw. **199**, 108418 (2021). https://doi.org/10.1016/j.comnet.2021.108418
4. Cormen, T.H., Leiserson, C.E. Rivest, R.L., Stein, C.: Introduction to Algorithms, 3rd edn. The MIT Press, Cambridge, pp. 594–623 (2009)
5. The official node software that runs on the IOTA Mainnet and Devnet. https://github.com/iotaledger/iri. Accessed 30 Dec 2021
6. IOTA - the tip selection process. https://legacy.docs.iota.org/docs/getting-started/0.1/network/tip-selection. Accessed 30 Dec 2021
7. Popov, S.: The tangle, April 30, 2018. Version 1.4.3. https://assets.ctfassets.net/r1dr6vzfxhev/2t4uxvsIqk0EUau6g2sw0g/45eae33637ca92f85dd9f4a3a218e1ec/iota1_4_3.pdf. Accessed 30 Dec 2021
8. Chrysalis (Path towards IOTA 1.5). https://blog.iota.org/chrysalis-b9906ec9d2de/. Accessed 5 Feb 2022
9. Towards full decentralization with IOTA 2.0, 2021. https://blog.iota.org/path-towards-full-decentralization-with-iota-2-0/. Accessed 5 Feb 2022
10. Van Hijfte, S.: Blockchain Platforms: A Look at the Underbelly of Distributed Platforms. Morgan & Claypool Publishers, San Rafael, Synthesis Lectures on Computer Science (2020)
11. Ferenczi, A.: Sample implementation of DFS-based Algorithm for the CWC Problem in IOTA Distributed Ledger. https://github.com/andrasfe/iota-research. Accessed 5 Feb 2022
12. Wang, Q., Yu, J., Chen, S., Xiang, Y.: SoK: diving into DAG-based Blockchain Systems, CoRR, abs/2012.06128, 2020, arXiv:2012.06128. Accessed 20 Mar 2022

Locally Differentially Private Quantile Summary Aggregation in Wireless Sensor Networks

Aishah Aseeri[1] and Rui Zhang[2]

[1] King Abdulaziz University, Jeddah 21589, Kingdom of Saudi Arabia
aaaseeri@kau.edu.sa
[2] University of Delaware, Newark, DE 19716, USA
ruizhang@udel.edu

Abstract. Privacy-preserving data aggregation has been widely recognized as a key enabling functionality in wireless sensor networks to allow the base station to learn valuable statistics of the sensed data while protecting individual sensor node's data privacy. Existing privacy-preserving data aggregation schemes all target simple statistic functions such as SUM, COUNT, and MAX/MIN. In contrast, a quantile summary allows a base station to extract the ϕ-quantile for any $0 < \phi < 1$ of all the sensor readings in the network and can thus provide a more accurate characterization of the data distribution. Unfortunately, how to realize privacy-preserving quantile summary aggregation remains an open challenge. In this paper, we introduce the design and evaluation of PrivQSA, a novel privacy-preserving quantile summary aggregation scheme for wireless sensor networks, which enables efficient quantile summary aggregation while guaranteeing ϵ-Local Differential Privacy for individual sensors. Detailed simulation studies confirm the efficacy and efficiency of the proposed protocol.

Keywords: Data aggregation · Wireless sensor network · Local differential privacy · Quantile summary

1 Introduction

Data aggregation [16] has been widely recognized as a key technique for reducing energy consumption and prolonging network lifetime by allowing sensed data to be aggregated by intermediate nodes along the route to the base station to eliminate possible redundancy. Data privacy is a key concern in many applications. For example, data generated by sensor nodes in an IoT-based smart-home system may contain a variety of sensitive information about users such as appliance usage and home occupancy. Since directly submitting such information to a base station would reveal sensitive information, there is strong need for privacy-preserving data aggregation solutions that can allow the base station to learn valuable statistic of the data generated in the network while ensuring the data privacy of individual sensor nodes.

© The Author(s), under exclusive license to Springer Nature Switzerland AG 2022
N. T. Nguyen et al. (Eds.): ACIIDS 2022, LNAI 13757, pp. 364–376, 2022.
https://doi.org/10.1007/978-3-031-21743-2_29

Privacy-preserving data aggregation has received significant attentions over the past decade. Existing privacy-preserving data aggregation schemes [7,9,11, 14,15,18,21] all target simple statistic functions such as SUM, COUNT, and MAX/MIN. In contrast, a quantile summary allows a base station to learn a more accurate distribution of the sensed data than simple statistics functions [5,6,10,19]. Specifically, a quantile summary allows the base station to retrieve the ϕ-quantile for any $0 \leq \phi \leq 1$, which can provide a much better characterization of the distribution of data generated by a wireless sensor network. Unfortunately, how to realize privacy-preserving quantile summary aggregation remains an open challenge.

In this paper, we introduce the design and evaluation of PrivQSA, a novel privacy-preserving quantile summary aggregation scheme for wireless sensor networks. Specifically, we design PrivQSA to satisfy ϵ-Local Differential Privacy (LDP), which is a model widely considered as the gold standard for data privacy. Under PrivQSA, every sensor node randomly perturbs its set of readings to ensure ϵ-LDP. All the nodes then participate in the quantile summary aggregation to allow the base station to obtain a quantile summary of the perturbed readings. The base station then estimates a quantile summary of the original sensed data based on the perturbation mechanism followed by individual sensor nodes. Our contributions in this paper can be summarized as follows.

- We are the first to study privacy-preserving quantile summary aggregation in wireless sensor networks.
- We introduce PrivQSA, a novel privacy-preserving quantile summary aggregation scheme that can allow the base station to learn a quantile summary of the sensed data while ensuring ϵ-LDP for individual sensor nodes.
- We confirm the efficacy and efficiency of PrivQSA via both theoretical analysis and detailed simulation studies, which demonstrate significant advantages over other baseline solutions.

The rest of this paper is structured as follows. Section 2 discusses the related work. Section 3 introduces the network model and some preliminaries. Section 4 introduces the design of PrivQSA. Section 5 evaluates PrivQSA via both theoretical analysis and simulation studies. Section 6 finally concludes this paper.

2   Related Work

In this section, we review some related work in quantile summary aggregation, privacy preserving data aggregation in WSNs.

Quantile summary aggregation in wireless sensor networks has been discussed in a number of articles throughout literature. Greenwald et. al. [5] studied quantile summary aggregation in wireless sensor networks. Shrivastava *et al.* [19] proposed a quantile digest summary structure to facilitate quantile aggregation. In [6], the authors designed a distributed algorithm to realize an ϵ-approximate quantile summary of all the sensor nodes data. Later, Huang *et al.* [10] introduced an improvement to the forementioned algorithm with the aim to reduce the maximum per node communication cost. Also, several efficient gossip distributed algorithms were proposed in [8] to compute the exact and approximate quantiles. However, none of the discussed quantile aggregation schemes accounts for privacy restrictions and therefore cannot be adapted to our proposed problem.

Privacy-preserving data aggregation in sensor networks has received a lot of attention over the past two decades [2–4,9,11,18,20]. Generally speaking, existing solutions for privacy preserving data aggregation can be classified into two categories. The first category uses encryption techniques such as homomorphic encryption [2,3,11,18,20,22], secure multiparty computation [12], and modulo addition-based encryption [1]. All these solutions target simple statistic functions such as SUM, COUNT, and MAX/MIN and cannot be adopted to support quantile summary aggregation. The second category uses random perturbation [4,9,17], in which each sensor node randomly perturbs its data according to a suitable probability distribution before participating in data aggregation, and the base station can still infer valuable statistics from the perturbed data. To the best of our knowledge, there is no prior work tackling privacy-preserving quantile summary aggregation.

3   Problem Formulation

In this section, we first introduce the network model and a background on quantile summary. We then provide the definition of Local Differential Privacy.

3.1   Network Model

We consider a wireless sensor network model consisting of a base station and n sensor nodes that form an aggregation tree. Let $R = \{1, \ldots, d\}$ be the range of possible readings. We assume that every sensor node i has a set of m readings $V_i = \{v_{i,1}, \ldots, v_{i,m}\}$, where every reading $v_{i,j} \in R$ for all $1 \leq i \leq n$ and $1 \leq j \leq m$. The set of all the sensed data generated in the sensor network is then $V = \bigcup_{i=1}^{n} V_i$. The base station aims to obtain a quantile summary of V.

3.2   Quantile Summary

A quantitle summary is a subset of readings along with their (estimated) global ranks which can support *value-to-rank* query over any $v \in R$ as well as ϕ-quantile queries for any $0 < \phi < 1$. Specifically, given a set of N distinct data values with a total order, the ϕ-quantile is the value v with rank $r(v) = \lfloor \phi N \rfloor$ in the set, where $r(v)$ is the number of values in the set smaller than v. Since a quantile summary that can provide the exact quantiles must contain the all N values in the worst case, an ϵ'-approximate ϕ-quantile is a value with rank between $(\phi - \epsilon')N$ and $(\phi + \epsilon')N$.

3.3   Local Differential Privacy (LDP)

Local Differential Privacy is a strong privacy notion widely considered as the gold standard for data privacy, which ensures that an adversary cannot differentiate two inputs based on the output he observe beyond certain predefined threshold. We give the definition of ϵ-Local Differential Privacy below.

Definition 1. *(ε-Local Differential Privacy). A randomized mechanism $\mathcal{M}$ satisfies ε-local differential privacy if and only if*

$$\frac{Pr[\mathcal{M}(x) = y]}{Pr[\mathcal{M}(x') = y]} \leq e^{\epsilon}$$

for any two inputs $x, x' \in X$ and any output $y \subseteq Range(\mathcal{M})$, where X is the domain of the input, $Range(\mathcal{M})$ is the domain of the output, and ϵ is commonly referred to as the privacy budget.

3.4   Design Goals

We seek to design a privacy-preserving quantile summary aggregation scheme with the following goals in mind.

- *Local Differential Privacy.* The scheme should satisfy ε-LDP for individual sensor nodes.
- *High accuracy.* The quantile summary obtained by the base station should be able to answer value-to-rank queries with high accuracy.
- *Communication efficiency.* The scheme should incur low communication overhead.

4   PrivQSA: Quantile Summary Aggregation with LDP

In this section, we first give an overview of PrivQSA and then detail its design.

4.1   Overview

We design PrivQSA by exploring the inherent connection between a quantile summary and a histogram. Specifically, a quantile summary can be viewed as an equi-depth histogram in which every bucket has the same number of values, and all the buckets in a standard histogram have the same width but different number of values. In addition, a quantile summary can be converted into a standard histogram under moderate assumptions, and vice versa. Based on this observation, we first let every sensor node randomly perturb its set of readings to generate a set of perturbed readings to ensure ε-LDP. All the sensor nodes then participate in quantile summary aggregation to allow the base station to receive a quantile summary, i.e., an equi-depth histogram, of the perturbed readings. The base station can then convert the quantile summary of the perturbed readings into a an equi-width histogram whereby to estimate an equi-width histogram of the original readings. Finally, the base station can convert the estimated equi-width histogram of the original readings into a quantile summary of the original readings whereby to answer any value-to-rank and percentile queries. In what follows, we detail the design of PrivQSA.

4.2   Detailed Design

PrivQSA consists of the following six steps.

Perturbation at Individual Sensor Nodes. Each sensor node i randomly perturbs its set of readings $V_i = \{v_{i,1}, \dots, v_{i,m}\}$ into a set of n perturbed readings S_i' via the exponential mechanism to ensure ϵ-LDP. The exponential mechanism [13] is a classical technique to provide differential privacy via outcome randomization. The key idea is to associate every pair of input x and candidate outcome o with a real-value quality score $q(x, o)$, where a higher quality score indicates higher utility of the outcome. Given the output space O, a score function $q(\cdot, \cdot)$, and the privacy budget ϵ, the exponential mechanism randomly selects an outcome $o \in O$ with probability proportional to $\exp(\epsilon q(x, o))$.

Here the input is a set $V_i \subseteq R$ of m readings, and the outcome $\tilde{V}_i$ of the exponential mechanism is also a subset of R with m elements. For every pair of possible input set V_i and output set $\tilde{V}_i$, we define the quality score function as

$$q(V_i, \tilde{V}_i) = \frac{|V_i \bigcap \tilde{V}_i|}{m}, \tag{1}$$

which is the ratio of their common elements. For example, if $V_i = \tilde{V}_i$, then the quality score is one. Under the quality score function $q(\cdot, \cdot)$, each node i then randomly chooses an m-element set $\tilde{V}_i \subset R$ with probability proportional to $\exp(\frac{\epsilon|V_i \bigcap \tilde{V}_i|}{m})$.

Data Augmentation. Since all existing quantile summary aggregation schemes including Huang $et\ al.$'s protocol [10] requires every data value to be distinct, every sensor node augments its perturbed readings by its node ID. Let $\tilde{V}_i = \{\tilde{v}_{i,1}, \dots, \tilde{v}_{i,m}\}$ be node i's set of perturbed readings. Each node i augments each perturbed reading $\tilde{v}_{i,j}$ as $\hat{v}_{i,j} = \tilde{v}_{i,j}||i$ for all $1 \leq j \leq m$, where $||$ denotes concatenation, and node ID i is encoded by $\gamma = \lceil \log_2 n \rceil$ bits. Doing so can ensure that every reading generated in the network is unique. We hereafter denote by $\hat{V}_i$ the set of perturbed and augmented readings of node i for all $1 \leq i \leq n$.

Quantile Summary Aggregation. Every node first generates a local quantile summary of $\hat{V}_i$. Specifically, each node i randomly samples each perturbed reading $\hat{v}_{i,j} \in \hat{V}_i$ independently with probability h to obtain a subset of perturbed readings $S_i \subseteq \hat{V}_i$, where $h \in (0, 1]$ is a system parameter. Node i's local quantile summary of $\hat{V}_i$ is then

$$Q_i = \{(\hat{v}_{i,j}, j)|\hat{v}_{i,j} \in S_i\}, \tag{2}$$

where j is the perturbed reading $\hat{v}_{i,j}$'s local rank within $\hat{V}_i$. The set $\hat{V}_i$ is commonly referred to as the *ground set* of local quantile summary Q_i

All the sensor nodes then participate in quantile summary aggregation according to Huang $et\ al.$'s protocol [10]. During the aggregation process, intermediate nodes may merge multiple local quantile summaries into one via random resampling to reduce the maximum per node communication cost. We refer readers to [10] for details of the merging process.

The base station performs value-to-rank query on every possible perturbed value to learn the distribution of the perturbed readings. Assume that the base station receive n' local quantile summaries $Q'_1, \ldots, Q'_{n'}$ at the end of the aggregation process, where each Q'_i corresponds to a ground set $\hat{V}'_i$ and $n' \leq n$ due to possible merging by intermediate sensor nodes. It is easy to see that $\bigcup_{i=1}^{n} \hat{V}_i = \bigcup_{i=1}^{n'} \hat{V}'_i$. For every possible augmented and perturbed value $\hat{v} = v\|i$ where $v \in R$ and $i \in \{1, \ldots, n\}$, the base station estimates its global rank within $\bigcup_{i=1}^{n} \hat{V}_i$ as

$$\hat{r}(\hat{v}) = \sum_{i=1}^{n'} \hat{r}(\hat{v}, \hat{V}'_i) , \tag{3}$$

where

$$\hat{r}(\hat{v}, \hat{V}'_i) = \begin{cases} r(\mathsf{pred}(\hat{v}|Q'_i), \hat{V}'_i) + 1/h, & \text{if } \mathsf{pred}(\hat{v}|Q'_i) \text{ exists}; \\ 0 & \text{otherwise}, \end{cases} \tag{4}$$

where $\mathsf{pred}(\hat{v}|Q_i)$ denotes the predecessor of value $\hat{v}$ in Q'_i. It has been shown in [10] that $\hat{r}(\hat{v})$ is an unbiased estimator of $r(\hat{v}, \bigcup_{i=1}^{n} \hat{V}'_i) = r(\hat{v}, \bigcup_{i=1}^{n} \hat{V}_i)$.

Next, the base station computes the global rank of each possible value $v \in R$ by removing the augmented node ID from the perturbed readings. In particular, for each pair of perturbed value and estimated rank $(\hat{v}, \hat{r}(\hat{v}))$, the base station updates its value as

$$\tilde{v} = \hat{v} \mod 2^\gamma \tag{5}$$

and records the pair $\langle \tilde{v}, \hat{r}(\hat{v}) \rangle$.

After removing the augmented IDs from all perturbed readings, the base station obtains one or more estimated global ranks for each possible perturbed value $\tilde{v} \in R$. Without loss of generality, let $r^-(\tilde{v})$ and $r^+(\tilde{v})$ be the lowest and highest estimated global ranks, respectively, of value $\tilde{v}$ for all $\tilde{v} \in R$. If value $\tilde{v}$ has only a unique estimated global rank, then $r^-(\tilde{v}) = r^+(\tilde{v})$.

Estimating Histogram of Perturbed Readings. The base station then constructs a histogram of the perturbed readings from the received quantile summaries by estimating the frequency of each perturbed value $\tilde{v} \in R$. We formulate the histogram construction as an optimization problem. In particular, let $f_{\tilde{v}}$ be the frequency of perturbed value $\tilde{v}$ for all $\tilde{v} \in R$. It follows that value 1 is ranked from the 1st to the f_1th, and value $\tilde{v}$ is ranked from $(\sum_{i=1}^{\tilde{v}-1} f_i + 1)$th to $(\sum_{i=1}^{\tilde{v}} f_i)$th for all $1 \leq \tilde{v} \leq d$. We can then formulate the estimation of $f_1, \ldots, f_d$

as the following optimization problem

$$\min_{(f_1,\ldots,f_d)\in\mathbb{N}^d} \sum_{\tilde{v}\in R}(\sum_{i=1}^{\tilde{v}-1} f_i + 1 - r^-(\tilde{v}))^2 + (\sum_{i=1}^{\tilde{v}} f_i - r^+(\tilde{v}))^2,$$

$$\text{such that} \quad \sum_{\tilde{v}=1}^{d} f_{\tilde{v}} = nm,$$

$$\sum_{i=1}^{\tilde{v}-1} f_i + 1 \le r^-(\tilde{v}), \forall \tilde{v} \in R,$$

$$\sum_{i=1}^{\tilde{v}} f_i \ge r^+(\tilde{v}), \forall \tilde{v} \in R,$$

(6)

where we seek to minimize the total square errors between the two boundaries and the corresponding lowest and highest estimated global ranks. In the above optimization problem, the first constraint indicates that the sum of all the values' frequencies should be nm, the second and third constraints guarantee that the lowest and highest estimated global ranks of a value $\tilde{v}$, i.e., $r^-(\tilde{v})$ and $r^+(\tilde{v})$, should fall in the range $[\sum_{i=1}^{\tilde{v}-1} f_i + 1, \sum_{i=1}^{\tilde{v}} f_i]$.

The solution of the above optimization problem is given by

$$\begin{cases} f_1 = \frac{r^+(1)+r^-(2)-1}{2}, \\ f_i = \frac{r^+(i)+r^-(i+1)-r^+(i-1)-r^-(i)}{2}, & \forall 2 \le i \le d-1, \\ f_d = nm - \frac{r^+(d-1)+r^-(d)-1}{2}. \end{cases}$$

(7)

Estimating Histogram of Original Readings. Given the estimated histogram of perturbed readings $f_1, \ldots, f_d$ obtained above, the base station further estimates the histogram of original readings based on the perturbation mechanism used by individual sensor nodes.

Denote by g_v be the frequency of value v among the original readings $\bigcup_{i=1}^{n} V_i$ for all $v \in R$. Consider any value $v \in R$ and sensor node i's original reading set V_i. Under the random perturbation mechanism, if $v \in V_i$, the probability that v shows up in the perturbed set $\tilde{V}_i$ is given by

$$Pr[v \in \tilde{V}_i | v \in V_i] = Pr[v \in V_i \bigcap \tilde{V}_i | v \in V_i]$$

$$= \sum_{k=1}^{m} Pr[v \in V_i \bigcap \tilde{V}_i | v \in V_i, |V_i \bigcap \tilde{V}_i| = k] \cdot Pr[|V_i \bigcap \tilde{V}_i| = k].$$

(8)

Moreover, we have $Pr[v \in V_i \bigcap \tilde{V}_i | v \in V_i, |V_i \bigcap \tilde{V}_i| = k] = \frac{k}{m}$ and $Pr[|V_i \bigcap \tilde{V}_i| = k] = \frac{c_k \exp(\frac{\epsilon k}{m})}{\sum_{j=0}^{m} c_j \cdot \exp(\frac{\epsilon j}{m})}$, where $c_k = \binom{m}{k} \cdot \binom{d-m}{m-k}$ is the number of m-element subsets of R that shared k common elements with V_i for all $k = 0, \ldots, m$. It follows that

$$Pr[v \in \tilde{V}_i | v \in V_i] = \sum_{k=1}^{m} \frac{k}{m} \cdot \frac{c_k \exp(\frac{\epsilon k}{m})}{\sum_{j=0}^{m} c_j \cdot \exp(\frac{\epsilon j}{m})} . \tag{9}$$

Now let us analyze the probability that v shows up in the perturbed set $\tilde{V}_i$ given that $v \notin V_i$. Following the similar analysis, we can derive

$$Pr[v \in \tilde{V}_i | v \notin V_i] = Pr[v \in \tilde{V}_i \setminus V_i | v \notin V_i]$$

$$= \sum_{k=0}^{m-1} Pr[v \in \tilde{V}_i \setminus V_i | v \notin V_i, |V_i \bigcap \tilde{V}_i| = k] \cdot Pr[|V_i \bigcap \tilde{V}_i| = k]$$

$$= \sum_{k=0}^{m-1} \frac{m-k}{d-m} \cdot \frac{c_k \exp(\frac{\epsilon k}{m})}{\sum_{j=0}^{m} c_j \cdot \exp(\frac{\epsilon j}{m})} . \tag{10}$$

Assume that value v appears in g_v sensor nodes' original reading sets. It follows that $n - g_v$ sets do not contain value v. The expected number of perturbed sets that include value v can be estimated as

$$E[f_v] = g_v \cdot Pr[v \in \tilde{V}_i | v \in V_i] + (n - g_v) \cdot Pr[v \in \tilde{V}_i | v \notin V_i] . \tag{11}$$

We therefore estimate the number of copies of v in $\bigcup_{i=1}^{n} V_i$ as

$$\hat{g}_v = \frac{f_v - n \cdot Pr[v \in \tilde{V}_i | v \notin V_i]}{Pr[v \in \tilde{V}_i | v \in V_i] - Pr[v \in \tilde{V}_i | v \notin V_i]} , \tag{12}$$

where $Pr[v \in \tilde{V}_i | v \in V_i]$ and $Pr[v \in \tilde{V}_i | v \notin V_i]$ are given in Eqs. (9) and (10), respectively.

Final Quantile Summary Construction. Given the estimated histogram of the original readings, the base station then constructs a final quantile summary of $\bigcup_{i=1}^{n} V_i$, which is equivalent to estimating the rank for every value $v \in R$ and answering ϕ-quantile query for all $0 < \phi < 1$.

To answer a value-to-rank query over value $v \in R$, the base station can simply return the median rank of value v as

$$r(v) = \lfloor \frac{r^-(v) + r^+(v)}{2} \rfloor , \tag{13}$$

where $r^-(v) = \sum_{i=0}^{v-1} \hat{g}_i + 1$ and $r^+(v) = \sum_{i=0}^{v} \hat{g}_i$. Moreover, to answer a ϕ-quantile query where $0 < \phi < 1$, the base station returns value v such that $r^-(v) \leq nm\phi \leq r^+(v)$.

5   Performance Evaluation

5.1   Theoretical Analysis

We first have the following theorem regarding the privacy guarantee of PrivQSA.

Theorem 1. *PrivQSA satisfies ϵ-LDP.*

Proof. Let V_i and V_j be two arbitrary sets of readings such that $|V_i| = |V_j| = m$. Let $\mathcal{M}$ denote the randomized perturbation mechanism used by each individual sensor node. Consider any possible output $\tilde{V}$ of $\mathcal{M}$. Since $0 \leq |\tilde{V} \cap V_i| \leq m$ and $0 \leq |\tilde{V} \cap V_j| \leq m$ for any V_i and V_j, we have

$$\frac{\Pr[\mathcal{M}(V_i) = \tilde{V}]}{\Pr[\mathcal{M}(V_j) = \tilde{V}]} = \frac{\exp(\epsilon \cdot \frac{|\tilde{V} \cap V_i|}{m})}{\exp(\epsilon \cdot \frac{|\tilde{V} \cap V_j|}{m})} \leq \frac{\exp(\epsilon \cdot \frac{m}{m})}{\exp(\epsilon \cdot \frac{0}{m})} \leq \exp(\epsilon). \qquad (14)$$

The theorem is thus proved.   □

5.2   Simulation Settings

We simulate a wireless sensor network consisting of $n = 1022$ sensor nodes. We assume that each node has $m = 10$ readings and every original reading is in the range $R = \{1, \ldots, 100\}$. Table 1 summarizes our default settings unless mentioned otherwise.

Table 1. Default simulation settings

Para.	Val.	Description.
ϵ	50	The privacy budget
h	0.5	The sampling probability
n	1022	The number of sensor nodes
m	10	The size of user value set
d	100	The maximum number in the range of user values

Since there is no prior solution for private quantile summary aggregation, we compare PrivQSA with the following two baseline schemes.

- *Baseline 1*: Every node randomly perturbs its reading set as in PrivQSA and independently samples its perturbed readings with probability h. It then submits only the sampled perturbed readings without rank information to the base station. The base station estimates the original distribution and the final quantile summary using the same method as in the last two steps of PrivQSA. Baseline 1 satisfies ϵ-LDP and does not involve any quantile summary aggregation.
- *Baseline 2*: Every node participates in quantile summary aggregation according to Huang *et al.* [10] without any privacy guarantee.

We use two metrics to measure the accuracy of the final quantile summary at the base station. Let $r(v)$ and $\hat{r}(v)$ be the true rank and estimated rank of a value v, respectively, for all $v \in \{1, \ldots, d\}$. Also let $r_{\max} = nm$ be the maximum global rank in the network which is the total number of readings in the network. The normalized average rank error (ARE) is defined as

$$\text{ARE} = \frac{\sum_{v=1}^{d} |\hat{r}(v) - r(v)|}{r_{\max} d}, \tag{15}$$

and the maximum rank error (MRE) is defined as

$$\text{MRE} = \frac{\max_{v=\{1,\ldots,d\}} (|\hat{r}(v) - r(v)|)}{r_{\max}}. \tag{16}$$

In addition, we also use total communication cost to compare PrivQSA and the two baseline solutions.

5.3 Simulation Results

Impact of Sampling Probability h. Figs. 1a to 1c compare the ARE, MRE, and total communication cost of PrivQSA and the two baseline solutions, respectively, with sampling probability varying from 0.1 to 1.0. We can see from Fig. 1a and 1b that both the ARE and MRE decrease as the sampling probability h increases under all three schemes. This is expected as the more readings we sample, the more accurate the value-to-rank query results, and vice versa. Moreover, we can see that the ARE and MRE of Baseline 2 is the lowest among the three because it does not involve any random perturbation. PrivQSA comes in the second place with a small difference compared to Baseline 2 which is the cost of providing ϵ-LDP. Finally, Baseline 1 incurs the largest rank errors because its does not make use of any rank information in estimating the original distribution. On the other hand, Fig. 1c shows that the total communication cost increases as the sampling probability increases under all three schemes, which is anticipated. Moreover, we can see that PrivQSA and Baseline 2 have the same communication cost, which is larger than Baseline 1's communication cost. This is because under both PrivQSA and Baseline 2 every sensor node needs to send the rank information besides the sampled values whereas under Baseline 1 only sampled readings need to be sent.

Impact of Privacy Budget ϵ. Figures 2a and 2b compare the ARE and MRE under PrivQSA and Baseline 2 with the privacy budget ϵ varying from 10 to 100, where those under Baseline 2 are plotted for reference only. We can see from both figures that both ARE and MRE decrease as the privacy budget ϵ increases both under PrivQSA and Baseline 1. This is because the larger the ϵ, the closer the perturbed reading set to the original reading set, the more accurate the estimated original distribution, the more accurate the value-to-rank query results, and vice versa. In addition, PrivQSA achieves significantly lower ARE and MRE than Baseline 1 due to the rank information included in the quantile summary aggregation.

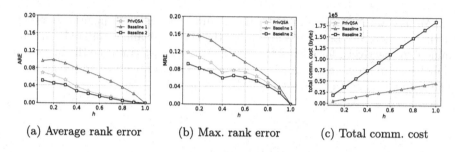

(a) Average rank error   (b) Max. rank error   (c) Total comm. cost

Fig. 1. Sampling probability h varying from 0.1 to 1.0.

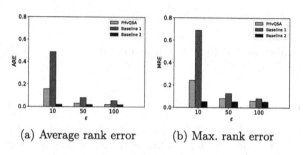

(a) Average rank error   (b) Max. rank error

Fig. 2. Privacy budget ϵ varying from 10 to 100.

Impact of m. Figures 3a to 3c compare the ARE, MRE, and total communi-cation cost of the three schemes with m varying from 10 to 100. As we can see from Figs. 3a and 3b, both ARE and MRE decrease as the number of reading per node increases under Baseline 2. The reason is that the more readings each node has, the more sampled readings, the more accurate the value-to-rank query results, and vice versa. In contrast, both ARE and MRE increase as the number of values per node increases under PrivQSA and Baseline 1. This is because perturbing a larger set of readings with the same privacy budget results in more noise added to the perturbed reading set and thus larger rank errors. Again, Baseline 2 has the lowest ARE and MRE, which is followed by PrivQSA and Baseline 1 for the same reasons mentioned earlier. Moreover, Fig. 3c shows that the total communication cost under all three schemes increase as m increases, which is expected. Moreover, Baseline 1 incurs the lowest communication cost, and PrivQSA and Baseline 2 incur the same communication cost.

Impact of d. Figures 4a and 4b compare the ARE and MRE of all three schemes with the size of reading domain d varying from 100 to 500. As we can see, Baseline 2 shows a slight increase in both ARE and MRE as d increases. This is expected as the larger the domain range, the more values that need to have their ranks estimated, the higher the ARE and MRE under a fixed sampling probability. For the same reason, we can see that PrivQSA and Baseline 1 incur higher ARE and MRE which also increase faster in comparison with Basline 2

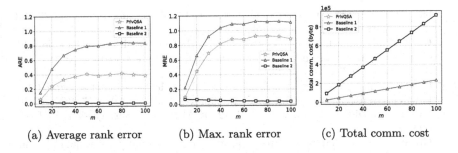

(a) Average rank error   (b) Max. rank error   (c) Total comm. cost

Fig. 3. Number of readings per node varying from 10 to 100.

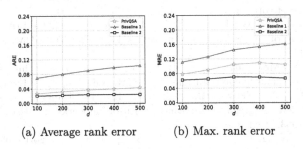

(a) Average rank error   (b) Max. rank error

Fig. 4. Size of reading domain d varying from 100 to 500.

as the size of domain range increases. The reason is that the larger the value domain, the fewer common elements between the original reading set and the perturbed reading set after perturbation, and the larger the rank estimation errors, and vice versa.

6 Conclusion

In this paper, we have introduced the design of PrivQSA, the first locally differentially private quantile summary aggregation protocol for wireless sensor networks, which can guarantee ϵ-LDP for individual sensor node's readings. We have confirmed the significant advantages of PrivQSA over baseline solutions via detailed simulation studies.

Acknowledgements. The authors would like to thank the anonymous reviewers for their constructive comments and helpful advice. This work was supported in part by the US National Science Foundation under grants CNS-1651954 (CAREER), CNS-1933047, and CNS-1718078.

References

1. Ács, G., Castelluccia, C.: I have a DREAM! (DiffeRentially privatE smArt Metering). In: Filler, T., Pevný, T., Craver, S., Ker, A. (eds.) IH 2011. LNCS, vol.

6958, pp. 118–132. Springer, Heidelberg (2011). https://doi.org/10.1007/978-3-642-24178-9_9

2. Castelluccia, C., Chan, A.C., Mykletun, E., Tsudik, G.: Efficient and provably secure aggregation of encrypted data in wireless sensor networks. ACM Trans. Sens. Netw. (TOSN) 5(3), 1–36 (2009)

3. Castelluccia, C., Mykletun, E., Tsudik, G.: Efficient aggregation of encrypted data in wireless sensor networks. In: IEEE Mobiquitous 2005, pp. 109–117. IEEE (2005)

4. Dwork, C.: Differential privacy. p. 1–12. ICALP 2006, Springer-Verlag, Berlin (2006)

5. Greenwald, M., Khanna, S.: Space-efficient online computation of quantile summaries. In: ACM SIGMOD 2001, p. 58–66. Santa Barbara, CA (2001)

6. Greenwald, M.B., Khanna, S.: Power-conserving computation of order-statistics over sensor networks. In: ACM PODS, pp. 275–285 (2004)

7. Groat, M.M., Hey, W., Forrest, S.: Kipda: k-indistinguishable privacy-preserving data aggregation in wireless sensor networks. In: IEEE INFOCOM, pp. 2024–2032. IEEE (2011)

8. Haeupler, B., Mohapatra, J., Su, H.H.: Optimal gossip algorithms for exact and approximate quantile computations. In: ACM PODC, pp. 179–188 (2018)

9. He, W., Liu, X., Nguyen, H., Nahrstedt, K., Abdelzaher, T.: PDA: privacy-preserving data aggregation in wireless sensor networks. In: IEEE INFOCOM, pp. 2045–2053. IEEE (2007)

10. Huang, Z., Wang, L., Yi, K., Liu, Y.: Sampling based algorithms for quantile computation in sensor networks. In: ACM SIGMOD, pp. 745–756 (2011)

11. Li, Q., Cao, G.: Efficient and privacy-preserving data aggregation in mobile sensing. In: 2012 20th IEEE ICNP, pp. 1–10. IEEE (2012)

12. Lindell, Y., Pinkas, B.: Secure multiparty computation for privacy-preserving data mining. J. Priv. Confid. 1(1) (2009)

13. McSherry, F., Talwar, K.: Mechanism design via differential privacy. In: IEEE FOCS (2007)

14. Naranjo, J.A., Casado, L.G., Jelasity, M.: Asynchronous privacy-preserving iterative computation on peer-to-peer networks. Computing 94(8–10), 763–782 (2012)

15. Ozdemir, S., Xiao, Y.: Secure data aggregation in wireless sensor networks: a comprehensive overview. Comput. Netw. 53(12), 2022–2037 (2009)

16. Rajagopalan, R., Varshney, P.K.: Data-aggregation techniques in sensor networks: a survey. IEEE Commun. Surv. Tutorials 8(4), 48–63 (2006)

17. Sun, J., Zhang, R., Zhang, Y.: PriStream: privacy-preserving distributed stream monitoring of thresholded PERCENTILE statistics. In: IEEE INFOCOM, pp. 1–9 (2016)

18. Shi, J., Zhang, R., Liu, Y., Zhang, Y.: Prisense: privacy-preserving data aggregation in people-centric urban sensing systems. In: IEEE INFOCOM, pp. 1–9. (2010)

19. Shrivastava, N., Buragohain, C., Agrawal, D., Suri, S.: Medians and beyond: new aggregation techniques for sensor networks. In: SenSys, pp. 239–249 (2004)

20. Westhoff, D., Girao, J., Acharya, M.: Concealed data aggregation for reverse multicast traffic in sensor networks: Encryption, key distribution, and routing adaptation. IEEE Trans. Mob. Comput. 5(10), 1417–1431 (2006)

21. Xue, M., Papadimitriou, P., Raïssi, C., Kalnis, P., Pung, H.K.: Distributed privacy preserving data collection. In: Yu, J.X., Kim, M.H., Unland, R. (eds.) DASFAA 2011. LNCS, vol. 6587, pp. 93–107. Springer, Heidelberg (2011). https://doi.org/10.1007/978-3-642-20149-3_9

22. Zhang, K., Han, Q., Cai, Z., Yin, G.: Rippas: a ring-based privacy-preserving aggregation scheme in wireless sensor networks. Sensors 17(2), 300 (2017)

XLMRQA: Open-Domain Question Answering on Vietnamese Wikipedia-Based Textual Knowledge Source

Kiet Van Nguyen[1,2(✉)], Phong Nguyen-Thuan Do[2], Nhat Duy Nguyen[2], Tin Van Huynh[1,2], Anh Gia-Tuan Nguyen[1,2], and Ngan Luu-Thuy Nguyen[1,2]

[1] Faculty of Information Science and Engineering, University of Information Technology, Ho Chi Minh, Vietnam
{kietnv,tinhv,anhngt,ngannlt}@uit.edu.vn
[2] Vietnam National University, Ho Chi Minh City, Vietnam
{18520126,18520118}@gm.uit.edu.vn

Abstract. Question answering (QA) is a natural language understanding task within the fields of information retrieval and information extraction that has attracted much attention from the computational linguistics and artificial intelligence research community in recent years because of the strong development of machine reading comprehension-based models. A reader-based QA system is a high-level search engine that can find correct answers to queries or questions in open-domain or domain-specific texts using machine reading comprehension (MRC) techniques. The majority of advancements in data resources and machine-learning approaches in the MRC and QA systems especially are developed significantly in two resource-rich languages such as English and Chinese. A low-resource language like Vietnamese has witnessed a scarcity of research on QA systems. This paper presents XLMRQA, the first Vietnamese QA system using a supervised transformer-based reader on the Wikipedia-based textual knowledge source (using the UIT-ViQuAD corpus), outperforming the two robust QA systems using deep neural network models: DrQA and BERTserini with 24.46% and 6.28%, respectively. From the results obtained on the three systems, we analyze the influence of question types on the performance of the QA systems.

Keywords: Question answering · Transformer · BERT · XLM-R · Transfer learning · Machine reading comprehension

1 Introduction

In recent years, the rapid development of social media has led to an explosion of data and information. People need to find information and knowledge through the support of machine question answering applications like Google, Siri, and Alexa. QA systems assist people in accessing information and knowledge faster without taking much time and effort. QA-based tasks are of interest to the Vietnamese natural language processing and computational linguistics

N. T. Nguyen et al. (Eds.): ACIIDS 2022, LNAI 13757, pp. 377–389, 2022.
https://doi.org/10.1007/978-3-031-21743-2_30

community. Machine reading comprehension-based QA systems [2] have gained much attention in recent years. Although several machine reading comprehension corpora are released for developing QA systems such as UIT-ViQuAD [15], UIT-ViWikiQA [6], and UIT-ViNewsQA [22], there is no reader-based QA system for Vietnamese yet.

Along with the strong development of machine learning, QA systems have been explored in various corpora and methods. In recent years, QA systems have followed two Retriever-Reader QA systems such as DrQA [2] and BERTserini [25], respectively. The input of QA systems is a question and a collection of passages or documents, and the output is a predicted answer extracted from a relevant document. Figure 1 shows the input and output of a QA system on the Vietnamese Wikipedia.

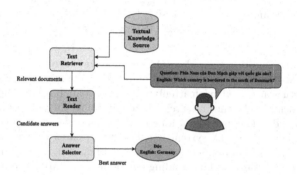

Fig. 1. An example of a retriever-reader-selector QA system on the wikipedia.

Our three main contributions are described as follows.

- We re-implement state-of-the-art QA systems on the Vietnamese Wikipedia knowledge texts: DrQA [2] and BERTserini [25] as baseline systems. The first experiments were performed on the retriever-reader-based QA model on Vietnamese texts.
- We propose XLMRQA, a retriever-reader-selector QA system for the Vietnamese language, outperforming the F1-score and exact match (EM) of two other SOTA systems: DrQA with the multi-layer recurrent neural network-based reader and BERTserini with the BERT based reader.
- We analyze the impacts of question types on the proposed QA system XLM-RQA for Vietnamese, which helps researchers improve the performance of the QA systems in future work.

2   Related Work

Building a QA system requires integrating two main parts: the text retriever and the text reader, to generate the whole QA system. In this section, we briefly review the techniques related to our work.

Text Retriever: In this study, we used two popular and effective techniques: TF-IDF and Pyserini. Term Frequency-Inverse Document Frequency (TF-IDF) is the count-based method that reveals the importance of a token or word to a document or text in a corpus [19]. Text retrieval models [10], text summarization models [3], and question answering models [2] have all employed TF-IDF. This approach is used as the initial baseline method for text retriever to compare with other techniques. Pyserini is a simple Python package that aids researchers in reproducing their findings by offering excellent first-component document retrieval for multi-component rating systems [13]. Because Pyserini is simple yet effective, it was chosen to be implemented for the text retriever as a component of the QA system.

Automatic Reader: The DrQA reader is a multi-layer neural network model that has been trained to identify answers from a document as input. In English [2], and Vietnamese [15,22], this reader is generated from machine reading comprehension tasks. Text segments from the Pyserini retriever are passed to the BERT reader [25]. These models were widely applied in automatic reading comprehension tasks in English [2,5] and Vietnamese [15,22]. Recently, there are more complex models that integrate additional linguistic knowledge into the models [18,24]. However, Vietnamese is a language with few resources and does not yet NLP pipeline tools really well to recognize linguistic knowledge to be associated with these language models.

Full QA System: Different from the previous QA systems [7,14,16] without readers, DrQA [2] is a full QA system combining a bigram hash-based TF-IDF retriever with a multi-layer iterative neural network reader trained to predict answers in the passage. BERTserini [25] is a QA system that combines the BERT-based reader and the open-source Anserini toolkit for text retriever. The system receives a small set of documents as input. In an end-to-end approach, the system combines best practices from document retrieval with a BERT-based reader to determine answers from a large-scale corpus of English Wikipedia articles. For the Vietnamese language, there are still not any QA systems based on the Retriever-Reader mechanism, mainly focusing on the traditional QA system [1, 11]. Therefore, we would like to develop this system as a starting point of the mechanism for Vietnamese QA.

3   UIT-ViQuAD: Vietnamese Wikipedia-Based Textual Knowledge Resource

In this paper, we use the UIT-ViQuAD corpus (abbreviated as ViQuAD) [15] as a Wikipedia-based textual knowledge source, which is the main resource to build the text retriever, reader, and the full QA system for Vietnamese.

ViQuAD is a Vietnamese corpus for assessing question answering, machine reading comprehension, and question generation models. Table 1 summarizes the statistics on the training (Train), development (Dev), and test (Test) sets of this corpus. ViQuAD comprises over 23,000 triples, and each triple includes a question, its answer, and a passage containing the answer. The numbers of passages and articles, the average question and answer lengths, and lexical unit sizes are presented in Table 1.

Table 1. Overview of the ViQuAD corpus.

Corpus	#Article	#Passage	#Questions	Average length			Vocabulary size
				Passage	Question	Answer	
Train	138	4,101	18,579	153.9	12.2	8.1	36,174
Dev	18	515	2,285	147.9	11.9	8.4	9,184
Test	18	493	2,210	155.0	12.2	8.9	9,792
Full	174	5,109	23,074	153.4	12.2	8.2	41,773

Besides, Nguyen et al. [15] also analyzed the distribution of seven types of questions: Who, What, When, Where, Why, How, and Others in the ViQuAD corpus. The most common type of question is What, accounting for 49.97% of all questions, Where questions have the lowest proportion of 5.64%, and other types of questions contribute to proportions of between 7% and 10%.

The question words in Vietnamese for each question type are diverse. The UIT-ViWiKiQA [6] corpus is a reading comprehension corpus that is automatically converted from the ViQuAD corpus's question-answer pairs. Do et al. [6] analyzed diverse Vietnamese question words to pose in each question type. Figure 2 shows the proportions of types of questions on the Dev and Test sets.

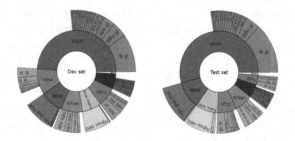

Fig. 2. Question types and question words statistics in the corpus [6].

The linguistic phenomenon in Vietnamese questions has various question words in different question types: What, Who, When, Why, and Where. What questions have the most variety of question words compared to other question types. Only the question words with a high frequency of occurrence are presented in Fig. 2 (approximately 0.7% or more). For example, "là gì", "điều", "làm gì", and "cái gì" are question words in "What questions" with "là gì" having the most significant rate (24.42% in the Test set and 20.69% in the Dev set). The question words in the How and How many question types are usually not diversified. In the Dev set, "điều" is the most popular question word reaching at 87.79%, whereas it accounts for 95.02% in the Test set. According to the previous investigation results [15], the What question type contains the most significant proportion of question words and the most remarkable diversity of question terms. The remaining types of questions make up a small percentage of the total, particularly the How and How many question categories, which include few question words.

4   XMLRQA: Retriever-Reader-Selector Question Answering System for the Vietnamese Language

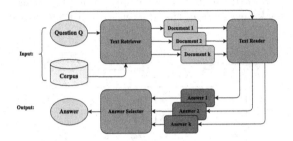

Fig. 3. Overview of the retriever-reader-selector QA system for Vietnamese.

4.1   Overview of QA System Architecture

Inspired by the DrQA system [2], we present an overview of the QA architecture using a supervised reader for the Vietnamese language. Figure 3 presents a QA system with three components: text retriever, text reader, and answer selector. The text retriever finds texts or documents and passes them to the text reader to find the candidate answers. The answer selector finds the answer that best matches the question from the candidate answers predicted by the text reader. In particular, we describe the QA system and its components as follows.

4.2   Text Retriever

We apply a basic retriever to find k passages that answer the input question, using the question as a bag-of-words question. The text retriever finds passages or documents related to the question from a set of 5,109 passages extracted from the ViQuAD corpus [15]. This corpus was built by aggregating all the passages from the Train, Dev, and Test sets of the ViQuAD corpus consisting of 4,101, 515, and 493 passages, respectively. This paper assesses two different text retrievers, including TF-IDF and the Anserini. To optimize the performance of QA systems, we apply word segmentation to the text retrievers.

382   K. V. Nguyen et al.

4.3   Text Reader

The retrieved passages are passed to the reader to extract the candidate answers. The questions are combined with their passages to generate k question-passage pairs that enter the reader to predict k candidate answers.

For each question-passage pair that enters the reader, the model reads the question with M tokens $[q_1, q_2, ..., q_M]$ and then reads all N tokens in the passage $[d_1, d_2, d_N]$. The model performs two probabilities for each token d_i in the document, with $Pstart_i$ being the score when d_i is the starting answer position and $Pend_i$ being the score when d_i is the ending answer position. The reader selects the best answer span from the $Pstart$ and $Pend$ scores list with the highest score from the document. After processing k passages by the reader, the reader obtains k answers and the answer-score list.

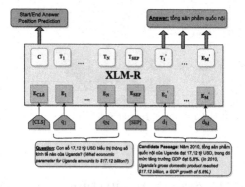

Fig. 4. Overview of the XLM-R based reader for the Vietnamese language.

XLM-RoBERTa (XLM-R) [4] is a multilingual language model trained on a large-scale dataset with 100 languages. XLM-R is used as a pre-trained language model for many tasks such as natural language inference and machine reading comprehension, which achieves state-of-the-art performances. In this paper, we use XLM-R to build a reader as the main component of the XLMRQA system to extract candidate answers before transferring them into the answer selector. Figure 4 shows an overview of the XLM-R-based Reader for the Vietnamese language.

4.4   Answer Selector

The candidate answer list, the reading score list, and the retrieving score list are fed into this component. Each score in the two score lists corresponds to each answer in the answer list. Following Yang et al. [25], we then combine the reading score with the retrieving score through linear interpolation to estimate the score for each answer and find the answer with the highest score.

$$Score_{answer} = alpha * Score_{Reader} + (1 - alpha) * Score_{Retriever} \quad (1)$$

where alpha is a hyper-parameter whose value is in the range [0,1], and alpha is found by tuning with a thousand question-answer pairs extracted randomly from the Train set.

5   Experimental Evaluation

5.1   Baseline Systems

For baseline systems, we re-implement two QA systems that have achieved state-of-the-art performance on English and Chinese corpora, including DrQA [2] and BERTserini [25]. Also, we compare the two QA systems with our proposed system using a powerful XLM-R-based reader, which obtains the best performance on the machine reading comprehension task on the ViQuAD corpus [15].

5.2   Experimental Settings

We used a single NVIDIA Tesla P100 GPU on the Google Collaboratory server[1] for all our experiments. We use sqlite3[2] to store the aggregated corpus from the ViQuAD corpus [15]. We set up our experiments described as follows.

Text Retriever: We implement two models for text retrievers, including TF-IDF and Anserini. The TF-IDF model is based on the DrQA model with the bi-gram language model. The Anserini model is a Python-compatible version called Pyserini[3]. In addition, in analyzing the influence of word splitting on text retrievers, we use pyvi[4] as a word segmentation tool.

Text Reader: We implement DrQA reader based on the DrQA system [2] as text reader as the first model and trained through 30 epochs with batch-size = 32. The pre-trained word embedding model when implementing the DrQA model is ELMO [17,23]. We use powerful pre-trained language models: BERT [5] and XLM-R [4] as text readers, of which the XLM-R model with two versions: large and base. The BERT and XLM-R models are fine-tuned with the baseline configurations provided by Huggingface[5] and we set them with a number of epochs = 2, a maximum string length = 384, and the learning rate = 2e–5.

5.3   Experimental Results

We assess the performance of each component, the end-to-end QA system and analyze the experimental results.

---

[1] https://colab.research.google.com/.
[2] https://docs.python.org/3/library/sqlite3.html.
[3] https://github.com/castorini/pyserini.
[4] https://pypi.org/project/pyvi/.
[5] https://huggingface.co/.

5.3.1   Experimental Results of Text Retriever

First and foremost, all the passages in the corpus are indexed. The text retriever then provides a score value to each passage, representing the likelihood that the passage includes the answer. Text retriever selects k passages with the highest score. P@k is used to assess the text retriever and is defined as the ratio of ranked passages that contained the answers to the top K relevant passages. Given $Q = \{q_1, q_2, ..., q_n\}$ as a set of questions, the answer of $q \in Q$ is a_q, $P \in Ps$ is the passage containing a_q where Ps if the set of passages in the corpus, and $P_k^*(q) \subset Ps$ is the set of the top k passages predicted by the text retriever. P@k is calculated using Formula 2.

$$P@k = \frac{1}{|Q|} \sum_1^n \begin{cases} 1 & \text{if } P \in P_k^*(q), \\ 0 & \text{Otherwise.} \end{cases} \tag{2}$$

Table 2 shows the P@k of two text retrievers, and it shows the effect of word segmentation for two text retrievers. Adding a word segmentation makes the TF-IDF model less efficient as its P@k decreases overall instances of k, decreasing by 5.60% on average. Pyserini model becomes more efficient when combined with pyvi as a word segmentation tool. Overall, Pyserini increases accuracy on all instances of k, increasing by 1.04% on average. Thus, when building the QA systems, we use the word segmentation for Pyserini and not for TF-IDF.

Table 2. P@k of the text retriever on the Test set of the ViQuAD corpus.

	TF-IDF	TF-IDF+Pyvi	Pyserini	Pyserini+Pyvi
P@1	64.39	58.14 (−6.25)	64.89	68.51 (+3.62)
P@5	84.34	78.42 (−5.92)	85.52	86.47 (+0.95)
P@10	90.68	84.89 (−5.79)	89.95	91.22 (+1.27)
P@15	92.99	87.96 (−5.03)	92.08	93.08 (+1.00)
P@20	94.39	88.82 (−5.57)	93.76	93.98 (+0.22)
P@25	95.38	90.09 (−5.29)	94.71	94.76 (+0.05)
P@30	96.06	90.72 (−5.34)	95.25	95.45 (+0.20)

5.3.2   Experimental Results of Text Reader

Text reader is used for extracting candidate answers from questions and their passages obtained by text retrievers. The text reader receives a question and a passage as input of the reader. For the reading component, exact match (EM) and F1 (following the evaluation of the ViQuAD corpus [15]) are the two assessment measures employed to estimate the performance of the text reader.

The performance of the text readers on the Dev and Test sets of the ViQuAD corpus is shown in Table 3. The BERT and XLM-R models outperform the DrQA Reader model in terms of overall performance. On the Test set of the ViQuAD corpus, the $XLMR_{Large}$ model outperforms the other models in both evaluation metrics, with an F1 of 86.86% and an EM of 70.29%.

Table 3. Evaluation of text readers on the ViQuAD corpus.

	Dev		Test	
	EM	F1	EM	F1
DrQA reader	44.60	65.99	39.10	62.92
mBERT	63.30	80.69	61.59	80.87
XLM-R$_{Base}$	63.60	81.95	63.87	82.56
XLM-R$_{Large}$	**73.23**	**88.36**	**70.29**	**86.86**

5.3.3 Experimental Results of Full QA Systems

The QA system is a complete system with three modules: text retriever, text reader, and answer selector. We implement three QA systems, including DrQA [2], BERTserini [25], and XLMRQA, where DrQA's text retriever is TF-IDF, and the other two are Pyserini. We use the DrQA reader for the DrQA system, BERT [5] for the reader of BERTserini, and XLM-R$_{Large}$ [4] for the reader of XLMRQA. We use the two assessment metrics including EM and F1 scores to measure QA systems' performance.

Table 4. Performances of the Vietnamese QA systems with different k-values.

k	DrQA				BERTserini				XLMRQA			
	Dev		Test		Dev		Test		Dev		Test	
	EM	F1	EM	F1	EM	F1	EM	F1	EM	F1	EM	F1
1	18.42	37.17	17.87	37.37	38.07	53.89	36.52	55.55	47.96	60.39	47.96	61.83
5	19.17	38.05	18.37	37.86	**41.84**	**57.50**	39.46	**58.30**	51.99	**64.10**	**51.94**	**64.99**
10	**19.17**	**38.05**	**18.42**	**37.86**	41.75	57.21	39.41	57.98	51.77	63.79	51.36	64.49
15	19.17	38.04	18.42	37.86	41.71	57.10	39.50	58.09	51.77	63.79	51.36	64.49
20	19.17	38.05	18.42	37.86	41.71	57.09	39.46	58.03	51.77	63.79	51.36	64.49
25	19.17	38.05	18.42	37.86	41.71	57.08	39.50	58.00	51.77	63.79	51.36	64.49
30	19.17	38.05	18.42	37.86	41.71	57.08	39.50	57.99	51.77	63.79	51.36	64.49

The evaluation of the QA systems is shown in Table 4. The three systems in ascending order of results are DrQA, BERTserini, and XLMRQA. The XLM-RQA QA system achieves the highest performance with EM and F1 set to the maximum at k = 5 with an F1 of 64.99% and an EM of 51.94% on the Test set. Similar to XLMRQA, the BERTserini system achieves the best performance at k = 5 with an F1 of 58.30% and an EM of 39.46% on the ViQUAD Test set. The DrQA system achieves the best performance with $k \geq 10$, and these are equal. Nevertheless, we have chosen k = 10 as the official value for the DrQA system because it achieves good performance in terms of time. At k = 10, the DrQA system achieved an F1 of 37.86% and EM of 18.42% on the Test set.

386   K. V. Nguyen et al.

5.4   Result Analysis

The majority of the predicted answers are focused on the first five passages, as seen in Fig. 5. For example, the XLMRQA system shows that 83.40% of the questions obtain predicted answers correctly when retrieving the passage with k=1, and all questions with correct answers appear in the top five passages. This explains the results in Table 4 that when $k \geq 10$, the results on the Test set are almost unchanged on both F1 and EM. The same is true for DrQA and BERTserini systems. Up to 99.90% of the answers appear the first to fifth passages in the DrQA system. This makes the results on the Test set of the DrQA system unchanged with $k \geq 10$.

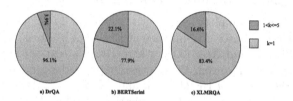

Fig. 5. Distribution of where predicted answers appear in the set of k passages.

Figure 6 shows the performance of QA systems for different types of questions. Overall, the XLMRQA QA system achieved the highest performance based on EM and F1 for all question types. The second-highest performing system is BERTserini, and the last is DrQA. Our analysis shows that the amount and diversity of question words and the complexity of the question impacted the performance of QA systems for each question type. Because Why and How are challenging to answer because they demand that the system comprehends the question and the retrieved passage, QA systems do not perform well. When questions include a limited number of questions but a large variety of question words (see Fig. 2) makes it difficult for QA systems, resulting in poor performance. Although the What type of question accounts for a vast number (nearly 50%) but has the highest diversity of words to ask, the QA system is problematic in recognizing and extracting answers. As for the How many question type, there is a shallow diversity of words to ask (see Fig. 2), so the performance of QA systems is higher. Analyzing the performance of QA systems based on question types to assess the difficulty level of Vietnamese questions for the QA task in ViQuAD helps researchers explore better models for Vietnamese in the future.

Figure 7 shows the average answer length for each question type. We found that the answers to the three types of What, Why, and How questions have the largest length. Thus, the performances of the QA systems on these three question types achieve low EM scores but high F1 scores. Especially with the Why question type, although QA systems achieve deficient performance on EM, QA systems achieve high performance on F1 because the answers to this type of question are long. This shows that the evaluation of the QA task based on F1 is only relative because it is easily dependent on the answer length.

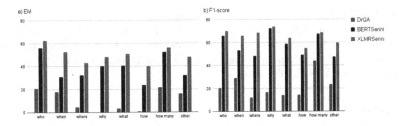

Fig. 6. Evaluation of QA systems on question types of the test set of ViQuAD.

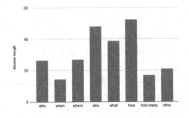

Fig. 7. Average predicted answer length for each question type.

How Many Languages will our Proposed System Implement? A powerful pre-trained language model XLM-RoBERTa using transformer architecture supports 100 different languages (including low-resource languages). Based on our guidelines for building QA systems with the Retriever-Reader-Selector mechanism in this paper, QA systems for other languages (especially low-resource languages) can be easily adapted and re-implemented as baseline QA systems. This proposed system can be extended to different datasets on other languages such as KorQuAD (for Korean) [12], SberQuAD (for Russian) [9], JaQuAD (for Japanese) [21], and FQuAD (for French) [8] in the near future.

6 Conclusion and Future Work

In this paper, we introduced XLMRQA, a QA system based on the retriever-reader-selector mechanism for Vietnamese open-domain texts, which outperformed two state-of-the-art question answering systems, DrQA [2], and BERTserini [25], on the ViQuAD corpus [15]. For assessing the performance of three QA systems on the ViQuAD corpus, we achieved the highest performance with the XLMRQA system: EM of 51.94% and F1 of 64.99%. Analysis of the performance of QA systems was performed on different types of questions. The results of our analysis indicate that the types of challenging questions to be addressed in future studies are How, Why, Where, and What. In the future, several future directions are recommended: (1) integrating diverse question words as linguistic features into the QA systems can boost their performances; (2) finding out a method to leverage the power of monolingual and multilingual BERTology-based language models [20]; and (3) expanding our QA system to other low-resource languages.

Acknowledgement. This research is funded by University of Information Technology-Vietnam National University HoChiMinh City under grant number D1-2022-13. Kiet Van Nguyen was funded by Vingroup JSC and supported by the Master, PhD Scholarship Programme of Vingroup Innovation Foundation (VINIF), Institute of Big Data, code VINIF.2021.TS.026.

References

1. Bach, N.X., Thanh, P.D., Oanh, T.T.: Question analysis towards a Vietnamese question answering system in the education domain. Cybern. Inf. Technol. **20**(1), 112–128 (2020)
2. Chen, D., Fisch, A., Weston, J., Bordes, A.: Reading Wikipedia to answer open-domain questions. In: Proceedings of ACL 2017 (Volume 1: Long Papers), pp. 1870–1879. Vancouver, Canada, Association for Computational Linguistics, July 2017
3. Christian, H, Agus, M.P., Suhartono, D.: Single document automatic text summarization using term frequency-inverse document frequency (TF-IDF). ComTech: Comput. Math. Eng. Appl. **7**(4), 285–294 (2016)
4. Conneau, A.: Unsupervised cross-lingual representation learning at scale. In: Proceedings of ACL 2020, pp. 8440–8451, Association for Computational Linguistics, July 2020
5. Devlin, J., Chang, M.W., Lee, K., Toutanova, K.: Bert: pre-training of deep bidirectional transformers for language understanding. In: Proceedings of NAACL 2019, Volume 1 (Long and Short Papers), pp. 4171–4186 (2019)
6. Do, P.N.-T., Nguyen, N.D., Van Huynh, T., Van Nguyen, K., Nguyen, A.G.-T., Nguyen, N.L.-T.: Sentence extraction-based machine reading comprehension for Vietnamese. In: Qiu, H., Zhang, C., Fei, Z., Qiu, M., Kung, S.-Y. (eds.) KSEM 2021. LNCS (LNAI), vol. 12816, pp. 511–523. Springer, Cham (2021). https://doi.org/10.1007/978-3-030-82147-0_42
7. Duong, P.H., Nguyen, H.T., Nguyen, D.D., Do, H.T.: A hybrid approach to answer selection in question answering systems. In: Huynh, V.-N., Inuiguchi, M., Tran, D.H., Denoeux, T. (eds.) IUKM 2018. LNCS (LNAI), vol. 10758, pp. 191–202. Springer, Cham (2018). https://doi.org/10.1007/978-3-319-75429-1_16
8. d'Hoffschmidt, M., Belblidia, W., Heinrich, Q., Brendlé, T., Vidal, M.: FQuAD: French question answering dataset. In: Findings of the Association for Computational Linguistics: EMNLP, vol. 2020, pp. 1193–1208 (2020)
9. Efimov, P., Chertok, A., Boytsov, L., Braslavski, P.: SberQuAD – Russian reading comprehension dataset: description and analysis. In: Arampatzis, A., et al. (eds.) CLEF 2020. LNCS, vol. 12260, pp. 3–15. Springer, Cham (2020). https://doi.org/10.1007/978-3-030-58219-7_1
10. Hiemstra, D.: A probabilistic justification for using tf×idf term weighting in information retrieval. Int. J. Digit. Libr. **3**(2), 131–139 (2000)
11. Le-Hong, P., Bui, D.-T.: A factoid question answering system for Vietnamese. In: Companion Proceedings of the The Web Conference, vol. 2018, pp. 1049–1055 (2018)
12. Lim, S., Kim, M., Lee, J.: Korquad1. 0: Korean QA dataset for machine reading comprehension. arXiv preprint arXiv:1909.07005 (2019)
13. Lin, J., Ma, X., Lin, S.C., Yang, J.H., Pradeep, R. and Nogueira, R.: Pyserini: an easy-to-use python toolkit to support replicable IR research with sparse and dense representations. arXiv preprint arXiv:2102.10073, 2021

14. Moldovan, D., et al.: The structure and performance of an open-domain question answering system. In: Proceedings of ACL, vol. 2000, pp. 563–570 (2000)
15. Nguyen, K., Nguyen, V., Nguyen, A., Nguyen, N.: A Vietnamese dataset for evaluating machine reading comprehension. In: Proceedings of the 28th International Conference on Computational Linguistics, pp. 2595–2605, Barcelona, Spain, International Committee on Computational Linguistics, December 2020
16. Nguyen, V.-T., Le, A.-C.: Deep neural network-based models for ranking question - answering pairs in community question answering systems. In: Huynh, V.-N., Inuiguchi, M., Tran, D.H., Denoeux, T. (eds.) IUKM 2018. LNCS (LNAI), vol. 10758, pp. 179–190. Springer, Cham (2018). https://doi.org/10.1007/978-3-319-75429-1_15
17. Matthew, E.P.: Deep contextualized word representations. In: Proceedings of NAACL 2018, Volume 1 (Long Papers), pp. 2227–2237, New Orleans, Louisiana, ACL, June 2018
18. Ponti, E.M., Aralikatte, R., Shrivastava, D., Reddy, S., Søgaard, A.: Minimax and neyman-pearson meta-learning for outlier languages. In: Findings of ACL: ACL-IJCNLP 2021, pp. 1245–1260 (2021)
19. Rajaraman, A., Ullman, J.D.: Mining of Massive Datasets. Cambridge University Press, Cambridge (2011)
20. Rogers, A., Kovaleva, O., Rumshisky, A.: A primer in bertology: What we know about how bert works. TACL **8**, 842–866 (2020)
21. So, B., Byun, K., Kang, K., Cho, S.: Jaquad: Japanese question answering dataset for machine reading comprehension. arXiv preprint arXiv:2202.01764 (2022)
22. Van Nguyen, K., Van Huynh, T., Nguyen, D.V., Nguyen, A.G.T. and Nguyen, N.L.T.: New Vietnamese corpus for machine reading comprehension of health news articles. arXiv preprint arXiv:2006.11138 (2020)
23. Vu, X.S., Vu, T., Tran, S.N., Jiang, L.: Etnlp: a visual-aided systematic approach to select pre-trained embeddings for a downstream task. arXiv preprint arXiv:1903.04433 (2019)
24. Wang, R., et al.: K-adapter: infusing knowledge into pre-trained models with adapters. In: Findings of ACL: ACL-IJCNLP, vol. 2021, pp. 1405–1418 (2021)
25. Wei, Y., et al.: End-to-end open-domain question answering with BERTserini. In Proceedings of NAACL 2019 (Demonstrations), pp. 72–77, Minneapolis, Minnesota, Association for Computational Linguistics, June 2019

On Verified Automated Reasoning
in Propositional Logic

Simon Tobias Lund and Jørgen Villadsen[(✉)][iD]

Technical University of Denmark, Kongens Lyngby, Denmark
jovi@dtu.dk

Abstract. As the complexity of software systems is ever increasing, so is the need for practical tools for formal verification. Among these are automatic theorem provers, capable of solving various reasoning problems automatically, and proof assistants, capable of deriving more complex results when guided by a mathematician/programmer. In this paper we consider using the latter to build the former. In the proof assistant Isabelle/HOL we combine functional programming and logical program verification to build a theorem prover for propositional logic. Finally, we consider how such a prover can be used to solve a reasoning task without much mental labor.

Keywords: Logic · Automated reasoning · Isabelle proof assistant

1 Introduction

Today's information systems must be not only intelligent but also trustworthy. Our contribution is two-fold. First, we program a basic theorem prover and discuss the solution to a riddle. Second, we verify the prover with the purpose of presenting modern automated reasoning tools to a wider audience.

The formal verification of a modern prover for full first-order logic is a major undertaking [16]. However, many natural language arguments can be handled in classical propositional logic, and using the Isabelle/HOL proof assistant [11] we have previously formally verified a number of such provers for formulations based on various sets of logical operators [20]. Recently we have obtained short formal soundness, completeness and termination proofs using the NAND or NOR operators [7]. However, the necessary translations of natural language statements into the modest logical languages of NAND or NOR are problematic for efficiency as well as explainability reasons.

In the present work we consider a more traditional formulation of propositional logic based on falsity and implication while still obtaining short formal proofs similar to the formal proofs obtained using the NAND or NOR operators. The prover can be automatically exported to Haskell and several other functional programming languages. The prover implements a deterministic algorithm for the sequent calculus proof system [10]. In order to obtain self-contained functional programs we code a few auxiliary list programs ourselves. We rely on a deep embedding of propositional logic such that we can reason in Isabelle/HOL about the syntax and semantics of propositional logic.

N. T. Nguyen et al. (Eds.): ACIIDS 2022, LNAI 13757, pp. 390–402, 2022.
https://doi.org/10.1007/978-3-031-21743-2_31

The aim of the prover is not to be as efficient as possible. We could, for example, decrease the computational complexity of checking whether a sequent is a tautology in the last step of the algorithm by representing the lists of atomic propositions as binary search trees. However, the aim of our development is clarity and quick development, to showcase how Isabelle can be used for prototyping and achieving very dependable results without much extra work.

We have developed the prover and its verification in such a way that a skilled programmer without knowledge of Isabelle and with only a basic knowledge of proof theory can understand the development described in this paper. Related to this, we have recently found the tool useful when teaching computational logic and formal methods to BSc and MSc students in computer science [22].

The entire prover as well as the solution to a riddle have been formally verified in Isabelle/HOL [11] and the formalizations are available online:

$$\text{https://hol.compute.dtu.dk/Scratch.thy}$$

Table 1 gives an overview of the file. We explain the name *Scratch* in the beginning of Sect. 4 and the riddle is considered in Sect. 6. Logically there is no difference in Isabelle between a theorem, corollary, lemma or proposition. Normally one would add a name to all results but we have left a few unnamed.

Table 1. Overview of the contents of the *Scratch.thy* file.

Name		Description
'a form	Data type	Propositional formulas with atomic propositions of type *'a*
semantics	Function	Evaluates a formula under an interpretation
sc	Function	Evaluates a sequent under an interpretation
member	Function	Checks membership of element in list
member-iff	Lemma	Equality between *member* and set-membership
common	Function	Checks if lists share common element
common-iff	Lemma	Equality between *common* and non-empty set-intersection
mp	Function	Definition of the micro prover for use on a sequent
main	Theorem	The sequent micro prover is sound and complete
prover	Function	Abbreviates the micro prover for use on a single formula
(unnamed)	Corollary	The formula micro prover is sound and complete
neg	Function	Abbreviates negation
con	Function	Abbreviates conjunction
bii	Function	Abbreviates bi-implication
one	Function	Abbreviates ternary exclusive disjunction
people	Data type	The type of atomic propositions in the riddle
riddle	Formula	The propositional formula describing the riddle
(unnamed)	Proposition	Ann is shown to be a knave
(unnamed)	Proposition	Cat is shown to be a knight
(unnamed)	Proposition	Bob cannot be shown to be a knave
(unnamed)	Proposition	Bob cannot be shown to be a knight

The paper is organized as follows. We first discuss related work (Sect. 2) and present the sequent calculus underlying the prover (Sect. 3). After this follows the definition of the prover, focused on functional programming in Isabelle (Sect. 4). Then comes the verification of the prover, focused on proving in Isabelle (Sect. 5). With the prover implemented and verified, we show how it can be used to solve a riddle (Sect. 6). Finally, we conclude with reflections (Sect. 7).

2   Related Work

Our work focuses on a formally verified micro prover based on the operator set containing falsity and implication. We have elsewhere implemented and verified several micro provers based on different operator sets [20]. Most recently, we made two based on just NAND and NOR, respectively. The verification of the prover in this paper is more succinct than the other provers, except the verification of the NAND and NOR provers. There are multiple reasons to prefer the operator set of implication and falsity. First, it is easier to understand the underlying sequent calculus as reasoning with implication and falsity is more intuitive than reasoning with NAND and NOR. Second, the prover in this paper is much more efficient for many problems than the NAND and NOR provers, since the translation of a natural language problem into a formula based on implication and falsity is likely much smaller than the translations into formulas based on NAND or NOR.

Shankar [17] and Michaelis and Nipkow [10] have made and verified provers for propositional logic, but for other operator sets. We focus on reasoning tools and development assisted by theorem provers and proof assistant. When it comes to theorem provers for propositional logic in general, we should mention SAT solving [2], which is much more efficient than our prover.

Blanchette [3] has made an overview of provers in Isabelle. This and our previous work is part of the IsaFoL (Isabelle Formalization of Logic) project for organizing and comparing various logics, proof systems and provers in Isabelle. For first-order logic there is leanTaP [1,6]: a small, unverified prover in Prolog. Larger and verified provers for first-order logic also exist [15,16,19,21,23].

3   Sequent Calculus for Propositional Logic

We start with a summary of the sequent calculus for propositional logic. In the following sections we show how to program and verify a prover based on the sequent calculus.

Formulas p, q, ... in classical propositional logic are built from propositional symbols, falsity ($\bot$) and implications ($p \to q$).

We introduce the rest of the usual operators as abbreviations:

$$\neg p \equiv p \to \bot \qquad p \wedge q \equiv \neg(p \to \neg q) \qquad p \vee q \equiv \neg p \to q$$

$$p \leftrightarrow q \equiv (p \to q) \wedge (q \to p)$$

We will use some terminology that might not be widely known with regards to the calculus. An interpretation is an assignment of a truth value to each of the atomic propositions. A formula is valid if it evaluates to true under every interpretation. A sequent is two sets of formulas written as $\Gamma \vdash \Delta$. Informally, a sequent expresses that if we assume all the formulas in Γ then at least one of the formulas in Δ is true. If this property holds for a sequent, then we call it valid. In $\Gamma \vdash \Delta$, we refer to Γ as the antecedents and Δ as the consequents. Assuming the antecedents means that if we consider some interpretation where a formula in the antecedents is false, then the sequent becomes true by default. Using the rules that follow we can construct valid sequents from other valid sequents using rules of inference. In the rules there are a few formulas changed in the transformation happening from premise to conclusion, and the rest of the formulas remain unaffected. We will call these unaffected formulas "passive".

The sequent calculus consists of two axiom schemas and two inference rules: Let Γ and Δ be finite sets of formulas.

The axiom schemas of the sequent calculus are of the form:

$$\Gamma \cup \{p\} \vdash \Delta \cup \{p\} \qquad \Gamma \cup \{\bot\} \vdash \Delta$$

The rules of the sequent calculus are the left and right introduction rules:

$$\frac{\Gamma \vdash \Delta \cup \{p\} \qquad \Gamma \cup \{q\} \vdash \Delta}{\Gamma \cup \{p \to q\} \vdash \Delta} \qquad \frac{\Gamma \cup \{p\} \vdash \Delta \cup \{q\}}{\Gamma \vdash \Delta \cup \{p \to q\}}$$

In the following explanation of the inferences, we write about the affected formulas as if they were the only ones present in the sequents. The explanations and arguments we provide for this special case also generalize to situations where there are passive formulas. For example, if a sequent is made true in some interpretation because the formula p is an antecedent and its truth value is false, then the sequent will also be made true by $p \to q$ being a consequent. If, on the other hand, a sequent is made true in an interpretation because of the truth value of a passive formula, then that passive formula will still make the derived sequent true. Similar arguments can be applied for the impossibility of introducing invalid sequents from valid ones.

The right introduction rule follows directly from the semantics (the truth of all formulas on the left-hand side imply the truth of at least one of the formulas on the right-hand side). If we can show q assuming p, then p implies q.

The left introduction rule is slightly harder to interpret. Consider that if $\Gamma \vdash \Delta$ can be shown, then for any interpretation either a formula in Γ is false or a formula in Δ is true. Showing the validity of a sequent with the empty set for Δ therefore amounts to showing that Γ is contradictory. Thus, if we can show that p is valid and that q is contradictory, we can also show that $p \to q$ is contradictory.

4 Programming the Prover

To begin the work in Isabelle we need to declare the name of our theory, import the other theories we will use, and create an Isabelle environment. The name of the theory needs to match the name of the file we are working in. Because of that we have given our theory the name *Scratch* so it can be copied and pasted into the default file displayed when starting Isabelle. *Main* is the core theory of Isabelle/HOL and contains everything necessary for implementing and verifying the prover.

theory *Scratch* **imports** *Main* **begin**

We start by defining the type of formulas. This takes a type variable, *'a*, which represents the type of atomic propositions (strings, natural numbers, and so on). A formula is one of three things: an atomic proposition of type *'a*, falsity, or an implication from one formula to another. We can observe that formulas are thus represented by binary trees, where the leafs are either propositions or falsity and parent nodes are implications from the left sub-tree to the right.

datatype *'a form*
    = *Pro 'a* (⟨ · ⟩) | *Falsity* (⟨ ⊥ ⟩) | *Imp* ⟨ *'a form* ⟩ ⟨ *'a form* ⟩ (**infixr** ⟨ → ⟩ *0*)

The interpretation of a formula is defined as a function from the type of atomic propositions to booleans. We can define the semantics of a formula by a function taking a formula and an interpretation, and returning the truth value of the formula under that interpretation. The function is defined by pattern matching on the formula: an atomic proposition gives its value in the interpretation, falsity always gives *False*, and implication is defined in terms of Isabelle's built-in implication.

primrec *semantics* **where**
    ⟨ *semantics i* (·*n*) = *i n* ⟩ |
    ⟨ *semantics* - ⊥ = *False* ⟩ |
    ⟨ *semantics i* (*p* → *q*) = (*semantics i p* ⟶ *semantics i q*) ⟩

We can convert lists to sets and use logical quantifiers on their elements. This allows us to express the semantics of sequents through the following one-line definition. It expresses that a sequent is true under an interpretation if all the antecedents imply at least one of the consequents.

abbreviation
    ⟨ *sc X Y i* ≡ (∀ *p* ∈ *set X. semantics i p*) ⟶ (∃ *q* ∈ *set Y. semantics i q*) ⟩

So that our theory is self-contained to the largest extent possible, and so that a translation of the prover to another programming language is as easy as possible, we define some functions for set operations on lists. These could be defined directly by translating the lists to sets, but then the program would not be directly translatable. Alternatively, we could have used the operations provided by Isabelle, but then the prover would not be as self-contained.

The first such function we define is *member*, which returns true if a given element exists in a list:

```
primrec member where
  ⟨ member - [] = False ⟩ |
  ⟨ member m (n # A) = (m = n ∨ member m A) ⟩
```

We can then prove that it is equivalent to the set-membership operator ∈. First, we define the lemma we want to show, then we give the proof. The proof is done by induction on the list. This is done by applying *induct A* to the proof state. This then creates two proof obligations; we need to show that the lemma holds for the empty list and for $x \# A$ (an arbitrary element x added to the front of A) if we assume it holds for A. By performing induction on A we obtain proof obligations matching the cases of the function. Since the function calls reduce nicely and the lemma is simple, we can solve both proof obligations with Isabelle's simple prover for logical rewriting: *simp*. To apply *simp* to both goals in the proof state we need to use *simp-all*.

```
lemma member-iff [iff]: ⟨ member m A ⟷ m ∈ set A ⟩
  by (induct A) simp-all
```

Using *member*, we can define a function for checking whether two lists contain a common element:

```
primrec common where
  ⟨ common - [] = False ⟩ |
  ⟨ common A (m # B) = (member m A ∨ common A B) ⟩
```

The desired property can be expressed using set operators as the intersection of the two lists not being the empty set:

```
lemma common-iff [iff]: ⟨ common A B ⟷ set A ∩ set B ≠ {} ⟩
  by (induct B) simp-all
```

In the later proofs Isabelle will automatically use the above translations to sets since we annotated the lemmas with [*iff*]. There is significant advantage to having the later proofs rely on these translations, as opposed to the actual definitions of *member* or *common*. We might at some point in the development change the implementation of these functions, for example by having the lists be binary search trees. It would then only be necessary to change the proofs of these two lemmas; the rest of the theory would still succeed. Thus, one can do modular development in Isabelle, with lemmas of the main properties of the different functions providing the static interfaces.

The following function (*mp*) is the core of the prover. It works by splitting implication-formulas up into their left- and right-sides at the appropriate sides of the sequents. Atomic propositions are moved into separate lists (A and B), so we can terminate when the lists containing non-atomic formulas (C and D) are empty. The inferences of the calculus are implemented quite directly, taking a conclusion and attempting to show the premise. The left implication rule, which uses two premises, calls the prover on both premise-sequents and conjuncts the result. Thus, the prover constructs the proof tree by recursive calls on the premises which could show the given sequent. If all of these calls lead to axioms, the proof succeeds. We attempt to use the propositional axiom (utilizing the *common* function from above) when all formulas are decomposed, while the falsity axiom can be used immediately when we obtain falsity on the left-hand side of a sequent. Falsity on the right-hand side of a sequent can never be used to show validity, so it is just removed in the recursive call.

function *mp* **where**
⟨ *mp A B* (·*n # C*) [] = *mp* (*n # A*) *B C* [] ⟩ |
⟨ *mp A B C* (·*n # D*) = *mp A* (*n # B*) *C D* ⟩ |
⟨ *mp* - - (⊥ # -) [] = *True* ⟩ |
⟨ *mp A B C* (⊥ # *D*) = *mp A B C D* ⟩ |
⟨ *mp A B* ((*p → q*) # *C*) [] = (*mp A B C* [*p*] ∧ *mp A B* (*q # C*) [])) ⟩ |
⟨ *mp A B C* ((*p → q*) # *D*) = *mp A B* (*p # C*) (*q # D*) ⟩ |
⟨ *mp A B* [] [] = *common A B* ⟩
by *pat-completeness simp-all*

To be able to perform structural induction on the function, and because it is a nice result in itself, we show termination of the prover. We do this by defining a notion of size of a sequent which decreases in each recursive call. A notion of size that works is the combined size of formulas in *C* and *D*. This obviously works for the recursive calls involving atomic propositions and falsity, as they are removed from either *C* or *D*. In the calls involving implication we always remove an operator, thereby reducing the size of a formula. With this size definition, the goal produced for each recursive call can be solved by *simp*. Thus, we finish the proof by applying *simp-all*.

termination
   by (*relation* ⟨ *measure* (λ(-, -, *C, D*). ∑ *p* ← *C* @ *D*. *size p*) ⟩) *simp-all*

Showing termination also makes automatic export to languages like Haskell possible through Isabelle's code generation (see end of Sect. 6).

5   Verifying the Prover

After constructing the prover, the big question is whether we did so correctly. In regards to this, there are really two properties that are important. Arguably the most important is soundness. We want to be sure that if our prover can construct a proof of a formula then the formula must also be valid. A property that is often harder to prove (and unattainable for some logics) is completeness, which expresses that any valid formula can be proven. The following theorem, which is the main result of our theory, shows both soundness and completeness. We do this by proving that a sequent is valid with regards to the semantics – i.e. true under every interpretation – if and only if our prover returns true.

The proof is done by structural induction on the cases of the function. This is initiated by *(induct rule: mp.induct)* which tells Isabelle to transform the main proof goal (the theorem itself) into inductive proof goals matching the cases of *mp*. Each of these goals are harder to prove than what we have considered up until this point, but still within the grasp of Isabelle's proof methods. We apply the proof methods *simp*, *blast*, *meson*, and *fast*, which is enough to solve all goals. Finding the right proof methods for a problem might seem like a challenge, but it is easy enough with Isabelle's proof-finding tool *Sledgehammer* [4]. When applying *Sledgehammer* to the proof state, it tries various theorem provers and returns the method capable of solving the current goal if it finds one. Thus, the proof methods can be found one by one, and then compacted to the one-line style we have applied.

theorem *main*: ⟨ (∀ *i*. *sc* (*map* · *A* @ *C*) (*map* · *B* @ *D*) *i*) ⟷ *mp A B C D* ⟩
by (*induct rule*: *mp.induct*) (*simp-all*, *blast*, *meson*, *fast*)

When we have a sound and complete prover for sequents, we can derive one for formulas. This is done using the fact that a formula can be proven by proving the sequent containing it on the right-hand side and nothing else.

definition ⟨ *prover p* ≡ *mp* [] [] [] [*p*] ⟩

We can then show soundness and completeness of the prover for formulas by using the soundness and completeness of the prover for sequents plus *simp*.

corollary ⟨ *prover p* ⟷ (∀ *i*. *semantics i p*) ⟩
unfolding *prover-def* **by** (*simp flip*: *main*)

Showing completeness is in general a daunting task. One can show soundness by induction on the structure of proofs, which means that the rules being independently sound is sufficient. Completeness is achieved when all the rules together are enough to prove any valid formula. This implication from the validity of a formula to the existence of a proof cannot be done by naive induction, as not all valid formulas are constructed from other valid formulas (as is the case for proofs).

In Isabelle we are able to show both soundness and completeness in one line! This is partly because of Isabelle's intelligent proof methods, which are able to prove many theorems automatically, and partly because we couple the calculus with a proof technique by defining it as a function which can be executed. Technically, this works because completeness is now achievable by induction. In the function, the relation between premise and conclusion in each rule is equality. Thus, completeness can be shown by induction on the implications from conclusions to premises. Equivalently, we need only prove that any formula for which our prover returns true is valid, and that any formula for which our prover returns false is not valid. In this case completeness is no more complicated than soundness.

6 Using the Prover

We will use the prover to solve a riddle based on one from *What is the Name of this Book* [18] by Raymond Smullyan, which itself is a (more interesting) variation of an older riddle. It is as follows:

You find yourself on a desert island inhabited by knights and knaves. Knights only ever tell the truth and knaves only ever tell lies. In a clearing, you come upon three islanders: Ann, Bob, and Cat. You ask Bob how many knights there are among them. Bob answers something in a foreign language. You then ask Ann what Bob said. Ann answers: "He said there is one knight among us." Cat then reacts: "Don't listen to Ann, for she is a knave!" Who, if any, among them are knights?

We translate the riddle to propositional logic. First, we create the following atomic propositions: A for "Ann is a knight", B for "Bob is a knight", and C for "Cat is a knight". Next we observe that a statement is true if and only if the person who said it is a knight. It becomes clear that it would be useful to have more propositional constructs than $\rightarrow$ and $\bot$. Thus, we define the operators from the start of Sect. 3 plus the ternary exclusive disjunction **one**(p, q, r), which is true when exactly one of the three arguments is true. A discussion on how to construct these operators comes later. We can sum up what we learn from Bob's and Ann's statements as "if Ann is a knight, then Bob is a knight if and only if there is exactly one knight", or in propositional logic: $A \rightarrow (B \leftrightarrow$ **one**$(A, B, C))$. We can see from the structure that we will not learn anything if Ann is a knave, which makes sense since we will then not know what Bob said. Cat's statement can be captured by "Cat is a knight if and only if Ann is a knave", or in propositional logic: $C \leftrightarrow \neg A$.

To utilize our prover we first need the operators used above. We implement them as abbreviations in Isabelle. The definition of $\neg$ (*neg*) is straightforward:

abbreviation ⟨ *neg a* ≡ (*a* → ⊥) ⟩

$p \rightarrow \neg q$ is true exactly when either p or q is false. Thus, $p \wedge q$ can be defined as $\neg(p \rightarrow \neg q)$:

abbreviation ⟨ *con a b* ≡ *neg* (*a* → *neg b*) ⟩

$p \leftrightarrow q$ is also straightforward:

abbreviation ⟨ *bii a b* ≡ *con* (*a* → *b*) (*b* → *a*) ⟩

one (*one*) is defined as three bi-implications, where each one expresses that a formula is true if and only if the other two are not:

abbreviation ⟨ *one a b c* ≡
   con
    (*bii a* (*con* (*neg b*) (*neg c*)))
    (*con*
     (*bii b* (*con* (*neg a*) (*neg c*)))
     (*bii c* (*con* (*neg a*) (*neg b*)))) ⟩

It is now possible to define the riddle itself.

First, we make a *people* data type, which is a union type of *Ann* (A), *Bob* (B), or *Cat* (C). This will be the type of atomic propositions in the formulas.

datatype *people = Ann | Bob | Cat*

We then put the facts learned from the islanders in the abbreviation *riddle*. It contains the formulas described previously, connected by a conjunction.

abbreviation ⟨ *riddle* ≡ (
con
 (·*Ann* → *bii* (·*Bob*) (*one* (·*Ann*) (·*Bob*) (·*Cat*)))
 (*bii* (·*Ann*) (*neg* (·*Cat*)))) ⟩

Now we use the prover to solve the riddle. In the following lines we write statements about the riddle, and prove them using the proof method *eval*. This will export the prover to Standard ML and execute it with the given formula as input. As this relies on the correctness of the code generator and the Standard ML compiler, the result is less trust-worthy than those obtained purely by Isabelle. We could use *code-simp* instead of *eval* to make sure all evaluation stays within Isabelle, but this comes at a considerable cost to performance.

First, we can show that Ann is a knave, as *riddle* implying her not being a knight is shown valid by the prover:

proposition ⟨ *prover* (*riddle* → *neg* (·*Ann*)) ⟩ **by** *eval*

Next, we show that Cat is a knight by proving that *riddle* implies *Cat*:

proposition ⟨ *prover* (*riddle* → ·*Cat*) ⟩ **by** *eval*

Finally, we show that Bob can be either knight or knave. This follows quite trivially from Ann being a knave, as it means Bob might have said anything (except for "there is one knight"). For Bob, both knighthood and knavehood is unprovable. Since we have shown that the prover is complete, one of these would have succeeded if we could be sure of either.

proposition ⟨ ¬*prover* (*riddle* → *neg* (·*Bob*)) ⟩ **by** *eval*

proposition ⟨ ¬*prover* (*riddle* → ·*Bob*) ⟩ **by** *eval*

We finalize the theory in Isabelle:

end

The lines examining the riddle could also be done in the environment of another programming language by exporting the prover using Isabelle's code generation. The supported languages are Haskell, OCaml, Scala and Standard ML. In this way Isabelle's code generation can be used to incorporate the verified prover into a larger project in one of these programming languages. In general, one can have a project where only important core elements are verified. Thus, verification does not need to happen all at once; one can start with the most important or difficult parts.

7 Concluding Remarks

We have developed a tool for automatic reasoning in propositional logic and verified it in Isabelle/HOL. We have also shown how to use it to solve a problem of reasoning, in our case a riddle. Many problems can be modelled in propositional logic, and many more in the higher-order logic of Isabelle/HOL. Each part of our development was accompanied by machine-checked proofs, which provide very high confidence while remaining easy to write up and quick to verify. We believe that most algorithmic challenges are solvable within the framework of proof assistants, with clear advantages gained from the confidence of machine-checked proofs, and often with not much added labor.

The abbreviation feature in Isabelle/HOL is convenient, but our prover solving the riddle works on the expressions of the formula data type, cf. Table 2. Nevertheless, the execution time in Isabelle/HOL for the file is around a second.

Table 2. Counting atomic propositions and operators in a formula.

	Count for formula: $riddle \rightarrow \neg Ann$
Proposition Ann	16
Proposition Bob	14
Proposition Cat	14
Falsity ($\bot$)	77
Implication ($\rightarrow$)	120

One of the benefits of working with a proof assistant like Isabelle/HOL is that appropriate changes can be made and the theory is automatically verified again. For example, the soundness and completeness theorems can be verified in Isabelle/HOL if falsity is omitted from the syntax of propositional logic, in the semantics and in the sequent calculus. Of course we can no longer define negation, conjunction, disjunction and the riddle. In fact we obtain the so-called implicational propositional logic [5,14,24].

We should ask the question: How can we trust Isabelle? After all, it is a program and could have bugs. This has been a challenge for more than 50 years since the first proof assistant SAM (Semi-Automated Mathematics) [9]. For decades, proof assistants like HOL4 and Isabelle have relied on the so-called LCF approach: a relatively small proof kernel, using abstract data types in the programming language Standard ML to ensure soundness, only assuming that the proof kernel is soundly implemented [8,12,13]. In addition, proof assistants are thoroughly tested and they are more and more considered practical tools for programmers, mathematicians and scientists in general.

Acknowledgements. We would like to thank Agnes Moesgård Eschen, Asta Halkjær From, Frederik Krogsdal Jacobsen and Alexander Birch Jensen for comments on drafts. We would also like to thank the anonymous reviewers for their suggestions. Following our riddle example, we can indeed dream about the system translating a text from a fragment of natural language to formulas in propositional logic, use our logical approach and next translate back to a conclusion expressed in natural language.

References

1. Beckert, B., Posegga, J.: leanTaP: lean tableau-based deduction. J. Autom. Reasoning **15**(3), 339–358 (1995)
2. Biere, A., Heule, M., van Maaren, H.: Handbook of Satisfiability. IOS Press, Amsterdam (2009)
3. Blanchette, J.C.: Formalizing the metatheory of logical calculi and automatic provers in Isabelle/HOL (invited talk). In: Mahboubi, A., Myreen, M.O. (eds.) Proceedings of the 8th ACM SIGPLAN International Conference on Certified Programs and Proofs, CPP 2019, pp. 1–13. ACM (2019)
4. Blanchette, J.C., Böhme, S., Paulson, L.C.: Extending sledgehammer with SMT solvers. J. Autom. Reasoning **51**(1), 109–128 (2013). https://doi.org/10.1007/s10817-013-9278-5
5. Church, A.: Introduction to Mathematical Logic. Princeton Mathematical Series, Princeton University Press, Princeton (1956)
6. Fitting, M.: leanTAP revisited. J. Logic Comput. **8**(1), 33–47 (1998)
7. From, A.H., Lund, S.T., Villadsen, J.: A Case Study in Computer-Assisted Metareasoning. In: González, S.R., et al. (eds.) DCAI 2021. LNNS, vol. 332, pp. 53–63. Springer, Cham (2022). https://doi.org/10.1007/978-3-030-86887-1_5
8. Gordon, M.: From LCF to HOL: a short history. In: Proof, Language, and Interaction: Essays in Honour of Robin Milner, pp. 169–185. MIT Press, Cambridge, MA, USA (2000)
9. Guard, J.R., Oglesby, F.C., Bennett, J.H., Settle, L.G.: Semi-automated mathematics. J. ACM **16**(1), 49–62 (1969). https://doi.org/10.1145/321495.321500
10. Michaelis, J., Nipkow, T.: Formalized proof systems for propositional logic. In: Abel, A., Forsberg, F.N., Kaposi, A. (eds.) 23rd International Conference on Types for Proofs and Programs (TYPES 2017). LIPIcs, vol. 104, pp. 6:1–6:16. Schloss Dagstuhl - Leibniz-Zentrum fuer Informatik (2018)
11. Nipkow, T., Wenzel, M., Paulson, L.C. (eds.): : 5. The Rules of the Game. In: Isabelle/HOL. LNCS, vol. 2283, pp. 67–104. Springer, Heidelberg (2002). https://doi.org/10.1007/3-540-45949-9_5
12. Paulson, L.C.: computational logic: its origins and applications. Proc. R. Soc. A. **474** 20170872 2210 (2018). https://doi.org/10.1098/rspa.2017.0872
13. Paulson, L.C., Nipkow, T., Wenzel, M.: From LCF to Isabelle/HOL. Formal Aspects Comput. **31**(6), 675–698 (2019). https://doi.org/10.1007/s00165-019-00492-1
14. Pfenning, F.: Single axioms in the implicational propositional calculus. In: Lusk, E., Overbeek, R. (eds.) CADE 1988. LNCS, vol. 310, pp. 710–713. Springer, Heidelberg (1988). https://doi.org/10.1007/BFb0012869
15. Ridge, T., Margetson, J.: A mechanically verified, sound and complete theorem prover for first order logic. In: Theorem Proving in Higher Order Logics, 18th International Conference, TPHOLs 2005, Oxford, UK, 22–25 August 2005, Proceedings, pp. 294–309 (2005)

16. Schlichtkrull, A., Blanchette, J.C., Traytel, D.: A verified prover based on ordered resolution. In: Proceedings of the 8th ACM SIGPLAN International Conference on Certified Programs and Proofs, CPP 2019, pp. 152–165. Association for Computing Machinery (2019). https://doi.org/10.1145/3293880.3294100

17. Shankar, N.: Towards mechanical metamathematics. J. Autom. Reasoning **1**(4), 407–434 (1985)

18. Smullyan, R.: What Is the Name of This Book? Prentice-Hall, Inc. Hoboken (1978)

19. Tourret, S., Blanchette, J.: A modular Isabelle framework for verifying saturation provers. In: Hritcu, C., Popescu, A. (eds.) CPP 2021: 10th ACM SIGPLAN International Conference on Certified Programs and Proofs, Virtual Event, Denmark, 17–19 January 2021, pp. 224–237. ACM (2021). https://doi.org/10.1145/3437992.3439912

20. Villadsen, J.: Tautology checkers in Isabelle and Haskell. In: Calimeri, F., Perri, S., Zumpano, E. (eds.) Proceedings of the 35th Edition of the Italian Conference on Computational Logic (CILC 2020), vol. 2710, pp. 327–341. CEUR-WS.org (2020). http://ceur-ws.org/Vol-2710/paper-21.pdf

21. Villadsen, J., Schlichtkrull, A., From, A.H.: A verified simple prover for first-order logic. In: Konev, B., Urban, J., Rümmer, P. (eds.) Proceedings of the 6th Workshop on Practical Aspects of Automated Reasoning co-located with Federated Logic Conference 2018 (FLoC 2018), Oxford, UK, 19th July 2018. CEUR Workshop Proceedings, vol. 2162, pp. 88–104. CEUR-WS.org (2018)

22. Villadsen, J., Jacobsen, F.K.: Using Isabelle in two courses on logic and automated reasoning. In: Ferreira, J.F., Mendes, A., Menghi, C. (eds.) FMTea 2021. LNCS, vol. 13122, pp. 117–132. Springer, Cham (2021). https://doi.org/10.1007/978-3-030-91550-6_9

23. Waldmann, U., Tourret, S., Robillard, S., Blanchette, J.: A comprehensive framework for saturation theorem proving. In: Peltier, N., Sofronie-Stokkermans, V. (eds.) IJCAR 2020. LNCS (LNAI), vol. 12166, pp. 316–334. Springer, Cham (2020). https://doi.org/10.1007/978-3-030-51074-9_18

24. Łukasiewicz, J.: The shortest axiom of the implicational calculus of propositions. In: Proceedings of the Royal Irish Academy. Section A: Mathematical and Physical Sciences, vol. 52, pp. 25–33 (1948)

Embedding and Integrating Literals to the HypER Model for Link Prediction on Knowledge Graphs

Thanh Le[1,2]($\boxtimes$) , Tuan Tran[1,2] , and Bac Le[1,2]

[1] Faculty of Information Technology, University of Science,
Ho Chi Minh City, Vietnam
{lnthanh,lhbac}@fit.hcmus.edu.vn, tttuan18@clc.fitus.edu.vn
[2] Vietnam National University, Ho Chi Minh City, Vietnam

Abstract. The link prediction on knowledge graphs is now one of the challenges that are gaining a lot of interest from the academic community. The leading solution for this problem is based on graph embedding. Recently, in embedding approaches, convolutional neural networks (CNN) have produced promising results, especially the HypER model. The HypER model outperforms the preceding approaches to maximize the quantity of information from the source entities and relations. However, HypER and other CNN-based methods only focus on retaining information (i.e., structure) of knowledge graphs in low dimension embedding spaces while ignoring literals of the entities. However, the literals can also have a significant impact on relation construction. As a result, this paper proposes an improved model called HypERLit, which is based on the HypER model and incorporates literals. Experiments prove that the role of literals significantly influences the accuracy of the prediction model on the benchmark datasets, including FB15k, FB15k-237, and YAGO3-10. Furthermore, our model outperforms the HypER and other CNN-based models on almost standard metrics.

Keywords: Knowledge graph embedding · Link prediction · Literals · Convolution neural network

1 Introduction

Knowledge is incredibly significant in today's society since it allows us to discover things never existed or been known before. However, knowledge is the accumulation of experience during human evolution that has resulted in a large amount of information. A structure called the "Knowledge Graphs" (KGs) is utilized to store them efficiently. Google Knowledge Graph [12], YAGO [13] and Wikipedia Knowledge Graph [11] are examples of popular KGs. KGs are frequently represented as triples (source entity, relation, target entity), implying that "the source entity has a relationship with the target entity." For example, a triple

N. T. Nguyen et al. (Eds.): ACIIDS 2022, LNAI 13757, pp. 403–415, 2022.
https://doi.org/10.1007/978-3-031-21743-2_32

(Biden, president_of, United_States) is understood as "Biden is the president of the United States."

Because KGs are fundamentally a graph, they can be directed or undirected. Directed graphs are used when swapping the source entity and the target entity impacts the meaning. For example, with the triple (A, student_of, class B), if the source and target entities are swapped, the meaning of the triple is incorrect since class B cannot have the relation "is a student of" with A. On the other hand, undirected graphs are utilized when the swapping of the source and target entities does not affect the meaning of the triple. For the triple (A, friend_of, B), if the source and target entities are swapped, the meaning of this triple stays the same.

One problem is that the data in KGs may be incomplete, even with large KGs. It means that some pairs of source and target entities may have a relationship between them, but there is no connecting edge between them in the graph. Because most of the data today is contributed by the community, it is difficult to avoid this problem. Therefore, researchers in KGs field have devised the link prediction problem. Its main task is to predict relations between the source and target entities in the triples. The input of the link prediction model is an incomplete KGs, and the output is a KGs containing the most likely new relations based on the model's assessment criteria.

The link prediction problem has helped to solve many practical problems such as anomaly detection, recommendation system, as well as a friend suggestion on social networks [7]. For example, we could consider users as nodes and links representing relationships in a social network. In a friend suggestion system, link prediction is the task that indicates the possibility of a connection between two people who do not know each other based on factors such as their mutual friends, residence, workplace, etc.

One of the typical research branches of the link prediction problem (LP) is knowledge graph embedding (KGE), which has recently achieved good results [15]. Embedding a knowledge graph is the process of finding a way to map entities and relations in a high-dimensional space of size n (usually equal to the number of entities) into a low-dimensional space of size k ($k \ll n$) so that information must be retained as much as possible. Using embedded entities and relations in low-dimensional space helps simplify the calculation and achieve better speed. It is one of the reasons why embedding-based models are more interested in LP than other models such as probabilistic models, rule-based models [8]. Embedding-based models are divided into three main research directions: translational models, implicit feature models (semantic matching models), and neural network-based models [5]. Among these methods, convolutional neural networks (CNNs) based models are a recently emerging research direction due to high accuracy and good feature extraction. So in this paper, we focus on CNN to make some improvements.

The survey found that current embedding models focus on keeping information of entities and relationships when converting from high embedding dimension to low embedding dimension. It means that literals in the knowledge graph are often ignored during embedding, even though they strongly impact creating relationships between entities. Figure 1 shows an example graph with additional literals for entities. In this example, *Tuan* and *Huy* both graduated from *Ho Chi Minh University of Science* with *Computer Science* major. If we only based on that information to conclude that *Tuan* is a friend of *Huy*, it is uncertain. By considering the literal as the birth year of *Tuan* and *Huy*, we can be more confident in predicting the friendship since they both attend the same school year.

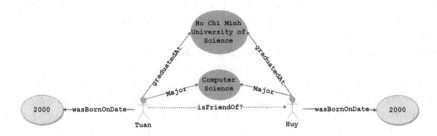

Fig. 1. An example of the influence of literals in the link prediction on KGs

The example in Fig. 1 shows that embedding-based models can give predictive results for a pair of entities, but the lack of literals can make the model uncertain on the relation or even give incorrect results. Therefore, counting literals to the link prediction model can help improve efficiency. In this paper, we choose the HypER model to implement this idea because it is currently one of the CNN-based models with robust and complete embedding representation. The literals are integrated with entity embeddings via a gate, then input to the HypER model. We call our model as HypERLit. Details are presented in the following sections.

2   Related Work

Convolutional neural network (CNN) has proven their success in many fields such as computer vision with recognition problems or in another area that can be said to be closer to KGs is natural language processing with classification problems. With that success, CNN has also been applied to problems on KGs, including embedding and link prediction. Therefore, this section analyzes the strengths and weaknesses of the important essential models in the CNN-based link prediction problem. Thereby, we give reasons for choosing the HypER model as the baseline model for our improvement proposal.

ConvE [3] can be said to be the pioneer model in applying CNN to KGE, and it also lays the foundation for other models to propose improvements. ConvE model performs convolution on a stacked 2D matrix of entity and relation embeddings. It helps to extract features between source entities and relations. As a result, they increase the interaction between source entities and relations as well as make the expressiveness of the model more efficient. In addition, using calculations on the target entity matrix saves training time compared to taking each target embedding vector in turn for computation. However, the limitation also lies in the concatenation of source and relation embeddings. It makes the features extracted at the convolution step not completely possess the properties of both source and relation embeddings.

InteractE [14] is a model developed based on ConvE. InteractE's steps are similar to ConvE's, but few changes. First, instead of stacking the 2D matrix of source and relation embeddings, the model arranges them in two different ways. They include (i) interleaving the rows of the source and relation embedding matrix; (ii) arranging the embeddings so that the source embedding is only adjacent to the relation embedding. As a result, these ways help to increase the interaction between the source entities and relations. Second, InteractE replaces the convolution with circular convolution and proves its efficiency in extracting interactions. These two changes have helped the InteractE model overcome the inherent disadvantage of the ConvE model, which is the poor interoperability between input embeddings. However, the InteractE model runs slower than ConvE because the 2D matrix must be transformed before proceeding to the later steps.

ConvKB [10] is a model that inherits the idea of using ConvE's 2D convolution layer. However, instead of using only source and relation embeddings to create a 2D matrix for convolution, it uses triples to perform. In this way, the features extracted in the convolution have the properties of all three embedding vectors, thus creating maximum interaction for both source entities, relation, and target entities. That is a considerable improvement compared to the ConvE model. Nonetheless, it is also a disadvantage because concatenating the embeddings in ConvKB makes the execution time slower than ConvE, especially in large datasets like YAGO3-10.

ConvR [5] is a model that uses relation embeddings to create filters instead of taking them into forming a 2D matrix for the convolution step like ConvE. Specifically, the relation embedding is divided into equal segments and then reshaped into 2D filters. Using filters to convolution on the source embedding matrix makes all embeddings interact together. Consequently, they increased the representation for embeddings. However, ConvR generates filters directly from the relation embeddings without considering the possibility that they can also significantly affect the result of the convolution.

HypER [1] is similar to ConvR when using relation embeddings to form filters. However, in the HypER model, the filters are generated from a fully connected layer instead of directly from relation embeddings like ConvR. Hence, the model can adjust the filters to have good results. It can be said that the improvements in the HypER model overcome most of the problems encountered by the previous CNN-based models. Experiments also show that HypER has high accuracy in benchmarking datasets. That is also why we choose the HypER model as the baseline model to integrate literals.

3   The Proposed Method

Before we describe the proposed model, we outline some significant underlying models, including HypER [1] and LiteralE [6]. HypER is a convolutional network model that was recently produced good results. Unlike ConvE employs randomly initialized filters, and ConvR uses filters formed directly from reshaping embedded relational vectors, HypER's filters are generated by a hypernetwork [4]. When performing convolution, this hypernetwork has the effect of tweaking the filters to extract the features of source embeddings efficiently. As a result, HypER may optimize the quantity of information gathered from source entities and represent them fully. Otherwise, the LiteralE model efficiently inserts literals into source entities. This efficiency is proved when substituting the source embedding in ConvE with the results of the LiteralE model. Building on the strengths of these two models, we designed a model named HypERLit by integrating literals into the filters and finding a way to attach literal information to entity embeddings. Hence, the model has more information to predict relations between entities.

The HypERLit model aggregates information from the source entity with literals by passing them through a Gate and does the same for all entities through an entity matrix whose each row is an entity vector in KGs. The related information is then transmitted via the hypernetwork in order to generate filters for convolution with the information accumulated through the previous Gate. Hence, we obtain the features retrieved from the source entity, literals, and relation. These features are projected into a d-dimensional space, and then a nonlinear function is applied to them. The acquired results are multiplied by the matrix of entities integrated with literals in the first stage and sent through the sigmoid function to generate the score of each triple in this matrix corresponding to each target entity. The whole procedure is depicted in Fig. 2.

To clarify in detail how the HypERLit model works, we introduce some related notations. Let E be the set containing entities of size n_e; R is a set of relations of size n_r; N is the set containing the numeric literals of each entity of size n_N; T is the set of text literals of each entity of size n_T; d is the number of dimensions of the entity and relation embedding vectors; dn and dt are the numbers of dimensions of the numeric literal and textual literal vectors, respectively.

First, the model uses the Gate structure of the LiteralE model to incorporate the source entity and the remaining entities with their literals. The Gate consists of three layers f, u, and h defined as follows:

$$\mathbf{f} = \sigma(\mathbf{W_e e} + \mathbf{W_n n} + \mathbf{W_t t} + \mathbf{b}) \tag{1}$$

$$\mathbf{u} = \theta(\mathbf{W_u}[\mathbf{e, n, t}]) \tag{2}$$

$$\mathbf{h} = (1 - \mathbf{f}) \odot \mathbf{e} + \mathbf{f} \odot \mathbf{u} \tag{3}$$

408   T. Le et al.

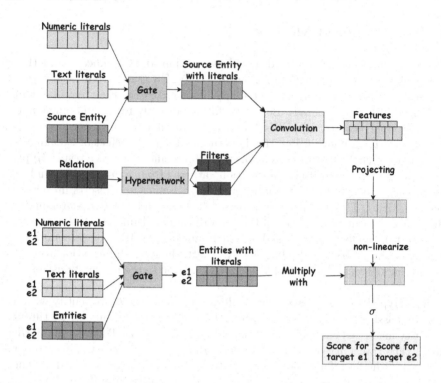

Fig. 2. The process of our HypERLit model

where $\mathbf{e} \in \mathbb{R}^d$; $\mathbf{n} \in \mathbb{R}^{dn}$; $\mathbf{t} \in \mathbb{R}^{dt}$; $[\mathbf{e}, \mathbf{n}, \mathbf{t}] \in \mathbb{R}^{d+dn+dt}$ is the operation of stacking entities and literals; $\mathbf{W_e}, \mathbf{W_n}, \mathbf{W_t}, \mathbf{W_u}$ are weights used for projecting $\mathbf{e}, \mathbf{n}, \mathbf{t}$ and $[\mathbf{e}, \mathbf{n}, \mathbf{t}]$ into $\mathbb{R}^d$ space; σ is sigmoid function and θ is nonlinear function as ReLU, tanh; $\odot$ is Hadamard product.

The $\mathbf{u}$ layer is called "update gate" that generates information which is a combination of entities by stacking them. This way saves costs because only one set of weights is used to project the stack into $\mathbb{R}^d$ space. Meanwhile, the $\mathbf{f}$ layer is the "forget gate" which decides what integration content to keep. This $\mathbf{f}$ layer does not use stacking as $\mathbf{u}$ layer, but instead uses an addition operation. The main reason is that the $\mathbf{f}$ layer wants the combined information between entities and literals, which significantly affects the decision to keep the information, should be preserved as much as possible. Next, the $\mathbf{h}$ layer performs the final output aggregation based on the "forget gate" decision. If the back values of "forget gate" have many 1s, the original entity information is almost ignored. Otherwise, if there are many 0 s, the information combined in the "update gate" is ignored. According to LiteralE, this Gate has a small number of parameters and low complexity, so it optimizes the model's execution time. In addition, the integration of more literals helps the entity be more specific, increasing the certainty in suggesting the relationship of these entities (similar to the case described in Fig. 1).

Next, HyperERLit employs a structure known as a hypernetwork. Hypernetwork is a fully connected layer, and the weights it initializes are sharable throughout subsequent layers. In our model, weights for relation embeddings are generated and used for the convolutional layer through filters as follows:

$$\mathbf{F} = vec^{-1}(\mathbf{Wr} + \mathbf{b}) \tag{4}$$

where $\mathbf{r} \in \mathbb{R}^d$ is relation embeddings; $\mathbf{W}$ and $\mathbf{b}$ are weights that are used to transform relation embeddings in hypernetwork; function vec^{-1} reshapes the 1D vector to the 2D matrix.

Sharing weights for the convolution layer is very important to our model. It can help the model learn how to tune the filters through the weights generated in the hypernetwork. Thus, the information extracted from the source entity with the associated literal is gradually better captured.

The filters generated from the hypernetwork continue convolution with the source embeddings with integrated literals by the Gate according to the following formula:

$$\mathbf{features} = vec(\mathbf{h} * \mathbf{F}) \tag{5}$$

where function vec reshaped the 2D matrix to the 1D vector; $*$ is the convolution operation.

The feature vectors extracted from Eq. 5 are projected into the $\mathbb{R}^d$ space. We then apply the nonlinear function to this projection as follows:

$$\mathbf{projection} = \theta(\mathbf{W_p}\mathbf{features} + \mathbf{b}) \tag{6}$$

Finally, the project from Eq. 6 is multiplied by a matrix whose each row is an entity vector with integrated literals, and then using the sigmoid function to get the score of each triple:

$$\mathbf{scores} = \sigma(\mathbf{ME} \cdot \mathbf{projection}) \tag{7}$$

where $\mathbf{ME} \in \mathbb{R}^{n_e \times d}$; scores={score of the first target entity is e_1,..., score of the nth target entity is e_{n_e}}.

The loss function we used in training is binary cross-entropy as follows:

$$Loss = -\frac{1}{n_e} \sum_{i=1}^{n_e} [label_i \times log(scores_i)] + [(1 - label_i) \times log(1 - scores_i)] \tag{8}$$

where $label_i$ is the label of the triple of the target entity e_i. If $label_i = 1$, then the triple of the target entity e_i is true and otherwise for $label_i = 0$.

4 Experiments and Result Analysis

4.1 Datasets

In this section, we evaluate the model on three benchmark datasets used in most link prediction problems including FB15k [15], FB15k-237 [15], and YAGO3-10 [3]. FB15k is a subset of the Freebase dataset [2], and it contains symmetrically related triples. FB15k-237 is a subset of the FB15k dataset with a smaller number of triplets due to the removal of symmetric triplets. Besides, YAGO3-10 is a subset of the YAGO3 [9] dataset containing triples taken from multiple Wikipedias, with entities and relationships mainly related to cities and people.

WN18 and WN18RR [1] are also well-known datasets. However, we do not use them in our experiment because literals of entities in these two datasets are mainly in the form of long text. Therefore, they require additional preprocessing steps in NLP to get better entity descriptions.

In addition, we also use literals of the datasets FB15k, FB15k-237, and YAGO3-10 which extracted and processed by [6] to add more information to the entities in experimental datasets. These literals include numeric literals such as year of birth, population, and textual literals with descriptions of the entity. Table 1 describes information about these datasets.

Table 1. Datasets' properties

	Num of entites	Num of relations	Num of triples			Num of literal triples	
			Train	Valid	Test	Numeric	Text
FB15k	14,951	1345	483,142	50,000	59,071	70,257	15,633
FB15k-237	14,541	237	272,115	17,535	20,466	70,257	15,633
YAGO3-10	123,182	37	1,079,040	5000	5000	111,406	107,326

4.2 Metrics

Two metrics commonly used to evaluate link prediction models are Mean Reciprocal Rank (MRR) and Hits@K. Assuming there are N correctly predicted links (in other words, triples that actually exist in KGs), two metrics are defined as follows:

$$MRR = \frac{1}{N} \sum_{i=1}^{N} \frac{1}{rank_i} \qquad (9)$$

$$Hits@K = \frac{|rank_i < K|}{N} \qquad (10)$$

where $rank_i$ is the rank of the correctly predicted i^{th} triple, $|rank_i < K|$ is the total number of correctly predicted triples that have rank less than a given K. Values of MRR and Hits@K are in the range [0, 1]. The closer these values are to 1, the better the prediction is.

4.3 Hyperparameters and Experimental Setup

The hyperparameters in the model include input dropout, hidden dropout, feature dropout, embedding dimension, learning rate, and kernel size. The embedding size is 200, which is the same as the baseline models for the empirical part of all datasets. For kernel size, it is 1×9 for all datasets as HypER's experiment. Similarly, we fixed the feature dropout value to 0.2. For the remaining hyperparameters, we set up with each dataset as follows: in the FB15k dataset, the input dropout is in $0.1, 0.2$, the hidden dropout is in $0.1, 0.2$, and the learning rate is 0.005; in the FB15k-237 dataset, the input dropout is in $0.1, 0.3, 0.4$, the hidden dropout is in $0.4, 0.5, 0.6$, and the learning rate is 0.001; in the YAGO3-10 dataset, the input dropout is 0.2, the hidden dropout is in $0.3, 0.5$, and the learning rate is in $0.001, 0.005$.

4.4 Results

To evaluate the effectiveness of the HypERLit model, we run experiments on the mentioned datasets in Sect. 4.1 and compare the results with the algorithms that have high results in link prediction on KGs. Since HypERLit is a CNN-based model, we prefer to choose the models in this group to perform the evaluation to ensure objectivity. These algorithms have been mentioned in Sect. 2 including ConvE [3], ConvKB [10], ConvR [5] and HypER [1]. In addition, we also selected the model with integrated literals such as ConvE-LiteralE [6] to compare the effectiveness of literals to our approach. These models' results are taken from the papers themself. Table 2, Table 3, and Table 4 describe the experimental results on the FB15k, FB15k-237, and YAGO3-10 datasets, respectively.

Table 2. The results on FB15k

	MRR	Hits@10	Hits@3	Hits@1
ConvE	0.745	0.873	0.801	0.670
ConvKB	0.211	0.408	–	0.114
ConvR	0.782	_0.887_	0.826	0.720
HypER	_0.790_	0.885	_0.829_	_0.734_
ConvE-LiteralE	0.733	0.863	0.785	0.656
HypERLit (our model)	**0.794**	**0.888**	**0.835**	**0.736**

In general, our model outperforms on all metrics in Table 2. Specifically, on MRR metric, we have a greater result ranging from 0.004 to 0.583. In addition, the model performs better on Hits@k metrics, with values ranging from 0.001 to 0.622. We keep the distance from the second-ranked model to 0.006, especially in Hits@3. Despite the fact that the FB15k dataset is no longer routinely employed in the assessment of later models due to data leakage issues, the comparison reveals that HypERLit is capable of doing well in this dataset.

The results when running on the FB15k-237 dataset are shown in Table 3. Although compared to ConvKB, we have worse results on the MRR and Hits@1

Table 3. The results on FB15k-237

	MRR	Hits@10	Hits@3	Hits@1
ConvE	0.316	0.491	0.350	0.239
ConvKB	**0.354**	0.535	–	**0.263**
ConvR	0.350	0.528	0.385	0.261
HypER	0.341	0.520	0.376	0.252
ConvE-LiteralE	0.303	0.471	0.330	0.219
HypERLit (our model)	0.353	**0.536**	**0.394**	0.259

metrics, the deviation is only between 0.001 and 0.004. Because this dataset does not contain inverse triples, it is more suitable for ConvKB's triple training process. For the other models, HypERLit is better. In particular, the models we inherited HypER and ConvE-LiteralE—are even much lower than when compared on the FB15k dataset.

Table 4. The results on YAGO3-10

	MRR	Hits@10	Hits@3	Hits@1
ConvE	0.440	0.620	0.490	0.350
HypER	0.533	**0.678**	0.580	0.455
ConvE-LiteralE	0.525	0.659	0.572	0.448
HypERLit (our model)	**0.538**	0.675	**0.586**	**0.461**

Because of the large size of the YAGO3-10 dataset with about 1M triples, only a few algorithms with fast processing speed such as ConvE, HypER and ConvE-LiteralE can conduct testing on it. Based on the results in Table 4, the HypERLit model has a better result from 0.005 to 0.098 on most metrics except for Hits@10 which is only about 0.003 worse than HypER. However, the HypERLit model is not too far apart from the baseline models. Because on the YAGO3-10 dataset, the information about the entities becomes more, the predictions can easily achieve equivalent accuracy even without the literal involved.

Next, we continue to evaluate the convergence of the model on the datasets. Figure 3 shows loss values in 1000 epochs for the FB15k, FB15k-237 and YAGO3-10 datasets. The chart shows that loss values of YAGO3-10 is the lowest, followed by FB15k-237 and the highest is FB15k. In general, the loss values of the datasets converge very quickly, which is only between epoch 1 and epoch 100. After that time, the loss values do not change significantly, especially YAGO3-10. In addition, the loss values on YAGO3-10 are also more stable than the other two datasets. It means that the information extracted from the graph is almost complete to predict the relations accurately.

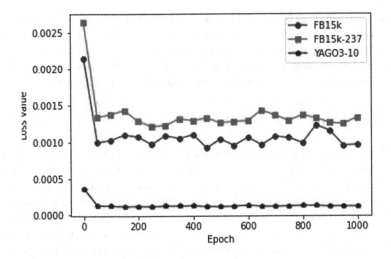

Fig. 3. Loss values on 1000 epochs

4.5   The Influence of Hyperparameters

Among the hyperparameters mentioned in Sect. 4.3, we find that input dropout and hidden dropout are the two hyperparameters that most influence the model's accuracy. Figure 4 depicts the resulting change when updating these hyperparameters. Specifically, the input dropout parameter is chosen for illustration includes the values 0.1, 0.3, and 0.4. We also tested other values and other hyperparameters, but we did not include them in this presentation due to the limitation of the paper length. Similarly, the hidden dropout parameter is selected with the values 0.3, 0.4, 0.6.

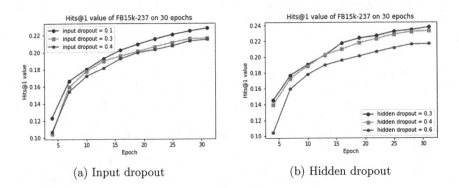

(a) Input dropout   (b) Hidden dropout

Fig. 4. Hit@1 of FB15k-237 dataset on input dropout and hidden dropout values

When considering the first 30 epochs with the Hits@1 metric, input_dropout = 0.1 gives the highest results; next is input_dropout = 0.3, and lowest is input_dropout = 0.4. Similarly, hidden_dropout = 0.3 gives the best results; next is hidden_dropout = 0.4, and the lowest is hidden_dropout = 0.6. In this way, we can determine the best parameters of the HypERLit model.

5 Conclusion

This paper proposes the HypERLit model based on HypER and LiteralE model to solve LP on KGs. The HypERLit model starts with passing literals along with the source entities through a gate so that the convolution can be performed with the filters generated by the HypER model. Consequently, the model can extract important features of the source entities with combined literals. In addition, the literals through the gate also form a matrix containing the complete information of entities and their description. This matrix is combined with the projection of the features to rank the predicted entities. Experiments on three standard datasets, including FB15k, FB15k-237, and YAGO3-10, have shown that the proposed model achieves better results than other CNN-based models on metrics MRR, Hits@K $(1, 3, 10)$. Particularly, for the FB15k dataset, the HypERLit model is higher than other models, about $0.001 - 0.622$ on the Hits@k measure and $0.004 - 0.583$ on the MRR measure. Similarly, on the FB15k-237 dataset, our model reaches a higher Hit@k and MRR result of about $0.001 - 0.009$. For the YAGO3-10 dataset, the result of HypERLit is about $0.005 - 0.098$ higher. It demonstrates that the inclusion of literals provides additional information for deciding which relation is most likely. We also evaluate the influence of hyperparameters on the model, especially input dropout and hidden dropout. In the future, we will apply other information aggregation structures from NLP to better integrate literals into entities.

Acknowledgements. This research is funded by the University of Science, VNU-HCM, Vietnam under grant number CNTT 2022-02 and Advanced Program in Computer Science.

References

1. Balažević, I., Allen, C., Hospedales, T.M.: Hypernetwork Knowledge Graph Embeddings. In: Tetko, I.V., Kůrková, V., Karpov, P., Theis, F. (eds.) ICANN 2019. LNCS, vol. 11731, pp. 553–565. Springer, Cham (2019). https://doi.org/10.1007/978-3-030-30493-5_52
2. Bollacker, K., Evans, C., Paritosh, P., Sturge, T., Taylor, J.: Freebase: a collaboratively created graph database for structuring human knowledge. In: Proceedings of the 2008 ACM SIGMOD International Conference on Management of Data, pp. 1247–1250 (2008)
3. Dettmers, T., Minervini, P., Stenetorp, P., Riedel, S.: Convolutional 2d knowledge graph embeddings. In: Thirty-Second AAAI Conference on Artificial Intelligence (2018)

4. Ha, D., Dai, A., Le, Q.V.: Hypernetworks. arXiv preprint arXiv:1609.09106 (2016)
5. Jiang, X., Wang, Q., Wang, B.: Adaptive convolution for multi-relational learning. In: Proceedings of the 2019 Conference of the North American Chapter of the Association for Computational Linguistics: Human Language Technologies, Volume 1 (Long and Short Papers), pp. 978–987 (2019)
6. Kristiadi, A., Khan, M.A., Lukovnikov, D., Lehmann, J., Fischer, A.: Incorporating literals into knowledge graph embeddings. In: Ghidini, C., Hartig, O., Maleshkova, M., Svátek, V., Cruz, I., Hogan, A., Song, J., Lefrançois, M., Gandon, F. (eds.) ISWC 2019. LNCS, vol. 11778, pp. 347–363. Springer, Cham (2019). https://doi.org/10.1007/978-3-030-30793-6_20
7. Liben-Nowell, D., Kleinberg, J.: The link-prediction problem for social networks. J. Am. Soc. Inform. Sci. Technol. **58**(7), 1019–1031 (2007)
8. Lü, L., Zhou, T.: Link prediction in complex networks: a survey. Physica A **390**(6), 1150–1170 (2011)
9. Mahdisoltani, F., Biega, J., Suchanek, F.: Yago3: a knowledge base from multilingual Wikipedias. In: 7th Biennial Conference on Innovative Data Systems Research. CIDR Conference (2014)
10. Nguyen, D.Q., Nguyen, T.D., Nguyen, D.Q., Phung, D.: A novel embedding model for knowledge base completion based on convolutional neural network. arXiv preprint arXiv:1712.02121 (2017)
11. Sarker, M.K., et al.: Wikipedia knowledge graph for explainable ai. In: Second Iberoamerican Knowledge Graphs and Semantic Web Conference (KGSWC) (11/2020 2020)
12. Singhal, A.: Introducing the knowledge graph: Things, not strings, May 2012. https://blog.google/products/search/introducing-knowledge-graph-things-not
13. Suchanek, F.M., Kasneci, G., Weikum, G.: Yago: a core of semantic knowledge. In: Proceedings of the 16th International Conference on World Wide Web, pp. 697–706 (2007)
14. Vashishth, S., Sanyal, S., Nitin, V., Agrawal, N., Talukdar, P.: Interacte: Improving convolution-based knowledge graph embeddings by increasing feature interactions. In: Proceedings of the AAAI Conference on Artificial Intelligence, vol. 34, pp. 3009–3016 (2020)
15. Wang, M., Qiu, L., Wang, X.: A survey on knowledge graph embeddings for link prediction. Symmetry **13**(3), 485 (2021)

A Semantic-Based Approach for Keyphrase Extraction from Vietnamese Documents Using Thematic Vector

Linh Viet Le and Tho Thi Ngoc Le[(✉)]

HUTECH University, Ho Chi Minh City, Vietnam
levietlinh@gmail.com, ltn.tho@hutech.edu.vn

Abstract. Keyphrase extraction plays an important role in many applications of Natural Language Processing. There are many effective proposals for English, but those approaches are not completely applicable for low resources languages such as Vietnamese. In this paper, we propose a Semantic-based Approach for Keyphrase Extraction (SAKE), which improved the TextRank algorithm [1]. In SAKE, we apply semantic to the phrases and incorporates the semantic to the ranking process. Technically, a document is represented as a graph, in which vertices are words and edges are relations among words. In each document, we get a representative thematic vector by computing the average of word embedding vectors. Each vertex has a similarity score to the thematic vector and this score will be involved to the scoring in the ranking process. The important vertices are highly weighted not only by their relationships to other vertices but also by the similarity to the document theme. We experimented our proposed method on Vietnamese news articles. The result shows that our SAKE improved TextRank for Vietnamese text by achieving 1.8% higher of F1-score.

Keywords: Keyphrase extraction · Thematic vector · TextRank · Vietnamese text · Word embedding

1 Introduction

Due to the increasing of information on the Internet, many unstructured texts are required to be efficiently stored and retrieved. One supportive information for these tasks is the usage of keyphrases for indexing the documents. Keyphrases are the terms that express the main ideas of the documents [2]. In practice, we can observe the keyphrase being indexed in many document systems. For examples, (i) in digital libraries, scientific documents are tagged with keyphrases by authors for indexing the documents; (ii) In news editorial department, documents that published in Internet are also tagged with keyphrases, so that documents are easier to be searched by search engine. It is the results of indexing the documents in which keyphrases were set as the key that represent to the document. Since keyphrases express the key information of documents, the more accurate keyphrases we choose the more quality search we achieve. That is one of the

N. T. Nguyen et al. (Eds.): ACIIDS 2022, LNAI 13757, pp. 416–427, 2022.
https://doi.org/10.1007/978-3-031-21743-2_33

reasons why search engine applies keyphrase extraction to automatically get the main ideas of the documents.

Keyphrase extraction is one of important tasks in Natural Language Processing (NLP). Formally, this task takes input as text documents and gives output as a list of keyphrases representing for the main contents of the documents. The task of automatic keyphrase extraction contributorily supports to many other tasks in NLP such as automatic indexing, automatic summarization, automatic classification, clustering, and filtering becomes incredibly popular in application of social media, tag of article in online news.

Many approaches for automatic keyphrase extraction have been proposed and implemented. There are two main methods: supervised methods and unsupervised methods. Supervised approaches treat keyphrase extraction as a classification task, which tries to leverage if the word/phrases are important or not. The features for these approaches are usually parts-of-speech, word scores such as TF-IDF and position information [2, 3]. Unsupervised approaches mainly find the highly weighted phrases as important ones. The methods for weighting the phrases include ranking on graphs [1, 4, 5], clustering for exemplars [6], calculating TF-IDF [7].

In this paper, we focus on extracting keyphrases from Vietnamese news articles. Due to the nature of the language, existing approaches for English are not fully applicable to Vietnamese. There are some previous works for Vietnamese keyphrase extraction [8–10]. However, most of them are supervised approaches or require ontology resource. Therefore, we motivate to adapt the general unsupervised approaches to extract keyphrases from Vietnamese text.

In unsupervised keyphrase extraction methods, TextRank [1] is a simple but effective algorithm. This algorithm has inspired a series of ranking algorithm for extracting keyphrases [11]. Most of variant algorithms are on the graph construction and the scores in the ranking process. In this paper, we propose a Semantic based Approach for Keyphrase Extraction (SAKE). In SAKE approach, we utilize a thematic vector to represent the theme for each document and exploit this vector to compute the score of candidates in ranking process. First, a document is represented as a graph, in which a vertex is a word/phrase, and an edge is the relation between a pair of vertices. Two vertices have an edge if two corresponding words appear in a fixed size window. Second, we find the thematic vector for each document by the average of word embedding vectors in the document. Then, each vertex in the graph is equipped an additional information, i.e., the semantic similarity between its word and the document thematic vector. Third, we rank the graph with the incorporating semantic similarity. Fourth, the highly ranked vertices are extracted as candidates for keyphrases. Finally, we post-process the candidates for keyphrases.

We experimented our proposed approach on 13,149 Vietnamese news articles. The experiment results show that document thematic vector is beneficial to the scores of candidates and therefore improve the performance of keyphrase extraction by 1.8% higher on F1-score.

In this paper, our contributions include:

1. Introducing SAKE approach for extracting keyphrases from Vietnamese news articles. In which, our approach considers the thematic vector to the weights of vertices.
2. Collecting the keyphrases dataset from Vietnamese news articles.
3. Evaluating the performance of our proposed method with the original approach.

The following of this paper will be organized as follows: Sect. 2 outlines the related work for keyphrases extraction in general and for Vietnamese text; Sect. 3 describes our approaches for extracting keyphrases from Vietnamese news articles; Sect. 4 presents the experiments and evaluation; Sect. 5 presents the comparison between other algorithms and SAKE. Finally, Sect. 6 concludes this paper.

2 Related Work

Keyphrase extraction is researched in recent years using two main approaches: supervised and unsupervised studies. In supervised studies, the task becomes classification using linguistic and statistical features extracted from a text of a document, such as Parts of Speech (POS) tags, TF-IDF scores and position information. The model is trained by Naive Bayes [3, 12, 13], CRFs [14], Bi-LSTM-CRF [15]. The broader view of supervised approaches for keyphrase extraction can be found in some recent surveys [11, 16, 17].

Since we focus on unsupervised keyphrase extraction in this paper, we briefly outline the most popular unsupervised approaches. On the other hand, as our target language is Vietnamese, we also summarize methods proposed for Vietnamese keyphrase extraction.

Frequency Based Solution
The simplest solution for keyphrase extraction assumes that the more frequency a word w is in a document d, the higher probability that word w is important in document d. This is known as TF (Term Frequency) which is simply counted each term appearing in text. When considering words are more frequent than others, it does not find any out of ordinary because it only considers inside a document. That is why people came up to refer a more common method. That is TF-IDF (Term Frequency-Inverse Document Frequency)

$$w_d = f_{w,d} \times \log(\frac{|D|}{f_{w,D}}) \tag{1}$$

In this equation, the words are generally frequent with f_w is also high in corpus D will be less frequent. This simple formular give a competitive result and it proved the effective until now.

Graph Based Solution
In 2004, Mihalcea and Tarau proposed TextRank algorithm [1], inspiring from the idea of PageRank [18]. TextRank can be considered the baseline for this category. TextRank represents text as graph, in which a vertex is a noun or an adjective, an edge is a relation

between two terms co-occur within a context window W of N words. When the graph is constructed, each vertex is set by an initial value, e.g., 1, and the whole graph is ranked until coverage. The node is voted by in and out edges, therefore if it has more score, it has more probability to be top ranked. The top T words is selected where T is chosen by one third of the number of vertices in the graph. However, there are still some important words with low frequency and therefore in low ranked, TextRank still need to be improved.

To compensate this semantic loss, many methods was proposed. In 2015, Danesh et al. proposed a hybrid statistic and graph-based approach called SGRank [4]. This algorithm computed an initial weight for each phrase based on TF-IDF score and position of first occurrence in document. Then the phrases are modeled as graph and recompute the weights using weighted PageRank centrality measurement, to get top ranked keyphrase. In 2017, Florescu và Caragea proposed another PageRank applied solution, called PositionRank [5] where words that is more frequent and occur in the beginning of a document have a higher initial weight.

Ensemble Solution
In 2019, Ye and Wang proposed an ensemble solution [19] where TF-IDF and TextRank are combined. At first each method is adopted separately to get 5 keyphrases. After that two results are taken the union to remove the duplicate phrases.

Vietnamese Keyphrase Extraction
In 2009, Nguyen et al. apply supervised classification for extracting keyphrases from Vietnamese text [8]. In this work, the authors treat keyphrase extraction as classification problem. The features for training are the words and theirs corresponding POS tags. The corpus for training is built by labelling each word as one of three classes I (In) – O (Out) – B(Begin) of a keyphrase. Then, classifying model such as SVM is applied to learn a classifier.

In the same year, Nguyen and Phan [10] utilize ontology to extract keyphrases. In this work, the authors use the Vietnamese Wikipedia as an ontology and some specific characteristics of the Vietnamese language for the key phrase selection step. They focus on the advanced candidate phrases recognition phase and POS tagging.

Recently, Hung et al. [9] use deep learning techniques to learn the keyphrases from Vietnamese text. In this work, the authors use hybrid the superior features of CNN and LSTM models to learn for keyphrases.

3 Semantic-Based Approach for Keyphrase Exaction (SAKE)

In this paper, we propose a Semantic-based Approach for Keyphrase Extraction (SAKE). This work is motivated from work by Key2Vec [20], proposed by Mahata et al. in 2018. Key2Vec is a method for training phrase embeddings which are used to represent for candidate keyphrases. However, this method is not immediately applicable to Vietnamese dataset due to the lack of Vietnamese phrase dataset to train model. Therefore, we are not able to implement this study for Vietnamese documents. Key2Vec motivates us on using semantic concept to extract phrases in Vietnamese news documents.

In a nutshell, our proposed SAKE algorithm includes the following steps:

- Pre-process the text for cleaning, only get words that are potential candidates
- Choose words from first sentences to create average vector that represents the semantic context of the text.
- Build a graph from words in text and from top words ranking, combining the single words to phrases that appears in text.

  o By using combination of single words to generate phrase, we can collect the potential phrases.

- Build a graph with vertices are single word and potential phrases that we generated, apply weighted graph-based algorithm to rank the top keys.

We chose Vietnamese news document as the main target and we collected dataset from online news website. Our solution includes 3 steps. Step 1 is preprocessing to tokenize words, get POS. Step 2 is using graph-based method to select candidates that includes keyword and keyphrase. Step 3 is basing on word-embedding, we score each candidate vector with a group of keyword's average vector. The score is used as cosine similarity. Using graph-based algorithm's ranking for getting top T keyphrases, we ranked the candidate.

3.1 Step 1: Preprocessing Documents

Document is processed the sentence segmentation by VnCoreNLP [21] tool, Vietnamese's tokenization by UETSegmenter [22] tool and make clean:

- Remove stop words
- Remove determiner
- Remove Unicode punctuation
- Remove numeric
- Only get parts-of-speech: Noun, Verb, Adjective

3.2 Step 2: Selecting Candidates

To get all phrases for candidate, we use TextRank to get the combination of single words. TextRank is graph based solution that proved efficiently in getting keywords. We selected keyphrases to join with tokenized words in Step 1.

3.3 Step 3: Scoring Candidates

We consider the document as graph of words and phrases. This is different to original TextRank which builds a graph of words. We built a directed graph $G = (V, E)$. In which, V is the set of vertices representing for words/phrases and E is set of edges representing the link between two vertices. The link between two vertices is set up when both corresponding words/phrases occur in a co-occurrence window of size W.

Calculate Weight of Edges: Word embedding techniques represent a word by a numeric vector. We can generate a vector that averages the words from the title and the top sentences (in our experiment, we used the first 3 sentences). This vector is considered representative thematic vector of the document d_i. When considering a word that is represented by a vertex $v_j^{d_i}$ and another co-occurred word that is represented by a vertex $v_k^{d_i}$, the vertex $v_k^{d_i}$ is the neighbor vertex of $v_j^{d_i}$.

Similarity score of a word that is represented by embedding vector is calculated as Eq. (2).

$$simsr\left(v_j^{d_i}, v_k^{d_i}\right) = \frac{1}{1 - cosine(v_j^{d_i}, v_k^{d_i})} \quad (2)$$

$cosine\left(v_j^{d_i}, v_k^{d_i}\right)$ is the cosine between 2 embedding vectors: $v_j^{d_i}$ and the neighbor $v_k^{d_i}$. It can be calculated by Eq. (3):

$$cosine\left(v_j^{d_i}, v_k^{d_i}\right) = \frac{\vec{v_j^{d_i}} \cdot \vec{v_k^{d_i}}}{\|\vec{v_j^{d_i}}\| \cdot \|\vec{v_k^{d_i}}\|} \quad (3)$$

Given a graph G, if $\varepsilon(v_j^{d_i})$ is a set of edges that incident vertex $v_j^{d_i}$, $w_{v_j}^{d_i}$ is weight of vertex $v_j^{d_i}$ and weight is calculated by Eq. (4) as follows:

$$w_{v_j}^{d_i} = cosine(v_j^{d_i}, d_i) \quad (4)$$

$cosine(v_j^{d_i}, d_i)$ is the cosine between two embedding vectors: the vertex $v_j^{d_i}$ and the document d_i.

Embedding vector of d_i is an average vector of embedding vectors for words and phrases in document d_i.

We used weighted personalized PageRank to calculate score of each vertex. The score is ranged from 0 to 1. The score 1 means the keyphrase is very near to content of the document. The score is calculated by Eq. (5) as follows:

$$S\left(v_j^{d_i}\right) = (1 - d)w_{v_j}^{d_i} + d \times \sum_{v_k^{d_i} \in \varepsilon(v_j^{d_i})} \left(\frac{simsr\left(v_j^{d_i}, v_k^{d_i}\right)}{\left|out\left(v_k^{d_i}\right)\right|}\right) S(v_k^{d_i}) \quad (5)$$

We sorted the score of all vertices and get top keyphrases as the result of extraction.

4 Experiments

4.1 Dataset

We use articles that have keyword tags as data for the experiment. We chose newspaper websites that lists keywords or tags inside the articles. Since the keywords in news articles are assigned manually by the news journalists and approved by the news editors, we can consider that the keywords are qualified keyphrases for news documents.

We collected news articles from online news websites by implementing a tool to crawl articles from three online newspapers (Thanh Nien[1], Phu Nu[2], Lao Dong[3]). These newspaper sites include content and keyword sections, and we collected these data to create a corpus of 13,149 articles with keywords. Data belongs to seven categories as shown in Table 1. For each news article, we collected author, sentences, link, description, time, title, keyword. Data for input of keyphrase extraction is JSON files that include a list of sentences, title, and a list of keywords.

Table 1. The number of articles and categories of article for experiment.

Categories	Thanh Nien	Phu Nu	Lao Dong
Politic	1,565	435	843
Technology	1,589	0	0
Finance	1,007	0	581
World	1,600	471	768
Sport	0	0	209
Culture	0	431	0
Other	1,192	1,051	1,407

In news, author usually applied 5W1H (who, what, where, when, why and how) to describe the information contents. We also observe that, the longer article is, the more difficulty to get correct keyphrases. We conduct a survey to examine where the keyphrases are usually distributed in the news articles.

Table 2. The percentage of keywords appearing in top N sentences of documents.

N	1	2	3	4	5	6	7	8	9	10
#keys (%)	79.7	61.1	53.4	49.8	49.2	47.4	46.1	45.9	45.3	42.3

Table 2 shows the percentage of keywords appear in the top N sentences of the news article. We found that the keywords tend to appear in the beginning of the documents. The rate of founding keyword from the 11th sentence is decreased. Hence, we limited the length of the news articles and get the first 10 sentences of each article to extract keyphrases.

---

[1] https://thanhnien.vn/

[2] https://phunuonline.com.vn/

[3] https://laodong.vn/

4.2 Evaluation

To evaluate, we used Precision, Recall, F1-score measurement. Precision P is defined as the number of correct keyphrases in the extracted keyphrases.

$$P = \frac{\#correct\ keyprhases}{\#extracted\ keyphrases} \quad (6)$$

Recall R is defined as the number of correct keyphrases being extracted comparing to the annotated keyphrases.

$$R = \frac{\#correct\ keyphrases}{\#annotated\ keyphrases} \quad (7)$$

F1-score is the harmony of Precision and Recall.

$$F1 = 2 \times \frac{P \times R}{P + R} \quad (8)$$

4.3 Results

We used FastText [23] as word embedding and the model that has been pre-trained[4] for creating vectors. We used TextRank as baseline for comparing with our solution. The experiment was run on Vietnamese news and result as in Table 3.

Table 3. Experiment result shows the number of terms in extract, true and evaluation.

W	TextRank					SAKE				
	#extract	#true	P(%)	R(%)	F1(%)	#extract	#true	P(%)	R(%)	F1(%)
2	59,195	6,483	11.0	17.6	13.5	59,195	6,930	11.7	18.8	14.0
3	59,195	6,623	11.2	18.0	13.8	59,195	7,497	12.7	20.3	15.6
4	59,195	6,688	11.3	18.1	13.9	59,195	7,572	12.8	20.5	**15.8**
5	59,195	6,705	11.3	18.2	**14.0**	59,195	7,596	12.8	20.6	**15.8**

The result in Table 3 shows that comparing to TextRank, SAKE is 2.3% higher on the Recall and 1.8% higher on the F1-score (in W = 3). In this example, we found that with SAKE, the keyphrase that has meaning which is nearest to the theme document is listed in the result of first 5 keyphrases.

In the article of "5 mối nguy_hiểm hết_sức nghiêm_trọng của **mỡ bụng**" (5 Dangers of **Belly Fat** That Doctors Warn Us About), we can realize that the keyphrase of "mỡ_bụng"(belly fat) is the importance term.

From the above sample, when scoring a node in graph, the relation between words/phrases is important but the relation of words/phrases to the theme of document is also important, so that we can use to extract more valued keyphrases.

---

[4] https://fasttext.cc/docs/en/crawl-vectors.html.

Table 4. An example shows the different result of top 5 keyphrase extraction.

Document	Extracted by	
	TextRank	SAKE
Bất_kể chất_béo xuất_hiện ở đâu trên cơ_thể, chúng_ta luôn lo_lắng, đặc_biệt khi nó ở dạng phình ra tại vòng 2 của bạn. Mỡ bụng luôn là nỗi sợ_hãi, không_chỉ đối_với chị_em phụ_nữ, mà cả với cánh mày râu. Nếu bạn là phụ_nữ với kích_thước vòng eo trên 90cm, hoặc nếu là nam_giới với kích_thước vòng eo trên 101cm, rất có_thể bạn đang thuộc nhóm đối_tượng có lượng mỡ bụng lớn. Theo các bác_sĩ, điều này có_thể gây nguy_hiểm cho sức khỏe. Nguy_cơ phát_triển bệnh tiểu_đường cao Khi gan của bạn bị bao_phủ bởi các mô mỡ do mỡ bụng dư_thừa, nó không_thể xử_lý đủ lượng đường trong máu. Lúc này mỡ bụng sẽ giữ đường trong máu của bạn thay_vì đường được xử_lý bởi gan. Điều này có_thể dẫn đến việc bạn có lượng đường trong máu cao hơn và có_thể gây ra bệnh tiểu_đường. Nguy_cơ mắc hội_chứng chuyển hóa cao	dạng phình, phụ_nữ, mỡ, phình, xuất_hiện	bụng, mỡ, **mỡ_bụng**, eo, mắc
{translated into English} No matter where fat shows up on our bodies, we always worry, especially when it's in the form of a bulge in our tummies. **Belly fat** is not only the haunting worry for woman, but also for man. If you're a lady and your waist size measures above 90 cm, or if you're a man and your waist size measures above 101 cm, chances are you have a large chunk of **belly fat**. According to doctors, this might be dangerous for your health Higher risk of developing diabetes When your liver is covered with fatty tissue as a result of excess **belly fat**, it fails to process blood sugar well enough. This, in turn, traps sugar in your bloodstream instead of the sugar being processed by the liver. This might lead to you having a higher blood sugar level which can cause diabetes	bulge form, woman, fat, bulge, show up	belly, fat, **belly fat**, waist, have

5   Discussions

SAKE is semantic-based ranking algorithm that is inspired by the popular TextRank algorithm. This paper considers the domain of Vietnamese news articles. Due to the nature of language, existing approaches for English are not fully applicable to Vietnamese such as Key2Vec, KeyBert. As the performance shown in Table 3 and Table 4, there is room to improve for keyphrase extraction for Vietnamese.

We compare our SAKE with Key2Vec and TextRank which are the baselines and inspiration of our algorithm in Table 5 and Table 6.

Table 5. Comparison of Key2Vec and SAKE.

Key2Vec	SAKE
– Use phrases in the vertices of graph – Use top sentences of document to get thematic vector – Use PageRank to rank candidates – Use FastText model to get word/phrase embedding	– Use phrases in the vertices of graph – Use top sentences of document to get thematic vector – Use PageRank to rank candidates – Use FastText model to get word/phrase embedding
Phrases are extracted by another tool	Phrases are extracted by applying TextRank
Thematic vector is calculated by summing thematic phrase's vectors	Thematic vector is calculated by averaging thematic phrase's vectors
Use Pointwise Mutual Information (PMI) in scoring candidates	Not use PMI in scoring candidates

Table 6. Comparison of TextRank and SAKE.

TextRank	SAKE
A graph-based algorithm	A graph-based algorithm
Topic of document is not yet concerned	Topic of document is concerned
Use words as vertices of graph	Use words and phrases as vertices of graph
Graph is undirected and weighted	Graph is bi-directed and weighted
Score of each vertex is initially set as 1	Score of each vertex is initial set by cosine similarity between vector of word or phrase and thematic vector

6 Conclusions

In this paper, we introduced a solution for keyphrase extraction where thematic vector is used to calculate score of candidates. Our SAKE is inspired by previous work: TextRank and Key2Vec. Candidates words and phrases that are selected by TextRank models a graph and weighted personalized PageRank is applied to rank the phrases.

Phrases and words in document become candidates for SAKE approach. Firstly, each document is represented as a graph in which a vertex is a word/phrase, and an edge is the relation between a pair of vertices. Two vertices have an edge if two corresponding words appear in a fixed size window. Secondly, we compute thematic vector for each document by calculating the average of word embedding vectors in the document. Thirdly, each vertex in the graph is equipped an additional information, i.e., the semantic similarity between its word and the document thematic vector. Fourthly, the score of each vertex is calculated using semantic similarity of corresponding word. PageRank's equation is modified using weight and similarity score. The score is then sorted to assure that the highest score of vertices will be the first element in list of keyphrases. We have evaluated our proposed approach on Vietnamese news articles. The result shows that the keyphrases from SAKE are matched more words than the baseline solutions. It indicates that semantic vector can enrich the graph representation of document for ranking the keyphrases.

References

1. Mihalcea, R., Tarau, P.: Textrank: Bringing order into text. In: Proceedings of the 2004 Conference on Empirical Methods in Natural Language Processing, pp. 404–411. Association for Computational Linguistics (2004)
2. Turney, P.D.: Learning algorithms for keyphrase extraction. Inf. Retrieval 2, 303–336 (2000). https://doi.org/10.1023/A:1009976227802
3. Witten, I.H., Paynter, G.W., Frank, E., Gutwin, C., Nevill-Manning, C.G.: KEA: practical automatic keyphrase extraction. In: Proceedings of the Fourth ACM conference on Digital Libraries, pp. 254–255 (1999). https://doi.org/10.1145/313238.313437
4. Danesh, S., Sumner, T., Martin, J. H.: SGRank: combining statistical and graphical methods to improve the state of the art in unsupervised keyphrase extraction. In: Proceedings of the Fourth Joint Conference on Lexical and Computational Semantics, pp. 117–126. Association for Computational Linguistics (2015). https://doi.org/10.18653/v1/S15-1013
5. Florescu, C., Caragea, C.: PositionRank: an unsupervised approach to keyphrase extraction from scholarly documents. In: Proceedings of the 55th Annual Meeting of the Association for Computational Linguistics, pp. 1105–1115. Association for Computational Linguistics (2017). https://doi.org/10.18653/v1/P17-1102
6. Liu, J., Li, P., Zheng, Y., Sun M.: Clustering to find exemplar terms for keyphrase extraction. In: Proceedings of the 2009 Conference on Empirical Methods in Natural Language Processing, pp. 257–266. Association for Computational Linguistics (2009). https://doi.org/10.3115/1699510.1699544
7. Le, T.T.N., Nguyen, M.L., Shimazu, A.: Unsupervised keyphrase extraction: introducing new kinds of words to keyphrases. In: Kang, B.H., Bai, Q. (eds.) AI 2016. LNCS (LNAI), vol. 9992, pp. 665–671. Springer, Cham (2016). https://doi.org/10.1007/978-3-319-50127-7_58
8. Nguyen, C.Q., Hong, L.T., Phan, T.T.: A support vector machines approach to Vietnamese key phrase extraction. In: Proceedings of IEEE-RIVF International Conference on Computing and Communication Technologies, pp. 1–5 (2009). https://doi.org/10.1109/RIVF.2009.5174613
9. Hung, B.T.: Vietnamese keyword extraction using hybrid deep learning methods. In: 2018 5th NAFOSTED Conference on Information and Computer Science (NICS), pp. 412–417 (2018). https://doi.org/10.1109/NICS.2018.8606906
10. Nguyen, C.Q., Phan, T.T.: An ontology–based approach for key phrase extraction. In: Proceedings of the ACL-IJCNLP 2009 Conference Short Papers, pp. 181–184. Association for Computational Linguistics (2009). https://doi.org/10.3115/1667583.1667639
11. AlamiMerrouni, Z., Frikh, B., Ouhbi, B.: Automatic keyphrase extraction: a survey and trends. J. Intell. Inf. Syst. 54(2), 391–424 (2019). https://doi.org/10.1007/s10844-019-00558-9
12. Caragea, C., Bulgarov, F., Godea, A., Gollapalli, S.D.: Citation-enhanced keyphrase extraction from research papers: a supervised approach. In: Proceedings of the 2014 Conference on Empirical Methods in Natural Language Processing (EMNLP), pp. 1435–1446. Association for Computational Linguistics (2014). https://doi.org/10.3115/v1/D14-1150
13. Nguyen, T.D., Luong, M.T.: WINGNUS: keyphrase extraction utilizing document logical structure. In: Proceedings of the 5th International Workshop on Semantic Evaluation, pp. 166–169. Association for Computational Linguistics (2010)
14. Gollapalli, S.D., Li, X.L., Yang, P.: Incorporating expert knowledge into keyphrase extraction. In: AAAI 2017: Proceedings of the Thirty-First AAAI Conference on Artificial Intelligence, pp. 3180–3187 (2017)

15. Al-Zaidy, R., Caragea, C., Giles, C.L.: Bi-LSTM-CRF sequence labeling for keyphrase extraction from scholarly documents. In: WWW 2019: The World Wide Web Conference, New York, pp. 2551–2557. Association for Computing Machinery (2019). https://doi.org/10.1145/330 8558.3313642

16. Hasan, K.S., Ng, V.: Automatic Keyphrase extraction: a survey of the state of the art. In: Proceedings of the 52nd Annual Meeting of the Association for Computational Linguistics, pp. 1262–1273. Association for Computational Linguistics (2014). https://doi.org/10.3115/v1/P14-1119

17. Papagiannopoulou, E., Tsoumakas, G.: A review of keyphrase extraction. In: Wiley Interdisciplinary Reviews: Data Mining and Knowledge Discovery, vol. 10, no. 2 (2020). https://doi.org/10.1002/widm.1339

18. Page, L., Brin, S., Motwani, R., Winograd, T.: The PageRank Citation Ranking: Bringing Order to the Web. Technical Report, Stanford InfoLab (1999). http://ilpubs.stanford.edu:8090/422/

19. Ye, H., Wang, L.: Semi-Supervised Learning for Neural Keyphrase Generation. arXiv preprint, p. arXiv:1808.06773 (2019). https://doi.org/10.48550/arXiv.1808.06773

20. Mahata, D., Kuriakose, J., Shah, R.R., Zimmermann, R.: Key2Vec: automatic ranked keyphrase extraction from scientific articles using phrase embeddings. In: Proceedings of the 2018 Conference of the North American Chapter of the Association for Computational Linguistics: Human Language Technologies, Volume 2 (Short Papers), pp. 634–639. Association for Computational Linguistics (2018). https://doi.org/10.18653/v1/N18-2100

21. Thanh, V., Dat Quoc, N., Dai Quoc, N., Dras, M., Johnson, M.: VnCoreNLP: a Vietnamese natural language processing toolkit. In: Proceedings of the 2018 Conference of the North American Chapter of the Association for Computational Linguistics: Demonstrations, pp. 56–60. Association for Computational Linguistics (2018). https://doi.org/10.18653/v1/N18-5012

22. Tuan-Phong, N., Anh-Cuong, L.: A hybrid approach to Vietnamese word segmentation. In: 2016 IEEE RIVF International Conference on Computing Communication Technologies, Research, Innovation, and Vision for the Future (RIVF), pp. 114–119 (2016). https://doi.org/10.1109/RIVF.2016.7800279

23. Bojanowski, P., Grave, E., Joulin, A., Mikolov, T.: Enriching word vectors with subword information. In: Transactions of the Association for Computational Linguistics, vol. 5, pp. 135–146 (2017). https://doi.org/10.1162/tacl_a_00051

Mixed Multi-relational Representation Learning for Low-Dimensional Knowledge Graph Embedding

Thanh Le[1,2]([✉])[iD], Chi Tran[1,2][iD], and Bac Le[1,2][iD]

[1] Faculty of Information Technology, University of Science, Ho Chi Minh City, Vietnam
{lnthanh,lhbac}@fit.hcmus.edu.vn, tdchi18@clc.fitus.edu.vn
[2] Vietnam National University, Ho Chi Minh City, Vietnam

Abstract. Hyperbolic embeddings have recently received attention in machine learning because of their better ability to handle hierarchical data than Euclidean embeddings. Moreover, Hyperbolic models are also being developed for multi-relational knowledge graphs which contain multiple hierarchical relationships and have achieved promising results, such as MuRP, HyperKG, and ROTH. However, not all data is hierarchical. We also found that most of the geometry models were trained to attain good results on high-dimensional embeddings and low-dimensional embeddings are often of little interest. Besides, neural networks and graph networks models have had an impressive performance. However, they also require a relatively high-dimensional embedding to achieve good results, making them limited to use in large-scale knowledge graphs. To address these issues, in this paper, we introduce a new model named **MuREL** (**Mu**lti-**R**elational **E**uclidean **L**orentzian), which learns a mixed embedding between two spaces, Euclidean and Lorentzian. They are not only suitable for a variety of data types but also work well on low-dimensional embeddings. Experiments on standard benchmark datasets from the task of link prediction show that our model outperforms existing Euclidean and Hyperbolic models, especially at lower dimensionality.

Keywords: Knowledge graph embedding · Link prediction · Multi-relational graph · Mixed-curvature spaces

1 Introduction

Since knowledge graph (KG) was introduced by Google in 2012, it has played an important role in many applications such as search engine, recommender systems and question answering. KG can be seen as a data structure that describes real-world objects and the relationships between them in terms of entities and relations. Thereby we can easily extract and analyze multi-dimensional information. Among the problems on KGs, we are focusing our attention on the link prediction problem [4,10] (LP). The goal of this problem is to predict new links

N. T. Nguyen et al. (Eds.): ACIIDS 2022, LNAI 13757, pp. 428–441, 2022.
https://doi.org/10.1007/978-3-031-21743-2_34

between nodes in KGs. However, real-world KGs are highly complex, this task has many challenges. If it is well solved, information query, anomaly detection, and information security systems will be more effective in practice.

To tackle this challenge, graph embedding can be viewed as an effective approach that shows better performance than other methods such as similarity-based and statistical probability-based methods. The main idea of this approach is to map entities/relations to a low-dimensional embedding space while preserving the semantics and inherent structures of the graph. Although there are many attempts to embed entities in a geometry space such as Euclidean [1,3,16], it seems that this method has reached its limitation. Recent works have taken advantage of Hyperbolic space [1,3] to solve hierarchical data which requires relatively low-dimensional embeddings while still achieving promising results.

In KGs, hierarchical data types often exist. As shown in Fig. 1, Food is a parent node of Fruit, Meat and Vegetable, so it is on at a higher level which forms a hierarchy. Although theories of Hyperbolic geometry have been widely introduced for representation learning, there are still very few models that embed a multi-relational graph data in this space because it is not easy to represent entities to form a kind of hierarchical tree under different relations. Recently, several models that use Hyperbolic space have been introduced including MuRP [1], HyperKG [23], and ROTH [3] which have succeeded in solving this difficulty by embedding entities and relationships in a Poincaré ball model of Hyperbolic space. However, we find that not all data is hierarchical and Euclidean models are better able to handle non-hierarchical data types than Hyperbolic models, suggesting a model that can handle both hierarchical and non-hierarchical data types well even on low-dimensional embeddings.

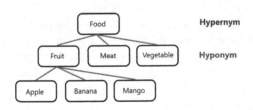

Fig. 1. An illustration of hierarchical structure.

In this paper, we propose MuREL, which embeds a static multi-relational knowledge graph in a mixture space between Euclidean and Lorentzian to handle a wide variety of data types well and get a good performance on low-dimensional embeddings. The model outperforms existing Euclidean and Hyperbolic models on link prediction tasks in the low-dimensional embeddings and has competitive results compared to state-of-the-art geometry-based models on high-dimensional embeddings.

2 Related Works

2.1 Euclidean Embeddings

In 2013, Border et al. proposed TransE [2] that can be seen as the first translation-based model and led to a large number of researchers joining the study of Trans series, in which the representative improved models include TransH [19], TransR [11], and TransD [6].

To summarize, the methods based on the translation model primarily utilizes the relationship between entities, the semantic information contained in the entity and the relation of KG and the structured information. They are simple, have a small number of parameters and can be used on large KGs. However, translations cause these models to fail to capture critical relational components such as symmetry/anti-symmetric, inversion and composition in KGs.

2.2 Complex Embeddings

In 2016, Trouillon et al. introduced ComplEx [17] to solve the problem of link prediction through latent factorization. In ComplEx, complex-valued embeddings are used because the composition of Complex embeddings can deal with many types of binary relations, such as symmetric and antisymmetric.

Based on the idea of Euler's identity $e^{i\theta} = \cos(\theta) + i\sin(\theta)$, Sun et al. proposed RotatE [16] that can model and infer various relationship patterns, including symmetric/antisymmetric, inversion and composition. The model represents entities as complex vectors and relations as rotations in the complex vector space.

Moreover, instead of using traditional complex-valued representations, Zhang et al. introduced QuatE [22] where hypercomplex representations are introduced with more expressive to model entities and relations.

The models in the Complex space often have bad results when running on low-dimensional embeddings compared to the Euclidean and Hyperbolic spaces. However, when run on a high-dimensional embeddings, they will show impressive results, but this will require a long training time and lead to high memory costs.

2.3 Hyperbolic Embeddings

Besides non-hierarchical data, recently, hierarchical data is also getting more attention. Based on recent works, MuRP can be seen as the first method that learns KG embeddings in Hyperbolic space to process hierarchical data by utilizing the Hyperbolic geometry of the Poincaré ball.

However, MuRP is still a translation-based model and it fails to capture logical properties of relations in KGs. Furthermore, using a fixed curvature for Hyperbolic space can lead to suboptimal results. Therefore, based on the advantages of rotation used in Complex space, Chami et al. proposed ROTH, a combination of the Poincaré model in MuRP and rotation operation in RotatE. By this way, learned embeddings were able to capture logical relational patterns better, and the trainable curvature has yielded more optimal results which is a flexible solution to different hierarchical structures.

3 Background

3.1 Riemannian Manifolds

Riemannian manifolds are smooth manifolds M which are equipped with variable inner products g, denoted (M, g) on tangent spaces T_pM (where $p \in M$ is a point). A Riemannian metric assigns to each point p a positive inner product $g_p: T_pM \times T_pM \to \mathbb{R}$, along with a norm $| \cdot |_p: T_pM \to \mathbb{R}$ defined by $|v|_p = \sqrt{g_p(v,v)}$ (where g is a geodesic distance between two points on the manifold).

3.2 Constant-Curvature Spaces

A constant curvature space M_K^d with curvature K and dimension $d \geq 2$ indicates deviation from a straight line or a plane such that $M_K^d = x \in \mathbb{R}^d :< x,x >_n = 1/K$ (where n is the inner product of a specific space). The Riemannian manifolds of constant curvature include Euclidean geometry $\mathbb{E}$, Elliptic geometry $\mathbb{S}_K$ and Hyperbolic geometry $\mathbb{H}_K$. These constant curvature spaces are defined in Eq 1.

$$M = \begin{cases} \mathbb{S}_K^d = x \in \mathbb{R}^{d+1} :< x,x >_2 = 1/K, & \text{for } K > 0, \\ \mathbb{E}^d = \mathbb{R}^d, & \text{for } K = 0, \\ \mathbb{H}_K^d = x \in \mathbb{R}^{d+1} :< x,x >_L = 1/K, & K < 0 \end{cases} \quad (1)$$

where $< x,y >_2 = x_1y_1 + \dots + x_ny_n, \forall x,y \in \mathbb{R}^{d+1}$ is the Euclidean inner product and $< x,y >_L = -x_1y_1 + \sum_{n=2}^{d+1} x_ny_n, \forall x,y \in \mathbb{R}^{d+1}$ is the Lorentz inner product.

3.3 Problem Formulation

A KG is a collection of triples (h, r, t) (where h, t $\in \mathcal{E}$ are head and tail entities and r $\in \mathcal{R}$ is a relation between these entities). If (h, r, t) holds, the head entity h is related to the tail entity t via the relation r. Our goal is to project entities e and relations r into entity embeddings **e** and relation embeddings **r** of a specific space $\mathcal{U}$ while preserving their semantic meaning. Learned KG embeddings are used to predict a head entity (?, r, t) or a tail entity (h, r, ?) of a given triple. Specifically, a score function $\phi : \mathcal{E} \times \mathcal{R} \times \mathcal{E} \to \mathbb{R}$ is defined to learn and compute a score for each triple and the higher the score of the triple, the more likely it is to form a new link. A non-linearity function such as sigmoid is applied to change the score to a binary predicted label $p = \sigma(s) \in [0,1]$ of the triple being true.

3.4 Lorentzian Distance

The squared Lorentzian distance d_L [8] for two points a, b $\in \mathcal{H}_\beta^d$ (where d is the embedding dimension and $\beta \in [10^{-1}, 1]$ is the curvature) is a distance function which has been used in Hyperbolic representation learning. Since we want to create a model that can be used on various of data types of different dimensional embeddings, we will take advantage of learning the diverse representation of the squared Lorentzian distance to combine with the Euclidean distance.

The formula of squared Lorentzian distance is defined:

$$d_L^2(a, b) = \|a - b\|_L^2 = -2\beta - 2 < a, b >_L \qquad (2)$$

where $<a, b>_L = -a_0 b_0 + \sum_{n=1}^{d} a_i b_i \leq -\beta$ is the squared Lorentzian norm, a $= (a_0...a_d) \in \mathcal{H}_\beta^d$, b $= (b_0...b_d) \in \mathcal{H}_\beta^d$ and $<a, b>_L = -\beta$ iff a $=$ b and otherwise.

In addition, given three points x, y, z in a space, a distance function between these points should be considered to satisfy the following axioms:

(a) $d(x, y) \geq 0$ (nonnegativity)
(b) $d(x, y) = 0$ if $x = y$ (nondegeneracy)
(c) $d(x, y) = d(y, x)$ (symmetry)
(d) $d(x, z) \leq d(x, y) + d(y, z)$ (triangle inequality)

For Euclidean distance, it satisfies all these axioms for a metric on $\mathbb{R}^d$, but Lorentzian distance just satisfies the first three axioms for a metric on $\mathbb{H}^d$ [12]. Based on this information, we believe that combining the Euclidean and squared Lorentzian distance functions can create a good enough mixture embedding space for our model to satisfy these axioms to work well on various data types.

4   Methodology

An entity can have many different relations and in each relation, there are many other relations and thus form hierarchical relations, making embedding models difficult to find suitable embedding spaces or scoring functions. For example, in Fig. 1, we have entity Food as a parent node to different types of food (e.g. Fruit, Meat, Vegetable) under the relation Hypernym and Fruit is a parent node to different fruit types (e.g. Apple, Banana, Mango) under the relation Hyponym. Our work aims to create a model to learn a mixed multi-relational embedding that can capture all hierarchies simultaneously and even work well on low-dimensional embeddings. Specifically, we propose MuREL, which combines two distance functions, Euclidean and squared Lorentzian. By this way, we expect our model to have two main benefits: taking advantage of Euclidean distance to handle non-hierarchical datasets well and representing different data types in squared Lorentzian distance to improve the model on hierarchical datasets.

4.1   MuREL

In this work, we build our model based on the form of Eq. 3 mentioned by [1]. Given a triple (h, r, t), the general scoring function is defined as follow:

$$\phi(h, r, t) = -d(\mathbf{h}, \mathbf{t})^2 + b_h + b_t \qquad (3)$$

where $d(\cdot, \cdot)$ is a distance function, $\mathbf{h}, \mathbf{t} \in \mathbb{R}^d$ are embeddings and $b_h, b_t \in \mathbb{R}$ are scalar biases of head and tail entities.

In Euclidean space, Eq. 3 can be rewritten as follow:

$$
\begin{aligned}
\phi_E(e_h, e_r, e_t) &= -d_E(e_h^{(r)}, e_t^{(r)})^2 + b_h + b_t \\
&= -d_E(\exp_0(\mathbf{R}\log_0(e_h)), e_t + e_r)^2 + b_h + b_t
\end{aligned}
\tag{4}
$$

where e_h, $e_t \in \mathbb{E}^d$ are Euclidean embeddings of head and tail entities, $R \in \mathbb{R}^{d \times d}$ is a diagonal relation matrix, $e_r \in \mathbb{E}^d$ is a translation vector of relation e_r, $\exp_0(x) = x$ and $\log_0(x) = x$ are exponential map zero and logarithm map zero in Euclidean space and d_E is the Euclidean distance.

The benefit of using $\mathbf{R}$ is that since it is a diagonal matrix, the number of parameters of the model increases linearly with the number of entities and relations in KGs, allowing it can be applied to large KGs. However, the scoring function in Eq. 4 is limited to handle hierarchical relations. To address this problem, instead of using the Poincaré ball model as previous works did, we apply squared Lorentzian distance to handle hierarchical relations because it has been proven to be a better choice than the Poincaré distance [8].

Based on equation Eq. 2, we define a scoring function for Lorentzian space by rewriting Eq. 3 as follow:

$$
\begin{aligned}
\phi_L(l_h, l_r, l_t) &= -d_L(l_h^{(r)}, l_t^{(r)})^2 + b_h + b_t \\
&= -d_L(\exp_0(\mathbf{R}\log_0(l_h)), l_t + l_r)^2 + b_h + b_t
\end{aligned}
\tag{5}
$$

where l_h, $l_t \in \mathbb{L}^d$ are Lorentzian embeddings of head and tail entities and $l_r \in \mathbb{L}^d$ is a translation vector of relation l_r and d_L is the squared Lorentzian distance.

As mentioned in the background section, the combination of Euclidean and square Lorentzian distances is expected to satisfy four axioms of metrics and be able to handle a variety of different data types. Thus, we combine Eq. 4 and Eq. 5 to produce mixed embeddings for MuREL and the scoring function of it is designed as follows:

$$
\begin{aligned}
\phi(h, r, t) &= -(d_E(e_h^{(r)}, e_t^{(r)})^2 + d_L(l_h^{(r)}, l_t^{(r)})^2) + b_h + b_t \\
&= -(d_E(\exp_0(\mathbf{R}\log_0(e_h)), e_t + e_r)^2 + d_L(\exp_0(\mathbf{R}\log_0(l_h)), l_t + l_r)^2) + b_h + b_t
\end{aligned}
\tag{6}
$$

To get the predicted probabilities of true triples, a non-linearity activation function such as logistic sigmoid is applied to the score function, i.e. $\sigma(\phi(h, r, t))$.

4.2 Training and Optimization

To train MuREL, we use the reciprocal negative sampling technique mentioned in [7] to increase the quantity of data points. First, reciprocal relations (t, r^{-1}, h) are added for every triple (h, r, t). Then, k negative samples are created for each true triple (h, r, t) by corrupting either the tail (h, r, t') or the head (t, r^{-1}, h') entity with a randomly chosen new entity from the training set $\mathcal{E}$. Finally, the model is trained to minimize the Bernoulli negative log-likelihood loss:

$$\mathcal{L}(y,p) = -\frac{1}{N}\sum_{i=1}^{N}(y^{(i)}log(p^{(i)}) + (1 - y^{(i)})log(1 - p^{(i)}))\qquad(7)$$

where p ∈ [0, 1] denotes the predicted probability, y ∈ {0, 1} denotes the binary label of a sample is positive or negative and N is the number of training samples.

For the optimization, instead of using Stochastic Gradient Descent or Riemannian Stochastic Gradient Descent [24], we use Adam and Adagrad because we have realized that the training time and the convergence speed using these two optimization functions are better than SGD and RSGD. Depending on the properties of each dataset, the Adam function can be slightly better than the Adagrad and vice versa.

5   Experiments

5.1   Datasets

To evaluate MuREL, we test its performance for both low and high-dimensional embeddings on the LP task by using two standard benchmark datasets WN18RR [9] and FB15k-237 [9].

Moreover, to demonstrate the effectiveness of this model on hierarchical datasets, we follow Balaževic [1] and use four datasets which are the subsets of NELL-995 [20] (containing 75.492 entities and 200 relations with about 22% of them being hierarchical) with different ratios of hierarchical and non-hierarchical relations. Specifically, Nell-995-h{100, 75, 50, 25} contain 100% (43 hierarchical relations and 0 non-hierarchical relations), 75% (43 and 14), 50% (43 and 43) and 25% (43 and 129) non-hierarchical relations, respectively. The summary of all these datasets are listed in Table 1.

Table 1. Dataset summary.

Dataset	Train	Valid	Test	Entities	Relations
FB15k-237	272.115	17.535	20.466	14.541	237
WN18RR	86.835	3.034	3.134	40.943	11
NELL-995-h100	50.314	3.763	3.746	21.010	43
NELL-995-h75	59.135	4.441	4.389	26.219	57
NELL-995-h50	72.767	5.440	5.393	32.277	86
NELL-995-h25	122.618	9.194	9.187	64.634	172

5.2   Evaluation Metrics

Similar to previous works, we use two ranking-based metrics mean reciprocal rank (MRR) and Hits@k (k ∈ {1, 3, 10}) to evaluate our model. MRR is the average of the reciprocal ranks assigned to the true triple over all evaluation triples. Hits@k measures the probability of correct triples in the top k candidate triples. The filtered setting [2] is also applied as standard evaluation method.

5.3 Settings and Hyperparameters

We conducted all experiments for our model in Pytorch version 1.0.1 on NVIDIA Tesla P100 GPUs. We report the best hyperparameters in both low and high-dimensional settings, including embedding dimension, learning rate, optimizer, batch size, negative samples and epochs for each dataset as follows: {WN18RR: 32, 0.001, Adam, 128, 50, 60}, {WN18RR: 200, 0.001, Adam, 128, 50, 50}, {FB15k-237: 32, 0.05, Adagrad, 128, 50, 250}, {FB15k-237: 200, 0.05, Adagrad, 128, 50, 150}, {NELL-995-h{100, 75, 50, 25}, 40/200, 0.001, Adam, 128, 50, 50}.

5.4 Low-Dimensional Embedding Results

Table 2. Link prediction results on WN18RR and FB15k-237 (d = 32). Best results in bold and second best in underlined. Except our model, all results are taken from [3].

Model	WN18RR				FB15k-237			
	MRR	Hits@1	Hits@3	Hits@10	MRR	Hits@1	Hits@3	Hits@10
RotatE [16]	0.387	0.330	0.417	0.491	0.290	0.208	0.316	0.458
MuRE [1]	0.458	0.421	0.471	0.525	0.313	0.226	0.340	0.489
ComplEx-N3 [7]	0.420	0.390	0.420	0.460	0.294	0.211	0.322	0.463
MuRP [1]	0.465	0.420	0.484	0.544	0.323	0.235	0.353	0.501
REFE [3]	0.455	0.419	0.470	0.521	0.302	0.216	0.330	0.474
ROTE [3]	0.463	0.426	0.477	0.529	0.307	0.220	0.337	0.482
REFH [3]	0.447	0.408	0.464	0.518	0.312	0.224	0.342	0.489
ROTH [3]	0.472	0.428	**0.490**	**0.553**	0.314	0.223	0.346	0.497
MuREL	0.472	**0.433**	0.483	0.552	**0.333**	**0.243**	**0.366**	**0.512**
% Improvement	3.057%	2.850%	2.548%	5.143%	6.390%	7.522%	7.647%	4.703%

Table 2 indicates the link prediction results for WN18RR and FB15k-237 in the low-dimensional embedding. In WN18RR, MuREL obtains the best performance on most metrics, except Hits@3 at 48.3% and Hits@10 at 55.2%. Compared with the baseline model MuRE, MuREL outperforms MuRE for all metrics, the percentage improvement ranges from 2.548% to 5.143%. This proves the effectiveness of the squared Lorentzian distance function when it is employed in a dataset with multiple hierarchies such as WN18RR. MuREL also has quite competitive results compared to ROTH and is only slightly behind in two metrics, hits@3 and hits@10, but much better than RotatE. With the exception of Hits@3 on WN18RR, MuREL outperforms MuRP which is a model using Poincaré distances in Hyperbolic space. This confirms that the Lorentzian squared distance is somewhat better at handling different data types than the Poincaré distance.

In FB15k-237, MuREL is better than the rest of the models for all metrics, which proved that the hypothesis of the combination of two different spaces would provide a mixed space suitable for processing a variety of data types is correct. Compared with MuRE, MuREL has slightly better results on FB15k-237 than on WN18RR, ranging from 4.703% to 7.647%. This indicates that although Euclidean space in MuRE has advantages in dealing with less hierarchical datasets, in order to achieve optimal results, it still needs a sufficiently large embedding dimension that require more time for training.

Table 3 shows the link prediction results for NELL-995-h{100, 75, 50, 25} in the low-dimensional embedding. As we expected, when the amount of non-hierarchical data is increased, the performance between MuRE and MuRP gradually narrows while MuREL retains good results in most datasets.

Table 3. Link prediction results on NELL-995-h{100, 75, 50, 25} (d = 40). Except our model, all results are taken from [1].

Dataset	Model	MRR	Hits@1	Hits@3	Hits@10
NELL-995-h100	MuRE	0.330	0.245	0.366	0.502
	MuRP	0.334	0.261	0.383	0.511
	MuREL	**0.356**	**0.268**	**0.396**	**0.528**
NELL-995-h75	MuRE	0.330	0.246	0.368	0.497
	MuRP	0.345	0.263	0.382	0.506
	MuREL	**0.354**	**0.268**	**0.396**	**0.519**
NELL-995-h50	MuRE	0.342	0.256	0.383	0.510
	MuRP	0.356	0.271	0.399	0.519
	MuREL	**0.363**	**0.276**	**0.406**	**0.532**
NELL-995-h25	MuRE	0.337	0.259	0.374	0.489
	MuRP	0.343	0.266	0.379	0.494
	MuREL	**0.357**	**0.281**	**0.392**	**0.504**

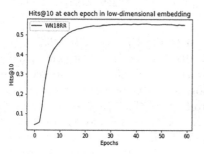

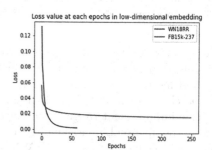

(a) Hits@10 after 60 epochs on WN18RR.

(b) Loss value after 60 epochs on WN18RR and 250 epochs on FB15k-237.

Fig. 2. Hits@10 and loss value in the low-dimensional embedding.

Figure 2a shows the results of Hits@10 after 60 epochs on WN18RR in the low-dimensional embedding. During the first 20 epochs, Hits@10 increases very quickly and achieves a pretty good result. However, in the later epochs, it only increased slightly and tended to start to level off. At this stage, other metrics such as Hits@1, Hits@3 and MRR started to increase more than Hits@10. Figure 2b shows the loss values after 60 epochs on WN18RR and 250 on FB15k-237 in the low-dimensional embedding. In general, the loss value of WN18RR at the early stage is almost twice compared with FB15k-237, but it has a relatively fast convergence speed. In contrast, the loss value of the FB15k-237 at the early stage is relatively good, but it has a slow convergence speed, and it takes until epoch 250 onwards to tend to slow down.

5.5 High-Dimensional Embedding Results

Table 4 indicates the link prediction results for WN18RR and FB15k-237 in the high-dimensional embedding. In WN18RR, MuREL achieves the best performance for all metrics in FB15k-237, but it only get the second best on WN18RR. Compared with MuRE, MuREL outperforms MuRE for all metrics in WN18RR, the percentage improvement ranges from 0.229% to 3.791%. This shows that when the dimensional embedding is large enough, MuRE has narrowed the results significantly more than the low dimensional embedding. For MuRP, although MuREL has better results on all metrics, the performance difference is not significant which shows that the Poincaré distance has quite competitive results with the squared Lorentzian distance on the high dimensional embedding. MuREL is only outperformed by ROTE although it has better results than ROTE on the low dimensional embedding. To account for this, the nature of MuREL is a translation model, so it fails to encode some logical properties of relations such as symmetry, inversion and composition in KGs. The use of rotation operation in Euclidean space of ROTE with a large enough dimensional embedding has overcome this weakness. In addition, we also find that although RotatE is slightly worse than MuREL, its results have increased significantly over the low dimensional embedding. This implies that the rotation

Table 4. Link prediction results on WN18RR and FB15k-237 (d = 200). Except our model, all results are taken from [1,3].

Model	WN18RR				FB15k-237			
	MRR	Hits@1	Hits@3	Hits@10	MRR	Hits@1	Hits@3	Hits@10
TransE [2]	0.226	–	–	0.501	0.294	–	–	0.465
M-WALK [15]	0.437	0.414	0.445	–	–	–	–	–
DistMult [21]	0.430	0.390	0.440	0.490	0.241	0.155	0.263	0.419
ComplEx [17]	0.440	0.410	0.460	0.510	0.247	0.158	0.275	0.428
ConvE [5]	0.430	0.400	0.440	0.520	0.325	0.237	0.356	0.501
MuRE [1]	0.475	0.436	0.487	0.554	0.336	0.245	0.370	0.521
MuRP [1]	0.481	0.440	0.495	0.566	0.335	0.243	0.367	0.518
RotatE [3]	0.476	0.428	0.492	0.571	0.338	0.241	0.375	0.533
ROTE [3]	**0.494**	**0.446**	**0.512**	**0.585**	0.346	0.251	0.381	**0.538**
MuREL	0.482	0.437	0.498	0.575	**0.351**	**0.258**	**0.386**	**0.538**
% Improvement	**1.474%**	**0.229%**	**2.259%**	**3.791%**	**4.464%**	**5.306%**	**4.324%**	**3.263%**

operation can handle different relational patterns well but it depends on the dimensional embedding and the geometry space.

In FB15k-237, MuREL is better than the rest of the models for all metrics and compared with MuRE, MuREL's improvement is still more significant than on WN18RR, ranging from 3.263% to 5.306%. MuRP has worse results than MuRE and ROTE which indicate the limitation of Hyperbolic space when used on datasets with little hierarchy and large number of relations such as FB15k-237. Although both ROTE and RotatE use the same rotation operator, the results of ROTE are slightly better than RotatE. Similarly, ComplEx also has lower results than TransE. From this observation, we can conclude that Euclidean space is still better than Complex and Hyperbolic space on datasets with multiple relations.

Table 5. Link prediction results on NELL-995-h{100, 75, 50, 25} (d = 200). Except our model, all results are taken from [1].

Dataset	Model	MRR	Hits@1	Hits@3	Hits@10
NELL-995-h100	MuRE	0.355	0.266	0.398	0.527
	MuRP	0.360	**0.274**	0.401	0.529
	MuREL	**0.362**	**0.274**	**0.405**	**0.534**
NELL-995-h75	MuRE	0.356	0.269	0.396	0.526
	MuRP	0.359	0.275	0.401	0.524
	MuREL	**0.364**	**0.278**	**0.404**	**0.531**
NELL-995-h50	MuRE	0.372	0.284	0.415	**0.544**
	MuRP	0.371	0.284	0.415	0.539
	MuREL	**0.373**	**0.285**	**0.416**	0.543
NELL-995-h25	MuRE	0.365	0.287	**0.404**	0.515
	MuRP	0.359	0.282	0.397	0.507
	MuREL	**0.366**	**0.288**	**0.404**	**0.517**

Table 5 shows the link prediction results for NELL-995-h{100, 75, 50, 25} in the high-dimensional embedding. We can see that for datasets with none or few non-hierarchical relations such as NELL-995-h100 and NELL-995-h75, MuRP has competitive results with MuREL. However, for the remaining two datasets where the amount of non-hierarchical relations is increased, MuRE, which uses embeddings in Euclidean space, has started to outperform MuRP on some metrics and has competitive results with MuREL.

Figure 3a shows the results of Hits@10 after 50 epochs on WN18RR in the high-dimensional embedding. During the first 20 epochs, Hits@10 also increases very quickly and achieves better results than the low-dimensional embedding. It begins to slow down in later stages, and the remaining metrics will tend to increase faster. Figure 3b shows the loss values after 50 epochs on WN18RR and 150 epochs on FB15k-237 in the high-dimensional embedding. Compared to

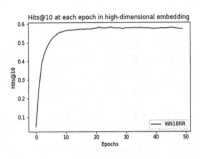

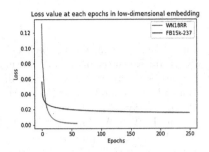

(a) Hits@10 after 50 epochs on WN18RR.

(b) Loss value after 50 epochs on WN18RR and 250 epochs on FB15k-237.

Fig. 3. Hits@10 and loss value in the high-dimensional embedding.

the low-dimensional embedding, the loss value at the early stage on the high-dimensional embedding for both datasets is slightly lower. The loss value at the early stage on the WN18RR is twice compared to FB15k-237, and it still has a much faster convergence rate. However, the main difference lies in the fact that the convergence speed is now faster in both datasets than running on the low-dimensional embedding.

6   Conclusion and Future Research Directions

In this paper, we introduced MuREL, a mixed model that utilizes two distance Euclidean and squared Lorentzian to better handle hierarchical datasets for both low and high-dimensional embeddings. We observe that the rotation operation in two models RotatE and ROTE just give good results when it is used with high-dimensional embeddings and also depends on the geometry space used. Therefore, the model that we have presented through the experiments on the low-dimensional embedding has shown quite promising results compared with the previous state-of-the-art models and imply that the combination of two distance functions belonging to two different geometric spaces has successfully handled a variety of data types in a relatively low-dimensional embedding.

Future directions for this work may use other geometric operations instead of rotation operation, and recently, Graph Convolutional Networks have shown quite impressive results in the link prediction task with some representative models such as R-GCN [13], SACN [14] and CompGCN [18]. Therefore, we can look into incorporating it into geometric-based models to achieve better results.

Acknowledgement. This research is funded by the University of Science, VNU-HCM, Vietnam under grant number CNTT 2022-2 and Advanced Program in Computer Science.

References

1. Balazevic, I., Allen, C., Hospedales, T.: Multi-relational poincaré graph embeddings. In: Advances in Neural Information Processing Systems, 32 (2019)
2. Bordes, A., Usunier, N., Garcia-Duran, A., Weston, J., Yakhnenko, O.: Translating embeddings for modeling multi-relational data. In: Advances in Neural Information Processing Systems, 26 (2013)
3. Chami, I., Wolf, A., Juan, D. C., Sala, F., Ravi, S., Ré, C.: Low-dimensional hyperbolic knowledge graph embeddings. arXiv preprint arXiv:2005.00545 (2020)
4. Chen, Z., Wang, Y., Zhao, B., Cheng, J., Zhao, X., Duan, Z.: Knowledge graph completion: a review. IEEE Access **8**, 192435–192456 (2020)
5. Dettmers, T., Minervini, P., Stenetorp, P., Riedel, S.: Convolutional 2d knowledge graph embeddings. In Proceedings of the AAAI Conference on Artificial Intelligence, vol. 32, No. 1, April 2018
6. Ji, G., He, S., Xu, L., Liu, K., Zhao, J.: Knowledge graph embedding via dynamic mapping matrix. In: Proceedings of the 53rd Annual Meeting of the Association for Computational Linguistics and the 7th International Joint Conference on Natural Language Processing, volume 1: Long papers), pp. 687–696, July 2015
7. Lacroix, T., Usunier, N., Obozinski, G.: Canonical tensor decomposition for knowledge base completion. In: International Conference on Machine Learning, pp. 2863–2872. PMLR, July 2018
8. Law, M., Liao, R., Snell, J., Zemel, R.: Lorentzian distance learning for hyperbolic representations. In: International Conference on Machine Learning, pp. 3672–3681. PMLR, May 2019
9. Le, T., Huynh, N., Le, B.: RotatHS: rotation embedding on the hyperplane with soft constraints for link prediction on knowledge graph. In: Nguyen, N.T., Iliadis, L., Maglogiannis, I., Trawiński, B. (eds.) ICCCI 2021. LNCS (LNAI), vol. 12876, pp. 29–41. Springer, Cham (2021). https://doi.org/10.1007/978-3-030-88081-1_3
10. Le, T., Nguyen, H., Le, B.: A survey of the link prediction on static and temporal knowledge graph. J. Inf. Technol. Commun. **2021**(2), 51–84 (2021)
11. Lin, Y., Liu, Z., Sun, M., Liu, Y., Zhu, X.: Learning entity and relation embeddings for knowledge graph completion. In: Twenty-Ninth AAAI Conference on Artificial Intelligence, February 2015
12. Ratcliffe, J. G., Axler, S., Ribet, K.A.:. Foundations of hyperbolic manifolds, vol. 149, pp. xii+-747. Springer, New York (1994)
13. Schlichtkrull, M., Kipf, T.N., Bloem, P., van den Berg, R., Titov, I., Welling, M.: Modeling relational data with graph convolutional networks. In: Gangemi, A., Navigli, R., Vidal, M.-E., Hitzler, P., Troncy, R., Hollink, L., Tordai, A., Alam, M. (eds.) ESWC 2018. LNCS, vol. 10843, pp. 593–607. Springer, Cham (2018). https://doi.org/10.1007/978-3-319-93417-4_38
14. Shang, C., Tang, Y., Huang, J., Bi, J., He, X., Zhou, B.: End-to-end structure-aware convolutional networks for knowledge base completion. In: Proceedings of the AAAI Conference on Artificial Intelligence, vol. 33, No. 01, pp. 3060–3067, July 2019
15. Shen, Y., Chen, J., Huang, P. S., Guo, Y., Gao, J.: M-walk: Learning to walk over graphs using monte carlo tree search. Advances in Neural Information Processing Systems, 31 (2018)
16. Sun, Z., Deng, Z. H., Nie, J. Y., Tang, J.: Rotate: Knowledge graph embedding by relational rotation in complex space. arXiv preprint: arXiv:1902.10197 (2019)

17. Trouillon, T., Welbl, J., Riedel, S., Gaussier, É., Bouchard, G.: Complex embeddings for simple link prediction. In: International Conference on Machine Learning, pp. 2071–2080. PMLR, June 2016
18. Vashishth, S., Sanyal, S., Nitin, V., Talukdar, P.: Composition-based multi-relational graph convolutional networks. arXiv preprint arXiv: 1911.03082 (2019)
19. Wang, Z., Zhang, J., Feng, J., Chen, Z.: Knowledge graph embedding by translating on hyperplanes. In: Proceedings of the AAAI Conference on Artificial Intelligence, vol. 28, no. 1, June 2014
20. Xiong, W., Hoang, T., Wang, W.Y.: Deeppath: a reinforcement learning method for knowledge graph reasoning. arXiv preprint arXiv: 1707.06690 (2017)
21. Yang, B., Yih, W.T., He, X., Gao, J., Deng, L.: Embedding entities and relations for learning and inference in knowledge bases. arXiv preprint arXiv: 1412.6575 (2014)
22. Zhang, S., Tay, Y., Yao, L., Liu, Q.: Quaternion knowledge graph embeddings. Advances in neural information processing systems, 32 (2019)
23. Kolyvakis, P., Kalousis, A., Kiritsis, D.: Hyperbolic knowledge graph embeddings for knowledge base completion. In: Harth, A., Kirrane, S., Ngonga Ngomo, A.-C., Paulheim, H., Rula, A., Gentile, A.L., Haase, P., Cochez, M. (eds.) ESWC 2020. LNCS, vol. 12123, pp. 199–214. Springer, Cham (2020). https://doi.org/10.1007/978-3-030-49461-2_12
24. Bonnabel, S.: Stochastic gradient descent on Riemannian manifolds. IEEE Trans. Autom. Control **58**(9), 2217–2229 (2013)

Learning to Map the GDPR to Logic Representation on DAPRECO-KB

Minh-Phuong Nguyen[1]([⊠]), Thi-Thu-Trang Nguyen[1], Vu Tran[2],
Ha-Thanh Nguyen[3], Le-Minh Nguyen[1], and Ken Satoh[3]

[1] Japan Advanced Institute of Science and Technology, Nomi, Japan
{phuongnm,trangttn,nguyenml}@jaist.ac.jp
[2] The Institute of Statistical Mathematics, Tachikawa, Japan
vutran@ism.ac.jp
[3] National Institute of Informaticsi, Tokyo, Japan
{nguyenhathanh,ksatoh}@nii.ac.jp

Abstract. General Data Protection Regulation (GDPR) is an important framework for data protection that applies to all European Union countries. Recently, DAPRECO knowledge base (KB) which is a repository of if-then rules written in LegalRuleML as a formal logic representation of GDPR has been introduced to assist compliance checking. DAPRECO KB is, however, constructed manually and the current version does not cover all the articles in GDPR. Looking for an automated method, we present our machine translation approach to obtain a semantic parser translating the regulations in GDPR to their logic representation on DAPRECO KB. We also propose a new version of GDPR Semantic Parsing data by splitting each complex regulation into simple subparagraph-like units and re-annotating them based on published data from DAPRECO project. Besides, to improve the performance of our semantic parser, we propose two mechanisms: *Sub-expression intersection* and *PRESEG*. The former deals with the problem of duplicate subexpressions while the latter distills knowledge from pre-trained language model BERT. Using these mechanisms, our semantic parser obtained a performance of 60.49% F1 in sub-expression level, which outperforms the baseline model by 5.68%.

Keywords: GDPR · Semantic parsing · Logic representation

1 Introduction

General Data Protection Regulation[1] is the regulation on the protection of EU citizens regarding the processing of personal data on the free flow within the European Union and on the transfer to third countries and international organizations. GDPR introduces a number of obligations that public administrations,

---

[1] https://gdpr-info.eu/.

M.-P. Nguyen and T.-T.-T. Nguyen—These authors contributed equally to this work.

N. T. Nguyen et al. (Eds.): ACIIDS 2022, LNAI 13757, pp. 442–454, 2022.
https://doi.org/10.1007/978-3-031-21743-2_35

enterprises and non-profit organizations need to observe when processing personal data. Because manual legal compliance checking is a time-consuming task, there has been an increasing interest in research on legal reasoning tools to automate the check.

When GDPR was first issued, there is a lack of logic representation for this document that can suffice to automate legal reasoning. Filling that lack, the DAPRECO knowledge base [12] which is a repository of if-then rules representing the regulations in GDPR has been introduced. DAPRECO KB uses the Privacy Ontology (PrOnto) [9] which models legal concepts in GDPR and also provides additional concepts which are needed to represent the semantics of the legal rules in GDPR. Following the Input/Output framework for legal reasoning [14], an ordinary legal rule in DAPRECO KB is usually represented by one constitutive norm (Entailment) and one regulative norm (Obligation or Permission) while a complex rule may have more norms of one or both types. The current version of DAPRECO KB is constructed manually and does not cover all articles in GDPR. This paper presents a machine translation approach to build a semantic parser which can automatically convert the regulations in GDPR to their corresponding logic representation on DAPRECO KB.

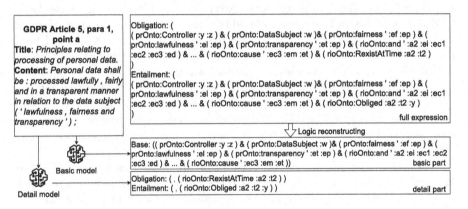

Fig. 1. Overview of Logic mapping GDPR on DAPRECO KB using sub-expression intersection mechanism.

The challenge of constructing a semantic parser for logic representation on DAPRECO KB comes from its constraints in the legal domain. For example, for reasoning, the explicit logic representation of time conditions is a frequently mandatory requirement in legal terms [12]. These conditions are repeated in constituent formulae (e.g. Obligation and Entailment formulae in Fig. 3) of GDPR expression in DAPRECO KB.

It is difficult to map directly a complex GDPR rule into its original logic expression consisting of multiple logic formulae. To approach the challenging task, we split a complex GDPR statement into simple legal rules and then build a model to generate logical representation of these simpler rules, inspired by the research on Semantic Parsing and Question Answering dealing with complex sentences [6,18]. As a result, we constructed two versions of the GDRP Semantic Parsing dataset. The first version of the dataset (**Original data**) consisting of 275 samples is constructed from the current version of the DAPRECO KB. One sample is a pair of a GDPR statement and its logic expression. For example, the logic representation of Article 5, paragraph 1, point a is a full expression (Fig. 1) consisting of 2 logic formulae (Obligation and Entailment). Each logic formula is an if-then rule which is a combination of sub-expressions (e.g., (prOnto:Controller :y :z) is a sub-expression). Similar to DAPRECO KB, a complex GDPR statement in the Original data is represented by more than two logic formulae and the number of these formulae can vary. To assist in solving the task of mapping a complex GDPR rule into its logic expression, we constructed a second version of dataset called **Relaxation data**. In this version, a complex sample is split into simple subparagraph-like units.

Concerning our semantic parser, we use machine translation approach and propose two mechanisms to improve the performance. Based on our observation on the GDPR expressions, for one GDPR statement there are a lot of duplicate sub-logic expressions in its logic formulae. To avoid them, we propose Sub-expression intersection mechanism. We separate shared sub-logic expressions (basic part) from the remaining (detail part) collected from the logic formulae of GDPR statements and use them to build two Transformer-based Neural Machine Translation (NMT) models [15] to generate the basic part and the detail part, respectively. For the basic part, we propose PRESEG (*i.e.*, **P**redicate **RE**trieval & **S**ub-**E**xpression **G**eneration) mechanism which consists of two steps. First, we utilize the power of the pre-trained language model BERT [3] to retrieve well-relevant predicates. After that, we apply a Transformer-based NMT model to generate sub-expressions for each predicate instead of generating the logic representation for the whole GDPR statement, which results in a more correct syntax of logic representation.

We evaluate our model on two versions of the dataset as mentioned above. We performed five experiment runs as the progress of developing our semantic parser. The proposed model achieves a performance of 60.49% F1 in the sub-expression level, which outperforms its baseline model.

2   Related Work

GDPR text extension (GDPRtEXT) [10] provides a hierarchy of concepts present in the GDPR. For example, identified data types such as personal data and anonymous data are defined as sub-classes of the common term Data. However, the GDPRtEXT does not really model the norms and legal axioms (e.g., the actions performed by the processor, the obligations of the controller and the rights of the data subject). Moreover, GDPRtEXT does not foster FRBR information for managing the versioning of the legal text over time and consequently the changes of the legal concepts due to modifications in the legal system.

GDPRov [11] is an OWL2 ontology for describing the provenance of data and consent life-cycles in the light of the linked open data principles such as fairness and trust. It extends the existing linked open data provenance ontologies - PROV ontology and ontology for Provenance and Plans. GDPRvo uses these provenance ontologies to express a data-flow model that can trace how consent and data are used by using GDPR terminology.

The Open Digital Rights Language (ODRL)[2] provides predicates and classes for managing obligations, permissions, prohibitions, but several parts of the deontic logic are missing (e.g., right and penalty classes). ODRL is good for modeling simple policies, but it is quite limited to manage the complex organization of the legal rules (e.g., exceptions in the constitutive rules or in prescriptive rules).

Privacy Ontology (PrOnto) [9] is designed in such a way that it models the essential legal concepts in the GDPR. PrOnto has been developed following thorough ontology development methodology called MeLON. PrOnto reuses existing ontologies: A Light Legal Ontology on TLCs (ALLOT), the FRBR (Functional Requirements for Bibliographic Records) ontology, the LKIF (Legal Knowledge Interchange Format) core, the Publishing Workflow Ontology, Time-indexed Value in Context and Time Interval. However, using the PrOnto alone only allows for basic reasoning, it is not sufficient to assess compliance checking.

In 2020, Livio Robaldo et al. [12] introduced the DAPRECO knowledge base, which is a repository of rules codified in LegalRuleML (an XML standard for representing legal documents) [8]. The rules represent the provisions of the GDPR. The DAPRECO knowledge base was built upon the PrOnto and added additional constraints in the form of if-then rules formalized in reified Input/Output logic [13]. To date, the DAPRECO knowledge base is the biggest knowledge base in LegalRuleMl, which allows complicated legal reasoning and suffice to check compliance. The DAPRECO knowledge base is used in this work for this particular reason.

Concerning automated GDPR compliance checking, Mousavi et al. [7] proposed a tool (KnIGHT) for mapping privacy policies to relevant GDPR articles. In 2021, Aberkane et al. [1] presented a systematic mapping study to explore the existing approaches for automating GDPR compliance in requirements engineering.

3   Methodology

To deal with the task of mapping a GDPR statement into its logic representation on DAPRECO KB, we apply the solution of the Semantic Parsing task in Natural Language Processing (NLP) [2,4,5,16,17]. With the approach using Intent Classification and Slot Filling [2,16], each logic representation is considered a semantic frame with the defined set of intent and slot information. This method requires annotated data to contain the label of slot information and intent type for each sample, which is difficult to extract from the GDPR data. A more flexible approach is using Neural Machine Translation (NMT) [4,5]. By

---

[2] https://www.w3.org/TR/odrl-model/.

considering source sentences and logic representations as source and target languages in the machine translation system, the semantic parser can be adapted to any logic representation syntax. Using this method, we build a semantic parser that can map GDPR statements into their logic representations on DAPRECO KB. Besides, we propose *PRESEG* mechanism which incorporates a pre-trained language model (e.g. BERT) and a NMT model to utilize the advantages of both: *knowledge distillation* capacity and *flexible generation*.

3.1   Baseline NMT Model

We use Transformer architecture [15] as our strong baseline model because this model is shown to be effective in learning long-range dependency, which is appropriate for a long document. In this architecture, the input is a sequence of words in a GDPR statement ($\mathbf{x} = [x_1, x_2, ..., x_{|S|}]$ where $|S|$ is the sentence length). This input is embedded by an Embedding Layer, and feed-forward via $N\times$ Transformer Encoder stacked layers to get the final vector representation. After that, the Transformer decoder based on the attention mechanism is used to decode each token in the expression ($\mathbf{y} = [y_1, y_2, ..., y_{|E|}]$ where $|E|$ is the number of tokens in the expression).

3.2   Sub-expression Intersection Mechanism

Based on our observation on the GDPR expressions, for one GDPR statement there are a lot of duplicate sub-expressions in its logic formulae (sub-formulae). For example, the GDPR expression in Table 1, all sub-formulae *f1* to *f4* share the sub-expression (prOnto:DataSubject :w). To avoid them, we propose *Sub-expression Intersection* mechanism to split a GDPR expression into 2 parts: the basic part contains common sub-expressions among logic formulae and the detail part contains the remaining sub-expressions. By using this mechanism, the GDPR expression is shortened but still preserves all semantic information.

3.3   PRESEG Mechanism

In this mechanism, we utilize the power of the pre-trained language model BERT [3] to support the expression parsing process. The parsing process is split into two steps (Fig. 2):

- *Predicate Retrieval.* This step uses a BERT retrieval model to generate a set of predicates related to an input GDPR statement ($\boldsymbol{x}$). In detail, we construct a vocabulary of predicates ($\mathcal{V}^{predicate} = \{p_i\}$) from the training data then fine-tune the pre-trained BERT model to predict the relation between text input and each predicate $\langle \boldsymbol{x}, p_i \rangle$.
- *Sub-expression Generation.* With each predicate generated from the previous step, we concatenate it with the GDPR statement to generate corresponding sub-expressions using the NMT model. After that, all generated sub-expressions are combined to present the final expression.

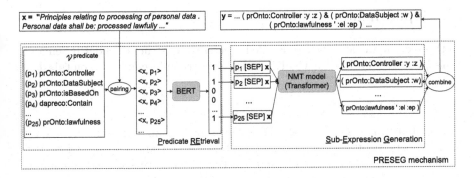

Fig. 2. PRESEG mechanism on GDPR Article 5, para 1, point a.

Predicate Retrieval. Given the predicate vocabulary ($\mathcal{V}^{predicate} = \{p_i\}$), we retrieve all predicates that are relevant to an input GDPR statement ($\mathbf{x}$). Following the task next-sentence-prediction (NSP) [3], for each pair $\langle x, p_i \rangle$, we generate an input string by a template *"[CLS] sentence1 [SEP] sentence2 [SEP]"* where *sentence1, sentence2* are the predicate (p_i) and the GDPR statement (x), respectively. We use the pre-trained BERT embedding to obtain the representation for this input. Next, we forward the hidden representation of *[CLS]* token ($\mathbf{h}^{[CLS]}$) to a Linear layer. Finally, we use a softmax function to calculate the probability of how relevant the predicate is to the GDPR statement. Mathematically, we use the formulae as follows:

$$\mathbf{h}^{[CLS]}, \mathbf{h}^{others} = \text{BERT}(\langle p_i, \mathbf{x} \rangle) \tag{1}$$

$$\mathbf{h}^{out} = \mathbf{W}^{out}\mathbf{h}^{[CLS]} + \mathbf{b}^{out} \tag{2}$$

$$P(\langle p_i, \mathbf{x} \rangle) \propto \text{softmax}(\mathbf{h}^{out}) \tag{3}$$

where $\mathbf{W}^{out}$, $\mathbf{b}^{out}$ are learnable parameters. The loss function is *Cross-Entropy*.

Sub-expression Generation. In this step, we construct an NMT model to generate sub-expressions for each predicate generated from the previous step based on the GDPR statement. For example (Fig. 2), the text input ($\mathbf{x}$ - the GDPR Article 5, para 1, point a) and the generated predicate ($p_1 = $ prOnto:Controller) are concatenated by *"[SEP]"* token. The NMT model generates the corresponding sub-expression (prOnto:Controller :y :z). The architecture of the NMT model in this step is the same as the baseline NMT model based on Transformer's architecture. Compared with the baseline model, the NMT model in this step is trained to generate sub-expression instead of the full expression. In this way, we got two advantages: the number of generated samples is 10 times larger than that of the baseline model, and the syntax of output expression is more correct. Finally, we unite all sub-expression to recover the full expression.

4   Experiments

In this section, we describe the detailed process to construct the GDPR Semantic Parsing dataset on DAPRECO KB and the experiments conducted on this data.

4.1   Datasets

We created two versions of the GDPR Semantic Parsing dataset: *Original data* - this version contains pairs of the GDPR statement and its expression (logic formulae) from [12]; *Relaxation data* - in this version, we split and re-annotate the complex samples to improve the consistency in all samples. By using the Relaxation data, although the semantic parser misses the target automatically mapping the whole GDPR statement into its expression, it has a meaning in verifying the improvement capacity of decomposing complex GDPR statement approach for the futures works or building a suggestion system for logic annotators on DAPRECO KB with higher accuracy.

Original Data. To construct the dataset, we crawl the content of the articles (GDPR statements) from General Data Protection Regulation homepage and the logic representations of the regulation in GDPR from the DAPRECO repository[3] [12]. Then we map each structural item in the GDPR article including the paragraph, the point in the paragraph to the corresponding logic formulae in the DAPRECO knowledge base. The mapping process is shown in Table 1. In detail, the mapping process between the GDPR statement and its expression is based on metadata information. For example, the paragraph 3 of article 37 is mapped to ref ID="GDPR:art_37__para_3" in published DAPRECO repository. Although the logic representations in fact for DAPRECO KB are written in XML syntax [12], we use the text version of these expressions because they have exactly the same semantic meaning and can be trivially converted back and forth. The final dataset has 275 samples: 198 normal samples and 97 complex samples.

Relaxation Data. The original data contains 97 complex samples. A complex one is defined as having more than two logic formulae in its logic expression. For example, the GDPR expression of Article 37, paragraph 3 (Table 1), has 4 logic formulae ($f1$ - $f4$). Another example, the GDPR expression of Article 35, paragraph 3, subparagraph 1, point b, has 20 logic formulae. Because the number of logic formulae for each GDPR statement varies, it is difficult for a semantic parser can generalize this inconsistency with limited samples like the original data. To assist in solving the task of mapping a complex GDPR rule into its logic expression, we constructed a *Relaxation* version of this data in which the complex sample is split into simple subparagraph-like units and re-annotated. For example, the complex sample in Table 1 is split into two new samples: the former refers to the controller while the latter refers to the processor. Finally,

---

[3] https://github.com/dapreco/daprecokb/blob/master/gdpr/rioKB_GDPR.xml.

Table 1. GDPR Mapping Example. This table is split into 2 parts, the upper part contains metadata information of each GDPR statement and its corresponding expression, the lower part shows their contents.

GDPR statement	GDPR expression
Article: 37	
Paragraph: 3	`<lrml:LegalReference refersTo="gdprC4S4A37P3-`
Sub-para: None	`ref" refID="GDPR:art_37__para_3" />`
Point: None	
Title: Art. 37 GDPR Designation of the data protection officer	*(f1)* Permission: (... (prOnto:DataSubject :w) & (prOnto:Controller :y2 :z) , (rioOnto:RexistAtTime :a2 :t2) ...)
Content: Where the controller or the processor is a public authority or body, a single data protection officer may be designated for several such authorities or bodies, taking account of their organisational structure and size.	*(f2)* Entailment: (... (prOnto:DataSubject :w) & (prOnto:Controller :y2 :z) , (rioOnto:Permitted :a2 :t2 :w) ...)
	(f3) Permission: (... (prOnto:DataSubject :w) & (prOnto:Processor :x1) , (rioOnto:RexistAtTime :a2 :t2) ...)
	(f4) Entailment: (... (prOnto:DataSubject :w) & (prOnto:Processor :x1) , (rioOnto:Permitted :a2 :t2 :w) ...)

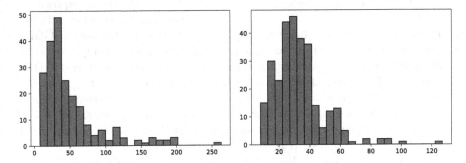

Fig. 3. Histogram comparison of number of sub-expressions on GDPR Semantic Parsing dataset between two versions: Original (left) and Relaxation (right).

the new dataset (relaxation version) consists of 390 samples: 198 initial ordinary samples and 192 new samples by splitting 97 complex ones. We split randomly 390 samples into training and testing datasets (the split ratio is 80:20). In Fig. 3, we show the histogram comparison of the number of sub-expressions between Original and Relaxation versions. In the original data, there are many complex samples having numbers of sub-expressions larger than 100, while the number of sub-expressions in the relaxation version is focused in a range less than 60.

Table 2 shows the statistic of the two versions of the datasets. The statistic shows that the GDPR semantic parsing version relaxation has 40% more samples than the original version with the shorter target expression.

Table 2. GDPR semantic parsing data analysis

Number of samples		Original	Relaxation
		275	390
GDPR statement	Average number of words	89	74
	Number of words	[17 - 1130]	[17 - 1130]
GDPR expression	Average number of tokens	401	275
	Number of tokens	[58 - 2023]	[18 - 1371]

4.2   Experimental Settings

For all experiments below, we set hyperparameter values according to the best setting of IWSLT 14 English-German NMT dataset [15], with 3 Transformer layers (because the size of this dataset is small), hidden size is 512, 4 heads, and 200 training epochs. We conduct five experiments as follows:

- **Setting 1**: (Baseline Model) We employed a single NMT model to map the GDPR statement into its expression. In this setting, all logic formulae for each GDPR statement are concatenated to present the target GDPR expression.
- **Setting 2**: In this setting, we aim to evaluate the effectiveness of *Sub-expression intersection* mechanism. We applied this mechanism on GDPR expressions to get the basic and the detailed parts. Then we employed a single NMT model for parsing. For each GDPR statement, the target GDPR expression is the concatenation of its basic part and detail part.
- **Setting 3**: In this experiment, we aim to evaluate the effectiveness of *sub-expression intersection* by generating separately two logic parts. Instead of employing a single NMT model as experiment 2, we employed two NMT models to learn the basic part and detail part separately.
- **Setting 4** (Our Semantic Parser): In this experiment, we aim to evaluate the effectiveness of *PRESEG* mechanism. We used *Sub-expression intersection* mechanism similar to experiment 3. Then we employed *PRESEG* mechanism for the basic part.

4.3   Experimental Results and Discussion

Main Result. Table 3 shows the performance on the Original test set and the Relaxation test set for our experiments, respectively. On the Original data, the performance of employing a single NMT model to learn full logic representation directly is average, $F1 = 37.16\%$ (setting 1). When the complex data samples were split and re-annotated (relaxation version), the performance increased by

17.39% ($F1 = 54.55\%$ in setting 1). Similar to experiment 1, the performance of remaining experiments when learning logic representation on Relaxation data also increased compared to on Original data. These improvements show that the GDPR expressions in Relaxation data are more consistent than the original version, which makes the model more generalizable. These results show the effectiveness of decomposing the complex GDPR statement into simple ones.

Using the *Sub-expression intersection* mechanism, on the full expression of Original data, the performance in experiments 2 and 3 increased respectively by 14.25% and 3.27% compared to the baseline model (setting 1). Similar to the setting on the original version, the F1-score on Relaxation data increased by at least 3% compared to the baseline model. We argue that this mechanism filters the duplicate sub-expressions in the GDPR sub-formulae; using this mechanism can reduce the complexity of the GDPR logic representation but still preserve the original semantic information. Besides, the performance in setting 2 when using this mechanism with a single NMT model (end-to-end model) is better than using separately two NMT models for the basic and detail parts because the end-to-end model utilizes the relation between these parts to improve parsing logic representation.

Using the *PRESEG* mechanism on Relaxation data (setting 4 Table 3), our proposed model outperforms all previous experiments with $F1 = 64.22\%$ in the basic part (increased by 1.92% compared to experiment 2 and increased by 2.9% when compared to setting 3). It boosts the performance on full logic representation to 60.23%, F1 increased by 2.47% compared to experiment 2 and increased by 2.93% compared to setting 3. However, this mechanism did not show improvement on the original version. The reason is that the GDPR expressions in the Original data are inconsistent.

Error Analysis. An analysis of mispredicted logic representation in the test set showed three main causes of generating errors relating to variables in sub-expressions. The variable names, which are named by human annotators, are usually not meaningful names. For example, in the GDPR expression of Article 5, paragraph 1, point a, x, ep are variable names for the controller, and the predicate `PersonalDataProcessing`, respectively. With limited data, that is not easy for a model to learn the way of naming variables if the annotators do not annotate variable names consistently. In addition, errors also occur in predicting the position of predicates. Instead of correctly predicting that a predicate belongs to the if statement, the model sometimes predicted that the predicate belongs in the then statement, and vice versa. Moreover, the model could not identify the predicates which rarely appear in logic representations.

The Need for Correct Variables in Sub-expression Component. One natural question is how often our semantic parser fail to generate the correct variable. To answer this, we conducted an oracle experiment (see Table 4). For oracle variable name, the decoder is forced to generate the correct variable name at each sub-logic expression. Likewise, oracle variable forces the decoder to choose the correct number of variables and the variable names in each sub-logic expression.

Table 3. Result of our experiments on GDPR Semantic Parsing data. The notation "n/a" indicates that the measurement method is not applicable.

		Setting 1	Setting 2	Setting 3	Setting 4
Single NMT model		✓	✓		
Multi NMT models				✓	✓
+ Sub-expression Intersection			✓	✓	✓
+ PRESEG					✓
Original data	Full Expression	37.16	**51.41**	40.43	38.59
	Basic Part	n/a	**68.41**	47.20	44.73
	Detail Part	n/a	**14.69**	1.82	1.82
Relaxation data	Full Expression	54.55	57.76	57.30	**60.23**
	Basic Part	n/a	62.30	61.32	**64.22**
	Detail Part	n/a	**32.27**	31.21	31.21

Table 4. F1 on the test set given an oracle providing correct number of variables and variable names in each sub-logic expression

	Basic part	Detail part	Full expression
Our semantic parser	64.22	31.21	60.23
Oracle variable name	65.21 (+0.99)	69.82 (+38.61)	64.19 (+3.96)
Oracle variable	69.05 (+4.83)	69.82 (+38.61)	67.72 (+7.49)

With just oracle "variable names", the F1 for the basic part is 65.21%, for the detail part is 69.82%, and for full logic representation is 64.19%. With oracle "variable" (containing oracle "number of variables" and oracle "variable names"), we observed a F1 of 69.05% for the basic part, 69.82% for the detail part, and 67.72% for full expression. This verifies that if the model can learn well the constraints between variables in each sub-logic expression according to each predicate, the performance of model can increase a lot. Therefore the problem of generating correct constraint between variable and predicate requires important future work.

5   Conclusion and Future Work

In this paper, we propose an effective semantic parser for mapping GDPR to corresponding logic representation on DAPRECO KB. Firstly, we create Relaxation data for this task by splitting and re-annotating the complex regulation. Secondly, we introduce Sub-expression intersection mechanism to solve the problem of generation of duplicate sub-logic expressions. Last but not least, we demonstrate how PRESEG mechanism utilized the power of the pre-trained language model BERT and the Transformer-based NMT model to generate the basic part in the logic representations. Empirically, our proposed model allows us to gain significant improvement on mapping the GDPR statement to its logic representation when compared to baseline model. Our semantic parser will be beneficial in tasks such as mapping other legal rules to logic representations.

In the future, we look forward to improving the architecture design by considering the constraint between the variable and predicate in each sub-expression.

Acknowledgment.. This work was supported by JSPS Kakenhi Grant Number 20H04295, 20K20406, and 20K20625.

References

1. Aberkane, A.-J., Poels, G., Broucke, S.V.: Exploring automated GDPR-compliance in requirements engineering: a systematic mapping study. IEEE Access **9**, 66542–66559 (2021)
2. Chen, Q., Zhuo, Z., Wang, W.: Bert for joint intent classification and slot filling. arXiv preprint arXiv:1902.10909 (2019)
3. Devlin, J., Chang, M.-W., Lee, K., Toutanova, K.: BERT: pre-training of deep bidirectional transformers for language understanding. In: Proceedings of the 2019 Conference of the North American Chapter of the Association for Computational Linguistics: Human Language Technologies, Volume 1 (Long and Short Papers), Minneapolis, Minnesota, June 2019, pp. 4171–4186. Association for Computational Linguistics (2019)
4. Dong, L., Lapata, M.: Coarse-to-fine decoding for neural semantic parsing. In: Proceedings of the 56th Annual Meeting of the Association for Computational Linguistics (Volume 1: Long Papers), Melbourne, Australia, July 2018, pp. 731–742. Association for Computational Linguistics (2018)
5. Jia, R., Liang, P.: Data recombination for neural semantic parsing. In: Proceedings of the 54th Annual Meeting of the Association for Computational Linguistics (Volume 1: Long Papers), Berlin, Germany, August 2016, pp. 12–22. Association for Computational Linguistics (2016)
6. Min, S., Zhong, V., Zettlemoyer, L., Hajishirzi, H.: Multi-hop reading comprehension through question decomposition and rescoring. In: Proceedings of the 57th Annual Meeting of the Association for Computational Linguistics, Florence, Italy, July 2019, pp. 6097–6109. Association for Computational Linguistics (2019)
7. Mousavi, N., Scerri, S., Lehman, J.: Knight: mapping privacy policies to GDPR, August 2018
8. Palmirani, M., Governatori, G., Rotolo, A., Tabet, S., Boley, H., Paschke, A.: LegalRuleML: XML-based rules and norms. In: Olken, F., Palmirani, M., Sottara, D. (eds.) RuleML 2011. LNCS, vol. 7018, pp. 298–312. Springer, Heidelberg (2011). https://doi.org/10.1007/978-3-642-24908-2_30
9. Palmirani, M., Martoni, M., Rossi, A., Bartolini, C., Robaldo, L.: PrOnto: privacy ontology for legal reasoning. In: Kő, A., Francesconi, E. (eds.) EGOVIS 2018. LNCS, vol. 11032, pp. 139–152. Springer, Cham (2018). https://doi.org/10.1007/978-3-319-98349-3_11
10. Pandit, H.J., Fatema, K., O'Sullivan, D., Lewis, D.: Gdprtext - GDPR as a linked data resource. In: ESWC (2018)
11. Pandit, H.J., Lewis, D.: Modelling provenance for GDPR compliance using linked open data vocabularies. In: PrivOn@ISWC (2017)

12. Robaldo, L., Bartolini, C., Palmirani, M., Rossi, A., Martoni, M., Lenzini, G.: Formalizing GDPR provisions in reified I/O logic: the DAPRECO knowledge base. J. Log. Lang. Inf. **29**(4), 401–449 (2020)
13. Robaldo, L., Sun, X.: Reified input/output logic: combining input/output logic and reification to represent norms coming from existing legislation. J. Log. Comput. **27**(8), 2471–2503 (2017)
14. Sun, X., van der Torre, L.: Combining constitutive and regulative norms in input/output logic. In: Cariani, F., Grossi, D., Meheus, J., Parent, X. (eds.) DEON 2014. LNCS (LNAI), vol. 8554, pp. 241–257. Springer, Cham (2014). https://doi.org/10.1007/978-3-319-08615-6_18
15. Vaswani, A., et al.: Attention is all you need. In: Guyon, I., et al. (eds.) Advances in Neural Information Processing Systems, vol. 30. Curran Associates Inc (2017)
16. Wang, B., Shin, R., Liu, X., Polozov, O., Richardson, M.: RAT-SQL: relation-aware schema encoding and linking for text-to-SQL parsers. In: Proceedings of the 58th Annual Meeting of the Association for Computational Linguistics, Online, July 2020, pp. 7567–7578. Association for Computational Linguistics (2020)
17. Wang, Y., Berant, J., Liang, P.: Building a semantic parser overnight. In: Proceedings of the 53rd Annual Meeting of the Association for Computational Linguistics and the 7th International Joint Conference on Natural Language Processing (Volume 1: Long Papers), Beijing, China, July 2015, pp. 1332–1342. Association for Computational Linguistics (2015)
18. Zhang, H., Cai, J., Xu, J., Wang, J.: Complex question decomposition for semantic parsing. In: Proceedings of the 57th Annual Meeting of the Association for Computational Linguistics, Florence, Italy, July 2019, pp. 4477–4486. Association for Computational Linguistics (2019)

Semantic Relationship-Based Image Retrieval Using KD-Tree Structure

Nguyen Thi Dinh[1,2], Thanh The Van[3], and Thanh Manh Le[1(✉)]

[1] University of Science - Hue University, Hue, Vietnam
dinhnt@hufi.edu.vn, lmthanh@hueuni.edu.vn
[2] University of Food Industry, HoChiMinh city, Vietnam
[3] HCMC University of Education, HoChiMinh city, Vietnam
thanhvt@hcmue.edu.vn

Abstract. The semantic relationship of visual objects plays an important role in determining the context and semantics of an image. In this paper, a method of classifying semantic relationships between objects on image is proposed and applied to a semantic-based image retrieval system. Firstly, the visual objects on an input image are extracted and classified using the R-CNN network model. Secondly, a semantic description of the image is determined by the KD-Tree structure. From that, a model of classifying semantic relationships and extracting semantic descriptions for an input image is proposed to retrieve a set of similar images by semantics. To prove the correctness of the proposed theoretical basis, an experiment was built on the COCO and Flickr image data sets with an average image retrieval performance of 0.6972, and 0.7794, respectively. Experimental results are compared with other works on the same data set to demonstrate the effectiveness of our proposed method and can be applied to multi-object image data sets.

Keywords: R-CNN · KD-Tree · Semantic relationship · Image retrieval · Similar images

1 Introduction

Nowadays, multimedia data is the main research object in the field of computer vision. Multi-object image data sets rapidly increasing in number and diversity in categories are opportunities and challenges for the problem of identifying semantic relationships of objects on an image and searching for images based on this semantic relationship. In addition, the analysis and extraction of semantic relationships of visual objects on the image are performed by many different methods to improve the performance of the semantic-based image retrieval system. In this paper, each input image is segmented into object regions by the R-CNN (*Region Convolution Neural Network*) network model [1]; Then a semantic description of the image is determined by the KD-Tree structure to create a semantic relationship description for the input image. Based on the triplet of describing the semantic and contextual relationships of an image, a SPARQL query is extracted to retrieve a set of similar images by semantics based on ontology [2].

The contributions of this article are (1) Extracting and classifying visual objects of an image using R-CNN; (2) Building a KD-Tree structure to determine a semantic relationship between objects of an input image; (3) Proposing a model to classify objects and extracting semantic relationships between objects; (4) Building experimental and proving the feasibility and correctness of the proposed method using the COCO [3], Flickr [4] image data sets. The contribution of this work is to build a triplet describing the relationship of the objects using the KD-Tree. This is a possible application method for multi-object image sets.

2   Related Works

A method of semantic relationship-based image retrieval is performed in several stages includes: extracting visual objects; classifying objects; defining the semantic relationships and retrieving a set of similar images based on describing semantic relationship of an input image. In this paper, several works are surveyed on R-CNN object extraction; image classification using the KD-Tree structure; semantic-based image retrieval using ontology.

Alina Kuznetsova et al. (2020) [5] proposed a method of analyzing the semantics of images. The authors have presented a method classification, object detection, and visual relationship detection. In this work, firstly, objects are detected and classified by an R-CNN model. Secondly, visual relationships are built by this set of images to train the process of building visual relationships between objects of the input image. The proposed method is evaluated the feasibility and effectiveness, applied to different sets of images.

Yuqian Zhang et al. (2016) [6] used a method of image classification for face detection by the k-NN algorithm combined with the KD-Tree structure. In this work, the KD-Tree structure was built according to the method of image classification. This method has demonstrated the feasibility and efficiency of the image classification problem. On this basis, a KD-Tree structure built according to the image data classification method published by us [2] proved the feasibility and efficiency of the image classification problem by the KD-Tree structure. Based on inheriting the KD-Tree [2] structure, in this paper, an improvement for the construction process to store data at the leaf, training the classifier vector set on a KD-Tree structure. Semantic relationship classification of objects on images after segmenting and classifying input images.

Zihao Wang et al. (2019) [7] performed an image retrieval method based on the results of object image segmentation to describe the image by text. Based on text content, each query image has extracted a set of similar images according to the Cross-modal Gated Fusion (CAMP) recommendation method. The authors demonstrate the effectiveness of our approach through experiments and analysis of benchmarks. The experiment is evaluated as effective and better than other methods on COCO and Flickr image data sets.

Nhi N T U et al. (2021) [8] built an ontology framework and SPARQL query to retrieve images by semantics. In this work, an ontology framework is built capable of enriching many sets of images and extending the semantic relationship of images. On this basis, a SPARQL query is extracted to retrieve a set of similar images by semantics based on ontology. This work is evaluated as feasible and effective for semantic image retrieval based on ontology.

The above works show the feasibility of the object extraction problem and classification by R-CNN; image classification by KD-Tree; semantic-based image retrieval on ontology. However, the combination of the R-CNN technique with KD-Tree and ontology for a semantic-based image retrieval system is limited in number. Therefore, it is necessary to propose a classification model and semantic relationships extraction for semantic-based image retrieval problem.

3   Object Detection Using R-CNN Network Model

3.1   R-CNN Network Model

Object detection on an image is performed by many methods such as using R-CNN, Fast R-CNN, and Faster R-CNN networks. The result of the object defect process is to classify, recognize an object, and find relationships between visual objects on images. Therefore, the R-CNN network is an advanced technique used in published works widely in recent years [5]. The architecture of R-CNN consists of 3 components including: (1) region proposal extraction creates and extracts feature regions bounded by bounding boxes; (2) feature extractor performs low-level feature extraction to help identify images from region proposals through deep convolutional neural networks; (3) a classifier classifies images containing appropriately labeled objects based on input features [9,16,17].

3.2   Object Detection and Classification Using R-CNN Network

Based on the R-CNN network structure presented in Sect. 3.1. In this section, the experiment of object extraction and classification on images by the R-CNN is performed. To extract features using the R-CNN, each object is defined by the region proposal, and the proposed regions containing the object bounded are extracted by bounding boxes. Based on the feature extracted bounding boxes and object classification by assigning labels to each extracted image region using the Mask R-CNN network model, this process is illustrated in Fig. 1. On this basis, each region of the image is extracted features by area, perimeter, color, shape, and texture consisting of 81 components [10].

Fig. 1. Illustration of object detection using Mask R-CNN

4 Classification of Semantic Relationship Based on KD-Tree

4.1 Triples Describe a Semantic Relationship

Triples describe a semantic relationship between objects is determined by the properties of the COCO and Flickr image data sets. The words expressing the relationship are extracted from the image data including On, In, At, Before, Behind, Under, Above, Left; Right, etc. [11]. In this paper, triples describe semantic relationship has selected a set of triples to be annotated. Each triple has the form of < *object1* + *semantic relationship* + *object2*>, e.g. <*person on chair*>. Each image consists of many objects, and they are related to each other according to triples after segmenting objects by the R-CNN network. In the work [10], the process of labeling leaves is performed by the set of labels on the image data sets. On this basis, the set of words describing the relationship between objects on each image data set (In, On, At, ...) are stored on the leaves to determine the relationship of the objects on each image. This is a proposal and improvement in this work. Figure 2 illustrates triples on COCO data set.

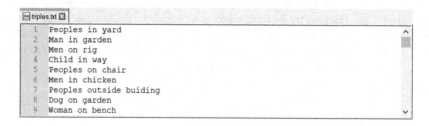

Fig. 2. Illustrtion of triples describe semantic relationship

4.2 Building a KD-Tree Structure for Semantic Relationship Classification

Each image region after detection and classification by the R-CNN model is extracted into a vector. The vectors of an input image on KD-Tree are combined into a vector by: $f_i(x_{i1},..,x_{in}) = f_{ij}(x_{ij1},..,x_{im}) + ...+ f_{ik}(x_{ik1},...,x_{ikm})$, where $m = 81$; $n = k*m$; k is a number of object on image.

Based on the original KD-Tree [12,15], in this paper, a KD-Tree structure is built as a balanced multi-branch tree at the node that stores a weight vector and is adjusted in training to allocate efficient relationships at the leaves. Initially, each randomly initialized leaf contains a word indicating the semantic relationship defined in Sect. 4.1. Each leaf stores a word indicating the relationship between objects on the image. For each image after segmentation and extraction feature vector is inserted into the KD-Tree from the root to the leaf. The building process of a KD-Tree structure according to the following steps is illustrated in Fig. 3. The process of building KD-Tree structure for semantic relationship is presented by Algorithm 1.

(1) Each region of the object is extracted by an m-dimensional vector feature $f_{ik} = (f_{ik1}, f_{ik2},..., f_{ikm})$

(2) The node being different from the leaf stores a 2*m-dimensional weight vector, initially initialized to be random

(3) Each vector $f_{ik} = (f_{ik1}, f_{ik2},..., f_{ikn})$ is inserted to the KD-Tree from the root to find the storage location at the $leaf_k$ on the KD-Tree by using the Sigmoid transfer function at the node

(4) $Leaf_k$ stores a word describe semantic relationship

(5) Based on the word describe semantic relationship at $leaf_k$, extracting a semantic relationship between the objects belonging to the input image.

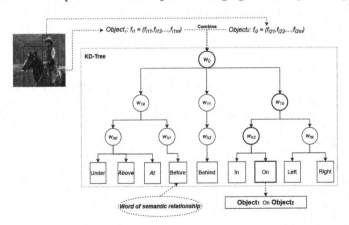

Fig. 3. Illustration of KD-Tree for semantic relationship

When building a KD-Tree structure has a height h needs to be added n elements. The KD-Tree is a balanced tree, so when adding elements to the tree, every element must be traversed from the root to the leaf. So, the complexity of the **1** algorithm is $O(n*h)$.

4.3   Training Weight Vector Process on KD-Tree

Initially, the KD-Tree structure is built with each leaf containing a randomly allocated relationship word as illustrated in Fig. 3. After inserting the image region feature vectors into KD-Tree, the allocation efficiency is calculated. On this basis, a process of adjusting a set of classifier vectors at the nodes is performed to improve the classification efficiency on KD-Tree as follows:

Algorithm 1: Building KD-Tree to classify semantic relationship

---

1 **Input**: Vector set $F = \{f_i : f_i = (x_{i0}, x_{i1}, ..., x_{in}); i = 1...k\}$;
2 h height; n maximum number of branches;
3 set of initialization weights $W_{kt} = \{W_i : W_i = (x_{i_0}, x_{i_1}, ..., x_{i_n}); i = 0..h - 1\}$;
4 **Output**: KD-Tree
5 **Function BKDT**(F, W_kt, h, n)
6 **begin**
7 | initialize parameters h, n; KD-Tree $= \phi$;
8 | **for** $i = 0$ to $h - 1$ **do**
9 | | Call function initialize parameter random InitWeight ();
10 | | Assign Wi.left = Epsilon and Wi.right = Epsilon;
11 | **end**
12 | **foreach** *(fi in F)* **do**
13 | | **if** $KD\text{-}Tree = \phi$ **then**
14 | | | create KD-Tree has one branch;
15 | | **end**
16 | | **else**
17 | | | **for** $j = 0$ to $h - 1$ **do**
18 | | | | **if** $Sigmoid(f_i, w_j) < w_j.left$ **then**
19 | | | | | Creat left branch;
20 | | | | **end**
21 | | | | **if** $Sigmoid(f_i, w_j) > w_j.right$ **then**
22 | | | | | Creat right branch;
23 | | | | **end**
24 | | | | **if** $w_j.left <= Sigmoid(f_i, w_j) and Sigmoid(f_i, w_j) <= w_j.right$ **then**
25 | | | | | Choose best branch;
26 | | | | **end**
27 | | | **end**
28 | | **end**
29 | **end**
30 | **foreach** *(f_i)* **do**
31 | | **if** *(f_i go to Leaf_k)* **then**
32 | | | Insert f_i into $Leaf_k$
33 | | **end**
34 | **end**
35 | Return KD-Tree;
36 **end**

---

(1) Calculating the correct classification performance for the set of image regions with the initial set of weight vectors.

(2) Finding $node_i$ is wrong location.

(3) Adjusting the vector at $node_i$ so that vector f_i at the correct leaf.

(4) Comparing the classifier performance after adjusting the weight vector and the classifier performance when using the set of random weight vectors.

(5) This process repeats until a given number of iterations is reached or the target performance is reached.

Algorithm 2

Call p is the number of times the weight vector is adjusted; h is the tree height (h is a constant); m is the number of elements participating in the tree construction process. The process of training the classifier model according to the KD-Tree structure is done by updating the weights to build KD-Tree. So, the complexity of the **2** algorithm is O(p*m).

4.4 Extraction Semantic Relationship of an Image Using KD-Tree

Currently, there are many different methods to build a triple describing the relationships of objects on multi-object images such as Graph Convolutional Networks, Knowledge Graph, etc. In this work, the semantics of an image is defined by a set of triples built and trained on KD-Tree. Each image is segmented into object image regions. Each object image region is extracted as a feature vector and combined into a feature vector for the input image to store at the leaf. On this basis, the semantic relationship is extracted based on the KD-Tree structure. Therefore, the algorithm for classification and extraction of semantic relationship is presented as Algorithm 3.

Algorithm 2: Algorithm train a set of classifier vector on KD-Tree

1 **Input:** Init vector $W_{kt} = W_i : W_i = (x_{i0}, x_{i1}, ..., x_{in}); i = 0...h - 1$;
2 **Output:** Trainvector;
3 **Function TKDT** $(W_{kt}, KD - TRee)$
4 **begin**
5   Vector $= W_{kt}$; Init M is target performance;
       repeat
          Create KD-Tree with W_{kt} set, height h, n branches;
          P_i = Calculate performance of semantic relationship classification with W_{kt};
          Determine the wrong path of the f_k vector on the KD-Tree;
          Determine the right path of the f_k vector on the KD-Tree;
          Find the position where Nodei misplaces the f_k vector;
          Adjust the vector stored at $node_i$ to f_k the right path;
          Recreate the KD-Tree according to the adjusted vector set;
          P_j = Calculatte the classifier performance after vector tuning at $node_i$;
       until $(P_j > P_i)$ *or* $(P_j > M)$;
       Vector = set of vector after adjustment;
       Return Vector;
6 **end**

Algorithm 3
Call h is the height of the KD-Tree (h is a constant), k is the maximum number of branches at $node_i$ (k is a constant). The **4** algorithm in turn traverses the layers of the KD-Tree from the root to the leaf with height h. Let N = h*k, N is a constant. Therefore, the complexity of the **3** algorithm is O(N).

5   A Model of Object Classification and Semantic Relationship Extraction

5.1   A Proposed Model

In the work [2,18], the authors performed the semantic-based image retrieval problem using an ontology. On the basis of inheriting ontology is built and enriched for COCO, and Flickr image data sets. In this work, a triple describing a semantic relationship between objects on an image and perform to retrieve by SPARQL query on the ontology (Fig. 4).

Algorithm 3: Classification and extraction semantic relationship on KD-Tree

1 **Input:** The vector f_I of the image I, KD-Tree
2 **Output:** Semantic relationship of image I;
3 **Function SRKD**($f_I, KD - Tree$)
4 **begin**
5     $SR_I = \emptyset$;
6   **for** $i = 0$ to $h - 1$ **do**
7   Finding the $(Left, Right)$ of w_i;
8   Finding a set of child nodes of w_i;
9   Calculating S = Sigmoid(Product(w_i,f_j));
10   **foreach** $(k$ in $(Left, Right)$ of $w_i)$ **do**
11   d(S,k) = caculate_distance(S,k);
12   **if** $d(S,k) = d\_min$ **then**
13   **SRKD**($f_I, KD - Tree.node_k$);
14   **end**
15   **end**
16   **end**
17   **if** $(f_I \in Leaf_k)$ **then**
18   Assign $SW_I = leaf\_k.$(word describe semantic relationship);
19   Assign $SR_I = Object\_i.lable + SW_I + Object\_j.lable$;
20   **end**
21   Return SR_I is semantic relationship of image I;
22 **end**

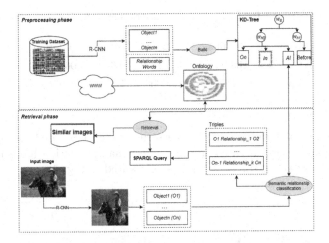

Fig. 4. A model of extracting objects semantic relationship on image

A model of semantic relationship-based image retrieval consists of two phases: the KD-Tree training phase and the image retrieval phase, including the following steps: (1) image segmentation and object classification using the R-CNN network model; (2) building a KD-Tree structure to store images and classify semantic relationships of images; (3) extracting triplets describing the semantic relationship of input images based on KD-Tree; (4) creating a SPARQL query; (5) retrieving on ontology to extract a set of similar images.

5.2 Multi-object Image Retrieval Based on Semantic Relationships and Ontology

Each input image, after extracting the triplet describing the image semantic relationship, is used as the basis for building the SPARQL query. The process of building a SPARQL query is inherited from the work [2,18]. The process of semantic-based image retrieval using ontology according to Algorithm 4.

The **4** algorithm creates a SPARQL query to retrieve the ontology to extract a set of similar images. So the complexity of the **4** algorithm is $O(c*l)$. Where c is the number of classifiers of the input image I; l is the classifier number of the ontology-constructed image data set. Since c, l are constants, set $N = c*l$, so the complexity of the **4** algorithm is $O(N)$.

5.3 Experiment and Evaluate the Results

The SR-KDT system is built on the .NET Framework 4.5 platform the $C\#$ programming language. The graphs are built on MathLab. The SR-KDT system is performed on computers with Intel(R) Core^{TM} i7-5200U processors, CPU 2.70 GHz, RAM 16 GB, and Windows 10 Professional operating systems. Description of experimental data in Table 1.

Table 1. Describe experimental data

Data Set	No. images	Training	Testing	Validation
COCO	163,957	118,287	40,670	5,000
Flickr	31,783	29,000	1,783	1,000

The process of building a KD-Tree structure is illustrated in Fig. 5. The process of image retrieval based on the image semantic relationship is illustrated in Fig. 6. The result of similar images is illustrated in Fig. 7.

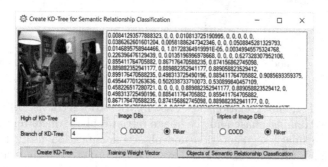

Fig. 5. Illustration of building and training KD-Tree structure

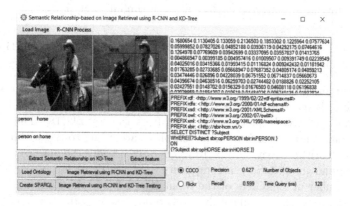

Fig. 6. Illustration of semantic relationship-based image retrieval **SR-KDT**

Algorithm 4: Multi-object image retrieval based on semantic description and ontology

---

1 **Input**: input image I
2 **Output:** *SI* is a set of similar images of image I
3 **Function SRS**()
4 **begin**
5   Build a SPARQL query from semantic relationship of image I ;
6   Retrieving on ontology used to SPARQL query;
7   *SI=* Extract a set similar image;
8   **return** *SI*;
9 **end**

Fig. 7. Illustration a set of similar images (000000183675.jpg COCO)

Figure 7 illustrates a set of similar images of the input image 000000183675.jpg (COCO) that has the semantic relationship between the objects as "person on horse" (Table 2).

Table 2. Performance of semantic relationship-based image retrieval system SR-KDT

Data set	Avg. precision	Avg. recall	Avg. query time (ms)
COCO	0.6972	0.6636	215
Flickr	0.7794	0.6788	126

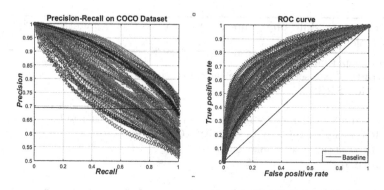

Fig. 8. Precision, Recall and ROC curve on COCO image data set

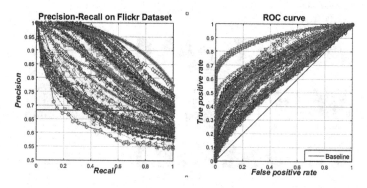

Fig. 9. Precision, Recall and ROC curve on Flickr image data set

Experimental results of the model evaluation are performed on COCO and Flickr image data sets. The average precision is defined by TopK (k = 5). These results of the **SR-KDT** system are illustrated by the ROC curve graph in Figs. 8 and 9 (Table 3).

Table 3. Compare the performance of SR-KDT system with other works

Method	Data set	No. images	AVG. precision
CN MAX, TopK = 5, [14]	COCO	5,000	0.3910
CAMP, TopK = 5, [7]	COCO	5,000	0.6890
R-CNN-KDT (Ours)	**COCO**	**5,000**	**0.6972**
IRSGS-GCN, TopK = 5, [13]	Flickr	1,000	0.5670
CAMP, TopK = 5, [7]	Flickr	1,000	0.7710
R-CNN-KDT (Ours)	**Flickr**	**1,000**	**0.7794**

The results of system semantic relationship-based image retrieval using the KD-Tree structure is higher than other works, because of: (1) SR-KDT system using R-CNN network for image segmentation and object classification, so the classification performance is high; (2) The training process of weight vector training to classify semantic relationships on KD-Tree should be highly effective.

6 Conclusion

In this paper, a model of semantic relationships-based image retrieval has been implemented using KD-Tree, R-CNN, and ontology. Based on the semantic relationship of an input image, the SPARQL query is built to retrieve a set of similar images by semantics on ontology. Experiments are built on the COCO, and Flickr image data sets to demonstrate the feasibility of our proposed method. The average image retrieval accuracy results for each COCO and Flickr image data sets were 0.6972 and 0.7794, respectively. Experimental results have proved that a method of combining the R-CNN, KD-Tree structure, and ontology for semantic-based image retrieval is feasible and effective. In the next development direction, we will combine semantic relationships for KD-Tree Random Forest with ontology to improve performance retrieval.

Acknowledgment. The authors would like to thank the Faculty of Information Technology, Univer-sity of Science - Hue University for their professional advice for this study. We would also like to thank HCMC University of Food Industry, HCMC University of Education, and research group SBIR-HCM, which are sponsors of this re-search.

References

1. Chen, S., Li, Z., Tang, Z.: Relation R-CNN A graph based relation-aware network for object detection. IEEE Signal Process. Lett. **27**, 1680–1684 (2020)
2. Dinh, N.T., Van Thanh, T., Thanh, L.M.: A method of image classification base on kd-tree structure for semantic-based image retrieval system. In: Proceedings of the National Conference on Basic Research and IT Applications (FAIR21) (2021). ISBN 978-604-9988-60-8
3. https://cocodataset.org/download
4. https://www.kaggle.com/datasets/hsankesara/flickr-image-datas
5. Kuznetsova, A., et al.: The open images dataset v4. Int. J. Comput. Vision **128**(7), 1956–1981 (2020)
6. Zhang, Y., et al.: Fast face sketch synthesis via kd-tree search. In: European Conference on Computer Vision. Springer, Cham (2016)
7. Wang, Z., Liu, X., Li, H., Sheng, L., Yan, J., Wang, X., Shao, J.: Camp Cross-modal adaptive message passing for text-image retrieval. In: Proceedings of the IEEE/CVF International Conference on Computer Vision, pp. 5764–5773 (2019)
8. Nhi, N.T.U., Le, T.M.: A model of semantic-based image retrieval using C-tree and neighbor graph. Int. J. Semant. Web Inf. Syst. (IJSWIS) **18**(1), 1–23 (2022)
9. He, K., Gkioxari, G., Dollár, P., Girshick, R.: Mask R-CNN. In: Proceedings of the IEEE International Conference on Computer Vision, pp. 2961–2969 (2017)

10. Nguyen, T.D., Van, T.T., Le, M.T.: Image classification using KdTree for image retrieval problem. J. Res. Dev. Inf. Coommun. Technol. **1**, 33–44 (2021)
11. Schroeder, B., Tripathi, S.: Structured query-based image retrieval using scene graphs. In: Proceedings of the IEEE/CVF Conference on Computer Vision and Pattern Recognition Workshops, pp. 178–179 (2020)
12. Bentley Jon Louis: Multidimensional binary search trees used for associative searching. Commun. ACM **18**(9), 509–517 (1975)
13. Yoon, S., et al.: Image-to-image retrieval by learning similarity between scene graphs, arXiv preprint arXiv, 4322, 2012.14700 (2020)
14. Icarte, R.T., Baier, J.A., Ruz, C., Soto, A.: How a general-purpose commonsense ontology can improve performance of learning-based image retrieval. arXiv preprint arXiv, 1705–08844 (2017)
15. Hou, W., Li, D., Xu, C., Zhang, H., Li, T.: An advanced k nearest neighbor classification algorithm based on KD-tree. In: IEEE International Conference of Safety Produce Informatization (IICSPI) (2018)
16. Lee, H., Eum, S., Kwon, H.: Me R-CNN: multi-expert R-CNN for object detection. IEEE Trans. Image Process. **29**, 1030–1044 (2019)
17. Cai, Z., Vasconcelos, N.: Cascade R-CNN: delving into high quality object detection. In Proceedings of the IEEE Conference on Computer Vision and Pattern Recognition, pp. 6154–6162 (2018)
18. Dinh, N. T., and Le, T. M.: An Improvement Method of Kd-Tree Using k-Means and k-NN for Semantic-Based Image Retrieval System. In: World Conference on Information Systems and Technologies, pp. 177–187. Springer, Cham (2022). https://doi.org/10.1007/978-3-031-04819-7_19

Preliminary Study on Video Codec Optimization Using VMAF

Syed Uddin$^{(\boxtimes)}$, Mikołaj Leszczuk, and Michal Grega

AGH University of Science and Technology, Krakow, Poland
{syed.uddin,mikolaj.leszczuk,michal.grega}@agh.edu.pl

Abstract. The growth in video streaming has been an exponential one for the last decade or so. High-resolution videos require high bandwidth to transport the videos over the network. There has been a growing demand for compression technologies to compress videos while simultaneously maintaining quality. Video codecs are used to encode and decode video streams. These codecs have been developed by MPEG, Google, Microsoft, and Apple Inc. There are many encoding parameters that affect bitrate and video quality. These performance parameters must be exploited, evaluated, and modeled to find the best possible solutions. This paper demonstrates some preliminary results for video coding sets with selected bitrates. The objective video multimethod assessment fusion (VMAF) metric is calculated for the encoded video versions. In this study, the quality of the encoded videos was evaluated and estimated using VMAF. The results confirm a strong relationship between bitrate and VMAF estimates. This study shows the impact of coding parameters on the VMAF values and provides the foundation for building robust models in the field of video quality analysis.

Keywords: Video coding · Video quality · MPEG · Codecs · VMAF

1 Introduction

Online video services have grown exponentially since their inception. It was predicted that, by 2022, 82% of all global internet traffic per year will be video content. According to forecasts, this figure will continue to increase in the following years [1], as also mentioned in Statista Research [2]. As the volume of video data increases, it becomes increasingly challenging to maintain a video service to the satisfaction of users. Due to various technological limitations, visible artifacts can occur at any stage of the video delivery cycle [3]. This, in turn, can lead to very low user satisfaction. As a result, online video providers are keen to find ways to measure and predict user satisfaction with videos so that they can optimize their video delivery chains.

The amount of video content and the transmission of that content in various resolutions over the Internet to viewers is increasing [4]. Video streaming over the Internet is possible due to video compression technology, also known as video codecs. Video compression and codecs offer better quality for streaming and minimize the storage complexity, transmission, and processing of the video frame sequence. Video codecs

© The Author(s), under exclusive license to Springer Nature Switzerland AG 2022
N. T. Nguyen et al. (Eds.): ACIIDS 2022, LNAI 13757, pp. 469–480, 2022.
https://doi.org/10.1007/978-3-031-21743-2_37

are essential for many applications such as video conferencing, HDTV video, digital cameras, and smartphones [5, 6].

Video streaming providers, such as Netflix and YouTube have more viewers and their number is increasing. These companies are investing a lot of budgets in research and development projects. The challenge with video streaming is that high bandwidth is required for content variability, and to satisfy video resolution constraints by storage capabilities. There are methods that can be used to compress videos and compactly optimize them. When videos are compressed, they can be stored or transmitted to viewers in an efficient manner [7]. There are a variety of codecs and international standards, currently in practice. The codecs used in practice are from ISO, Google, Microsoft, and Apple. The most widely used codecs are Moving Picture Experts Group (MPEG) ones [8–10].

The video sequence is encoded at different bit rates. These data rates are the constant bit rate (CBR) and the variable bit rate (VBR). The use of one of these rate control modes depends on video streaming. The application of these data rates has an impact on file size, encoding time, and video quality. In addition to data rates, there are also quantization parameters that also affect video quality. Video multimethod assessment fusion (VMAF) is also an objective metric for estimating the video quality affected by video compression. The main goal of this work is to encode the video stream with different video codecs. Moreover, the VMAF between the original video and the distorted version was calculated. Based on the results, we can compare different codecs and perform an analysis.

The paper is structured as follows - Sect. 2 describes the main work related to the study. Section 3 discusses the research methodology. Section 4 describes the study's conclusion. Section 4 is also dedicated to future work.

2 Related Work

Video streaming requires a high bandwidth to transmit videos. There is a need for compression technologies to compress videos. The compression techniques need to compress a video at different bitrates and resolutions to save bits, maintain the quality of the video, and stream it in an appropriate manner so that no information is lost.

Previous research has provided valuable information on video codec compression and the optimization of video sequences. The parameters investigated potentially affect video quality, bitrate, and encoding time. The development of video compression techniques is continuously increasing as researchers investigate from MPEG-2, the basic international video coding standard [11], to the latest codecs such as advanced video coding (H.264/AVC) [12].

There is a recent addition to compression technology known as Versatile Video Coding (VVC) [13]. VVC compresses videos by reducing bit rates by 30 to 40% compared to the High Efficiency Video Coding (HEVC) standard [14]. The Alliance for Open Media (AOMedia) developed open-source codec to compete with MPEG standards. Research shows that the AOMedia Video 1 (AV1) codec [15] outperforms its predecessor VP9 [16].

The H.264/MPEG-4-AVC codec [12] is still the most prolific video coding standard, although the H.265/HEVC [14] standard offers better coding performance. A next-generation video coding standard, Versatile Video Coding (VVC), is emerging and targets the coding gain over H.265/HEVC. The VVC standard supports immersive formats *(360° videos)* and higher resolutions, for example, 16K video.

The Alliance for Open Media (AOMedia) develops open-source video codecs. Google developed the VP9 standard [16] to compete with MPEG and formed the basis for the AV1 standard [15]. The AV1 standard is expected to compete with the MPEG standard in the context of video streaming applications.

The performance of video coding algorithms is evaluated by comparing their rate distortion (RD) or rate quality (RQ) performance on a variety of test sequences. Objective quality metrics or subjective ratings are used to evaluate compressed video quality. The difference in RD and RQ performance between codecs can be calculated using objective quality measurements [17] or SCENIC [17] (subjective ratings).

To compare video codecs and optimize performance, quality and rate, convex hull rate distortion for adaptive streaming applications is optimized [18].

3   Methodology

In this section, the experimental setup and procedure are described in detail. The experimental results and analysis are also presented in this section.

3.1   Experiment Setup

In this section, the experimental setup is presented in detail. The experimental procedure consists of two parts: a) the video sequence used was from Big Buck Bunny (BBB). The video was in the YUV format (raw format). This was the main reference video file used in the experiment. The resolution was 1920 × 1080, which corresponds to the Full high definition (FHD). The 720 frames were extracted from the main video using FFMPEG. Several different video frames (screenshots) of the main reference file are shown in Fig. 1. Information on the main file can be found in Table 1.

Table 1. Main reference file configuration

Video	Dimension	FPS	Number of frames
Big buck bunny	1920 × 1080	24	720

First, the original video sequence (reference video) is encoded with different codecs and configurations. The sample list of codecs is provided in Table 2.

To get an idea of this, the original video sequence was encoded at two (2) different bitrates. In the first, the rate was set to 5M (megabits) and in the second, the input was encoded to 1M (megabits).

Table 2. Sample codecs

Codec family	Codec name
ISO/IEC/ITU-T	MPEG-4 MPEG-4/AVC HEVC/H.265 (MPEG-H Part 2)
Google	VP 8 VP 9
Microsoft	Windows Media Video (WMV)
Adobe flash	Sorenson Spark
Xiph.org	Theora

Fig. 1. Reference video scenes

3.2 Results

This section presents the preliminary results of the experiment. The results are based on experiment part 1 (a).

As shown in Table 3, the influence of the input bit rate is demonstrated. The initial results show that the HEVC/H.265 codec achieves a high VMAF score. VMAF 90 (or >90) means that the encoded video sequence has not lost quality [20]. The size of the encoded file is also promising (19.9 MB). The Sorenson Spark codec shows a weak relationship between the input bitstream and the VMAF. The results show a drastic degradation of video quality when encoded.

As evidenced by the study results, the outputs with the MPEG-4, AVC, HEVC, and VP9 codecs are the most promising. The codecs that reduce the quality of video when

Table 3. Encoded video with 5 Mbit/s

Codec name	Bitrate kbps	Frame rate f/sec	File size MB	VMAF
MPEG-4	5,573	24	19.9	91.02
MPEG-4/AVC	5,490	24	19.6	96.7
HEVC/H.265	5,579	24	19.9	97.56
VP 8	5,043	24	18	80.76
VP 9	6,673	24	23.8	96.49
Windows Media Video (WMV)	4,901	30	18.2	71.92
Sorenson Spark	5,075	24	18.1	70.65
Theora	5,014	24	17.9	86.02

encoded are Windows Media Video (WMV) and Sorenson Spark. A look at the results shows that the codecs that minimize file size and have VMAF ($\geq$90) give a promising output.

The VP9 codec shows interesting results, but on the other hand, the file size is somewhat large compared to MPEG codecs. At the input bitrate provided (5 Mbit/sec), the codec VP9 outputs a bit rate (6,673 bits) that is larger than that set in the input stream. This relationship is shown in Figs. 2 and 3.

Table 4. Encoded video with 1 Mbit/sec

Codec name	Bitrate kbps	Frame rate f/sec	File size MB	VMAF
MPEG-4	1,760	24	6.6	54
MPEG-4/AVC	1,085	24	4.06	82
HEVC/H.265	1,066	24	4.01	89
VP 8	1,078	24	4.04	62
VP 9	1,872	24	7.02	87
Windows Media Video (WMV)	980	30	93	82
Sorenson Spark	2,134	24	8.0	51
Theora	1,069	24	4.01	50

Table 4 shows the results of the codecs when encoded with an input stream (1 Mbit/s). Compared to the results shown in Table 3, these results are not very promising.

If the video sequence was encoded on an input stream (1 Mbit/s). The file size was minimized. However, on the other hand, the quality deteriorates. Here, none of the codecs shows promising output results. This means that these codecs show poor performance when operated with a minimal input stream.

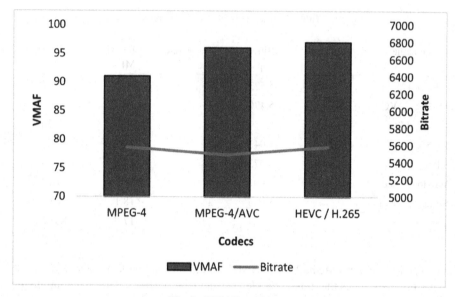

Fig. 2. VMAF and bitrate

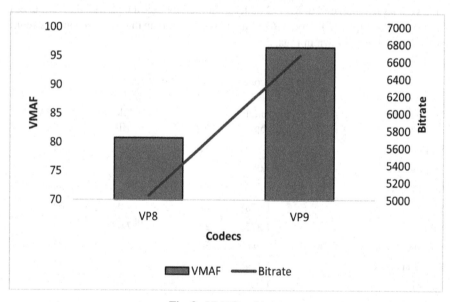

Fig. 3. VMAF vs bitrate

The following results are based on experiment setup part 2 (b). In this part of the study, elements were developed to determine the level of occurrence of Just a Noticeable Difference (JND) [19]. The VMAF model is used to determine JND. The evaluation is performed on the H.264/MPEG-4 AVC codec. FFMPEG is used to encode the video sequence. The input rate is set accordingly, and the output bitrate is obtained according to the input bitrate. Both the input and output bitrates demonstrate significant differences. The trend between the two rates is shown in Fig. 4.

The output bit rate is used as a reference in the following steps. First, the bitrate cutoff point will be determined, which is assumed to yield a VMAF of 90. VMAF measurements were made in the input bit rate range of 8.0 to 13.0 Mb/s. These values correspond to an output bit rate of 8.6 Mb/s to 14.0 Mb/s. The VMAF acceptability interval is assumed to start at 70 (corresponding to a MOS value of 3.5), the condition under which the VMAF value was lowered was not considered. The results of the measurements are shown in Fig. 5.

As can be seen in the graph, the VMAF level occurs at an output bitrate of 11.3 Mbit/s. The level determined in this way was taken as the starting point for further consideration. The maximum VMAF results are shown at an output bit rate of 14.0 Mbit/s.

Furthermore, three compression parameters were selected. These parameters can potentially affect the quality of the video stream while maintaining the same output bitrate. The parameters selected are the size of -keyint_min (Minimum GOP – Group of pictures) size, -g (set GOP size) and -preset (set x264 encoding preset). The impact of all these parameters on video quality is described in Fig. 6, Fig. 7, and Fig. 8, respectively.

As can be seen in the diagram (Fig. 6), this parameter has no influence on the quality expressed by VMAF in the tested case. It shows no fluctuations and remains stable at the level of 90 (VMAF), although an attempt was made to use a wide range of parameters.

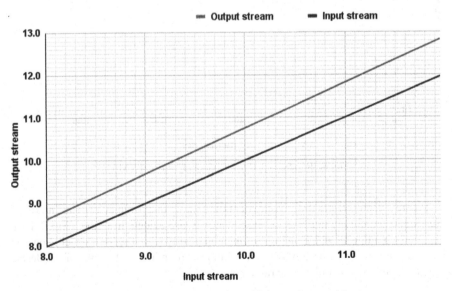

Fig. 4. Relationship between input bitrate and output bitrate

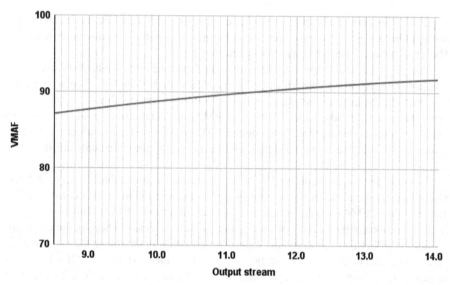

Fig. 5. Relationship between VMAF and output stream

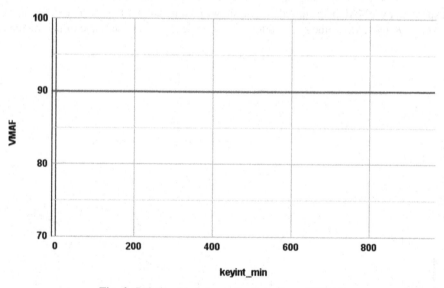

Fig. 6. Relationship between keyint_min and VMAF

Figure 7 demonstrates the measurements and shows that parameter (g) had no significant impact on quality in the tested case, which is also evident from the VMAF. There were two situations in which an improvement in quality was observed, which, however, practically remained invariably within the limits of rounding to the level of 90, despite attempts to use a wide range of parameters.

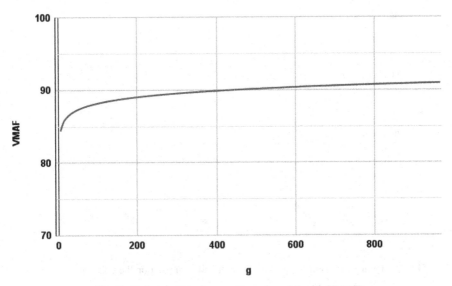

Fig. 7. Relationship of parameter values (g) with VMAF

Figure 8 shows the measurements of the parameter (-preset). This parameter is in practice, a meta-parameter that changes many other compression parameters simultaneously. The option specifies a range of options, from 'ultrafast' to 'placebo'. After testing the sequence by setting this parameter, we could see a very clear influence of the parameter on quality. On the other hand, this influence was observed at the cost of greater computational complexity. Computational complexity refers to the measure of the amount of computing resources (e.g. time). The time is taken into the consideration while running each condition of 'preset' values.

Moving on to testing, we examined how much output bitrate could be achieved by assigning different values to the -preset (slower than the 'fast'). The sequence was tested, and the parameter value was assigned to the 'veryslow'. The 'placebo' value is ignored, as it provides only a minimal increase in quality at the expense of much slower coding. Figure 9 shows the results obtained from the experiment.

If we look at Fig. 9, we can see, that for the conditions presented, increasing the output bit rate does not have a great effect on the VMAF value. It can be assumed that increasing the output bit rate by 0.5 Mb/s results in a yield of only 1 unit on the VMAF scale. On the other hand, each optimization, such as changing the preset parameter, causing the VMAF value to be increased by 1 unit, results in savings of as much as 0.5 Mb /s of the stream. Here, the point to be noted is that VMAF levels for the same bit rate may have different values due to the characteristics of the contact that need to be compressed.

478   S. Uddin et al.

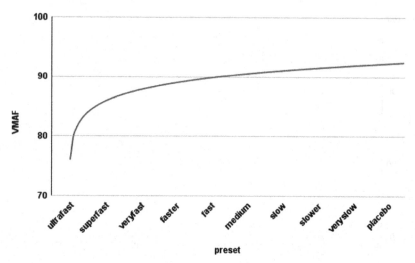

Fig. 8. The measurement of parameter preset and its impact on the video quality score

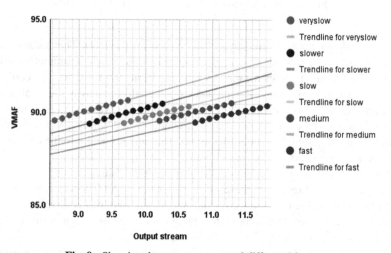

Fig. 9. Showing the output stream and different bitrates

4   Conclusion and Future Work

In this paper, the video sequence was encoded with different codecs. Different input bitrates were used to investigate the effect on the output bitrate, file size, frame rate, and VMAF model. The purpose of this paper is to address the issues of using different bitrates and how much they affect video quality. First, the video sequence was encoded with 5 Mbit/s as the input. In the second test, 1 Mbit/s was used as the input stream. Both tests gave completely different results. The initial idea is that a high value should be used for the input bitrate. Using minimal bitrate values did not show significant results.

The preliminary results demonstrate that the MPEG and VP9 codecs are potential codecs for better compression under constrained conditions. The HEVC/H.265 codec is a better codec, as evident by the encoding settings and VMAF.

As this is a preliminary study, we evaluated the quality of the video. This study should be extended, and a model should be derived to measure video quality using machine learning algorithms. Since very few parameters are considered. The list of parameters needs to be expanded, and the most promising parameters that affect bitrate reduction and video quality retention need to be discovered. In the next stage, the factor of computational complexity will also be considered. The current study is based on objective measures. This study can be repeated with a collection of subjective measures. Then, objective and subjective measurements can be compared to verify whether there is a significant difference.

Acknowledgement. This research received funding from the Polish National Center for Research and Development (POIR.01.01.01-00-1896/20).

References

1. Cisco: Annual Internet Report (2018–2023) White Paper (2020). https://www.cisco.com/c/en/us/solutions/collateral/executive-perspectives/annual-internet-report/white-paper-c11-741490.html
2. Lindlahr, S.: Forecast of video-on-demand revenue by segment in the United States 2017–2025. https://www.statista.com/forecasts/459396/video-on-demand-revenue-in-the-united-states-forecast
3. Pérez-Castilla, A., García-Ramos, A., Padial, P., Morales-Artacho, A.J., Feriche, B.: Effect of different velocity loss thresholds during a power-oriented resistance training program on the mechanical capacities of lower body muscles. J. Sports Sci. **36**, 1331–1339 (2018)
4. Tian, L., et al.: Understanding user behavior at scale in a mobile video chat application. In: UbiComp 2013 - Proceedings of the 2013 ACM International Joint Conference on Pervasive and Ubiquitous Computing, pp. 647–656 (2013). https://doi.org/10.1145/2493432.2493488
5. Guna, S.: The evolution of video codecs and the future (2017)
6. Li, Z., Drew, M.S., Liu, J.: Modern video coding standards: H.264, H.265, and H.266. In: Li, Z., Drew, M.S., Liu, J. (eds.) Fundamentals of Multimedia. Texts in Computer Science, pp. 423–478. Springer, Cham (2021). https://doi.org/10.1007/978-3-030-62124-7_12
7. Pullareddi, M., Fathima, A.: A review of image and video compression standards. Asian J. Pharmac. Clin. Res. **10**, 373 (2017). https://doi.org/10.22159/ajpcr.2017.v10s1.19760
8. Karwowski, D., Grajek, T., Klimaszewski, K., Stankiewicz, O., Stankowski, J., Wegner, K.: 20 years of progress in video compression – from MPEG-1 to MPEG-H HEVC. general view on the path of video coding development. In: Choraś, R.S. (ed.) IP&C 2016. AISC, vol. 525, pp. 3–15. Springer, Cham (2017). https://doi.org/10.1007/978-3-319-47274-4_1
9. Ahmed, R., Islam, M., Uddin, J.: Optimizing the apple lossless audio codec algorithm using NVIDIA CUDA architecture. Int. J. Electr. Comput. Eng. (IJECE) **8**, 70 (2018). https://doi.org/10.11591/ijece.v8i1.pp70-75
10. Goyal, D., Hemrajani, D.: Comparative analysis of performance of WMV & MPEG formats video streaming in a cloud. Int. J. Digit. Appl. Contemp. Res. **2**, 7 (2014)
11. ITU-T. H.262, information technology - generic coding of moving pictures and associated audio information: video, ITU-T std (2012). https://www.itu.int/rec/T-REC-H.262

12. ITU-T. H.264, advanced video coding for generic audiovisual services (2005). https://www. itu.int/rec/T-REC-H.264

13. Im, S.-K., Chan, K.-H.: More probability estimators for CABAC in versatile video coding, pp. 366–370 (2020). https://doi.org/10.1109/ICSIP49896.2020.9339374

14. ITU-T. H.265: High-efficiency video coding (2014). http://www.itu.int/rec/T-REC-H.265

15. Sethuraman, S., Rajan, C., Patankar, K.: Analysis of the emerging AOMedia AV1 video coding format for OTT Use cases. SMPTE Motion Imaging J. 127, 44–50 (2018). https://doi. org/10.5594/JMI.2018.2821419

16. Chen, C., Inguva, S., Rankin, A., Kokaram, A.: A subjective study for the design of multiresolution ABR video streams with the VP9 codec. Electron. Imaging 2016(2), 1–5 (2016)

17. Hanhart, P., Ebrahimi, T.: Calculation of the average coding efficiency based on subjective quality scores. J. Vis. Comun. Image Representation 25(3), 555–564 (2014)

18. Katsavounidis, I., Guo, L.: Video codec comparison using the dynamic optimizer framework, p. 26 (2018). https://doi.org/10.1117/12.2322118

19. Wu, W., Song, B.: Just-noticeable-distortion-based fast coding unit size decision algorithm for high-efficiency video coding. Electron. Lett. 50, 443–444 (2014)

20. Li, Z., Bampis, C.: VMAF: the journey continues. Netflix Technology Blog | Netflix TechBlog. https://netflixtechblog.com/vmaf-the-journey-continues-44b51ee9ed12

Semantic-Based Image Retrieval Using R^S-Tree and Knowledge Graph

Le Thi Vinh Thanh[1,2], Thanh The Van[3], and Thanh Manh Le[1(✉)]

[1] University of Science - Hue University, Hue, Vietnam
lmthanh@hueuni.edu.vn
[2] Ba Ria – Vung Tau University, Vũng Tàu, Vietnam
[3] HCMC University of Education, Ho Chi Minh City, Vietnam
thanhltv@bvu.edu.vn, thanhvt@hcmue.edu.vn

Abstract. High-level semantic image retrieval is the problem which has applied in many different fields. In this paper, we approached a method of similarity image retrieval and image semantic extraction based on the combination of R^S-Tree and knowledge graph. The main tasks include: (1) building a knowledge graph to store objects, attributes, and relationships of objects on images; (2) searching a set of similar images based on the low-level features stored in R^S-Tree; (3) extracting a scene graph of image using Visual Genome dataset; (4) generating SPARQL query based on the scene graph to extract high-level semantic of the image from the built knowledge graph. In order to evaluate the efficiency as well as compare the accuracy of R^S-Tree, the COREL and Wang image datasets are used. On the basis of the proposed method, a knowledge graph is built based on the Visual Genome dataset. The experimental results are compared with related works to demonstrate the effectiveness of the proposed method. Therefore, our proposed method is feasible in semantic-based image retrieval systems.

Keywords: R^S-Tree · SBIR · SPARQL · Scene graph · Knowledge graph

1 Introduction

The problem of retrieving a set of similar images in large databases is interested in many researchers and is applied in many fields [1, 2]. Many image retrieval methods have been proposed, in which content-based image retrieval (CBIR) is one of the effective methods. However, the major challenge of the CBIR is the semantic gap between the low-level features and the high-level semantic concepts [3]. In recent years, many works have approached the knowledge graph and the scene graph to describe semantics of images [4, 5]. Then, objects, attributes, and relationships of the objects on the image must be extracted and represented [6].

In this paper, we propose a model of semantic-based image retrieval using R^S-Tree and knowledge graph (KG). In this model, there are two main tasks including (1) retrieving a set of similar images based on low-level visual features using R^S-Tree; (2) extracting high-level semantics and relationships of objects on images using KG. Firstly,

© The Author(s), under exclusive license to Springer Nature Switzerland AG 2022
N. T. Nguyen et al. (Eds.): ACIIDS 2022, LNAI 13757, pp. 481–495, 2022.
https://doi.org/10.1007/978-3-031-21743-2_38

the low-level features of images in the dataset are extracted such as color, position, shape... Then, the set of similar images is retrieved using R^S-Tree. Secondly, a KG is built based on Visual Genome (VG) dataset and WordNet. To perform this, we extract the objects, attributes, and relationships on the image from VG dataset. The KG is built by RDF/XML triple language, which is including a set of vertices as the object labels on the image and edges as the relationships between objects. The vertices on the knowledge graph consist of three types (1) the object label; (2) the object identifier concept; (3) the image identifier. The relationships represent the context of objects in the image include spatial relationships, action relationships, descriptive verb relationships, and comparative relationships. With a query image, the system extracts a set of similar images and the scene graphs based on VG dataset. After that, the SPARQL query is generated from the scene graphs. Then, the high-level semantic concepts and the semantic descriptions for the image are extracted from KG.

2   Related Works

The scene graph contains objects, attributes, and relationships between them to represent the image [6]. Scene graph generation consists of two main tasks as follows: extracting visual objects in images and creating relationships between them [7, 8].

Johnson J. et al. [6] proposed a semantic-based image retrieval framework using the concept of the scene graph. The authors used scene graphs to retrieve semantically related images. In the work, a conditional random field model is de-signed to reason about possible groundings of scene graphs. The likelihoods of these groundings are used as ranking scores for retrieval. The experiment of this work is built on 5000 scene graphs to evaluate query performance. The experimental results show that the proposed method is more effective than the methods using low-level features.

Wang S. et al. [10] introduced an image retrieval model using scene graphs including visual scene graphs (VSG) and text scene graphs (TSG). Image-text retrieval tasks are formulated as cross-modal scene graph matching. In the process of image retrieval, object features and relationships are used to evaluate image similarity. The authors carry out on the Flickr30k and MS COCO dataset, the experimental results show that the graph match-based approach to image retrieval is effective.

Schroeder B. et al. [11] presented a method that uses scene graph embedding as the basis for an approach to image retrieval. The authors examine how visual relationships, derived from scene graphs, can be used as structured queries. The visual relationships are directed subgraphs of the scene graph with a subject and object as nodes connected by a predicate relationship. The authors used six relationships including left of, right of, above, below, inside, and surrounding. This work conducts on the COCO-Stuff dataset. The results show that the recall increased by 10% in the best case.

Yoon S. et al. [12] proposed an approach for image-to-image retrieval using scene graph by graph neural networks. In this work, graph neural networks were trained to predict the image relevance measure, computed from human-annotated captions using a pre-trained sentence similarity model. The authors collected the dataset for image relevance measured by human annotators to evaluate retrieval algorithms. The experiment is implemented on two image datasets involving VG-COCO and Flickr30. The experimental results show that the proposed method was more effective than others.

Qi M. et al. [13] proposed a new framework for online cross-modal scene retrieval based on binary representations and semantic graphs. This approach is also used to perform text-based image retrieval. The authors adopt the cross-modal hashing based on the quantized correlation and measure the semantic agreement and similarity of the semantic graph for each instance. This work used four main components including multimodal binary representation, semantic graph, common objective function, and online update method. The test results on four data sets show the effectiveness of the proposed method compared to the existing method.

Ramnath et al. [14] introduced a neural-symbolic approach for a one-shot retrieval of images from a large-scale catalog, given the caption description. To facilitate this, they represent the catalog and caption as scene graphs and model the retrieval task as a learnable graph matching problem, trained end-to-end with a reinforce algorithm. The work used the CLEVR dataset for their experiments. This work meets with a retrieval accuracy of 99.9% when the full scene graph is used, it drops to 89% and subsequently 75% when 20% and 30% of the nodes are dropped from the query.

Quinn M. H. et al. [15] described a novel architecture for retrieving instances of a query visual situation in a collection of images. This architecture, called Situ-ate, is a combination of object localization models based on visual features with probabilistic models that represent learned multi-object relationships. With a query image, the system uses these models to create a bounding box that localizes relevant objects and relationships in the image. In addition, the system uses the resulting grounding to calculate a score for images containing an instance of the situation. This work experimented on 500 images taken from the VG dataset.

These recent approaches have focused on methods automatically scene graph generation. The results of those works show that applying scene graphs for the problem of semantic-based image retrieval is feasible. However, these works have not built knowledge graphs to describe the semantics of images. In addition, the works have not combined content-based image retrieval and semantic-based image retrieval. On the other hand, the works have not performed queries based on SPARQL query language to query scene graphs on knowledge graphs. Based on the findings and limits of the existing studies, we propose a model of semantic-based image retrieval using R^S-Tree and knowledge graph to improve image retrieval performance.

3   Architecture of the Semantic-Based Image Retrieval System

In this section, we describe a model of semantic-based image retrieval using R^S-Tree and KG, named SBIR-RSTKG, illustrated in Fig. 1. The model SBIR-RSTKG consists of two phases: (1) KG framework is built for a description of images semantic based on VG dataset and WordNet; the feature vectors of the image dataset are extracted and stored on R^S-Tree; (2) the low-level features of the query image is extracted; at the same times, its scene graph is extracted from VG dataset; the process of retrieving image is performed on the R^S-Tree; SPARQL queries are created from the scene graph and high-level semantics of images is extracted based on the built KG.

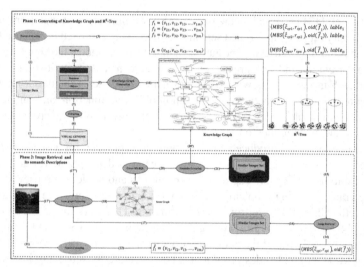

Fig. 1. The model of semantic-based image retrieval using R^S-Tree and KG

The process of semantic-based image retrieval in the SBIR-RSTKG model is described as follows:

Pre-processing phase: The process of building R^S-Tree and knowledge graph using VG dataset includes two tasks as follows:

i) Building the R^S-Tree structure

Step 1. Extract the feature vectors of VG image dataset (1–3).

Step 2. Represent the feature vector of image dataset in the form of a sphere (4).

Step 3. Build R^S-Tree structure based on the proposed similar measure (5).

ii) Generating a scene graph

Step 1: Extract the objects, attributes, and relationships in VG (6, 7).

Step 2: Extract semantics, URIs, and metadata from the WordNet (8).

Step 3: Build a knowledge graph using RDF/XML triple language (9–10).

Image retrieval phase: The process of semantic-based image retrieval is per-formed on R^S-Tree and the high-level semantics is extracted on KG. This process includes two stages as follows:

i) Retrieving similar images on R^S-Tree

Step 1. Extract feature vector of the query image and represent in the form of a sphere (11, 12, 13).

Step 2. Retrieve similar images using R^S-Tree (14, 15).

Step 3. Return set of content-based similar images (16).

ii) Extracting image semantics based on KG

Step 1. Extract scene graph of query image and set of similar images using VG dataset (17, 18).

Step 2. Create a SPARQL query from scene graphs (19).

Step 3. Extract high-level semantic concepts and semantic descriptions of objects in the image based on KG (20).

Step 4. Return set of similar images and their semantic descriptions (21).

4 Image Retrieval System

4.1 Description of R^S-Tree Structure

R^S-Tree [23] is built based on SS-Tree structure. The center vector of the sphere of the leaf and node on R^S-Tree is similar to SS-Tree. The improvements of R^S-Tree including (1) creating a sphere structure to store feature vector of an image; (2) improving the node splitting process based on the difference measure; (3) proposing a theta threshold to cluster similar data; (4) combining nearest neighbor and spatial region queries.

R^S-Tree is a balanced multi-branch tree used for similar image retrieval. The process of data clustering is performed on each node of R^S-Tree based on the Euclidean distance and threshold θ. This process creates a balanced multi-branch clustering tree to enhance the accuracy of the retrieval system and reduce the retrieval time. R^S-Tree is a data partition structure including a root, a set of nodes, and a set of leaves.

Internal node denoted S_{node} is of the form $\langle MBS, p \rangle$. Where MBS is a sphere that has center denoted $\overrightarrow{c}_{node}$, and radius r_{node}, p is the link pointer to the child nodes. This sphere covers the spheres of nodes in each sub branch of the tree. Each S_{node} has a minimum element of 2 and a maximum of N.

The leaf node denoted S_{leaf} is of the form $\langle MBS, element \rangle$. Where MBS is a sphere that has center denoted $\overrightarrow{c}_{leaf}$ and radius r_{leaf} contains a set of elements. Each element $spED$ is of the form $\langle MBS, oid \rangle$. Where MBS is a sphere that has center denoted $\overrightarrow{c}_{sp}$, and radius r_{sp}, contains the object space, oid is identifier $\overrightarrow{f} = (v_1, v_2, v_3, ..., v_d)$. Each leaf node S_{leaf} has the maximum number of elements called M and the minimum $m(1 < m < M/2)$.

4.2 Knowledge Graph Construction

In this section, we present a method of building KG to describe semantics for images from Visual Genome dataset and WordNet. The VG dataset includes the following main components: region descriptions, objects, attributes, relationships, region graphs, and scene graphs. We extract class labels of objects, attributes, and the most frequent relationships between them. From there, a knowledge graph is generated from these components using RDF/XML triple language. In addition, we build a dictionary based on WordNet to describe high-level semantics for image objects.

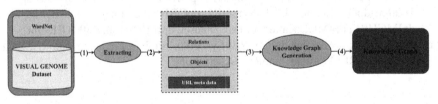

Fig. 2. The process of building KG

A knowledge graph is a graph $G = <O, A, R>$ illustrated in Fig. 2, where $O = \{o_1, o_2, \ldots, o_n\}$ is a set of objects that are vertices on the graph, o_i is an object class label, concept, or instance; $A = \{a_1, a_2, \ldots, a_n\}$ is the set of attributes, a_i is the attribute of the concept or instance; $R = \{r_1, r_2, \ldots, r_n\}$ is the set of relationships that are the edges of the graph, r_i is a relationship between concepts and instances, or the relationship between concepts and attributes, or relationship between instances and attributes. A data model and visualized structure of KG is illustrated in Fig. 3 and Fig. 4.

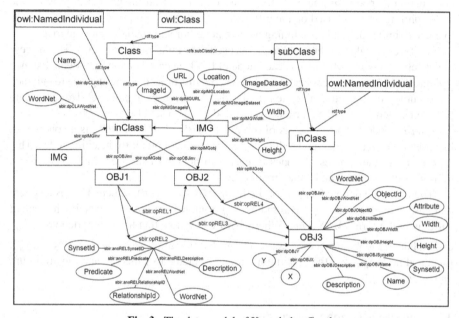

Fig. 3. The data model of Knowledge Graph

Compoments in KG structure include:

(1) Node Types: Classes (class), inClass (individual), Objects (individual), Images (individual);
(2) Relationships: Object Properties (opOBJinv, opIMGinv, opIMGobj), Relations between Objects (ON, IN, OF, WEAR, RIDE, …);
(3) Data Properties: Object (dpOBJWordNet, dpOBJObjectID, dpOBJAtribute, dpOBJWidth, dpOBJHeight, dpOBJSynsetID, dpOBJName) Image (dpIMGImageID, dpIMGURL, dpIMGLocation, dpIMGDataset, dpIMGWidth, dpIMGHeight);
(4) Annotations: Relationship (anoRELSynsetID, anoRELPredicate, anoRELRelationID, anoRELWordNet, anoRELDescription).

4.3   Scene Graph

A scene graph is a data structure that describes the contents of a sceneillustrated in Fig. 5. A scene graph presents object instances, attributes of objects, and relationships between

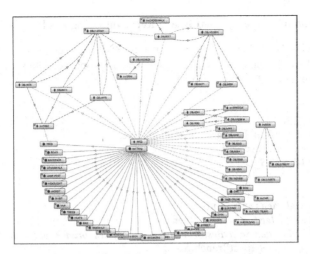

Fig. 4. Visualized structure of KG on Protégé.

objects [8]. Let C be a set of object classes, a set of at-tribute types as A, and a set of relationship types as R. A scene graph of an image is described as follows:

A scene graph$G = (O, E)$, where $O = \{o_1, ..., o_n\}$ is a set of objects and $E \subseteq O \times R \times O$ is a set of edges. Each object has the form $o_i = (c_i, A_i)$ where $c_i \in C$ is the class of the object and $A_i \subseteq A$ are the attributes of the object.

In this section, we perform the extraction of a sub-scene graph from the scene graph in the VG dataset. In the VG dataset, there are more than 2.3 million relationships, of which more than 50% of the relationships involve very small and unclear objects. Therefore, in this paper, we only select 9 objects with the largest region area to generate sub-scene graphs and describe semantics for these objects. Finally, a sub-scene graph is generated from the extracted objects, attributes, and relationships. The objects and attributes are vertices, and the relations between them are edges. A sub-scene graph for the image 2371376.jpg in the VG dataset is illustrated in Fig. 5.

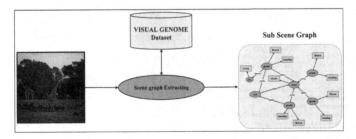

Fig. 5. A sub scene graph of image 2371376.jpg

4.4   Algorithm of Content-Based Image Retrieval Using R^S-Tree

From the spatial data where clustering structure R^S-Tree is created, an algorithm for content-based similar images retrieval on R^S-Tree is proposed. The process of similar image retrieval on R^S-Tree is described as in Algorithm 1

Algorithm 1. RSIR
Input: element $spED$ of the image retrieval, R^S-Tree.
Output: the similar image set SI
Function: $\textbf{RSIR}(S_{Nr}, S_{spED})$
begin
  $S_{Node} = S_{Nr}$;
  if $S_{Node} = null$ **then return** $null$;
  else if $(S_N.Child.leaf == false)$ **then**
          Use function SelectBestBranch to select the best branch follow similar measure
          $S_{Nk} = \text{SelectBestBranch}(S_{Node}, S_{spED})$;
          $\textbf{RSIR}(S_{Nk}, S_{spED})$;
      else $SI = S_{Node}.Child.spEC[i].ListSpED|i = 1..S_{Node}.size$;
      endif
  endif
  return $\{SI\}$
end

Call n as size of the dataset, M is the maximum number of the elements in a node of R^S-Tree. Algorithm RSIR browses from the root to the leaf. Each time of browsing, algorithm RSIR must be compared with M element of a node. M is constant. Therefore, the complexity of the algorithm RSIR is $O(\log n)$.

4.5   Creating of SPARQL Query

Based on the scene graph, the SPARQL command is automatically generated for retrieval on knowledge. The query result is a set of URIs with the image semantics and metadata of the similar image dataset [20]. Figure 6 illustrates the query generated from the scene graph.

```
1   PREFIX rdf: <http://www.w3.org/1999/02/22-rdf-syntax-ns#>
2   PREFIX owl: <http://www.w3.org/2002/07/owl#>
3   PREFIX rdfs: <http://www.w3.org/2000/01/rdf-schema#>
4   PREFIX xsd: <http://www.w3.org/2001/XMLSchema#>
5   PREFIX SBIR: <http://www.semanticweb.org/vinhthanh/ontologies/2022/3/untitled-ontology-195#>
6   SELECT ?has_Name?has_Synset?has_Att?has_H?has_objectID?has_W?has_X?has_Y
7   WHERE{ {{?semantics owl:annotatedSource SBIR:IMG1}
8       {?semantics owl:annotatedProperty SBIR:opBuildingR1}
9       {?semantics owl:annotatedTarget SBIR:inBuilding}
10      {?semantics SBIR:has_name ?has_Name}
11      {?semantics SBIR:hasSynsets ?has_Synset}
12      {?semantics SBIR:has_Attributes ?has_Att}
13      {?semantics SBIR:has_h ?has_H}
14      {?semantics SBIR:has_object-id ?has_objectID}
15      {?semantics SBIR:has_w ?has_W}
16      {?semantics SBIR:has_x ?has_X}
17      {?semantics SBIR:has_y?has_Y}}}
```

Fig. 6. Example of a semantic query using SPARQL

4.6   Algorithm of Image Semantic Retrieval

Relying on the scene graph SG of an image generated from the VG, the SPARQL query is created to execute the query on the built KG. Algorithm of image semantic retrieval is described as in Algorithm 2

Input: Scene graph SG of an image I
Output: Description semantics of image I
Function: IRSG(SG)
Begin
    Create Enum OBJ = {Name, ID}
    Create Enum REL = {ObjectID,SubjectID, RelID}
    1. Extracting name of object
    Foreach (object in SG.O) **do**
        OBJ_i.Name = object.Name; OBJ_i.ID = object.ID
    EndForeach
    2. Extracting relationship
    Foreach (relationship in SG.R) **do**
        REL_i.ObjectID = relationship.ObjectID;
        REL_i.SubjectID = relationship.SubjectID;
        REL_i.RelID = relationship.ID;
    EndForeach
    3. Extracting object semantics
    Foreach (obj in OBJ) **do**
        AttQuery = CreateSPARQL(obj.Name, "has-a", ?Attribute);
        SynQuery = CreateSPARQL(object.Name, "has-a", ?Synset);
        ExecuteQuery(AttQuery);   ExecuteQuery(SynQuery);
    EndForeach
    4. Extracting relationship semantics
    Foreach(rel in REL) **do**
        PreQuery = CreateSPARQL(rel.ObjectID,?Predicate, rel.subjectID);
        RelQuery = CreateSPARQL(rel.ID, "which is", ?Synset);
        ExecuteQuery(PreQuery);   ExecuteQuery(RelQuery);
    EndForeach
    End

5   Experiments and Discussions

5.1   Experimental Environment

In this paper, each image was extracted into low-level visual features describing content of the image such as color, location, shape. The feature vector had 292 values consisting of MPEG-7, SIFT, HOG, and Sobel.

    Pre-processing phase was performed on PC CPU 2.3 GHz 8-core 9th-generation Intel Core i9, 16 GB 2666 MHz memory, 1TB flash storage. The image retrieval phase is examined on PC CPU Intel Core i7-6500U CPU @ 2.50 GHz, 8.0 GB RAM, the operating system of Windows 10 Pro 64 bit.

We built a KG based on VG dataset. The VG consisted of 108,077 images which include an average of 35 objects, 50 regions, 26 attributes, and 21 relationships between object pairs per image. In the experiment of image retrieval, we used VG, COREL [21], and Wang [22] data sets. The experimental applications were illustrated in Figs. 7, 8, 9 and 10.

Figure 7 describes the process of image retrieving on R^S-Tree and a result of a similar image set. For each image, the objects with the largest region area were extracted. The properties of each object were extracted from the VG database and the descriptive semantics are automatically extracted from WordNet. Figure 8 descript process of creating KG. Figure 9 depicts the semantic relationship between the objects and the scene graph of an image. Figure 10 presents semantic descriptions of image.

Fig. 7. A result of image retrieval in R^S-Tree

Table 1 statistics the results of KG construction with information including data size, triple, creation time, memory processing, time loading data, storage memory to load data. The statistical data such as objects on the image, relationships, KG of classes, KG of images, KG of objects on the image, KG of the relationships of objects on the image, KG of the Visual Gennome dataset.

5.2 Experimental Evaluation

In this paper, the parameters M, m, N, θ were used in the process of building R^S-Tree to improve the retrieval precision. Experimental processing was conducted as follows:

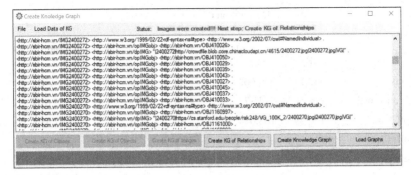

Fig. 8. Experimental process of creating KG

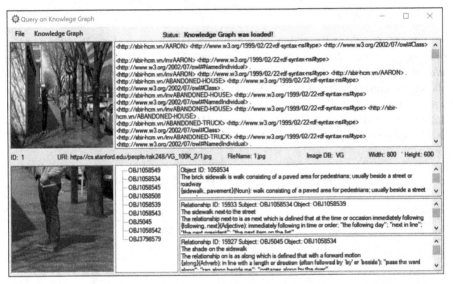

Fig. 9. Retrieving and extracting scene graph of image on KG

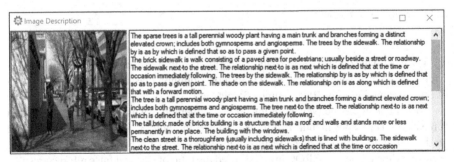

Fig. 10. Description of image semantic

Table 1. Description of the result creating KG

	Size (KB)	Tripple/records	Creating time	Loadingtime	Loadingmemory
ObjectsData	332,133	929,662	N/A	32 s	1.1 GB
RelationshipsData	111,508	584,373	N/A	7 s	403 MB
KGofClasses	13,449	108,084	2 s	3 s	262 MB
KGofImages	115,176	1,126,986	24 s	24 s	2.4 GB
KGofOjects	393,449	3,763,961	85 s	116 s	6.7 GB
KGofRelationships	192,330	1,753,119	34 s	47 s	4.0 GB
SumofpartialKG	714,404	6,752,150	145 s	190 s	13.362 GB

Let $nMax_{inclass}$ be the maximum number of elements in a class. We choose $M \in [nMax_{inclass} - \alpha, nMax_{inclass} + \alpha]$, α is constant; $N \in [2, M]$. Threshold θ is used to evaluate the similarity of elements belonging to a cluster. The number elements of each classifier estimate approximately 10% of the dataset. In the this paper, we experiment with the values $\theta \in [0.1 - \mu, 0.1 + \mu]$, $\mu \in [0, 0.05]$ to cluster the data depending on the dataset. Optimal parameters in the experimental process, the Top of retrieved elements and R^S-tree build time are presented in Table 2.

Table 2. Description of experimental parameters

Parameters	COREL	Wang	VG dataset
M	110	90	120
m	1	1	1
N	25	35	20
θ	0.09	0.12	0.09
topK	60	50	90
R^S-Tree building time (hours)	0.50	3.57	3.2

The test image dataset on VG dataset was divided into two subsets, including dataset 1 (consisting of images containing the 36 most common classes) dataset 2 (consisting of images containing the 2nd common 36 classes).

To evaluate the effectiveness of the proposed method, we used precision, recall, and F-measure. The experimental results was shown in Table 3, and Fig. 11. Each curve on the figure described retrieval results in the dataset of COREL, Wang, VG. Also, the corresponding curve in the ROC graph showed the ratio of true and false retrieval results. The area under this curve evaluates the correctness of retrieval results.

To evaluate the performance of the proposed model, we compare the results with previous works on the COREL, Wang dataset described in Table 4.

Table 3. Performance of retrieval systems on COREL, Wang, VG dataset

Dataset	Avg. Precision	Avg. Recall	Avg. F-measure	Avg. Query time (ms)
Dataset 1	0.7107	0.5975	0.6492	201.25
Dataset 2	0.6826	0.5824	0.6285	216.51
Average-VG	**0.6966**	**0.5900**	**0.6389**	**208.88**
COREL	**0.8029**	**0.6575**	**0.7229**	**15.39**
Wang	**0.7731**	**0.6036**	**0.6779**	**57.23**

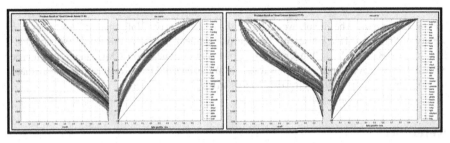

Fig. 11. Precision-Recall and curving ROC of VG dataset

Table 4. Comparison mean average precision (MAP) of methods

Methods	MAP	Dataset
L. Haldurai (2015) [17]	0.7388	COREL
Ahmed (2019) [18]	0.7210	COREL
SBIR-RSTKG	**0.8029**	**COREL**
P. Chhabra, 2020 [19]	0.6320	Wang
Lande, Milind V, 2014 [16]	0.6100	Wang
SBIR-RSTKG	**0.7731**	**Wang**

The comparison results showed the precision and effectiveness of the proposed model and algorithm. Therefore, the model SBIR-RSTKG can be developed to improve the efficiency of semantic image retrieval systems.

6 Conclusion

In this paper, we performed similar image retrieval and semantic extracting for those images. The retrieving of a similar image was performed on R^S-Tree in which nodes were spatial spheres to describe the set of images. The experiment was performed on COREL, Wang dataset to evaluate precision and demonstrate the feasibility of retrieving images using R^S-Tree. Then, the image set of the VG dataset was extracted features and stored on the R^S-Tree to perform similar image retrieval. We used the scene graphs of the images on the VG dataset to retrieve the semantics of those images on KG. There were two contributions in the paper including improving R^S-Tree and building KG. From there, the problem of similar image retrieval and its semantic extraction is conducted. To perform semantic extraction, we created a SPARQL query and execute the query on the KG using RDF triple language. The experiments were evaluated on Visual Genome, COREL, and Wang image datasets to demonstrate the effectiveness of our proposed method. In future work, we will combine the R-CNN network to classify objects on images and extract relationships between these objects based on spatial to generate scene graphs automatically.

Acknowledgment. The authors would like to thank the Faculty of Information Technology, University of Sciences - Hue University for their professional advice for this study. We would also like to thank Ba Ria - Vung Tau University, University of Education HCMC, and research group SBIR HCM, which are sponsors of this research. We also would like to express our sincere thanks to reviewers for their helpful comments on this article.

References

1. Li, X., Yang, J., Ma, J.: Recent developments of content-based image retrieval (CBIR). Neurocomputing **452**, 675–689 (2021)
2. Hameed, I.M., Abdulhussain, S.H., Mahmmod, B.M.: Content-based image retrieval: a review of recent trends. Cogent Eng. **8**(1), 1927469 (2021)
3. Bai, C., Chen, J.N., Huang, L., Kpalma, K., Chen, S.: Saliency-based multi-feature modeling for semantic image retrieval. J. Vis. Commun. Image Represent. **50**, 199–204 (2018)
4. Cui, P., Liu, S., Zhu, W.: General knowledge embedded image representation learning. IEEE Trans. Multim. **20**(1), 198–207 (2017)
5. Zareian, A., Karaman, S., Chang, S.-F.: Bridging knowledge graphs to generate scene graphs. In: Vedaldi, A., Bischof, H., Brox, T., Frahm, J.-M. (eds.) ECCV 2020. LNCS, vol. 12368, pp. 606–623. Springer, Cham (2020). https://doi.org/10.1007/978-3-030-58592-1_36
6. Johnson, J., et al.: Image retrieval using scene graphs. In: Proceedings of the IEEE Conference on Computer Vision and Pattern Recognition, pp. 3668–3678 (2015)
7. Ren, S., He, K., Girshick, R., Sun, J.: Faster R-CNN: towards real-time object detection with region proposal networks. Adv. Neural Inf. Process. Syst. **28** (2015)
8. Dai, B., Zhang, Y., Lin, D.: Detecting visual relationships with deep relational networks. In: Proceedings of the IEEE Conference on Computer Vision and Pattern Recognition, pp. 3076–3086 (2015)
9. Krishna, R., et al.: Visual genome: connecting language and vision using crowdsourced dense image annotations. Int. J. Comput. Vis. **123**(1), 32–73 (2017)
10. Wang, S., Wang, R., Yao, Z., Shan, S., Chen, X.: Cross-modal scene graph matching for relationship-aware image-text retrieval. In: Proceedings of the IEEE/CVF Winter Conference on Applications of Computer Vision, pp. 1508–1517 (2020)
11. Schroeder, B., Tripathi, S.: Structured query-based image retrieval using scene graphs. In: Proceedings of the IEEE/CVF Conference on Computer Vision and Pattern Recognition Workshops, pp. 178–179 (2020)
12. Yoon, S., et al.: Image-to-image retrieval by learning similarity between scene graphs. arXiv preprint arXiv:2012.1470.4322 (2020)

13. Qi, M., Wang, Y., Li, A.: Online cross-modal scene retrieval by binary representation and semantic graph. In: Proceedings of the 25th ACM International Conference on Multimedia, pp. 744–752, October 2017

14. Ramnath, S., Saha, A., Chakrabarti, S., Khapra, M.M.: Scene graph based image retrieval--a case study on the CLEVR dataset. arXiv preprint arXiv:1911.00850 (2019)

15. Quinn, M.H., Conser, E., Witte, J.M., Mitchell, M.: Semantic image retrieval via active grounding of visual situations. In: 2018 IEEE 12th International Conference on Semantic Computing (ICSC), pp. 172–179. IEEE, January 2018

16. Ahmed, K.T., Ummesafi, S., Iqbal, A.: Content based image retrieval using image features information fusion. Inf. Fus. **51**, 76–99 (2019)

17. Chhabra, P., Garg, N.K., Kumar, M.: Content-based image retrieval system using ORB and SIFT features. Neural Comput. Appl. **32**(7), 2725–2733 (2018). https://doi.org/10.1007/s00521-018-3677-9

18. Lande, M.V., Bhanodiya, P., Jain, P.: An effective content-based image retrieval using color, texture and shape feature. In: Mohapatra, D.P., Patnaik, S. (eds.) Intelligent Computing, Networking, and Informatics. AISC, vol. 243, pp. 1163–1170. Springer, New Delhi (2014). https://doi.org/10.1007/978-81-322-1665-0_119

19. Haldurai, L., Vinodhini, V.: Parallel indexing on color and texture feature extraction using r-tree for content based image retrieval. Int. J. Comput. Sci. Eng. **3**, 11–15 (2015)

20. Nhi, N.T.U., Le, T.M., Van, T.T.: A model of semantic-based image retrieval using C-Tree and neighbor graph. Int. J. Seman. Web Inf. Syst. **18**(1), 1–23 (2022)

21. Li, J., Wang, J.Z.: Automatic linguistic indexing of pictures by a statistical modeling approach. IEEE Trans. Pattern Anal. Mach. Intell. **25**(9), 1075–1088 (2003)

22. Wang, J.Z., Li, J., Wiederhold, G.: SIMPLIcity: semantics-sensitive integrated matching for picture libraries. IEEE Trans. Pattern Anal. Mach. Intell. **23**(9), 947–963 (2001)

23. Thanh, L.T.V., Thanh, L.M., Thanh, V.T.: Semantic-based image retrieval using RS-tree and neighbor graph. In: Rocha, A., Adeli, H., Dzemyda, G., Moreira, F. (eds.) Information Systems and Technologies. WorldCIST 2022. LNNS, vol. 469. Springer, Cham, pp. 165–176 (2022). https://doi.org/10.1007/978-3-031-04819-7_18

An Extension of Reciprocal Logic for Trust Reasoning: A Case Study in PKI

Sameera Basit[ID] and Yuichi Goto[(✉)][ID]

Department of Information and Computer Sciences, Saitama University,
Saitama 338-8570, Japan
{sameera,gotoh}@aise.ics.saitama-u.ac.jp

Abstract. Trust relationship is one of the kinds of reciprocal relationship and basis of communications among agents, especially in open and decentralized systems, e.g., public key infrastructure (PKI). In such systems, it is difficult to know whether an agent that is required to communicate with us can be trusted or not. Thus, it is indispensable to calculate the degree of trust of the target agent by using already known facts, hypotheses, and observed data. Trust reasoning is a process to calculate the degree of trust of the target agents. Although the current extension of reciprocal logic is an expectable candidate for a logic system underlying trust reasoning, it has a limitation when we deal with trust messages from other agents as a proposition. From the viewpoint of predicate logic, the current extension of reciprocal logic deals with messages from other agents as countable objects and are represented as individual constants. However, Demolombe represents messages from other agents as a proposition. From the viewpoint of expressive power, Demolombe's approach is better and the current extension of reciprocal logic is not enough. Following the Demolombe's approach, we introduced modal operators *Bel* and *Inf* and add several axioms to the current extension of reciprocal logic and a case study of trust reasoning based on the proposed extension in PKI is also presented.

Keywords: Trust reasoning · Strong relevant logic · Reciprocal logic

1 Introduction

A trust relationship is one of the important reciprocal relationships in our society and cyberspace. There are many reciprocal relationships that must concern two parties, e.g., parent-child relationship, relative relationship, friendship, cooperative relationship, complementary relationship, trade relationship, buying and selling relationship, and so on [6]. Especially, the trust relationship is the basis of communications among agents (human to human, human to system, and system to system), and the basis of the decision-making of the agents.

Trust reasoning is an indispensable process for establishing trustworthy and secure communication under open and decentralized systems that include multi-agents. In open and decentralized systems, although it is difficult to know

N. T. Nguyen et al. (Eds.): ACIIDS 2022, LNAI 13757, pp. 496–506, 2022.
https://doi.org/10.1007/978-3-031-21743-2_39

whether an agent that is required to communicate with us can be trusted or not before communication with it, we want to know whether the agent is trusted or not to establish trustworthy and secure communication, e.g., public key infrastructure (PKI). Thus, we should calculate the degree of trust of the target agent by using already known facts, hypotheses, and observed data. Trust reasoning is a process to calculate the degree of trust of the target agents and messages that come from other agents.

Although reciprocal logic [6] is an expectable candidate for a logic system underlying trust reasoning, current reciprocal logic cannot deal with several trust properties. Cheng proposed reciprocal logic [6] as a logic system underlying reasoning for such reciprocal relationships and formalized trust relationships between agents and agents, and agents and organizations in the logic. On the other hand, there are various trust properties for trust relationships, i.e., "an agent α trusts another agent β about a message from β in *property*" where *property* are sincerity, validity, vigilance, credibility, cooperativity, completeness, and so on [10,13–16]. The trust properties focus on not only trust relationships between agents and agents but also trust relationships between agents and messages that are informed by other agents. We cannot describe the trust relationship between agents and messages in current reciprocal logic.

An extension of reciprocal logic is demanded to deal with trust relationships between agents and messages which includes trust properties. We proposed an extension of reciprocal logic [4]. In the extension, messages that come from other agents are regarded as countable objects and are represented as individual constants from the viewpoint of predicate logic. On the other hand, Demolombe [8] defined several trust properties and formalized them. He regarded messages from agents as the beliefs of the agents and represented them as propositions (logical formulas) from the viewpoint of predicate logic. From the viewpoint of expressive power, Demolombe's approach is better than the approach of our last extension, so the last extension is not enough.

This paper presents a new extension of reciprocal logic that can deal with trust properties and shows a case study of trust reasoning based on the proposed extension in PKI. We introduced two modal operators *Bel* and *Inf* into reciprocal logic to represent trust relationships between agents and messages from other agents according to Demolombe's approach. We also add several axioms into reciprocal logic. One of the reasons why we want to know trustworthy agents is to reduce the process of whether the messages from the target agents are correct or not. In other words, the reason is to filter messages. Under this consideration, we defined axioms representing how agent α deals with a message from agent β if α trusts β in some trust property. Finally, we conducted a case study of trust reasoning based on the proposed extension in PKI. The case study shows that the proposed extension can deal with several trust properties. Thus, we can conclude that the proposed extension is an expectable candidate for a logic system underlying trust reasoning. The rest of the paper is organized as follows: Sect. 2 presents a summary of survey results concerning trust relationships and trust properties, the limitations of reciprocal logic, and the last extension. Section 3 presents the new extension of reciprocal logic. In Sect. 4, we showed a case study of trust reasoning based on the new extension in PKI. Some concluding remarks are given in Sect. 5.

2   Related Works

2.1   Trust Relationship and Trust Properties

Trust is a common phenomenon, and it is an essential element in a relationship that concerns two parties. These two parties are usually regarded as a trustor and a trustee when we consider a trust relationship, i.e., a trustee provides trustworthy messages to make the trustor trust the trustee. Trust relationships between parties are more tractable with the aid of trust properties.

Many previous works focus on trust properties. Several authors in [9,11] target a certain property and focus on one dimension only whereas other authors in [10,16] deal with trust in the reliability, credibility, and collectively. The authors in [14,15] provide a classification of trust properties from the viewpoint of the trustor and trustee and regard them as essential in the establishment of a trust relationship.

In the context of trust not all the information from the other agent can be taken as a true message, i.e., an agent α trusts another agent β with respect to some property means that α believes that β satisfies this property. Demolombe [8] defined several trust properties. His definitions are as follows.

- *Sincerity*: An agent α trusts in the sincerity of an agent β if β informs α about a proposition p then β believes p.
- *Validity*: An agent α trusts in the validity of an agent β if β informs α about a proposition p then p is the case.
- *Completeness*: An agent α trusts in the completeness of an agent β if p is the case then β informs α about p.
- *Cooperativity*: An agent α trusts in the cooperativity of an agent β if β believes p then β informs $alpha$ about p.
- *Credibility*: An agent α trust in the credibility of an agent β if β believes p then p is the case.
- *Vigilance*: An agent α trust in the vigilance of an agent β if p is the case then β believes p.

Demolombe also provided a formal definition of the above properties. His formalization is based on classical mathematical logic.

2.2   Reciprocal Logic and Its Extension

Reciprocal logic was proposed by Cheng [6] as a logic system underlying reasoning for a reciprocal relationship. Classical mathematical logic and its various conservative extensions are not suitable for logic systems underlying reasoning because they have paradoxes of implication [2,3]. Strong relevant logic has rejected those paradoxes of implication and is considered the universal basis of various applied logic for knowledge representation and reasoning [5,7]. Thus, strong relevant logic and its conservative extensions are candidates for logic systems underlying reasoning. Reciprocal logic is one of the conservative extensions of strong relevant logic to deal with various reciprocal relationships. Reciprocal logic provides primitive predicates representing trust relationships between an agent and another agent, and between an agent and an organization, defined predicates based on the primitive predicates, and several axioms that include the predicates [6]. Let pe_1, pe_2, and pe_3 be individual variables representing agents, and let o_1 and o_2 be individual variables representing organizations. The primitive predicates are as follows

- $TR(pe_1, pe_2)$: pe_1 trusts pe_2.
- $B(pe_1, o_1)$: agent pe_1 belongs to organization o_1

Defined predicates based on the above primitive predicate are as follows.

- $NTR(pe_1, pe_2) =_{df} \neg(TR(pe_1, pe_2))$ ($NTR(pe_1, pe_2)$ means pe_1 does not trust pe_2).
- $TREO(pe_1, pe_2) =_{df} TR(pe_1, pe_2) \wedge (TR(pe_2, pe_1))$ ($TREO(pe_1, pe_2)$ means pe_1 and pe_2 trust each other.)
- $ITR(pe_1, pe_2, pe_3) =_{df} \neg(TR(pe_1, pe_2) \wedge TR(pe_1, pe_3))$ ($ITR(pe_1, pe_2, pe_3)$ means pe_1 does not trust both pe_2 and pe_3 (Incompatibility))
- $XTR(pe_1, pe_2, pe_3) =_{df} (TR(pe_1, pe_2) \vee TR(pe_1, pe_3)) \wedge (NTR(pe_1, pe_2) \vee NTR(pe_1, pe_3))$ ($XTR(pe_1, pe_2, pe_3)$ means pe_1 trusts either pe_2 or pe_3 but not both (exclusive disjunction)).
- $JTR(pe_1, pe_2, pe_3) =_{df} \neg(TR(pe_1, pe_2) \vee TR(pe_1, pe_3))$ ($JTR(pe_1, pe_2, pe_3)$ means pe_1 trusts neither pe_2 nor pe_3 (joint denial)).
- $TTR(pe_1, pe_2, pe_3) =_{df} (TR(pe_1, pe_2) \wedge TR(pe_1, pe_3)) \Rightarrow TR(pe_1, pe_3)$ ($TTR(pe_1, pe_2, pe_3)$ means pe_1 trusts pe_3 if pe_1 trusts pe_2 and pe_2 trusts pe_3).
- $CTR(pe_1, pe_2, pe_3) =_{df} (TR(pe_1, pe_3) \Rightarrow (TR(pe_2, pe_3))$ ($CTR(pe_1, pe_2, pe_3)$ means pe_2 trusts pe_3 if pe_1 trusts pe_3.)
- $NCTR(pe_1, pe_2, pe_3) =_{df} (\neg TR(pe_1, pe_3) \Rightarrow (TR(pe_2, pe_3))$ ($NCTR(pe_1, pe_2, pe_3)$ means pe_2 trusts pe_3 if pe_1 does not trusts pe_3)
- $CNTR(pe_1, pe_2, pe_3) =_{df} \neg(TR(pe_1, pe_3) \Rightarrow \neg(TR(pe_2, pe_3))$ ($CNTR(pe_1, pe_2, pe_3)$ means pe_2 does not trusts pe_3 if pe_1 does not trusts pe_3)
- $TRpo(pe_1, o_1) =_{df} \forall pe_2(B(pe_2, o_1) \wedge (TR(pe_1, pe_2))$ ($TRpo(pe_1, o_1)$ means pe_1 trusts o_1).
- $NTRpo(pe_1, o_1) =_{df} \forall pe_2(B(pe_2, o_1) \wedge (NTR(pe_1, pe_2))$ ($NTRpo(pe_1, o_1)$ means pe_1 does not trusts o_1).
- $TRop(o_1, pe_1) =_{df} \forall pe_2(B(pe_2, o_1) \wedge (TR(pe_2, pe_1))$ ($TRop(o_1, pe_1)$ means o_1 trusts pe_1).
- $NTRop(o_1, pe_1) =_{df} \forall pe_2(B(pe_2, o_1) \wedge (NTR(pe_2, pe_1))$ ($NTRop(o_1, pe_1)$ means o_1 does not trusts pe_1).
- $TRoo(o_1, o_2) =_{df} \forall pe_1 \forall pe_2(B(pe_1, o_1) \wedge (B(pe_2, o_2)) \wedge (TR(pe_1, pe_2))$ ($TRoo(o_1, o_2)$ means o_1 trusts o_2).

- $NTRoo(o_1, o_2) =_{df} \forall pe_1 \forall pe_2 (B(pe_1, o_1) \wedge (B(pe_2, o_2))) \wedge (NTR(pe_1, pe_2)$
  ($NTRoo(o_1, o_2)$ means o_1 does not trusts o_2).

Through the above definitions of predicates, we can consider that reciprocal logic focuses on only trust relationship between an agent and other agent, and between an agent and an organization.

Axioms of the reciprocal logic are as follows:

TR1: $\neg(\forall pe_1 \forall pe_2 (TR(pe_1, pe_2) \Rightarrow TR(pe_2, pe_1)))$
TR2: $\neg(\forall pe_1 \forall o_1 (TRpo(pe_1, o_1) \Rightarrow TRop(o_1, pe_1)))$
TR3: $\neg(\forall o_1 \forall pe_1 (TRop(o_1, pe_1) \Rightarrow TRpo(pe_1, o_1)))$
TR4: $\neg(\forall o_1 \forall o_2 (TRoo(o_1, o_2) \Rightarrow TRoo(o_2, o_1)))$
TR5: $\neg(\forall pe_1 \forall pe_2 \forall pe_2 (TR(pe_1, pe_2) \wedge TR(pe_2, pe_3) \Rightarrow TR(pe_1, pe_3)))$
TR6: $\neg(\forall pe_1 \forall pe_2 \forall o_1 (TRpo(pe_1, o_1) \wedge TRop(o_1, pe_2) \Rightarrow TR(pe_1, pe_2)))$
TR7: $\neg(\forall pe_1 \forall pe_2 \forall o_1 (TRop(o_1, pe_1) \wedge TR(pe_1, pe_2) \Rightarrow TRop(o_1, pe_2)))$
TR8: $\neg(\forall o_1 \forall o_2 \forall o_3 (TRoo(o_1, o_2) \wedge TRoo(o_2, pe_3) \Rightarrow TR(o_1, o_3)))$

$TrTcQ =_{df} TcQ + \{TR1, \dots, TR8\}$, $TrEcQ =_{df} EcQ + \{TR1, \dots, TR8\}$, and $TrRcQ =_{df} RcQ + \{TR1, \dots, TR8\}$ are the minimal logic systems of reciprocal logic where TcQ, EcQ, and RcQ are logic systems of the first order predicate strong relevant logics [6].

We proposed an extension of reciprocal logic to deal with trust properties [4]. Current reciprocal logic cannot deal with the trust properties explained in Sect. 2.1 because it does not provide a representation method of the trust relationship between an agent and a message that came from other agents. We introduced several predicates to represent the trust relationship between an agent and a message into reciprocal logic. In the extension, messages that come from other agents are regarded as countable objects, and are represented as individual constants from the viewpoint of predicate logic. However, the extension is not enough to represent the trust properties explained in Sect. 2.1.

3 A New Extension of Reciprocal Logic

Although we proposed an extension of reciprocal logic for trust reasoning [4], we regarded messages that come from agents as countable objects (individual constants). From the viewpoint of applications of trust reasoning, we should regard the messages from other agents as propositions like Demolombe's logic system [8]. Thus, we replaced the trust properties in the first extension with Demolombe's logic system like logical formulas.

At first, we add a predicate "$TR(pe_1, pe_2, PROP)$" where pe_1 and pe_2 are agents, and $PROP$ is individual constant that represents trust properties: sincerity, validity, completeness, cooperativity, credibility, and vigilance into reciprocal logic. For example, "$TR(pe_1, pe_2, sincerity)$" means "$pe_1$ trusts pe_2 in sincerity". Note that "$TR(pe_1, pe_2, all)$" means "pe_1 trusts pe_2 in all trust properties", i.e., "$TR(pe_1, pe_2)$" in reciprocal logic is as same as "$TR(pe_1, pe_2, all)$" in our new extension.

Secondly, we introduced two modal operators $Bel_i(A)$ and $Inf_{i,j}(A)$ used in Demolombe's logic system into reciprocal logic to represent the trust relationship between agents and information that comes from other agents. The two modal operators follow the KD systems of modal logic [8].

$Bel_i(A)$: an agent i believes that a proposition A is true.
$Inf_{i,j}(A)$: an agent i has informed an agent j about A.

Finally, we add new axioms into reciprocal logic.

ERcL1: $\forall i \forall j (TR(i, j, sincerity) \Rightarrow (Inf_{j,i}(A) \Rightarrow Bel_j(A)))$
ERcL2: $\forall i \forall j (TR(i, j, validity) \Rightarrow (Inf_{j,i}(A) \Rightarrow A))$
ERcL3: $\forall i \forall j (TR(i, j, vigilance) \Rightarrow (A \Rightarrow Bel_j(A)))$
ERcL4: $\forall i \forall j (TR(i, j, credibility) \Rightarrow (Bel_j(A) \Rightarrow A))$
ERcL5: $\forall i \forall j (TR(i, j, cooperativity) \Rightarrow (Bel_j(A) \Rightarrow Inf_{j,i}(A)))$
ERcL6: $\forall i \forall j (TR(i, j, completeness) \Rightarrow (A \Rightarrow Inf_{j,i}(A)))$
BEL: $\forall i (Bel_i(A \Rightarrow B) \Rightarrow (Bel_i(A) \Rightarrow Bel_i(B)))$.

We summarize our new extension of reciprocal logic. Let RcL be all axioms of reciprocal logic. Our new extension is $RcL \cup \{ERcL1, \ldots, ERcL6, BEL\}$.

4   A Case Study of Trust Reasoning Based on New Extension in PKI

4.1   Scenario

We present a simple scenario in PKI inspired from [12]. We have formalized the scenario and applied the trust reasoning process based on new extension in PKI.

Suppose that a certificate c_2 is signed by the subject of a certificate c_1 with the private key corresponding to the public key of c_1. Agent e_1 trusts the certificate c_1 because c_1 is informed by its parent agent. In PKI, we consider that every agent trusts its parent agent in its validity, i.e., $\forall e(TR(e, parent(e), validity))$. Moreover, agent e_2 informs agent e_1 about certificate c_2. We assume that agent e_1 trusts agent e_2 in its completeness, i.e., $TR(e_1, e_2, completeness)$. Agent e_1 does not trust the certificate but wishes to use certificate c_2. We need to know that whether certificate c_2 informed by agent e_2 is valid or not. From these two trust relationship: $TR(e_1, parent(e_1), validity)$ and $TR(e_1, e_2, completeness)$, we can conclude that certificate c_2 informed by agent e_2 is valid, i.e., $Inf_{e_2,e_1}(isValid(c_2))$.

4.2   Formalization

To formalize the above scenario, we defined following constants, functions, and predicates.

- Individual variables:
  - e: an agent

- c, c': certifications
- Individual constants:
    - e_1, e_2: agents
    - c_1, c_2: certifications
    - $today$: date of today
- Functions:
    - $I(c)$: Issuer of certification c.
    - $S(c)$: Subject of certification c.
    - $PK(c)$: Public key of c.
    - $SK(c)$: Share key of c.
    - $DS(c)$: Start date of c.
    - $DE(c)$: End date of c.
    - $Sig(c)$: Signature of c.
    - $parent(e)$: The parent of agent e.
- Predicates:
    - $inCRL(c)$: c is in certification revocation list.
    - $isValid(x)$: x is valid.
    - $isSigned(x, k)$: x is message signed by key k.
    - $x = y$: x is equal to y.
    - $x \leq y$: x is equal to or less than y.
    - $x < y$: x is less than y.

In PKI, we can assume following empirical theories.

PKI1: $\forall e(\, TR(e, parent(e), validity))$
(Any agent trusts its parent agent in validity).

PKI2: $\forall c(\exists c'((isValid(c'))) \land (I(c) = S(c')) \land (isSigned(c, PK(c')))) \Rightarrow isValid(Sig(c)))$

PKI3: $\forall c((isValid(Sig(c)) \land (DS(c) \leq today) \land (today < DE(c)) \land \neg inCRL(c)) \Rightarrow isValid(c))$
(PKI2 and PKI3 allows to verify the signature and certificate itself on the basis of another certificate whose validity has been proven).

From scenario, we can assume following logical formulas.

P1: $I(c_2) = S(c_1)$
(This observed facts are used as a premises in our reasoning process and it is true in this scenario only).

P2: $isSigned(c_2, PK(c_1))$
(A certificate c_2 is signed by the subject of certificate c_1 with the private key corresponding to the public key of c_1).

P3: $Inf_{parent(e_1), e_1}(isValid(c_1))$
(The parent agent of e_1 has informed e_1 about "certificate c_1 is valid").

P4: $TR(e_1, e_2, completeness)$
(our assumption)

P5: $DS(c_2) \leq today$ (our assumption)

P6: $today < DS(c_2)$ (our assumption)

P7: $\neg inCRL(c_2)$ (our assumption).

In the next section, we used inference rules of reciprocal logic and our new extension for trust reasoning. The inference rules are as follows.

$\Rightarrow$E: "from A and $A \Rightarrow B$ to infer B" (Modus Ponens)
$\wedge$I: "from A and B infer $A \wedge B$" (Adjunction).

4.3 Trust Reasoning Process

According to the above formalization, we can reason out the expected conclusion "$Inf_{e2,e1}(isValid(c_2))$". The reasoning process is as follows.

1. $Inf_{j,i}(A) \Rightarrow A$ [Deduced from PKI1 and ERcL2 with $\Rightarrow$E]
2. $isValid(c_1)$ [Deduced from P3 and 1 with $\Rightarrow$E]
3. $isValid(c_1) \wedge (I(c_2) = S(c_1)) \wedge isSigned(c_2, PK(c_1))$ [Deduced from 2, P1 and P2 with $\wedge$I]
4. $\exists c'((isValid(c')) \wedge (I(c_2) = S(c')) \wedge (isSigned(c_2, PK(c')))) \Rightarrow isValid(Sig(c_2))$ [Substitute c_2 for c in PKI2]
5. $isValid(Sig(c_2))$ [Deduced from 3 and 4 with $\Rightarrow$E]
6. $isValid(Sig(c_2)) \wedge (DS(c_2) \leq today) \wedge (today < DE(c_2)) \wedge \neg inCRL(c_2)$ [Deduced from 5 and P5 to P7 with $\wedge$I)
7. $(isValid(Sig(c_2)) \wedge (DS(c_2) \leq today) \wedge (today < DE(c_2)) \wedge \neg inCRL(c_2)) \Rightarrow isValid(c_2)$ [Subsutitute c_2 for c in PKI3]
8. $isValid(c_2)$ [Deduced from 6 and 7 with $\Rightarrow$E]
9. $A \Rightarrow Inf_{j,i}(A)$ [Deduced from P4 and ERcL6 with $\Rightarrow$E]
10. $Inf_{e2,e1}(isValid(c_2))$ [Deduced from 8 and 9 with $\Rightarrow$E].

Having completed the trust reasoning process, we can therefore have $Inf_{e2,e1}(isValid(c_2))$ derived from the fact $Inf_{parent(e_1),e_1}(isValid(c_1))$.

Instinctively, it represents a trust transfer. Agent e_1 trust in certificate c_2 informed by an agent e_2 is transferred from its trust in the validity of its parent entity. In PKI, agents can transfer their trust from where it exists to where it is needed, e.g., if you initially trust the authenticity of a public key and you verify a message signed by the corresponding private key, then you will also trust the authenticity of the message [18]. Our trust reasoning process enables agents to achieve trust transfer correctly because it includes trust relationships with trust properties. Various forms of trust transfer occur in PKI. Since these certificates and PKI's do not create trust, they just propagate it [18]. Therefore, first agents must trust something. We can call it initial trust.

One of the advantages of our trust reasoning process based on reciprocal logic is that it provides us with trust relationships and their properties and these trust relationships can be regarded as initial trust. In our PKI Scenario, trust relationship between agent e_1 and its parent entity $TR(e_1, parent(e_1), validity)$ is considered as an initial trust. Therefore, based on the initial trust agent e_1 believes that certificate c_1 informed by its parent entity is valid. Moreover, agent e_1 trust in completeness of agent e_2 $TR(e_1, e_2, completeness)$ but at this point only agent e_2 believes that certificate c_2 is valid and agent e_1 need to know whether the informed certificate c_2 is valid or not. Therefore, through initial trust and other known trust relationships, agents can reason out the desired beliefs by correctly achieving trust transfer through our trust reasoning process.

4.4   Discussion

In PKI (Public key infrastructure), trust relationships play an important role, especially in cases when an agent wants to know whether the certificate informed by another agent is valid or not. Usually, authors have focused on certification relationships [12,19] instead of trust relationships. Our trust reasoning process focuses on two trust relationships, i.e., validity, and completeness. Why are these two trust relationships essential? Because if an agent trusts another agent in its validity, it means that the agent believes that the other agent is a valid information source about both p and $\neg p$. In PKI, every agent has trust in the validity of its parent entity and an agent believes that the information provided by its parent entity, whether p and $\neg p$, is valid. Also, if an agent in PKI has complete information, e.g., about a certificate the agent should inform another agent about that certificate.

Such trust properties are essential in a trust relationship, especially when an agent deals with a message from another agent. Traditional reciprocal logic only deals with the trust relationships between agents. Trust reasoning without these trust properties does not provide us with room to deal with messages from other agents. For such a purpose, a new extension of reciprocal logic is introduced with two new modal operators Bel and Inf and axioms. Thus, our new extension is $RcL \cup \{ERcL1, \dots , ERcL6, BEL\}$.

Moreover, traditional reciprocal logic does not help us deal with complex situations, for example, when agents have trust relationships based on trust properties. $TR(pe_1, pe_3, validity)$ cannot be concluded from $TR(pe_1, pe_2, validity)$ and $TR(pe_2, pe_3, validity)$ because trust relationships with trust properties are not transitive. Also, some studies have discussed reasons why trust is not transitive [17]. However if we consider the trust transitive as pseudo-transitivity [11], i.e., if all agents in a trust relationship have similar trust properties for the same proposition p then we can say that agent pe_1 trust agent pe_3 in its validity, i.e., $\neg(\forall pe_1 \forall pe_2 \forall pe_3 (TR(pe_1, pe_2, validity) \wedge TR(pe_2, pe_3, validity) \Rightarrow TR(pe_1, pe_3, validity)))$. Therefore, we can conclude that in a case where there are three agents pe_1, pe_2, and pe_3, and two of them, pe_1 and pe_3, do not have a trust relationship. Thus, based on pseudo transitivity, a trust relationship may be derived if agents pe_1 and pe_2, as well as pe_2 and pe_3, have trust relationships with the same trust property and hold the same belief.

Alongside, there are cases when agents can have different trust relationships with different trust properties and agents believe in different propositions. This refers to trust transfer. We have already discussed in Sect. 4.3 how agents can reason out desired beliefs by correctly achieving trust transfer through our trust reasoning process. In the scope of the current paper, trust relationships in the current PKI scenario focus on validity and completeness trust property only. Because in the domain of PKI, validity and completeness are one of the essential properties when dealing with messages from other agents. Also, not all the axioms have been used in the current scope of this paper, but future studies include the application of a trust reasoning process based on these axioms in scenarios complex PKI scenarios and other areas.

We know that trust relationships change themselves over space and time, e.g., at time t if agent α believe p coming from β and at time $t + 1$ agent α believe $\neg p$ coming from β. This could cause complexity for agent α when trusting agent β. This problem provides us a new insight into maintaining and updating trust relationships an agent has with other agents as an agent view. Through a trust relationship, one can capture the beliefs of the agent about the message from other agents at a specific time or in a particular space. These captured beliefs can be added to agents view whom the agent would trust. These agent views containing trust relationships need to be maintained because two different agents may unequally trust any received message and may act differently. Also, these view needs to be updated when an agent makes new trust relationships with other agents. Maintaining and updating views will not only help deal with the future threats but also aid in making a decision. Further research is needed to establish such agents view in new extensions of reciprocal logic.

5 Concluding Remarks

We have proposed a new extension of reciprocal logic that can deal with trust properties. Two modal operators *Bel* and *Inf* have been introduced to represent trust relationships between agents and messages from other agents. A case study has been shown in PKI. Modal operators and new axioms aid in reasoning out new trust relationships in PKI. This we believe is an improvement our new extension. One of the advantages of our approach is generality. Trust reasoning based on a new extension of reciprocal logic is general in terms that not only trust relationships in PKI could be described as an empirical theory but also trust relationships in various complex scenarios can be described.

In the future, the aim is to provide a trust reasoning framework based on a new extension of reciprocal logic and its implementation in various areas and the ideas contained in this paper. Moreover, dealing with trust relationships with time-related constraints is also part of future works.

References

1. Amgoud, L., Demolombe, R.: An argumentation-based approach for reasoning about trust in information sources. Argument Comput. **5**(2–3), 191–215 (2014). https://doi.org/10.1080/19462166.2014.881417
2. Anderson, A.R., Belnap Jr., N.D.: Entailment: The Logic of Relevance and Necessity, vol. 1. Princeton University Press, Princeton (1975)
3. Anderson, A.R., Belnap Jr., N.D., Dunn, J.M.: Entailment: The Logic of Relevance and Necessity, vol. 2. Princeton University Press, Princeton (1992)
4. Basit, S., Goto, Y.: An extension of reciprocal logics for trust reasoning. In: Nguyen, N.T., Jearanaitanakij, K., Selamat, A., Trawiński, B., Chittayasothorn, S. (eds.) ACIIDS 2020. LNCS (LNAI), vol. 12034, pp. 65–75. Springer, Cham (2020). https://doi.org/10.1007/978-3-030-42058-1_6

5. Cheng, J.: A strong relevant logic model of epistemic processes in scientific discovery. In Kawaguchi, E., Kangassalo, H., Jaakkola, H., Hamid, I.A. (eds.) Information Modeling and Knowledge Bases XI. Frontiers in Artificial Intelligence and Applications, vol. 61, pp. 136–159. Amsterdam, IOS Press (2000)

6. Cheng, J.: Reciprocal logic: logics for specifying, verifying, and reasoning about reciprocal relationships. In: Khosla, R., Howlett, R.J., Jain, L.C. (eds.) KES 2005. LNCS (LNAI), vol. 3682, pp. 437–445. Springer, Heidelberg (2005). https://doi.org/10.1007/11552451_58

7. Cheng, J.: Strong relevant logic as the universal basis of various applied logics for knowledge representation and reasoning. In Kiyoki, Y., Henno, J., Jaakkola, H., Kangassalo, H. (eds.) Information Modeling and Knowledge Bases XVII. Frontiers in Artificial Intelligence and Applications, vol. 136, pp. 310–320. IOS Press (2006)

8. Demolombe, R.: Reasoning about trust: a formal logical framework. In: Jensen, C., Poslad, S., Dimitrakos, T. (eds.) iTrust 2004. LNCS, vol. 2995, pp. 291–303. Springer, Heidelberg (2004). https://doi.org/10.1007/978-3-540-24747-0_22

9. Josang, A., Ismail, R., Boyd, C.: A survey of trust and reputation systems for online service provision. Decision Support Syst. **43**(2), 618–644 (2007). https://doi.org/10.1016/j.dss.2005.05.019

10. Koutrouli, E., Tsalgatidou, A.: Credibility enhanced reputation mechanism for distributed e-communities, In: Proceedings of the 2011 19th International Euromicro Conference on Parallel, Distributed and Network-Based Processing, pp. 627–634. IEEE Computer Society, Washington (2011). https://doi.org/10.1109/PDP.2011.68

11. Leturc, C., Bonnet, G.: A normal modal logic for trust in the sincerity. In: Proceedings of the 17th International Conference on Autonomous Agents and MultiAgent Systems, Stockholm, Sweden, pp. 175–183. International Foundation for Autonomous Agents and Multiagent Systems, Richland (2018)

12. Liu, C., Ozols, M., Cant, T.: An axiomatic basis for reasoning about trust in PKIs. In: Varadharajan, V., Mu, Y. (eds.) ACISP 2001. LNCS, vol. 2119, pp. 274–291. Springer, Heidelberg (2001). https://doi.org/10.1007/3-540-47719-5_23

13. Liau, C.: Belief, information acquisition, and trust in multi-agent systems: a modal logic formulation. Artif. Intell. **149**(1), 31–60 (2003). https://doi.org/10.1016/S0004-3702(03)00063-8

14. Namiluko, C.: An architectural approach for reasoning about trust properties. Ph.D. thesis, University of Oxford (2016)

15. Yan, Z., Zhang, P., Vasilakos, A.: A survey on trust management for Internet of Things. J. Netw. Comput. Appl. **42**, 120–134 (2014)

16. Zhao, H., Li, X.: A group trust management system for peer-to-peer desktop grid. J. Comput. Sci. Technol. **24**(5), 833–843 (2009). https://doi.org/10.1007/s11390-009-9275-7

17. Christianson, B., Harbison, W.S.: Why isn't trust transitive? In: Lomas, M. (ed.) Security Protocols 1996. LNCS, vol. 1189, pp. 171–176. Springer, Heidelberg (1997). https://doi.org/10.1007/3-540-62494-5_16

18. Jøsang, A., Pedersen, I.G., Povey, D.: PKI seeks a trusting relationship. In: Dawson, E.P., Clark, A., Boyd, C. (eds.) ACISP 2000. LNCS, vol. 1841, pp. 191–205. Springer, Heidelberg (2000). https://doi.org/10.1007/10718964_16

19. El Bakkali, H., Kaitouni, B.I. A logic-based reasoning about PKI trust model. In: Proceedings of Sixth IEEE Symposium on Computers and Communications, pp. 42–48 (2001). https://doi.org/10.1109/ISCC.2001.935353

Common Graph Representation
of Different XBRL Taxonomies

Artur Basiura[1,2](✉) ⓘ, Leszek Kotulski[2]ⓘ, and Dominik Ziembiński[1]

[1] BFT24.COM, ul. Chopina 9/11, 20-026 Lublin, Poland
{basiura,dominik.ziembinski}@bft24.com
[2] Department of Applied Computer Science, AGH University of Science
and Technology, Al. Mickiewicza 30, 30-059 Kraków, Poland
{abasiura,kotulski}@agh.edu.pl

Abstract. Information nowadays plays a critical role in our lives, and its misinterpretation or lack of data, makes decisions wrong. It is important to systematize it, not only in the local context but globally. Finance is one of the key areas where standardization and normalization are attempted. One of the attempts is the XBRL format which is widely used in finance. However, the problem is the nature of local implementations. There are many different taxonomies that are implemented independently by countries and organizations. Currently there are no attempts to combine them and create a single standard.

The paper presents a formal model for storing data in graph structures and the concept of using graph grammar to search financial indicators in big data storage. It provides a basis for the future construction of a common graph representation and thus the accumulation of cross-cutting knowledge

Keywords: Graph · Graph methods · Xbrl · Similarity graph

1 Introduction

Growing global market requires us to make quick and precise decisions. It is not possible without relying on correct and verified data. A perfect example of the misinterpretation of the data is the appearance of the real estate bubble, and consequently the financial crisis of 2007–2008. The reason was not the lack of information but the reliance on incorrect indicators and studies. The scale of the problem touched the globe. In one of the reports on financial crimes [4] it was indicated that reliance on incorrect financial data may be one of the cyber security threats. As a result, access to raw financial data is becoming more and more important. Data that should be verified and available for wider analysis. A popular format that has been implemented in many different countries is XBRL (eXtesible Business Reporting Language) [15]. This format was introduced in 2005 by the SEC (U.S. Securities and Exchange Commission) in order to standardize the reporting structures. Currently, reporting in this format is carried

© The Author(s), under exclusive license to Springer Nature Switzerland AG 2022
N. T. Nguyen et al. (Eds.): ACIIDS 2022, LNAI 13757, pp. 507–515, 2022.
https://doi.org/10.1007/978-3-031-21743-2_40

out not only in the United States but also in Japan, and since the last year in the European Union. The format which in theory unifies the method of settlements at the level of one country does not provide an opportunity to look at the economy globally. It is implemented differently in the United States and in the European Union. In the United States the applicable taxonomy is US GAAP and in the European Union the ESEF. The two taxonomies differ, among other things, with the names of the tags and the way the presentation and clustering layer is organized. The problem is complicated by the fact that it is possible to extend the taxonomy to include own metrics. It is not possible to easily transpose items from one report to another.

This article introduces formal graph notation that is dedicated to the storage of data from the XBRL format. The notation is independent of taxonomy. Importing reports to a graph database gives us the opportunity to perform operations related to the identification of structures with the same similarity [1,5,6,8,12]. It can be used in practice for very fast data retrieval. The last part shows a practical representation of the report parts in the US GAAP taxonomy. The compilation and storing those data in graph structure allow to search quickly for the needed data. Not only based on one taxonomy.

2   XBRL Format and Taxonomies

XBRL (eXtensible Business Reporting Language) is a format used to exchange information between business systems. The language enables the semantic expression of financial and business metrics. This data is stored in an organized manner using XML (Extensible Markup Language) format. Related technologies, such as XML Schema which contains definitions of report components and Namespaces were used to extend the basic structure in order to be able to comprehend the relationships between the report elements and to organize the references. XBRL can be used to store information related to various financial areas or various forms of reporting. To organize reporting issues and provide a single point of reference various taxonomies are defined.

Taxonomy is a system that can be used to identify and structure information that are used to identify financial metrics. It provides information not only about type of structures but also how factors and metrics should be organized (Fig. 1).

To make the XBRL approach consistent, it introduces several ordering layers:

- structure definitions - define what metrics and values may appear in the final report, and what is their meaning. Usually the description comes down to the name of the tag and a short description of the value it represents and the size in which it is expressed (including price tag, currency),
- description layer - is an extension of the definition with information on how a given tag is to be presented in the context of a specific report, or in the context of the language in which the report is created,
- presentation layer - that is, the definition of what types of statements may be included in the final report definition, and how they should be presented. The main purpose is to be able to present reports in a consistent manner.

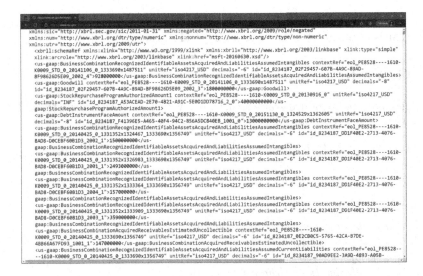

Fig. 1. Example of the fragment of the MICROSOFT report for 2016 in XBRL format [9]

- calculation layer - defines the basic rules for verifying the values that are presented on the report. The layer introduces the calculation rules. Apart from validation they can be used to count the fields that were not included in the report. Usually it has a simple form based on the operation of adding up several indicators.
- formula layer - describes more complex relationships between elements that cannot be described by simple arithmetic operations. Despite the attempt to formalize reports and statements in an orderly manner, it is common to extend the basic definitions introduced by taxonomies by companies that provide reporting. This creates a mess and cannot be easily compared.

All the layers described above are linked to the actual data that is stored in the report file. Single financial data is stored in the form of XML tag in which we have its name, value and additional attributes. For a better understanding of the values and their meaning, a new format has been introduced. InLine XBRL is a format that is based on XBRL tags but allows them to be organized in the form of an unstructured document (HTML/XHTML format) which contain references to XBRL tags. On the one hand these reports are more readable; on the other hand they still contain XBRL fields that can be analyzed by external tools. However, the structure may not be as well defined by the taxonomy. Finding specific information in such data and creating reports can be difficult.

It is also noticeable that each of the above representations can be represented in a graph structure. For example, the calculation rules associated with calculating simple values represent the relationship between tags can take the role of nodes while edges can denote the relationship and the weight with which elements are associated. The next step is to define the data storage model in a way that is optimized for data processing and retrieval.

3   Graph Representation

It is necessary to introduce graph structure in which we store information related to XBRL raport data files. It is referred to as a *XBRL graph*.

Definition 1. *XBRL Graph* *(abbrev. XG) is a graph of the form:*

$$XG = (V, E, \Sigma, \Gamma, type, attr),$$

where:

- *V is a finite, non-empty set of graph nodes,*
- *E is a finite set of edges,*
- *Σ is a set of node types,*
- *Γ is a set of edge types, where $\Sigma \cap \Gamma = \emptyset$,*
- *type : $V \cup E \rightarrow \Sigma \cup \Gamma$ is a function that returns the type of a given node/edge: type(V) = Σ, type(E) = Γ,*
- *attr is a function that returns a set of attribute types for a given node/edge type.*

We will use the graph to store information from both a single report and a group of reports. Therefore, the following node types have been defined (Σ):

- *FINT (Financial Indicator Node Type)* - Set of attributes associated with a node are: atr(FINT) = {indicator, value, metric}
- *SNT (Statement Node Type)* - The report includes various statements/types of reports (Profit and Loss Account, Balance Sheet and others). The top type determines the type of report that is included in the report. A set of attributes associated with a node: atr(SNT) = {name}.
- *DNT (Document Node Type)* - node type specifying the type of document that represents the report. A set of attributes associated with this type of node: atr(DocId) = {period, report type, submission date}.
- *CNT (Company Node Type)* - the type of node that identifies reporting institution. Set of at- tributes associated with a node are : atr(CNT) = {fullname, symbol}.
- *PNT (Period Node Type)* - the type of node specifying the period and date associated with the report or indicator contained in the report
- *TNT (Taxonomy Node Type)* - the type of node that identifies taxonomy which was used to present financial indicators,

This list of node is not complete and can be extended to include specific types depending on the type of analysis to be performed. Proposed graph structure is designed to optimize the search for similar information by year and company type.

We use the following edge types for analysis (Γ):

– *INCL (Include)* - used to create hierarchical structures. It shows the relationships between the elements of a graph.
– *REL (Related)* - used to model the relationships between elements, the relationships may for example show the calculation rules and how values are related to them.

As with node types, this list is not complete and may be expanded in the future.

Organizing a series of documents using the introduced graph structure gives the possibility to define custom graph grammars and transformations or use mechanisms introduced in other areas. As an example we use graph methods and formal grammars which are used in lighting optimization [1,7,10,11,13].

4 Practical Examples

To better illustrate the concepts, the following is an example of a graph representation of one of the US GAAP taxonomy rules that are represented in XBRL graph form . The income statement is the most common statement in financial statements. One of the representations presented in the US GAAP taxonomy is the calculation rules, which can be written in the form of relationships. The figure visualizes the rules and tag names in an Excel file (shown on Fig 2).

name	label	dept	orde	priorit	weigl
GrossProfit	Gross Profit	9	10,0	0	1,0
Revenues	Revenues	10	10,0	0	1,0
SalesRevenueNet	Revenue, Net	11	10,0	0	1,0
FinancialServicesRevenue	Financial Services Revenue	11	20,0	0	1,0
NetInvestmentIncome	Net Investment Income	11	30,0	0	1,0
RealizedInvestmentGainsLosses	Realized Investment Gains (Losses)	11	40,0	0	1,0
RevenuesExcludingInterestAndDiv	Revenues, Excluding Interest and Dividends	11	50,0	0	1,0
InvestmentBankingRevenue	Investment Banking Revenue	11	60,0	0	1,0
UnderwritingIncomeLoss	Underwriting Income (Loss)	11	70,0	0	1,0
MarketDataRevenue	Market Data Revenue	11	80,0	0	1,0
OtherOperatingIncome	Other Operating Income	11	90,0	0	1,0
OtherIncome	Other Income	11	100,0	0	1,0
CostOfRevenue	Cost of Revenue	10	20,0	0	-1,0
CostOfGoodsAndServicesSold	Cost of Goods and Services Sold	11	10,0	0	1,0
FinancialServicesCosts	Financial Services Costs	11	20,0	0	1,0
LiabilityForFuturePolicyBenefitsPer	Liability for Future Policy Benefits, Period Expens	11	30,0	0	1,0
InterestCreditedToPolicyholdersAc	Interest Credited to Policyholders Account Balan	11	40,0	0	1,0
PolicyholderDividends	Policyholder Dividends, Expense	11	50,0	0	1,0
DeferredSalesInducementsAmortiz	Deferred Sales Inducement Cost, Amortization Ex	11	60,0	0	1,0
PresentValueOfFutureInsurancePr	Present Value of Future Insurance Profits, Amorti	11	70,0	0	1,0
AmortizationOfMortgageServicingR	Amortization of Mortgage Servicing Rights (MSRs	11	80,0	0	1,0
DeferredPolicyAcquisitionCostAmc	Deferred Policy Acquisition Costs, Amortization E	11	90,0	0	1,0
InsuranceTax	Insurance Tax	11	100,0	0	1,0
AmortizationOfValueOfBusinessAc	Amortization of Value of Business Acquired (VOB	11	110,0	0	1,0
OtherCostOfOperatingRevenue	Other Cost of Operating Revenue	11	120,0	0	1,0

http://fasb.org/us-gaap/role/statement/StatementOfIncome
124000 - Statement - Statement of Income (Including Gross Margin)

Fig. 2. Part of rules for the Revenue tag with calculation base

For example, we can see that statement 124000 defines a GrossProfit indicator, which is dependent on Revenues and CostOfReveues. Further analysis shows that Revenues depends on a number of different indicators that are extended relative to the underlying taxonomy.

These structures can be represented using XBRL graph, in a hierarchical manner. Thus, all the tags that are in the Microsoft report for the year, are represented as FINT nodes (Financial Indicator Nodes) associated with SNT nodes (Statements), which identify the company and period (shown on Fig 3).

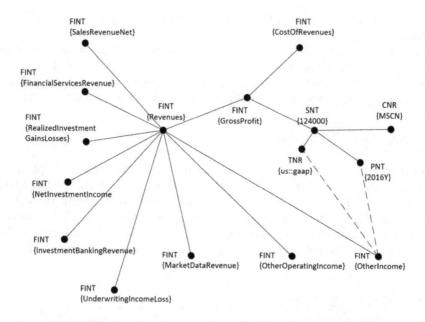

Fig. 3. Part of rules for the Revenue tag with calculation base

On such a graph, it is possible to use graph grammars which are used to divide graphs into smaller graphs to implement agent processing [3,7,11,13]. The created graph can be split to form balanced graphs on which independent functions can be executed.

5   Searching XBRL Graph Database

For further analysis, a database was created consisting of reports from 2007 to 2017 containing 46,737 statements, which occupied a space of over 1 Gb of data. From this data, an XBRL graph was created containing 72 million nodes. The current set of reports posted on the SEC website [14] contains over 1 Tb of data. To download data sets, tools provided by BFT24 were used, with which allow to download selected XBRL report files [2].

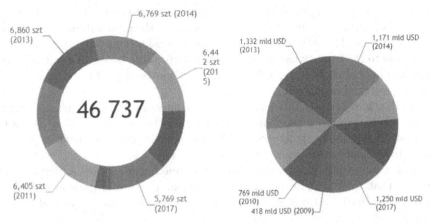

(a) Number of reports process by year

(b) One of output net revenues of processed companies divided into years

Fig. 4. Results of processing searching Revenues financial indicator

The created database was used to search the set and create simple graph rules, which were aimed at finding the value of revenues and presenting them on a pie chart. To compare the effectiveness, different approaches have been used:

- data processing based on native XBRL sets (XML files has been processed),
- creating XBRL Graph, and then using them to search Net Revenues values,
- creating an XBRL chart, dividing it into 10 subcharts and parallel data search.

The results from the data search were summarized in tabular form, shown in Table 1.

Table 1. Time to calculate revenue values per year for the entire dataset

Data storage structure	Time to create a full report
XBRL raw data	557 min
XBRL graph	15.5 min
XBRL graph divided into 10 subgraphs	130 s

There is a very high acceleration in searching for data based on a structured structure, this time can be accelerated when an agent environment is used.

6   Conclusions

In this paper, we present a formal model which can be used for storing XBRL reports and case study of processing a group of 46,736 individual XBRL reports.

With the proposed approach it took about 130 s to create the report and in comparison with the standard method we reduced the preparation time by more than 10 times. The XG-based method using well-known graph grammars also allows us to perform "what if" analyses. It is possible to analyze datasets and estimate indicators that are not available in reports or to determine groups of errors generated in reports.

Taking into account different types of taxonomies and different methodologies of their creation a common format can also be an element that allows for separation of similar structures and better translation of elements. This will allow to estimate in which set the parameter value will occur.

The concept presented here is also an outline for developing an agent-based system that would offer even faster estimation. The use of parallel processing would allow initial estimates to be obtained in time comparable to real time.

References

1. Basiura, A., Sędziwy, A., Komnata, K.: Similarity and conformity graphs in lighting optimization and assessment. In: International Conference on Computational Science, pp. 145–157 (2021)
2. BFT24: blockchain financial tools (XBRL processing). https://prod.bft24.com
3. Flasiński, M., Kotulski, L.: On the use of graph grammars for the control of a distributed software allocation. Comput. J. **25**, 167–175 (1992)
4. Hasham, S., Joshi, S., Mikkelsen, D.: Financial crime and fraud in the age of cybersecurity. https://mckinsey.com/business-functions/risk-and-resilience/our-insights/financial-crime-and-fraud-in-the-age-of-cybersecurity (2019). Accessed 06 Apr 2022
5. Komnata, K., Basiura, A., Kotulski, L.: Graph-based street similarity comparing method. In: Zamojski, W., Mazurkiewicz, J., Sugier, J., Walkowiak, T., Kacprzyk, J. (eds.) DepCoS-RELCOMEX 2020. AISC, vol. 1173, pp. 366–377. Springer, Cham (2020). https://doi.org/10.1007/978-3-030-48256-5_36
6. Kotulski, L.: Rozproszone transformacje grafowe: teoria i zastosowania. Wyd, Naukowe AGH (2013)
7. Kotulski, L., Sedziwy, A.: GRADIS - the multiagent environment supported by graph transformations. Simul. Model. Pract. Theory **18**(10), 1515–1525 (2010)
8. Kotulski, L., Sedziwy, A.: Parallel graph transformations supported by replicated complementary graph. In: ICANNGA, pp. 254–264 (2011)
9. Microsoft: XBRL Raport Data File. https://sec.gov/Archives/edgar/data/789019/000119312516662209/msft-20160630.xml (2016). Accessed 06 Apr 2022
10. Sédziwy, A., Kotulski, L., Basiura, A.: Enhancing energy efficiency of adaptive lighting control. In: Nguyen, N.T., Tojo, S., Nguyen, L.M., Trawiński, B. (eds.) ACIIDS 2017. LNCS (LNAI), vol. 10192, pp. 487–496. Springer, Cham (2017). https://doi.org/10.1007/978-3-319-54430-4_47
11. Sédziwy, A., Kotulski, L., Basiura, A.: Multi-agent support for street lighting modernization planning. In: Nguyen, N.T., Gaol, F.L., Hong, T.-P., Trawiński, B. (eds.) ACIIDS 2019. LNCS (LNAI), vol. 11431, pp. 442–452. Springer, Cham (2019). https://doi.org/10.1007/978-3-030-14799-0_38
12. Sedziwy, A.: Effective graph representation supporting multi-agent distributed computing. Int. J. Innovative Comput. Inf. Control **10**(1), 101–113 (2014)

13. Sedziwy, A., Basiura, A.: Energy reduction in roadway lighting achieved with novel design approach and leds. LEUKOS: J. Illum. Eng. Soc. North Am. **14**(1), 45–51 (2017). https://doi.org/10.1080/15502724.2017.133015
14. U.S. Securities and Exchange Commision(SEC): Electronic Data Gathering, Analysis, and Retrieval system data. https://www.sec.gov/Archives/edgar/full-index/. Accessed 06 Apr 2022
15. U.S. Securities and Exchange Commision(SEC): Structured Disclosure at the SEC: History and Rulemaking. https://www.sec.gov/page/osdhistoryandrulemaking. Accessed 06 Apr 2022

Natural Language Processing

Development of CRF and CTC Based End-To-End Kazakh Speech Recognition System

Dina Oralbekova[1,2]([✉]) [iD], Orken Mamyrbayev[2] [iD], Mohamed Othman[3] [iD],
Keylan Alimhan[4] [iD], Bagashar Zhumazhanov[2] [iD], and Bulbul Nuranbayeva[5] [iD]

[1] Satbayev University, Almaty, Kazakhstan
`dinaoral@mail.ru`
[2] Institute of Information and Computational Technologies, Almaty, Kazakhstan
[3] Universiti Putra Malaysia, Kuala Lumpur, Malaysia
[4] L.N. Gumilyov, Eurasian National University, Nur-Sultan, Kazakhstan
[5] Caspian University, Almaty, Kazakhstan

Abstract. Architecture end-to-ends are commonly used methods in many areas of machine learning, namely speech recognition. The end-to-end structure represents the system as one whole element, in contrast to the traditional one, which has several independent elements. The end-to-end system provides a direct mapping of acoustic signals in a sequence of labels without intermediate states, without the need for post-processing at the output, making it easy to implement. Combining several end-to-end method types perform better results than applying them separately. Inspired by this issue, in this work we have realized a method for using CRF and CTC together to recognize a low-resource language like the Kazakh language. In this work, architectures of a recurrent neural network and a ResNet network were applied to build a model using language models. The results of experimental studies showed that the proposed approach based on the ResNet architecture with the RNN language model achieved the best CER result with a value of 9.86% compared to other network architectures for the Kazakh language.

Keywords: Automatic speech recognition · End-to-end · Connectionist temporal classification · Conditional random fields · ResNet

1 Introduction

It's no secret that some of the processes processed by humans are not so easy to implement at the machine level, this concerns speech recognition systems. Today there are many speech recognition applications and software. Companies like Google and Yandex have achieved a fairly good level of accurate word recognition in speech for languages with a large body of training data. However, the data quality of the systems requires improvements in decoding sequences from speech.

Conventional automatic speech recognition systems were built on the basis of independent components, this is an acoustic model, a language model and a vocabulary, which were tuned and trained separately. The context-dependent states of phonemes are

© The Author(s), under exclusive license to Springer Nature Switzerland AG 2022
N. T. Nguyen et al. (Eds.): ACIIDS 2022, LNAI 13757, pp. 519–531, 2022.
https://doi.org/10.1007/978-3-031-21743-2_41

forecasted by the acoustic model, the language model and lexicon determine the most possible sequences of spoken phrases.

In traditional speech recognition systems, a Hidden Markov Model (HMM) model with a Gaussian Mixture Model (GMM) has been widely used. With the help of HMM, statistical models of words were built, and GMM represents the unit of pronunciation, i.e. distribution of signals within a certain period of time [1]. The development of deep learning technologies has contributed to the improvement of other scientific areas, which includes speech recognition. Deep neural networks began to be used for acoustic modeling instead of GMM, which led to improved results [2]. Thus, the HMM-DNN architecture, i.e. a combination of these methods is becoming popular in the field of continuous speech recognition.

In the future, the end-to-end model became widespread, which trains the components of the traditional model simultaneously without isolating individual elements, representing the system as a single neural network. In research papers [3–5] DNN was used to develop an acoustic model, language models and dictionaries were implemented by RNN. In addition, convolutional neural networks were used to extract features from the original signal [6]. And the application of these modifications has led to an enhancement in the efficiency of speech recognition systems. Different ANN architectures can be used at all stages of recognition, and this makes it effective in terms of performance compared to other popular systems. This approach is end-to-end. The end-to-end structure represents the system as one whole element, in contrast to the traditional one, which has several independent elements. The end-to-end system provides a direct mapping of acoustic signals in a sequence of labels without intermediate states, without the need for post-processing at the output, making it easy to implement.

For the first time the end-to-end model was mentioned in the work of Alex Graves [7], he presented the model based on the Connectionist temporal classification. But the application of the E2E model was implemented after 2013 [8]. This was due to the absence of computing power. After the emergence of powerful techniques with support for parallel computing, end-to-end models began to be used in speech recognition in popular languages that have a fairly large amount of training data. Also, the appearance of other models of end-to-end architecture influenced the result of indicators of speech recognition systems. Thus, to build an end-to-end model, it is enough to observe three basic rules: 1) the presence of high-performance technology, a supercomputer; 2) the presence of a large speech corpus for several thousand hours of speech, which will consist of audio data and their transcriptions, 3) the selected architecture of the E2E model with the appropriate settings that is suitable for speech recognition of the desired language or a group of languages of the same structure (inflectional, agglutinative, etc.).

The E2E system realizes direct reflection of acoustic indicators in a sequence of marks without intermediate states, which does not require further processing at the output. These processes make the system easy to implement. There are several basic types of E2E models, such as connectionist temporal classification (CTC) [7, 8], encoder-decoder with attention models [9, 10], and Conditional Random Fields (CRF) [11]. In the CTC-based model, there is unnecessary frame-level alignment between acoustics and transcription, since a special token is allocated, like an empty "label", which defines the start and end of one phoneme. In encoder models based on the attention mechanism, the encoder is an acoustic model (AM), the decoder is similar to language model (LM) – it works autoregressive, predicting each output token depending on previous predictions [9, 10]. The CRF-based model allows you to combine local information to predict the global probabilistic model by sequences [11]. These models greatly simplify the speech recognition process. The increase in training data refined the quality of the ASR systems compared to the HMM [12]. But data reduction increases recognition errors. However, E2E models have the flexibility to combine them to mitigate their specific disadvantages.

E2E models require a large amount of speech data for training, which is problematic for languages with limited training data. And one of these languages is the Kazakh language. The Kazakh language has an agglutinative character, in which the dominant type of inflection is agglutination [13], opposite to the inflectional one. Some research works [14–16] have shown that the combined use of E2E models like CTC and attention can be trained from start to finish, while this combination gave a very good result, which almost came close to the accuracy of the human level [16]. Based on these studies, we study the end-to-end system of the joint CTC and CRF models. Until now, systems have been developed on E2E CTC and Transformer models [17, 18] for the recognition of Kazakh speech with different sets of training data. At the moment, E2E CRF based systems for the recognition of Kazakh speech have not been investigated, and we decided to build this model. The effectiveness of this model has been realized in natural language processing.

In this research work, we have built a hybrid model based on two end-to-end methods, CRF and CTC for Kazakh speech recognition.

The structure of the research work is given in the following order: Sect. 2 provides a brief analytical review on scientific topics. Section 3 shows the principle of operation of models based on CTC and CRF. Further in Sect. 4 our experimental data, speech corpus and model settings are described, the results obtained are analyzed. The final section summarizes the findings.

2 Literature Review

Conditional Random Fields (CRF) is a model that allows you to combine local information to predict the global probabilistic model from sequences. This model is considered to be a kind of Markov random field. This model was first proposed in [11] for speech recognition. J. Lafferty et al. (2001) proposed an algorithm for estimating parameters for conditional random fields and showed that CRF has a greater advantage over HMM and MEMM (maximum entropy Markov models) for natural language data. In [19, 20], the CRF model was applied to assess the measurement of accuracy in the problem of phonetic recognition, as well as the accuracy of detecting boundaries between them. The results show that when using transition functions in the CRF-based recognition structure, recognition performance is significantly improved by reducing the number of phoneme deletions. Show that when using transition functions in the recognition structure based on CRF, recognition performance is significantly improved by decreasing the number of phone deletions. Bounding efficiency is also improved, mainly for transitions between the phonetic classes of silence, stop, and clicks. In addition, the CRF model gives a lower error rate than the HMM and Maxent models in the task of detecting the boundaries of sentences in speech.

Currently, the most common in speech recognition are linear and segmental CRF (linear chain & segmental CRF) models. This model is most often used to solve the problems of marking and segmenting sequences.

Keyu An et al. (2019) [21] demonstrated a new CAT toolkit that represents an implementation of CTC-CRF E2E models. For the experiment, Chinese and English tests, such as Switchboard and Aishell, were applied, thus obtaining the most modern results among the existing end-to-end models with fewer parameters and competitive in comparison with the hybrid models DNN-HMM. In addition, the same authors in [22] (2020) proposed a new technique called contextualized soft forgetting that allows the CAT tool to perform streaming ASR without compromising accuracy and has performed well with limited datasets compared to existing ones.

In [23] (2017), an end-to-end model was built based on segmented conditional random fields (SCRF) and connection time classification (CTC). SCRF uses a globally normalized joint label and segment length model, and CTC classifies each frame as either an output symbol. Through experimentation with the TIMIT dataset, a multitasking approach to training improved the recognition accuracy of CTC and SCRF models. In addition, it has been illustrated that CTC can be used to pre-train an RNN encoder, accelerating the training of a collaborative model.

Hongyu Xiang et al. [24] (2019) developed a single-stage acoustic simulation based on a CRF with a state topology based on CTC. Evaluation experiments were conducted with WSJ, Switchboard and LibriSpeech datasets. In direct comparison, the CTC-CRF model using simple bidirectional LSTMs consistently outperformed SS-LF-MMI (lattice-free maximum-mutual-information) on all three benchmark datasets and in both monophones and mono symbols. And it was revealed that the CTC-CRF model avoids some special operations in SS-LF-MMI.

In [25] (2021), methods were investigated to apply the newly developed text word modeling modules and Conformer neural networks in the CTC-CRF. Research observations are conducted on Switchboard and LibriSpeech, and the German CommonVoice dataset. Experimental results show that Conformer can significantly improve recognition quality; verbal systems perform slightly worse than telephone systems for a target language with a low graphemphoneme match, while both systems can perform equally well when that match is high for the target language.

Yang, Li et al. (2019) [26] propose a text processing model after Chinese speech recognition that combines a bidirectional long-term short-term memory (LSTM) network with a conditional random field (CRF) model. The need to process the text after recognition is associated with the appearance of a problem with the dialect and accent, since it is necessary to correct the text after speech recognition before displaying it. The objective is divided into two steps: detecting text errors and editing text errors. In this article, a bi-directional long-term short-term memory (Bi-LSTM) network and a conditional random field are used in two stages of text error detection and text error correction, respectively. Through validation and system testing of the SIGHAN 2013 Chinese Spelling Check (CSC) dataset, experimental results show that the model can effectively improve the accuracy of text after speech recognition.

The reviewed works shows us that the joint use of E2E models developed the productivity of the ASR system than using them separately.

Unfortunately, there is very little new research on this topic specifically in the study of speech recognition. But we tried to consider the publications that were in the public domain to the maximum.

3 Methodology of E2E Models

This section describes the CTC and CRF models and their joint model.

3.1 Connectionist Temporal Classification (CTC)

The CTC function is used to train a neural network in sequence recognition. Let's say we have an output sequence $y = S_w(x)$. Let each element of this sequence have a probability distribution vector for each symbol V' at time t. Therefore, we must define y_k^t, which is the probability of pronouncing the character k from the alphabet V' at time t. If μ is a sequence of symbols and a "space" for the given input x, then the probability $P(\mu|x)$ can be defined as follows (1):

$$P(\mu|x) = \prod_t y_{\mu t}^t \tag{1}$$

From the above equation, you can see that the components of the output sequence are independent of each other. To align data, you need to add an auxiliary character that will remove duplicate letters and spaces. Let's denote it as B. Thus, the total probability of the output sequence can be expressed by the following (2):

$$P(y|x) = \sum_{\mu \in B^{-1}(y)} P(\mu|x) \tag{2}$$

The given above equation determines the sum over all alignments using dynamic programming, and helps to train the ANN on unlabeled data (3):

$$CTC(x) = -log P(y|x) \tag{3}$$

From the above it follows that ANN can be trained on any gradient optimized algorithm. In the CTC architecture, any kind of ANN can be used as an encoder, such as LSTM and BLSTM.

To decode the CTC-model, the assumption was presented in [7] (4):

$$argmax P(y|x) \approx B(\mu^\circ) \tag{4}$$

where $\mu^\circ = argmax P(y|x)$.

CTC eliminates the need for data alignment and allows for quite a few layers, a simple network structure to implement a model that maps audio to sequence of utterances.

3.2 Conditional Random Fields (CRF)

Conditional random fields (CRF) are a class of statistical modeling techniques that are usually used in machine learning and are used for structured prediction. While the classifier predicts a label for one sample without considering "adjacent" samples, the CRF can take context into account. For this, the forecast is modeled as a graphical model that implements the relations between the forecasts.

Other examples of the use of CRF are: labeling or analysis of sequential data for natural language or biological sequence processing, POS marking, and object recognition and image segmentation in computer vision [27].

Conditional Random Field (CRF) is a discriminative undirected probabilistic graphical model [28]. This method, in contrast to the Markov model of maximum entropy, does not have a label bias problem [29, 30]. CRF and its various modifications have found applications in areas such as natural language processing, computer vision, speech recognition, etc. (see Fig. 1).

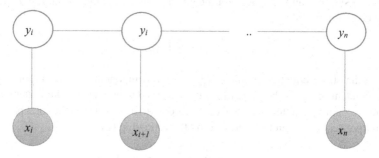

Fig. 1. General structure of the model CRF

Here $X = (x_i, ..., x_n)$ is the input sequence to be recognized, and $Y = (y_i, ..., y_n)$ is the recognized set of character set labels, where $i = 1$.

In the following paragraphs, the widely used linear chain and segmental CRF will be discussed. These models are successfully used to solve machine learning problems, where the precedent is a sequence of random variables with labels assigned to them. These are the so-called sequence marking and segmentation tasks. Therefore, this model is popular in areas that are characterized by a sequence of variables, for example, in natural language processing tasks.

Linear-Chain CRF. The linear CRF model is a discriminative model, and this is where it resembles the popular Maximum entropy Markov model (MEMM). In contrast to CRF, the maximum entropy model has a disadvantage called label bias [29, 30]. The essence of the problem is that, due to the peculiarities of learning, the maximum entropy model tends to give preference to those latent states that have a lower entropy of distribution of subsequent states.

The linear CRF model put in to solve many natural language processing problems, mostly for the English language, for example, to the problem of POS-tagging, to the problem of surface parsing, or to the problem of resolving anaphora [31].

The disadvantage of the linear CRF model is the used complex algorithm of the analysis of the training sample, that makes it hard to constantly update the model when new training data arrives [32].

Segmental Conditional Random Fields. Segmental CRFs are built on the basis of segmented recurrent neural networks, namely, it is semi-Markov CRF [33].

SCRF generalizes conditional random fields to operate at the segment level rather than the traditional frame level. Basically, each segment is labeled with a word. Features are then extracted, each measuring some form of consistency between the underlying sound and the word hypothesis for the segment. They are combined in a log-linear model to obtain the posterior probability of a sequence of words, including sound.

In the context of speech recognition, taking into account the sequence of input vectors X and the corresponding sequence of output labels Y CRF of the linear chain of zero order determines the conditional probability at the level of sequence (5) using auxiliary segments in the following way

$$P(y, s|x) = \frac{1}{Z(x)} \prod_{i=1}^{n} exp f(y_i, s_i, x_i) \qquad (5)$$

where Z (X) is the normalization term and the complete calculation of this value is given in [34], where the segmental level functions were studied using RNN and the segmental RNN (SRNN) model was applied as an implementation of the SCRF acoustic model for multitasking learning.

In [34] SRNNs were applied for speech recognition. On the TIMIT speech corpus, a PER score of 17.3% was obtained. Contemporaneously, the implemented model did not use LM.

3.3 Joint Training Models for Kazakh Speech Recognition

Joint training of E2E models can be expressed as follows (6):

$$L_{CTC/Att} = \tau L_{CTC}(x) + (1 - \tau)L_{SCRF}(x) \qquad (6)$$

where τ is an adjustable parameter and satisfies the condition $- 0 \leq \tau \leq 1$. Neural networks with short and long-term memory LSTM and bidirectional LSTM and ResNet architecture were used as neural networks.

In the CTC model, it uses monotonic alignment between speech and tag sequences and trains the network quickly. And besides, the proposed model will be effective in recognizing speech in long sequences if training took place in short training data. In addition, CTC helps speed up the process of assessing the desired alignment without the help of rough estimates of this process, which is labor-intensive and time-consuming.

4 Experiments and Results

This section contains descriptions of the case, preliminary model settings, experimental data, as well as the results obtained and a comparative analysis of the data obtained.

4.1 Data Preparation

The corpus of speech for the Kazakh language was implemented by the researchers of the laboratory "Computer Engineering of Intelligent Systems" of the Institute for Computer Engineering of the Ministry of Education and Science of the Republic of Kazakhstan [35]. This corpus consists of pure speech and speech of telephone conversations.

For the recording, 250 speakers took part (55% of them were men and 45% were women). The corps mainly includes young and middle-aged people. Thus, the group of speakers has relatively small changes in age, profession and education. The recording was made in an office environment: the windows and doors were closed to avoid any external noise. Headphones with a noise canceling microphone were used for recording. For efficiency, we have chosen phonetically rich words in which consonants dominate vowels. The base includes the read text, consisting of 94 267 words in 1200 sentences. All speech files were named with a unique identification code.

Given the recent advances in speech recognition, it is necessary to add data recorded under various conditions. Thus, we added to our corpus the recordings of telephone conversations, which were transcribed by young volunteers who were involved from the Higher School - these are undergraduates, undergraduates and doctoral students of Almaty national universities.

As a result, the total volume of Kazakh-language speech data was 300 h of speech, with 90% of the audio data used for training, and 10% for model validation.

All audio materials were in.wav format. All audio data has been converted to single channel. The PCM method was used to convert the data into digital form. Discrete 44.1 kHz, 16-bit.

4.2 Presetting Models

End-to-end models with different variations of neural networks were implemented, as well as the implementation of models separately and jointly.

Feature extraction is an obligatory part of the system, and in this part, convolutional neural networks were used, since they have a stronger anti-interference ability than the use of cepstral coefficients with a mel-frequency frequency [36]. The activation function ReLU was applied for convolutional layers [37]. Next, a maxpooling layer was added to filter the low frequencies of speech. In addition, layers of compression and normalization of the extracted features were additionally introduced.

During training, the gain of the weight coefficient of the CTC model was set: $\lambda = 0.2$. Also, the external language model was leveraged.

To the basic CTC model, the RNN varieties were applied, such as LSTM and BLSTM with five layers containing 1024 cells for each layer, the ResNet network - a method for eliminating the gradient fading effect for CNN. ResNet consists of 8 convolutional layers and a max-pooling layer with batch normalization [38].

The initial learning rate coefficient was set to 0.001. Dropout was used for each output of the recurrent layer as a regularization and is equal to 0.5. For our model, we used a gradient descent optimization algorithm based on Adam [39].

To measure the quality of the Kazakh speech recognition system based on E2E models, the CER metric was used - the number of incorrectly recognized characters, as well as on the basis of the word error rate (WER), which is calculated using the Levenshtein distance [40].

4.3   Results and Analysis of Experimental Studies

Audio recordings with transcription from news sites in the Kazakh language (https://qazaqstan.tv/live, https://24.kz/kz/), audiobook sites (https://kitap.kz/, https://e-history.kz/ru/audio/) and which is 1 600 separate phrases of different speakers and none of the speakers was used simultaneously in both parts.

The end-to-end model, when using the CTC function without a language model, reached a CER of 17.45% and a WER of 29.01% (see Fig. 2). Integration of the external language model into the CTC end-to-end system improved the CER and WER indicators by 13% and 18%, respectively. Our joint model of CTC and CRF with ResNet architecture showed good results without the use of LM, and CER reached 11.57% and WER - 18.32%.

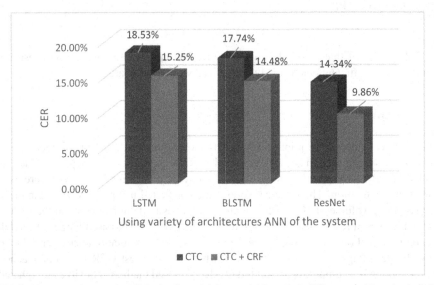

Fig. 2. Comparative graphs of the basic and joint model based on different architectures of the NN with LM in terms of CER

And after adding LM, the model slightly improved its quality, by almost 1.5%. The gain values are shown in Table 1.

Table 1. Results of experiments on the use of E2E models.

Model	WER, %	CER, %
CTC without LM	34,48	21,56
LSTM	32,59	20,56
BLSTM	29,01	17,45
ResNet		
CTC with LM	26,63	18,53
LSTM	24,31	17,74
BLSTM	23,65	14,34
ResNet		
Without LM	31,92	18,20
CTC (LSTM) + CRF	27,19	16,82
CTC (BLSTM) + CRF	18,32	11,57
CTC (ResNet) + CRF		
With LM	26,35	15,25
CTC (LSTM) + CRF	23,61	14,48
CTC (BLSTM) + CRF	16,75	9,86
CTC (ResNet) + CRF		

Table 1 shows that the models using the ResNet network showed the best result in terms of the coefficients of correctly recognized words and characters.

5   Conclusion

The work considered the joint end-to-end models CTC and CRF for the recognition of Kazakh speech. To implement this model, RNN variations were applied, such as LSTM and BiLSTM, as well as the ResNet. Convolutional neural networks were used for feature extraction. The practice works were conducted using the Kazakh language corpus with a volume of 300 speech hours, and the result demonstrated that the system can achieve high results using the ResNet and the use of RNN-based language model. Decoding based on these models does not increase the computational cost, and due to this, the decoding speed does not slow down. Thus, the best CER indicator reached 9.86%, which is a competitive result today. The proposed method is flexible enough and does not require conditional independence of variables. In addition, we can realize that proposed model can be used to recognize other languages with limited training data, which are part of the Turkic languages.

Now, we target to study insertion-based models for recognizing agglutinative languages.

Acknowledgement. This research has been funded by the Science Committee of the Ministry of Education and Science of the Republic Kazakhstan (Grant No. AP08855743).

References

1. Gales, M., Young, S.: 2007. The application of hidden Markov models in speech recognition. Found. Trends Signal Process. 1(3), 195–304 (2008). https://doi.org/10.1561/2000000004
2. Hinton, G., et al.: Deep neural networks for acoustic modeling in speech recognition: the shared views of four research groups. IEEE Signal Process. Mag. 29(6), 82–97, (2012). https://doi.org/10.1109/MSP.2012.2205597
3. Maas, A., Qi, P., Xie, Z., Hannun, A., Lengerich, C., Jurafsky, D., Ng, A.: Building DNN acoustic models for large vocabulary speech recognition. Comput Speech Lang. 41 (2016). https://doi.org/10.1016/j.csl.2016.06.007
4. Fohr, D., Mella, O., Illina. I.:New Paradigm in speech recognition: deep neural networks. In: IEEE International Conference on Information Systems and Economic Intelligence, Marrakech, Morocco. ffhal-01484447f (2017)
5. Shi, Y., Zhang, WQ., Liu, J., et al.: RNN language model with word clustering and class-based output layer. J. Audio Speech Music Proc. 22 (2013). https://doi.org/10.1186/1687-4722-201 3-22
6. Huang, S., Tang, J., Dai, J., Wang, Y.: Signal status recognition based on 1DCNN and its feature extraction mechanism analysis. Sensors (Basel) 19(9) (2018). https://doi.org/10.3390/s19092018
7. Graves, A., Fernández, S., Gomez, F., Schmidhuber, J.: Connectionist temporal classification: Labelling unsegmented sequence data with recurrent neural 'networks. In: ICML 2006 - Proceedings of the 23rd International Conference on Machine Learning, pp. 369–376 (2006). https://doi.org/10.1145/1143844.1143891
8. Mamyrbayev, O., Oralbekova, D.: Modern trends in the development of speech recognition systems. News Nat. Acad. Sci. Republic of Kazakhstan, 4(32), 42 – 51 (2020). https://doi.org/10.32014/2020.2518-1726.64
9. Chan, W., Jaitly, N., Le, Q.V., Vinyals, O.L.: Attend and Spell. ArXiv, abs/1508.01211. (data of request: 14.09.2021) (2015)
10. Bahdanau, D., Chorowski, J., Serdyuk, D., Brakel, P., Bengio, Y.: End-to-end attention-based large vocabulary speech recognition. In: Proceedings of the 2016 IEEE International Conference on Acoustics, Speech and Signal Processing (ICASSP), Shanghai, China, pp. 4945–4949 (2016)
11. Lafferty, J., McCallum, A., Pereira, F.: Conditional random fields: Probabilistic models for segmenting and labeling sequence data. In: Proceedings of the International Conference on Machine Learning (ICML 2001), Williamstown, MA, USA, pp. 282–289 (2001)
12. Garcia-Moral, A., Solera-Ureña, R., Peláez-Moreno, C., Díaz-de-María, F.: Data balancing for efficient training of hybrid ANN/HMM automatic speech recognition systems. IEEE Trans. Audio Speech Lang. Process. 19. 468 - 481 (2011). https://doi.org/10.1109/TASL.2010.2050513
13. Agglutinating language - http://www.glottopedia.org/index.php/Agglutinating_language, (data of request: 27 Sep 2021)
14. Hori, T., Watanabe, S., Zhang, Y., Chan, W.: Advances in Joint CTC-Attention based End-to-End Speech Recognition with a Deep CNN Encoder and RNN-LM (2017)
15. Kim, S., Hori, T., Watanabe, S.: Joint CTC-attention based end-to-end speech recognition using multi-task learning (2016)
16. Mamyrbayev, O., Alimhan, K., Oralbekova, D., Bekarystankyzy, A., Zhumazhanov, B.: Identifying the influence of transfer learning method in developing an end-to-end automatic speech recognition system with a low data level. Eastern-Euro. J. Enter. Technol. 1(9(115), 84–92 (2022). https://doi.org/10.15587/1729-4061.2022.252801

17. Mamyrbayev, O., Kydyrbekova, A., Alimhan, K., Oralbekova, D., Zhumazhanov, B., Nuran-bayeva, B.: Development of security systems using DNN and i & x-vector classifiers. Eastern-Euro. J. Enter. Technol. **4** (9 (112)), 32–45 (2021). https://doi.org/10.15587/1729-4061.2021. 239186
18. Orken, M., Dina, O., Keylan, A., Tolganay, T., Mohamed, O.: A study of transformer-based end-to-end speech recognition system for Kazakh language. Sci Rep **12**, 8337 (2022). https://doi.org/10.1038/s41598-022-12260-y
19. Dimopoulos, S., Fosler-Lussier, E., Lee, C., Potamianos, A.: Transition features for CRF-based speech recognition and boundary detection. In: 2009 IEEE Workshop on Automatic Speech Recognisip & Understanding, pp. 99–102 (2009). https://doi.org/10.1109/ASRU. 2009.5373287
20. Liu, Y., Stolcke, A., Shriberg, E., Harper, M.: Using Conditional Random Fields for Sentence Boundary Detection in Speech (2005). https://doi.org/10.3115/1219840.1219896
21. An, K., Xiang, H., Ou, Z.: CAT: CRF-based ASR Toolkit. arXiv: abs/1911.08747, https://arxiv.org/abs/1911.08747 (2019)
22. An, K., et al.: CAT: A CTC-CRF based ASR Toolkit Bridging the Hybrid and the End-to-end Approaches towards Data Efficiency and Low Latency. In: NTERSPEECH (2020)
23. Lu, L., Kong, L., Dyer, C., Smith, N.A.:Multitask Learning with CTC and Segmental CRF for Speech Recognition In: Interspeech (2017)
24. Xiang, H., Ou, Z.: CRF-based single-stage acoustic modeling with CTC topology. In: 2019 IEEE International Conference on Acoustics, Speech and Signal Processing (ICASSP), pp. 5676–5680 (2019)
25. An, K., Xiang, H., Ou, Z.: CAT: A CTC-CRF based ASR Toolkit Bridging the Hybrid and the End-to-end Approaches towards Data Efficiency and Low Latency. In: INTERSPEECH (2020)
26. Yang, L., Li, Y., Wang, J., Tang, Z.: Post Text Processing of Chinese Speech Recognition Based on Bidirectional LSTM Networks and CRF. Electronics **8**(11) 1248 (2019). https://doi.org/10.3390/electronics8111248
27. Abney S.: Parsing by chunks. In: Berwick, R., Abney, S., Tenny, C., (eds.) Principle-based Parsing. Kluwer Academic Publishers, pp. 257–279 (1991)
28. Sutton, C., McCallum, A.: An Introduction to Conditional Random Fields for Relational Learning. MIT Press (2006)
29. Lafferty, J., McCallum, A., Pereira, F.: Conditional random fields: Probabilistic models for segmenting and labeling sequence data. In: Proceedings of the 18th International Conference on Machine Learning, Williamstown, Massachusetts, pp. 282–289 (2001)
30. Bottou, L.: Une approche theorique de l'apprentissage connexionniste: Applications a la reconnaissance de la parole. Doctoral dissertation, Universite de Paris XI (1991)
31. Culotta, A., Wick, M., Hall R., McCallum, A.: First-order probabilistic models for coreference resolution. In: Proc. of HLT-NAACL (2007)
32. Markovnikov, N.M., Kipyatkova, I.S.: An analytic survey of end-to-end speech recognition systems. Tr. SPIIRAN **58**, 77–110 (2018)
33. Kong, L., Dyer C., Smith, N.A.: Segmental recurrent neural networks. arXiv: 1511.06018, https://arxiv.org/abs/1511.06018. (Accessed 02 Oct 2021) (2015)
34. Lu, L., Kong, L., Dyer, C., Smith, N., Renals, S.: Segmental recurrent neural networks for end-to-end speech recognition. In: Proc. INTERSPEECH (2016)
35. Laboratory of computer engineering of intelligent systems – https://iict.kz/laboratory-of-computer-engineering-of-intelligent-systems/ (data of request: 02 Aug 2021)
36. Li, F., et al.: Feature extraction and classification of heart sound using 1D convolutional neural networks. EURASIP J. Adv. Signal Process. **2019**(1), 1–11 (2019). https://doi.org/10.1186/s13634-019-0651-3

37. Zhao, G., Zhang, Z., Guan, H., Tang, P., Wang, J.: Rethinking ReLU to Train Better CNNs. 603–608 (2018). https://doi.org/10.1109/ICPR.2018.8545612
38. Ioffe, S., Szegedy, C.: Proceedings of the 32nd International Conference on Machine Learning, PMLR, vol. 37, pp. 448–456 (2015)
39. Kingma D. P., Ba J. Adam: A method for stochastic optimization. http://arxiv.org/abs/1412. 6980 (data of request: 01.11.2021) (2014)
40. Levenshtein, V.I.: Binary codes capable of correcting deletions, insertions, and reversals. Soviet Phys. Doklady **10**, 707–710 (1996)

A Survey of Abstractive Text Summarization Utilising Pretrained Language Models

Ayesha Ayub Syed[1]([✉]) [iD], Ford Lumban Gaol[1] [iD], Alfred Boediman[2],
Tokuro Matsuo[3,4], and Widodo Budiharto[5] [iD]

[1] Department of Doctor of Computer Science – BINUS Graduate Program, Bina Nusantara
University, Jakarta, Indonesia
ayeshaayubsyed@yahoo.com

[2] Department of Econometrics and Statistics, The University of Chicago, Booth School of
Business, Chicago, USA

[3] Graduate School of Industrial Technology, Advanced Institute of Industrial Technology,
Tokyo 140-0011, Japan

[4] Department of M-Commerce and Multimedia Applications, Asia University, Taichung 41354,
Taiwan

[5] Computer Science Department, School of Computer Science, Bina Nusantara University,
Jakarta 11480, Indonesia

Abstract. We live in a digital era - an era of technology, artificial intelligence,
big data, and information. The data and information on which we depend to fulfil
several daily tasks and decision-making can become overwhelming to deal with
and requires effective processing. This can be achieved by designing improved and
robust automatic text summarization systems. These systems reduce the size of
text document while retaining the salient information. The resurgence of deep
learning and its progress from the Recurrent Neural Networks to deep transformer based Pretrained Language Models (PLM) with huge parameters and ample
world and common-sense knowledge have opened the doors for huge success and
improvement of the Natural Language Processing tasks including Abstractive Text
Summarization (ATS). This work surveys the scientific literature to explore and
analyze recent research on pre-trained language models and abstractive text summarization utilizing these models. The pretrained language models on abstractive
summarization tasks have been analyzed quantitatively based on ROUGE scores
on four standard datasets while the analysis of state-of-the-art ATS models has
been conducted qualitatively to identify some issues and challenges encountered
on finetuning large PLMs on downstream datasets for abstractive summarization.
The survey further highlights some techniques that can help boost the performance
of these systems. The findings in terms of performance improvement reveal that
the models with better performance use either one or a combination of these strategies: (1) Domain Adaptation, (2) Model Augmentation, (3) Stable finetuning, and
(4) Data Augmentation.

Keywords: Abstractive text summarization · Performance improvement ·
Pretrained language models

© The Author(s), under exclusive license to Springer Nature Switzerland AG 2022
N. T. Nguyen et al. (Eds.): ACIIDS 2022, LNAI 13757, pp. 532–544, 2022.
https://doi.org/10.1007/978-3-031-21743-2_42

1 Introduction

In the current digital era, data and information are generated at a pace greater than the effective processing capacity of machines. This rapid overflow of information owes to the availability of largescale storage and low-cost reproduction facilities for digital textual data [1]. Therefore, automated systems are required to capture and filter the most important information, to better utilize it for other purposes like decision making, research, journalism, etc. These automated systems well known as automatic summarization systems produce a condensed version of the lengthy document that retains the salient information and detains the redundant details.

Automatic text summarization is classified into two broad categories in the literature: extractive and abstractive methods. Extractive methods generate the summary by extracting key phrases or sentences from the input text and concatenating them [2]. Abstractive techniques provide a shortened and paraphrased version of the original material. The generated summary is human-like in appearance, expressing the most important information utilizing both source text and innovative words. There are also hybrid models that utilize extractive techniques to extract key sentences and then abstractive techniques to construct the summary. Automatic summarization, whether extractive, abstractive, or hybrid, has been advancing at a breakneck pace thanks to the pretraining paradigm. Pretrained Language Models have been used to significantly increase performance on a variety of NLP tasks. The concept is to train models on large-scale corpora and then fine-tune them for downstream applications. The current generation of PLMs is built on the notion of transfer learning and makes use of the Transformer architecture [3] in conjunction with self-supervised learning objectives [4]. The transformer architecture is based entirely on attention mechanism eliminating the need for recurrence or convolution. A transformer model takes a sequence as an input, processes it through the 6-layer encoder and decoder stacks and produces a sequence as an output. Transformer architecture trains large deep neural network models while self-supervised learning eliminates the need for human supervision/manual annotation and leverages intrinsic text correlations to learn. For example, hiding or masking a word in a sentence and then using the remaining words to predict the masked word is achieved by self-supervised learning [2].

Abstractive text summarization has many variations: single/multi-document summarization, mono/multilingual summarization, generic/query-focussed summarization, and short-text/long document summarization. [3] contributed XL-Sum, a diverse multilingual dataset covering 44 languages. The pretrained multilingual language model mT5 was finetuned on this dataset. The results were competitive to that of results for other rich resource languages like English.

We conduct an up-to-date review of the literature on state-of-the-art pre-trained language models for abstractive text summarization and compare their performance on four standard datasets. Second, we examine recent work in abstractive text summarization that makes use of these models and discuss some of the concerns and obstacles, as well as ways for improving the performance of existing ATS systems. The paper's second section presents a brief overview of neural abstractive text summarization. Section 3 summarizes the state-of-the-art PLMs in the context of abstractive summarization. Section 4 discusses various current ATS models that make use of PLMs. Section 5 summarizes the survey's analysis and findings. This article finishes with Sect. 6.

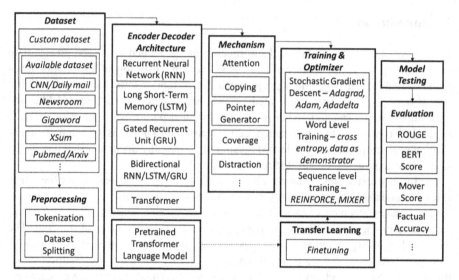

Fig. 1. Key architectural elements and process of Abstractive text summarization

2 Background

2.1 Neural Abstractive Text Summarization

Neural Abstractive Text Summarization is based on sequence-to-sequence modeling. Models known as seq2seq convert an input sequence (source text) to an output sequence (target summary). Abstractive sentence summarization was initially accomplished using local attention in neural networks. The attention aided the model in acquiring a soft alignment on the input text in order to familiarize the output. Later models were built on top of the RNN encoder-decoder architecture and adjusted to solve other difficulties such as keyword capture, document structure recognition, and out-of-vocabulary terms via the generator-pointer switch. The neural abstractive text summarization paradigm based on RNN, LSTM, and GRU rapidly evolved with the advent of Transformer [3].

The Transformer Architecture. It consists of a stack of encoders and a stack of decoders. Both the encoder and decoder stacks have their embedding layer. The input to the encoder stack is the input sequence while the input to the decoder stack is the target sequence. At the end of the decoder stack, there is an output layer that generates the transformer's output. Inside a single encoder block, there is a self-attention layer and a feed-forward layer. Also, there are residual skip connections and normalization layers around both the self-attention layer and the feed-forward layer. A single decoder block also has a self-attention layer and a feed-forward layer but in between these two layers, there is another layer of attention known as encoder-decoder attention. Both the encoder and decoder have a separate set of weights.

Numerous works have used transformer model to improve abstractive summarization performance in a variety of contexts, including multi-document summarization, multi-lingual summarization, length-controllable summary generation, and abstractive meeting summarization. The state-of-the-art in abstractive summarization is to improve performance through the use of pretrained language models.

2.2 Elements of Abstractive Text Summarization Systems

The dataset, the encoder-decoder architecture, the attention mechanism, training and optimization, and evaluation are the fundamental components of neural abstractive text summarization systems. Figure 1 depicts these features as well as the modeling approach for abstract text summarization. The reader is directed to [5] for a more extensive discussion of these aspects. The design of abstractive text summarization models begins with the training dataset, which can be a newly gathered dataset or an existing publically available dataset such as CNN/Daily Mail, Newsroom, XSum, and so on. The document-summary pairs are typically found in large quantities in the collection. The dataset must be prepared and pre-processed before it can be used. Preproceesing cleans up the dataset by removing punctuation and stop words, lowercasing, stemming, lemmatizing, and converting the dataset into tokens to prepare it for further processing. The encoder-decoder model architecture is then designed. Certain mechanisms, such as attention, pointer-generator, and coverage, can be used to offer additional functionality. The training and optimization processes are carried out utilizing the selected training method and optimizer. If a pretrained encoder-decoder language model is utilized, it must be finetuned via transfer learning on the target dataset/task. Finally, the model is tested and evaluated with metrics such as ROUGE, Bert Score, and so on.

ROUGE. ROUGE (Recall Oriented Understudy for Gisting Evaluation) is the most commonly used metric to evaluate summarization models. It measures the number of overlapping n-grams (words) between the automatic summary and the reference summary. The evaluation is conducted using three variations of ROUGE metric: ROUGE-1 (measures unigrams overlap), ROUGE-2 (measures bigrams overlap), and ROUGE-L (measures longest common subsequence) between the automatic and reference summary. ROUGE is computed in terms of recall and precision and then f-score.

$$ROUGE_{recall} = \frac{overlapping\ words\ between\ the\ reference\ summary\ \&\ automatic\ summary}{total\ words\ in\ reference\ summary}$$

(1)

$$ROUGE_{prec} = \frac{overlapping\ words\ between\ the\ reference\ summary\ \&\ automatic\ summary}{total\ words\ in\ automatic\ summary}$$

(2)

3   Pretrained Language Models and Abstractive Text Summarization

The performance of abstractive text summarization has been considerably influenced by the pretrained language models. The process is based on the concept of Transfer learfning where the pretrained model is finetuned on a downstream summarization task/dataset and the knowledge from the source is used to improve the target function. During fine-tuning, the model is initialized with the learned weights from the pretrained language model and then these weights are optimized according to the target task. Abstractive text summarization is a natural language generation task where the objective is to generate a summary conditioned on the input text. Unlike encoder-only or decoder-only PLMs like BERT [6] and GPT [7], language generation tasks are handled effectively by PLMs with an encoder-decoder based seq2seq architecture [8] like, UNILM [9], BART [10], MASS [8], PEGASUS [11], T5 [12], LED [13] etc. However, BERT and GPT have been effectively utilized as generalizing either the encoder or the decoder component for other architectures.

The encoder-decoder-based PLMs unleashed the potential of pretraining for language generation tasks. UNILM [9] was proposed as an improvement to BERT. Due to its bidirectional nature, BERT is difficult to adapt to the language generation tasks while UNILM with its left to right, right to left, bidirectional, and seq2seq language modelling objectives combined in a single model, can be applied to both language understanding and the generation tasks. MASS [8] used fragment masking as the seq2seq pretraining objective. Fragment masking is different from BERT's masked language modelling (single token masking) and GPT's standard language modelling. In fragment masking, fragment refers to a consecutive sequence of tokens in the input sentence. BART [10] model with its bidirectional encoder and an autoregressive decoder can also be directly finetuned on sequence generation activities like abstractive summarization. However, BART proposes several novel noising approaches as presented in Table 2. Among several schemes, BART with text infilling approach performed effectively superseding the previous models across various tasks. Google's PEGASUS [11] proposed new pretraining objectives exclusively for abstractive text summarization that are self-supervised and are known as Gap Sentence Generation (GSG) and Masked Language Modelling (MLM). The model has been evaluated on a diverse range of downstream datasets including but not limited to news domain, scientific articles, email, patents, etc., and has achieved the state-of-the-art summarization performance across many domains.

Transformer based models have limitations in processing lengthy sequences because the self-attention mechanism increases quadratically on increasing the sequence length causing the problem of quadratic complexity (denoted as $O(n^2)$) in terms of computation and memory usage. Pretrained models like Big Bird [14] and Long former [13] have been proposed to address this issue. Both architectures work with modified attention patterns to reduce the quadratic complexity to linear $O(n)$. Big Bird [14] solves the problem by using sparse attention instead of full attention to lower the quadratic dependency to linear. For abstractive summarization, both encoder and decoder suffer from this issue. Big Bird [14] researchers modified the self-attention to sparse attention only on the encoder side arguing that it is more efficient in cases where the input sequence is lengthy, and the output sequence is short as in the case of text summarization. Long former [13] also proposed an attention pattern based on the sliding window attention plus global attention (sparse). The idea behind both Big Bird and Long former architectures is local–global attention. However, the difference between the two lies in the way global–local attention is applied. Both these architectures are particularly effective for handling tasks requiring longer contexts like abstractive summarization and question answering. The standard transformer can roughly handle an input sequence length of up to 512 tokens while Big Bird scales this input tokens number up to 8x (8 × 512 = 4,096 tokens).

4 State-of-the-Art Models for Abstractive Text Summarization

This Section presents some novel architectures for abstractive text summarization utilizing pre-trained transformers under two modelling paradigms: fully abstractive and extractive-abstractive.

4.1 Fully Abstractive

[15] used locality modelling strategy to improve the performance of abstractive summarization. Localty modelling enhances the language encoding capability of the model by introducing convolutions in the transformer attention. The work used BERT language model as the encoder part and two variations of decoder, the transformer decoder and the BERT transformer decoder. The model was evaluated on English and German datasets (CNN/Daily mail and Swiss Text). [16] explored transfer learning for abstractive text summarization on BBC News dataset. The pretrained transformer T5 (Text-to-Text Transfer Transformer) was used in this work. It basically compares the performance of T5 with the previous attention-based neural network models on abstractive summarization and reveals that T5 outperforms the previous models. [17] investigated the use of pretrained model for short text abstractive summary generation on a Chinese dataset. The summaries were generated based on the idea of keyword templates that were automatically created from the training dataset. The resulting generated summaries were of high quality as compared to previous models on the Chinese dataset. [18] proposed the use of source embeddings and domain adaptive training to utilize pretrained language models efficiently for the task of abstractve summarization. The improvements were more prominent on abstractive datasets as compared to extractive.

4.2 Extractive-Abstractive

[19] explored abstractive summarization to produce simple lay summaries of the scientific articles. This work employs a combination of extractive and abstractive methods along with the readability metric. It utilized PEGASUS language model for generating the abstractive summary while BERT for extractive summary. The output was evaluated using both ROUGE and readability metrics. The work of [20] implemented long document abstractive summarization in a low resource setting. The training dataset only had 102 document-summary pairs. The work is based on extractive-abstractive paradigm. The extractive model works as a salience classifier for the input document. The salient sentences from the output of extractive module are then passed onto the abstractive module that generates the final summary using the pretrained BART generative model. The model was evaluated using ROUGE scores. [21] worked on modelling content importance using information theory for text summarization with pretrained language models. The experiments were conducted for three types of pretrained language models; Autoregressive (GPT-2), masked language models (BERT), and permutation language

models (XLNET) using CNN/Daily mail dataset and the New York Times corpus. The results indicated better importance modelling as compared to the previous approaches. [22] worked on long document abstractive summarization using the tansformer language models. The important information extraction task was done prior to generating the abstractive summaries. The transformer language model was conditioned with the relevant information collected during the extractive step. The results proved the model to be better at producing abstractive summaries.

5 Analysis and Findings

This Section presents the qualitative and quantitative analysis of abstractive text summarization using pretrained language models. The quantitative analysis has been conducted using ROUGE score while the qualitative analysis has been done by extracting the data of interest and tabulating it. The qualitative analysis has been used to identify the characteristics of pretrained language models, to identify the challenges and issues and to highlight the performance boosting techniques. Table 1 presents several characteristics for UNILM [9], BART [10], MASS [8], PEGASUS [11], T5 [12], LED [13], and BIGBIRD [14].

Table 1. Characteristics of various pretrained language models

Model	Pretraining objective	Variants	Pretraining Dataset
UNILM [9]	Cloze task (context-based prediction of the masked word), next sentence prediction	UNILM	English Wikipedia, Book Corpus
BART [10]	Token masking, Token deletion, Text infilling, Sentence permutation, Document rotation	BART (base), BART (large)	Books and Wikipedia data
MASS [8]	Sequence fragment masking	MASS	WMT News Crawl Datasets
PEGASUS [11]	Masked language modeling, gap sentence generation	PEGASUS (base), PEGASUS (large)	C4 and Huge News corpora
T5 [12]	Prefix language modeling, masked language modeling, Deshuffling	T5 (base), T5 (small), T5 (large)	C4 (Colossal Clean Crawled Corpus)
LED [13]	Masked language modeling	LED (base), LED (large)	Books corpus, English Wikipedia, RealNews dataset

(continued)

Table 1. (*continued*)

Model	Pretraining objective	Variants	Pretraining Dataset
BigBird [14]	Masked language modeling	BigBird-RoBERTa, BigBird-PEGASUS	Books, CC-News, Stories, Wikipedia

Table 2. Performance comparison on Abstractive Text Summarization based on ROUGE

Dataset	Pretrained Language Model	R1/R2/RL
CNN/Daily mail	BART [10]	44.16/21.28/40.90
	UniLM [9]	43.33/20.21/40.51
	T5 – 11B [12]	43.52/21.55/40.69
	PEGASUS (large) [11]	**44.17/21.47/41.11**
XSum	BART [10]	45.14/22.27/37.25
	PEGASUS (large) [11]	**47.21/24.56/39.25**
Gigaowrd	MASS [8]	38.73/19.71/35.96
	UniLM [9]	38.45/19.45/35.75
	PEGASUS (large) [11]	**39.12/19.86/36.24**
ArXiv	LED (large) [13]	**46.63/19.62/41.83**
	PEGASUS (large) [11]	44.67/17.18/25.73
	Big Bird – PEGASUS [14]	46.63/19.02/41.77

Table 2 provides ROUGE score-based performance comparison between pretrained language models on abstractive text summarization across four standard datasets: CNN/Daily mail, XSum, Gigaword, and ArXiv. The results indicate that BART and PEGASUS performed competitively on CNN/Dailymail dataset, PEGASUS gives high performance on both XSum and Gigaword datasets while, BIGBIRD and LED perform equally well in terms of R1 on the ArXiv dataset. The low performance of PEGASUS on ArXiv dataset in terms of RL might be due to the input limitation of the PEGASUS model. The input limit of PEGASUS is 1024 tokens while average input length of the ArXiv dataset is beyond 1024 tokens.

5.1 Issues and Challenges in Finetuning the PLMs

This subsection defines several issues that arise during the finetuning of pretrained language models on the target tasks. *Catastrophic Forgetting* is a phenomena that arises in sequential learning. During the finetuning process, when the weights of the pretrained language model are altered to meet the objectives of another task, the model tends to lose some of the knowledge and language ability gained during the pretraining stage [23]. *Representational Collapse* is the deterioration of the quality of the generalizable representations from the pretrained language model during the finetuning stage [24].

Overfitting/Underfitting problem occurs during finetuning process when the encoder part of the neural network is a pretrained language model and the decoder part is a standard transformer and needs to be trained from scratch [25]. This unstability issue can also arise when the pretrained language model is finetuned on a low resource target dataset. *Few Shot Domain Transfer* becomes a challenge when the abstractive summarization model is unsupervised and the training dataset is based on a few examples like 100 or 200 examples [26]. Finetuning a pretrained language model using cross entropy loss results in less robust and *low quality generalization* ability of the neural network model [27].

5.2   Performance Improvement Techniques

Domain Adaptation. Pretrained models possess a great deal of language understanding. It has been studied that taking these models through the second phase of domain-related pretraining results in performance improvement of the task under both high and low resource scenarios [28]. In supervised domain adaptation, there is labelled data (supervision signals) from the target domain while in unsupervised domain adaptation, data from the target domain is unlabelled. The performance of domain adaptation depends on the divergence between the data distributions of the source and target domains [29]. [23] investigated domain adaptation for abstractive summarization across multiple domains in a low-resource setting. The researchers performed a second phase of pretraining on BART and investigated the results under three different scenarios: Source Domain Pre-Training (SDPT), Domain Adaptive Pre-Training (DAPT) and, Task Adaptive Pretraining (TAPT). During SDPT, with the News domain as the selected source domain, the training samples from XSum dataset were utilized to support the fast adaptation in the target domains. For DAPT, an unlabelled domain related corpus is leveraged to help introduce the domain knowledge in the BART. With TAPT, pretraining continues with the unlabelled documents from the target domain. The results indicate that the SDPT approach performed better than DAPT and TAPT in terms of ROUGE-1 metric. Although helpful, the second phase of pretraining makes the model vulnerable to catastrophic forgetting where the language model partially loses/forgets the language understanding learned during the initial pretraining. To avoid the catastrophic forgetting, RecAdam (Recall Adam) optimizer has been used by some researchers. [18] proposed domain adaptive pretraining on Newswire dataset as a solution to efficiently adapt GPT-2 for abstractive summarization. The approach was tested empirically across three datasets: CNN/Daily mail, XSum, and Newsroom. The results revealed more concise and clear summaries compared to the baselines.

Model Augmentation. Model augmentations have been reported to enhance the functionality and improve the performance of abstractive text summarization systems [30]. Certain types of model augmentations have been proposed for neural abstractive text summarization including copying mechanism [31], pointer-generator implementation and coverage method [32], distraction-based summarization [33], and Reinforcement learning-based methods [34, 35]. However, all these approaches have not experimented with transformer-based summarization models. [15] applied locality modelling and language model conditioning to abstractive text summarization. Locality modelling is the way to restrict calculations to the local context and is achieved by introducing 2-dimensional convolutional self-attention in the initial layers of the transformer encoder. This approach enabled better encoding of the local dependencies between the tokens improving the language encoding capability of the model. A pointer-generator is used on the top of the decoder network. The transformer-based encoder decoder model is conditioned with the BERT language model. A method for BERT windowing is also proposed to process the texts longer than the BERT window size. Training is done on CNN/daily mail and Swiss text dataset (German language). Ref. [36] augmented the abstractive document summarization model with an explicit information selection layer that performs two functions: (1) remove the redundant or unnecessary information by using gated global information filtering, (2) select sentences with salient information using local sentence selection. The approach led to summaries that are more informative as well as concise.

Better Finetuning. The standard finetuning suffers from the issue of representation collapse - "the degradation of generalizable representations of pre-trained models during the fine-tuning stage" [24]. Also, there exist cases in abstractive summarization where the encoder is pre-trained while the decoder is a standard transformer. This situation causes the finetuning procedure to be unbalanced due to the encoder overfitting and decoder underfitting. [25] proposed an effective finetuning approach with separate optimizers for both encoder and the decoder that helps to stabilize the process. The same work also demonstrated that a two-stage finetuning procedure with the encoder first finetuned on an extractive summarization task and then on abstractive summarization task results in high-quality summaries because now the model can leverage the information that is shared between the two tasks without modifying the model architecture. Based on Trust Region Theory, [24] proposed R3F (Robust Representations through Regularized Finetuning) and R4F (Robust Representations through Regularized and Reparametrized Finetuning) methods addressing the problem of representational collapse. The results from their probing experiment revealed that with standard finetuning, the model diverges from the source task much more than the proposed R3F and R4F. The R3F approach was evaluated on language generation tasks with various benchmark datasets including CNN/Daily mail, Gigaword, Reddit TIFU. BART combined with R3F becomes the new SOTA across the abstractive text summarization datasets with R1/R2/RL of 44.38/21.53/41.17 across CNN/Daily mail, 40.45/20.69/36.56 across Gigaword, and rouge scores equal to 30.31/10.98/24.74 across the Reddit TIFU dataset. [37] proposed on-manifold and off-manifold regularization techniques to address and mitigate the problem of miscalibration that commonly occurs due to over-parameterization while finetuning pre-trained language models on limited-sized datasets. A supervised contrastive learning (SCL) objective for finetuning is proposed in the work of [27]. SCL loss combined with the cross-entropy loss results in significant performance gains for multiple datasets of the GLUE benchmark.

Data Augmentation. It is a technique that increases the amount of training data by applying some form of transformation to the available dataset. The result is a new synthetic large sized training dataset. The meanings and patterns in the augmented dataset remain the same as that of the original dataset. Data augmentation has been found to be effective in boosting the performance of abstractive text summarization [26]. There are multiple techniques for data augmentation that has been applied to abstractive summarization. For example, [26] applied round-trip translation to generate the paraphrases of the documents and summaries. In round-trip translation or backtranslation, the original data is passed through a translation model say, English to French (another language) and then translated back from French to English to create an enriched paraphrased version. The inclusion of words from another language enriches the dataset. Another work [38] used part-of-speech tagger to obtain artificial data for French text summarization of financial reports. The tagger reads the reports and assigns parts of speech to each word, then the modifiers are masked and predicted with the pretrained language model, resulting in several new reports.

6   Summary and Conclusions

This article presents a comprehensive overview on neural abstractive text summarization. The emphasis is on the latest trends i-e, the use of pretrained language models for the task of abstractive text summarization. For this purpose, several encoder-decoder based PLMs have been compared on standard datasets for abstractive summarization in terms of ROUGE scores. The results of comparison give some clues on which PLMs perform better with what types of datasets. For example, every PLM has constraints on the input and output length. The length of the input and output documents in the dataset should always be considered while choosing the model otherwise the model performance might be suboptimal. Also, it has been seen that some PLMs like PEGASUS performs better on fully abstractive datasets like XSum as compared to datasets like CNN/Daily mail that are not very abstractive.

This survey has also highlighted some issues in the finetuning process of pretrained language models. The problems include catastrophic forgetting, representational collapse, model overfitting/underfitting, few-shot domain transfer, and low quality generalization. Future research efforts may be directed towards tackling these challenges effectively. That would help improve the performance of the abstractive summarization as well as other related tasks in NLP that utilize finetuning pretrained language models. Another open research area in abstractive text summarization with PLMs is zero-shot (no training examples) abstractive summarization. Currently, most PLMs give an extractive output summary in zero-shot settings. Narrating an abstractive summary in zero-shot setting is still a challenge.

One of the outcomes of this survey include strategies for improvement of current abstractive summarization systems. Domain adaptation can improve model performance under both high and low resource settings, model augmentation can improve performance by enhancing model functionality, stable finetuning leads to optimal performance by reducing the representation degradation and collapse, and data augmentation can help when the available labelled dataset is small in size.

We expect this survey would provide the abstractive summarization research community with the latest trends, useful insights, and future research directions to design and develop better abstractive summarization models.

References

1. Klymenko, O., Braun, D., Matthes, F.: Automatic Text Summarization: a State-of-the-Art Review, vol. 1, pp. 648–655 (2020)
2. Han, X., et al.: Pre-Trained Models: Past, Present and Future. AI Open (2021). https://doi. org/10.1016/j.aiopen.2021.08.002
3. Hasan, T., et al.: XL-Sum: large-scale multilingual abstractive summarization for 44 languages. In: ACL-IJCNLP, pp. 4693–4703 (2021)
4. Cao, Y., Wan, X., Yao, J., Yu, D.: MultiSumm: Towards a Unified Model for Multi-Lingual Abstractive Summarization. Proc. AAAI **34**(01), 11–18 (2020). https://doi.org/10.1609/aaai. v34i01.5328
5. Syed, A.A., Gaol, F.L., Matsuo, T.: A survey of the state-of-the-art models in neural abstractive text summarization. IEEE Access **9**, 13248–13265 (2021)
6. Devlin, J., et al.: BERT: Pre-training of deep bidirectional transformers for language understanding. NAACL-HLT **1**, 4171–4186 (2019)
7. Peters, M.E., et al.: Improving language understanding by generative pre-training. In: OpenAI, pp. 1–10 (2018)
8. K. Song, X. Tan, T. Qin, J. Lu, and T. Y. Liu, "MASS: Masked sequence to sequence pre-training for language generation," 36th ICML, pp. 10384–10394, 2019.
9. Dong, L., et al.: Unified language model pre-training for natural language understanding and generation. In: NIPS., vol. 32 (2019)
10. Lewis, M., et al.: BART: denoising sequence-to-sequence pre-training for natural language generation, translation, and comprehension. In: ACL, pp. 7871–7880 (2020)
11. Zhang, J., Zhao, Y., Saleh, M., Liu, P. J.: PEGASUS: pre-training with extracted gap-sentences for abstractive summarization. In: ICML, pp. 11328–11339 (2020)
12. Raffel, C., et al.: Exploring the limits of transfer learning with a unified text-to-text transformer. J. Mach. Learn. Res. **21**, 1–67 (2020)
13. Beltagy, I., et al.: Longformer: The Long-Document Transformer (2020)
14. Zaheer, M., et al.: "Big bird: Transformers for longer sequences," 2020.
15. Aksenov, D., et al.: Abstractive text summarization based on language model conditioning and locality modeling. In: 12th International Conference on Language Resources and Evaluation, pp. 6680–6689 (2020)
16. Zolotareva, E., et al.: Abstractive text summarization using transfer learning. CEUR Workshop Proc. **2718**, 75–80 (2020)
17. Zhao, S., You, F., Liu, Z.: Leveraging pre-trained language model for summary generation on short text. IEEE Access 1–6 (2020)
18. Hoang, A., et al.: Efficient adaptation of pretrained transformers for abstractive summarization (2019)
19. Kim, S.: Using pre-Trained Transformer for Better Lay Summarization, pp. 328–335 (2020). https://doi.org/10.18653/v1/2020.sdp-1.38.
20. Bajaj, A., et al.: Long Document Summarization in a Low Resource Setting using Pretrained Language Models, pp. 71–80 (2021). https://doi.org/10.18653/v1/2021.acl-srw.7.
21. Xiao, L., Wang, L., He, H.: Modeling Content Importance for Summarization with Pre-trained Language Models, pp. 3606–3611 (2020)
22. Pilault, J., Li, R., Subramanian, S., Pal, C.: On Extractive and Abstractive Neural Document Summarization with Transformer Language Models, pp. 9308–9319 (2020)
23. Yu, T., Liu, Z., Fung, P.: AdaptSum: Towards Low-Resource Domain Adaptation for Abstractive Summarization, pp. 5892–5904 (2021)
24. Aghajanyan, A., et al.: Better Fine-tuning by Reducing Representational Collapse (2021)
25. Liu, Y., Lapata, M: Text Summarization with Pretrained Encoders (2019)

26. Fabbri, A., et al.: Improving zero and few-shot abstractive summarization with intermediate fine-tuning and data augmentation. In: NAACL, pp. 704–717 (2021)
27. Gunel, B., Du, J., Conneau, A., Stoyanov, V.: Supervised contrastive learning for pre-trained language model fine-tuning. In: ICLR, pp. 1–21 (2021)
28. Gururangan, S., et al.: Don't stop pretraining: adapt language models to domains and tasks. In: ACL, pp. 8342–8360 (2020)
29. Guo, H., et al.: Multi-source domain adaptation for text classification via DistanceNet-bandits. In: 34th AAAI Conference on Artifical Intelligence, pp. 7830–7838 (2020)
30. Khandelwal, U., Clark, K., Jurafsky, D., Brain, G.: Sample Efficient Text Summarization Using a Single Pre-Trained Transformer (2018)
31. Gu, J., Lu, Z., Li, H., Li, V.O.K.: Incorporating copying mechanism in sequence-to-sequence learning. ACL 3, 1631–1640 (2016)
32. See, A., Liu, P.J., Manning, C.D.: Get to the point: Summarization with pointer-generator networks. ACL 1, 1073–1083 (2017)
33. Chen, Q., et al.: Distraction-Based Neural Networks for Document Summarization (2016)
34. Li, S., Lei, D., Qin, P., Wang, W.Y.: Deep reinforcement learning with distributional semantic rewards for abstractive summarization. In: EMNLP, pp. 6038–6044 (2020)
35. Paulus, R., Xiong, C., Socher, R.: A deep reinforced model for abstractive summarization. In: 6th ICLR, pp. 1–12 (2018)
36. Li, W.: et al.: Improving neural abstractive document summarization with explicit information selection modeling. In: EMNLP, pp. 1787–1796 (2020)
37. Kong, L., Jiang, H., Zhuang, Y., Lyu, J., Zhao, T., Zhang, C.: Calibrated Language Model Fine-Tuning for In- and Out-of-Distribution Data, pp. 1326–1340 (2020). https://doi.org/10.18653/v1/2020.emnlp-main.102.
38. Laifa, A., et al.: Data augmentation impact on domain-specific text summarization Data augmentation impact on domain-specific text summarization (2021)

A Combination of BERT and Transformer for Vietnamese Spelling Correction

Trung Hieu Ngo$^{(\boxtimes)}$, Ham Duong Tran, Tin Huynh, and Kiem Hoang

The Saigon International University, Ho Chi Minh City, Vietnam
{ngotrunghieu,tranhamduong,huynhngoctin,hoangkiem}@siu.edu.vn

Abstract. Recently, many studies have shown the efficiency of using Bidirectional Encoder Representations from Transformers (BERT) in various Natural Language Processing (NLP) tasks. Specifically, English spelling correction task that uses Encoder-Decoder architecture and takes advantage of BERT has achieved state-of-the-art result. However, to our knowledge, there is no implementation in Vietnamese yet. Therefore, in this study, a combination of Transformer architecture (state-of-the-art for Encoder-Decoder model) and BERT was proposed to deal with Vietnamese spelling correction. The experiment results have shown that our model outperforms other approaches as well as the Google Docs Spell Checking tool, achieves an 86.24 BLEU score on this task.

Keywords: Vietnamese spelling correction · BERT · Transformer

1 Introduction

A spelling error is a word written in a wrong spelling standard, including various forms: homophone, acronym, uppercase, or fairly the phenomenon of wrong-written words. Usually, there are many groups of origination causing the spelling errors to happen in Vietnamese: typing, semantic confusion, local pronunciation, rules, and standards in the written text, not mastery in grammar and influence of social network language, etc. [29].

Spelling correction is a Natural Language Processing task that focuses on correcting spelling errors in text or a document. The spelling correction task keeps a critical role in enhancing the user typing experience and guarantees the information integrity of Vietnamese. Besides, one primary application is the ability to incorporate with other tasks. For example, when using the spelling correction attached to the last phase of the Scene Text Detection / Optical Character Recognition (OCR) task, the results are improved significantly [1,11]. Consider the chatbot task, if spelling correction is applied to preprocess user inputs, the chatbot will have better accuracy and performance in understanding the user requests [27].

Frequently, the spelling correction can be divided into two steps, including spell checking and spell correcting [7]. In the first phase, mistakes are investigated if there are any in the given input and then try to transform the wrong words into corrected words in the second phase.

© The Author(s), under exclusive license to Springer Nature Switzerland AG 2022
N. T. Nguyen et al. (Eds.): ACIIDS 2022, LNAI 13757, pp. 545–558, 2022.
https://doi.org/10.1007/978-3-031-21743-2_43

Contrary to English and other languages, the Vietnamese possess up to six complex diacritic marks and uses them as a discrimination sign. Therefore, a word that combines with different diacritic marks can create up to six written forms, and each of them also has independent meaning and usage. For instance, the word "ma" (ghost) can be written in 5 more ways with 5 different diacritic marks: "má" (mother), "mả" (nevertheless), "mả" (tomb), "mã" (code), "mạ" (rice seedlings). All the originations and elements described above have made the Vietnamese spelling correction problem a very challenging task.

There are many initial approaches to the Vietnamese Spelling Correction task that has been carried out such as applying rule-based methods [10], using edit-distance Algorithm [20], collating with dictionaries, using n-gram/big-gram language model [19], etc. However, most of these approaches neither adapted to out-of-vocabulary words nor did they take the contextualized word embeddings into account. In order to deal with these gaps, many deep learning models using Recurrent Neural Network (RNN) or Long Short-term Memory (LSTM) networks have been proposed and achieved impressive performance [21].

Recently, spelling correction studies that took advantage of the Encoder-Decoder model have attracted much attention and achieved state-of-the-art in the English spelling correction task [14,30]. This is a novel approach, which is notably potential because of the optimal utilization of the parallelism calculation ability and the strength of powerful pre-trained language models. One of the most attention is the usage of the Transformer architecture [28] with the language model known as BERT [2]. Despite its success in English [31], there is still no implementation in Vietnamese that can be used in practice. Therefore, this paper aims to apply these architectures and techniques to improve the performance of correcting Vietnamese spelling errors. The experiment results show that the proposed solution achieves considerable efficiency and is able to integrate with practical services. The main contributions of this study could be summarized as follows:

- Applying the Transformer architecture and leveraging the pre-trained BERT to provide a solution to the Vietnamese spelling correction problem.
- Constructing a large and creditable dataset based on the most common practical Vietnamese spelling errors. The evaluation dataset is published for the Vietnamese NLP community using in related works.

The remainder of this paper is organized as follows. In Sect. 2, related works are presented and discussed. Section 3 describes the proposed methods in detail. The dataset, experimental results, and discussion are provided in Sect. 4. Section 5 summarizes, concludes and gives future orientation.

2   Related Works

Spelling correction is not a new problem in NLP tasks. Earlier there have been many approaches for this problem, from straightforward approach using probability, such as implementing the Naive Bayes algorithm (Peter Norvig[1]). The

---

large N-gam-based language modeling approach of both left and right side has improved the performance of spelling correction tasks [19]. After training with a large corpus, this model can predict the probabilities of multiple N-gram candidates for correcting words. Large N-gram LM is a pure probability approach. It expects high memory resources to store all pre-calculated probabilities of N-gram pairs and can not handle a not-pre-trained error, which leads to all of the probability of N-gram pairs to zero.

The advantages of contextual embedding in word presentation model, likes Word2Vec [16], Glove [24], etc., is being taken into the spelling correction task [4]. An edit-distance algorithm generates the candidates, then each candidate's score is calculated by the cosine similarity between the candidate vector and the target word vector, the highest score ranking candidate will be selected. This method has shown significant results in the spelling correction task and is suitable for many languages, especially in Vietnamese, English, etc. On the other hand, this approach requires many resources to represent the rich context embeddings of a language accurately. Also, out-of-vocabulary (OOV) is a large major problem to the ranking system.

Another approach to using deep learning has been developed through the use of LSTM network [18]. A LSTM network [8] is constructed that encodes the input sequence and then decodes it to the expected correct output sequence, respectively. The accuracy of their model makes a significant gap compared to the current state-of-the-art model [19]. Studies have reported that spelling correction can be beneficial from Encoder-Decoder architecture [9,25]. A state-of-the-art approach in English is implemented by the Encoder-Decoder architecture [26] and also makes use of the powerful pretrained BERT model [9]. They first fine-tuned the BERT model and then used its last hidden presentation output as additional features to the an error correction model, which is a customized Transformers [28] architecture. A similar method for Vietnamese grammatical error correction using OpenNMT framework [13] instead of the Transformer architecture [25]. This method, respectively, depends on using the Microsoft Office spelling tool to check and detect the incorrect text before the correction step.

Through previous related works, deep learning approaches to spelling correction are our focus. The approach is receiving much concern, gains state-of-the-art performance is the Encoder-Decoder architecture with prominent pre-trained MLM. Both well-known pre-trained Google Multilingual [2] and vinai/phobert [17] are used to extract hidden presentations and implement the transformer architecture into a specific Vietnamese spelling correction task.

3 Our Approach

3.1 Introduction to the Vietnamese Language

This section briefly presents the characteristics and differences from English of the Vietnamese language. Unlike neighbor countries, the Vietnamese does not use hieroglyphic letters, but a modified Latin (Roman) alphabet. The Vietnamese alphabet uses 29 letters, unlike the English alphabet, it does not use 4 letters 'w', 'f', 'j', 'z' and uses 6 more vowel letters (with special mark): 'ă', 'â', 'ê', 'ô', 'ơ', 'ê', and the letter 'đ' [3,6]. Along with the above 6 types of diacritics, it forms up to 67 separate letters (nearly 3 times larger than the number of letters in English). Therefore, spelling mistakes are much more common in Vietnamese than in English.

3.2 Analyzing of Vietnamese Common Spelling Error

In this section, the concept of common error type in the Vietnamese language are presented. Due to the lacks of scientific public research or national survey constructed on this topic, various types of Vietnamese error type from previous related work [18–20] are summarized and divided them into six groups:

- **Abbreviation:** There are a wide variety of abbreviation for common words in Vietnamese writing. Despite its convenience, this style of writing may raise misunderstanding, make the writing less formal and not accepted by most people. To determine this error cases, a list of most common abbreviation substitutions in Vietnamese is compiled from the Internet.
- **Region:** The region error type is the most complicated type to analyze owing to its variety of happening contexts. The region error type comes from different region pronunciation across the Vietnam territory. When people tend to write a word the same way they pronounce it, this error occurs. Many adults may mistake this type of error if not a native speaker or do not have enough knowledge of the Vietnamese language. An incorrect word with region type stands alone, may still have meaning. Some examples of region error type are described in Table 1.

Table 1. Some examples of region error type

Original	Usually mistake for	Original	Usually mistake for
ch-	tr-	-nh	-n
tr-	ch	c-	k-
-n	-ng	k-	c-
-ng	-n	ngh-	ng-
g-	gi-	gi-	g-
...	...	...	...

- **Teencode:** Teencode (or Teen-code) is a method of writing used by teenagers on social media or through messaging. Those teenagers put words into special encryption so the adults can not understand.
- **Telex:** Telex is a convention for encoding Vietnamese text in plain ASCII characters, used initially for transmitting Vietnamese text over telex systems. Forgetting to turn on the language encoder or entering the wrong Vietnamese Telex rules leads to this type of error.

- **Fat Finger** Fat Finger, also known as the clumsy finger, means when typing through a cell phone or computer keyboard, the user's finger mistypes the surrounding key instead of the target key, causing the wrong words.
- **Edit Distance** Edit Distance is a pseudo error generation strategy in which several characters equal to a 'distance' to the original are randomly replaces. Although this error rarely happens logically, a low percentage amount is still generated in our data set.

For the convenience of observation, a list of examples corresponding to the type of error is presented in Table 2.

Table 2. A summary about error types

Error type	When it happends	Examples	
		Correct	Incorrect
Abbreviation	To make writing faster and more convenient, people use abbreviation instead of full words.	Không (No) Mọi người (Everyone) Bình thường (Normal)	Kg Mn Bt
TeenTeencode	When teenagers try to encode their text in order to hide information.	Ví dụ (Example) Chồng (Husband) Điện thoại (Phone)	Vj du Cho'ng Dj3n tk04j
Fat-Finger	While typing with a virtual keyboard, the user's finger mistypes the surroundings instead of the target key.	Xin chào (Hello) Trường học (School) Điện thoại (Phone)	Xim chào Trường hịc Điện thoak
Telex	Forgetting to turn on the language encoder or entering the wrong Vietnamese Telex rules.	Xin chào (Hello) Trường học (School) Điện thoại (Phone)	Xin chafo Truowng hojc Ddieejn thoaji
Region	Different region pronunciation across the Vietnam (as people tend to write the same as as they pronounce).	Tranh (Painting) Lạnh (Cold) Nghỉ (Rest)	Chanh (Lemon) Nạnh Nghỉ(Think)
Edit-Distance	A common pseudo misspelling error generation strategy.	Tranh (Painting) Lạnh (Cold) Mưa (Rain)	Thanh (Bar) Lhoạnh Mtưa

3.3 BERT

BERT is a language representation model based on multi-layer bidirectional transformers encoder architecture. There is a wide variety of challenging natural language tasks that BERT can handle and achieve state-of-the-art performance, from classification, question answering, and sequence-to-sequence learning task, etc. BERT can represent sentence effectively by its encoding mechanism, including various embedding step as token embeddings, sentence embeddings, and transformers positional embeddings. Then, this BERT's last hidden presentation output from BERT is used as the input into the transformers architecture.

In this study, two well pre-trained BERT on the Vietnamese: the Google Multilingual BERT[2] (bert-base-multilingual-cased) and VinAI/phoBERT[3] are

---

[2] Github: https://github.com/google-research/bert.
[3] Github: https://github.com/VinAIResearch/PhoBERT.

considered. They are both trained with extensive Vietnamese corpus, while the multilingual BERT is the BERT base model, and phoBERT is a RoBERTa model [15] (which is a modified version from the base model).

3.4 Transformers

Before the transformers architecture, Encoder-Decoder architecture using RNN/ LSTM/ GRU (Gated Recurrent Unit) cell is used widely in machine translation and sequence-to-sequence tasks. This Encoder-Decoder architecture, also known as the Seq2Seq model, uses several RNN cells to encode the input tokens to hidden states and then sum all hidden states up before sending them to the decoder. Thanks to this hidden state, the decoders receive all previously encoded information and use it for the output token prediction task. Despite large capacities in handling sequence-to-sequence tasks, the Seq2Seq decoder may fail to fully capture the meaning and context of the last hidden presentation from the encoder. That means the more extended and more complex the input sequence, the less effective the hidden presentation can represent, which is known as the bottleneck problem.

While the attention mechanism takes several inputs simultaneously, construct weight matrices captured from each hidden presentation input to calculate a weighted sum of all the past encoder states. The decoder will then take the inputs and the provided attention weights, and through that, the decoder knows how to 'pay much attention' to which hidden presentation and vice versa. Another limitation of Seq2Seq architecture is that it handles the input sequentially, which means to compute for the current token at time t, we need the previous hidden state t-1 and so on. Therefore, especially in the spelling correction task, if many erroneous tokens stick together, the correction of the last tokens can be poorly affected. The transformer architecture and attention mechanism come to cross those boundaries of those previous architectures. Based on the Encoder-Decoder architecture [26], the transformers architecture [28] uses stacked multi-head self-attention and fully connected layers. The transformer is designed to allow parallel computation and reduce the drop in performance due to long dependencies. It uses positional embeddings and multi-head self-attention to encode more information about the position of a token and the relation between each token.

As Shun has provided an advanced insights to incorporating pseudo data into the spelling correction task [12]. Consequently, a vast pseudo training dataset can be generated so that not only can the transformer maximize its parallelization ability but also BERT can represent its rich contextual embedding vectors.

3.5 Incorporate BERT into Transformers

As mentioned in Sect. 3.3, BERT is capable of deep language understanding by capturing contextual embedding of different words in a sequence. In addition, the Transformer model has been proved to be more efficient than popular Encoder-Decoder architectures, especially in the Machine Translation problem.

Some recent studies have treated the spelling correction problem as a machine translation job where the error sentence is the source sequence and the corrected sentence is the target sequence. And the combination of BERT and Transformer achieves state-of-the-art results for the English spelling correction task [9]. However, to the best of our knowledge, there has not been any research combining BERT and Transformer for Vietnamese spelling correction problem. The combination can be briefly summarized in the following steps:

- **Step 1**: Let the input sentence notated as $X = (x_1, ..., x_n)$, where n is the number of its tokens; x_i is the i-th token in X. BERT receives the input sequence tokens, and through its layers, BERT extracts them to hidden presentations notated as $H_B = (h_1, h_2, ..., h_n)$, where H_B is the output of the last layers in BERT.
- **Step 2**: The Encoder will take H_B from the previous step and encode the representation of each l layer H_E^l. The final contextual representation of the last encoder layer H_E^L is the output of the Encoder. The Encoder components consist of the multi-head self-attention mechanism, position-wise fully connected feed-forward network. A residual connection around each of the two sub-layers, followed by layer normalization. A Multi-head Attention is a component allowing the model to jointly attend to information from different representations and helps the encoder look at other words in the input sentence as it encodes a specific word for better-capturing contextual embedding.
- **Step 3**: The Decoder receives the representation H_E^L from the Encoder and decodes through its layers into final representation H_D^L. Similar to the Encoder, the Decoder possessed the same components of the Encoder. These Encoder H_E^L are to be used by each Decoder in its "encoder-decoder attention" layer which helps the Decoder focus on appropriate places in the input sequence.
- **Step 4**: Finally, the Decoder final representation H_D^L is mapped via a linear transformation and softmax to get the t-th predicted word $\hat{y}$. The decoding process continues until meeting the end-of-sentence token.

An illustration of our proposed method is shown in Fig. 1.

4 Experimental Evaluation

This section includes dataset description, evaluation method, model hyperparameter setting as well as experimental results of applying the Transformer architecture and BERT to Vietnamese spelling correction.

4.1 Experimental Dataset

This section describes the process of creating our training and testing set based on the Binhvq News Corpus[4] which contains 14,896,998 Vietnamese news

---

[4] Github: https://github.com/binhvq/news-corpus.

crawled from the Internet and preprocessed, including steps like HTML tag removal, duplicate removal, NFC standardization, and sentence segmentation. The corpus is gathered from reputable news and media sites in Vietnam, so the data is very reliable in terms of spelling. For the purpose of training and evaluating spelling correction, our newly constructed dataset must consist of two fields that can be described as a pair of correct and incorrect spelling sentences.

To the best of our knowledge, there is no specific survey as well as assessment on the rate of error types appearing in Vietnamese. However, Vietnamese often has common spelling mistakes: Region, FatFinger, Telex. Besides, some other types of errors are concerned, such as Edit-Distance, Abbreviation, Teencode, but rarely happened in practice. Therefore, error rate is reproduced based on our experience. Details of the rates of error types in the generated data set are listed in Table 3.

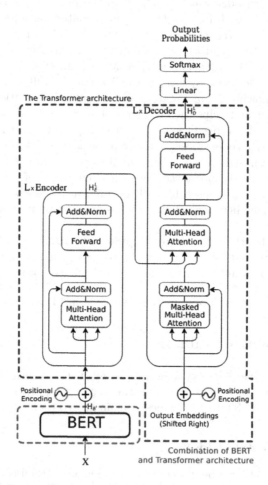

Fig. 1. Proposed combination between BERT and Transformer

Table 3. Training and testing sets

Error type	Error ratio (%)
Acronym	3.0
Teencode sets	3.0
Edit-Distance	3.0
FatFinger	30.0
Telex	30.0
Region	31.0

The training set is composed by randomly selecting 4,000,000 sentences from the Binhvq corpus and then apply the pseudo error generator to these correct ones. All the sentences must have an average word count between 50–60 words per sentence. Similarly, validating set and testing set are generated with the number of correct sentences chosen from the above corpus 20,000 and 6,000 respectively. Details of the dataset can be summarized in Table 4. The testing set is public and can be downloaded at the following link[5]

Table 4. Training and testing sets

Dataset	Size (#Pair of sentence)	Avg. Length per sentence (#token)
Training sets	4,000,000	60
Validating sets	20,000	60
Testing sets	6,000	60

4.2 Evaluating Metric

In the perspective of a spelling correction task, many traditional approaches used Accuracy, Precision, Recall, and F1 for evaluation [18,19]. These metrics require the predictions' words to have the same length as labels' words. Recently, BLEU score is chosen, especially in deep learning models because of its ability to adapt to different prediction lengths [9,31]. Therefore, BLUE [23] is selected for the evaluating task. BLEU, or the Bilingual Evaluation Understudy, is a score for comparing a candidate translation of text to one or more reference translations. Although developed for translation, it can be used to evaluate text generated for a suite of natural language processing tasks. Our BLEU configuration uses four n-grams settings because the spelling correction task critically requires the order of words in the sentence.

$$BP = \begin{cases} 1 & \text{if} \quad r > c \\ e^{(1-r/c)} & \text{if} \quad r \le c \end{cases} \tag{1}$$

---

[5] Github: https://github.com/tranhamduong/Vietnamese-Spelling-Correction-testset.

$$BLEU = BP \cdot \exp \left(\sum_{n=1}^{N} w_n \log p_n \right) \tag{2}$$

where BP stands for Brevity Penalty. c, r is the length of the predictions and labels, respectively. BP will penalty cases where the model failed to propose correction, or the change happens more than allowed (as the number of words need to be corrected must be equal to the actual corrected). p_n stands for modified n-gram precision, using n-grams up to length N and positive weights w_n summing to one. The n-gram precision can be simply understands as 'the number of corrected words which occur in reference sentence (ground-truth)' divided by 'the number of words after sentence transformed'. Therefore, the BLEU metrics has potential to not only to keep track strictly of word ordering by measuring n-gram (up-to-4) overlapping but also evaluate how a sentence has been corrected from the original despite the action (remove, edit, add more words).

4.3   Model Settings

Our models are implemented by fairseq toolkit [22] which is an re-implementation on the base Transformer architecture [28]. To find the appropriate hyperparameters for the proposed model, experiments with multiple model designations has been reviewed and the configuration of Jinhua work [31] are selected. Training details with hyperparameter settings are in the Table 5:

Table 5. Hyperparameters of transformer model

BERT model	Bert-base-multilingual-cased Vinai/phobert-base
Number of epochs	100
Dropout	0.3
Loss function	Labeled smoothed cross-entropy
Optimizer	Adam(0.9,0.98)
Learning rate	0.005
Label smoothing	0.1
Weight decay	0.0001
Beam search	5
Max tokens	1280

4.4   Experimental Results and Discussion

In this phase, we compared with the Google Docs spell checking tool[6] and other methods. From the results showed in Fig. 6, two versions of our model, Trans-

---

[6] The tool can be found on the Google Docs website (https://docs.google.com/). We collected samples by using a web browser behavior simulator based on Selenium framework that manipulate the Google spell checking tool to correct all of its possible suggestions.

former+vinai/ phoBERT and Transformer+BERT-multi-cased, achieved better results than the previous methods. This partly reinforces our hypothesis that using a pre-trained language model BERT brings two benefits to the spelling correction problem: being applicable in the spelling correction task and taking advantage of contextualized word embeddings. Firstly, as mentioned in the BERT paper, tasks such as Text Classification, Question and Answering, Sentence Tagging, etc., are recommended to be used in the fined-tuning phase but the spelling correction task. Due to our modification, at the first step, BERT is verified to be beneficial for the correction task. Secondly, when correcting an error word, the action of choosing a suitable candidate based on context words is the main characteristic of the spelling correction problem. Concretely, BERT produces contextualized word embedding (the same word for different contexts have different embeddings) helps the models to better utilize word embedding at correcting phase. Besides, the pre-training of BERT on a huge data set also makes fine-tuning for our model easier because of no need to re-train from the beginning, taking advantage of the knowledge from the language model.

Table 6. BLEU scores on models

Model	BLEU score
Google Docs spellchecking tool	0.6829
Transformer + vinai/phobert-base	0.8027
Word2Vec	0.8222
Transformer + bert-multi-cased	**0.8624**

Transformer + vinai/phobert-base: The proposed model based on the Transformer architecture and PhoBERT [17].

Word2Vec: The reimplementation of the Word2Vec approach in spelling correction [5].

Transformer + bert-multi-cased: The proposed model based on the Transformer architecture and BERT multilingual model [2].

For the objective of comparison and practical application, there are a few patterns that our excellent model gain out performance: telex and edit-distance error types, compared to the google docs spellchecking tool. This happened partially because we designated more of these types of error to achieve our goal. More tuning is needed in future work on the error type distribution to improve performance for other types of errors.

The google docs spellchecking tool has another advantage over our model is the ability to restrict unnecessary correction. Additionally, the emergence of proper nouns also makes our model ineffective. When it comes to a proper noun, especially Vietnamese proper names, our model tends to correct them, which should not be the case. To overcome this weakness, some supporting components can be developed to the proposed architecture: Applying a name entity recognition component or an independent spellchecker to determine to correct a word or not.

5 Conclusion

In this paper, a combination of BERT and Transformer architecture is implemented for the Vietnamese spell correction task. The experimental results show that our model outperforms other approaches with a 0.86 BLUE score and can be used in real-world applications. Besides, a dataset is constructed for related works based on a breakdown of the spelling correction problem to define which errors commonly happened and need more attention.

To our concern, despite the improvement in the model's performance, there are late inferences due to large and complex architecture. In addition, due to the different distribution of data in the pre-trained model compared to data for the spelling correction task, we can not fully utilize the representation of pre-trained words, resulting in the model sometimes try to correct the unwanted words.

In the future, our architecture will be experimented with other existing pre-trained language models to see how well the compatibility they are. Moreover, we also evaluate our model's accuracy on a bigger dataset. Finally, investigating and analyzing errors that may happen in practices is our priority in order to create a better error pseudo generator.

References

1. Bassil, Y., Alwani, M.: Ocr post-processing error correction algorithm using google online spelling suggestion. J. Emerging Trends Comput. Inf. Sci. (2012)
2. Devlin, J., Chang, M., Lee, K., Toutanova, K.: BERT: pre-training of deep bidirectional transformers for language understanding, pp. 4171–4186 (2019)
3. of Education Vietnam M: Sách Giáo khoa tiếng Việt 1 (Tập Một) Ministry of Education Publisher (2002)
4. Fivez, P., Šuster, S., Daelemans, W.: Unsupervised context-sensitive spelling correction of clinical free-text with word and character n-gram embeddings. In: BioNLP 2017, pp. 143–148. Association for Computational Linguistics, Vancouver, Canada Aug 2017
5. Fivez, P., Suster, S., Daelemans, W.: Unsupervised context-sensitive spelling correction of english and dutch clinical free-text with word and character n-gram embeddings (2017)
6. Hao, C.X.: Tiếng Việt, văn Việt, người Việt Youth Publisher (2003)
7. Hladek, D., Staš, J., Pleva, M.: Survey of automatic spelling correction. Electronics 9, 1670 (2020)
8. Hochreiter, S., Schmidhuber, J.: Long short-term memory. Neural Comput. 9, 1735–80 (1997)
9. Kaneko, M., Mita, M., Kiyono, S., Suzuki, J., Inui, K.: Encoder-decoder models can benefit from pre-trained masked language models in grammatical error correction. In: Proceedings of the 58th Annual Meeting of the Association for Computational Linguistics, pp. 4248–4254. Association for Computational Linguistics (2020)

10. Khanh, P.H.: Good spelling of vietnamese texts, one aspect of computational linguistics in vietnam. In: Proceedings of the 38th Annual Meeting on Association for Computational Linguistics, ACL 2000 p. 1–2. Association for Computational Linguistics, USA (2000)

11. Kissos, I., Dershowitz, N.: Ocr error correction using character correction and feature-based word classification. In: 2016 12th IAPR Workshop on Document Analysis Systems (DAS), pp. 198–203. IEEE (2016)

12. Kiyono, S., Suzuki, J., Mita, M., Mizumoto, T., Inui, K.: An empirical study of incorporating pseudo data into grammatical error correction. In: Proceedings of the 2019 Conference on Empirical Methods in Natural Language Processing and the 9th International Joint Conference on Natural Language Processing (EMNLP-IJCNLP), pp. 1236–1242. Association for Computational Linguistics, Hong Kong, China, Nov 2019

13. Klein, G., Kim, Y., Deng, Y., Senellart, J., Rush, A.: OpenNMT: Open-source toolkit for neural machine translation. In: Proceedings of ACL 2017, System Demonstrations, pp. 67–72. Association for Computational Linguistics, Vancouver, Canada (Jul 2017)

14. Liu, J., Cheng, F., Wang, Y., Shindo, H., Matsumoto, Y.: Automatic error correction on Japanese functional expressions using character-based neural machine translation. In: Proceedings of the 32nd Pacific Asia Conference on Language, Information and Computation. Association for Computational Linguistics, Hong Kong, 1–3 Dec 2018

15. Liu, Y., et al.: Roberta: A robustly optimized bert pretraining approach. arXiv preprint arXiv:1907.11692 (2019)

16. Mikolov, T., Chen, K., Corrado, G., Dean, J.: Efficient estimation of word representations in vector space. In: 1st International Conference on Learning Representations, ICLR 2013, Scottsdale, Arizona, USA, 2–4 May 2013. Workshop Track Proceedings (2013)

17. Nguyen, D.Q., Nguyen, A.T.: PhoBERT: Pre-trained language models for Vietnamese. In: Findings of the Association for Computational Linguistics: EMNLP 2020, pp. 1037–1042 (2020)

18. Nguyen, H.T., Dang, T.B., Nguyen, L.M.: Deep learning approach for vietnamese consonant misspell correction. In: Nguyen, L.-M., Phan, X.-H., Hasida, K., Tojo, S. (eds.) PACLING 2019. CCIS, vol. 1215, pp. 497–504. Springer, Singapore (2020). https://doi.org/10.1007/978-981-15-6168-9_40

19. Nguyen, H., Dang, T., Nguyen, T.T., Le, C.: Using large n-gram for vietnamese spell checking. Adv. Intell. Syst. Comput. **326**, 617–627 (2015)

20. Nguyen, P.H., Ngo, T.D., Phan, D.A., Dinh, T.P., Huynh, T.Q.: Vietnamese spelling detection and correction using bi-gram, minimum edit distance, soundex algorithms with some additional heuristics. In: 2008 IEEE International Conference on Research, Innovation and Vision for the Future in Computing and Communication Technologies, pp. 96–102. IEEE (2008)

21. Nguyen, Q.D., Le, D.A., Zelinka, I.: Ocr error correction for unconstrained vietnamese handwritten text, pp. 132–138 (12 2019)

22. Ott, M., et al.: fairseq: A fast, extensible toolkit for sequence modeling. In: Proceedings of the 2019 Conference of the North American Chapter of the Association for Computational Linguistics (Demonstrations). pp. 48–53. Association for Computational Linguistics, Minneapolis, Minnesota (Jun 2019)

23. Papineni, K., Roukos, S., Ward, T., Zhu, W.J.: Bleu: a method for automatic evaluation of machine translation. In: Proceedings of the 40th Annual Meeting

of the Association for Computational Linguistics, pp. 311–318. Association for Computational Linguistics, Philadelphia, Pennsylvania, USA (Jul 2002)

24. Pennington, J., Socher, R., Manning, C.: GloVe: Global vectors for word representation. In: Proceedings of the 2014 Conference on Empirical Methods in Natural Language Processing (EMNLP), pp. 1532–1543. Association for Computational Linguistics, Doha, Qatar (Oct 2014)

25. Pham, N.L., Nguyen, T.H., Nguyen, V.V.: Grammatical error correction for vietnamese using machine translation. In: Nguyen, L.-M., Phan, X.-H., Hasida, K., Tojo, S. (eds.) PACLING 2019. CCIS, vol. 1215, pp. 505–512. Springer, Singapore (2020). https://doi.org/10.1007/978-981-15-6168-9_41

26. Sutskever, I., Vinyals, O., Le, Q.V.: Sequence to sequence learning with neural networks. In: NIPS 2014, pp. 3104–3112. MIT Press, Cambridge, MA, USA (2014)

27. Tedjopranoto, M., Wijaya, A., Santoso, L., Suhartono, D.: Correcting typographical error and understanding user intention in chatbot by combining n-gram and machine learning using schema matching technique. Int. J. Mach. Learn. Comput. **9**, 471–476 (2019)

28. Vaswani, A., Shazeer, N., Parmar, N., Uszkoreit, J., Jones, L., Gomez, A.N., Kaiser, Ł., Polosukhin, I.: Attention is all you need. In: Advances in Neural Information Processing Systems, pp. 5998–6008 (2017)

29. Xuan, P.: Solutions to spelling mistakes in written vietnamese. VNU J. Sci. Educ. Research **33**(2) (2017)

30. Yuan, Z., Briscoe, T.: Grammatical error correction using neural machine translation. In: Proceedings of the 2016 Conference of the North American Chapter of the Association for Computational Linguistics: Human Language Technologies, pp. 380–386. Association for Computational Linguistics (Jun 2016)

31. Zhu, J., Xia, Y., Wu, L., He, D., Qin, T., Zhou, W., Li, H., Liu, T.: Incorporating BERT into neural machine translation. In: Eighth International Conference on Learning Representations (2020)

Enhancing Vietnamese Question Generation with Reinforcement Learning

Nguyen Vu[1,2] and Kiet Van Nguyen[1,2(✉)]

[1] Faculty of Information Science and Engineering, University of Information Technology,
Ho Chi Minh City, Vietnam
18520323@gm.uit.edu.vn, kietnv@uit.edu.vn
[2] Vietnam National University, Ho Chi Minh City, Vietnam

Abstract. Along with automatic question answering and machine reading comprehension, question generation has become a popular yet challenging task of natural language understanding in recent years. However, as far as we are concerned, there was no study being conducted with a concentration on method for question generation in Vietnamese known as a low-resource language. In this paper, we evaluate different powerful question generation systems in two benchmark Vietnamese datasets: UIT-ViNewsQA and UIT-ViQuAD. First, we conduct experiments on deep neural network and sequence-to-sequence approaches, based on a context and an answer to generate a question. In addition, in order to investigate several powerful approaches, we utilize two strong language models (LM): the monolingual language model PhoBERT and a massively multilingual pretrained language model mT5. To obtain higher performance, we enhance LM-based methods with reinforcement learning during the decoding process. Our experiments show that the best model achieves the BLEU 4 scores of 19.77 on UIT-ViNewsQA and 20.43 on UIT-ViQuAD.

Keywords: Question generation · BERT · Transformer · UIT-ViQuAD · UIT-ViNewsQA

1 Introduction

Human beings tend to ask questions to find out more information about things said or done, especially those that interest them. Asking questions can help gain insight into others' point of view, knowledge and perspective. The majority of the knowledge people want to learn is in written form. When striving to learn new knowledge, people often ask themselves questions, for instance, "What is this text about? What is the main element of this text?". This paper builds an automatic system to generate questions for computers based on texts and questions, more specifically, based on the Vietnamese Corpus for Machine Reading Comprehension (MRC), UIT-ViNewsQA [20] and UIT-ViQuAD [13].

Based on the Question Answering (QA) - predicting the answer by text and question has been attracted the natural language processing community on the world, many datasets, and many approaches to the task of reading comprehension. However, the

N. T. Nguyen et al. (Eds.): ACIIDS 2022, LNAI 13757, pp. 559–570, 2022.
https://doi.org/10.1007/978-3-031-21743-2_45

Question Generation (QG) depends on the text and answer has not really resulted in many methods and many datasets. Especially in a language with few data sources like Vietnamese, there have been a limited number of publications on question generation until now.

Instead of a regular-standard MRC, we will teach machines how to question text in this task. In a paragraph, many elements can knock off the centre of the text. Asking questions about the main issues in the passage will help stimulate the readers' curiosity and ability to answer questions. Compared to the standard problem, making questions can be more complex and varied, as the machine can generate a sentence that may not appear in the text. Another motivation when doing this problem is that asking questions on text bring many applications, such as: creating questions for data sets to serve the MRC afterwards, asking about the main issues of the text to cope with summarizing the text, and developing the ability to read and answer human inquiries.

We approach and build the system for the task as the predecessors, creating questions based on the text. However, this paper explicitly constructs a baseline for the Vietnamese corpus, a language that is difficult to handle because of its semantic complexity. We build the model based on the Seq2seq model. Before this study, there were many tasks about QG, typically the one building Seq2seq [19] models on the SQuAD dataset [16] (e.g. Yuan et al. [25], Du et al. [4], and Kim et al. [7]). Next, we propose some pre-trained language models for this study. Based on which Chan and Fan [1] was investigated BERT-based for question generation task, we proposed a system using PhoBERT [12], which is the SOTA model in Vietnamese. Besides that, we use a multilingual text-to-text transformer model named mT5 [24] to generate questions. Moreover, we combine both reinforcement learning [23] and cross-entropy losses to ensure the generation closer to the human question generated.

2   Related Work

The Question Generation is a challenging task which has attracted the computational linguistics community to explore datasets and machine-learning methods. As a result, many neural network models built on SQuAD have been released. The most prevalent technique for question generation was built on a sequence-to-sequence model, using the sentence as context, e.g. Duan et al. [5]; Du et al. [4]; Yuan et al. [25]; Kim et al. [7]. Another one used paragraph as context, e.g. Zhao et al. [26].

The appearance and growth of the Transformer-based models have also improved the result for QG. For example, the task using BERT pre-trained improved the result for sentence-level, e.g. Kriangchaivech and Wangperawong [8]; Chan and Fan [1]. Furthermore, the most recent is a method of Lopez et al. [9], built end-to-end model based on Transformer. In recent years, with Transformer, many methods could do dual-task as QA and QG, performed NER task in preprocessing and marking words are used to generate questions. Therefore, manually defining the answer to be used as inputs like the predecessors is unnecessary anymore.

To our knowledge, QG in the Vietnamese dataset has not been publicly released until now. In this paper, we present the method for the UIT-ViNewsQA dataset [20] and the UIT-ViQuAD dataset [13], a machine reading comprehension dataset for the low-resource language of Vietnamese. There have been many MRC and QA studies [3, 14, 21] but not any research on question generation.

3 Task Definition

Given a context, a context could be an article or a sentence split by our system from a text, then highlight an answer in a context. We use a context with an answer highlighted as input. The question generation models aim to generate a question based on the answer in the context.

The input is presented as follows.

$$X = ([\text{CLS}]\ c_1, c_2, ..., [\text{HL}]\ a_1, a_2, ..., a_A\ [\text{HL}]\ c_i, ..., c_C\ [\text{SEP}])$$

where c is context with length C, a is an answer with length A defined in context. Based on Chan and Fan work, [CLS], [SEP] tokens were auto-generated through our data processing; it represents start and end positions of the context. The [HL] token aims for the answer highlighted in the context. By highlighting the answer, the models could learn how to generate questions without duplicating the answers.

In this task, we generate question Y from X. Our system embedding the input X. Then, the input representation is sent into other layers to perform the question prediction. Table 1 shows sample output generated by human and our system.

Table 1. Questions are generated by human and our best performance system. Bold texts in the context represent the answers as inputs.

Context
Ông Trần Ngọc Tụ, Chi cục trưởng An toàn Vệ sinh Thực phẩm Hà Nội cho biết hiện ghi nhận số người bị ngộ độc thực phẩm tại trường mầm non Xuân Nộn (Đông Anh, Hà Nội) tuần trước lên đến **225 người**. Trong số này có 3 giáo viên, còn lại là các bé học sinh. Sau ba ngày điều trị, vẫn còn **hơn 100 bệnh nhân** phải lưu lại hai bệnh viện Đa khoa huyện Đông Anh, Đa khoa Bắc Thăng Long để theo dõi. (*Mr. Tran Ngoc Tu, Director of Hanoi Food Safety and Hygiene Department, said the record of some people getting food poisoning at Xuan Non kindergarten (Dong Anh, Hanoi) last week was up to 225 people. Among these are 3 teachers, the rest are small students. After three days of treatment, more than 100 patients have to stay at two hospitals in Dong Anh district, North Thang Long General Hospital for monitoring.*)
Question generation by human and our system
Question 1)
Human: Số ca bị ngộ độc thực phẩm ở trường mầm non Xuân Nộn là bao nhiêu? (*What is the number of cases of food poisoning at Xuan Non kindergarten?*)
Our system: Bao nhiêu ca bị ngộ độc thực phẩm? (*How many cases of food poisoning?*)
Question 2)
Human: Hiện nay còn bao nhiêu bệnh nhân vẫn đang nằm điều trị tại bệnh viện đa khoa huyện Đông Anh? (*How many patients are still being treated at Dong Anh district general hospital?*)
Our system: Bao nhiêu bệnh nhân đang điều trị? (*How many patients are still being treated?*)

In this study, we define two levels of text, the "sentence-level" and "text-level". At the "sentence level", we only give the sentence combine the answer in it. Other sentences in the paragraph are removed and not to be used as input. Besides, the text level means we give the whole passage is input. We expected the text-level helps the machine read the context more precisely, then generate the question with nearly meaning to human-generated.

4   Methods

In this paper, a question generation system based on PhoBERT and reinforcement learning is proposed for Vietnamese. Figure 1 shows PhoBERT architecture with reinforcement learning.

4.1   Baseline Models

Seq2seq. Inspired by the RNN encoder-decoder architecture first outline by Cho et al. [2], we build the seq2seq-based system for Vietnamese question generation. In fact, to generate the question more relevant and focus on the context per question, we take the target answer as input and use pre-trained embedding word representations for initialization. Furthermore, an attention mechanism at decoding can be used to tackle question generation.

- Embedding: Basically, we teach the machine to learn words and embedding it into vectors. Moreover, we decide to choose word embedding of 300 dimensions and use pre-trained word vectors for Vietnamese, which distributed by Grave et al. [6] to enhance the performance. In addition, PhoW2V pre-trained word-level embedding is also used for Vietnamese by Nguyen er al. [11].
- Encoder - Decoder: For this study, we employ an RNN encoder-decoder architecture [2]. The input sequence is represented by the encoder as a learned word vector. Then, utilizing the encoder's output, the decoder produces an output. It is common practice to decode predictions using beam search. In this study, we use beam search decoding to assess our system. We evaluate our model with a beam width of 3.
- Attention Mechanism: The central issue surrounding the seq2seq model can be solved by the Attention Mechanism; Thus, by using an attention mechanism proposed by Luong et al. [10] we can ensure the model focus on common elemenst of the input during the decoding process.

mT5. The recent text-to-text transfer transformer (T5) was defined as pre-trained on a new Common Crawl-based dataset including over a hundred of languages and covering between 300 million and 13 billion parameters and also known principle that Multilingual pre-trained text-to-text transformers (mT5) [24] based on. In terms of both design and approach, the architecture and training of the mT5 model are quite similar to those of T5. We employ the mT5-small pre-trained transformer model for this study. By utilizing a highlight input sequence, we are fine-tuning the model's parameters to handle Vietnamese question generating.

PhoBERT. We use the VnCoreNLP [22] to perform word segmentation in processing. Then, the Vietnamese words can be easier to process.

PhoBERT pre-trained model by Nguyen and Nguyen [12] is built on the RoBERTa model. Using BERT or any models built on BERT's language masking strategy and modifies key hyper-parameters in BERT, the input must be aligned as the BERT-based specific input sequence. In Sect. 3, we presented the input in BERT's format. In general, a [CLS] token is inserted as the first token of the input string. To distinguish between other information, a unique token [SEP] is added between two sentence strings.

Technically, PhoBERT pre-trained model by Nguyen and Nguyen [12] is constructed on the RoBERTa model. In fact, the input must be aligned as the BERT-based specific input sequence which is the compulsory condition for using BERT or any models built on BERT's language masking strategy and modifies key hyper-parameters in BERT. In general, a unique token [SEP] is added between two sentence strings to discriminate between other information; plus, we use a [CLS] token for the first token of the input string. All of this input in BERT's format is presented in the Sect. 3.

Inspired by Chan and Fan [1], we implement a BERT-based approach for automatic question generation using a highlight sequence model for the input sequence. Using this method, the question generated has improved quality when processing lengthy context. Besides, it helps to avoid an answer appear multiple times in the input sequence.

Specifically, the question generating is based on the fill-mask task of the BERT model and at the end of the input sequence for this method, special token [MASK] is given. After that, by embedding layers and the transfers forward into the transformer model PhoBERT, X_i then is represented. Next, taking the final state hidden for [MASK] in the input sequence, we compute the predicted word and appended the new predicted token into X. Until the [SEP] token is predicted, the process must be repeated with the new X.

Furthermore, to improve quality, we use an attention mechanism at decoding. Reinforcement Learning is also applied to fine-tune the model to get higher performance. Figure 1 shows the overall architecture of our proposed approach using PhoBERT enhanced with reinforcement learning.

4.2   Question Generation Enhanced with Reinforcement Learning

Reinforcement learning is applied to achieve higher performances on all methods.

Using regular cross-entropy loss, we train our system, defined as:

$$\mathcal{L}_{ce} = -\sum_t log P(y_t|X, y_{<t})$$

where y_t can be considered as the t-th word of the sample output. Based on the current word, our system predict what the next word is. Therefore, use the cross-entropy loss for sequence learning, the generator could generate the question unclear. We following the Reinforcement algorithm [23], which makes the prediction to be more effective during decoding. As result, the calculated contrast between generated sequence Y and corresponding sample sequence Y^s is the definition of the reward:

$$r = \alpha f_{eval} + (1 - \alpha) \cos(\theta)$$

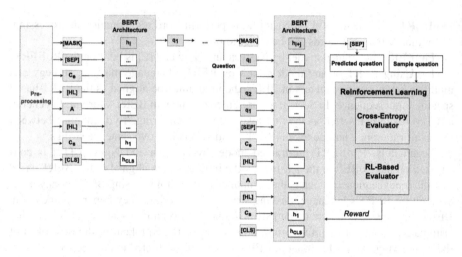

Fig. 1. Question generation architecture with Reinforcement Learning

where α is a scalar. Then, we define the last reward by scale the reward:

$$r = \frac{r + 1 - \alpha}{2 - \alpha}$$

BLEU 4 [15] was used as an evaluation metric and a reward function f_{eval}. We follow the effective self-critical sequence training algorithm (called SCST) [17] to optimize the evaluation metric directly, obtained by the greedy search. The loss function is defined as follows.

$$\mathcal{L}_{rl} = (r(Y) - r(Y^s)) \sum_t logP(y_t^s | X, y_{<t}^s)$$

To compute the cross-entropy loss function with reward, we define the loss funtion $\mathcal{L}_t$ as:

$$\mathcal{L}_t = (1 - r)\mathcal{L}_{ce}$$

The models are fine-tuned by optimizing a mixed objection function combining both cross-entropy loss function with reward and the Reinforcement Learning loss, defined as:

$$\mathcal{L} = \gamma\mathcal{L}_{rl} + (1 - \gamma)\mathcal{L}_t$$

where γ is scale factor to control the take-off between cross-entropy loss with reward $\mathcal{L}_t$ and Reinforcement learning loss $\mathcal{L}_{rl}$. A beam size of 5 is used in the beam search algorithm for the final prediction.

5   Experiments and Results

5.1   Datasets

Two benchmark datasets: UIT-ViQuAD [13] and UIT-ViNewsQA [20] are used for evaluating Vietnamese automatic reading comprehension, question answering, and question generation models. However, in this paper, we use them to evaluate the question generation for Vietnamese. While UIT-ViQuAD is an open domain dataset on Wikipedia data, UIT-ViNewsQA is a closed domain dataset with health news articles. They are consistent with our goal of evaluating question generation on both open and closed domains. Table 2 shows overview statistics of the two benchmark datasets for Vietnamese question generation.

Table 2. Overview statistics of the two benchmark datasets: UIT-ViNewsQA and UIT-ViQuAD.

Dataset	Type	Train	Dev	Test	all
UIT-ViNewsQA	Articles	3,517	500	399	4,416
	Passage	–	–	–	-
	Pairs	17,568	2,497	1,992	22,057
	Avg. length	342.9	323.9	360.4	342.4
UIT-ViQuAD	Articles	138	18	18	174
	Passages	4,101	515	493	5,109
	Pairs	18,579	2,285	2,210	23,074
	Avg. length	153.9	147.9	155.0	153.4

5.2   Evaluation Metrics

Following the previous study [15], we use BLEU as a automatic evaluation metric from the evaluation package provided by Sharma et al. [18]. In this paper, we use four BLEU evaluation metrics comprising BLEU 1, BLEU 2, BLEU 3, and BLEU 4. BLEU measures the average n-gram precision on a set of reference sentences. BLEU-n is a BLEU score variant that uses up to n-grams for counting co-occurrences. In our study, the BLEU 4 is used as a primary evaluation metric.

5.3   Implementation Details

For the seq2seq model, we use the Pytorch library to train our system. First, in the pre-trained embedding, we choose 300 dimensions, we use and compare the fastText and PhoW2V pre-trained words representation. The encoder with a vocab of 45,000 words, and the decoder with 20,000 words. The stochastic gradient descent (SGD) optimizer is applied during the training process with an initial learning rate of 1e–4. We train for 15 epochs, and the batch size set to 32. We set the LSTM hidden size to 600 and set the number of layers to 2. Dropout probability is set to 0.3 between vertical LSTM stacks. Our system is training on a single GPU.

Next, for the PhoBERT model, we also use the Pytorch version to train our system. We use the RoBERTa Config (12 layers, hidden size of 768, 12 attention head) with a vocab of 64,001. The AdamW optimizer is applied with an initial learning rate of 5e–5. A single GPU for 3 epochs is used for the training phase. Dropout probability is set to 0.1 between transformer layers. Also, for sequence decoding, we use an attention mechanism and the Beam search strategy for sequence decoding with a beam size of 3.

For mT5, we also use the Pytorch version to train our system. We train on a single GPU for 5 epochs, and the batch size set to 32. We use the base optimizer with an initial learning rate of 1e–4. The gradient accumulation set to 8 and the logging step of 50. We use mT5-small pre-trained for this task. During decoding, we also use the Beam search strategy with a beam size of 4.

5.4 Model Comparison

In this paper, we compare our proposed system with the baseline model. For that, we choose the seq-2-seq and mT5 as the baseline systems. The detail of the baseline model is mentioned in Sects. 4.1 and 4.2.

5.5 Experimental Results

To our knowledge, we propose the first system to solve automatic question generation in Vietnamese. In this section, we compare our systems to each other.

Table 3 shows the comparison result using standard metric BLEU. On both the UIT-ViNewsQA dataset and the UIT-ViQuAD dataset, PhoBERT enhanced with Reinforcement Learning (RL) has the best performance for the Question Generation task using the sentence-level context. Besides, our systems have the performance approximately each other. The RNN encoder-decoder architecture with seq2seq model has lowest performance. The difference about performance around 2–3% on BLEU 4 metric.

Table 3. Experimental results of different systems using the sentence-level context.

Dataset	System	BLEU 1	BLEU 2	BLEU 3	BLEU 4
UIT-ViNewsQA	seq2seq	64.46	46.01	21.55	17.26
	seq2seq + RL	**66.21** (↑ **1.75**)	**47.97** (↑ **1.96**)	**22.70** (↑ **1.15**)	**17.41** (↑ **0.15**)
	mT5-small	42.29	31.77	23.91	18.14
	mT5-small + RL	**44.09** (↑ **1.80**)	**33.12** (↑ **1.35**)	**24.70** (↑ **0.79**)	**18.44** (↑ **0.30**)
	PhoBERT	44.52	34.86	24.16	19.32
	PhoBERT + RL	**48.02** (↑ **3.50**)	**36.11** (↑ **1.25**)	**25.86** (↑ **1.70**)	**19.77** (↑ **0.45**)
UIT-ViQuAD	seq2seq	66.43	49.89	22.15	17.64
	seq2seq + RL	**67.89** (↑ **1.46**)	**50.01** (↑ **0.12**)	**22.32** (↑ **0.17**)	**18.09** (↑ **0.45**)
	mT5-small	43.08	34.43	25.79	18.92
	mT5-small + RL	**44.87** (↑ **1.79**)	**35.93** (↑ **1.50**)	**26.71** (↑ **0.92**)	**19.55** (↑ **0.63**)
	PhoBERT	48.21	30.35	26.52	20.21
	PhoBERT + RL	**51.46** (↑ **3.25**)	**33.01** (↑ **2.66**)	**27.14** (↑ **0.62**)	**20.43** (↑ **0.22**)

Not only the comparison using the sentence-level context, but we also compare text-level as input. The PhoBERT outperform seq2seq and mT5 model on both datasets. But for this level, mT5 has the lowest performance. Besides that, the seq2seq model has a big difference between datasets. The method on the UIT-ViQuAD dataset has a higher performance than the UIT-ViNewsQA dataset. The difference belongs to the average context length. The average context length of the UIT-ViQuAD dataset has less than the UIT-ViNewsQA dataset. The comparison result using text-level context texts is presented in Table 4.

Table 4. Experimental results of different systems using the text-level context.

Dataset	System	BLEU 1	BLEU 2	BLEU 3	BLEU 4
UIT-ViNewsQA	seq2seq	57.57	38.25	23.15	10.98
	seq2seq + RL	**60.11** (↑ **2.44**)	**40.90** (↑ **2.65**)	**25.43** (↑ **2.32**)	**13.45** (↑ **2.47**)
	mT5-small	40.92	28.75	19.20	9.94
	mT5-small + RL	**41.86**(↑ **0.94**)	**28.95** (↑ **0.20**)	**19.63** (↑ **0.43**)	**11.03** (↑ **1.09**)
	PhoBERT	43.76	34.18	23.50	17.31
	PhoBERT + RL	**46.50** (↑ **2.74**)	**36.10** (↑ **1.92**)	**25.46** (↑ **1.96**)	**19.22** (↑ **1.91**)
UIT-ViQuAD	seq2seq	63.73	55.20	22.64	16.52
	seq2seq + RL	**65.66** (↑ **1.93**)	**57.14** (↑ **1.94**)	**24.33** (↑ **1.69**)	**16.91** (↑ **1.39**)
	mT5-small	39.27	24.14	18.66	11.21
	mT5-small + RL	**39.65** (↑ **0.38**)	**24.58** (↑ **0.44**)	**19.24** (↑ **0.58**)	**11.36** (↑ **0.15**)
	PhoBERT	43.98	32.65	23.02	16.83
	PhoBERT + RL	**44.54** (↑ **0.56**)	**33.76** (↑ **1.11**)	**24.13** (↑ **1.11**)	**17.19** (↑ **0.36**)

For all methods, using reinforcement learning at the decoder help the predicted questions closely. Following both Table 3 and table 4, we can see that methods enhanced with Reinforcement Learning reached higher performance on the evaluation metrics.

5.6 Question Type Prediction Analysis

After evaluating the question generation models, we aim to analyze question types predicted on the test set. We group questions into 8 categories: "What" *(cái gì, là gì)*, "Why" *(tại sao, vì sao)*, "How" *(như thế nào)*, "When" *(khi nào)*, "Where" *(ở đâu)*, "Who" *(ai)*, "Which" *(nào)* and Others.

Through this evaluation, the question types including "Why" and "How" present in big magnitude on the UIT-ViNewsQA dataset. The UIT-ViNewsQA dataset consists of health news. Most of the answers are about new things in healthcare, the reason for the health problem, the way to take health care and the quantity (How many, How much) - that why most of the predicted questions are "What", "Why" and "How" types. The result of this evaluation has presented in Table 5.

For the UIT-ViQuAD dataset, the question types including "What" and "Others" have a high proportion of correct prediction. The UIT-ViQuAD dataset consists of Wikipedia articles. Most of the articles are about new things, such as bringing new

Table 5. Question type predictions on UIT-ViNewsQA dataset (%).

	What	Why	How	When	Where	Who	Which	Others
seq2seq	24.68	35.52	22.31	1.26	1.83	4.60	0.76	9.04
PhoBERT	15.77	39.86	19.42	2.02	1.80	3.11	0.97	17.05
mT5-small	23.21	46.74	14.39	0.71	0.83	2.15	0.52	11.45

knowledge to readers, so most predicted questions are "What" type. "Others" type like "Yes/No" question, "Will", "Can", or tag question also have the high prediction on this dataset. Table 6 shows the result of question type prediction on the UIT-ViQuAD dataset.

Table 6. Question type predictions on UIT-ViQuAD dataset (%).

	What	Why	How	When	Where	Who	Which	Others
seq2seq	38.23	7.03	8.12	3.68	3.75	9.61	2.08	27.5
PhoBERT	41.10	6.19	8.53	2.97	2.65	10.29	1.11	27.16
mT5-small	33.25	9.43	8.77	4.32	4.54	8.22	0.96	30.51

Our systems predicted the question type mostly inclined to two or three types on each dataset. On both datasets, the prediction of question type "Which" is very low. It demonstrated many question types "Which" like "Which year", "Which place" and "Which person" maybe generate as "When", "Where", and "Who".

6   Conclusion and Future Work

In this study, we evaluate different powerful models in natural language processing that generate any question automatically from the given context (which is a sentence or a text) and the answer. In particular, three different models: seq2seq based on RNN encoder-decoder architecture, PhoBERT and mT5 based on pre-trained transformer model were investigated. Especially, we proposed a transformer-based question generation approach enhanced with reinforcement learning for predicted questions closely to humans. Currently, our approach using the sentence-level context has a better performance than text-level context across both the two benchmark datasets: UIT-ViQuAD and UIT-ViNewsQA in terms of the BLEU 4 metric.

We would like to improve the performance of our proposed approach on both sentence-level and text-level. Moreover, with the appearance and growth of the transformer model, we want to use more and more pre-trained models to generate questions. Besides that, we are also interested in further that the standard QA models can answer the automatically generated questions and the multi-task machine reading comprehension as automatically extracted answers, then generate questions on the context and answers, and then the machine can answer questions.

Acknowledgement. This research was supported by The VNUHCM-University of Information Technology's Scientific Research Support Fund.

References

1. Chan, Y.-H., Fan, Y.-C.: A recurrent BERT-based model for question generation. In: Proceedings of the 2nd Workshop on Machine Reading for Question Answering, pp. 154–162 (2019)
2. Cho, K., et al.: Learning phrase representations using RNN encoderdecoder for statistical machine translation. In: arXiv preprint arXiv:1406.1078 (2014)
3. Do, P.N.-T., Nguyen, N.D., Van Huynh, T., Van Nguyen, K., Nguyen, A.G.-T., Nguyen, N.L.-T.: Sentence extraction-based machine reading comprehension for Vietnamese. In: Qiu, H., Zhang, C., Fei, Z., Qiu, M., Kung, S.-Y. (eds.) KSEM 2021. LNCS (LNAI), vol. 12816, pp. 511–523. Springer, Cham (2021). https://doi.org/10.1007/978-3-030-82147-0_42
4. Du, X., Shao, J., Cardie, C.: Learning to ask: Neural question generation for reading comprehension. In: arXiv preprint arXiv:1705.00106 (2017)
5. Duan, N.: et al.: Question generation for question answering. In: Proceedings of the 2017 Conference on Empirical Methods in Natural Language Processing, pp. 866–874 (2017)
6. Grave, E., et al.: Learning word vectors for 157 languages. In: arXiv preprint arXiv:1802.06893 (2018)
7. Kim, Y., et al.: Improving neural question generation using answer separation. In: Proceedings of the AAAI Conference on Artificial Intelligence. Vol. 33, pp. 6602–6609, January 2019
8. Kriangchaivech, K., Wangperawong, A.:Question generation by transformers. In: arXiv preprint arXiv:1909.05017 (2019)
9. Lopez, L.E., et al.: Simplifying paragraph-level question generation via transformer language models. In: arXiv preprint arXiv:2005.01107 (2020)
10. Luong, M.-T., Pham, H., Manning, C.D.: Effective approaches to attention-based neural machine translation". In: arXiv preprint arXiv:1508.04025 (2015)
11. Nguyen, A.T., Dao, M.H., Nguyen, D.Q.: A pilot study of textto- SQL semantic parsing for Vietnamese". In: arXiv preprint arXiv:2010018a91 (2020)
12. Nguyen, D.Q., Nguyen, A.T.: PhoBERT: pre-trained language models for Vietnamese. In: arXiv preprint arXiv:2003.00744 (2020)
13. Nguyen, K., et al.: A Vietnamese dataset for evaluating machine reading comprehension. In: Proceedings of the 28th International Conference on Computational Linguistics, pp. 2595–2605 (2020)
14. Van Nguyen, K., et al.: XLMRserini: open-domain question answering on Vietnamese Wikipedia-based textual knowledge source. In: 14th Asian Conference on Intelligent Information and Database Systems (2022)
15. Papineni, K., et al.: BLEU: a method for automatic evaluation of machine translation. In: Proceedings of the 40th annual meeting of the Association for Computational Linguistics, pp. 311–318 (2002)
16. Rajpurkar, P., et al.: SQUAD: 100,000+ questions for machine comprehension of text. In: arXiv preprint arXiv:1606.05250 (2016)
17. Rennie, S.J., et al.: Self-critical sequence training for image captioning. In: Proceedings of the IEEE conference on Computer Vision and Pattern Recognition, pp. 7008–7024 (2017)
18. Sharma, S., et al.: Relevance of unsupervised metrics in task-oriented dialogue for evaluating natural language generation. In: arXiv preprint arXiv:1706.09799 (2017)
19. Sutskever, I., Vinyals, O., Le, Q.V.: Sequence to sequence learning with neural networks. In: Advances in Neural Information Processing Systems, pp. 3104–3112 (2014)

20. Van Nguyen, K., et al.: New Vietnamese corpus for machine reading comprehension of health news articles. In: ACM Trans. Asian Low-Resour. Lang. Inf. Process. (2022). Just Accepted. ISSN: 2375–4699. https://doi.org/10.1145/3527631.https://doi.org/10.1145/3527631

21. Van Nguyena, K., et al.: Vireader: a Wikipedia-based Vietnamese reading comprehension system using transfer learning. J. Intell. Fuzzy Syst. **1**, 1–5 (2021)

22. Vu, T., et al.: VnCoreNLP: a Vietnamese natural language processing toolkit. In: arXiv preprint arXiv:1801.01331 (2018)

23. Williams, R.J.: Simple statistical gradient-following algorithms for connectionist reinforcement learning. Mach. Learn. **8**(3), 229–256 (1992)

24. Xue, L., et al.: MT5: a massively multilingual pre-trained text-to-text transformer. In: arXiv preprint arXiv:2010.11934 (2020)

25. Yuan, X., et al.: Machine comprehension by text-to-text neural question generation. In: arXiv preprint arXiv:1705.02012 (2017)

26. Zhao, Y., et al.: Paragraph-level neural question generation with maxout pointer and gated self-attention networks. In: Proceedings of the 2018 Conference on Empirical Methods in Natural Language Processing, pp. 3901–3910 (2018)

A Practical Method for Occupational Skills Detection in Vietnamese Job Listings

Viet-Trung Tran[✉], Hai-Nam Cao, and Tuan-Dung Cao

Hanoi University of Science and Technology, Hanoi, Vietnam
{trungtv,namch,dungct}@soict.hust.edu.vn

Abstract. Vietnamese labor market has been under an imbalanced development. The number of university graduates is growing, but so is the unemployment rate. This situation is often caused by the lack of accurate and timely labor market information, which leads to skill miss-matches between worker supply and the actual market demands. To build a data monitoring and analytic platform for the labor market, one of the main challenges is to be able to automatically detect occupational skills from labor-related data, such as resumes and job listings. Traditional approaches rely on existing taxonomy and/or large annotated data to build Named Entity Recognition (NER) models. They are expensive and require huge manual efforts. In this paper, we propose a practical methodology for skill detection in Vietnamese job listings. Rather than viewing the task as a NER task, we consider the task as a ranking problem. We propose a pipeline in which phrases are first extracted and ranked in semantic similarity with the phrases' contexts. Then we employ a final classification to detect skill phrases. We collected three datasets and conducted extensive experiments. The results demonstrated that our methodology achieved better performance than a NER model in scarce datasets.

Keywords: Skill extraction · Named entity recognition · Text embedding · Text ranking

1 Introduction

Labor market is the foundation and key driver for economic growth. To achieve high efficiency, labor market needs information. Policymakers rely on labor supply and demand relationships to chart economic and social policies. Educators need to align curriculum development with employers' demand, especially in fast-changing sectors. Job seekers need information on skill requirements, company profiles, work conditions, and growth trajectories. In an increasingly digitized economy, labor market policymakers need to continuously maintain an up-to-date vision, focusing on growing skills that are less likely to be replaced by automation [6].

N. T. Nguyen et al. (Eds.): ACIIDS 2022, LNAI 13757, pp. 571–581, 2022.
https://doi.org/10.1007/978-3-031-21743-2_46

It is clear that regulators of Vietnam labor market lack updated and relevant information to make market-driven decisions. Usually, the labor force surveys, which conducted quarterly by General Statistics Office of Vietnam (GSO), face six key challenges: freshness, accuracy, coverage, analysis, usability, and cost. In consequence, Vietnamese labor market has been under imbalance development for many years. Although the number of university graduates is growing, but so is the unemployment rate. In the first quarter of 2018 [10], the number of people with intermediate and college degrees that found jobs was 79.1% and 72.9% respectively; meanwhile, only 55.6% of university graduates have jobs.

Besides, Vietnamese job portals have been considered as an important bridge between recruitment managers and job seekers. Over the years, these portals have accumulated a growing amount of digital labor-related market data such as job listings and applicants' resumes. However, the exploitation of these data is limited as these portals only provide job categories and keyword-based search functionality.

To enable advanced analysis, it is imperative to have a model that can automatically detect skills from labor market-related data. The model can benefit advanced labor market analysis and ultimately facilitate orienting workforce training and re-skilling programs. Various approaches [11,13,17,18] consider this skill detection task as a Named Entity Recognition (NER) task in natural language processing. They have a common drawback: a large number of labeled sentences is needed to train the NER models in a supervised setting. Other approaches detect skills from a given document by performing a direct match between n-gram sequences and terms in the target taxonomy [2,7,9]. These approaches, however, do not work for Vietnamese language as there is no such a taxonomy yet.

In Vietnamese job listing websites, a job opening usually has a common semi-structural format. Each job opening has the following sections:

- **Title.** A short, one sentence highlighting for the job to attract job seekers. The title often mentions job position, job level, and salary range.
- **Description.** One paragraph or a list that describes the job characteristics: What and how the work will be carried on.
- **Compensation.** One paragraph or a list that shows salary range and benefits paid to employees in exchange for the services they provide.
- **Requirements.** One paragraph or a list that contains experiences, qualifications, and skills necessary for the candidates to be considered for a role.
- **About the company.** Brief introduction to the company and its environment.
- **Contact point.** An email address and a phone number to submit and question the application.

The order of those sections may vary, however, most skill mentions will be within the requirement section. In this paper, we present a practical approach for skill detection in Vietnamese job listings. Rather than viewing the task as a NER task, we model the task as a ranking problem. Our approach exploits the structural property of a job description: any skill mention found in a requirement section will have a high semantic similarity score with the section itself.

The rest of this paper is organized as follows: we start in Sect. 2 by outlining the main steps of the proposed method. In Sect. 3, we describe in detail the implementation of the tasks in the previous section: embedding, phrase mining, term ranking, and term classification. In Sect. 4, we carry out a comprehensive experimental study to validate the proposed method. We conclude with a summary of results and future work in Sect. 5.

2 Methodology

Our method is depicted in Fig. 1. In comparison to the traditional NER approach, our methodology is more practical and less expensive in terms of manual efforts. It is a pipeline composed of 4 layers:

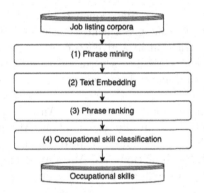

Fig. 1. The pipeline of our proposed method

1. **Phrase mining.** Occupational skill mentions can be multi-word or single-word phrases (e.g., "Java", "data mining"). Thus, a crucial step in our pipeline is phrase mining, which aims at extracting high-quality phrases in a given document. The output of this layer is considered as skill mention candidates for the next ranking layer. To reduce manual effort, we leverage a semi-supervised, weak supervision approach for this layer.

2. **Text embedding.** This layer is responsible to output the corresponding embedding vectors, given a word, a phrase, a sentence, and a paragraph. The output of the embedding layer will be used to compute ranking similarity scores. For this layer, we can leverage powerful embedding methods such as SIF [1], and BERT [4]. Thus, this layer requires no manual labeling effort.

3. **Phrase ranking.** This layer ranks the importance of a phrase w.r.t. its outer context, the parental requirement section in the job description. Apparently, it can be observed from practical experiments that skill-related phrases often achieve high rankings, while low-ranked phrases are undoubtedly not occupational skills. Therefore, ranking important phrases are capable of discarding

clusters of words that are irrelevant to the occupational topic. This layer does not require labeling effort.

4. **Occupational skill classification.** Generally, there are cases where many extracted phrases in the previous steps are not occupational skill terms. Therefore, this layer is necessary as a binary classification model to identify truly occupational skill terms. This layer requires a labeled dataset. Beginning with a subset of the output phrases in the ranking layer, our labeling workers are required to prepare two subsets: skill and non-skill ones. In contrast, NER labeling workers must carefully select the phrase spans and assign the corresponding labels. Thus, the dataset construction effort is generally cheaper and faster in our methodology than that of the NER approach.

3   Implementation

3.1   Phrase Mining

A phrase is defined as a consecutive sequence of words in the text, forming a complete semantic unit in that given context. Phrase mining is the process of high-quality phrases extraction in a given corpus. Phrases can be multi-words and single-word phrases. There are several existing methods for phrase mining, but they suffer from domain dependence, human labeling, and a variety of languages.

In search of appropriate methods, we select AutoPhrase [16] method for its proven significant effectiveness in different domains. Autophrase only requires a collection of "good" phrases which are cheap to construct from external sources such as Wikipedia or existing dictionaries. Autophrase emphasizes the following features for the detection of high-quality multi-word phrases:

- **Popularity.** High-quality phrases should be popular. Autophrase uses the probability of occurrence $p(v)$ as the popularity score of v in a given dataset d. For each word or phrase u in a given dataset d, the probability of occurrence $p(u)$ is defined as:

$$p(u) = \frac{f_u}{\sum_{u' \in d} f_{u'}}$$

  where f_u is the raw frequency, that is the raw count of u in the dataset d.
- **Concordance.** For each high-quality phrase, the collocation of its constituent words should occur with a higher probability than what is expected due to chance. Given a phrase v, let u_l, u_r denote the two most likely left and right parts of v such that the concatenation of u_l, u_r is equal to v and the Pointwise Mutual Information (PMI) of u_l, u_r is minimized:

$$u_l, u_r = argmin_{u_l + u_r = v} PMI(u_l, u_r)$$

  where the operation $u_l + u_r$ is the concatenation of u_l and u_r; $PMI(u_l, u_r)$ is the Pointwise Mutual Information score. PMI is calculated as follows:

$$PMI(u_l, u_r) = log \frac{p(v)}{p(u_l)p(u_r)}$$

where $p(v)$, $p(u_l)$, $p(u_r)$ are the probability of occurrence of v, u_l, u_r respectively.

Another concordance feature can be used is the Pointwise Kullback-Leibler divergence (PKL):

$$PKL(v|| < u_l, u_r >) = p(v)log\frac{p(v)}{p(u_l)p(u_r)}$$

- **Informativeness** refers to the possibility of a high-quality phrase being indicative of a specific topic or concept. Informativeness feature includes: (1) Whether stop-words are located within the phrase candidate; (2) Average inverse document frequency (IDF); (3) Probabilities of a phrase in quotes, brackets, or capitalized.
- **Completeness** refers to the possibility of a high-quality phrase to express a complete semantic unit in a given document context.

Leveraging these features, Autophrase method of extracting high-quality phrases consists of two main steps:

1. Estimate quality of phrases by positive-only distance training.
   - A collection of "good" phrases is collected from external sources such as Wikipedia or existing dictionaries. These "good" phrases are put on a positive pool.
   - Given the document collection, every $n-gram$ phrase is extracted. Those phrases that satisfy the popularity threshold are treated as candidate phrases. Phrases that are not in the positive pool are put on a negative pool.
   - Since training a classifier to distinguish "good" phrases and "bad" phrases directly from the whole noisy data is not effective. Autophrase proposed to use an ensemble method that composes of multiple independent classifiers, each is trained in a subset of mixture data from a positive pool and a negative pool. The output of the trained model is the phrase quality score.
2. Re-estimate quality of phrases based on phrasal segmentation to rectify the inaccurate phrase quality initially estimated.
   - This step aims to guarantee the completeness characteristic of detected phrases by exploiting the local context in a given document. Autophrase proposes a POS-guided phrasal segmentation to locate the boundaries of phrases more accurately.

In order to detect salient phrases in Vietnamese corpus, We adapt AutoPhrase to accept the Vietnamese tokenizer and POS tagger. Then, we train Autophrase on the Vietnamese job listing corpus. High-quality phrases are constructed by crawling Wikipedia and job listing websites.

3.2   Text Embedding: Universal Methods for Words, Phrases, Sentences, and Paragraphs

Embedding is one of the most popular representation that has radically transformed the natural language processing (NLP) landscape in recent years. Embedding mechanisms can encode pieces of text, such as word, phrase, sentence and paragraph, as fixed-sized vectors, namely embedding vectors, that allow capturing context, syntactic and semantic similarity.

Word2vec, Bert are the two most popular word embedding mechanisms. There are dozens of proposed methods to encode sentences, and phrases. One simple method is to average vectors of words in the sentence. Kiros et al. [8] trains an encoder-decoder architecture to reconstruct the surrounding sentences. InferSent [3], Sentence-Transformer [15] learn sentence embeddings by training siamese-based networks with labeled paraphrases.

In our methodology, we aim at being able to measure the similarity from words, phrases to sentences and paragraphs. So, we use the same embedding mechanisms that can encode various types of text pieces to the same multi-dimensional space. Specifically, we experiment with SIF [1], SimCSE [5], PhoBert [14], and SBERT [15] embeddings.

SIF Scheme. SIF (Smooth Inverse Frequency) [1] is a method of averaging weight to improve sentence embedding performance, with the weight being the inverse frequency smoothed. SIF relies on the assumption that frequent words in the whole dictionary pool are less important than the infrequent words in the computation of the sentence embedding.

The weights of words in SIF are computed as follows:

$$\text{Weight(w)} = \frac{a}{a + f_w}$$

where f_w is the raw count of the occurrence of the word w in the dataset, a is a hyper-parameter, whose values range from 10^{-3} to 10^{-4} . The embedding vector v_s of a sentence s is calculated as follows:

$$v_s = \frac{1}{|s|} \sum_{w \in s} \text{Weight}(w) v_w$$

where v_w is the word embedding vector of w.

SIF works well with most language models pre-trained. Despite the simple formula, SIF works surprisingly well on sentiment tasks, and the tasks of measuring semantic text similarity. The embedding vectors computed by SIF well reflect the intrinsic meaning of words in the sentences.

SimCSE Scheme. SimCSE [5] is a method to train sentence embeddings without having training data. In training, SimCSE encodes a sentence twice by using different dropout masks. A contrastive loss is used to minimize the distance between the two embedding outputs, while minimizing the distance to other embeddings of the other sentences in the same batch (these are negative examples).

SimCSE serves well as our embedding layer as there is currently no Vietnamese paraphrase datasets. Thus a supervised training is not possible. We leverage pre-trained PhoBert and fine-tune SimCE on a collection of 200K Vietnamese job listings. The trained model is used to predict embedding vectors for all types of text pieces such as words, phrases, sentences, and paragraphs.

3.3 Term Ranking

To rank a phrase to its container, the job requirement section, we measure the cosine similarity of their respective embedding vectors v_p, v_d.

$$SIM(v_p, v_d) = \cos(\theta) = \frac{v_p \cdot v_d}{\|v_p\| \|v_d\|}$$

After obtaining similarity scores of phrases, all phrases with scores lower than a defined threshold (e.g., 0.5) will be discarded since they are definitely irrelevant to skills.

3.4 Occupational Skill Classification

We can select any classification algorithm to implement this layer. In our experiment, we build a small multi-layer neural network for this purpose. The only feature we used for the classification model was term embeddings.

4 Experiments and Results

4.1 Data Collection

To demonstrate our methodology, we prepare multiple datasets as follows:

- **Vn_job_corpus.** Vn_job_copus is a collection of 1.122.159 unique sentences extracted from 370.770 Vietnamese-language job listings. The data had been crawled periodically by our crawler system since 2020 Oct. We tokenized the data by vi_spacy[1] toolkit.
- **Vn_requirements_NER.** Vn_requirements_NER consists of 625 requirement paragraphs of the job descriptions, with manual labeling of skill terms. The data is split in the ratio of 70:30 for train and test sets.
- **Vn_resumes_NER.** Vn_resumes_NER consists of 77284 sentences of the experience sections in the resumes. Due to the scarcity of available sources, these resumes belong only to the information technology sector. The dataset is manually labeled for skill terms. It is also split in the ratio of 70:30 for train and test sets.

---

[1] https://github.com/trungtv/vi_spacy.

4.2   Evaluation Methods

We compare our proposed methodology to the common NER approach for skill terms detection in Vn_requirements_NER and Vn_resumes_NER datasets. The models are compared in precision, recall, and F1 scores in complete match and partial match assessments.

The traditional evaluation requires a complete match of the predicted token and the ground-trust token in order to count a True Positive (TP) case. However, partial match assessment [12] is valuable in practice for several reasons: (1) A partial match reveals a significant piece of information: the NER model can predict the right entity, even if whose boundary does not perfectly match with that of the actual entity; (2) The ground-trust datasets are not always in high quality, especially for specific domains in crowd-sourcing data labeling.

4.3   Experiments

Since Vn_requirements_NER is more generic where Vn_resumes_NER clusters in the information technology sector, we trained two separated NER models on the Spacy framework[2].

We use the Vn_job_corpus to train the Autophrase component, the weighs of words in SIF embedding and SimCSE embedding. We also experiment with two more pretrained embedding models such as PhoBert [14] and SBERT [15]. For each embedding method, we have the corresponding ranking component and the corresponding classification component as they rely only on the embedding layers. The training data for the classification component is extracted from the NER dataset. Concretely, positive samples are labeled entities whereas negative samples are randomly selected in the corpus.

To evaluate the effectiveness of different components in our methodology we measure 4 settings:

- **M1: Autophrase.** In this setting, we only measure the performance of the Autophrase component. We expect this component to result in a high recall score so that it does not filter out good skill terms.
- **M2: Autophrase → Ranking.** In this setting, the Autophrase output is used as the input to the Ranking component. We expect the ranking component can boost the precision score while maintaining the high recall score.
- **M3: Autophrase → Classification.** In this setting, the model consists of Autophrase and Classification components. This setting aims to compare to the last setting in order to clarify the role of our ranking component.
- **M4: Autophrase → Ranking → Classification.** This setting is the full pipeline of our methodology. In this setting, the output of Autophrase is passed to the ranking component, then the output of the ranking component will be used as the input for the final classification model.

---

[2] https://spacy.io/api/architectures.

4.4 Results

Table 1. Performance evaluation on the Vn_resumes_NER

Method	Full match			Partial match			Embedding
	Precision	Recall	F1_score	Precision	Recall	F1_score	
M1	0.28	**0.72**	0.40	0.35	**0.91**	0.51	
M2	0.32	0.66	0.43	0.40	0.83	0.54	
M3	0.58	0.58	**0.58**	0.67	0.67	**0.67**	
M4	**0.59**	0.53	0.56	**0.68**	0.62	0.65	SIF
M2	0.30	0.64	0.41	0.38	0.81	0.52	
M3	0.58	0.58	**0.58**	0.67	0.67	**0.67**	
M4	0.58	0.52	0.55	**0.68**	0.60	0.64	PhoBert
M2	0.29	0.60	0.39	0.36	0.77	0.49	
M3	0.58	0.58	**0.58**	0.67	0.67	**0.67**	
M4	0.58	0.49	0.53	**0.68**	0.57	0.62	SimCSE
M2	0.27	0.57	0.36	0.34	0.72	0.46	
M3	0.49	0.68	0.57	0.58	0.80	0.67	
M4	0.48	0.55	0.51	0.56	0.64	0.60	SBERT
NER	0.65	0.7	0.67	0.7	0.75	0.72	

The experimental results on the Vn_resumes_NER and the Vn_requirements_NER are depicted in Fig. 1 and Fig. 2 respectively. Overall, our methodology demonstrates superior performance in comparison to the NER model in the Vn_requirements_NER dataset. This result explained that the NER model cannot learn from small labeled data. In the Vn_resumes_NER dataset, our methodology achieves lower performance in full match but comparable performance in partial match assessment.

The experimental results shown that despite simple formula, SIF embedding is the best embedding scheme for our embedding component. In Vn_requirement_NER dataset, our ranking component based on SIF embedding could filter negative terms so that it boosted up the overall performance of the M4: Autophrase → Ranking → Classification pipeline.

Table 2. Performance evaluation on the Vn_requirements_NER

Method	Full match			Partial match			Embedding
	Precision	Recall	F1_score	Precision	Recall	F1_score	
M1	0.19	**0.74**	0.31	0.21	**0.81**	0.33	
M2	0.28	0.73	0.41	0.30	0.79	0.44	
M3	0.68	0.63	0.65	0.70	0.65	0.67	
M4	**0.72**	0.62	**0.67**	**0.74**	0.64	**0.68**	SIF
M2	0.21	0.71	0.33	0.23	0.78	0.36	
M3	0.68	0.63	0.65	0.70	0.65	0.67	
M4	0.68	0.61	0.65	0.70	0.63	0.66	PhoBert
M2	0.20	0.66	0.30	0.21	0.72	0.33	
M3	0.69	0.61	0.65	0.71	0.62	0.66	
M4	0.71	0.54	0.61	0.72	0.56	0.63	SimCSE
M2	0.20	0.68	0.31	0.22	0.74	0.34	
M3	0.67	0.60	0.63	0.69	0.62	0.65	
M4	0.68	0.56	0.62	0.70	0.58	0.63	SBERT
NER	0.32	0.24	0.27	0.37	0.27	0.31	

5   Conclusions and Future Work

In this paper, we have presented a practical methodology for occupational skills detection in Vietnamese job listings. Our methodology exploits the structural property of a job description: most of the skill mentions are within the requirement section and the skill mentions have high semantic similarity scores with the section itself. According to the best of our knowledge, we are the first to propose such a methodology for skill extraction from Vietnamese text.

Our methodology is practical so that it does not require expensive manual labeling datasets. Skill mentions are first detected through an automated phrase detection component that relies on limited positive only terms. Then a ranking component based on text embeddings is used to filter out non-related skill terms. The remaining terms are fed to a classification model to finalize the skill detection pipeline.

Our methodology can achieve comparable performance to a popular industrial-ready NER model in the Vn_resumes_NER dataset while being superior in the smaller Vn_requirements_NER dataset.

The future scope of this work is to detect skill synonyms for skill normalization and skill taxonomy construction. This will benefit further analytic works such as job recommendations and labor market analysis.

Acknowledgments. This research is funded by NAVER corporation.

References

1. Arora, S., Liang, Y., Ma, T.: A simple but tough-to-beat baseline for sentence embeddings. In: Proceedings of the 5th ICLR International Conference on Learning Representations (2016)
2. Bastian, M., et al.: Linkedin skills: large-scale topic extraction and inference. In: Proceedings of the 8th ACM Conference on Recommender Systems, pp. 1–8 (2014)
3. Conneau, A., Kiela, D., Schwenk, H., Barrault, L., Bordes, A.: Supervised learning of universal sentence representations from natural language inference data. arXiv preprint arXiv:1705.02364 (2017)
4. Devlin, J., Chang, M.W., Lee, K., Toutanova, K.: Bert: Pre-training of deep bidirectional transformers for language understanding. arXiv preprint arXiv:1810.04805 (2018)
5. Gao, T., Yao, X., Chen, D.: SimCSE: Simple Contrastive Learning of Sentence Embeddings. arXiv preprint arXiv:2104.08821 (2021)
6. Group, W.B.: World Development Report 2016: Digital Dividends. World Bank Publications (2016)
7. Javed, F., Hoang, P., Mahoney, T., McNair, M.: Large-scale occupational skills normalization for online recruitment. In: Twenty-Ninth IAAI Conference (2017)
8. Kiros, R., et al.: Skip-thought vectors. In: Advances in Neural Information Processing Systems, pp. 3294–3302 (2015)
9. Kivimäki, I., et al.: A graph-based approach to skill extraction from text. In: Proceedings of TextGraphs-8 Graph-based Methods for Natural Language Processing, pp. 79–87 (2013)
10. Le, Q.T.T., Doan, T.H.D., Nguyen, Q.L.H.T.T., Nguyen, D.T.P.: Competency gap in the labor market: evidence from Vietnam. J. Asian Finan. Econ. Bus. 7(9), 697–706 (2020)
11. Li, J., Arya, D., Ha-Thuc, V., Sinha, S.: How to get them a dream job? Entity-aware features for personalized job search ranking. In: Proceedings of the 22nd ACM SIGKDD International Conference on Knowledge Discovery and Data Mining, pp. 501–510 (2016)
12. Nadeau, D.: Semi-supervised named entity recognition: learning to recognize 100 entity types with little supervision. Ph.D. thesis, University of Ottawa (2007)
13. Nadeau, D., Sekine, S.: A survey of named entity recognition and classification. Lingvisticae Investigationes 30(1), 3–26 (2007)
14. Nguyen, D.Q., Nguyen, A.T.: PhoBERT: pre-trained language models for Vietnamese. arXiv preprint arXiv:2003.00744 (2020)
15. Reimers, N., Gurevych, I.: Sentence-bert: sentence embeddings using siamese bert-networks. arXiv preprint arXiv:1908.10084 (2019)
16. Shang, J., Liu, J., Jiang, M., Ren, X., Voss, C.R., Han, J.: Automated phrase mining from massive text corpora. IEEE Trans. Knowl. Data Eng. 30(10), 1825–1837 (2018)
17. Vasudevan, S., et al.: Estimating fungibility between skills by combining skill similarities obtained from multiple data sources. Data Sci. Eng. 3(3), 248–262 (2018)
18. Zhao, M., Javed, F., Jacob, F., McNair, M.: SKILL: a system for skill identification and normalization. In: Twenty-Seventh IAAI Conference (2015)

Neural Inverse Text Normalization with Numerical Recognition for Low Resource Scenarios

Tuan Anh Phan[1,2], Ngoc Dung Nguyen[1], Huong Le Thanh[2],
and Khac-Hoai Nam Bui[1(✉)] 🆔

[1] Viettel Cyberspace Center, Viettel Group, Hanoi, Vietnam
{anhpt161,dungnn7,nambkh}@viettel.com.vn
[2] Hanoi University of Science and Technology, Hanoi, Vietnam
huonglt@soict.hust.edu.vn

Abstract. Neural inverse text normalization (ITN) has recently become an emerging approach for automatic speech recognition in terms of post-processing for readability. In particular, leveraging ITN by using neural network models has achieved remarkable results instead of relying on the accuracy of manual rules. However, ITN is a highly language-dependent task that is especially tricky in ambiguous languages. In this study, we focus on improving the performance of ITN tasks by adopting the combination of neural network models and rule-based systems. Specifically, we first use a seq2seq model to detect numerical segments (e.g., cardinals, ordinals, and date) of input sentences. Then, detected segments are converted into the written form using rule-based systems. Technically, a major difference in our method is that we only use neural network models to detect numerical segments, which is able to deal with the low resource and ambiguous scenarios of target languages. Regarding the experiment, we evaluate different languages in order to indicate the advantages of the proposed method. Accordingly, empirical evaluations provide promising results for our method compared with state-of-the-art models in this research field, especially in the case of low resource scenarios.

Keywords: Inverse text normalization · Automatic speech recognition · Neural network models · Rule based systems

1 Introduction

Inverse text normalization is a natural language processing (NLP) task of converting a token sequence in spoken form (source sentence) to the corresponding written form (target sentence), which is applied to most speech recognition systems. Figure 1 depicts the pipeline of a spoken dialogue system with ITN. ITN is the inverse problem of text normalization (TN), which transforms the written form into spoken form. However, different from the exploitation of promising methods for TN problem in recent years [6], there are not many remarkable

© The Author(s), under exclusive license to Springer Nature Switzerland AG 2022
N. T. Nguyen et al. (Eds.): ACIIDS 2022, LNAI 13757, pp. 582–594, 2022.
https://doi.org/10.1007/978-3-031-21743-2_47

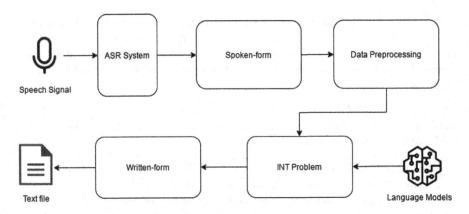

Fig. 1. ITN in spoken dialogue systems.

achievements for the ITN problem, which is regarded as one of the most challenging NLP tasks.

The conventional approach for addressing ITN is rule-based systems, for instance, finite state transducer (FST) based models [2], which have proved the competitive results [15]. However, the major problem with this approach is the scalability problem, which requires complex accurate transformation rules [14]. Recently, Neural ITN has become an emerging issue in this research field, by exploiting the power of neural networks (NN) for ITN tasks. Specifically, NN-based models, typically seq2seq, have achieved high performances and become state-of-the-art models for the ITN problem [3,7,10]. Nevertheless, as we mention above, ITN is a highly language-dependent task and requires linguistic knowledge. In this regard, the data-hungry problem (i.e., low resource scenarios) is an open issue that needs to take into account for improving performance. Furthermore, due to the significant difference between written and spoken forms, handling numbers with minimal error is a central problem in this research field. In particular, to be able to read the numeric values, the models should be worked on both consecutive tasks such as recognizing the parts that belong to numeric values and combining those parts to precise numbers. Specifically, in the shortage-data situation, models might lack information for the training stage in order to recognize and transform numerical segments, which is the cause of the bad performance of ITN. Using pre-trained embedding models (e.g., BERT, GPT3,...) can improve performance. However, for real-world systems, this issue is inefficient due to the cost of memory and time calculation.

In this study, we take an investigation to improve the performance of Neural ITN in terms of low resources and ambiguous scenarios. Particularly, for formatting number problems, conventional seq2seq models might fail to generate sequentially character by character digit, which often appears in long numbers (e.g., phone numbers, big cardinal). For example, the number 'one billion and eight' must be converted to 10 sequential character: '1 0 0 0 0 0 0 0 0 8'. Moreover, the poverty of data in the training process can cause this issue more worse in case the considered languages have lots of ambiguous semantics between numbers and words. For instance, in Vietnamese, the word 'không' (English translate: no) can be a digit '0', but also used to indicate a negative opinion. Table 1 illustrates several examples of the ambiguous semantic problem in the Vietnamese language.

Table 1. Examples of the ambiguous semantic problem in Vietnamese language. The bold parts are ambiguous words.

Spoken form (English translation)	Number	Word
tôi **không** thích cái bánh này (I do not like this cake)		✓
không là số tự nhiên nhỏ nhất -(zero is the smallest natural number)	✓	
năm một nghìn chín trăm chín bảy (nineteen ninety-seven)		✓
năm mươi nghìn (fifty thousand)	✓	
chín qủa táo (nine apples)	✓	
quả táo **chín** (a ripe apple)		✓

In this paper, the proposed framework includes two stages: i) In the first stage, we use a neural network model to detect numerically segments in each sentence; ii) Then, the output of the first phase is converted into the written form by using a set of rules. Accordingly, the main difference compared with previous works is that we use the neural network to detect numerical segments in each sentence as the first phase. Reading number is processed in the second stage by a set of rules, which is able to supply substantial information for the system without requiring much data to learn as end-to-end models. Generally, the main contributions of our method are threefold as follows:

- We propose a novel hybrid approach by combining a neural network with a rule-based system, which is able to deal with ITN problems in terms of low resources and ambiguous scenarios.
- We evaluate the proposed methods in two different languages such as English and Vietnamese with promising results. Specifically, our method can extend easily to other languages without requiring any linguistics knowledge.
- We present a pipeline to build an ITN model for Vietnamese. To the best of our knowledge, this is the first study of the ITN problem for Vietnamese speech systems.

The rest of the paper is organized as: Section 2 presents the literature review and background of our study. Our methodology is proposed in Sect. 3. We report and analyze the evaluated results in Sect. 4. Section 5 concludes our work and discusses the future work regarding this study.

2 Literature Review

The research on ITN is closely related to the TN problem in which recent works can be classified into three approaches, which are sequentially described as follows:

2.1 Rule-Based Systems

Most ASR and Text-to-Speech (TTS) systems are based on Weighted FST grammar for the TN problem. Kestrel, a component of the Google TTS synthesis system [2], has achieved around 99.9% on the Google TN test dataset. For ITN, the set of rules in [7] achieves 99% on internal data from a virtual assistant application. In order to develop products, Zhang et al. introduce an open-source python WFST-based library [15], which includes two-stage normalization pipeline that first detects semiotic tokens (classification) and then converts these to written form (verbalization). Both stages consume a single WFST grammar. The major problem is that this approach requires significant effort and time to scale the system across languages, especially linguists experiences.

2.2 Neural Network Models

Recurrent Neural Network (RNN)-based seq2seq models [11], have been adopted for reducing manual processes. Sproat et al. [9] considers the TN problem as a machine translation task and develop an RNN-based seq2seq model trained on window-based data. Specifically, an input sentence is regarded as a sequence of characters, and the output sentence is a sequence of words. Furthermore, since the length of sequence input problem, they split a sentence into chunks with window size equals three for creating sample training in which normalized tokens are marked by distinctive begin tag <norm> and end tag </norm>. In this regard, this approach is able to limit the number of input and output nodes to something reasonable. Their architecture neural network follows closely that of [1]. Sequentially, Sevinj et al. [13] proposed a novel end-to-end Convolutional Neural Network (CNN) architecture with residual connections for the TN task. Particularly, they consider the TN problem as the classification problem which includes two stages: i) First, the input sentence is segmented into chunks, which is similar to [9] and use a CNN-based model to label each chunk into corresponding class based on scores of soft-max function; ii) After the classification stage, they apply rule-based methods depending on each class.

Recently, Transformer is a new seq2seq structure, which has achieved high performance for most NLP tasks [12]. Accordingly, the work in [10] has shown the advantage of Transformer compared with RNN-based models in the ITN problem. Figure 2 depicts a general architecture of Transformer.

Particularly, the model handles the input into three kinds of vectors (i.e., Key, Value, and Query vectors), which are driven by multiplying the input embedding with three matrices that we trained during the training process. Furthermore,

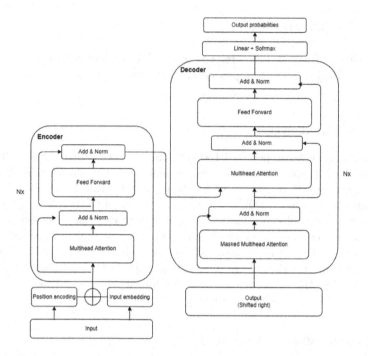

Fig. 2. The general architecture of Transformer. The encoder layer is built by 6 identical components which contain one multi-head attention layer with a fully connected network. These two sub-layers are equipped with residual connection as well as layer normalization. The decoder layer is more complicated and also includes 6 components stacked. Each component includes three connected sub-layers in which two sub-layers of multi-head self-attention and one sub-layer of fully-connected neural network.

Positional Encoding is adopted to the model for the sequence order, and Self-Attention with multi-head is applied for allowing the model to jointly attend to inform different representation sub-spaces at different positions. Sequentially, the log conditional probability can be interpreted as follows:

$$logP(y|x) = \sum_{t=1}^{t=N} logP(y_t|y_{<t}, x) \qquad (1)$$

Furthermore, although there is no out-of-vocab (OOV) problem of the input by using the window-based seq2seq model, however, it is able to occur in the decode. In this regard, Courtney et al. [5] propose a new method for directly translating the input written-form to output spoken form without tagger or window-based segment. Accordingly, in order to handle OOV on both sides, the study uses the subword method [8] to decompose words into subparts, which have been proved the capability in open-vocabulary speech recognition and machine translation tasks.

2.3   Hybrid Models

Employing a weak covering grammar to filter and correct the misreading of NN models. Pusateri et al. [7] presents a data-driven approach for ITN problems by a set of simple rules and a few hand- crafted grammars to cast ITN as a labeling problem. Then, a bi-directional LSTM model is adopted for solving the classification problem. [10] propose a neural solution for ITN by combining transformer-based seq2seq models and FST-based text normalization techniques for data preparation. Specifically, similar to [5], they implement a word and subword-based seq2seq neural network models except reversing the order of input sentence and output sentence. Accordingly, integrating Neural ITN with an FST is able to overcome common recoverable errors in production environments.

3   Methodology

3.1   General Framework

In this paper, we propose a novel hybrid model Neural ITN problem using seq2seq neural network and rule base systems by considering the ITN task as the Machine Translation (MT) task. Figure 3 describe general our framework, which includes two main stages.

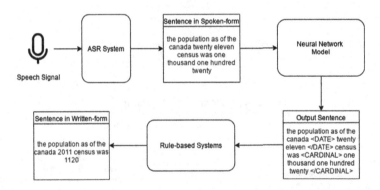

Fig. 3. The general framework of the proposed method for the Neural ITN approach

Specifically, in the first stage, each sentence is put into a transformer-based seq2seq model for detecting numerical segments by using tag <n> and </n>, in which n represents a numerical classes (e.g., *DATE, CARDINAL, ORDINAL*). Then, a set of rules is employed to convert tokens, which be wrapped by tag to number, into the written form. Otherwise, all parts of a sentence, which are not in the tag are conserved. Particularly, instead of using a neural network to directly translate numbers [9], we only use NN to detect numerical segments. Essentially, NN is only utilized for distinguishing which is in the number and which is not. After that, when the model has candidates for numbers, they are transformed to the correct form by the set of rules. Consequentially, the model is able to read accurately numbers in sentences.

For example, if we have input spoken sentence: *'the population as of the canada twenty eleven census was one thousand one hundred twenty'*. After pass through this sequence to NN model, the output will be formalized as: *'the population as of the canada <DATE> twenty eleven </DATE> census was <CARDINAL> one thousand one hundred twenty </CARDINAL>'*. As result, numerical phrases *'twenty eleven'* and *'one thousand one hundred twenty'* are wrapped, and transformed to written form by rules: '27' and '1120' based on two class DATE and CARDINAL. A set of rules, which are utilized can be considered as the replacement for a great deal of knowledge in the training process.

3.2 Data Creation Process

Since there is no publicly available data for training ITN, following the work in [10], we employ a novel data generation pipeline for ITN using the TTS system. Figure 4 shows the main steps of our data creation process, which are sequentially described as follows:

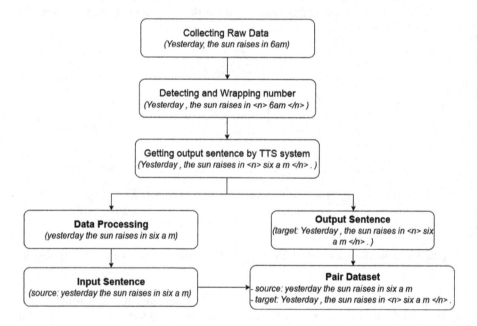

Fig. 4. Data creation process for training seq2seq model

- **Step 1**: Crawling/downloading raw data from published websites.
- **Step 2**: Cleaning raw data (e.g., removing HTML, CSS, and so on), and removing noise documents.

- **Step 3**: Detecting non-standard words, for instance, alphabet (e.g 'David') or number (e.g '2017'), in sentence. For handling alphabet words, we split them into characters (extremely subword) and bound them by using tag <oov> and </oov>. For handling numerical words, we split them into sequence digits and bound them by couple tag <n> and </n>.
- **Step 4**: Passing output sentences through TTS systems (e.g., Google TTS).
- **Step 5**: The output sentences from TTS systems are used as target sentences for NN models. For creating source sentences, we remove punctuation, tag <n>, lower all tokens, and only preserve tag <oov> similar to the spoken form.
- **Step 6**: Saving data to file, simultaneously.

3.3   Training Model

As we mention above, we model the ITN as an MT problem where the source and target are the output of spoken form and the detected segments of text, respectively. For NN models, we implement two training models which are RNN-based and transformer-based seq2seq models. Specifically, for the RNN model, we employ a stacked bi-direction long short term memory network (bi-LSTM) as encoder and a stacked long short term memory network (LSTM) as decoder, respectively [1]. Regarding the non-recurrent model, we implement a transformer model based on the work in [12]. Additionally, the source and target sentences are segmented into subword sequences [5].

3.4   Rule-Based Systems

The output of NN models with detected segments is transformed into the written form using a set of rules. Table 2 demonstrates an example with the input is the output sequence of the first stage and final outputs of this process.

Table 2. An example output in second stage. Tokens labeled by NONE is conserved, while tokens are bounded by numerical tag (eg. CARDINAL) will be rewritten under using rules.

Spoken-form	Label	After label	Written-form
He	NONE	he	he
Collected	NONE	collected	collected
Four	CARDINAL start	$\langle$ CARDINAL$\rangle$	400000
Hundred	CARDINAL in	four hundred thousand	
Thousand	CARDINAL end	$\langle$ /CARDINAL$\rangle$	
Records	NONE	records	records

Specifically, the input sentence is segmented and specified by labels using post-processing grammar. Sequentially, with each numerical token, we used the word2number1 python package[1] for converting spoken numbers into written numbers. In particular, since the tool works only for positive numbers and the largest value is limited to 999,999,999,999, we extended the tool in order to handle negative cardinals and larger numbers. Moreover, we also construct the python modules for reading the number, which belongs to other classes such as MEASURE, DATE, PHONE, TIME... and so on based on the aforementioned extended tool.

4 Experiment

4.1 Dataset

Regarding evaluation datasets, we test our method on two different language datasets such as English and Vietnamese, which are extracted from publicly available data sources as follows:

- **English Dataset**: the original version consists of 1.1 billion words of English text from Wikipedia, divided across 100 files. The normalized text is obtained by running through Google TTS system's Kestrel text normalization system [2]. In this study, we use the first file which contains approximately 4 million samples for our experiments. We split randomly the file into two parts in which the first part includes 1 million sentences for training data, and the second part contains 50,000 sentences for testing. For the ITN dataset, we conserved all tokens of sentences and swapped input and output.
- **Vietnamese Dataset**: the raw dataset is extracted from the largely published source[2]. After that, we decode them as UTF-8 and remove all sentences including tags (e.g., Html and CSS). We also extracted 1 million and 50,000 sentences for training and testing data, respectively. Sequentially, we execute the dataset through Viettel TTS System[3]. For formatting into the ITN dataset, we construct the data following the pipeline in Sect. 3.2.

Since our study focuses on low resource scenarios, we divide the training data into various numbers of samples such as 100 k, 200 k, 500 k, and 1000 k, respectively, in order to evaluate the advantage of the proposed method. Specifically, Table 3 illustrates the size of training, validation, and vocab size of the training datasets.

4.2 Baseline Models and Hyperparameter Configuration

Regarding baseline models, we implement two well-known NN models such as RNN and Transformer. Specifically, the Neural ITN is regarded as the MT problem. Furthermore, all models are implemented with the subword approach, which has proven the better performance [5]. Particularly, the baseline models are sequentially configured as follows:

---

[1] https://pypi.org/project/word2number/.
[2] https://github.com/binhvq/news-corpus.
[3] https://viettelgroup.ai/.

Table 3. Size of training datasets with 20% for validation.

Language	Dataset	Training	Validation	Vocabulary size
English	100 k	80 k	20 k	35903
	200 k	160 k	40 k	47328
	500 k	400 k	100 k	65079
	1000 k	800 k	200 k	80585
Vietnamese	100 k	80 k	20 k	7223
	200 k	160 k	40 k	8225
	500 k	400 k	100 k	10400
	1000 k	800 k	200 k	11657

- **RNN Model**: For the recurrent seq2seq baseline model, we use Encoder-Decoder architecture with RNN-Encoder consisting of two bi-directional long short-term memory (Bi-LSTM) layers and two LSTM layers for the decoder. Both encoder and decoder contain 512 hidden states. Global attention mechanisms [4] is implemented in Decoder.
- **Transformer Model**: For the non-recurrent seq2seq baseline model, we implement the architecture, which is similar to [12]. Specifically, we employ the subword-transformer model with 6 layers for both Encoder and Decoder. Each sub-layer block contains a 512-dimension vector hidden state. Furthermore, the number of multi-head self-attention is set to 8.

Consequentially, we execute our proposed method with two versions by adopting two aforementioned baseline models for the first phase of segment detection, respectively. For the hyperparameter configuration, we use Adam optimizer with learning rate annealing and the initial value is 0.01, and dropout is set to 0.1. All models are trained with 100k timesteps, early stop is used on validation loss.

4.3   Evaluation Metrics

Bi Lingual Evaluation Understudy (BLEU) is the popular way to measure the performance of a machine translation system, which is calculated as:

$$BLEU = BP\, exp \left(\sum_{n=1}^{n-grams} w_n\, logp_n \right) \qquad (2)$$

where P_n is the modified n-$grams$ precision and w_n denotes the uniform weight. BP refers to the brevity, which can be calculated as follows:

$$BP = \begin{cases} 1, & \text{if } c > r. \\ e^{1-\frac{r}{c}}, & \text{otherwise.} \end{cases} \qquad (3)$$

where c and r refers the candidate and reference sentences, respectively.

Word Error Rate (WER) is a common metric of the performance of speech recognition or machine translation system, which is calculated as:

$$WER = \frac{S + D + I}{S + D + C} \tag{4}$$

where S, D, I, C are the number of substitutions, deletions, insertions, and correct words, respectively.

4.4   Result Analysis

Table 4 shows the comparison results of our experiments on the test set, which bold parts are the best results.

Table 4. Comparison of models on test set with BLEU and WER scores. Bold texts indicate the best results.

Lang	Training data	Metric	RNN	Trans	Our(RNN)	Our(Trans)
English	100k	BLEU	0.7405	0.7048	**0.8334**	0.741
		WER	0.1467	0.1561	**0.1017**	0.152
	200k	BLEU	0.7909	0.7919	**0.8353**	0.8188
		WER	0.1129	0.1033	0.112	**0.1003**
	500k	BLEU	0.7706	**0.8558**	0.8377	0.8394
		WER	0.128	**0.0708**	0.1009	0.0874
	1000k	BLEU	0.7959	**0.9138**	0.848	0.8933
		WER	0.1087	**0.0405**	0.0953	0.0517
Vietnamese	100 k	BLEU	0.6422	0.6755	**0.7019**	0.6774
		WER	0.2495	0.2447	**0.1897**	0.2029
	200 k	BLEU	0.6794	0.7201	0.7144	**0.7286**
		WER	0.2111	0.2101	0.184	**0.1774**
	500 k	BLEU	0.6921	0.7447	0.718	**0.7667**
		WER	0.2016	0.1767	0.1784	**0.1327**
	1000 k	BLEU	0.6777	0.7594	0.711	**0.7885**
		WER	0.2103	0.1821	0.1817	**0.1199**

Note that for the BLEU score, the higher is the better, contrary to the case of WER score. Accordingly, there are several issues we can summarize based on the results as follows:

– Our method outperforms baseline models in the case of low resource scenarios (i.e., 100 k and 200 k) and is able to achieve competitive results in the case of higher resources (i.e., 500 k and 1000 k) with the English language. The experiment results indicate that when the number of the training sample is low, using two stages is able to cover several prediction errors in end-to-end models.

- For the Vietnamese language, our method is able to achieve the best results in all cases. The reason is that different from English, the numerical classes in Vietnamese are more ambiguous in Spoken-form (as shown in Table 1), in which the end-to-end modes are not able to distinguish the ambiguous problem between numerical classes in Vietnamese.
- Recurrent-based seq2seq models with attention achieve better performance compared with Transformer in the case of low resource scenarios. Meanwhile, Transformer-based models are able to achieve the best results by increasing the number of training samples. Therefore, combining two methods (hybrid models) is able to improve the performance. We take this issue as our future work regarding this study.

5   Conclusion

In this study, we introduce a new method for the neural ITN approach. Specifically, the difference from previous works, we divide the neural ITN problem into two stages. Particularly, in the first stage, neural models are used to detect numerical segments. Sequentially, the written form is extracted based on a set of rules in the second stage. In this regard, our method is able to deal with the low resource scenarios, where there is not much available data for training. Furthermore, we showed that our method can be easily extended to other languages without linguistic knowledge requirements. The evaluation of two different language datasets (i.e., English and Vietnamese) with different sizes of training samples (i.e., 100 k, 200 k, 500 k, and 1000 k) indicates that our method is able to achieve comparable results in the English language, and the highest results in Vietnamese languages. Regarding the future work of this study, we focus on extending the model with deeper architectures, for instance, 12 layers of Transformer, and combine models, including pre-trained models for improving the performance of Neural ITN problem.

References

1. Chan, W., Jaitly, N., Le, Q.V., Vinyals, O.: Listen, attend and spell: a neural network for large vocabulary conversational speech recognition. In: Proceeding of the 41st International Conference on Acoustics, Speech and Signal Processing (ICASSP), pp. 4960–4964. IEEE (2016). https://doi.org/10.1109/ICASSP.2016.7472621
2. Ebden, P., Sproat, R.: The kestrel TTS text normalization system. Nat. Lang. Eng. **21**(3), 333–353 (2015). https://doi.org/10.1017/S1351324914000175
3. Ihori, M., Takashima, A., Masumura, R.: Large-context pointer-generator networks for spoken-to-written style conversion. In: Proceeding of the 45th International Conference on Acoustics, Speech and Signal Processing (ICASSP), pp. 8189–8193. IEEE (2020). https://doi.org/10.1109/ICASSP40776.2020.9053930
4. Luong, T., Pham, H., Manning, C.D.: Effective approaches to attention-based neural machine translation. In: Proceeding of the 2015 Conference on Empirical Methods in Natural Language Processing (EMNLP), pp. 1412–1421. The Association for Computational Linguistics (2015). https://doi.org/10.18653/v1/d15-1166

5. Mansfield, C., Sun, M., Liu, Y., Gandhe, A., Hoffmeister, B.: Neural text normalization with subword units. In: Proceedings of the 17th Annual Conference of the North American Chapter of the Association for Computational Linguistics: Human Language Technologies (NAACL-HLT), pp. 190–196. Association for Computational Linguistics (2019). https://doi.org/10.18653/v1/N19-2024

6. Pramanik, S., Hussain, A.: Text normalization using memory augmented neural networks. Speech Commun. **109**, 15–23 (2019). https://doi.org/10.1016/j.specom.2019.02.003

7. Pusateri, E., Ambati, B.R., Brooks, E., Plátek, O., McAllaster, D., Nagesha, V.: A mostly data-driven approach to inverse text normalization. In: Proceeding of the 18th Annual Conference of the International Speech Communication Association (Interspeech), pp. 2784–2788. ISCA (2017)

8. Sennrich, R., Haddow, B., Birch, A.: Neural machine translation of rare words with subword units. In: Proceedings of the 54th Annual Meeting of the Association for Computational Linguistics (ACL), pp. 1715–1725. Association for Computational Linguistics (2016). https://doi.org/10.18653/v1/P16-1162

9. Sproat, R., Jaitly, N.: An RNN model of text normalization. In: Proceeding of the 18th Annual Conference of the International Speech Communication Association (Interspeech), pp. 754–758. ISCA (2017)

10. Sunkara, M., Shivade, C., Bodapati, S., Kirchhoff, K.: Neural inverse text normalization. In: Proceeding of the 46th International Conference on Acoustics, Speech and Signal Processing (ICASSP), pp. 7573–7577. IEEE (2021). https://doi.org/10.1109/ICASSP39728.2021.9414912

11. Sutskever, I., Vinyals, O., Le, Q.V.: Sequence to sequence learning with neural networks. In: Proceeding of the Advances in Neural Information Processing Systems 27: Annual Conference on Neural Information Processing System (NeurIPS), pp. 3104–3112 (2014)

12. Vaswani, A., et al.: Attention is all you need. In: Proceeding of the Advances in Neural Information Processing Systems 30: Annual Conference on Neural Information Processing System (NeurIPS), pp. 5998–6008 (2017)

13. Yolchuyeva, S., Németh, G., Gyires-Tóth, B.: Text normalization with convolutional neural networks. Int. J. Speech Technol. **21**(3), 589–600 (2018). https://doi.org/10.1007/s10772-018-9521-x

14. Zhang, H., et al.: Neural models of text normalization for speech applications. Comput. Linguist. **45**(2), 293–337 (2019). https://doi.org/10.1162/coli_a_00349

15. Zhang, Y., Bakhturina, E., Gorman, K., Ginsburg, B.: Nemo inverse text normalization: from development to production. CoRR abs/2104.05055 (2021)

Detecting Spam Reviews on Vietnamese E-Commerce Websites

Co Van Dinh[1,2], Son T. Luu[1,2(✉)], and Anh Gia-Tuan Nguyen[1,2]

[1] University of Information Technology,
Ho Chi Minh City, Vietnam
19521293@gm.uit.edu.vn, {sonlt,anhngt}@uit.edu.vn
[2] Vietnam National University, Ho Chi Minh City, Vietnam

Abstract. The reviews of customers play an essential role in online shopping. People often refer to reviews or comments of previous customers to decide whether to buy a new product. Catching up with this behavior, some people create untruths and illegitimate reviews to hoax customers about the fake quality of products. These are called spam reviews, confusing consumers on online shopping platforms and negatively affecting online shopping behaviors. We propose the dataset called ViSpamReviews, which has a strict annotation procedure for detecting spam reviews on e-commerce platforms. Our dataset consists of two tasks: the binary classification task for detecting whether a review is spam or not and the multi-class classification task for identifying the type of spam. The PhoBERT obtained the highest results on both tasks, 86.89%, and 72.17%, respectively, by macro average F1 score.

Keywords: Spam reviews · Text classification · Deep neural models · Transformer models · Dataset · Annotation guidelines

1 Introduction

Vietnam has witnessed strong growth in e-commerce in recent years. Many Vietnamese online trading platforms are constructed and attract consumers. Online shopping is now popular in people's daily routines because of its convenience and flexibility. However, besides the advantages of online shopping, the rise of fake products and fraud qualifications in online trading concerns customers and shop owners.

Customer reviews play an essential part in the behaviors of consumers in online shopping. Customer reviews express their opinions, emotions, and attitudes about products, and these opinions affect other customers in deciding whether to buy a product or not. If a customer wants to buy a product in an online shop, they tend to refer to reviews of previous customers about that product. Catching up with this behavior, some people create spam reviews, which are illegitimate means and untruth facts about the actual quality of products to confuse the consumers to boost the financial business or fame of an individual or

N. T. Nguyen et al. (Eds.): ACIIDS 2022, LNAI 13757, pp. 595–607, 2022.
https://doi.org/10.1007/978-3-031-21743-2_48

organization [9]. Therefore, detecting these spam reviews will protect both sellers and customers from the risk of low-quality products and preserve the reputation of the sellers.

Our purpose in this paper is to propose a method to detect spam reviews about products on online shopping platforms. First, we constructed a corpus for spam detection from users' reviews by texts. Second, we use machine learning approaches to build classification models for detecting spam comments and evaluate classification models' performances on the constructed dataset.

The paper is structured as follows. Section 1 introduces our works. Section 2 takes a survey about relevant research on the problem of online spam detection. Section 3 describes the data creation process and gives an overview of our dataset. Section 4 introduces our approaches for the online spam reviews detection problem by machine learning and deep learning. Section 5 displays our empirical results and analysis of the performances of classification models. Finally, Sect. 6 concludes our research and proposes future works.

2   Related Works

Preliminary research about spam reviews showed how to construct a classification model for detecting whether a user's reviews are spam or not and the difficulty in detecting spam reviews from both the content of the review and the reviewers [9]. The challenge of identifying spam reviews comes from reviewers' behavior when they try to create spam content just like other innocent reviewers. Therefore, [10] proposes three aspects of spam reviews to claim the problem of spam reviews.

Many approaches are applied for the task of spam reviews detection, including patterns and rules [13,14,30], machine learning and deep learning approaches [24, 26], linguistic features [15,22]. Overall, the dataset is the key point for training and evaluating the classification models applied in the opinion spam detection task. Available datasets for detecting spam reviews are introduced by [25].

In Vietnamese, there are several datasets about user reviews on e-commerce platforms, such as the dataset about phone and restaurant reviews [18,29], the smartphone feedback datasets [16,19], and the complaining detection on e-commerce websites dataset [20]. However, there is no particular dataset for spam review detection on Vietnamese e-commerce websites yet. Hence, our motivation is to construct a dataset for detecting spam reviews on Vietnamese E-commerce platforms.

3   The Dataset

3.1   Dataset Creation Process

We collected data from leading online shopping platforms in Vietnam. Then, we select some of the most recent selling products for each product category and collect up to 15 reviews per product. After collecting, we get a dataset of 19,868 product reviews which contains the number of star reviews, comments about the product, and the link to that product. Subsequently, we construct the annotate guideline and annotate for the corpus.

Our data annotating process is impressed from the MATTER framework [23]. The annotation process consists of two phases. The first is the training phase for the annotators. The annotators read a guideline describing the meaning of labels, a sample review, and some examples of specific cases. The annotators will read the guideline and annotate 300 random samples in the dataset. Then, we calculate and evaluate the average inter-annotator agreement. If the inter-annotator agreement is satisfied, we move to the second phase, which is the annotation phase. In contrast, we re-train the annotators and update the annotation guidelines. According to [4], the inter-annotator agreement calculated by Cohen's Kappa index [2] is acceptable when higher than 0.5.

In the second phase, the annotators will be provided with a complete dataset and annotate on this dataset. There are all three annotators during the annotation phases, and the final label of the dataset will be decided by voting for the most assigned label. To ensure objectivity when annotating the dataset, we keep annotators annotating independently.

3.2 Annotation Guidelines

Our dataset comprises two tasks. The first task determines whether the reviews are spam or not spam (**Task 1**), and the second task indicates the types of spam reviews (**Task 2**). The dataset contains two labels: SPAM and NO-SPAM. For each spam review, we label one of three types of spam labels [10]. The label of a review is described as follows:

NO-SPAM: Reviews labeled with this label are regular reviews, true to the product's reality. Reviews like these provide helpful information for buyers to get an overview of the product before deciding whether to buy it or not.

SPAM: Reviews labeled with this label are reviews that are entirely or partially untrue about products sold on e-commerce sites. Reviews like these often make it easier to sell products or hurt the sales and reputation of stores and provide inaccurate or unhelpful information. According to [10], we divided the labels for the reviews as spam into three labels:

- **SPAM-1 (fake review):** These reviews mislead customers by giving negative review comments to the product to damage the reputation of the store selling the product or giving an excellent review to the product in order to attract customers for the product and the shop even though the product is not relevant.
- **SPAM-2 (review on brand only):** These reviews do not comment specifically on the product but only on the brand, manufacturer, or seller of the product. Although these reviews can be informative for product buyers, they are often negative and considered spam.
- **SPAM-3 (non-review):** These are reviews whose commentary is not about the product or anything related to the product. These reviews tend to promote another product, get commissions from an e-commerce site, or have no purpose.

Table 1. Several example reviews and instruction for annotating labels

Comments	Type	Stars	Explanation
Gia Vị Lẩu Haidilao (Hải Để Lao) Chắc hẳn ai đi Trung Quốc cũng 1 lần ghé ăn nhà hàng lẩu Haidelao siêu siêu ngon với vị lẩu tuyệt hảo. Độ ngon độ hot của thương hiệu này thì không phải bàn cãi nữa, chỉ cần 1 gói gia vị lẩu chế được 1-2 nồi nước ngon ngọt siêu thơm. Sản phẩm gia vị độc đáo dùng. (*English: Haidilao Hot Pot Seasoning - Surely everyone who goes to China will visit once at the super delicious Haidelao hotpot restaurant with excellent hotpot taste. The deliciousness and hotness of this brand is not in dispute anymore, just 1 pack of hot pot spices can make 1-2 pots of super delicious. This is a unique seasoning product.*)	SPAM-1	5	This comment was created in order to advertise for the brand named Haidilao. It does not mention to any aspect of the product.
JBL thương hiệu có tiếng về loa. Cảm ơn Tiki đã giao hàng thành công, mặc dù chậm hơn 1 ngày so với lịch hẹn nhưng mình hiểu vì lý do thời tiết trời mưa. Chúc Tiki thành công và phát triển mạnh. (*English: JBL brand is famous for speakers. Thank you Tiki for successful delivery, although 1 day later than scheduled, I understand because of the rainy weather. Wish Tiki success and thrive.*)	SPAM-2	5	This review only focuses on the specific brand, which does not give any evaluation of the product.
ầy đẹp lắm shopppppppppppppppppppp. Chất đẹp mịn sang. Nhận hàng ưng cực kì. Giá cả hợp lýyyyyyyyyyyyyyyyyyyyyyyyyyyyyyyyy. Sẽ ủng hộ shopppppp dài dài ạ (*English: very beatiful, shopppppppppppppppppppppp. Quality fabric is beautiful, smooth, luxurious. Received very satisfied. Reasonable priceeeeeeeeeeeeeeeeeeeeee. Will support shopppppp for a long time*)	SPAM-3	5	This comment is not relevant to any aspect about the product. In this comment, the users use the word with duplicated characters to improve the comment length. However, this is not used to give any valuable opinion about the product quality.
Phấn mịn, bám da tốt. Mình sử dụng màu Peach hơi trắng hơn da mình. Không biết mấy màu khác thì sao. (*English: The powder is smooth and adheres well to the skin. I use Peach color which is slightly whiter than my skin. I don't know about other colors*)	NO SPAM	5	This comment express the opinion of the users about the product quality.
Sản phẩm rất tệ, có mùi mật ong nhưng nhỏ trực tiếp lên đường đi của kiến để cả ngày chẳng thấy con kiến nào bu vào. Mà chúng còn làm thêm đường khác để đi. Phải rải thêm 1 ít sữa bột thì kiến mới bu vào. Nhưng cuối cùng kiến vẫn đi khắp nhà. Mình sử dụng dũng mấy ngày kquả rất bực nên mới đánh gía (*English: The product is very bad, has a smell of honey but drops directly onto the ant's path so that no ants can come in all day. But ants also make another way to go. Have to sprinkle a little more milk powder to get the ants in. But in the end ants still go around the house. I used it for a few days and the results were so frustrating that I gave a review*)	NO SPAM	1	This comment gives the opinion of the consumer about the specific product. It can be seen that the attitude of the consumer with the product is matched with the rating stars he/she gave. Hence, this is a good review of the product.

In addition, Table 1 describes several examples of reviews from users and the explanation for choosing the label for each review. For each review, annotators choose one suitable label.

3.3   Inter-annotators Agreement and Discussion

We have three different annotators to annotate the dataset. We let those three annotators work independently on the sample to measure the inter-annotator agreement in the training phase. Then, we calculate the inter-annotator agreement in pairs of annotators by the Cohen's Kappa index [2].

Table 2. Inter-annotator agreement of three annotators A1, A2 and A3 on the two tasks. Annotators are working independently

	Task 1			Task 2		
	A1	A2	A3	A1	A2	A3
A1	–	0.36	0.34	–	0.41	0.37
A2	–	–	0.61	–	–	0.57
A3	–	–	–	–	–	–
Average κ	0.43			0.44		

Table 3. Inter-annotator agreement of three annotators A1, A2 and A3 on the two tasks after re-training with the updated annotation guidelines

	Task 1			Task 2		
	A1	A2	A3	A1	A2	A3
A1	–	0.93	0.59	–	0.87	0.57
A2	–	–	0.65	–	–	0.60
A3	–	–	–	–	–	–
Average κ	0.72			0.68		

According to Table 2, the average inter-annotator agreements of two tasks are lower than 0.5, which does not satisfy the minimum agreement level, according to [4]. Therefore, we update the current annotation guidelines with more explanations and examples and re-train the annotators to boost the quality of the annotation guidelines. As illustrated in Table 3, the inter-annotator agreement is improved on both tasks. The final inter-annotator agreement of our dataset is 0.72 and 0.68, which is the substantial level according to [12].

Table 4. Confusion matrix between three pairs of annotators when annotating the samples. We calculate the values by taking the average of three pairs

	NO SPAM	SPAM-1	SPAM-2	SPAM-3
NO SPAM	**170.33**	2.00	4.33	1.33
SPAM-1	7.33	**8.67**	1.33	3.67
SPAM-2	12.00	1.33	**12.00**	0.00
SPAM-3	11.00	1.00	5.00	**58.67**

In addition, Table 4 describes the number of annotated comments by three different annotators. It can be seen that the number of disagreement data fell into the case of determining whether a comment belongs to a specific spam type and comment is not a spam review. Therefore, we attach the original links of products to the reviews for annotators to reference. This way, annotators can identify the reviews' context, then give the accrue labels. However, as shown in Table 5, two annotators are disparity when annotating Reviews #1, because this review contains the user's opinion about not only the product but also the brands of the providers, which is categorized as SPAM-2 according to the annotation guidelines. Besides, Comments #2 and Comments #3 give the opinion about the product's quality and design but do not mention the product carefully, which confuses the annotators for deciding these comments as non-review (SPAM-3).

In general, the challenge of annotating for this task is identifying whether or not the reviews are spam. Despite detailed annotation guidelines and the information about the products relevant to the reviews, annotators are still misunderstood because consumers' opinions are diverse, and the stylistics of users' reviews are unclear. Therefore, to guarantee the objectives, we let three annotators annotate the entire dataset, then take the final label by major voting.

Table 5. Several sample reviews that contain disagreement labels between annotators

No.	Reviews	Annotator 1	Annotator 2
1	Tiki giao hàng cực nhanh dù trong mùa dịch. Nivea xài tốt xưa giờ rồi. Tuy có hơi rít nhẹ, xài một thời gian da mềm hơn, sáng hơn. (*English: Tiki delivered very quickly even during the epidemic season. Nivea used it very well from the past to now. Although there is a slight hiss, after using it for a while, the skin is softer and brighter.*)	SPAM	NO SPAM
2	Hàng đúng loại, giá rẻ (*English: Right product, cheap price*)	NO SPAM	SPAM
3	Đẹp nha :)) sang xịn mịn (*English: Beautiful :)) luxurious and smooth*)	NO SPAM	SPAM

3.4 Dataset Overview

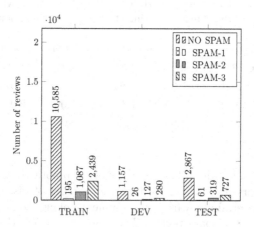

Fig. 1. The distributions of three labels on the train, development, and test sets

After annotating the dataset, we have nearly 19,000 reviews from users, in which each review is categorized as spam or not spam. If the reviews are spam, they consist of the types of spam. Then, we divided the dataset into train, development, and test sets with proportions 7-1-2. The overall information about the dataset is illustrated in Table 6.

In addition, Fig. 1 shows the distribution of reviews by each label on the train, development, and test sets. The reviews which are annotated as not spam account for the highest proportion. For the spam types, the SPAM-3 reviews are more than two remaining types. The data distribution on the training, development, and test set is similar.

Table 6. Overview about the dataset. The vocabulary size is computed on syllable level

	Train	Development	Test
Num. reviews	14,306	1,590	3,974
Vocabulary	19,677	5,046	9,040

Table 7. The distribution of spam and non-spam reviews based on the rating stars of users

Stars	1	2	3	4	5
NO-SPAM	213	93	193	292	9,793
SPAM	3	1	1	0	190

Besides, according to Table 7, most spam reviews have 5 stars rated by users. For no spam reviews, although the distribution is more uniform from 1 to 4, most reviews are rated as 5 stars. Hence, the rating star for the product is not reliable information for expressing the opinions of users about the quality of the product.

4 Methodologies

4.1 Task Definition

The problem of spam review detection is categorized as the text classification task. This problem comprises two tasks: Task 1 is the binary classification task for classifying whether a review is spam or not spam, and Task 2 is the multi-class classification task for identifying the type of spam, which are one of three types as mentioned in Sect. 3.

4.2 Word Embedding

Word embedding is a vector space used to represent text data that can describe the relationship, semantic similarity, the context of data. In natural language processing, word representation plays a vital role in many downstream tasks, such as classification tasks. On the task of text classification, the empirical results from [8] showed that the fastText pre-trained embedding provided by [5] obtained robust results when integrating with various deep neural networks on social media texts. Therefore, we choose the fastText word embedding[1] for our empirical results.

---

[1] https://fasttext.cc/docs/en/crawl-vectors.html.

4.3   Deep Neural Network Models

Text-CNN [11]: Convolutional neural network (CNN) is a model combined with many different layers. CNN is often applied in computer vision to extract features of images for image classification and has achieved high performance than traditional approaches. In addition, CNN is also applied in natural language processing problems, typically text classification with the Text-CNN model. This model is based on convolutional architecture to extract valuable features from natural texts.

LSTM [6]: Long Short Term Memory (LSTM) is an improved model from Recurrent Neural Network (RNN). LSTM helps the model remember the previous information for a long time, which is a restriction faced by the RNN model. LSTM comprises three gates: input gate, output gate, and forget gate. The input gate selects information to add to the context, the output gate decides whether the input is necessary for the present, and the forget gate is used to remove information from the context when it is no longer needed. This model helps the model classify the text better because it can capture contextual information in the entire text.

GRU [1]: Gated Recurrent Unit (GRU) is a variant of the LSTM model. This model has lower complexity than the LSTM. While the LSTM has three gates, the GRU has only two gates: the update and reset gates. The update gate determines if any past information is retained and used in the future, and the reset gate decides that past information should be kept and any information forgotten. The advantage of GRU is using fewer parameters during training and therefore uses less memory, and training time is faster than LSTM.

Transformers is an architecture that has been proposed in recent years and is currently in widespread use. The appearance of BERT [3] helps many downstream tasks in NLP attain high-performance results while training on a small dataset. BERT and its variances become the baseline approaches in many NLP tasks, which is called BERTology [27].

In the Vietnamese language, there are two kinds of BERTology approaches: multilingual and monolingual models [7,28]. As a result, the monolingual obtained better results than the multilingual models for the text classification task [28], and sequence-to-sequence task [7]. Therefore, we applied two monolingual BERT models, including PhoBERT [17] and BERT4News [21] for our problem of detecting spam reviews.

5   Empirical Results

5.1   Baseline Results

We implement two tasks for the spam detection problem. First, we implement a binary classifier to classify reviews as spam or no spam. Second, we construct a classification model to determine spam types for the review. We adapt the Text CNN, LSTM, GRU models, and transformers with PhoBERT and BERT4News for our tasks. Finally, we will use the Accuracy and macro-averaged F1-score metrics to evaluate the performance of baseline models.

Table 8. The empirical results of classification models on the dataset

Model	Accuracy (%)		F1-macro (%)	
	Task 1	Task 2	Task 1	Task 2
Text-CNN	84.18	83.42	77.89	64.74
LSTM	82.97	83.35	77.24	66.58
GRU	83.50	82.84	77.67	66.51
PhoBERT	**90.01**	**88.93**	**86.89**	**72.17**
BERT4News	86.39	86.20	86.16	62.62

On classifying reviews as spam or not, the Text CNN model results on the test set with Accuracy and macro-averaged F1-score of 84.18% and 77.89%, better than the two models LSTM and GRU. Also, on this task, the PhoBERT achieved better results than BERT4News with Accuracy and macro-averaged F1-scores of 90.01% and 86.89%, respectively. For the spam type detection task, the PhoBERT model obtains 88.93% for Accuracy and 72.17% for macro-averaged F1-score, higher than BERT4News. The results of the evaluation between the models are described in Table 8.

Based on the results from the training and evaluation of the models, we can see that the classification of the evaluations using the Transformer models gives better performance than the deep neural network models. The PhoBERT model obtained the best performance on the two tasks. As for detecting spam types, the results of Accuracy and F1-score are more different than classifying reviews are spam or not because of the imbalance between the data labels. Besides, classifying the reviews as spam or not is less complicated than detecting the types of spam, so the results of this task are higher than the task of spam types detection.

5.2 Error Analysis

According to Fig. 2, the error prediction of SPAM and NO SPAM is not too much. In contrast, the error predictions are significantly different on the second task. Most of the SPAM-2 reviews are predicted as NO SPAM, and the number of wrong prediction (predicted as NO SPAM is 177) are higher than the accurate prediction (predicted as SPAM-2 is 135). The proportion of the wrong prediction of SPAM-1 reviews is also very high, in which most SPAM-1 reviews are predicted as no spam. However, this error prediction is not too much in the whole test set. The reviews with type SPAM-3 are the same as type SPAM-1. In general, most wrong predictions are caused by the doubt between NO SPAM label and other labels. Thus, the challenge of classification models on our dataset for this task is to determine whether the reviews are spam or not and to identify the type of spam reviews.

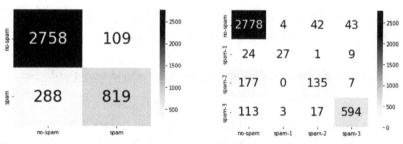

Task 1: Detecting spam or no spam   Task 2: Identifying type of spams

Fig. 2. Confusion matrices of PhoBERT model on two tasks. The confusion matrices are created by using the sklearn library

Table 9. Several wrong predictions of reviews with type SPAM-2 to NO SPAM label

No.	Reviews	True labels	Predicted labels
1	"Cảm ơn Tefal về chất lượng sản phẩm và cảm ơn Tiki về chất lượng dịch vụ. Chất lượng sản phẩm Tefal luôn vượt qua sự mong đợi của mình". (*English: "Thank you Tefal for product quality and thank you Tiki for service quality. Tefal product quality always exceeds my expectations".*)	SPAM-2	NO-SPAM
2	hợp với giá tiền.giao hàng nhanh (*English: suitable for the price.fast delivery*)	SPAM-2	NO-SPAM
3	Giao hàng nhanh, giá ổn (*English: Fast delivery, good price*)	SPAM-2	NO-SPAM
4	3 túi nước ariel 3,2 kg đã nhận.... chất lượng thì mình giặt đồ xong sẽ đánh giá sau. Thanhk Tiki và shop... (*English: 3 bags of Ariel 3.2 kg laundry detergent received.... the quality will be evaluated after I wash the clothes. Thanks Tiki and shop...*)	SPAM-2	NO-SPAM
5	Pop It là đồ chơi xả stress theo lời con gái khoe. Shop giao hàng nhanh, trao đổi nhiệt tình (*English: Pop It is a stress reliever according to what the girl said. Shop delivered products quickly, enthusiastic exchange*)	SPAM-2	NO-SPAM

To study the wrong prediction in SPAM-2 labels, we take several random examples with predicted labels by the highest classification model and compare them with the real label. Those examples are described in Table 9. According to Table 9, there are two main reasons for the wrong predictions. The first reason is the identification of the praise for the brand of the product and the brand of the retailer or provider, and the reviews about the quality of products. For example, reviews No #1, No #3, and No #4 mention the opinion about Tefal, Ariel (the two famous consumer goods brands in Vietnam), and brand Tiki (the online shopping service provider and retailer). However, those reviews do not focus on product quality, only express the thank to the retailer and the brands. The classification model cannot discriminate between the praise of the brands and the opinion of customers about the quality of products. The second reason is the short reviews of users, which are not giving any information about the products or services provider, such as reviews No #2 and No #3.

In general, the cons of current classification models on the dataset are the perplexity of the spam and no-spam reviews, in which the reviews about products must directly focus on the characteristic of the product, the quality of the products, and their services. To overcome this problem, the model should integrate extra information about the product, such as the information page about the product and the previous reviews of users, to enhance the classification ability.

6 Conclusion

This paper provided the ViSpamReviews - a dataset for spam reviews detection on Vietnamese online shopping websites with more than 19,000 reviews annotated by humans. The dataset follows the strict annotation process, and annotators are provided with detailed annotation guidelines for labeling the dataset. The final inter-annotator agreements are $\kappa = 0.72$ for the task of determining whether a review is spam or not and $\kappa = 0.68$ for the task of detecting the types of spam reviews. Besides, we also applied robust classification models to the dataset, and the PhoBERT model obtained the highest result with 86.89% by F1-score for the spam classification task and 72.17% by F1-score for the spam types detection task. From the error analysis, we found that it is necessary to integrate extra metadata about the product as well as the previous reviews to boost up the classification models.

Our next study is to extend the dataset for detecting spans of spam in the reviews and identify the opinion of users on the specific characteristic of products and their relevant services. Finally, based on the current results, the dataset can be used for developing an application to help shop owners filter spam reviews from users.

Acknowledgements. We would like to thank the annotators for their contribution to this work. This research was supported by The VNUHCM-University of Information Technology's Scientific Research Support Fund.

References

1. Cho, K., van Merriënboer, B., Bahdanau, D., Bengio, Y.: On the properties of neural machine translation: Encoder-decoder approaches. In: Proceedings of SSST-8, Eighth Workshop on Syntax, Semantics and Structure in Statistical Translation, pp. 103–111. Association for Computational Linguistics, Doha (2014)
2. Cohen, J.: A coefficient of agreement for nominal scales. Educ. Psychol. Measur. **20**(1), 37–46 (1960)
3. Devlin, J., Chang, M.W., Lee, K., Toutanova, K.: BERT: pre-training of deep bidirectional transformers for language understanding. In: Proceedings of the 2019 Conference of the North American Chapter of the Association for Computational Linguistics: Human Language Technologies, Volume 1 (Long and Short Papers), pp. 4171–4186. Association for Computational Linguistics, Minneapolis (2019)

4. Di Eugenio, B.: On the usage of kappa to evaluate agreement on coding tasks. In: Proceedings of the Second International Conference on Language Resources and Evaluation (LREC 2000). European Language Resources Association (ELRA), Athens (2000)

5. Grave, E., Bojanowski, P., Gupta, P., Joulin, A., Mikolov, T.: Learning word vectors for 157 languages. In: Proceedings of the Eleventh International Conference on Language Resources and Evaluation (LREC 2018). European Language Resources Association (ELRA), Miyazaki (2018)

6. Hochreiter, S., Schmidhuber, J.: Long short-term memory. Neural Comput. 9(8), 1735–1780 (1997)

7. To, N.N.L.T.H., Quoc, K.N.V., Gia-Tuan, A.N.: Monolingual vs multilingual BERTology for Vietnamese extractive multi-document summarization. In: Proceedings of the 35th Pacific Asia Conference on Language, Information and Computation, pp. 210–217. Association for Computational Lingustics, Shanghai (2021)

8. Huynh, H.D., Do, H.T.T., Nguyen, K.V., Nguyen, N.T.L.: A simple and efficient ensemble classifier combining multiple neural network models on social media datasets in Vietnamese. In: Proceedings of the 34th Pacific Asia Conference on Language, Information and Computation, pp. 420–429. Association for Computational Linguistics, Hanoi (2020)

9. Jindal, N., Liu, B.: Review spam detection. In: Proceedings of the 16th International Conference on World Wide Web, pp. 1189–1190 (2007)

10. Jindal, N., Liu, B.: Opinion spam and analysis. In: Proceedings of the 2008 International Conference on Web Search and Data Mining, pp. 219–230 (2008)

11. Kim, Y.: Convolutional neural networks for sentence classification. In: Proceedings of the 2014 Conference on Empirical Methods in Natural Language Processing (EMNLP), pp. 1746–1751. Association for Computational Linguistics, Doha (2014)

12. Landis, J.R., Koch, G.G.: The measurement of observer agreement for categorical data. Biometrics 159–174 (1977)

13. Li, H., Chen, Z., Mukherjee, A., Liu, B., Shao, J.: Analyzing and detecting opinion spam on a large-scale dataset via temporal and spatial patterns. In: Ninth International AAAI Conference on Web and Social Media (2015)

14. Li, H., et al.: Bimodal distribution and co-bursting in review spam detection. In: Proceedings of the 26th International Conference on World Wide Web, pp. 1063–1072 (2017)

15. Li, J., Ott, M., Cardie, C., Hovy, E.: Towards a general rule for identifying deceptive opinion spam. In: Proceedings of the 52nd Annual Meeting of the Association for Computational Linguistics (Volume 1: Long Papers), pp. 1566–1576. Association for Computational Linguistics, Baltimore (2014)

16. Luc Phan, L., et al.: SA2SL: from aspect-based sentiment analysis to social listening system for business intelligence. In: Qiu, H., Zhang, C., Fei, Z., Qiu, M., Kung, S.-Y. (eds.) KSEM 2021. LNCS (LNAI), vol. 12816, pp. 647–658. Springer, Cham (2021). https://doi.org/10.1007/978-3-030-82147-0_53

17. Nguyen, D.Q., Tuan Nguyen, A.: PhoBERT: pre-trained language models for Vietnamese. In: Findings of the Association for Computational Linguistics: EMNLP 2020, pp. 1037–1042. Association for Computational Linguistics (2020)

18. Nguyen, H.T., et al.: VLSP shared task: sentiment analysis. J. Comput. Sci. Cybern. 34(4), 295–310 (2018)

19. Nguyen, K.T.T., Huynh, S.K., Phan, L.L., Pham, P.H., Nguyen, D.V., Van Nguyen, K.: Span detection for aspect-based sentiment analysis in Vietnamese. arXiv preprint arXiv:2110.07833 (2021)

20. Nguyen, N.T.H., Phan, P.H.D., Nguyen, L.T., Van Nguyen, K., Nguyen, N.L.T.: Vietnamese open-domain complaint detection in e-commerce websites. arXiv preprint arXiv:2104.11969 (2021)
21. Nguyen, T.C., Nguyen, V.N.: NLPBK at VLSP-2020 shared task: compose transformer pretrained models for reliable intelligence identification on social network. arXiv preprint arXiv:2101.12672 (2021)
22. Ott, M., Choi, Y., Cardie, C., Hancock, J.T.: Finding deceptive opinion spam by any stretch of the imagination. In: Proceedings of the 49th Annual Meeting of the Association for Computational Linguistics: Human Language Technologies, pp. 309–319. Association for Computational Linguistics, Portland (2011)
23. Pustejovsky, J., Stubbs, A.: Natural Language Annotation for Machine Learning: A Guide to Corpus-Building for Applications. O'Reilly Media, Inc. (2012)
24. Radovanović, D., Krstajić, B.: Review spam detection using machine learning. In: 2018 23rd International Scientific-Professional Conference on Information Technology (IT), pp. 1–4 (2018)
25. Ren, Y., Ji, D.: Learning to detect deceptive opinion spam: a survey. IEEE Access 7, 42934–42945 (2019)
26. Ren, Y., Zhang, Y.: Deceptive opinion spam detection using neural network. In: Proceedings of COLING 2016, the 26th International Conference on Computational Linguistics: Technical Papers, pp. 140–150 (2016)
27. Rogers, A., Kovaleva, O., Rumshisky, A.: A primer in BERTology: what we know about how BERT works. Trans. Assoc. Comput. Linguist. 8, 842–866 (2020)
28. Van Thin, D., Le, L.S., Hoang, V.X., Nguyen, N.L.T.: Investigating monolingual and multilingual BERT models for Vietnamese aspect category detection. arXiv preprint arXiv:2103.09519 (2021)
29. Van Thin, D., Nguyen, N.L.T., Truong, T.M., Le, L.S., Vo, D.T.: Two new large corpora for Vietnamese aspect-based sentiment analysis at sentence level. ACM Trans. Asian Low-Resour. Lang. Inf. Process. 20(4), 1-22 (2021)
30. Xie, S., Wang, G., Lin, S., Yu, P.S.: Review spam detection via temporal pattern discovery. In: Proceedings of the 18th ACM SIGKDD International Conference on Knowledge Discovery and Data Mining, pp. 823–831 (2012)

v3MFND: A Deep Multi-domain Multimodal Fake News Detection Model for Vietnamese

Cam-Van Nguyen Thi$^{(\boxtimes)}$, Thanh-Toan Vuong, Duc-Trong Le, and Quang-Thuy Ha

Vietnam National University (VNU), University of Engineering and Technology (UET), Hanoi, Vietnam
{vanntc,18021279,trongld,thuyhq}@vnu.edu.vn

Abstract. Fake news become a critical problem on the Internet, especially social media. During the worldwide COVID-19 epidemic, social networking sites (SNSs) are primary sources to spread false news, which are incredibly difficult to detect and regulate them since they rapidly grow everyday. With multimedia technology advances, the content of social media news now is manifested via various modalities, such as text, photos, and videos. Approaches that learn the multimodal representation for detecting fake news have evolved in recent years. Additionally, there exist diverse content domains in news platforms. Exploiting data from these domains potentially solve the data sparsity problem as well as simultaneously boosting overall performance. In this paper, we propose an effective Deep Multi-domain Multimodal Fake News Detection model for Vietnamese, **v3MFND** for short. Extensive experiments on a real-life dataset reveal that **v3MFND** improves the performance of multi-domain multimodal fake news detection for Vietnamese considerably. An ablation study is also carried out to evaluate the role of each individual modality in the multimodal model.

Keywords: Vietnamese fake news detection · Multimodal · Multi-domain

1 Introduction

Fake news has been confounded with phrases, e.g., rumor, false news, and misinformation, however there is no an universal definition [15]. The dissemination of fake news through the Internet has grown into a scourge of our time, especially through social networks in the news media. According to the Digital 2021 research[1], there are around 72 million Vietnamese users on board with the blazing-fast surge of SNSs, e.g., Facebook, Zalo, or Lotus, with a growth of 7 million accounts over the same period in 2020, equating to a penetration rate of 73.7%. SNSs are also an ideal platforms to spread fake news during the outbreak of the global Coronavirus (COVID-19) pandemic. It is extremely difficult to

---

[1] https://datareportal.com/reports/digital-2021-global-overview-report.

© The Author(s), under exclusive license to Springer Nature Switzerland AG 2022
N. T. Nguyen et al. (Eds.): ACIIDS 2022, LNAI 13757, pp. 608–620, 2022.
https://doi.org/10.1007/978-3-031-21743-2_49

detect and control these fake news when they grow exponentially everyday [11]. Once fake news can also cause extremely unpredictable consequences for society in general and people in particular, fake news detection becomes an important and urgent problem that needs to be addressed.

With the modern multimedia technology, one trend that cannot be overlooked is that more and more social media news now includes information in several modalities, such as text, images, and videos. Multiple information modalities provide greater proof of news events occurring and more attractive to readers, hence this kind of fake news is harder to detect and widespread easier. Methods for detecting fake news have steadily progressed from unimodal to multimodal techniques in recent years. The question of how to learn a joint representation that includes multimodal information has sparked a lot of interest in the scientific community. As an example, Shivangi et al. [14] introduce SpotFake, a multimodal framework for fake news detection using BERT [1]. As a variant improved from SpotFake, they propose another novel transfer learning-based fake news detection method named SpotFake+ [13], which is mainly based on XLNet [20]. In [4], Khattar et al. employs a bimodal auto-encoder in conjunction with a conditional classifier to build the MAVE model. In order to extract text and visual representation, they exploit bi-directional LSTMs and VGG-19 respectively. Probabilistic latent variable models are trained via maximizing the marginal probability of observed data. Song et al. [15] leverage an attention method to fuse a number of word embeddings and one image embedding to create fused features, and then extract essential features as a joint representation of the fused features. Generally, the modeling multiple modalities is efficient to improve the fake news detection rate.

In real-world scenarios, news platforms provide a variety of news in many disciplines, i.e., domains, on a daily basis. Potentially, levering data from these domains may help alleviate the data sparsity problem while also improving the performance of all domains. This raises the notion of multi-domain fake news detection (MFND).However, major domain shift and lack of labeled data in certain domains are two serious problems that make MFND becomes more challenging [19]. As an effort to solve MFND, Quiong Nan et al. [9] propose a model named MDFEND, whereby uses domain gate to combine different representations retrieved by mixture-of-experts in order to cope with multi-domain transfer and isolation. The limitation of MDFEND is just use the text feature while ignoring the connected images in each news.

Overall, these mentioned points motivate us to exploit the fake news detection problem in the direction of multi-domain multimodal methods. Because there is a dearth of multi-domain multimodal dataset for Vietnamese in fake news detection problems, it is critical to build an appropriate dataset. In this paper, we proposed a model named **v3MFND: A Deep Multi-domain MultiModal Fake News Dectection Model for Vietnamese.** Our main contributions are as follows:

1. Construct a multi-domain multimodal dataset for Vietnamese fake news detection up on the ReINTEL dataset [6], a fake news dataset on Vietnamese SNSs.

2. Propose a multi-domain multimodal automatic fake news detection model for Vietnamese using advanced deep learning methods for text features and image features.
3. Conduct extensive experiments to assess the role of each individual modality in multimodal model.
4. Provide useful insight into several aspects of our approach for future research.

The rest of the paper is organized as follows. Section 2 delves into the details of the labeled Vietnamese multidomain multimodal fake news detection dataset. This is followed by a discussion of the proposed model in Sect. 3. In Sect. 4, we describe how we set up our experiment and go through the results in depth. Finally, we conclude the paper with Sect. 6.

2   M2-ReINTEL: Vietnamese Multi-domain Multimodal Dataset

ReINTEL dataset[2] is a fake news dataset collected from Vietnamese SNSs [6]. The most straightforward information, i.e., news content, is gathered for each item of news, as well as different modalities, i.e., photos, sequential signals, i.e., timestamp, and social context, i.e., number of likes, number of shares. Table 1 summarises the characteristics of this dataset. It is worth noting that, in addition to the text of the news gathered, the urls associated with the photos used in the news is also provided; however, not all news feature photos.

Table 1. Example of common features in ReINTEL dataset

Feature	Example	Explaination
id	1952_public	unique id for a news post on SNSs
uid	7d31f701ce1abfc1fc 0b7b311debc99d	the anonymized id of the owner
text	The State Bank of Vietnam reduced a series of operating interest rates from March 17. (Ngân hàng Nhà nước giảm hàng loạt lãi suất điều hành từ 17/3)	the text content of the news
timestamp	1584336550	the time when the news is posted
image_links	_	image urls associated with the news
nb_likes	4	the number of likes
nb_comments	0	the number of comment
nb_shares	0	the number of shares
label	0	unreliable news are labeled 1, reliable news are labeled 0

[2] https://vlsp.org.vn/vlsp2020.

The dataset includes 4825 news, which only 3583 news have visual features. All of news are manually labeled with ten domains namely *science, health, politics, education, economics, disaster, military, sports, entertainment, society.* We have two experts who independently label each record, then do a match between these two results. For labels that do not agree, we review the record together and come up with a final label. However, because the content of the news is quite long, the domains covered in those news come from many different domains, it is quite difficult to assign a label. We selected the more popular domain among the domains considered. In the future, multi-label label and simultaneous identification may be considered for implementation.

Because this data was gathered during the COVID-19 epidemic in Vietnam, there are a lot of linked articles, which is why the disaster domain is so popular. A simple statistic is shown in Table 2. To make identification and future referencing easier, we gave this multi-domain multimodal dataset for fake news detection the name **M2-ReINTEL**.

Table 2. The statistics of dataset M2-ReINTEL

Domain		Science	Health	Politics	Education	Economic
Real	no_image	66	62	92	92	223
	has_image	6	30	15	54	33
Fake	no_image	2	14	31	7	18
	has_image	0	4	10	1	5
Total		74	110	148	154	279
Domain		Disaster	Military	Sport	Entertainment	Society
Real	no_image	902	19	62	35	1443
	has_image	433	3	35	30	371
Fake	no_image	274	3	0	1	237
	has_image	107	1	1	6	97
All		1716	26	98	72	2148

We can quickly identify a severe imbalance in this dataset by looking at the data statistics table. The difference in fake/real labels, the quantity of articles between domains, and the heterogeneity in the number of photos all contribute to this imbalance. Only popular domains have a large number of related articles such as: *society, disaster, economics, politics*. This is also true in practice, when a large amount of fake news that confuses public opinion originates from ordinary social concerns, breaking news related to the economic and political situation, and especially news about epidemics and disasters.

3 v3MFND: A Deep Multi-domain Multimodal Fake News Dectection Model for Vietnamese

In this section, we present our novel model named **v3MFND**: A Deep Multi-domain MultiModal Fake News Dectection Model for Vietnamese. Figure 1 depicts the overall architecture of **v3MFND**. The model mainly consists of four components: pre-processing phase, modalities representation learning, fusion layer, and a classification layer.

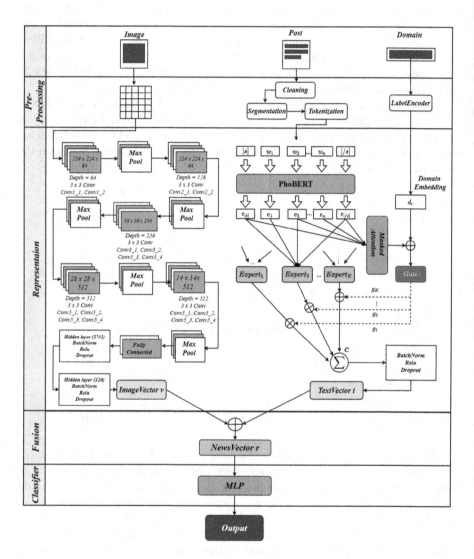

Fig. 1. An overview of v3MFND architecture.

Given a dataset $\mathcal{D}$, with input consisting of two modalities: $\mathcal{C}$ containing the contents of n news items C_i, $\mathcal{V}$ consists of l news' photos V_{ij}, $1 \leq i \leq n$, $= 1 \leq j \leq l$. The input dataset is fed through a separate pre-processing phase for each modality. Text data will be cleaned, normalized, then segmented using VNCoreNLP [18] and tokenized using PhoBERT Tokenizer [10]. Because one news may have zero or many images, a consistent procedure is applied for synchronization: (1) randomly select an image to be attached as the featured image in the post, (2) for posts without images, use a black image (value 0) as the attacked image, and (3) the images are then resized to $224 \times 224 \times 3$ size. We are also performing the label encoder with the domain data.

3.1 Multimodal Representation Learning

Textual Feature Extractor: For each news content C_i, the text content after going through pre-processing phase is tokenized into a sequence of tokens denoted as $W = [w_1, w_2, \ldots, w_n]$, where n is the total number of tokens. To correctly interpret input, we additionally add special tokens $[<s>]$ and $[</s>]$ into W to obtain new sequence $W_s = [[<s>], w_1, w_2, \ldots, w_n, [</s>]]$.

Pre-trained language models, particularly BERT [1] based on the architecture of Transformer [17], have lately gained a lot of traction and have achieved state-of-the-art improvements on different NLP tasks. PhoBERT [10] is the best pre-trained BERT model for Vietnamese based on RoBERTa [7], which optimized the BERT pre-training strategy for more robust results. Thus, instead of a vector representation of the text, we leveraged PhoBERT to extract word embeddings contain textual information, which is denoted as $E = [e_{[<s>]}, e_1, e_2, \ldots, e_n, e_{[</s>]}]$. Each word embedding e_i is a textual feature since it provides information about the entire text content.

In MDFEND [9], the authors leverage advantage of Mixture-of-Expert [3,8] to extract the news' representations for multiple domains. Because each expert specializes in a specific domain, the news representation retrieved by a single expert may only contain incomplete information, and hence may not fully describe the features of news content. Thus, it motivates us to use multiple experts networks, e.g., TextCNN [5] in **v3MFND**. A representation extracted by an expert network denoted as follows:

$$p_i = \Phi_i(E, \gamma_i) \tag{1}$$

where Φ_i is an expert network, $1 \leq i \leq K$, γ_i represents the parameters to be learned and K is a hyper-parameter that indicates the number of TextCNN model.

MDFEND [9] also employs a domain gate with the domain embedding as well as sentence embedding as input to guide the selection process. In order to assist with domain-specific representation extraction. We construct a domain embedding $D = [d_1, d_2, \ldots, d_m]$ where m is the total number of domains. This selection method produces a vector v representing each expert's weight ratio as:

$$g = SoftMax(Gate(D \oplus S; \rho)) \tag{2}$$

where $Gate(D \oplus S; \rho)$ is a domain gate feed-forward network, S is sentence-level embedding obtained from mask-attention network and ρ is the parameters in the domain gate. Softmax funtion is used to nomalize the output of $Gate(.; \rho)$. The news' final textual representation vector is obtained via:

$$t = \sum_{i=1}^{K} g_i p_i \qquad (3)$$

Visual Feature Extractor: VGG19 (a.k.a., VGGNet-19), a pre-trained model on the ImageNet dataset [12], is a variant of the VGG model which in short consists of 19 layers including 16 convolution layers, 3 Fully connected layers, 5 MaxPool layers and 1 SoftMax layer. In **v3MFND**, the VGG-19 model is used to learn different visual features. We extract the output of the VGG-19 convolutional network's second last layer, denoted as o. The final visual representation v is acquired by passing o via a fully connected layer to reduce down to a final dimension of length 14 as follows:

$$v = \sigma(W.o) \qquad (4)$$

where W is the weight matrix of the fully connected layer in the visual feature extractor.

3.2   Multimodal Fusion

To obtain the desired news representation r, the two feature vectors received from separate modalities are fused together using a simple concatenation approach. It is not only combines different visual features, but also reflects the dependencies between textual features, visual features and meta-data in the same news.

$$r = t \oplus v \qquad (5)$$

Additionally, we also conducts experiments with some other pseudo-combination methods besides concat, which are addition and average.

- **Sum:** the vectors are added together to produce a representative vector representing the entire news by the following formula:

$$r = t + r \qquad (6)$$

- **Average:** the vectors are averaged to give a representative vector representing the entire news using the following formula:

$$r = \frac{t + r}{2} \qquad (7)$$

3.3   Learning

The final feature vector of the news is fed into the classifier, which is a Multi-layer Perception (MLP) network with a SoftMax function to make prediction as follows:

$$\hat{y} = SoftMax(r) \qquad (8)$$

The fake news detector's purpose is to determine whether or not the news is fake, the loss function is set to Binary Cross-Entropy as follows.

$$L(\theta) = -ylog(\hat{y} - (1-y)log(1-\hat{y})$$ (9)

where θ indicates all of the proposed model's learnable parameters, and $y \in \{0,1\}$ signifies the ground-truth label.

4 Experimental Setup

4.1 Dataset

The ReINTEL dataset [6], which was utilized in the evaluation campaign for the 7th International Workshop on Vietnamese Language and Speech Processing (VLSP 2020) on detecting fake news on Vietnamese SNSs, was used to evaluate our model with the new version named M2-ReINTEL. Details of M2-ReINTEL dataset and how we proceed to label to obtain the final experimental dataset are described in Sect. 2 above.

4.2 Baselines

To our best knowledge, there is no multimodal multi-domain model has been proposed up to now. In order to evaluate the effectiveness of our proposed model v3MFND, we seek to compare against two baseline models namely MDFEND [9] (multi-domain model) and SpotFake [14] (multimodal model).

The dataset also provides metadata, which is also mined by many competing teams [16], so we conduct more experiments to evaluate the role of metadata in the model. We also examine different fusion operations, such as average and sum, in addition to the concat given in the model.

4.3 Experimental Settings

After tuning the proposed model on various settings to find the optimal value, the final settings is shown in Table 3.

4.4 Evaluation Metrics

For each experiment, we report Area Under the ROC Curve (AUC-ROC) as the performance measure. The fundamental reason is that due to data imbalance, traditional binary classification model assessment metrics including accuracy, precision, recall, and F-1 score [2] do not appropriately assess the model's efficacy. The performance is rounded to four digits.

Table 3. Parameters configuration

Parameters	10 domains	5 domains
num "expert"	6	3
Optimizer	Adam	
Learning rate	0.01	0.001
Batch size	32	
MLP	1 dense layer (384 unit) + batch norm 1d + relu + drop_out = 0.4	
Resize image	dense layer (2742 unit) + batch norm 1d + relu + drop_out = 0.4 dense layer (320 unit) + batch norm 1d + relu + drop_out = 0.4	
Resize metadata	dense layer (320 unit) + batch norm 1d + relu + drop_out = 0.4	
Norm text data (fusion)	batch norm 1d + relu + drop_out = 0.4	
Max len of sentence	170	
Word embedding vector dimension	768	

5   Results

In this section, we will report and discuss the experimental results of the multimodal multi-domain deep learning model applied to Vietnamese (v3MFND) and baseline models on M2-ReINTEL.

We conducted experiments using dataset on all 10 domains and the 5 domains with the largest amount of labels due to the imbalance of data regarding domains and labels (disaster, social, economics, health and politics). Notice that, in the model/fusion type column, for example v3MFND - concat, it means the v3MFND model with the fusion type is concat. With a value of "−", it means that the result cannot be calculated. The result values in bold are the highest values by domain, underlined the highest of the fusion types.

With the results presented in Table 4 and Table 5, we can see that the proposed model v3MFND, with concat in the fusion phase, performs well across the board. The MDFEND model scored 0.9753 and 0.9548 in the Disaster and Social domains, respectively, whereas v3MFN scored 0.9294, which was lower

Table 4. Results on 10 data domains with measurement AUC-ROC

Model/Fusion type		Disasters	Education	Entertainment	Economic	Health
MDFEND		**0.9753**	*1.0*	–	**0.9710**	0.875
SpotFake		0.9216	*1.0*	–	1.0	1.0
v3MDFN_{meta}	concat	0.9597	*1.0*	–	**0.9710**	0.875
	mean	0.9534	*1.0*	–	0.9565	1.0
	sum	0.9516	*1.0*	–	0.8406	1.0
v3MDFN_{img+meta}	concat	0.8324	*1.0*	–	0.3623	1.0
	mean	0.7903	*1.0*	–	0.6377	0.5
	sum	0.7126	*1.0*	–	0.5362	1.0
v3MDFN_{ProposedModel}	concat	0.7781	*1.0*	–	0.7826	1.0
	mean	0.8988	*1.0*	–	0.942	1.0
	sum	0.8313	*1.0*	–	0.7681	1.0
Model/Fusion type		Military	Politics	Science	Society	Sport
MDFEND		–	0.6429	–	**0.9548**	–
SpotFake		–	0.7302	–	0.8359	–
v3MDFN_{meta}	concat	–	0.8036	–	0.9277	–
	mean	–	0.6786	–	0.9243	–
	sum	–	**0.8750**	–	0.9294	–
v3MDFN_{img+meta}	concat	–	0.7679	–	0.7049	–
	mean	–	0.7857	–	0.6794	–
	sum	–	0.7143	–	0.6085	–
v3MDFN_{ProposedModel}	concat	–	0.6250	–	0.6696	–
	mean	–	0.6250	–	0.7356	–
	sum	–	0.7679	–	0.8286	–

Table 5. Average results across 10 data domains with measurement AUC-ROC

Model/Fusion type		All
MDFEND		**0.9576**
SpotFake		0.8872
v3MDFN_{meta}	concat	0.7178
	mean	0.7046
	sum	0.6669
v3MDFN_{img+meta}	concat	0.7195
	mean	0.7641
	sum	0.8198
v3MDFN_{ProposedModel}	concat	0.9404
	mean	0.9364
	sum	0.9399

than MDFEND's 0.0254. (2.6 %). All models projected properly in the Health and Education domains, which is owing to a severe data scarcity that caused the model to overfit. The SpotFake model received the highest score of 1.0 in the Ecocomics domain. The rationale is similar to that of Education and Health.

The v3MFND model with fusion as 'concat' produced the best results of 1.0 in the Political domain. The AUC - ROC measure cannot be used to calculate the results in the domains of entertainment, military, science, and sports because the number of news news items in these domains is very small, and the majority of these news items are real news. As a result, when dividing the data set, these domains only have real news.

The models that deal with the multi-domain problem, MDFEND and v3MFND, give better overall results than the SpotFake model. In addition, with the use of more metadata in the v3MFND model, the model not only ineffective, but also reduced the performance of the model. Besides, in the v3MFND model, in most domains, the 'concat' fusion gives higher results than other fusion methods.

Table 6. Results on 5 data domains with measurement AUC-ROC

Model/Fusion type		Disasters	Finance	Health	Politics	Society	All
$MDFEND$		0.9662	**1.0**	**1.0**	**0.8333**	0.8724	0.9263
$SpotFake$		0.944	0.9697	0.9394	0.7407	0.805	0.8876
$v3MDFN_{meta}$	concat	0.8006	0.6875	0.6	0.875	0.7546	0.7748
	mean	0.6935	0.625	0.65	0.8333	0.7544	0.6673
	sum	0.7504	0.9062	0.45	0.8333	0.7357	0.7309
$v3MDFN_{img\_meta}$	concat	0.7134	0.8125	0.5	**0.9583**	0.6413	0.6812
	mean	0.8728	0.9062	0.9	0.75	0.7904	0.8266
	sum	0.7169	0.6875	1.0	0.6667	0.706	0.72435
$v3MDFN_{ProposedModel}$	concat	**0.9754**	**1.0**	**1.0**	0.625	**0.8858**	**0.9375**
	mean	0.9594	**1.0**	**1.0**	0.7917	0.8332	0.9080
	sum	0.9657	**1.0**	**1.0**	0.625	0.8493	0.9157

In Table 6, we can see that our proposed model v3MFND as concat fusion gives the best results on most domains and on the entire evaluation set. Specifically, on the Disaster, Social, Economics and Health domain, the v3MFND model achieved 0.9754, 0.8858, 1.0 and 1.0, respectively, with the AUC-ROC measure. As for the entire evaluation set, the v3MFND model reached 0.9375.

From the above results table, we can also see that MDFEND and v3MFND models with multi-domain problem handling give better results in most domains and overall better results than SpotFake model. In addition, with the use of more metadata on the v3MFND model, the overall performance of the model was reduced, but on some domains such as Politics, the $v3MFND_{metadata}$ model gave the highest result of 0.9583 with a precision of 0.9583.

6 Conclusion

In this study, we investigate the problem of detecting multi-domain, multi-modal fake news for Vietnamese. As the first contribution, we build a multi-domain, multi-method fake news dataset for Vietnamese up on the ReINTEL dataset, whereby assigns domain names to news items. Subsequently, we propose a multi-domain, multi-method fake news detection framework for Vietnamese named v3MFND using advanced deep learning models to extract features for multi-modal data (text, images) and multi-domain data problem solving. Furthermore, extensive experiments on the newly constructed dataset demonstrate the effectiveness of v3MFND compared against baselines for the fake new detection task.

As future works, the challenge with learning data is that the existing data set employed by the model is still not diverse and general, thus additional data (in particular, fake news items) on some domains, such as sports, entertainment, military, and science, is required. The imbalance of data between domains and the imbalance of labels (fake news, real news) are issues we haven't solved yet. These imbalances occur in tandem rather than their own. Furthermore, when labeling, we discovered that certain items belong not just to one news domain, but to a group of domains. Therefore, our next research will focus on resolving multi-domain data imbalance issues, including not only data disequilibrium between domains but also label imbalance.

Acknowledgment. .

References

1. Devlin, J., Chang, M.W., Lee, K., Toutanova, K.: Bert: pre-training of deep bidirectional transformers for language understanding. arXiv preprint arXiv:1810.04805 (2018)
2. Hossin, M., Sulaiman, M.N.: A review on evaluation metrics for data classification evaluations. Int. J. Data Mining Knowl. Manage. Process 5(2), 1 (2015)
3. Jacobs, R.A., Jordan, M.I., Nowlan, S.J., Hinton, G.E.: Adaptive mixtures of local experts. Neural Comput. 3(1), 79–87 (1991)
4. Khattar, D., Goud, J.S., Gupta, M., Varma, V.: MVAE: multimodal variational autoencoder for fake news detection. In: The World Wide Web Conference, pp. 2915–2921 (2019)
5. Kim, Y.: Convolutional neural networks for sentence classification. CoRR abs/1408.5882 (2014). https://arxiv.org/abs/1408.5882
6. Le, D.T., et al.: ReINTEL: a multimodal data challenge for responsible information identification on social network sites. In: Proceedings of the 7th International Workshop on Vietnamese Language and Speech Processing, pp. 84–91. Association for Computational Lingustics, Hanoi, Vietnam (2020). https://aclanthology.org/2020.vlsp-1.16
7. Liu, Y., et al.: Roberta: a robustly optimized bert pretraining approach. arXiv preprint arXiv:1907.11692 (2019)

8. Ma, J., Zhao, Z., Yi, X., Chen, J., Hong, L., Chi, E.H.: Modeling task relationships in multi-task learning with multi-gate mixture-of-experts. In: Proceedings of the 24th ACM SIGKDD International Conference on Knowledge Discovery & Data Mining, pp. 1930–1939 (2018)
9. Nan, Q., Cao, J., Zhu, Y., Wang, Y., Li, J.: Mdfend: multi-domain fake news detection. In: Proceedings of the 30th ACM International Conference on Information & Knowledge Management, pp. 3343–3347 (2021)
10. Nguyen, D.Q., Nguyen, A.T.: Phobert: pre-trained language models for vietnamese. arXiv preprint arXiv:2003.00744 (2020)
11. Shu, K., Wang, S., Liu, H.: Beyond news contents: the role of social context for fake news detection. In: Proceedings of the twelfth ACM International Conference on Web Search and Data Mining, pp. 312–320 (2019)
12. Simonyan, K., Zisserman, A.: Very deep convolutional networks for large-scale image recognition. arXiv preprint arXiv:1409.1556 (2014)
13. Singhal, S., Kabra, A., Sharma, M., Shah, R.R., Chakraborty, T., Kumaraguru, P.: Spotfake+: a multimodal framework for fake news detection via transfer learning (student abstract). In: Proceedings of the AAAI Conference on Artificial Intelligence, vol. 34, pp. 13915–13916 (2020)
14. Singhal, S., Shah, R.R., Chakraborty, T., Kumaraguru, P., Satoh, S.: Spotfake: a multi-modal framework for fake news detection. In: 2019 IEEE Fifth International Conference on Multimedia Big Data (BigMM), pp. 39–47. IEEE (2019)
15. Song, C., Ning, N., Zhang, Y., Wu, B.: A multimodal fake news detection model based on crossmodal attention residual and multichannel convolutional neural networks. Inf. Process. Manage. 58(1) (2021)
16. Tuan, N.M.D., Minh, P.Q.N.: Reintel challenge 2020: a multimodal ensemble model for detecting unreliable information on vietnamese sns. arXiv preprint arXiv:2012.10267 (2020)
17. Vaswani, A., et al.: Attention is all you need. Adv. Neural Inf. Process. Syst. 30 (2017)
18. Vu, T., Nguyen, D.Q., Nguyen, D.Q., Dras, M., Johnson, M.: Vncorenlp: a vietnamese natural language processing toolkit. arXiv preprint arXiv:1801.01331 (2018)
19. Weiss, K., Khoshgoftaar, T.M., Wang, D.: A survey of transfer learning. J. Big Data 3(1), 1–40 (2016)
20. Yang, Z., Dai, Z., Yang, Y., Carbonell, J., Salakhutdinov, R., Le, Q.V.: Xlnet: generalized autoregressive pretraining for language understanding (2019). https://doi.org/10.48550/ARXIV.1906.08237, https://arxiv.org/abs/1906.08237

Social Networks and Recommender Systems

Fast and Accurate Evaluation of Collaborative Filtering Recommendation Algorithms

Nikolaos Polatidis[1]([✉]), Stelios Kapetanakis[1], Elias Pimenidis[2],
and Yannis Manolopoulos[3]

[1] School of Architecture, Technology and Engineering, University of Brighton, Brighton BN2
4GJ, UK
{N.Polatidis,S.Kapetanakis}@Brighton.ac.uk
[2] Department of Computer Science and Creative Technologies,
University of the West of England, Bristol BS16 1QY, UK
Elias.Pimenidis@uwe.ac.uk
[3] Faculty of Pure and Applied Sciences, Open University of Cyprus, 2220 Nicosia, Cyprus
Yannis.Manolopoulos@Ouc.ac.cy

Abstract. Collaborative filtering are recommender systems algorithms that provide personalized recommendations to users in various online environments such as movies, music, books, jokes and others. There are many such recommendation algorithms and, regarding experimental evaluations to find which algorithm performs better a lengthy process needs to take place and the time required depends on the size of the dataset and the evaluation metrics used. In this paper we present a novel method that is based on a series of steps that include random subset selections, ensemble learning and the use of well-known evaluation metrics Mean Absolute Error and Precision to identify, in a fast and accurate way, which algorithm performs the best for a given dataset. The proposed method has been experimentally evaluated using two publicly available datasets with the experimental results showing that the time required for the evaluation is significantly reduced, while the results are accurate when compared to a full evaluation cycle.

Keywords: Recommender systems · Collaborative filtering · Evaluation · Mean absolute error · Precision

1 Introduction

Recommender systems are algorithms that are based on opinions of a community of users to provide personalized recommendations of items, such as movies, books, jokes and music among others to users [1, 2]. One of the most successful technologies to provide such recommendations to users is Collaborative Filtering (CF), an approach that is based on a history of common ratings between users. When common ratings exist between users then, in its basic approach, a distance is calculated between users to form a neighborhood of common users using distance metrics such as the Pearson Correlation Coefficient (PCC), the Cosine or the Jaccard similarities [1–3]. However, numerous CF

© The Author(s), under exclusive license to Springer Nature Switzerland AG 2022
N. T. Nguyen et al. (Eds.): ACIIDS 2022, LNAI 13757, pp. 623–634, 2022.
https://doi.org/10.1007/978-3-031-21743-2_50

algorithms have been developed in the past few years that improve the quality of the recommendations in one way or another, usually for a specific domain.

The challenge in this area comes in clearly identifying which is the best algorithm. This can take place using an online approach where users are requested to evaluate a system with the click rate being counted, or in an offline approach where evaluation metrics are being used [4]. In the later approach, which is the most common amongst researchers, there should be a suitable volume of data and the selected evaluation metrics should be appropriate for the task. Based on the volume of the data several experimental evaluation rounds need to take place to identify which algorithm is the best [1, 4]. Evaluating recommender systems using offline metrics is usually based on accuracy metrics such as the Mean Absolute Error (MAE) and the Root Mean Squared Error (RMSE), or on information retrieval metrics such as Precision, Recall, and F1 among others [4]. The drawback of this approach is that depending on (a) how many algorithms need to be tested, (b) the volume of the data, (c) how many metrics will be used, and (d) how many tests are considered enough, the process could prove very time-consuming. It might take from a few hours to several days to execute all the tests, collect the results and go through a manual comparison to conclude which algorithm performs best. To this extent we have developed a randomization-based method that is both practical and effective in recommending a ranked list of collaborative filtering algorithms in a time efficient way. The contributions of the paper are:

- A method for fast evaluation of collaborative filtering algorithms, based on random subset data selection is delivered.
- The proposed method has been evaluated using two publicly available datasets and well-known evaluation metrics.

The rest of the paper is structured as follows: Sect. 2 contains the related work, Sect. 3 delivers the proposed method, Sect. 4 presents the experimental evaluation and Sect. 5 is the conclusions.

2 Related Work

This section is divided into two parts. The first part presents several CF algorithms found in the literature and the second part explains the offline evaluation approaches available in the domain. In the literature there are numerous CF approaches and various evaluation metrics which usually results in a time-consuming procedure.

2.1 Collaborative Filtering Algorithms

The most traditional and widely used CF algorithms are the ones that form a neighborhood of similar users based on a history of common ratings using a distance metrics such as the PCC or Cosine similarity [1]. The method is usually referred as K nearest neighbors (KNN) where K is the number of neighbors assigned to each user. This traditional method has been extended by Polatidis and Georgiadis [2] where the similarity is divided into multiple levels according to the number of co-rated items and the actual similarity value provided by PCC, while the accuracy and the precision is improved. This work was extended in Polatidis and Georgiadis [5] to create the multiple levels dynamically using information such as the number of users, items, and ratings. The previous dynamic multi-level work has been extended by Shojaei and Saneifar [3] where the authors introduced a fuzzy model which outperforms the dynamic approach that it is based on.

Another work is the one from Anand and Bharadwaj [6] where the authors have utilized sparsity measures based on local and global similarities to improve recommendation quality. Other works include the one from Bobadilla et al. [7] where the authors proposed a new metric that can be used as an alternative to PCC or Cosine that considers both common and uncommon ratings. This method has been validated and it was shown that recommendation quality is improved in terms of accuracy and precision/recall. Another metric from Bobadilla et al. [8] uses singularities to improve the quality of the recommendations.

There are other methods in the literature that aim to improve recommendations in different ways. RF-Rec is such a method that is based on rating frequencies to provide fast and accurate recommendations, while it outperforms the traditional baselines [9]. The method from Liu et al. [10] analyses the disadvantages of traditional recommendation methods such as PCC and Cosine and proposes a metric that is based on Proximity, Impact and Popularity (PIP) which improves the provided recommendation list. Najafabadi et al. [11] proposed a similarity metric that is based on clustering and association rule mining to improve the accuracy.

CF algorithms work in different ways such as using ontologies and dimensionality reduction techniques as proposed by Nilashi et al., [12]. Sarwar et al. [13] introduced the concept of Incremental Singular Value Decomposition (SVD) algorithms that makes recommender systems scalable. HU-FCF is a hybrid user-based fuzzy collaborative filtering method that uses fuzzy logic to improve the accuracy [14]. A work of further interest is the one from Wang et al. [15] where the authors developed a new metric that utilizes entropy to improve recommendations. A more recent work is one that uses a neural network to deliver neural CF and improved quality in the recommendations [16]. Xiaojun [17] delivered an improved collaborative filtering recommendation algorithm that is based on clustering. Dimensionality reduction and clustering have also been used in [18]. AutoSVD ++ is a method where CF is delivered with the use of contractive auto-encoders which improves the quality of the recommendations [19].

2.2 Evaluation Methods

Recommender systems evaluation using offline evaluation metrics is usually divided into two parts, accuracy, and retrieval.

The first part is to calculate the error of the predictions made using either the Mean Absolute Error (MAE) or the Root Mean Squared Error (RMSE) evaluation metrics [1, 2, 4]. Both metrics work in a similar way: In MAE when a rating prediction is made for a user the predicted value is compared to the actual value and an error value is calculated. The smaller the error value is, the better the rating prediction algorithm. In RMSE the difference is that the error is squared, thus larger error values are more harshly punished.

The second part uses Information Retrieval (IR). Here metrics are being used to evaluate how good a recommendation list is. Metrics such as Precision, Recall, F1, Mean Average Precision (MAP) and the Mean Reciprocal Rank (MRR) are commonly used for evaluation. This case is different to the previous one, with higher values being better. Values are in the range of 0 to 1 or 0 to 100 when converted to a % scale [1, 2, 4].

During an evaluation process different metrics are typically used. Researchers perform experiments using different algorithms, they collect results, and evaluate them based on those metrics. Accuracy and IR metrics are well established in the recommender systems community and are used to evaluate algorithms in several rounds of experiments by using different settings for each. While the correct approach is to follow such an experimentation procedure the drawback here is that it is very time consuming, and the reproducibility of the results is often difficult since in published research papers settings used in algorithms and experiments are not explained in detail. The proposed method fills this gap by automating the evaluation process in a fast and accurate approach with default algorithmic and evaluation settings.

3 Proposed Method

The novelty of the proposed method is in the random selection of an N number of subsets from the dataset, evaluating and combining the results using an ensemble approach. Each of the N subsets selected is independent from any other, which means that the method does not run concurrently but procedurally, and each time a new execution is running steps 1 to 6 of the method as follows: The MAE value is calculated, converted to a 1 to 100% and then the accuracy is calculated by subtracting the error value from 100. Then the precision is calculated, and the two metrics are combined to form a new metric that gives a final output value between 0 and 1. This value is then used to rank CF algorithms and recommend a list of CF algorithms with the ones having higher values appearing higher in the list. The above steps run independently from each other for N times and at the final step shown in Eq. 7 an ensemble of ranking values is calculated using a soft voting ensemble approach. In the equation n is the number of recommended items, p is the predicted rating and r is the actual rating.

$$MAE = \frac{1}{n} \sum_{i=1}^{n} |p_i - r_i| \tag{1}$$

Convert MAE value to a 0 to 100 scale as follows:

$$MAE100 = \frac{100}{Max\ rating\ value} \tag{2}$$

At the next step we calculate the error in a % scale as follows:

$$Error100 = MAE * MAE100 \tag{3}$$

Following on, we calculate the accuracy of the algorithm in terms of rating prediction as follows:

$$Accuracy = 100 - Error100 \tag{4}$$

The precision is calculated using Eq. 5. Precision is a value from 0 to 100 which tells us how good a list of recommendations is. Good recommendations are usually provided to users that satisfy a minimum rating criterion such as 4 out of 5 and "all recommendations" are all the recommendations provided to each user.

$$Precision = \frac{Good\ recommendations}{All\ recommendations} \qquad (5)$$

The next step of the method involves the combination of the Accuracy (MAE) and Precision values now that both are in a positive 0 to 100% value as shown in Eq. 6.

$$Combine = 2 * \frac{Accuracy * Precision}{Accuracy + Precision} \qquad (6)$$

The value obtained from Eq. 6 is then used to rank each collaborative filtering algorithm. The higher the value the higher in the list the algorithm appears. Once this step has finished a voting ensemble algorithm is used as shown in Eq. 7. At this step the algorithm that gathers the highest ensemble value appears higher in the list, since a re-ranking process takes place.

$$Ensemble = \frac{Combine1 + Combine2 + \ldots \ldots \ldots Combine N}{N} \qquad (7)$$

4   Experimental Evaluation

This section explains the experiments and includes subsections with the settings, datasets, algorithms, and results. We found that a subset of about 10% of users includes ratings for many of the items which results to similar outputs using MAE and Precision compared to when evaluating with the full dataset and we used three subsets. Therefore, N = 3 with approximately 10% of users with all their ratings for each dataset and the values are rounded up or down to the closest value. For example, from the Epinions dataset 4000 of 40163 have been used and for the MovieTweetings dataset 2000 of 21645 users have been selected. It is shown that in each evaluation cycle of each of the datasets, even when a subset of users is used, all data items are being used in the process. The Java programming language has been used in an Intel Dual core i7 2.5 GHz with 8 gigabytes of RAM computer running Windows 10. 5-fold cross validation has been used in all experiments.

4.1   Datasets

We used two publicly available datasets the statistical information of which is presented in Table 1. The datasets were chosen based on their size and specific characteristics such as the number of users and items.

Epinions: This is a general commerce dataset with more items than users and a rating scale from 1 to 5 [20].

Table 1. Dataset statistics

Dataset	Users	Items	Ratings	Scales
Epinions	40163	139738	664823	[1, 5]
MovieTweetings	21645	12989	150000	[1, 10]

MovieTweetings: This is a movies dataset with the details about the users, items and ratings crawled from Twitter and it rating scale is from 1 to 10 [21].

Figures 1 shows 10 different selections of random users for the Epinions, and MovieTweetings datasets respectively. It is shown that each time a random subset of users is selected the number of ratings remains similar.

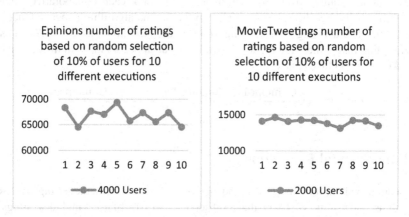

Fig. 1. Number of ratings is presented on the left side of each figure for 4000 users of the Epinions dataset and 2000 users of the MovieTweetings dataset for 10 random selections

4.2 Algorithms

Three algorithms have been used in the experiments and are described in detail below along with their settings.

KNN: This is the traditional user-user algorithm that forms a neighborhood of similar users using PCC with minimum similarity value of 0.0 and K number of neighbors equals 50 and a minimum overlap of 3 items.

Rf-Rec: This is an algorithm that generates predictions based on the counting and combination of different rating values [9].

Funk SVD: This is an SVD based algorithm that ignores missing values in the rating matrix [22].

4.3 Metrics

Three evaluation metrics have been used in the evaluation process. MAE, Precision and MRR.

MAE: This metric calculates the difference between an actual rating and a predicted rating. It has been defined in Eq. 1 since it is also used in the method.

Precision: This metric calculates the quality of the recommendations. It has been defined in Eq. 5 since it is also used in the method.

MRR: This is a metric that can be used to calculate the ordered probability of correctness. It is defined in Eq. 8 below, where Q is the number of queries executed and rank of "i" is the order in which the first most relevant answer appears within a list of ranked answers.

$$MRR = \frac{1}{|Q|} \sum_{1}^{|Q|} \frac{1}{rank_i} \tag{8}$$

MAE and Precision are used to calculate the accuracy and precision of the three algorithms against each of the datasets for 5 random 5-fold executions. Moreover, the same metrics are being used for 3 random executions of the proposed method to create a ranked list of the algorithms for each dataset. At the end, MRR is used to calculate if the best algorithm is ranked within the first place of the recommended list.

4.4 Results

Tables 2, 3, 4 and 5 present the results for the Epinions dataset. Tables 2 and 3 show the results for the whole dataset using MAE and Precision respectively for top-5 recommendations. Five different executions based on 5-fold cross validation and random data selection took place. The results are consistent, showing that, for MAE, Funk SVD performs the best among the three algorithms, RF-Rec second best and KNN third. For precision results RF-Rec performs the best, followed by Funk SVD and KNN. Tables 4 and 5 present three executions with k (4000 in this case) random users on each. Funk SVD is the best for MAE followed by Rf-Rec and KNN while for precision RF-Rec is the best followed by Funk SVD and KNN.

Tables 6, 7, 8 and 9 present the results for the MovieTweetings dataset. Tables 6 and 7 are the results for the whole dataset using MAE and Precision respectively for top-5 recommendations. Five different executions based on 5-fold cross validation and random data selection took place. The results are consistent, showing that, for MAE, KNN performs the best among the three algorithms, Funk SVD the second best, and RF-Rec the third. For precision results RF-Rec performs the best, followed by Funk SVD and KNN. Tables 8 and 9 present three executions with k (2000 in this case) random users each. KNN is the best for MAE followed by Funk SVD and RF-Rec. For precision RF-Rec is the best, followed by Funk SVD, and KNN.

Table 2. MAE results for Epinions

Methods	1st execution	2nd execution	3rd execution	4th execution	5th execution
KNN (PCC)	0.906	0.907	0.907	0.905	0.909
RF-Rec	0.867	0.867	0.868	0.867	0.867
Funk SVD	0.802	0.802	0.803	0.802	0.802

Table 3. Precision results for Epinions for top-5 recommendations

Methods	1st execution	2nd execution	3rd execution	4th execution	5th execution
KNN (PCC)	0.75	0.75	0.75	0.75	0.751
RF-Rec	0.809	0.808	0.808	0.809	0.808
Funk SVD	0.8	0.8	0.799	0.8	0.799

Table 4. MAE results for Epinions with 4000 users

Methods	1st execution	2nd execution	3rd execution
KNN (PCC)	0.966	0.946	0.945
RF-Rec	0.951	0.943	0.94
Funk SVD	0.88	0.855	0.852

Table 5. Precision results for Epinions with 4000 users for top-5 recommendations

Methods	1st execution	2nd execution	3rd execution
KNN (PCC)	0.765	0.759	0.776
RF-Rec	0.802	0.803	0.807
Funk SVD	0.792	0.793	0.799

Table 6. MAE results for MovieTweetings

Methods	1st execution	2nd execution	3rd execution	4th execution	5th execution
KNN (PCC)	2.526	2.526	2.525	2.528	2.524
RF-Rec	2.613	2.613	2.613	2.613	2.613
Funk SVD	2.569	2.57	2.57	2.57	2.569

Table 7. Precision results for MovieTweetings for top-5 recommendations

Methods	1st execution	2nd execution	3rd execution	4th execution	5th execution
KNN (PCC)	0.746	0.743	0.741	0.746	0.744
RF-Rec	0.85	0.849	0.849	0.848	0.849
Funk SVD	0.82	0.82	0.819	0.819	0.82

Table 8. MAE results for MovieTweetings with 2000 users

Methods	1st execution	2nd execution	3rd execution
KNN (PCC)	2.435	2.447	2.436
RF-Rec	2.587	2.597	2.644
Funk SVD	2.536	2.588	2.615

Table 9. Precision results for MovieTweetings with 2000 users for top-5 recommendations

Methods	1st execution	2nd execution	3rd execution
KNN (PCC)	0.737	0.749	0.737
RF-Rec	0.845	0.853	0.85
Funk SVD	0.817	0.826	0.823

Table 10 presents the Mean Reciprocal Rank results for the datasets followed by the overall value. The recommendation list is the list of the algorithms recommended by the proposed method. The best result represents which is the best algorithm for the corresponding dataset, followed by the actual rank and the reciprocal rank value. Initially, by manually observing the results, one can clearly assess which algorithm is better in terms of MAE and Precision; while all executions use the full dataset and for all datasets the executions are similar to each other. Furthermore, by observing the results using a subset of each dataset, and by comparing algorithms using this approach, it can be observed which algorithm is the best. By randomly choosing subsets the values are not very similar, thus several executions are necessary to get average MAE and Precision values.

Table 10 presents the results of the best performing algorithm (best result) as derived by manually observing the results using the whole dataset while the recommendation list is generated using the proposed method. The results of Table 10 are based on the Mean Reciprocal Rank and are 100% accurate. This metric calculates if the best result is in the top position of the recommendation list. Since the rank is 1 for all three the reciprocal rank is also 1 and the Mean Reciprocal Rank is calculated as shown below Table 10.

Mean Reciprocal Rank: $(1 + 1)/2 = 1$ (100%).

Table 10. Mean Reciprocal Rank results

Dataset	Ranked list based on full executions	Best result based on the proposed method	Rank	Reciprocal rank
Epinions	Funk SVD, Rf-Rec, KNN	Funk SVD	1	1
MovieTweetings	RF-Rec, Funk SVD, KNN	RF-Rec	1	1

Figure 2 presents the performance evaluation comparison results for the Epinions and MovieTweetings datasets, using MAE, Precision, and the KNN, RF-Rec and FunkSVD recommendation algorithms for top-5 recommendations. The times for the first part (Epinions dataset) of the figure are in hours and for the second (MovieTweetings dataset) in minutes while these are approximate, which means that these might slightly vary according to how many items will be retrieved through the randomization process, the settings, processing power, operating system and background processes running.

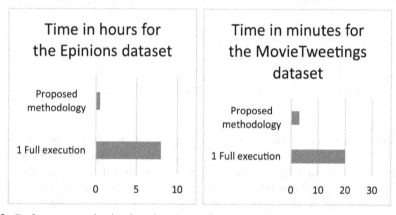

Fig. 2. Performance evaluation based on hours for the Epinions dataset and minutes for the MovieTweetings dataset

The results show that random selection of user subsets and combination of their results is a good approach when evaluating recommender systems. The evaluation and ranking of the algorithms with the use of subsets of the dataset delivers an accurate list of ranked and recommended CF algorithms, while the time required to do so is significantly reduced which is especially useful for larger datasets. For Epinions, a relatively large dataset, the proposed method takes about 30 min of processing time while a typical execution that uses the whole dataset takes about 8 h. For MovieTweetings, a smaller dataset, the proposed method requires about 3 min processing time whereas a typical round of evaluation using the whole dataset needs about 20 min. The proposed method can also be used as white box evaluation approach to automatically evaluate algorithms, but a limitation is the settings of the algorithms used.

5 Conclusions

Collaborative filtering has matured, and several algorithms can be found in the literature. Additionally, there are several evaluation metrics that can be used to evaluate such algorithms and are either related to accuracy or IR. CF is being used in various domains, and algorithms that are considered good in a domain might not be as good in another. To find out which algorithm is good a manual and time-consuming procedure of running experiments needs to take place. In this article we delivered an evaluation, ranking and recommendation method which can be used to evaluate CF algorithms using subsets of the dataset in a fast and accurate way. The proposed method has been tested on two publicly available datasets, the size of which is significantly different from each other, with the results indicating that the method is fast, accurate, while it is straightforward to use.

The proposed method is stable and can be used as a basis for a white box method which can automate the evaluation process and allow researchers to reproduce results without the worry of omitted settings and parameters. Thus, in the future we aim to investigate how the selection of subsets can assist in the evaluation of deep learning recommendation algorithms especially applying only IR metrics. In addition to that, we aim to deliver a white-box evaluation approach which researchers will be able to use for experimentation and will allow the experimental results to be reproduced in a straightforward way.

References

1. Bobadilla, J., Ortega, F., Hernando, A., Gutiérrez, A.: Recommender systems survey. Knowl. Based Syst. **46**, 109–132 (2013)
2. Polatidis, N., Georgiadis, C.K.: A multi-level collaborative filtering method that improves recommendations. Expert Syst. Appl. **48**, 100–110 (2016)
3. Shojaei, M., Saneifar, H.: MFSR: a novel multi-level fuzzy similarity measure for recommender systems. Expert Syst. Appl. **177**, 114969 (2021)
4. Herlocker, J.L., Konstan, J.A., Terveen, L.G., Riedl, J.T.: Evaluating collaborative filtering recommender systems. ACM Trans. Inf. Syst. **22**(1), 5–53 (2004)
5. Polatidis, N., Georgiadis, C.K.: A dynamic multi-level collaborative filtering method for improved recommendations. Comput. Stand. Interf. **51**, 14–21 (2017)
6. Anand, D., Bharadwaj, K.K.: Utilizing various sparsity measures for enhancing accuracy of collaborative recommender systems based on local and global similarities. Expert Syst. Appl. **38**(5), 5101–5109 (2011)
7. Bobadilla, J., Serradilla, F., Bernal, J.: A new collaborative filtering metric that improves the behavior of recommender systems. Knowl. Based Syst. **23**(6), 520–528 (2010)
8. Bobadilla, J., Ortega, F., Hernando, A.: A collaborative filtering similarity measure based on singularities. Inf. Process. Manage. **48**(2), 204–217 (2012)
9. Gedikli, F., Bagdat, F., Ge, M., Jannach, D.: RF-REC: Fast and accurate computation of recommendations based on rating frequencies. In: 2011 IEEE 13th Conference on Commerce and Enterprise Computing, pp. 50–57. IEEE (2011)
10. Liu, H., Hu, Z., Mian, A., Tian, H., Zhu, X.: A new user similarity model to improve the accuracy of collaborative filtering. Knowl. Based Syst. **56**, 156–166 (2014)

11. Najafabadi, M.K., Mahrin, M.N.R., Chuprat, S., Sarkan, H.M.: Improving the accuracy of collaborative filtering recommendations using clustering and association rules mining on implicit data. Comput. Hum. Behav. **67**, 113–128 (2017)
12. Nilashi, M., Ibrahim, O., Bagherifard, K.: A recommender system based on collaborative filtering using ontology and dimensionality reduction techniques. Expert Syst. Appl. **92**, 507–520 (2018)
13. Sarwar, B., Karypis, G., Konstan, J., Riedl, J.: Incremental singular value decomposition algorithms for highly scalable recommender systems. In: Fifth International Conference on Computer and Information Science, vol. 1, no. 012002, pp. 27–8 (2002)
14. Son, L.H.: HU-FCF: a hybrid user-based fuzzy collaborative filtering method in recommender systems. Expert Syst. Appl. Int. J. **41**(15), 6861–6870 (2014)
15. Wang, W., Zhang, G., Lu, J.: Collaborative filtering with entropy-driven user similarity in recommender systems. Int. J. Intell. Syst. **30**(8), 854–870 (2015)
16. Wang, X., He, X., Wang, M., Feng, F., Chua, T.S.: Neural graph collaborative filtering. In: Proceedings of the 42nd international ACM SIGIR conference on Research and development in Information Retrieval, pp. 165–174 (2019)
17. Xiaojun, L.: An improved clustering-based collaborative filtering recommendation algorithm. Clust. Comput. **20**(2), 1281–1288 (2017). https://doi.org/10.1007/s10586-017-0807-6
18. Zarzour, H., Al-Sharif, Z., Al-Ayyoub, M., Jararweh, Y.: A new collaborative filtering recommendation algorithm based on dimensionality reduction and clustering techniques. In: 2018 9th International Conference on Information and Communication Systems (ICICS), pp. 102–106. IEEE (2018)
19. Zhang, S., Yao, L., Xu, X.: AutoSVD++ an efficient hybrid collaborative filtering model via contractive auto-encoders. In: Proceedings of the 40th International ACM SIGIR conference on Research and Development in Information Retrieval, pp. 957–960 (2017)
20. Massa, P., Souren, K., Salvetti, M., Tomasoni, D.: Trustlet, open research on trust metrics. Scalable Comput. Pract. Exper. **9**(4) (2008)
21. Dooms, S., De Pessemier, T., Martens, L.: Movietweetings: a movie rating dataset collected from twitter. In: Workshop on Crowdsourcing and human computation for recommender systems. CrowdRec at RecSys, vol. 2013, p. 43 (2013)
22. Funk, S. (2006). https://sifter.org/~simon/journal/20061211.html

Improvement Graph Convolution Collaborative Filtering with Weighted Addition Input

Tin T. Tran[1,2]($\boxtimes$) and Václav Snasel[1]

[1] Faculty of Electrical Engineering and Computer Science, VŠB -Technical University of Ostrava, Ostrava-Poruba, Czech Republic
{trung.tin.tran.st,vaclav.snasel}@vsb.cz
[2] Faculty of Information Technology, Ton Duc Thang University,
Ho Chi Minh City, Vietnam
trantrungtin@tdtu.edu.vn

Abstract. Graph Neural Networks have been extensively applied in the field of machine learning to find features of graphs, and recommendation systems are no exception. The ratings of users on considered items can be represented by graphs which are input for many efficient models to find out the characteristics of the users and the items. From these insights, relevant items are recommended to users. However, user's decisions on the items have varying degrees of effects on different users, and this information should be learned so as not to be lost in the process of information mining.

In this publication, we propose to build an additional graph showing the recommended weight of an item to a target user to improve the accuracy of GNN models. Although the users' friendships were not recorded, their correlation was still evident through the commonalities in consumption behavior. We build a model WiGCN (Weighted input GCN) to describe and experiment on well-known datasets. Conclusions will be stated after comparing our results with state-of-the-art such as GCMC, NGCF and LightGCN. The source code is also included (https://github.com/trantin84/WiGCN).

Keywords: Recommender system · Collaborative signal · Graph neural network · Embedding propagation

1 Introduction

Recommendation systems are an important research area of Information Systems. They are widely used in e-commerce, advertising and social media to give suggestions to the target user, and thus, improve the user experience. Based on previous interactions, such as purchases or reviews, of multiple users on items, the system will look for features and point out similar users and similar items to offer a set of recommendations combine products and recommendations to target customers. Collaborative filtering (CF) is a method of assessing the similarity between users based on a series of past behavior that has been applied

© The Author(s), under exclusive license to Springer Nature Switzerland AG 2022
N. T. Nguyen et al. (Eds.): ACIIDS 2022, LNAI 13757, pp. 635–647, 2022.
https://doi.org/10.1007/978-3-031-21743-2_51

in [1, 2]. In this method, the user and item features are represented by vectors. The distance of vectors in the calculating space represent how similar the users or items is. By this method, the items do not need to specify the attributes nor need to collect the user's personal information.

The user-item relationship matrix is usually a sparse matrix, because the number of items a user has purchased or is interested in accounts for only a very small portion of all items. Reducing the number of matrix dimensions helps to create embedded matrices of much smaller size but without losing the features of users and items [3]. Furthermore, the Numpy [4] and TensorFlow libraries [5] can efficiently handle very large sparse matrices without taking up a lot of computer memory in execution.

In addition to traditional recommendations, social recommendations are systems that consider the social relation between users. Users can also be friends with each other in real life or on social networks, a friend's advice is always taken with higher trust. The methods outlined in [6, 7] mined a relational database of friends to enrich recommendation information. However, a friendship database is not always available. Social networking services exploit user relationships only, while e-commerce sites have item review data and treat users as independent entities.

On the other hand, the user and item relations can also be naturally represented by graphs, and can be exploited by Graph neural network [8]. The process of capturing collaborative signals can go through two stages, decomposer and combiner as suggested at [9], where decomposer will capture signals from graph and then combiner will combine them into one unified embedding. High-order connectivity is found in the NGCF model [10] where propagating embedding is used when mining the user-item relationship graph. Recently, the LightGCN model has removed the weights when receiving the signal from the graph into the embedded vectors but still retains the accuracy of the algorithm [11].

In this publication, we propose a model that uses Graph neural network to receive signals from two separate matrices, that is, the implicit matrix that records the user's attention on the item and the weight matrix number of interactions between users. Although there is no data on social relationships between users, mutual interest in a set of items also constitutes strong signals between users. We also implement this model and perform experiments on common data sets to evaluate and compare with state-of-the-art algorithms.

2   Related Works

2.1   Collaborative Filtering

From the beginning, recommendation systems exploited the characteristics of all items as well as the preferences of users to find suitable items and make recommendations to them [12]. For example, a movie can be categorized into action, comedy, romance, etc; a toy can have properties of color, shape, age, etc. With users, information such as age, gender, address of residence or education could also used to enrich the input data [13]. However, with a very large number of items and frequent additions, it is difficult to find out and give items attributes, the systems need additive filtering methods. Collaborative filtering approach algorithms were used to find similarity between users or items without attribute information from them [14].

Collaborative filtering can be implemented by memory-based models [15] or model-based models [16]. With memory-based model, the history of every user's rating on items will be recorded in a rating matrix. There are many ways to define a rating scale, which can be an integer value from 1 to 5, or an implicit rating. In a space made up of user set, item set, and rating values, each user is represented by a feature vector e_u, also as each item is represented by feature vector e_i. The algorithms will find out the distance between each pair of users (or pair of items for item-based algorithms) to find the neighbors and form the recommended results. The distance can be Cosine, Pearson, Euclidean similarity or Mean Squared Differences [17]. The weakness of these methods is that the input matrix is very large to be stored in the computer memory and the matrix is sparse because a user usually rates a few items.

In the model-based collaborative filtering systems, the algorithms will look for patterns in the learning data to create a model for future prediction [18]. Matrix analysis techniques can be applied to reduce the dimensional of the user rating data matrix on items. Rating matrix can be decomposed into user feature matrix and item feature matrix with smaller size; but still ensure the accuracy of distance between users as well as between items, and more accuracy on recommendation prediction results [19], scalability and easier data learning process.

The evaluative measure will be selected appropriately for each algorithm [20]. A commonly evaluating method is to extract part of the data set to make a test set, the rest as a training set. The algorithm is applied on the training set to make predictions that are evaluated on the test set. The difference between the learning result and the actual data value shows the accuracy of the algorithm. This difference can be expressed in mean absolute error (MAE) or root mean square error (RMSE). Besides accuracy, coverage, scalability, learning time, memory consumption or interpretability are also an important criteria in evaluating the recommended system.

$$MAE = (\frac{1}{n}) \sum_{i=1}^{n} |\widehat{y}_i - y_i| \quad \text{and} \quad RMSE = \sqrt{\frac{1}{n}\Sigma_{i=1}^{n}\left(\frac{\widehat{y}_i - y_i}{\sigma_i}\right)^2} \quad (1)$$

where $\widehat{y}_i$ is predicted rating while y_i is the value of that rating in the test set.

For implicit data, interactions between the user and the item are recorded binary, rather than rated as a specific value. Then, the algorithms will use the accuracy measure for the classification. The commonly used metrics are Precision and Recall [21,22]. Precision is the ratio between correct predictions on the test set, and Recall is the sensitivity of the algorithm, or the proportion of relational assertions that have been retrieved from the test set. The F1 score is used to balance accuracy and recall because the two are often opposite. Last but not least, discounted cumulative gain score (DCG) [23] assumes that judges have assigned labels to each result, and accumulates across the result vector a gain function G applied to the label of each result, scaled by a discount function D of the rank of the result, and it's normalized by the dividing DCG of an ideal result vector I.

$$Precision = \frac{TP}{TP + FP}$$
$$Recall = \frac{TP}{TP + FN}$$
$$F1 = \frac{2 * Precision * Recall}{Precision + Recall} \tag{2}$$
$$NDGC_u = \frac{DCG_u}{CDG_{max}}$$

where TP is true positive set of correctly predicting interaction exists between user and item. FP is false positive set of missing prediction of interaction while FN (False negative) shows that the predicted interaction doesn't exist on the test set.

2.2   Graph Neural Networks

Neural networks have had great success in setting up hidden layers to extract input data and identify features of the output data set [24]. A graph is a data structure that represents the relationship between entities, people, data, and attributes using a set of vertices V to represent an object and an set of edges E to represent the relationship. An edge of a graph can be undirected, representing only the relationship, or specifying the weight of the relationship, or might be a directed edge. Publication [25] have applied neural networks with graph input data and obtained positive results when vertex features are propagated and aggregated into each other during the learning process [26]. Inheriting from Neural Network algorithms, GNN also uses multiple propagation layers and applies different aggregation and update solutions. The properties of the neighbor vertex are updated and the target vertex by pooling or attention mechanisms [27].

GCN is a method of collecting repeated information [28]. For example, when there is a similar item of interest to many of the target user's friends, it should be recommended with a higher degree than other items. GraphSAGE [29] embedded inductive for each vertex of the graph and learned the topological structure of the graph as well as the effect of vertices on neighboring vertices. This method not only focus on feature-rich graphs but also make use of structural features of all vertices.

Multi Component Graph Convolutional [9] viewed a two-component input graph by separating the vertices representing the user as one group and the vertices representing the item as the other group. The features of the users are calculated to embed each other and so are the features of the items. The learning process is divided into two main stages: decomposer and combiner. In which, the decomposer component will obtain some M latent flows from the user-item relation matrix through M transformation matrices and aggregate into the user (or item) vertices. The combiner component will combine the feature vectors of the items on a particular user and those of the users on a particular item, and finally the system obtains a predictive rating on the output.

Assuming that the influence of the users is different depending on the distance between the vertices in the graph, Neural Graph Collaborative Filtering [10] generates hop-by-hop propagation classes in the input graph. The kth layer of propagation, called $k - order$ propagation, receives messages from items to a user as well as attention signals from that user's neighbors. The number of propagation layers k can be considered as the input parameter and the value $k = 3$ is considered to be the most optimal. The loss function is built from the difference of the predicted and actual evaluation values of the test set. The learning process to minimize the loss function takes place after a number of iterations depending on the learning rate, the size of the embedding matrices as well as the size of the learning data. In a subsequent version of NGCF, LightGCN [11] removed the characteristic transition weight matrices during propagation, and a non-linear activation function to remove negative effects on objects in the NGCF propagation process.

3 Proposed Method

In this chapter, we will present a recommendation model to learn the user and item characteristics at the input, and then predict the recommended outcome at the output. We also suggest the metric for the recommended results and how to build functions in the hidden layers as well as the loss function in the learning process.

3.1 Adjacent Graph and Weighted References Matrix

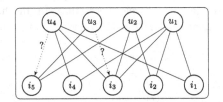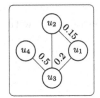

Fig. 1. Bi-parties users-items graph and graph of users' influences.

Due to history of interaction between users and items, we can model a bi-parties graph $G = \{V, E\}$, where set of vertices $V = \{U, I\}$ is union from set of users and set of items; the E contains edge $e_{i,j} = (u_i, i_j)$ if user u had interaction on item i. From the bi-parties graph G, we can define a matrix $R \subseteq U \times I$ and it's elements has a value of

$$R_{i,j} = \begin{cases} 1 & \text{user } u_i \text{ has an interaction on item } i_j \\ 0 & \text{otherwise} \end{cases} \tag{3}$$

The weighted user references matrix $W_U = R \times R^T$ shows how many common items that user u_i and u_j have interaction by value of $W_{U_{i,j}}$. Because the number of interactive items of each user are different, matrix W should by normalized by least absolute deviations. Similarly, the weighted item references matrix $W_I = R^T \times R$ indicates how many users the same two items were referenced by.

According to [10], the Laplacian matrices for user-item relation should be formed in propagation rule because all embedding vectors could be updated simultaneously. The first input Γ is a square matrix whose dimensions each are the sum of the number of users and the number of items.

$$\Gamma = D^{-\frac{1}{2}} A D^{-\frac{1}{2}} \quad \text{and} \quad A = \begin{bmatrix} 0 & R \\ R^\top & 0 \end{bmatrix} \tag{4}$$

where 0 is all-zero matrix and D is the diagonal degree matrix, the ith element $D_{i,i} = |N_i|$. The second input is matrix Δ, which has the same size with Γ, collects Weighted references values of both users and items.

$$\Delta = D^{-\frac{1}{2}} B D^{-\frac{1}{2}} \quad \text{and} \quad B = \begin{bmatrix} W_U & 0 \\ 0 & W_I \end{bmatrix} \tag{5}$$

3.2   Embedding Layers

In publication [30], authors used two vector for user's latent and item's latent, the size of the vectors for each user and each item can be taken as an input parameter. Thus, we define users embeddings and items embeddings as $e_u \in \mathbb{R}^d$ and $e_i \in \mathbb{R}^d$ with d is size of embeddings vector. In the first loop of propagation, embedding layer $E^{(0)}$ need an initial state of He normal weight initialization method which has good performance with Rectified Linear Unit activation function [31].

$$\begin{aligned} E_{user} &= [e_{u_1}, \dots, e_{u_n}] \\ E_{item} &= [e_{i_1}, \dots, e_{i_m}] \\ E &= [E_{user} | E_{item}] \end{aligned} \tag{6}$$

With input matrices in (4) and (5), the embedding should be trained though each round of propagation as

$$E^{(k)} = LeakyReLU\left[\Gamma E^{(k-1)} W_1^{(k-1)} + \Delta \Gamma E^{(k-1)} W_2^{(k-1)} + b^{(k-1)} \right] \tag{7}$$

where $W_1^{(k-1)}$, $W_2^{(k-1)}$ are the trainable transformation matrices and b^{k-1} is bias vector.

3.3 Propagation Process

Collaborative filtering method will capture signals inside the graphs' structure and training the embeddings of both users and items. As message-passing model of GNN [32,33], we extract the signals and then make an aggregation for the embedding at output.

Extract Signals. With a user-item pair, the signals sent out by an users and an item into target user is respectively

$$m_{u \leftarrow u} = W_1 e_u \tag{8}$$

$$m_{u \leftarrow i} = \frac{1}{\sqrt{|N_u||N_i|}} * \left(W_1 e_i + \Delta W_2 e_i \right) \tag{9}$$

where W_1 and W_2 are trainable weight matrices. Δ is the Weighted addition matrix at the input. $|N_i|$ and $|N_u|$ are is the force of the user and item set, respectively.

Aggregate Signal. With messages received from neighborhood of a user u, we aggregate all of them to refine representation of u.

$$e_u^{(1)} = LeakyReLU\big(m_{u \leftarrow u} + \sum_{i \in \aleph_u} m_{u \leftarrow i}\big) \tag{10}$$

where $e_u^{(1)}$ present embedded latent of user u after the first propagation layer. LeakyReLU is the activation function allows messages to encode both positive and a little negative signals [34]. The relation of users is taken into consideration by two sources that are $m_{u \leftarrow u}$ and matrix Δ.

3.4 Prediction and Optimization

After a number of propagation iterations, the embedding vector E^* will be acquired, the predicting score between user u_i on item i_j can be calculated by $\widehat{y}_{ui} = e_{u_i}^{*\top} e_{i_j}^*$. We build the loss function with Bayesian Personalized Ranking because it is the best suitable method from implicit feedback datasets [35]. With two pooling observable sets, we have Ω_{ui}^+ is observed relations and Ω_{uj}^- is unobserved relations. Thus,

$$Loss_{bpr} = \sum_{\Omega_{ui}^+} \sum_{\Omega_{uj}^-} -ln\sigma(\widehat{y}_{ui} - \widehat{y}_{uj}) + \lambda \parallel \Phi \parallel_2^2 \tag{11}$$

where Φ is embedding E^* and all of trainable weight W_1, W_2, bias b; $\sigma(.)$ is sigmoid function and λ controls the regularization.

We generalize the propagation process from the two input matrices, through the output embedding and stack layers to calculate a predicted rating value in Fig. 2.

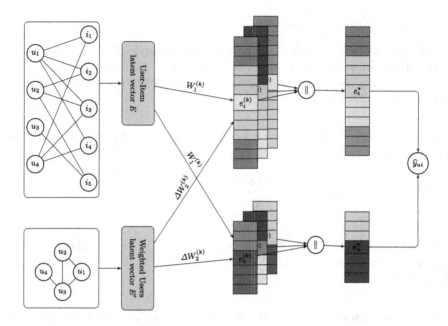

Fig. 2. The proposed framework with bi-parties graph and weighted influence users as additional input.

4 Experiments

We make experiments with our proposed model on three well-known datasets which are Gowalla, Amazon-book and Yelp2018; and compare the result with three state-of-the-art models which are GCMC, NGCF and LightGCN.

4.1 Datasets Description

We use setting 10-core with all datasets to ensure that each user has at least ten interactions. For each dataset, 20% of interactions were random selected as test set and the remaining (80% interaction) is used for training process. We present a summary of the data sets in Table 1.

- **Gowalla**: Gowalla was a location-based social networking service and users were able to check in at their current location.
- **Amazon-book**: This dataset contains product reviews and metadata from Amazon books.
- **Yelp2018**: Yelp is a popular online directory for discovering local businesses such are hotels, bars, restaurants, and cafes.

4.2 Experimental Settings

Setting Parameters. We configure our model with embedding fixed-size to 64 for all models and the embedding parameters and Weighted matrix W_1, W_2

Table 1. Statistic of the experiment datasets.

Dataset	#Users	#Items	#Relations	Density
Gowalla	29,858	40,981	1,027,370	0.000084
Amazon-book	52,643	91,599	2,984,108	0.00062
Yelp2018	31,668	38,048	1,561,406	0.00130

are initialized with the He normal weight method. The optimization process was done with Adam algorithm [36]. In the settings, $3 - layers$ give the best results.

Baseline. To demonstrate the result, we compare our proposed model with the following state-of-the-art methods:

- **GCMC**[9] designed with two modules, decomposer and combiner, this method distinguishes the underlying ranking motives underlying the clearly observed interactions between the user and the item. The first module decomposes rating values to extract the interactions and the second one combines all signals into a unified embeddings for prediction.
- **NGCF**[10] conducts propagation processes on embeddings with several iterations. The stacked embeddings on output contains high-order connectivity in interactions graph. The collaborative signal is encoded into the latent vectors and it make the model more sufficient.
- **LightGCN** [11]: focus on the neighborhood aggregation component for collaborative filtering. This model uses linearly propagating to learn users and items embeddings for interaction graph. The final embedding is weighted sum of all learned embeddings.

4.3 Results

The overall performance comparison shows in Table 2. Our proposed model gives out the better scores, both precision, recall and NGCD.

Table 2. Overall performance comparisons

Dataset	Gowalla			Amazon-book			Yelp2018		
	Precision	Recall	ndcg@20	Precision	Recall	ndcg@20	Precision	Recall	ndcg@20
GCMC	0.03690	0.11799	0.09042	0.01122	0.02539	0.02033	0.02320	0.05114	0.04141
NGCF	0.04080	0.13123	0.11149	0.01713	0.41160	0.03045	0.02192	0.04865	0.03917
LightGCN	0.03961	0.12637	0.11032	0.01181	0.02657	0.02114	0.01975	0.04262	0.03462
Our model	0.04167	0.13503	0.11648	0.01772	0.04335	0.031619	0.02225	0.04894	0.03954

The only worse case is when compared to GCMC with the Yelp2018 dataset. We suspect that Yelp2018 clustered users into categories database, which is an advantage for GCMC, as a cluster becomes a component in decomposition and combination. In this case, NGCF and LightGCN also give worse results than GCMC.

Figure. 3 shows the comparison of all methods in loss by epoch. Our proposed model reduces the loss faster others. It shows that the propagation process that has been effective recorded after each mini batch.

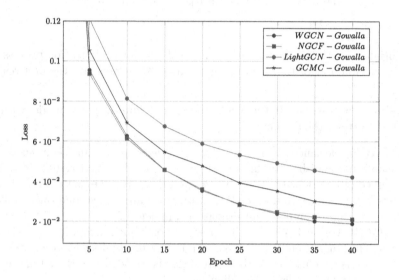

Fig. 3. Test performance of each epoch of MF and NGCF.

5   Conclusion and Future Works

5.1   Conclusion

In this work, we have proposed a method of weighting the influence among users, which can be considered equivalent to the Social Relation of Trust in the item recommendation problem. The input complement matrix has been calculated to enrich the interactions between the user and the item, and it serves as a moderator of the signals during the extraction and synthesis of messages. By contributing a machine learning model, we believe we have created an inspiration for future studies on the Recommender systems.

5.2   Future Works

Trust weight graphs between users , which were prepared and send into the proposed computational model, based on the interactions of many users on the same item. This is not a graph that represents real social relationships between users. Future works might be to find out if a social relation graph, by replace with (either collaborative filtering with) the weighted trust graph, will result in better recommendation results. The problem of clustering users according to interests, habits or behaviors also needs to be explored for more effective collaborative filtering.

Last but not least, the tuning parameters for the learning model are difficult to determine optimally, as well as the problem of initializing the embedded layer value at the first step that needs to be considered. We hope that the proposed WGCN model has provided a perspective on data preparation and learning methods from multiple sources of information.

References

1. Zhang, J., Lin, Z., Xiao, B., Zhang, C.: An optimized item-based collaborative filtering recommendation algorithm. In: 2009 IEEE International Conference on Network Infrastructure and Digital Content, pp. 414–418 (2009)
2. Tewari, A., Kumar, A., Barman, A.: Book recommendation system based on combine features of content based filtering, collaborative filtering and association rule mining. In: 2014 IEEE International Advance Computing Conference (IACC), pp. 500–503 (2014)
3. Shi, Y., Larson, M., Hanjalic, A.: Collaborative filtering beyond the user-item matrix: a survey of the state of the art and future challenges. ACM Comput. Surv. **47**(1), 1–45 (2014). https://doi.org/10.1145/2556270
4. Walt, S., Colbert, S., Varoquaux, G.: The numPy array: a structure for efficient numerical computation. Comput. Sci. Eng. **13**, 22–30 (2011)
5. Abadi, M., et al.: TensorFlow: a system for large-scale machine learning. In: 12th USENIX Symposium on Operating Systems Design and Implementation (OSDI 16), pp. 265–283 (2016). https://www.usenix.org/conference/osdi16/technical-sessions/presentation/abadi
6. Ma, H., Yang, H., Lyu, M., King, I.: SoRec: social recommendation using probabilistic matrix factorization. In: Proceedings of the 17th ACM Conference on Information and Knowledge Management, pp. 931–940 (2008). https://doi.org/10.1145/1458082.1458205
7. Ma, H., Zhou, D., Liu, C., Lyu, M., King, I.: Recommender systems with social regularization. In: WSDM, pp. 287–296 (2011)
8. Scarselli, F., Gori, M., Tsoi, A., Hagenbuchner, M., Monfardini, G.: The graph neural network model. IEEE Trans. Neural Netw. **20**, 61–80 (2009)
9. Wang, X., Wang, R., Shi, C., Song, G., Li, Q.: Multi-component graph convolutional collaborative filtering. In: Proceedings of the AAAI Conference on Artificial Intelligence, vol. 34, pp. 6267–6274 (2020)
10. Wang, X., He, X., Wang, M., Feng, F., Chua, T.: Neural Graph Collaborative Filtering. CoRR. abs/1905.08108 (2019). https://arxiv.org/abs/1905.08108
11. He, X., Deng, K., Wang, X., Li, Y., Zhang, Y., Wang, M.: Lightgcn: simplifying and powering graph convolution network for recommendation. In: Proceedings of the 43rd International ACM SIGIR Conference on Research and Development in Information Retrieval, pp. 639–648 (2020)
12. Jalili, M., Ahmadian, S., Izadi, M., Moradi, P., Salehi, M.: Evaluating collaborative filtering recommender algorithms: a survey. IEEE Access. **6**, 74003–74024 (2018)
13. Pazzani, M.: A framework for collaborative, content-based and demographic filtering. Artif. Intell. Rev. **13**, 393–408 (1999)

14. Adomavicius, G., Sankaranarayanan, R., Sen, S., Tuzhilin, A.: Incorporating contextual information in recommender systems using a multidimensional approach. ACM Trans. Inf. Syst. (TOIS). **23**, 103–145 (2005)
15. Bojnordi, E., Moradi, P.: A novel collaborative filtering model based on combination of correlation method with matrix completion technique. In: The 16th CSI International Symposium on Artificial Intelligence and Signal Processing (AISP 2012), pp. 191–194 (2012)
16. Goldberg, D., Nichols, D., Oki, B., Terry, D.: Using collaborative filtering to weave an information tapestry. Commun. ACM. **35**, 61–70 (1992). https://doi.org/10.1145/138859.138867
17. Candillier, L., Meyer, F., Boullé, M.: Comparing state-of-the-art collaborative filtering systems. In: Perner, P. (ed.) MLDM 2007. LNCS (LNAI), vol. 4571, pp. 548–562. Springer, Heidelberg (2007). https://doi.org/10.1007/978-3-540-73499-4_41
18. Yu, K., Xu, X., Tao, J., Ester, M., Kriegel, H. Instance selection techniques for memory-based collaborative filtering. In: Proceedings of the 2002 SIAM International Conference on Data Mining (SDM), pp. 59–74 (2002). https://epubs.siam.org/doi/abs/10.1137/1.9781611972726.4
19. Koren, Y., Bell, R., Volinsky, C.: Matrix factorization techniques for recommender systems. Computer **42**, 30–37 (2009)
20. Herlocker, J., Konstan, J., Terveen, L., Riedl, J. Evaluating collaborative filtering recommender systems. ACM Trans. Inf. Syst. **22**, 5–53 (2004). https://doi.org/10.1145/963770.963772
21. Sarwar, B., Karypis, G., Konstan, J., Riedl, J.: Application of Dimensionality Reduction in Recommender System - A Case Study. (2000,8)
22. Sarwar, B., Karypis, G., Konstan, J., Riedl, J.: Analysis of recommendation algorithms for E-commerce. In: Proceedings of the 2nd ACM Conference on Electronic Commerce, pp. 158–167 (2000). https://doi.org/10.1145/352871.352887
23. Najork, M., McSherry, F.: Computing information retrieval performance measures efficiently in the presence of tied scores. In: 30th European Conference on IR Research (ECIR). (2008). https://www.microsoft.com/en-us/research/publication/computing-information-retrieval-performance-measures-efficiently-in-the-presence-of-tied-scores/
24. Gao, Y., Li, Y., Lin, Y., Gao, H., Khan, L.: A survey, deep learning on knowledge graph for recommender system (2020)
25. Wu, S., Sun, F., Zhang, W., Cui, B.: Graph neural networks in recommender systems: a survey. ACM Comput. Surv. (CSUR) (2021)
26. Wu, Z., Pan, S., Chen, F., Long, G., Zhang, C., Yu, P.: A comprehensive survey on graph neural networks. IEEE Trans. Neural Netw. Learn. Syst. **32**, 4–24 (2021). https://dx.doi.org/10.1109/TNNLS.2020.2978386
27. Wang, X., et al.: Heterogeneous Graph Attention Network. CoRR. abs/1903.07293 (2019). https://arxiv.org/abs/1903.07293
28. Ying, R., He, R., Chen, K., Eksombatchai, P., Hamilton, W., Leskovec, J.: Graph convolutional neural networks for web-scale recommender systems. In: Proceedings of the 24th ACM SIGKDD International Conference on Knowledge Discovery Data Mining, pp. 974–983 (2018). https://doi.org/10.1145/3219819.3219890
29. Hamilton, W., Ying, R., Leskovec, J.: Inductive Representation Learning on Large Graphs. CoRR. abs/1706.02216 (2017). https://arxiv.org/abs/1706.02216
30. He, X., Liao, L., Zhang, H., Nie, L., Hu, X., Chua, T.: Neural collaborative filtering. In: Proceedings of the 26th International Conference on World Wide Web, pp. 173–182 (2017)

31. Datta, L.: A Survey on Activation Functions and their relation with Xavier and He Normal Initialization. CoRR. abs/2004.06632 (2020). https://arxiv.org/abs/2004.06632

32. Xu, K., Li, C., Tian, Y., Sonobe, T., Kawarabayashi, K., Jegelka, S.: Representation learning on graphs with jumping knowledge networks. In: Proceedings of the 35th International Conference on Machine Learning, vol. 80, pp. 5453–5462 (2018). https://proceedings.mlr.press/v80/xu18c.html

33. Hamilton, W., Ying, Z., Leskovec, J.: Inductive Representation learning on large graphs. Adv. Neural Inf. Process. Syst. **30** (2017). https://proceedings.neurips.cc/paper/2017/file/5dd9db5e033da9c6fb5ba83c7a7ebea9-Paper.pdf

34. Maas, A.: Rectifier nonlinearities improve neural network acoustic models. In: ICML Workshop on Deep Learning For Audio, Speech, and Language Processing (WDLASL), vol. 30, no. 1, p. 3 (2013)

35. Rendle, S., Freudenthaler, C., Gantner, Z., Schmidt-Thieme, L.: Bayesian personalized ranking from implicit feedback, BPR (2012)

36. Kingma, D., Ba, J.: A Method for Stochastic Optimization, Adam (2017)

Combining User Specific and Global News Features for Neural News Recommendation

Cuong Manh Nguyen[1,2], Ngo Xuan Bach[1(✉)], and Tu Minh Phuong[1]

[1] Department of Computer Science, Posts and Telecommunications Institute of
Technology, Hanoi, Vietnam
{bachnx,phuongtm}@ptit.edu.vn
[2] FPT Smart Cloud, Hanoi, Vietnam

Abstract. News recommendation engines play a vital role in online
news services. Such systems help users discover news articles they may
want to read, thus increasing user experience and engagement. Existing
news recommendation methods usually model users' interest from their
reading history by learning user and news representations. Those methods represent news by features specific to users to create personalized
user interest models and recommend news similar to user interest. However, in many cases, users may also read the news that is not in their usual
interest, for example, breaking news. Such news often has global characteristics that are independent of specific users. In this paper, we propose
a neural news recommendation method that combines user-specific news
models with global news models to better deal with those situations.
To create informative news models, we adopt an attentive neural approach that can learn representations of the news from different news
aspects such as title, category, and news contents. The attentive mechanism allows the model to focus on an aspect that attracts global interest
from different users. Experiments on a real-world dataset show that our
model achieves improved performance over the state-of-the-art news recommendation method and is able to recommend candidate news of both
types: those are relevant to a specific user and those are of common
interest for different users.

Keywords: Recommender systems · News recommendation · Deep
neural networks

1 Introduction

With the development of the Internet, newsreaders have access to various sources
of news online, which include online versions of traditional newspaper companies
or online news platforms such as Google News[1] and MSN News[2] that aggregate
news from other news sites [1]. Such sources generate a considerable number of

---

[1] https://news.google.com.

[2] https://www.msn.com/en-us/news.

© The Author(s), under exclusive license to Springer Nature Switzerland AG 2022
N. T. Nguyen et al. (Eds.): ACIIDS 2022, LNAI 13757, pp. 648–660, 2022.
https://doi.org/10.1007/978-3-031-21743-2_52

news articles every day. Together with a fast update cycle, such an abundance of information makes it difficult for readers to find news that is most interesting to them. Personalized news recommendation techniques have shown to be helpful in alleviating the effect of information overload and improve reader experience in such situations [2–5].

Recently, deep-learning-based methods have shown good performance for the news recommendation task [6,7]. Those methods use different neural network architectures to learn the representations of users and news. A common strategy is to learn news representations from the body part or titles of the news articles, and learn the user representations from news representations, usually in a coordinated manner [6,8,9]. In prediction, one can match the representations of candidate news against the representation of a user to find relevant news for the recommendation. Because such methods learn user representation from news representations, they tend to focus on news features that are user-specific, i.e., news features that better characterize the personal interests of users. While this is a benefit for making personalized recommendations, focusing on user-specific features may fail to recommend news that is of global interest, such as breaking news or trending news.

In this work, we propose to address this problem by learning both user-specific and global features of news, which we use to generate personalized and global recommendations. Our model has two different news encoders. In the first encoder, we learn user-specific features, from which we then create the representations of users. We create the models of users from the representations of news they clicked, which are the outputs of the first encoder. The second encoder focuses on global features of news, which characterize how well the news attracts the common interest of the users regardless of their personal reading taste. We combine two user encoders to make final predictions that promote both types of news articles.

In learning news representation, an important observation is that a news article often contains different attributes such as title, abstract, body, category, and topic, among others. Although previous work used only one type of attribute, for example, article body [6,10], recent work has shown that using more news attributes allows learning more informative news representations. In this work, we adopt a similar approach and learn the representations of the news from multiple attributes or components of news articles. Since different attributes may have different contributions to the representations of news, we follow the approach described in [8] and apply the attention mechanism to the news encoder. We also explicitly explore the contributions of each attribute to news representations in an ablation study.

Among news components, components such as title, abstract, and body contain essential information in the form of texts. In learning the representation of textual information, the context has an important influence on the meaning of words, which finally influences the news representations. To capture the context of words, we adopt the multi-head self-attention mechanism, which has shown the ability to model the interactions between words [11]. We then apply another attention layer to learn the representation of each textual component from important words.

We perform experiments on a real-world dataset to evaluate the effectiveness of the proposed method. The results show that our model can effectively improve the performance of news recommendations in terms of various metrics.

2   Related Work

News recommendation has been an active research area in recent years due to its broad applicability in online news services. Various methods have been developed for building news recommender systems. In this section, we give a brief review of news recommendation methods, with a focus on deep learning-based approaches.

2.1   Conventional Approaches

Traditional methods for building news recommender systems can be roughly classified as content-based filtering (CBF), collaborative filtering (CF), and hybrid approaches [12,13]. To generate recommendations, a CBF method is primarily based on user profiles and news contents (i.e., metadata), while a CF algorithm exploits the behaviors of users in the past. As its name implies, a hybrid method integrates both types of information in a single recommender system or combines the outputs of different CBF and CF systems. Raza and Ding show in their survey paper [14] that in the news domain, CBF is the most popular recommendation method, while the hybrid comes in the second place, and CF is the least popular among the three. It is reasonable because the content of news is the most important information to be considered when producing recommendations.

2.2   Deep Learning-Based Approaches

Deep learning has become the most powerful tool and the first choice in most research areas of computer science, such as computer vision, natural language processing, and recommender systems [15,16]. Compared with conventional approaches, deep learning-based methods offer more advanced solutions to issues in news recommendations [14]. First, deep learning models are very powerful in representation learning or feature learning tasks. Deep neural networks like Convolutional Neural Networks (CNNs) [17], Recurrent Neural Networks (RNNs) [18], and Transformers [11] can produce good representations for news metadata, a key component in content-based news recommendation. Second, deep learning models are able to learn rich user-item interactions from the news data, which has been shown to be outperformed traditional CF methods. Third, deep neural networks like RNNs and their variants provide a natural way for sequential modeling and therefore are effective in session-based news recommendation tasks. Finally, deep learning models can mitigate the cold-start and data sparsity problems by extracting valuable features from news and user data to improve user and item profiles.

Many studies have been conducted using different deep neural network architectures for the news recommendation tasks. Yu et al. [19] with Multi-Layer Perceptrons (MLPs), Cao et al. [20] and Okura et al. [6] with Autoencoders (AEs), An et al. [21], Wang et al. [7], Wu et al. [22], and Zhu et al. [23] with CNNs, Zhu et al. [23], An et al. [21], Zhang et al. [24], and de Souza Pereira Moreira [25] with RNNs and their variants Long Short-Term Memory (LSTM) and Gated Recurrent Units (GRUs), Lee et al. [26], Sheu and Li [27], and Ge et al. [28] with Graph Neural Networks (GNNs), Wang et al. [7], An et al. [21], Sun et al. [29], and Wu et al. [30] with attention networks and Transformers are some typical examples among the others. The most closely related work to ours is the one of Wu et al. [8], which presents a neural news recommendation model to learn representations of news and users by exploiting different types of news information, including title, abstract, body, and category. Our model, however, has two different news encoders for learning user-specific news features and global news features, respectively. Therefore, our recommendations consist of both personalized and global interest news.

3   Our Approach

This section describes our method for generating news recommendations. As illustrated in Fig. 1, our model consists of four major modules: 1) (User specific) News encoder; 2) User encoder; 3) Global news encoder; and 4) Click predictor. In the following sections, we introduce these modules in detail.

3.1   News Encoder

The news encoder module is used to learn news representations from different kinds of news information, such as titles, bodies, and topic categories. Figure 2 shows the architecture of the news encoder which comprises four main components: Title encoder, Body encoder, Category encoder, and Attentive pooling.

Title Encoder. The first component is the title encoder whose goal is to learn news representations from their titles. The title encoder again consists of three layers. The first layer is word embedding, which converts a news title from a word sequence into a sequence of low-dimensional semantic vectors. Let $[w_1^t, w_2^t, \ldots, w_M^t]$ denote the word sequence of a news title with M words. A word embedding lookup table $W_e \in R^{V \times D}$ is used to convert the word sequence in the title into a sequence of vectors $[e_1^t, e_2^t, \ldots, e_M^t]$, where V and D are the vocabulary size and word embedding dimension, respectively.

The second layer is an word-level multi-head self-attention network to learn contextual representations of words by capturing their interactions. The representation of the ith word learned by the kth attention head is computed as follows:

$$\alpha_{i,j}^k = \frac{\exp\left(\mathbf{e}_i^T \mathbf{Q}_k^w \mathbf{e}_j\right)}{\sum_{m=1}^{M} \exp\left(\mathbf{e}_i^T \mathbf{Q}_k^w \mathbf{e}_m\right)} \tag{1}$$

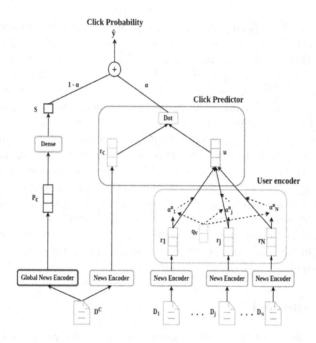

Fig. 1. The architecture of our approach for news recommendation.

$$\mathbf{h}_{i,k}^w = \mathbf{V}_k^w \left(\sum_{j=1}^M \alpha_{i,j}^k \mathbf{e}_j \right) \tag{2}$$

where Q_k^w and V_k^w are the projection parameters of the kth self-attention head, and $\alpha_{i,j}^k$ indicates the relative importance of the interaction between the ith and jth words. The multi-head representation h_i^w of the ith word is the concatenation of the representations produced by h separate self-attention heads, i.e., $h_i^w = [h_{i,1}^w; h_{i,2}^w; \ldots; h_{i,h}^w]$.

The third layer of the title encoder is an additive word attention network, where we use an attention mechanism to select important words in news titles for learning more informative news representations. The attention weight α_i^t of the ith word in a news title can be computed as follows:

$$a_i^t = \mathbf{q}_t^T \tanh(\mathbf{V}_t \times \mathbf{h}_i^w + \mathbf{v}_t) \tag{3}$$

$$\alpha_i^t = \frac{\exp\left(a_i^t\right)}{\sum_{j=1}^M \exp\left(a_j^t\right)} \tag{4}$$

where $\mathbf{V}_t$ and $\mathbf{v}_t$ are the projection parameters, and $\mathbf{q}_t$ denotes the attention query vector.

The final representation of a news title after going through three layers is the summation of the contextual representations of its words weighted by their attention weights:

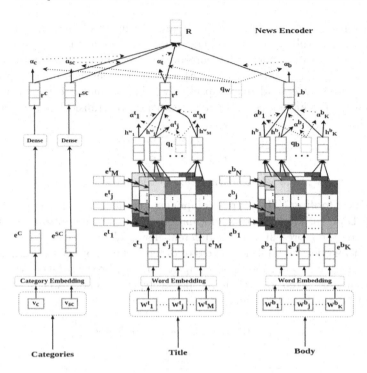

Fig. 2. The architecture of news encoder.

$$\mathbf{r}^t = \sum_{j=1}^{M} \alpha_j^t \mathbf{h}_j^w \tag{5}$$

Body Encoder. The second component of the news encoder is the body encoder. Similar to the title encoder, there are also three layers with the same function in the body encoder. The final representation of a news body is the summation of the contextual representations of its words weighted by their attention weights:

$$\mathbf{r}^b = \sum_{j=1}^{M} \alpha_j^b \mathbf{h}_j^b \tag{6}$$

Category Encoder. The third component is the category encoder, which is used to learn news representations from their categories. On many online news platforms such as Microsoft Start, Google News, news articles are usually labeled with topic categories (e.g., "health" and "finance") and subcategories (e.g., "medical" and "markets") for target user interests that contain essential information of the news articles. For example, if a user clicks on many news articles with the category of "sports", we can infer that this user may be interested in sports news. Thus, we propose to incorporate both the category and subcategory information in news representation learning.

The inputs of the category encoder are the index of the category v_c and the index of the subcategory v_{sc}. There are two layers in the category encoder. The first layer is a category index embedding layer, which converts the discrete indexes of categories and subcategories into low-dimensional dense representations (denoted by $\mathbf{e}_c$ and $\mathbf{e}_{sc}$, respectively). The second one is a dense layer used to learn the hidden category representations by transforming the category embeddings:

$$
\begin{aligned}
\mathbf{r}^c &= \mathrm{ReLU}\left(\mathbf{V}_c \times \mathbf{e}_c + \mathbf{v}_c\right) \\
\mathbf{r}^{sc} &= \mathrm{ReLU}\left(\mathbf{V}_s \times \mathbf{e}_{sc} + \mathbf{v}_s\right)
\end{aligned}
\tag{7}
$$

where $\mathbf{V}_c$, $\mathbf{v}_c$, $\mathbf{V}_s$, and $\mathbf{v}_s$ are parameters in the dense layers.

Attentive Pooling. The fourth component of the news encoder is attentive pooling. Different kinds of news information may have different informativeness for learning the representations of different news. We use a view-level attention network to model the informativeness of different kinds of news information for learning news representations. Let α_t, α_b, α_c, and α_{sc} be the title, body, category, and subcategory attention weights, respectively, the attention weight of the title view can be calculated as follows:

$$
a_t = \mathbf{q}_v^T \tanh\left(\mathbf{U}_v \times \mathbf{r}^t + \mathbf{u}_v\right), \text{and}
\tag{8}
$$

$$
\alpha_t = \frac{\exp\left(a_t\right)}{\exp\left(a_t\right) + \exp\left(a_b\right) + \exp\left(a_c\right) + \exp\left(a_{sc}\right)}
\tag{9}
$$

where $\mathbf{U}_v$ and $\mathbf{u}_v$ are projection parameters, $\mathbf{q}_v$ is the attention query vector. The attention weights of other types of news information, i.e., body, category, and subcategory can be computed in a similar way.

The final unified news representations learned by the news encoder module is the summation of the news representations from different views weighted by their attention weights:

$$
\mathbf{r} = \alpha_c \mathbf{r}^c + \alpha_{sc} \mathbf{r}^{sc} + \alpha_t \mathbf{r}^t + \alpha_b \mathbf{r}^b.
\tag{10}
$$

3.2   User Encoder

The user encoder module in our approach is used to learn the representations of users from the representations of their browsed news. As shown in Fig. 1, we apply a news attention network to learn more informative user representations by selecting important news in the same way as those of Wu et al. [8]. The attention weight α_i^n of the ith news browsed by a user is calculated as follows:

$$
a_i^n = \mathbf{q}_n^T \tanh\left(\mathbf{W}_n \times \mathbf{r}_i + \mathbf{b}_n\right)
\tag{11}
$$

$$
\alpha_i^n = \frac{\exp\left(a_i^n\right)}{\sum_{j=1}^{N} \exp\left(a_j^n\right)}
\tag{12}
$$

where $\mathbf{q}_n$, $\mathbf{W}_n$ and $\mathbf{b}_n$ are the parameters in the news attention network, and $\mathbf{r}_i$ denotes the representation of the ith news. The final representation of a user is the summation of the representations of news browsed by this user weighted by their attention weights:

$$\mathbf{u} = \sum_{i=1}^{N} \alpha_i^n \mathbf{r}_i \qquad (13)$$

where N is the number of browsed news.

3.3 Global News Encoder

News recommender systems usually use the relation between historical news articles browsed by a user to find news of his/her interest. However, the popularity or notability of a news article is also an essential piece of information to determine whether the news article is of interest or not. For example, a user who regularly reads sports news will also be attracted to news related to the Covid-19 epidemic in particular or hot news in general. Therefore, in addition to the news encoder, we propose to use a global news encoder to determine the notability of a news article. The global news encoder has the same architecture as the news encoder but has its own parameter set.

3.4 Click Predictor

The click predictor is used to estimate the probability of a user browsing a candidate news article based on their representations. Let $\mathbf{u}$ and $\mathbf{r}_c$ denote the representation of a user u and the representation of a candidate news D_c, respectively. We estimate the probability that user u clicks on candidate news D_c in three steps as follows:

- First, the click score y is calculated by the inner product of the representation vectors of the user and the news candidate:

$$y = \langle \mathbf{u}, \mathbf{r}_c \rangle \qquad (14)$$

- Second, the notability score of the news article is calculated. We pass it through the global news encoder to get the vector representation, then through the dense layer with softmax activation function to get the score value s, which in the range of $[0,1]$. The higher the score s is, the higher the notability of the candidate news is.
- The final click probability score $\hat{y}$ is the weighted sum of y and s:

$$\hat{y} = \alpha \times y + (1 - \alpha) \times s \qquad (15)$$

where $\alpha \in [0, 1]$ is a parameter to be learned.

Table 1. Statistics of MIND dataset for the news recommendation task

# Users	1,000,000	# News	161,013
# Train samples	2,232,748	# Test samples	376,471
# Click behavior	24,155,470	Avg. click history length	37.07
Avg. title length	11.52	Avg. body length	585.05

Table 2. Parameter/Hyperparameter setting

Parameter/Hyperparameter	Value
Word embedding dimension	300
Category embedding dimension	100
Sub-category embedding dimension	100
Number of self-attention heads	20
Number of CNN filters	400
Negative sampling ratio	4
Batch size	20
Number of epochs	20
Learning rate	0.0001
Dropout rate	20%

4 Experiments

4.1 Datasets and Experimental Settings

Our experiments were conducted on a real-world dataset, namely MIND [31], which was constructed from the user click logs of Microsoft News[3]. MIND contains 1 million users and more than 160k English news articles, each of which has rich textual content such as title, abstract, and category. The detailed statistics of the dataset are shown in Table 1. In our experiments, since the MIND dataset has no information about the article's main body, we used the abstract of the news article instead of the body. The performances of news recommendation models were evaluated using standard metrics, including AUC, MRR, nDCG@5, and nDCG@10.

To train the neural networks, we used Adam optimizer with the cross-entropy loss, Relu activation functions, and pre-trained Glove word embeddings [32]. We set the dimensions of word embeddings, category embeddings, and sub-category embeddings to 300, 100, and 100, respectively. We trained the models in 20 epochs with a batch size of 20 and a learning rate of 0.0001. To mitigate the overfitting problem, we used the dropout technique with the rate of 20%. The values of parameters/hyperparameters are summarized in Table 2.

---

[3] https://microsoftnews.msn.com/.

Table 3. Experimental results of different news recommendation methods

#	Method	AUC	MRR	nDCG@5	nDCG@10
1	LibFM	0.5993	0.2823	0.3005	0.3574
2	EBNR	0.6542	0.3124	0.3376	0.3947
3	DKN	0.6460	0.3132	0.3384	0.3948
4	NPA	0.6669	0.3224	0.3498	0.4068
5	NAML	0.6686	0.3249	0.3524	0.4091
6	LSTUR	0.6773	0.3277	0.3559	0.4134
7	NRMS	0.6776	0.3305	0.3594	0.4163
8	**Our method**	**0.6945**	**0.3340**	**0.3725**	**0.4372**

4.2 Performance Evaluation

We first conducted experiments to evaluate the performance of our approach by comparing it with several strong baseline methods as follows:

1. LibFM [33]: A recommendation model with factorization machines.
2. EBNR [6]: An embedding-based news recommendation model with auto-encoders and GRU network.
3. DKN [7]: A method using deep knowledge-aware network.
4. NPA [22]: A news recommendation model with personalized attention.
5. NAML [8]: A model with attentive multi-view learning.
6. LSTUR [21]: A news recommendation method with long and short-term user representations.
7. NRMS [10]: A recommendation method based on multi-head self-attention.
8. Our method: A model for combining user specific and global news features for neural news recommendation.

Experimental results of these methods are summarized in Table 3. Our first observation is that methods using a user encoder architecture combining multi-head self-attention and attentive multi-view learning, i.e., NAML, LSTUR, nRMS, and our method, got better results than the previous approaches. The second observation is that the proposed method outperformed the baselines by large margins in all metrics. We got scores of 0.6945, 0.3340, 0.3725, and 0.4372 in AUC, MRR, nDCG@5, and nDCG@10, respectively.

4.3 Effectiveness of Global News Encoder

Next, we conducted experiments to evaluate the effect of the global news encoder in our model by changing the value of α coefficient. With $\alpha = 1$, the output of the Click Predictor ignores the informations of the global news encoder. With $\alpha = 0$, the model only uses information from candidate news to determine whether a user clicked on that article, regardless of the user history. Table 4 shows experimental results with different values of α coefficient.

Table 4. Experimental results with different values of α coefficient

α value	AUC	MRR	nDCG@5	nDCG@10
0.0	0.5165	0.2382	0.2506	0.3124
0.1	0.6447	0.2887	0.3032	0.3862
0.2	0.6611	0.3009	0.3388	0.4069
0.3	0.6756	0.3129	0.3494	0.4133
0.4	0.6802	0.3188	0.3541	0.4176
0.5	0.6880	0.3258	0.3662	0.4286
0.6	0.6891	0.3270	0.3679	0.4299
0.7	0.6904	0.3284	0.3688	0.4313
0.8	0.6913	0.3307	0.3702	0.4338
0.9	0.6924	0.3318	0.3709	0.4349
1.0	0.6910	0.3298	0.3695	0.4314
0.875 (learned)	**0.6945**	**0.3340**	**0.3725**	**0.4372**

Table 5. Global news encoder with different views

Views	AUC	MRR	nDCG@5	nDCG@10
Title	0.6885	0.3269	0.3665	0.4308
Abstract	0.6883	0.3289	0.3667	0.4312
Abstract+Category	0.6912	0.3315	0.3699	0.4315
Title+Abstract	0.6896	0.3293	0.3691	0.4302
Title+Category	0.6887	0.3281	0.3679	0.4313
Title+Abstract+Category	**0.6945**	**0.3340**	**0.3725**	**0.4372**

4.4 Ablation Study

We also conducted an ablation study on the global news encoder with different views. Experimental results in Table 5 shows that: 1) our model got relatively good results when using news titles or abstracts only; 2) adding category information improved the performance of the news recommender system; and 3) the version with all three types of information achieved the best results. The results also demonstrate the important role of each type of information in modeling the notability of news articles.

5 Conclusion

We proposed in this paper a model to combine user-specific and global news features for neural news recommendation. The core of our model consists of a news encoder, a user encoder, and a global news encoder. In the news encoder, we apply multi-head self-attention to learn contextual word representations by modeling the interactions between words in the title and abstract. We also use a multi-view learning framework to learn unified news representations by incorporating categories as different news views. We also apply an attention mechanism to the news encoder to select important words and views for learning informative news representations. In the user encoder, we learn representations of users from their browsed news and apply a news attention network to select the important news for learning informative user representations. We propose a global news encoder to determine the popularity of an article and its effect on user click behavior. We also conducted experiments on a real-world dataset to show the effectiveness of our model in comparison with several strong baselines.

In each news reading session, a user is often interested in news articles in similar categories such as sports, health, or politics. In future work, we plan to use a graph neural network to determine what kind of news articles a user is likely to be interested in based on the articles he/she has read. This information can help us determine whether a user clicks on a news article.

Acknowledgements. We would like to thank FPT Smart Cloud for financial support.

References

1. Das, A.S., Datar, M., Garg, A., Rajaram, S.: Google news personalization: scalable online collaborative filtering. In: Proceedings of WWW, pp. 271–280 (2007)
2. Cleger-Tamayo, S., Fernández-Luna, J.M., Huete, J.F.: Top-n news recommendations in digital newspapers. Knowl.-Based Syst. **27**, 180–189 (2012)
3. Liu, J., Dolan, P., Pedersen, E.R.: Personalized news recommendation based on click behavior. In: Proceedings of IUI, pp. 31–40 (2010)
4. Kirshenbaum, E., Forman, G., Dugan, M.: A live comparison of methods for personalized article recommendation at Forbes.com. In: Flach, P.A., De Bie, T., Cristianini, N. (eds.) ECML PKDD 2012. LNCS (LNAI), vol. 7524, pp. 51–66. Springer, Heidelberg (2012). https://doi.org/10.1007/978-3-642-33486-3_4
5. Bach, N.X., Hai, N.D., Phuong, T.M.: Personalized recommendation of stories for commenting in forum-based social media. Inf. Sci. **352–353**, 48–60 (2016)
6. Okura, S., Tagami, Y., Ono, S., Tajima, A.: Embedding-based news recommendation for millions of users. In: Proceedings of KDD, pp. 1933–1942 (2017)
7. Wang, H., Zhang, F., Xie, X., Guo, M.: DKN: deep knowledge-aware network for news recommendation. In: Proceedings of WWW, pp. 1835–1844 (2018)
8. Wu, C., Wu, F., An, M., Huang, J., Huang, Y., Xie, X.: Neural news recommendation with attentive multi-view learning. In: Proceedings of IJCAI, pp. 3863–3869 (2019)
9. Wu, C., Wu, F., Huang, Y., Xie, X.: User-as-graph: user modeling with heterogeneous graph pooling for news recommendation. In: Proceedings of IJCAI, pp. 1624–1630 (2021)
10. Wu, C., Wu, F., Ge, S., Qi, T., Huang, Y., Xie, X.: Neural news recommendation with multi-head self-attention. In: Proceedings of EMNLP/IJCNLP, pp. 6388–6393 (2019)
11. Vaswani, A., et al.: Attention is all you need. In: Proceedings of NIPS, pp. 6000–6010 (2017)

12. Adomavicius, G., Tuzhilin, A.: Toward the next generation of recommender systems: a survey of the state-of-the-art and possible extensions. IEEE Trans. Knowl. Data Eng. **17**(6), 734–749 (2005)
13. Phuong, N.D., Thang, L.Q., Phuong, T.M.: A graph-based method for combining collaborative and content-based filtering. In: Ho, T.-B., Zhou, Z.-H. (eds.) PRICAI 2008. LNCS (LNAI), vol. 5351, pp. 859–869. Springer, Heidelberg (2008). https://doi.org/10.1007/978-3-540-89197-0_80
14. Raza, S., Ding, C.: News recommender system: a review of recent progress, challenges, and opportunities. Artif. Intell. Rev. (2021)
15. Kumar, V., Khattar, D., Gupta, S., Gupta, M., Varma, V.: Deep neural architecture for news recommendation. In: Proceedings of CLEF (2017)
16. Phuong, T.M., Thanh, T.C., Bach, N.X.: Neural session-aware recommendation. IEEE Access **7**, 86884–86896 (2019)
17. Kim, Y.: Convolutional neural networks for sentence classification. In: Proceedings of EMNLP, 1746–1751 (2014)
18. Elman, J.: Finding structure in time. Cogn. Sci. **14**(2), 179–211 (1990)
19. Yu, B., Shao, J., Cheng, Q., Yu, H., Li, G., Lu, S.: Multi-source news recommender system based on convolutional neural networks. In: Proceedings of ICIIP, pp. 17–23 (2018)
20. Cao, S., Yang, N., Liu, Z.: Online news recommender based on stacked autoencoder. In: Proceedings of ICIS, pp. 721–726 (2017)
21. An, M., Wu, F., Wu, C., Zhang, K., Liu, Z., Xie, X.: Neural news recommendation with long- and short-term user representations. In: Proceedings of ACL, pp. 336–345 (2019)
22. Wu, C., Wu, F., An, M., Huang, J., Huang, Y., Xie, X.: NPA: neural news recommendation with personalized attention. In: Proceedings of KDD, pp. 2576–2584 (2019)
23. Zhu, Q., Zhou, X., Song, Z., Tan, J., Guo, L.: DAN: deep attention neural network for news recommendation. In: Proceedings of AAAI, pp. 5973–5980 (2019)
24. Zhang, L., Liu, P., Gulla, J.A.: Dynamic attention-integrated neural network for session-based news recommendation. Mach. Learn. **108**(10), 1851–1875 (2019). https://doi.org/10.1007/s10994-018-05777-9
25. de Souza Pereira Moreira, G.: CHAMELEON: a deep learning meta-architecture for news recommender systems. In: Proceedings of RecSys, pp. 578–583 (2018)
26. Lee, D., Oh, B., Seo, S., Lee, K.: News recommendation with topic-enriched knowledge graphs. In: Proceedings of CIKM, pp. 695–704 (2020)
27. Sheu, H., Li, S.: Context-aware graph embedding for session-based news recommendation. In: Proceedings of RecSys, pp. 657–662 (2020)
28. Ge, S., Wu, C., Wu, F., Qi, T., Huang, Y.: Graph enhanced representation learning for news recommendation. In: Proceedings of the Web Conference, pp. 2863–2869 (2020)
29. Sun, F., et al.: BERT4Rec: sequential recommendation with bidirectional encoder representations from transformer. In: Proceedings of CIKM, pp. 1441–1450 (2019)
30. Wu, C., Wu, F., Yu, Y., Qi, T., Huang, Y., Liu, Q.: NewsBERT: distilling pretrained language model for intelligent news application. arXiv (2021)
31. Wu et al.: MIND: a large-scale dataset for news recommendation. In: Proceedings of ACL, pp. 3597–3606 (2020)
32. Pennington, J., Socher, R., Manning, C.: GloVe: global vectors for word representation. In: Proceedings of EMNLP, pp. 1532–1543 (2014)
33. Rendle, S.: Factorization machines with libFM. ACM Trans. Intell. Syst. Technol. **3**(3), 57:1–57:22 (2012)

Polarization in Personalized Recommendations: Balancing Safety and Accuracy

Zakaria El-Moutaouakkil$^{(\boxtimes)}$, Mohamed Lechiakh, and Alexandre Maurer

UM6P-CS, Mohammed VI Polytechnic University, Benguerir, Morocco
z.elmoutaouakkil@ieee.org, {mohamed.lechiakh,alexandre.maurer}@um6p.ma

Abstract. Recommender systems are beneficial to both service providers and users, but may also have unintended side effects. In this paper, we address the societal impact of recommender systemss on user polarization resulting from over-personalized recommendations. First, we model the user preference gap (uPG) as the distance between two timeseries, representing the user's consumption per content categories. Second, we map the uPG score onto the not-yet-rated items, and consider two approaches to minimize uPG per user, when they go beyond a certain threshold. In particular, we propose post and pre-constrained versions of the alternating least square algorithm that reduce the user's uPG score, hence avoiding to reinforce her polarization over time. Interestingly, these newly constrained algorithms still maintain a high level of recommendation accuracy. Our simulation results, derived from real datasets, show that our solutions enable personalization with a reasonable level of polarization.

Keywords: Recommender system · User polarization ·
Personalization · User preference · Alternating least square · Items
recommendation

1 Introduction

1.1 Motivation

Recommender systems (RSs) facilitate users' search for relevant items (e.g., books, videos, press articles, etc.) among a huge list of choices. Basically, a RS algorithm provides the user with a small set of items, often referred to as top-N items, supposed to best meet her preferences. For instance, collaborative filtering is a memory-based class of RS algorithms that learns similarity among users or items to generate personalized recommendations for each user. Given their central role in several businesses, numerous research works focused on enhancing the offline performance of RS algorithms [1], especially in terms of accuracy, diversity, novelty and serendipity, to name a few metrics. However, as users started relying more and more on the RS output, that in turn relies on historical data generated by other users, these algorithms are prone to societal

N. T. Nguyen et al. (Eds.): ACIIDS 2022, LNAI 13757, pp. 661–674, 2022.
https://doi.org/10.1007/978-3-031-21743-2_53

side effects such as bias propagation, user's preference amplification, filter bubbles and echo-chambers formation, as a result of over-personalization. Therefore, ensuring the safety of RSs against these effects, and making them beneficial to their users [2], is a crucial research direction today. In this paper, we focus on mitigating users' polarization by controlling their preference gap (uPG score) while interacting with the RS, and propose a simple way to avoid its widening throughout the user-RS interaction process.

1.2   Related Works

Recently, dealing with user polarization in RSs has attracted a lot of interest [3–8]. Prior to its minimization, some of these works [3–5] have modeled user polarization as an ongoing opinion formation process backed by DeGroot and Friedkin-Johsen's models in social science theory. In this context, a RS is often represented by a graph where vertices represent users, and the (weighted) edges represent the similarity between them. Alternatively, other very recent research works [6–8] considered the evolution of users' latent vectors derived from the matrix factorization of the RS's rating matrix. These vectors (or their euclidean norm) are assumed to encode the user's preference towards specific features of the items she has rated. Therefore, limiting their amplification over time is a potential solution to mitigate user polarization in RSs [9]. In this paper, we follow this line of reasoning, and conceptually classify works that are closely related to ours under three main sources of user polarization in RSs:

Item: Some controversial recommended contents can be polarizing. In an effort to handle polarization in the top-N items recommendation, [10–12] propose an approach where each item is assigned a polarization score, for the purpose of separating polarized from non-polarized items, as it is the case in [10,11] or in [12] for polarization measurements. Given a predefined parameter $\theta \in \,]0,1[$, this score is calculated as the ratio of ratings falling below $r_{min} + \theta\,(r_{max} - r_{min})$ and above $r_{max} - \theta\,(r_{max} - r_{min})$ to the total number of ratings assigned to an item by the users who rated it. Here, r_{max} and r_{min} are the maximum and minimum rating value of the considered rating scale. If this ratio (i.e., item polarization score) is high, implying that the item receives almost opposite ratings (forming a U-shaped histogram), then the item is classified as "polarizing". This approach has some limitations, as the polarization score of an item can be high, and yet, the form of its ratings histogram may not have a U shape. This case arises, for instance, when an item receives a small number of ratings. Thus, it becomes difficult to deduce whether an item is polarizing or not by only looking to its ratings histogram.

User: A *biased user* is a user who carries certain doctrines and rigid ideas on certain topics, thus preconditioned to be polarized even before interacting with the RS. Dandekar et al. introduced in [5] a framework that analyzes the effect of biased assimilation on user polarization in RSs, and concluded with this important result: whether a simplified RS algorithm (e.g., SALSA, ICF and PPR) becomes polarizing or not does largely depend on the user's bias. RSs can propagate and amplify different types of user biases, leading to the so-called bias disparity. In [13], the authors addressed this phenomenon via a simple re-ranking top-N recommendation algorithm. Very recently, [14] investigated the users' bias resulting from being exposed to the RS algorithm, called *exposure bias*.

RS Algorithm: The RS algorithm itself can be polarizing, as it encourages users, through recommendations, to adopt the views and opinions of like-minded users, and keeps them away from the views of opposite-minded users. This may result in the formation of echo-chambers, consisting of isolated and polarized sets of users sharing similar preferences, opinions and/or views. Some empirical studies have shown that increasing the diversity of recommendations or their novelty does not necessarily bridge the gap between these echo-chambers, and can even increase political polarization [16] as a counter-reaction. As user polarization caused by the RS algorithm develops over time [17], it is crucial to take control of the RS's degree of personalization in order to minimize its polarizing effect timely, i.e., in the early interaction stage between the user and the RS.

To sum up, linking users' polarization to their exposure to specific items, following [10–12], might be useful in detecting highly polarized items, thus preventing them from being recommended. However, this approach does not tell us whether a RS algorithm is polarizing or not. Furthermore, a biased user should not prevent the RS from moderating her bias and safely treating her preferences. Therefore, item- and user-based polarization detection techniques should be combined with a non-polarizing RS algorithm too, so that the user preference is not significantly amplified over time.

1.3 Contributions

We consider a RS algorithmic approach to minimize user polarization in RSs, and summarize our contributions in this paper as follows:

- We introduce a new model for minimizing user polarization by capturing her uPG between a predefined set of items' attributes (i.e. categories). Thus, at each moment, the RS supervises the dynamics of its users' preferences, and adapts its recommendations accordingly.
- Using our model, we measure the uPG scores based on hierarchical agglomerative clustering of item category timeseries, as the minimum euclidean distance between two timeseries (each one represents the user's consumption from a subset of item categories).
- We test our model on the MovieLens 100k and 1m datasets, and empirically illustrate that it can ensure the minimization of user polarization in RSs.
- We propose a user-item polarization mapping through assigning uPG scores to items, before being re-ranked accordingly. Thus, the top-N items recommended to the active user represent a balance between accuracy and user polarization reduction.
- Finally, using the above framework, we propose post- and pre-constrained versions of the alternating least square (ALS) algorithm that minimize the uPG while maintaining an acceptable accuracy, close to the unconstrained ALS and its counterparts.

The rest of the paper is organized as follows. In Sect. 2, we introduce our model for measuring uPG in RSs. Then, in Sect. 3 and 4, we develop our post-

and pre-constraining methods, respectively, to avoid uPG amplification over the user-RS interaction time. Finally, Sect. 5 concludes the paper.

2 Modeling User Preference Dynamics in RSs

By looking at the user preference dynamics and how they change over time, we aim to evaluate whether the user-RS interaction results in a user preference amplification (thus reinforcing her polarization) or not. To this end, we proceed by attributing categories to each item (e.g., book and movie genres in the case of book and movie recommendation). The categories are denoted by $C = \{C_1, \ldots, C_K\}$; at least one category is attributed to each item. Then, we evaluate the consumed proportions of each category by each user of the RS over time.

2.1 Modeling Items Consumption per Category as Timeseries

The set of items $\mathcal{I}_u(t)$ consumed by user u up to timestamp t is supposed to engender a set of consumed item categories

$$\mathcal{C}_u(t) : \mathcal{I}_u(t) \to C_1^{n_{u,1}(t)} \times \cdots \times C_K^{n_{u,K}(t)} \tag{1}$$

where $n_{u,k}$ is the number of items consumed by user u and belonging to category C_k. In (1), we assume, without loss of generality, that the cardinals $n_{u,1}(t), \ldots, n_{u,k}(t)$ are sorted in increasing order, i.e., $\forall k_1 \leq k_2 \in \mathbb{N}^*$, we have $n_{u,k_1}(t) \leq n_{u,k_2}(t)$. Then, we model the proportion of an item category C_k consumed by user u up to timestamp $t \in \{t_0, \ldots, t\}$ as a timeserie, expressed as

$$S_{u,k}(t) = \frac{n_{u,k}(t)}{\sum_{k=1}^{K} n_{u,k}(t)}. \tag{2}$$

For example, if the first item consumed at timestamp t_0 by user u belongs to two categories C_1 and C_2, then $S_{u,1}(t_0) = S_{u,2}(t_0) = 1/2$. Then, if, at timestamp t_1, the user has consumed an item belonging to a different category C_5, then $S_{u,5}(t_1) = 1/3$, while $S_{u,1}(t_1) = S_{u,2}(t_1) = 1/3$ (given that the user has rated only two items). Note that, at every timestamp t, we have

$$\sum_{k=1}^{K} S_{u,k}(t) = 1. \tag{3}$$

2.2 Hierarchical Agglomerative Clustering of Category Timeseries

To calculate the preference gap score for each user, we perform a *hierarchical agglomerative* (HA) clustering [15] of all category timeseries into two distinct major clusters $\mathcal{G}_{u,1} \subset C$ and $\mathcal{G}_{u,2} \subset C \setminus \mathcal{G}_{u,1}$, such that $\mathcal{G}_{u,1}$ represents the group of categories towards which the user is highly polarized, and $\mathcal{G}_{u,2}$ represent the subsequent group (including remaining categories). For each user, the cardinals of $\mathcal{G}_{u,2}$ and $\mathcal{G}_{u,1}$ vary depending on the item categories rated by user u. Finally, we model the user preference gap (uPG score) at timestamp t as the average normalized Euclidean distances between two category timeseries $k_1 \in \mathcal{G}_{u,1}$ and $k_2 \in \mathcal{G}_{u,2}$ (each one belonging to the two resultant groups of our item category clustering) as

$$\text{uPG}^e(t) = \sum_{k_1 \in \mathcal{G}_1, k_2 \in \mathcal{G}_2} \frac{\sqrt{\sum_{t=t_0}^{t} \left(S_{u,k_1}(t) - S_{u,k_2}(t) \right)^2}}{|\mathcal{G}_1| \, |\mathcal{G}_2| \, \sqrt{(t - t_0 + 1)}} \tag{4}$$

where the timestamp threshold $t_0 \in \{0, \ldots, t-1\}$ has been introduced to ensure that, for some $t > t_0$, user u has rated a certain minimum number of items. For comparison purposes, the proposed metric in (4) is compared with the following simplified version of user preference gap, defined as

$$\text{uPG}^s(t) = \frac{\max_{1 \leq k \leq K} n_{u,k}(t)}{\sum_{k=1}^{K} n_{u,k}(t)}. \tag{5}$$

It is worth noting that both definitions (4) and (5) represent a gap measuring the dispersion of the user consumption from the different item categories over time. However, while (4) is an averaged version of the uPG scores, (5) captures the peak (instantaneous) increase of the uPG score being observed. Furthermore, the fact that (4) takes into account the previously rated items, back to t_0, makes this definition of uPG more relevant in detecting the user preference tendencies towards all item categories. Clearly, the second definition is more sensitive to quick and pronounced variations of user preferences over time, also called user drifts. One advantage of the uPG model in (5) is that it can easily be implemented.

Taking into account the time dimension in both definitions (4) and (5), user polarization can be reinforced if the RS keeps recommending items that increase the user consumption from a specific item category (or group of categories), and thus, her uPG score. There are two possible causes to the increase of this score: either the user is biased, or the RS over-personalizes its recommendations to narrowly match the user preferences, thus becoming narrower over time. In the next section, we address the second cause by preventing user preference polarization from being amplified while the user is interacting with the RS.

2.3   Measurement and Analysis of the uPG Score

We experiment our proposed uPG models on the MovieLens dataset, in which an item category corresponds to a movie genre. Under the assumption that a user only consumes from her list of recommended items, we analyze the consumption history of a user from the different movie genres to anticipate her future movies' consumption per genre. To facilitate the reproducibility of our results, we used the ml-1m MovieLens dataset - available for download at http://grouplens.org/datasets - to generate Fig. 1 and Fig. 2.

- **Item categories HA clustering step:** Consider the users whose Ids are $u = 601$ and $u = 255$, among the 6040 users rating over 3706 items. As

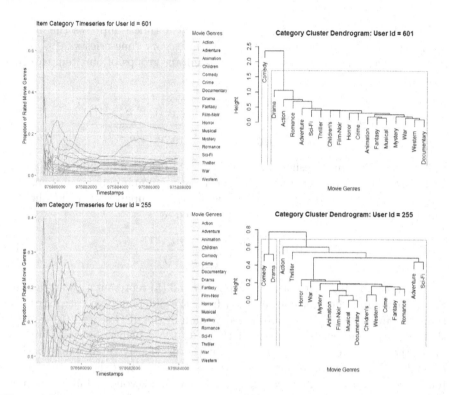

Fig. 1. Movie genre timeseries as expressed in (2) and the result of their HA clustering for user Id 601 and 255 in the ml-1m MovieLens dataset.

described in the previous subsection, prior to measuring each user's uPG at timestamp t, we cluster the consumed movie genres timeseries following our HA clustering, as shown in the upper right and lower right sides of Fig. 1, resulting in two distinct groups of movie genres: $\mathcal{G}_{601,1} = \{\text{Comedy}\}$ and $\mathcal{G}_{601,2} = \mathcal{C} \setminus \mathcal{G}_{601,1}$ for user $u = 601$; and $\mathcal{G}_{255,1} = \{\text{Drama, Comedy}\}$ and $\mathcal{G}_{255,2} = \mathcal{C} \setminus \mathcal{G}_{255,1}$ for user $u = 255$. Compared against the timeseries of the different movie genres shown in the upper left and lower left sides of Fig. 1, our HA clustering results accurately reflect the actual distances between the different movie genres timeseries. Note that the resulting clusters $\mathcal{G}_{601,1}$ and $\mathcal{G}_{601,2}$ will be used to evaluate the uPG score according to (4) and (5), while t_0 should carefully be set in both expressions.

- **uPG measurement step:** Fig. 2 shows the result of our uPG score modeling and measurement strategies for user $u = 255$. Initially, for both equation (4) and (5), the clusters $\mathcal{G}_{255,1}$ and $\mathcal{G}_{255,2}$ are identified at each timestamp following our item categories HA clustering step. Then, depending on the value of t_0 in (4), i.e., the size of the time interval $(t - t_0 + 1)$ during which the user has interacted with the RS, the uPG score is calculated. Importantly, tracking the historical changes of the user consumption per item category is crucial in summarizing her preferences over a large period of time. Thus,

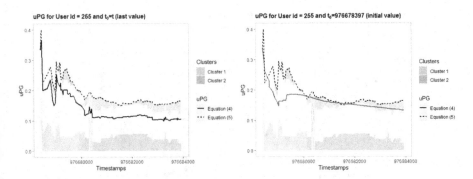

Fig. 2. uPG evolution over timestamps for user $u = 255$ following our models in (4) and (5). In the left figure, we took $t_0 = t$, i.e., only the last timestamp is taken into account when performing the HA clustering of item categories timeseries. In the right figure, t_0 is set to the first interaction time between the user and the RS.

compared to the left-hand side plot in Fig. 2 where $t_0 = t$, the plot on the right-hand side captures the past and present pattern of the user preference tendencies since her first interaction with the RS. As a result, the uPG score (in red) smoothly changes over time.

3   Post-constrained uPG for Polarization-Safe Top-N Items Recommendation

Based on our models of uPG, we introduce a low-complexity yet effective approach for the minimization of user polarization, as constraining the RS optimization problem is shown to be very complex to solve in practice. In particular, we directly operate on the output data of the RS model, and favor top-N items leading to less uPG scores if consumed by the user.

3.1   Unconstrained MF-Based RS Model and Initial Results

An unconstrained matrix factorization (MF)-driven RS predicts the missing ratings in its utility matrix, then recommends (to the active user) the top-N items with the highest predicted ratings, by solving the following unconstrained optimization problem [18]

$$\mathcal{OP}: \min_{(u_i, v_j)\,|\,r_{ij}\ \text{given}} \sum_{i,j} \left(r_{ij} - u_i^T v_j\right)^2 + \lambda \left(\sum_i n_{u_i} \|u_i\|_F^2 + \sum_j n_{v_j} \|v_j\|_F^2\right) \quad (6)$$

where r_{ij} is the observed rating given by user u_i to item v_j, and $\|u\|_F^2$ stands for the Frobenius norm of vector u. The λ-weighted term in (6) corresponds to the Tikhonov regularization [18], ensuring that (6) does not overfit the test

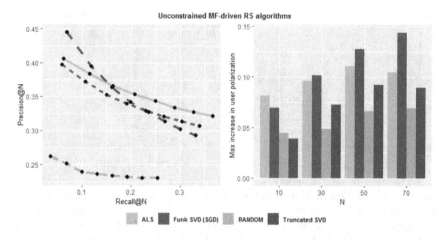

Fig. 3. Left: Precision vs Recall of different unconstrained MF-driven RS algorithms evaluated on the MovieLens 100k dataset - available for download at http://grouplens. org/datasets - for values of N ranging from $N = 10$ to $N = 80$, with an increment of 10; k, as well as the number of iterations of SGD and ALS, are fixed at 10. The only considered ratings in the training and test data are those above the threshold of 3 in the scale of 0–5 integer values. Right: the change in uPG defined in (5) after actually consuming a subset of items from the top-N recommendations.

data, while n_{u_i} (resp. n_{v_j}) is the number of items rated by user u_i (resp. the number of users rating an item v_j). The user and item latent vectors u_i and v_j, respectively, are assumed to embed k features.

To solve the unconstrained optimization problem (6), several MF algorithms, among which truncated singular value decomposition (SVD) and Funk SVD using stochastic gradient descent (SGD), have successfully been applied to locally minimize the loss function. Furthermore, alternating least squares (ALS) can potently solve (6) as a bi-convex loss function in (u_i, v_j). In other words, fixing either user or item latent vectors renders the loss function convex and quadratic, and therefore, its global minimizer can be derived in a closed form [18].

Using the precision-recall plot, Fig. 3 shows, on the left side, the results of running each of the aforementioned algorithms to minimize the unconstrained loss function (6). Apparently, ALS starts taking the lead in both precision and recall, starting from $N = 30$. On the right side of Fig. 3, each of these algorithms is assessed in terms of the maximum increase in uPG, as defined in (5). For each user in the test set, this increase is calculated as the difference in uPG before and after *actually* consuming the subset of items in the top-N recommendations, using one of the competing algorithms. Clearly, Funk SVD has a more polarizing effect than ALS and Truncated SVD as N grows larger. As a result, unconstrained personalization can lead to a reinforced user polarization over time. Also, as shown by Fig. 3 on the right side, the RANDOM algorithm picking just a randomized top-N recommendations has the lowest user polarization effect, due to the non-personalization of its generated content.

To generate the results of Fig. 3, we randomly split the MovieLens 100k dataset (excluding users who rated less than 50 items, and items receiving less than 100 ratings) into training and test sets (representing 90% and 10% of the users, respectively). In the test set, for each user, 15 rated items are also randomly selected and hold unknown, for the purpose of evaluating the RS accuracy. The same operation is repeated for 50 runs before averaging the obtained recall and precision. As a result, we already observe that each of the assessed RS algorithms has a different impact on the uPG. Although these MF algorithms are basic ones, our objective is to analyze their polarizing effect, and provide a way to minimize it that is applicable to more advanced RS models.

3.2   User-Item uPG Scores Mapping: Key Enabling Factor to Minimize Polarization in RSs

We now introduce our user-item polarization scores mapping function, as a key enabler to minimize user polarization while solving (6) independently of the chosen RS algorithm and uPG model, as suggested in Sect. 2.

Definition 1. *An item polarization score (iPS) is defined as the uPG score of the user after she presumably consumes it from its recommendation list, using a model of uPG between (4) and (5). We say that the user uPG is assigned or mapped on to the not-yet-rated items among the top-N.*

It follows from Definition 1 that each item in the top-N recommendations is assigned an iPS that is calculated based, for instance, on the euclidean-distance-based uPG model in (4). Depending now on the type of constraint put on iPS, the RS needs to balance the trade-off between maximizing its accuracy and minimizing the uPG scores of its users.

3.3   Solving The Constrained RS Optimization Problem is NP-hard

If we reformulate our RS model $\mathcal{OP}$ presented in Eq. (6) as an uPG-constrained RS optimization problem, the active user preference gap uPG(t) can be minimized in two regimes. In a strict one, $\mathcal{OP}_1$ does not allow the uPG to increase over time, and finds application when a specific user has already reached a high uPG score. It can be expressed as

$$\mathcal{OP}_1 : \quad \mathcal{OP} \quad w.t. \quad \text{uPG(t+1)} \leq \text{uPG(t)}. \tag{7}$$

On the other hand, in a tolerant uPG-constrained regime, $\mathcal{OP}_2$ can allow the uPG score to increase up to a certain pre-defined threshold $\overline{\text{uPG}}$ such that

$$\mathcal{OP}_2 : \quad \mathcal{OP} \quad w.t. \quad \text{uPG(t+1)} \leq \overline{\text{uPG}}. \tag{8}$$

In both variants of our optimization problems $\mathcal{OP}_1$ and $\mathcal{OP}_2$, the uPG score expression under constraint in (7) or (8) can be one of the models defined in (4) or (5). Following the user-item polarization mapping function introduced in

Definition 1, the top-N MF-driven RS models $\mathcal{OP}_1$ (resp. $\mathcal{OP}_2$) can equivalently be reformulated as

$$\widetilde{\mathcal{OP}}_1 : \quad \mathcal{OP} \quad s.t. \quad \max_{j^*} \left\{ u_a^T v_{j^*} \right\} < \gamma_N \qquad (9)$$

where γ_N is the smallest predicted rating of the top$-N$ recommended items satisfying the constraint in (7). That is, items j^* increasing the active user (with a latent vector u_a) uPG score are prevented from being recommended, by constraining their predicted ratings $u_a^T v_{j^*}$ to be below γ_N.

To solve $\widetilde{\mathcal{OP}}_1$, and, similarly, $\mathcal{OP}_2$, note that the constraint in (9) involves a sorting function to calculate γ_N that is not a constant but a term dependent on u_a and the non-polarizing items' latent vectors. This makes our optimization problems NP-hard, and thus prohibitively complex to solve, especially when the RS rating matrix becomes very large.

3.4   Cost-Effective Post-Constrained RS Model: Approach 1

Ideally, a safe RS should not recommend its users items that significantly amplify their preferences; yet, most importantly, the RS accuracy has to be maintained high. Therefore, we first propose a sub-optimal uPG-post-constrained approach to minimize user polarization in RSs, by dropping from the top$-N$ list items whose iPS scores are above the current uPG, or above some predefined (constant) uPG threshold. Finally, the recommended top-N items are re-ranked from highest-to-lowest predicted ratings, but now lowering or at least maintaining the user's uPG score below a certain threshold. As a result, this approach does not impact the RS algorithm nor its inputs, as it only operates on the output list of items being recommended to the user. This first simple and effective solution to minimize user polarization has been made possible thanks to the user-item uPG scores mapping detailed in Subsect. 3.2. In the next section, we introduce our second approach for the ALS algorithm, and provide comparison results based on the MovieLens dataset.

4   Pre-constrained ALS Algorithm and Comparative Empirical Results

Another simple but more involved approach to minimize uPG scores is to consider that the not-yet-rated items whose iPSs are high (i.e., polarizing items) have virtually been rated by the user, and fictitiously assign them a rating with a minimal value. Consequently, the RS model treats these ratings as training data, and will not recommend them to the active user. Obviously, assigning a low rating to all not-yet-rated items with high iPSs may result in accuracy loss for the RS, because not all of these items will necessarily be among the top-N.

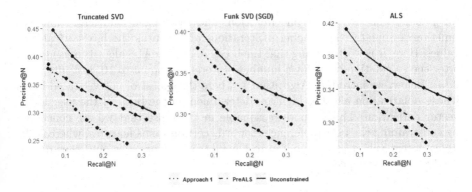

Fig. 4. Precision vs Recall for various algorithms, using our two simple approaches to minimize user polarization, evaluated on the MovieLens 100k dataset - available for download at http://grouplens.org/datasets - for values of N ranging from $N = 10$ to $N = 80$ with an increment of 10; k, as well as the number of iterations of SGD and ALS, are fixed at 10.

4.1 PreALS Algorithm For Mitigated User Polarization in RSs

We apply and test our second pre-constrained solution approach to the ALS algorithm, even though the same idea can be applied to other optimization algorithms to prevent uPG scores from being amplified while interacting with RSs. Before introducing our PreALS algorithm, let us recall both loss functions, that are alternatively minimized. We rewrite the loss function in (6) for fixed latent vectors v_js as $f(u_1, \ldots, u_U) = 2\sum_{i \in \mathcal{U}} f_i(u_i) + \varepsilon_1$, where $\varepsilon_1 = \lambda \sum_{j \in \mathcal{I}} n_{v_j} \|v_j\|_F^2$,

$$f_i(u_i) = \frac{1}{2}u_i^T \underbrace{\left(A_{1i}^T A_{1i} + \lambda n_{u_i}\mathrm{I}\right)}_{Q_{1i}} u_i + \underbrace{\left(-A_{1i}^T b_{1i}\right)^T}_{P_{1i}} u_i \qquad (10)$$

and $A_{1i} = [\ldots, v_j, \ldots]^T \in \mathbb{R}^{I \times k}$, where j belongs to the I-cardinal subset of items rated by user $i \in \mathcal{U}$ and $b_{1i} = [\ldots, r_{ij}, \ldots]^T \in \mathbb{R}^{I \times 1}$, where $\mathcal{U}$ and $\mathcal{I}$ are the users and items global sets. Alternating the u_i and v_j values, the loss function (6) (for fixed u_i values) can similarly be expressed and developed as $g(v_1, \ldots, v_I) = 2\sum_{j \in \mathcal{I}} g_j(v_j) + \varepsilon_2$, where $\varepsilon_2 = \lambda \sum_{i \in \mathcal{U}} n_{u_i} \|u_i\|_F^2$,

$$g_j(v_j) = \frac{1}{2}v_j^T \underbrace{\left(A_{2j}^T A_{2j} + \lambda n_{v_j}\mathrm{I}\right)}_{Q_{2j}} v_i + \underbrace{\left(-A_{2j}^T b_{2j}\right)^T}_{P_{2j}} v_j, \qquad (11)$$

and $A_{2j} = [\ldots, u_i, \ldots]^T \in \mathbb{R}^{U \times k}$, where i belongs to the U-cardinal subset of users rated items $j \in \mathcal{I}$, and $b_{2j} = [\ldots, r_{ij}, \ldots]^T$. In (10) and (11), ε_2 and ε_2 should be considered as constants. In our proposed PreALS solution, the RS performs its MF by iteratively solving (10), injecting its solution $u_i^\star = Q_{1i}^{-1}P_{1i}$ for $i \in \mathcal{U}$ into (11), then deriving the optimal solution of the latter as $v_j^\star = Q_{2j}^{-1}P_{2j}$

for $j \in \mathcal{I}$. This completes one iteration, and reaching a sufficiently low loss function terminates the number of iterations and returns the latest $u_i^{\star}$ and $v_j^{\star}$ values. Here, it is worth noting that the break in our PreALS algorithm is applied after the calculation of $u_i^{\star}$ for a number of iterations equal to 10, just after step 2.(a). Below, we introduce Algorithm 1, where the subset of not-yet-rated items whose polarization effect is high are determined on the basis of the user latent vector derivation. This is more reasonable in the sense that we are concerned by constraining the user preferences towards certain items features, whereas the items' latent vectors are supposed to have intrinsic properties, and thus do not need to be constrained.

Algorithm 1. Proposed PreALS Solution Algorithm.

---

1. **Intialize** A_{1i} for $i \in \mathcal{U}$, where the I elements of the first column are the average ratings per item in the rating matrix. The other elements of A_{1i} are set to zero.
2. **Iterate** until the 10th iteration:
   (a) **While** the top$-N$ recommended items (for each user in the test set) include a polarizing item, i.e., whose iPS score is such that $upG(t) \leq iPS$, **do**
   - Compute $u_i^{\star} = Q_{1i}^{-1} P_{1i}$ for $i \in \mathcal{U}$
   - Form an inclusive (i.e. growing with the iterations) black list $\mathcal{B}_a$ of items that are polarizing for the active user, based on $u_i^{\star}$
   - Set the ratings of items in $\mathcal{B}_a$ to 3 and go back to 2.(a)
   (b) **Mimimize** $v_j^{\star} = Q_{2j}^{-1} P_{2j}$ for each $j \in \mathcal{I}$
3. **Return** the top$-N$ items with the highest predicted ratings to the active user a

---

4.2   Empirical and Comparative Results

As shown in Fig. 4 on the right side, PreALS outperforms Approach 1, yet remarkably with variable gains achieved in terms of precision and recall, depending on the value of the pre-constrained items ratings (initially set to 3, as stated in step 2.(a) of Algorithm 1). In our simulations, it is realized that with a slight increase in this value, we obtained optimal performances for pre-constrained items ratings equal to 3.5. Clearly, PreALS shows better performance results than Approach 1 for all values of N.

Furthermore, PreALS achieves a close performance to the unconditional Truncated SVD and Funk SVD algorithms in terms of the precision and recall, and far better than their post-constrained variants following Approach 1, as shown in the left and middle plots of Fig. 4. Quite importantly, pre-constraining the input data of the RS algorithm in order to minimize user polarization does not necessarily lead to a reduced RS accuracy. Compared to the low implementation complexity of Approach 1 (described in subsection 3.3) relying on re-ranking items by dropping those whose polarization scores are high, PreALS allows a good trade-off: achieving a high RS accuracy, while maintaining a low user polarization over the user-RS interaction period.

5   Conclusion

In this paper, we addressed the problem of mitigating user preference amplification in RSs, through a new modeling and measurement of its evolution. To minimize its effect in the short and long runs, we introduced a simple uPG post-constraining approach to minimize user polarization based on our user-item scores mapping, while guaranteeing an acceptable RS accuracy. Using our proposed RS algorithm PreALS, our empirical results on the MovieLens datasets reveal promising gains in terms of user polarization reduction, that can be obtained by adopting the appropriate safety measurements in contemporary RSs.

References

1. Ricci, F., Rokach, L., Shapira, B.: Recommender Systems Handbook. Springer Publishing Company, Incorporated (2015)
2. Milano, S., Taddeo, M., Floridi, L.: Recommender systems and their ethical challenges. AI Soc. **35**, 957–967 (2020)
3. Donkers, T., Ziegler, J.: The dual echo chamber: modeling social media polarization for interventional recommending. In: Fifteenth ACM Conference on Recommender Systems, pp. 12–22 (2021)
4. Amendola, L., Marra, V., Quartin, M. The evolving perception of controversial movies. ArXiv.abs/1512.07893 (2015)
5. Dandekar, P., Goel, A., Lee, D.: Biased assimilation, homophily, and the dynamics of polarization. PNAS **110**, 5791–5796 (2013)
6. Kalimeris, D., Bhagat, S., Kalyanaraman, S., Weinsberg, U.: Preference amplification in recommender systems. In: Proceedings of ACM SIGKDD, pp. 805–815 (2021)
7. Stray, J. Designing Recommender Systems to Depolarize. ArXiv (2021)
8. Badami, M., Nasraoui, O.: PaRIS: polarization-aware recommender interactive system. In: Proceedings of the 2nd Workshop OHARS (2021)
9. Jiang, R., Chiappa, S., Lattimore, T., György, A., Kohli, P.: Degenerate feedback loops in recommender systems. In: ACM AIES, pp. 383–390 (2019)
10. Badami, M., Nasraoui, O., Sun, W., Shafto, P.: Detecting polarization in ratings: an automated pipeline and a preliminary quantification on several benchmark data sets. In: 2017 IEEE International Conference on Big Data, pp. 2682–2690 (2017)
11. Badami, M., Nasraoui, O., Shafto, P.: Pre-recommendation counter-polarization. In: KDIR, PrCP, pp. 280–287 (2018)
12. Keyes, R., Gillet, T., Manggala, P.: On measuring polarization using recommender system scores. Reveal Workshop, RECSYS (2018)
13. Tsintzou, V., Pitoura, E., Tsaparas, P.: Bias disparity in recommendation systems. ArXiv Preprint ArXiv:1811.01461 (2018)
14. Khenissi, S., Nasraoui, O.: Modeling and counteracting exposure bias in recommender systems. ArXiv Preprint ArXiv:2001.04832 (2020)
15. James, G., Witten, D., Hastie, T., Tibshirani, R.: An Introduction to Statistical Learning: with Applications in R. Springer (2013)
16. Bail, C., et al.: Exposure to opposing views on social media can increase political polarization. In: Proceedings of the NAS, vol. 115, no. 37, pp. 9216–9221 (2018)

17. Celis, L.E., Kapoor, S., Salehi, F., Vishnoi, N.K.: Controlling polarization in personalization: an algorithmic framework. In: Proceedings of ACM FAT, pp. 160–169 (2019)
18. Zhou, Y., Wilkinson, D., Schreiber, R., Pan, R.: Large-scale parallel collaborative filtering for the netflix prize. In: Fleischer, R., Xu, J. (eds.) AAIM 2008. LNCS, vol. 5034, pp. 337–348. Springer, Heidelberg (2008). https://doi.org/10.1007/978-3-540-68880-8_32

Social Multi-role Discovering with Hypergraph Embedding for Location-Based Social Networks

Minh Tam Pham[1], Thanh Dat Hoang[1], Minh Hieu Nguyen[1], Viet Hung Vu[1], Thanh Trung Huynh[2(✉)], and Quyet Thang Huynh[1]

[1] Hanoi University of Science and Technology, Hanoi , Vietnam
{tam.pm202708m,dat.ht202714m,hieu.nm2052511m,
hung.vv162050}@sis.hust.edu.vn, thanghq@soict.hust.edu.vn
[2] Griffith University, Queensland, Australia
thanh.huynh@epfl.ch

Abstract. Location-based social networks (LBSNs) have become more and more popular in the recent years. The typical LBSN platforms such as Foursquare, Facebook Local or Yelp allow the user to share their daily digital footprints in the form of check-ins with other people in different communities. The dynamic between users' social context and their mobility plays an important role in LBSN, e.g. users potentially participate with their friends in the same activity. The social interaction also demonstrate in the form of multi-role context, as each user may experience different activities with each particular community. Existing representation learning for LBSNs analysis often fails to fully capture such complex social pattern. In this paper, we propose a representation learning model in which the multi-role social interaction can be captured simultaneously with the mobility information. More specifically, the model first applies a "persona" decomposition process, where each user node is splitted into several pseudo nodes presenting for his social roles. The process then learns multiple presentations for each persona that reflect the corresponding role by maximizing the collocation of nodes sampling on input user-user edges (friendships) and user-time-POI-semantic hyperedges (check-ins). We conduct experiments on 5 real-world datasets with 7 state-of-the-art baselines to demonstrate the robustness of our model on downstream tasks such as friendship suggestion and location prediction.

Keywords: Location-based social network · Hypergraph embedding · Recommendation system

1 Introduction

Location-based social networks (LBSNs) such as Facebook Local, Foursquare, Yelp, Brightkite and Gowalla have emerged in the recent years [8]. LBSNs allow the users to share their daily experiences to other people by the check-ins, each includes a location (a.k.a point of interest (POI) such as university, plaza, supermarket), a specific timestamp, a semantic category (e.g. attending class, working

© The Author(s), under exclusive license to Springer Nature Switzerland AG 2022
N. T. Nguyen et al. (Eds.): ACIIDS 2022, LNAI 13757, pp. 675–687, 2022.
https://doi.org/10.1007/978-3-031-21743-2_54

out or shopping). LBSNs data contain rich socio-spatial properties of user activities. For instance, users who enjoy the same activity at the same place within overlapping period potentially meet and make friend with each other, due to a common presence and hobby. Also, members in a community have high chance to collaborate in the same activity (e.g. students attending a class, gym members working out at a gym).

Given the great benefit of LBSNs analysis, a rich body of researches has been proposed to model the social interaction to user mobility. Traditional techniques design hand-crafted features extracted from either user mobility data (e.g. co-location rates) or user friendships (e.g. Katz index [4]) to investigate the impact from one on the other. Such approaches often require significant human effort and domain knowledge as well as lack of generalizability to different applications. Recent techniques leverage the advances in graph representation learning [9] to embed the nodes into low-dimensional embedding spaces that automatically capture the users mobility and social context, based on the original graph topology and nodes' attribute.

However, the existing techniques have not fully exploited the multi-context nature of LBSNs. In LBSNs, the user nodes associate with other key modalities including spatial pattern (POI nodes), temporal pattern (time nodes), and semantic pattern (activity nodes) in the check-ins, results in a high-order modal. Thus, the LBSNs bring the nature of hypergraph, where the check-ins can be presented as hyperedges (each edge including four nodes) besides the classical friendship (user-user) edges. Most embedding-based approaches simplify the complexity of LBSNs by first breaking the hyperedges into smaller classical edges, then apply the existing representation learning technique for classical graph. Such transformation process might involve information loss and performance degradation [22,23].

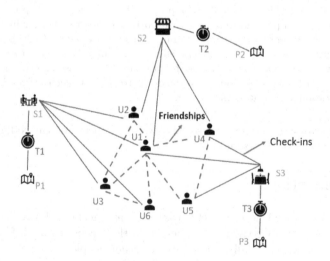

Fig. 1. Example of LBSNs

Another key property of LBSNs that the existing works fail to capture is the social multi-role context, which is often presented in the form of social groups or communities. Users may have different type of activities depending on different groups that they participate in (e.g. an university student, a gym member) [10, 15]. For example, one student often goes to class in university with his classmates or goes to the gym with other member of a club, but when he plan to go out with his family, they often go shopping at a supermarket near their home or eating out at the plaza. The usage of a single context for multi social roles of the user therefore might be imprudent. Also, the sampling process using in existing embedding-based works [22,23] often flattens out the community structures that the users participate in and considers only the close neighborhoodship.

In this work, we propose a Multi-Role Social Interaction aware embedding framework for **LBSN** (MR-LBSN), where we leverage hypergraph embedding to capture the at the same time multi-role social context and mobility dynamic of the users. Our approach models the holistic interactions between multi-role social and mobility context via hypergraph representation and persona decomposition. Our model first devise multiple presentations, so-called user personas, for each user; each persona corresponds to one social role of the user. Then, the friendship and check-in information are used together to learn the latent feature for each persona node. We summarize our contributions as below:

- We propose a multi-context hypergraph embedding method for LBSNs that captures simultaneously both friendship and check-ins information. We are the first to address the multi-role nature of users' social context by analyzing the overlapping structure of communities.
- We specially designed a persona decomposition algorithm that considers at the same time user context scope and check-in assignment for persona nodes. To this end, the each user social role is depicted by a friend community and their shared mobility pattern.
- We integrate user multi-role context and user mobility under the same modal by combining persona transition and check-in exploration simultaneously in the random walk process. The random walk is also designed to avoid the dominance of the number of check-ins over the friendships.
- We develop a social multi-role aware embedding that leverage the sampled random walks to integrate simultaneously the check-in and friendship information. For each hyperedge, we maintain the n-wise proximity that is reflected through the co-occurrence of the nodes.
- We conduct extensive experiments with five real-world datasets to justify the improvements over others embedding baselines in two essential LBSN applications: friend suggestion and POI recommendation.

2   Related Work

Recently, location-based social networks such as Facebook Local, Yelp, Foursquare or Gowalla have attracted million of users to share their daily experiences with others. These platforms contain great pool of information and mostly accessible to the public. To this end, many researches attempts to learn the underlying patterns of user behaviour in LBSNs [6]. The earlier techniques often leverage hand-crafted features such as daily routines [10,14], dispersion metric [1] and applies heuristics to retrieve the needed insight about the users. For example, Wang et al. [19] observes that friends often have shared communication activities, thus they define a threshold and determine that two people potentially be friends if the number of shared activity between them being greater than the threshold. Yang et al. [22] characterises user mobility based on two criteria: the total time-independent travel distance and the probability of returning to particular locations. Song et al. [14] uses a metric called mobility entropy to reflect users' daily mobility. Backstrom and Kleinberg [1] proposed dispersion metric to estimate the tie strength between the connected users and detect the couples and romantic partners by their strong social bond pattern. However, the feature engineering requires significant human effort and expert knowledge.

Recent advances in graph embedding techniques [2,5,12,16] enables the recent LBSN mining techniques [7,23] to learn automatically the underlying features, which helps to save the human efforts and the requirement of domain expert in feature engineering process. As the LBSN data can be naturally presented as a hypergraph, with users, POI, time and semantic being the nodes and the friendship and check-in hyperedge connecting them, it would be complicated to use the traditional adjacency matrix. To this end, the embedding technique allow to learn new representation, where the nodes are assigned to low-dimensional vectors such that the learnt latent space can reflect the topology of the original hypergraph. The learnt embeddings play the roles as the feature vector for the nodes and downstream applications such as friend suggestion and POI prediction can leverage this new representation to enhance the prediction ability. Since the existing graph embedding techniques were often developed for classical graph, the existing works often transform the original hypergraph to multiple classic graphs. For instance, [21] extracted the POI-POI graph where the edges demonstrate the POIs visit by the same user, which thus help the technique to emphasize on the sequential dynamic of the POIs. Xie et al. [20] break the original LBSN into four bipartite graphs, namely the POI-POI, POI-region, POI-time and POI-category graphs, respectively; then integrate the information in these graphs simultaneously using a multi-objective loss function. However, Yang et al. [22] found that the irreversible transformation of LBSN data into classical graphs causes information loss and thus may lead to performance degradation. They proposed a hypergraph embedding solution for LBSN, but ignore the multi-role of users, which we successfully address in this work.

3   Problem and Approach

3.1   Problem Statement

Before formulating the problem, we describe the following necessary definition, including the hypergraph and its elements:

Users and Friendship. In LBSNs, users, presented by u, play a central role. Users can establish relations with each other (e.g. friendship, followership), where each edge connects two user nodes. In this work, we call such edges as *friendship edge*. For example, in Fig. 1, $(u1, u2), (u3, u4)$ are user friendship edges.

Check-ins. Besides, in LBSNs, the user activities (check-ins) are specified in the form of quadruples *(user, semantic, time, POI)*, denoted by (u, s, t, p). A POI p is a specific site (e.g. a cinema or a plaza). A time slot t is the timestamp when the check-in occurs. An activity semantic s is the content of the activity (e.g. shopping or working out), and different activity semantics can occur at the same POI. The check-in tuples can be seen as hyperedges of size four in the LBSN hypergraph. In Fig. 1, (u_1, s_1, t_1, p_1) is a check-in by user u_1.

LBSN Hypergraph. Given all the elements, the LBSN hypergraph can be denoted by $G = (V, E)$. V contains four node types, including user nodes, POI nodes, semantic nodes and time nodes, whose subset is denoted by V_u, V_p, V_s and V_t respectively. E contains two edge types: user friendships and check-ins, as defined above.

Problem 1 *(Multi-role embedding for LBSN hypergraph).* *Given a LBSN hypergraph $G = (V, E)$, the problem is to learn the representations $Z_v = \{z_v^1, \ldots, z_v^n\}$ for each node $v \in V$, with n being the number of social roles if v is user node, otherwise $n = 1$. The learnt space has to capture the relationship of the original nodes: $\omega(u, v) = \sigma(Z_u, Z_v)$, with ω determines the similarity between the original hypergraph nodes and σ calculates the distance in the embedding space.*

In practice, ω can be the function which determines whether the two nodes co-occur in a hyperedge; and σ can be the cosine distance. This formulation allows each user to be represented by multiple embeddings w.r.t his social roles.

3.2 Framework Overview

Figure 2 demonstrates the pipeline of our proposed model. Given a LBSN, we identify the local overlapping subgraph cluster of user nodes to split the node into

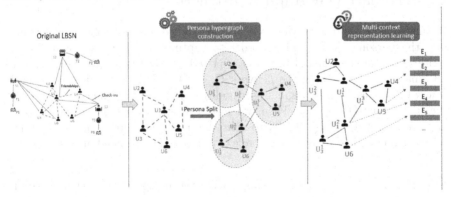

Fig. 2. Framework overview

multiple pseudo nodes (often referred as *personas*), results in a new structure namely *persona hypergraph*, to capture the social dynamics of users. With the obtained structure, we apply a biased random walk process that sample and balance the users' social context and mobility context. Finally, we learn the embeddings for all LBSN nodes by a skip-gram model that maximises the nodes co-occurrence in the same walk context. In summary, our model implements following functions:

Persona Hypergraph Construction. Given an LBSN hypergraph G, we first construct a persona hypergraph G_p where each user node u in G is represented by a series of replicas $\{u_1, u_2, \ldots, u_k\}$. Each replica node of the user, so-called *personas*, reflect a social role of that user in a corresponding community. This mechanism allows the model to capture the multi-context nature of social dynamics in the real world. We go beyond the state-of-the-art by proposing a novel splitting approach that includes two important constraints. The first constraint is an isolation constraint, while the second is an anti-fragmentation constraint. The constraints aim to prevent the exploration of the users multi-context from going too far from the original structure. The details can be found in Sect. 4.

Social Multi-role Aware Representation Learning. After persona decomposition, we perform a sampling process using random walks to determine the context of each user persona by its close relation nodes. As the number of check-ins is often outweighs that of friendships, we incorporate a bias weighting mechanism into the random walk process to balance the social and mobility contexts. From the obtained walks for each node, we apply the skip-gram model to generate the node embeddings that maximize the co-occurrence of the nodes within the same context but with some efficient twist. In more details, unlike the contexts in classic graphs that contains only two nodes in an edge, the hyperedge context might contain multiple nodes. To this end, we leverage the best-fit-line mechanism from linear regression on n-wise proximity of hyperedge to reflect the topology of the original hypergraph in the embedding space. The details can be found in Sect. 5.

4   Persona Hypergraph Construction

In this section, we discuss the process that transforms the original hypergraph into the persona hypergraph in order to capture the users' social multi-role contexts. From the original LBSN $G = (V, E)$, we first decouple the friendship information by extracting the network $G_u = (V_u, E_u)$, so-called *user network*, where the nodes are the original user nodes u_i from G; and the edges represent the friendships ($E_u \subset E$). Then, we inspire from the persona decomposition algorithm introduced by the [3] to split the user node into multiple personas, each persona node presents a social role of the user. The detailed process contains of the three steps as follows:

Step 1: For each user $u \in G_u$, we apply a clustering algorithm $\mathbb{A}_l$ that partitioning the local information to retrieve the ego-net of u. Mathematically, given the neighboring set of u as N_u and the corresponding constructed subgraph $G_u[N_u]$, we have $\mathbb{A}_l(G_u[N_u]) = \{N_u^1, N_u^2, \ldots, N_u^k\}$, with k being the number of the partitioning result.

Step 2: From the ego-net partitions $\mathbb{A}_l(G_u[N_u]) = \{N_u^1, N_u^2, \ldots, N_u^k\}$ for each user u, we assign k corresponding persona nodes for that user. We denote the collection of persona nodes for each user u by $Per(u)$.

Step 3: The user edges between the persona nodes are then formed by connecting the user personas that belong to the same community. In more details, if any two persona u_1^i, u_2^j of users u_1, u_2 satisfy that one belongs to the same group of the other: $u_1 \in N_{u_2}^j \bigwedge u_2 \in N_{u_1}^i)$; we connect them with an edge. The result is a persona-aware edge set E_p.

Note that the direct usage of this traditional ego-network analysis might lead to the three main problems. First, the persona graph is potentially quite different from the original graph, if the output of the local clustering algorithm $\mathbb{A}_l$ consists of too many subgraphs. This phenomenon should be avoided, since one person often has only a few social roles in the real-world. Second, the redundant persona split might create the disconnected components even if the original graph was connected. Third, the original persona decomposition only captures multi-role contexts, but lacks the interaction with high-order and user mobility contexts.

For the first issue, we add edges between any two personas of a user to make the whole persona graph connected, instead of leaving them unconnected as in the original traditional ego-net analysis [3]. To further mitigate the issue as well as to address the second issue, we enforce a *social scope constraint* that control the quality of the subgraphs generated by the local clustering algorithm $\mathbb{A}_l$, which helps to avoid creating too many personas for a user. For the third issue, we assigning all check-ins information of each user to its corresponding personas. More precisely, if a user u_i related to the POI p and p is assigned to the persona u_i^k; all the check-ins visited by u_i to that POI are added to E^{per}. The final persona hypergraph is denoted by $G^{per} = (V_u^{per}, E_u^{per})$.

5 Social Multi-role Aware Representation Learning

5.1 Biased Random Walk Based Sampling

After constructing the persona graph, we perform a biased random walk process to integrate at the same time the multi-role social context and the mobility of the users. In more details, starting from the persona nodes as the root, the friendship hyperedges and the check-in hyperedges of the persona graph are jointly visited by the walks to reflect the holistic interactions between the social relationship and the user mobility of the users. Also, as each original user node is now represented by set of persona nodes in the persona graph, our sampling scheme is designed to balance between the users internal information (related to personas belong to the same user) and external information (w.r.t personas belong to neighborhood users), so-called *persona transition*. Also, to efficiently explore the mobility of the users while sampling through the user personas, the random walk stays at each persona to explore the corresponding check-ins.

Persona Transition. As mentioned above, we introduce a persona transition mechanism to control the exploration between the user personas since the original user are transformed into multiple persona nodes. For any two persona nodes, there are three cases: (i) the two persona nodes are connected and belong to the different original users, (ii) the two persona nodes are connected and belong to the same original users, and (iii) they are disconnected.

For the first case, we calculate the transition weights between the personas using the proportion of the number of shared locations over all visited locations. Mathematically, for any two personas belonging to different original users i and j, the weight w_{ij} between them is calculated by the Jaccard distance between the visited POI set of i and j:

$$w_{ij} = Jac(loc(i), loc(j)) + 1 = \frac{\#(loc(i) \cap loc(j))}{\#(loc(i) \cup loc(j))} + 1 \qquad (1)$$

with Jac being the Jaccard distance between two location sets $loc(i)$ and $loc(j)$.

For the second case, if the two personas are connected and belong to the same original user, the transition weight is set to one. Otherwise, the transition weight is set to zero for any two disconnected persona nodes i and j. After assigning the transition weight, we apply the same biased random walk process as in [5].

Check-in Exploration. To capture the mobility dynamic along with the social context, we extend the random walk process to explore the check-ins information. During the transition through the persona nodes, the random walks also stop at each persona to explore the check-ins related to the personas. To this end, not only the social dynamic of the users are sampled through the co-occurrence of the personas but also the mobility pattern of the users are investigated through the check-in hyperedges, and both these information are integrated at the same time. As the number of check-ins often dominates the number of users, we use a check-in visit probability α as a hyperparameter to control and balance the exploration process. At each persona node, the process has the probability of α to explore its check-ins and $(1 - \alpha)$ to continue and move to another persona. Such mechanism allow the process to modulate the number of influential hyperedges of each type, so that α reflects the importance of user mobility and $(1 - \alpha)$ depicts that of social dynamic to the learned node embeddings. In practice, we set α as 0.1 due to the dominance of the number of check-ins.

5.2   Embedding Learning

From the sampled contexts, we employ an embedding strategy that aims to capture the co-occurrence of the nodes that appear in a context as well as to preserve the n-wise node proximity of the hyperedges of the LBSN. To this end, we embed the network nodes into the embedding space where the nodes proximity in the original context is reflected by the cosine similarity. In more details, the direct neighborhoods of the hyperedges of the graph and the nodes occuring in the same walk context are learnt by maximising the co-occurrence probability of all of such pairs of vertices v_i, v_j:

$$L = \sum_{v_i, v_j \in E} log(p(v_i, v_j)) = \sum_{v_i, v_j \in E} log(\sigma(z_i, z_j)) \quad (2)$$

where v_i, v_j are two any neighboring nodes. We model the co-occurrence probability of the nodes by the similarity function σ between their corresponding embeddings z_i, z_j, which is chosen by cosine similarity rather than the popular dot product function used in the existing works [5,17] since it is stable to the norm of the input vectors, which has a large variance due to the varying of the hyperedge size ($n = 2$ or 4 in our LBSN hypergraph).

A potential issue is the solution overfitting and the heavy computation because the large amount of user node pairs as well as their check-ins profile (a quadratic problem). To mitigate this problem, we employ a *negative sampling* strategy which allows us to only modify a small percentage of the weights, rather than all of them for each training sample:

$$L = \sum_{v_i, v_j \in E} log(\sigma(z_i, z_j)) + \sum_{k=1}^{K}(1 - log(\sigma(z_i, z_k^{neg}))) \quad (3)$$

where the former term represents the co-occurrence of the nodes and the latter term depicts K negative samples z_k^{neg}. As the LBSN hypergraph contains four different modalities, the negative samples for each node v_i are uniformly sampled with other nodes with the same type.

6 Empirical Evaluation

6.1 Experimental Setting

Datasets. We use five real-world datasets collected from Foursquare (details shown in Table 1). The first four datasets were collected from tourist cities from 2012 to 2014 [22] over four cities, including Jakarta (**JK**), Kuala Lumpur (**KL**), Sao Paulo (**SP**), Istanbul (**IST**), results in total of 27,148 users and 2,062,965 check-ins. The fifth dataset is collected from 114,508 Foursquare users all over the **USA** in the period from 2010 to 2011, results in total 1,434,668 check-ins.

Table 1. Statistics of the datasets

Dataset	SP	KL	JK	IST	USA
#User	3,954	6,432	6,395	10,367	3,707
#POI	6,286	10,817	8,826	12,693	13,105
#Checkins	249,839	526,405	378,559	908,162	255,557
#Friendships	16,969	35,001	18,641	45,969	16,256

Downstream Tasks and Metrics. We evaluate the performance of the techniques on two important downstream tasks: friend suggestion and POI recommendation. For the former task, we leverage the embedding similarity between the users to determine whether they are potential to make friend in the future, then evaluate the result using five metrics: precision, recall, F1-score and nDCG to the top-K predicted friendships [13]. For POI recommendation, for a set of (user, time, semantic) as the query, we choose the POI that having the closet embedding as the result, then apply Hit@K to evaluate the result [11].

Baselines. We compared our technique with several state-of-the-art baselines, including classic graph embedding techniques DeepWalk [12], Node2vec [5]; and hypergraph embedding techniques DHNE [18] and LBSN2vec [22,23]. Since most of these baseline methods cannot take directly the LBSN hypergraph as input, we used three settings to adapt the hypergraph to the techniques:

- (S): Only the social information is considered by keeping only the user nodes and their friendship edges. This setting can be applied to classical embedding techniques such as DeepWalk and Node2Vec.
- (M): Only the check-in hyperedges is kept. The result hypergraph contains only one type of hyperedge (check-ins), thus is compatible to homogeneous hypergraph embedding technique like DHNE.
- (S&M): Full LBSN hypergraph is considered. This setting can be directly applied to LBSN2Vec and our technique MR-LBSN.

Hyperparameter Setting. The hyperparameters for all experiences were set as follows: the random walk length and number of walks were set at ($l = 80$, $r = 10$), respectively; the mobility-social balance hyperparameter α was set at 0.1; and the node embedding dimension was set to 128.

6.2 Performance on Friendship Suggestion

Figure 3 demonstrates the performance of our technique and the baselines on friendship suggestion task. It can be seen that our technique *MR-LBSN* outperformed the others in all datasets. *MR-LBSN* achieved the gain of nearly 50% in Precision and F1-score compared to the best baseline, LBSN2Vec, in the *USA* dataset. This might be because that the dataset has the densest check-ins, which indicates a strong user-POI correlation. For the other datasets, our technique outperformed LBSN2Vec by almost 40%. When it comes to the Recall metric, *MR-LBSN* achieved slightly less significant improvement as for precision and F1-score, but still reached an average enhancement of 21.35%. For the nDCG@k metric, our techniques outperforms other baselines by 15–30% in all five datasets. This is a solid proof for the capability of our technique in recognizing the ranked candidate list to recommend for the users, which is an essential and popular use case in real-world application. Besides, it is worth noting that the gap between our technique and the baselines for precision@K and F1-score@K became smaller when K increases. This is because the number of friends in the test set was relatively small compared to the sample size K, which led to this decrease.

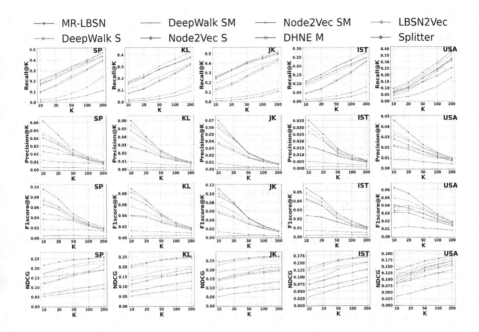

Fig. 3. Performance of the techniques on friend suggestion task

For the baseline methods, LBSN2Vec outperformed other baselines in all scenario. This is because this baseline successfully maintained the complex hypergraph nature of the LBSN data like our technique. However, the gain of our technique over LBSN2Vec justifies the need of analyzing user social relationship in different angles. For the other baselines, which required the irreversible transformation of the original hypergraph, it is unsurprising that their performances on the social only setting (S) was much higher than the mobility setting (M). This is because the S setting contains the social network information, which is more crucial for friendship suggestion task, while the M setting contains only mobility data. Another interesting finding is that including both social network and mobility information in the S&M setting did not lead to an improvement in the techniques' performance. This might be because these techniques did not consider the dominance of the number of check-ins like our technique, as the loss of information in the transformed graph hinder such ability.

6.3 Performance on POI Recommendation

Figure 4 shows the performance of the techniques in a different task - POI recommendation. For this task, we only consider the M and S&M setting, as the mobility of the users is lacking in the S setting. The result in Hit@3, Hit@5 and Hit@10 of all the techniques in the five datasets is reported. In overall, our technique was the clear winner in this task for all scenarios. *MR-LBSN* achieved an average improvements of 30.23% in Hit@5 and 28.46% in Hit@10 to the best

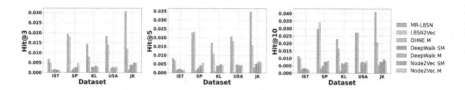

Fig. 4. End-to-end comparison for the location prediction task

baseline, *LSBN2Vec*. Similar to the friendship suggestion task, *MR-LBSN* and LBSN2vec, the hypergraph embedding techniques, achieved more robust result on this POI recommendation task comparing to the other embedding techniques, since the classic graphs are insufficient to represent the complex nature of LBSN data. Also, the same phenomenon in friendship suggestion can be seen in POI recommendation is that the inappropriate use of both social relationship and mobility information (S&M setting) was less effective that using only the mobility data (M setting). This might be because the dominance of the number of check-ins over that of friendship caused a degradation in the prediction ability. However, this degradation was less significant as in the friendship suggestion, since the number of check-ins still outweighs that of the number of friendships.

7   Conclusion

In this paper, we propose a LBSN hypergraph embedding technique that captures the holistic interactions of multi-role social context and user mobility. To this end, we develop a local overlapping clustering analysis to detect the potential social roles of the users, results in the set of pseudo nodes (persona) for each original users. Using the augmented LBSN hypergraph, we develop a biased random walk scheme that jointly samples the social patterns and mobility dynamic of the users. The random walk also balances the internal social multi-role and the interaction of the persona of neighborhood users by a persona transition scheme. From the sampled walks, the embeddings then are learnt by preserving the n-wise node proximity of the friendship and check-ins hyperedges. Extensive experiments on two downstream tasks in five real-world datasets justify the robustness of our technique against existing LBSN pattern extraction frameworks. In the future work, we would want to explore mobility dynamic patterns such as sequential effects, cyclic effects, which are crucial for LBSN analysis [8].

References

1. Backstrom, L., Kleinberg, J.: Romantic partnerships and the dispersion of social ties: a network analysis of relationship status on facebook. In: CSCW, pp. 831–841 (2014)
2. Du, C.X.T., Trung, H.T., Tam, P.M., Hung, N.Q.V., Jo, J., et al.: Efficient-frequency: a hybrid visual forensic framework for facial forgery detection. In: AusDM, pp. 707–712. IEEE (2020)

3. Epasto, A., Lattanzi, S., Paes Leme, R.: Ego-splitting framework: from non-overlapping to overlapping clusters. In: KDD, pp. 145–154 (2017)
4. Glückstad, F.K.: Terminological ontology and cognitive processes in translation. In: PACLIC, pp. 629–636 (2010)
5. Grover, A., Leskovec, J.: node2vec: Scalable feature learning for networks. In: KDD, pp. 855–864 (2016)
6. Huynh, T.T., et al.: Network alignment with holistic embeddings. TKDE **34**, 1–14 (2021)
7. Huynh, T.T., Van Tong, V., Nguyen, T.T., Jo, J., Yin, H., Nguyen, Q.V.H.: Learning holistic interactions in LBSNs with high-order, dynamic, and multi-role contexts. TKDE (2022)
8. Kefalas, P., Symeonidis, P., Manolopoulos, Y.: A graph-based taxonomy of recommendation algorithms and systems in LBSNs. TKDE **28**(3), 604–622 (2015)
9. Kipf, T.N., Welling, M.: Semi-supervised classification with graph convolutional networks. In: ICLR, pp. 1–14 (2017)
10. Lan, C., Yang, Y., Li, X., Luo, B., Huan, J.: Learning social circles in ego-networks based on multi-view network structure. TKDE **29**(8), 1681–1694 (2017)
11. Liben-Nowell, D., Kleinberg, J.: The link-prediction problem for social networks. JASIST **58**(7), 1019–1031 (2007)
12. Perozzi, B., Al-Rfou, R., Skiena, S.: Deepwalk: online learning of social representations. In: KDD, pp. 701–710 (2014)
13. Scellato, S., Noulas, A., Mascolo, C.: Exploiting place features in link prediction on location-based social networks. In: KDD, pp. 1046–1054 (2011)
14. Song, C., Qu, Z., Blumm, N., Barabási, A.L.: Limits of predictability in human mobility. Science **327**(5968), 1018–1021 (2010)
15. Tam, N.T., Weidlich, M., Zheng, B., Yin, H., Hung, N.Q.V., Stantic, B.: From anomaly detection to rumour detection using data streams of social platforms. In: PVLDB, pp. 1016–1029 (2019)
16. Trung, H.T., et al.: A comparative study on network alignment techniques. ESWA **140**, 112883 (2020)
17. Trung, H.T., Van Vinh, T., Tam, N.T., Yin, H., Weidlich, M., Hung, N.Q.V.: Adaptive network alignment with unsupervised and multi-order convolutional networks. In: ICDE, pp. 85–96 (2020)
18. Tu, K., Cui, P., Wang, X., Wang, F., Zhu, W.: Structural deep embedding for hyper-networks. In: AAAI, vol. 32, no. 1 (2017)
19. Wang, D., Pedreschi, D., Song, C., Giannotti, F., Barabasi, A.L.: Human mobility, social ties, and link prediction. In: KDD, pp. 1100–1108 (2011)
20. Xie, M., Yin, H., Wang, H., Xu, F., Chen, W., Wang, S.: Learning graph-based poi embedding for location-based recommendation. In: CIKM, pp. 15–24 (2016)
21. Xie, M., Yin, H., Xu, F., Wang, H., Zhou, X.: Graph-based metric embedding for next POI recommendation. In: Cellary, W., Mokbel, M.F., Wang, J., Wang, H., Zhou, R., Zhang, Y. (eds.) WISE 2016. LNCS, vol. 10042, pp. 207–222. Springer, Cham (2016). https://doi.org/10.1007/978-3-319-48743-4_17
22. Yang, D., Qu, B., Yang, J., Cudre-Mauroux, P.: Revisiting user mobility and social relationships in LBSNs: a hypergraph embedding approach. In: WWW, pp. 2147–2157 (2019)
23. Yang, D., Qu, B., Yang, J., Cudré-Mauroux, P.: Lbsn2vec++: heterogeneous hypergraph embedding for location-based social networks. TKDE (2020)

Multimedia Application for Analyzing Interdisciplinary Scientific Collaboration

Veslava Osińska[1]([⊠]) [iD], Konrad Poręba[2], Grzegorz Osiński[3] [iD], and Brett Buttliere[4] [iD]

[1] Institute of Information and Communication Research, Nicolaus Copernicus University, Toruń, Poland
wieo@umk.pl

[2] Faculty of Mathematics and Information Technologies, Jagiellonian University, Kraków, Poland
konrad.poreba@student.uj.edu.pl

[3] Institute of Computer Science, Academy of Social and Media Culture, Toruń, Poland
grzegorz.osinski@wsksim.edu.pl

[4] University Center of Excellence IMSERT, Nicolaus Copernicus University, Toruń, Poland
bbutliere@umk.pl

Abstract. Information about scientific research can be represented on interactive science maps, which can yield insight into scientific activity at different levels of organization: micro, meso and macro. There are many examples at the meso or macro levels, i.e., research fields. But the micro level – individual analysis - is rather rarely considered in academic writing or app design. The authors designed and tested a web application, Scientific Visualizer, to illustrate an individual's scientific activity, which became a basis for a wide-ranging analysis. The visualizations generated by software will aid researchers in management of their own career, as well as planning effective communication and collaboration. The authors also have discussed their own approach to constructing a disciplinary space of publications based on the current classification of science. They added to scholarship on software and its application in academic community.

Keywords: Science visualization · Science map · Interface application · Multidisciplinary space · Output portfolio

1 Introduction

Data science not only offers significant opportunities; it also affects the sharing, management and reusing of scientific data. The researchers who work with full bibliographic records (for example scientometricians, science sociologists and philosophers, librarians) often struggle to see the patterns in large sets of data extracted from a repository. As data become more complex, it is increasingly more difficult to find the most significant patterns. The importance of visual analysis in data science cannot be overestimated: the first scholar to note the importance of visualization in data analysis was statistician John Tukey [1]. The impact of visualization on big data analysis is visible in global systems

N. T. Nguyen et al. (Eds.): ACIIDS 2022, LNAI 13757, pp. 688–700, 2022.
https://doi.org/10.1007/978-3-031-21743-2_55

use, such as MOOC (Massively Open Online Courses), in which applied statistical analysis allows students, teachers and developers to improve learning dynamics, trajectories, and progress at the individual and mass level.

While developers of data visualization software prioritize macro analyses, i.e. data on big groups of users or global economic, politic processes, individual users' needs remain unnoticed. Scientists also employ big data analysis to analyze the patterns of collaboration between research groups at the (inter)national level [2, 3]. There is no opensource tool which would allow for the complete analysis of a single scholar's output and for the visual representation of the results.

To redress this lack, the authors first considered the individual researchers' activity, in many overlapping contexts. Basing on analysis results, they designed a Web application – Scientific Visualizer, dedicated to micro, i.e. individual analysis of scholars' output and for scholars.

There are many benefits to developing a tool which will analyse the data pertaining to a single user. The individual might examine the data held by the university and the government to ensure that they are accurate and representative of their work. This matters, because the process of collecting these data is imperfect, and often fails: it is difficult to translate a scholar's affiliations into data. Making the data available to individual users gives them a chance to make sure that their work will be recorded. The goal was not only to make the data available, but to also develop ways of visualizing them such that the scholars would understand the data in new ways, impossible in a traditional CSV format.

This paper provides a different perspective on data analysis in micro scale and discusses new insights yielded by visual maps, in particular the formation of multidisciplinary space.

2 Science Mapping and Inter- and Multidisciplinarity

The process of mapping science (SM – Science Mapping) aimed to provide analysis of social, intellectual and technological structure of science and of its evolution [2–4]. The typical product of such analysis is a visualization of literature in a given discipline, preceded by a quantitative analysis, classification or clusterisation. These visual representations of scholarly knowledge are called science maps. They give insight into patterns of activity performed by selected units, such as: authors, countries, articles, citations. Quickly developed in the last two decades, SM became an interdisciplinary field which draws on bibliometrics, computer science and particularly data science [3].

According to Katy Borner [2] we may distinguish five types of data analysis in relation to academic publishing. The basic analysis involves statistical description of working data; summarizing N and basic statistical measures of the data table. It answers the main research question **how**. If we want to see how the data changed, we present them on a timeline – an appropriate analysis concern **when** question. As data distribution may be tracked on maps, we may provide a geospatial description of data, answering the question **where**. Modern NLP (Natural Language Processing) methods offer new opportunities for analysing textual data, which should be appropriately cleaned and ordered – in this case topical analysis is undertaken which answers the question **what**.

The last type of analysis can be performed based on Social Network Analysis (SNA), which describes the relations between people, or groups of people or inanimate entities such as organizations, nations or continents [5]. It answers the research question with **whom**. All five types of data analysis might be performed on bibliographic metadata of academic articles. However, Web of Science, Scopus and other global databases implement such insight only partly; and the results are available to scholars for a fee.

The later SM works often focus on the comparison of visualizations generated by different algorithms and statistical methods, such as factor analysis, multi-dimensional scaling, cluster analysis, or Kohonen networks [4–7]. Maps can reveal not only the social structure of research but also disciplinary relationships between them. The major challenge in science mapping today is to give insight into research topics and fields extending beyond a single discipline, or institution, so as to track the evolution of different domains and disciplines in science [2]. Examining the overlap between certain research field and the gradual fading away of others, we develop a better understanding of the relations between disciplines [8].

Studies focused on these issues employ topological analysis algorithms. We can define interdisciplinarity as the integration of knowledge and methodologies specific to particular disciplines, used in the paradigm of particular scientific field by means of both synthesis and analysis [9]. In data analysis, a study of interdisciplinarity involves a large-scale consideration of knowledge and methodology. The analysis of the output of individual scholars becomes less important. Taking the multidisciplinary approach, we pay the greatest attention to the work of individual scientists who, while working on a specific project, bring their unique, field-specific knowledge to it [10]. This is why it is so important to quantify the value of potential possibilities of scientific activity of individual scientists in various fields of research. The best method in this case seems to be the use of algorithms calculating the "coverage area" for each scholar analyzed [11]. In order to estimate multidisciplinarity, the authors designed an appropriate science map with a given topology. For this purpose, a predefined list of academic journals was mapped onto 2D representation.

3 Data and Method of Mapping

Our source of data was a university bibliographic database called Expertus (https://bg.cm. umk.pl/expertus/umk/). Expertus is a platform used by the majority of university libraries in Poland. Expertus uses the Dublin Core standard for object description, as well as filtering and searching tasks. A single record contains 30 subfields. Apart from standard bibliographic metadata describing particular entities: authors (name, affiliation), publication (title, year, language, abstract and keywords), journal (title, IF) and publisher (name, open access policy), the record notes how many points the Ministry of Science and Higher Education has awarded to a journal, which is currently the most important indicator of the journal's prestige. The disciplines and science domains ascribed to particular journals, according to Polish classification of science (https://konstytucjadlan auki.gov.pl), are also listed.

Expertus allows scholars to download their records in different formats: Excell/OpenCalc, RTF, CSV and bibliometric specified BibTeX. Moreover they can select the amount of output data table (e.g. choose between full name of the journal, or an abbreviation) depending on their purposes.

We selected an RTF file for data input because it provided most complete information. We divided the file into different sections, representing individual publications and then matched them with various individual fields (eg. List of authors, the title of article, journals, formal and subject classification). To parse data we used the rtf-parser library (https://github.com/iarna/rtf-parser), custom string operations and regular expressions to retrieve desired information and build an array of record objects used for all visualizations. The array of publication data is stored in-memory for visualizing. All information which needs to be stored at the server side was saved using a NoSQL document database - MongoDB. The application is based on the web framework Express which allows the user to render templates and create all required endpoints for retrieving data or for routing between application components. For internationalization and localization of our application we used the i18n library which allowed us to define all translations in JSON files. For the visualizations, we mostly relied on the D3.js library which provides a powerful toolkit for data presentation. For manipulation of geocoding data in map visualization we used an open-source leaflet library (https://leafletjs.com/).

Current Polish classification lists 32,681 journals (both international and national), classified according to 46 disciplines and 8 knowledge domains. Each journal has its own score. We matched two tables: the Expertus list of journals which were established in the last five years, and the Ministry list. Thus, we obtained 23,635 items which became the basis for mapping academic publications. t-SNE (t-distributed stochastic neighbour embedding) was applied as an appropriate algorithm for science mapping when co-occurrences of categories are not dense. t-SNE mapping is characterized by qualitative clustering and local accuracy [6]. This is a machine-learning technique developed in 2010 as a variation on stochastic neighbour embedding [12]. Using this method, each high-dimensional object is modelled by a two- or three-dimensional point in such a way that similar objects are modelled by nearby points and dissimilar objects are modelled by distant points with high probability. As the creators mentioned „The similarity of datapoint x_j to datapoint x_i is the conditional probability $p_{j|i}$, that x_i would pick x_j as its neighbour if neighbors were picked in proportion to their probability density under a Gaussian centered x_i":

$$p_{j|i} = \frac{\exp\left(\frac{\|x_i - x_j\|^2}{2\sigma_i^2}\right)}{\sum_{k \neq i} \exp\left(\frac{\|x_i - x_j\|^2}{2\sigma_i^2}\right)}, \text{ for } i \neq j \tag{1}$$

where σ_i is Gauss distribution variance centered in x_i. Then probabilities p_{ij} that are proportional to the similarity of objects x_i and x_j in N-dimensional space can be defined as:

$$p_{ij} = \frac{p_{j|i} + p_{i|j}}{2N} \tag{2}$$

This algorithm successfully applies to high-dimensional data for visualization in a low-dimensional space of two or three dimensions. Therefore 46 disciplines assigned to journals in Expertus database might be arranged on a t-SNE map according to their disciplinary similarity where the common number of ascribed disciplines defines proximity. The data table for t-SNE mapping (23,635 rows and 44 columns) is coded binary: 1 if there is relation to a discipline and 0 if not. Thus, we obtained a map of disciplines that displays the similarities between different journals and the clustering of disciplines among 8 scientific domains.

4 Analysis Measures

Data of different types contained in bibliographic records might be used to visualize academic activity and to describe its qualities. For example, we may determine an academic's seniority considering when their work was published. Basic bibliometric indicators (IF, coauthors, ministerial points and so on) for each article allow us to compute yearly sums and averages.

Therefore, the visualization software should take into account such indexes of research activity as development, internationalization, collaborations [2]. We prepared the list of possible indexes:

- the first year of publishing
- sum of papers
- yearly paper count
- ministry total score (according to the last version of the List of scored journals [13])
- ministry yearly score
- Impact Factor of journal
- The name of the closest collaborator.

The main disadvantage of these indexes is that none are based on citation measures. Expertus and both local and national databases do not contain the complete data on citation, so Polish academics have to rely among global indexes on quick Google Scholar Citations [14]. The authors plan to gather the data from Google Scholar and incorporate them into their sample to make analysis more comprehensive and sophisticated. Citation data can provide constructing citation networks between scientists as well as between topics themselves. In presented web application all available parameters have been applied to create the appropriate measures (Table 1). New and valuable indicators were introduced in accordance with Open Access policies, such as Open Access publications ratio (OAP) and conferences papers ratio (CPR), which demonstrate the extent to which an academic's work is disseminated. Table 1 presents calculated normalized parameters of academic activity which the authors incorporated into the app. To reach wider audiences, Polish researchers should publish in English – an appropriate index (EPR) was computed.

Publishers' full descriptions provide information regarding works published nationally and internationally; two measures based on the relationships of these measures are introduced: RNCP and RIP. The CC10 and SFA measures represent the closest and most distant co-authors, respectively. The last measure, based on geometry, relates to t-SNE space of disciplines, where each of 46 disciplines is defined by two coordinates and one of eight domain colors (Fig. 3). If we position researchers' articles ascribed to particular disciplines in this space, it is possible to estimate the "coverage area" of a polygon formed by article nodes - MASI.

Table 1. Normalized parameters of a multifaceted researcher's activity.

Parameter	Description	Definition
CPR	Ratio of conference papers to all published papers	N_c/N, where N_c is conference proceedings count and N is overall number of works
OAP	Open Access papers proportional ratio to all published papers	N_{OA}/N, where N_{OA} is OA papers count and N is the overall number of works
EPR	English papers ratio to all published papers, where we assumed the polish language is dominating	N_{en}/N, where N_{en} is the number of English papers and N is the overall number of papers
RNCP	Range of National Cooperation within communication with polish publishers	N_{pl}/N, where N_{pl} is number of articles and chapters published in the country and N is overall number of published works
RICP	Range of International Cooperation in the scope of publishing regarding to national publishing	N_{int}/N_{pl}, where N_{int} is number of articles and chapters published abroad
CC10	Core Collaboration finds the contribution of the papers published with top ten collauthors to overall output	N_{10c}/N, where N_{10c} is number of articles published with 10 top coauthors (ten is contractual threshold)
SFA	Scattering Factor of Authors shows the contribution of extreme coauthors to overall output	N_{exct}/N, where N_{exc} is number of extreme coauthors who appears in external ring of coauthorship graph
MASI	Multidisciplinary Area of Scientific Interests is defined by disciplinary space of researchers' articles	S_{art}/S_{disc}, where $S_{art} = \oint_{Pol} \Gamma(x_{ij})$ is area of polygon constructed on disciplinary plan area and S_{disc} is an area created by the most external disciplines nodes

To estimate the coverage area 'MASI' we must first determine the location of the edge nodes for each discipline. The computational algorithm carries out this process by calculating the distances from the density center of each research paper published in a journal assigned to a given discipline(s). This allows us to determine the location of nodes in the two-dimensional topological space (x_v, y_v) illustrated on Fig. 3.

Then the topological center of reconstructed map will be defined as:

$$\begin{cases} x_v = \frac{\oiint x dS}{S} = \frac{\sum_{i=1}^n x_i s_i}{\sum_{i=1}^n s_i} \\ y_v = \frac{\oiint y dS}{S} = \frac{\sum_{i=1}^n y_i s_i}{\sum_{i=1}^n s_i} \end{cases} \tag{3}$$

The theoretical value of the area integral is calculated iteratively as the sum of the products of the position coordinate (x_i, y_i) of the individual values of the area assigned to the specific visualized value. Determining the MASI parameter requires the use of the appropriately adjusted Monte-Carlo algorithm [15] to calculate the area S_{art} within the polygon (see Table 1). Thus we can use the standard algorithm based on the image segmentation [16]. To calculate the coverage region between two neighboring regions R1, R2 as the minimum weight edge we should apply the following formula:

$$S(R_1, R_2) = v_i \in R_1, v_j \in R_1, \overset{min}{(v_i, v_j)} \in E \, w\big((v_i, v_j)\big) \tag{4}$$

where (v_i, v_j) relates to each node on visual representation.

Parametrization of the distance between the specific nodes makes it possible to estimate the value of the distance module, which normalizes the value of the surface integral needed to calculate the value of S_{art}.

5   Web Application Design

The web application was developed using JavaScript programming language in a NodeJS environment.

An express framework served as a skeleton of an entire app rendering layouts, routing between them, managing sessions, translating and creating HTTP endpoints which served to retrieve the data from the server. Most of the visualization scripts which use a D3.js library [17] run in the browser using data from the files uploaded by the user without the need to fetch any data from the server. The whole application has been containerized using Docker to make it platform-independent and to facilitate the deployment process on the destined server. Source code is published in a public repository under the following address https://gitlab.com/hotone/vis. Ultimately, the application will be placed on the university's website.

5.1   Communication

The interface contains a menu with the key information regarding the application and the developers' team, a module for the uploaded data, a measures panel and the visualizations area with six buttons which run appropriate visualizations. Hovering the courser over the buttons opens popup windows with general information about each chart. The application is protected with a password for the reasons of data security and archiving. The user has to first download an RTF file containing their bibliographic records from the Expertus database. The next step is uploading this file to the application.

All labels and descriptions are available in two language versions: Polish and English. An UTF-8 coding page was used in data communication between Expertus and the app; the information about the authors and journals is properly displayed. The users are able to download the final visualization (in PDF format) and the table of calculated scientometric measures. Visualizations based on the D3.js library are interactive: the scholars receive additional information about metadata in popup fields; they can also filter data by setting a threshold value in the appropriate panel.

5.2 Visual Layouts

Temporal analysis shows how the scholar's publishing habits changed over time on a stacked bar chart. It allows us to compare the different forms of published units, such as: articles, books and conference papers. Such information is essential, because the weight of a particular item can depend on the faculty or discipline. For example books are scored more highly for humanities, articles – for medicine, and conference papers, where the newest technologies are introduced - for engineers. The X scale is scaled to the dates range (Fig. 1A). Information about particular articles and books is displayed during hovering the cursor.

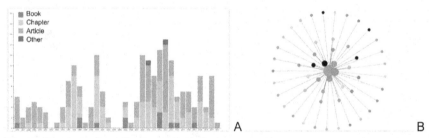

Fig. 1. A stacked bar chart for temporal analysis of publishing (A) and collaboration graph of scholar (B).

Collaboration analysis is an important part of planning future scientific research that will foster scientific cooperation at both national and international levels. This is key for interdisciplinary research, which requires a collaborative network whose participants' scientific interests are sufficiently varied. To an extent, the quality of the environment can be predicted on the basis of previous experience but our app provides a graphic representation of collaboration networks which might yield more information.

The visualization algorithm allows us to identify a number "collaboration circles" (from closest to most distant) that we can parameterize by entering a value for the core collaboration coefficient. The papers published to date were assumed as initial values of these ranges; for 'core collaboration authors' (CCA) - the closest ring of graph (Fig. 1B), a value of more than 10 was assumed (it was set arbitrarily and will need to be examined further). The 'potential collaboration class' consists of the coauthors with whom a researcher published less than 10 but more than 3 papers. All others will be placed in the external rings of the graph and should be referred to as the 'scattering factor authors' (SFA) class.

Similarly to other local databases, Expertus does not store information regarding coauthors' affiliation and countries, so it is not possible to extract information about the geographical range of cooperation from this database. However, we can access the information about the place (city and sometimes country) where the books and papers were published and perform **geospatial analysis**.

At this stage, the coordinates (longitude and latitude) of cities are established through searching city names in a static JSON file, but in the future we plan to retrieve these data by integrating it with an external API.

The RNCP and the RICP (see Table 1 and Fig. 2A) provide information regarding the geographical area within which the researcher publishes their work. The values of these coefficients, calculated on the basis of previous scientific activity, may direct the scholar's plans for research and publishing. They will be able to select collaborators from their country with appropriate RNCP values in relation to scientific fields that the scholar has not yet explored in their research. The comparison of these two parameters – the ratio of one to another - indicates both the degree of internationalization and the necessity of extending the circle of national and international collaborations.

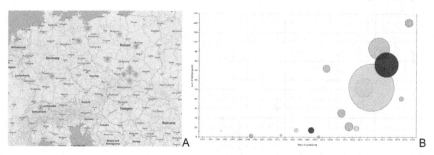

Fig. 2. Geospatial scope of works published (A), the cumulated ministry scores over time (B).

A bubble chart (Fig. 2B) presents indicators most important for Polish academics as they constitute the basis of their professional advancement [14]. Researchers can see how the **yearly ministry scores** change across years. The size of a bubble indicates the sum of pages of published works (the colour is random, but it can be used as a variable in future iterations of the app). The user can scale the diameter of all bubbles.

Every scholar is interested in a general view of their scientific output in relation to the traditions of their discipline. Only metadata such as keywords and disciplinary classification provide qualitative information, i.e. verified by experts, regarding the scholar's area of research. We used these metadata to make **topical analysis** and create semantic spaces relating to research, visualized through graphs of two types.

Word cloud or tag cloud is a very popular way of communicating scientific concepts to an audience who wont a quick overview of a subject, but do not need to understand it in depth. This simple technique displays how frequently words appear in a given body of text – the frequency is reflected by the size of a given word. The words are randomly arranged, depending on the spatial positioning algorithms employed.

Creating a **disciplines map** is usually very challenging for researchers because of the multidimensionality of data and the ambiguity of classification [4, 6]. We used a t-SNE map of disciplines as a basic layout for further mapping a scholar's classified articles (Fig. 3). It consists of coloured 8 areas and 44 circles indicating disciplines. For example, we can see three medical subdisciplines within the domain of medicine (red). Articles' nodes (black circles), define the area of the scholar's interests across the disciplines; the larger the area, the wider the scope of the researcher's interests. For a quantitative comparison of DASI the area of the researcher's interests has to be referenced to the entire disciplines space. MASI parameter is defined as a relation of the first area to the area extending between the most external nodes of disciplines (Fig. 3).

The D3 library allowed us to carry out calculations using functions such as *d3.polygonHull( )* to indicate the points in an array which act as the nodes of the biggest possible polygon which would include all other points and *d3.polygonArea( )* to calculate the area of that polygon.

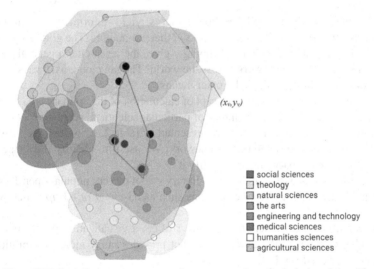

Fig. 3. The t-SNE disciplines map as a space for representation of scholar's articles. The colours represent 8 domains, coloured circles – 46 disciplines ascribed to their domain, dark small points – scholar' articles forming the space of disciplines (green inner polygon). The outlined area is used as a reference for the calculation of MASI – equals 0.11 for presented researcher. (Color figure online)

6 Summary and Conclusions

This paper presents a web application – Scientific Visualizer that visualizes and analyses an individual's scientific output as an alternative to commercial tools: Scival or InCites.

Aside from the basic statistical measures (the mean, the yearly count), we used new non-linear parameters including quantification of muldisciplinary scope. Our visualizations answer the questions "How many", "When?" "Where?" "What?" and "With Whom?" through quantitative, temporal, geospatial, topical, and network approaches.

The timeline charts allow the researchers to analyse the dynamics of their publishing.

Table 2. MASI values for selected researchers.

Scientist Id	Discipline(s)	MASI value
S1	history	0.1
S2	psychology, cognitive science	0.3
S3	art history, library science	0.2
S4	automation, computer science	0.4
S5	computer science, cognitive science, physics	0.8

The visualizations of collaboration patterns yield information regarding the researcher's network, as well as the networks of their co-authors. The graph of geospatial patterns – an innovation in science mapping – gives the researcher a general overview of the dissemination of their work across the world. The tag cloud, as always, provides an immediate insight unavailable by other means.

The article also introduces the concept of calculation of disciplinary area formed by researchers' interests. Several examples of particular scientists' scope of specialization were chosen; their computed values are presented in Table 2. As we can see, MASI can be a basic component of analytical framework. We use it to identify the shape of the researcher's interests space, and to quantify it.

The application is based on the D3.js library - a set of very popular open JS scripts. The D3.js library [16] has many extensions such as *d3-cloud* or *d3-tip* which we also used.

Our web application Scientific Visualizer offers a visualization framework which can resemble Google Analytics dashboard but provides many more possibilities than any commercial platform [18].

Main Conclusions
The application gives the researchers using the Expertus data base the means to produce a sophisticated visual representation of their scientific output in relation to their collaboration networks and in context of the disciplinary range of their research. The insight into collaboration networks allows the researchers to see their own connections to their co-authors, and the connections their co-authors might have to one another. The appropriately prepared visualizations help the researcher not only to determine their institutional reach (key for international collaboration), but also to situate their research in relation to classified disciplines. A careful study of these maps may direct the researcher's plans for future research, inspiring them to extend their collaboration networks, or to consider new research fields. This will result in a more efficient planning of future research and

of the researcher's career trajectory. Values of the RNCP and RICP parameters indicate whether the researcher should pursue international collaboration, or rather focus on the already established network, which perhaps remains underused – as indicated by the value of the CC10 parameter. As far as the disciplinary range is concerned, the application shows the researcher the "overlap area" of the work they have published so far, which then may suggest ways of expanding it. The visualizations also allow the researcher to evaluate the efficiency of their publishing activity, and indicate the best directions for future development according to the strategy employed: expanding or reducing the current collaboration network.

The application therefore may replace the previously popular scientometric ranking comparisons, which do not take into account the characteristics of individual research, offering in their place personalized parameters supported by visualization maps prepared for specific researchers. This customization will facilitate planning and personal evaluation of one's research output.

Limitations

The main disadvantage of the application is that it does not account for citations data, which are not stored in Expertus. This is a limitation of all Polish data bases. However the authors plan to gather the citations data from Google Scholar and integrate them with Expertus to ensure that the analyses will be comprehensive and advanced. Unfortunately, not all records in Expertus are complete, which means that not all visualizations are perfectly accurate. Lack of a modern API in systems such as Expertus, allowing for retrieving the list of all journals recorded in the database, is an obstacle to the development of the application.

Future Work

The authors consider further development of visualizations to allow for more extensive interactivity and new possibilities for analysis based on both existing datasets and new ones. They may gather new data from other national bibliographic databases and from Google Scholar. They plan to create a single integrated database which would illuminate all aspects of scientific activity, not only scientometrics.

Authors also intend to study the usability of Scientific Visualizer's interface by collecting feedback from end-users (i.e., researchers). They will use BCI and eye-tracking. One advantage of the proposed interface is the possibility of monitoring users' behaviour, which will yield insight into their information and visualization needs. The final step of the study will consist of surveys and interviews.

The authors are considering developing applications for disabled users, which should include a text reader and voice control functions.

Acknowledgments. The research is a part of project Bitscope supported by the National Science Centre, Poland within CHIST-ERA IV from the EU Horizon 2020 Research and Innovation Programme (Grant Agreement no 857925). Presented tool allowing these results was funded from IDUB IMSErt – the NCU center.

References

1. Tukey, J.: Visual Exploration of Data, 1st edn. Pearson, New York (1977)
2. Börner, K.: The Atlas of Knowledge. MIT Press, Cambridge (2016)
3. Chen, C.: CiteSpace II: detecting and visualizing emerging trends and transient patterns in scientific literature. J. Am. Soc. Inf. Sci. Technol. **57**(3), 359–377 (2006)
4. Chen, C., Song, M.: Representing Scientific Knowledge: The Role of Uncertainty. Springer, Heidelberg (2017). https://doi.org/10.1007/978-3-319-62543-0
5. Fortunato, S., et al.: Science of science. Science **359**, 6379 (2018). https://doi.org/10.1126/science.aao0185
6. Osinska, V.: A qualitative–quantitative study of science mapping by different algorithms: the Polish journals landscape. J. Inf. Sci. **47**(3), 359–372 (2021). https://doi.org/10.1177/016555 1520902738
7. Osiński, G, Malak, P. Osinska, V.: PCA algorithms in the visualization of big data from Polish digital libraries, In: Choroś, K., et al. (eds.) Proceedings on 11th International Conference on MISSI 2019, AISC, vol. 833, pp. 522–532. Springer, Wrocław (2019)
8. Wang, Q., Ahlgren, P. Measuring the interdisciplinarity of research topics. In: Costas, R., et al. (eds.) Proceedings of the 23rd International Conference on Science and Technology Indicators STI 2018, pp. 134–142. Leiden University (2018)
9. Stember, M.: Advancing the social sciences through the interdisciplinary enterprise. Soc. Sci. J. **28**(1), 1–14 (1991)
10. Choi, B.C., Pak, A.W.: Multidisciplinarity, interdisciplinarity and transdisciplinarity in health research, services, education and policy: definitions, objectives. Clin. Invest. Med. **29**(6), 351–364 (2006)
11. Walters, W.H.: Google Scholar coverage of a multidisciplinary field. Inf. Process. Manag. **43**(4), 1121–1132 (2007)
12. van der Maaten, L.J.P., Hinton, G.E.: Visualizing high-dimensional data using t-SNE. J. Mach. Learn. Res. **9**(11) (2008). https://www.jmlr.org/papers/volume9/vandermaaten08a/van dermaaten08a.pdf
13. New extended list of scientific journals and peer-reviewed materials from international conferences. Report, https://www.gov.pl/web/ [In Polish]
14. Dowgier, R., Zawadzka-Pąk, U.K.: Scientific publications as an element of evaluation of scientific output in Poland. Ann. Center Rev. **9**, 30–38 (2016). http://hdl.handle.net/11320/5630
15. Zhou, C., Cui, H.: Monte Carlo method of polygon intersection test. J. Comput. Inf. Syst. **9**(12), 4707–4713 (2013)
16. Peng, B., Zhang, L., Zhang, D.: Automatic image segmentation by dynamic region merging. IEEE Trans. Image Process. **20**(12), 3592–3605 (2011). https://doi.org/10.1109/TIP.2011.215 7512
17. Bostock, M., Ogievetsky, V., Heer, J.: D3: data-driven documents. Proc. IEEE Trans. Vis. Comput. Graph. **17**(12), 2301–2309 (2011). https://doi.org/10.1109/TVCG.2011.185
18. Hung, Y.-H., Liang, J.-B.: Developing a knowledge-based system for lean communications between designers and clients. In: Kurosu, M. (ed.) HCII 2021. LNCS, vol. 12764, pp. 34–48. Springer, Cham (2021). https://doi.org/10.1007/978-3-030-78468-3_3

CORDIS Partner Matching Algorithm for Recommender Systems

Dariusz Król[1]($\boxtimes$) (iD), Zuzanna Zborowska[2], Paweł Ropa[3], and Łukasz Kincel[3]

[1] Department of Applied Informatics, Wroclaw University of Science and Technology,
Wroclaw, Poland
`dariusz.krol@pwr.edu.pl`
[2] Faculty of Computer Science and Management, Wroclaw University of Science and
Technology, Wroclaw, Poland
[3] Software Development Agency Nomtek Ltd., Wroclaw, Poland
`{p.ropa,l.kincel}@nomtek.com`

Abstract. The purpose of this paper is developing a method for recommending business and scientific partners matchmaking with use of a deep learning model based on historical data on previously completed European Union projects. The paper starts with an introduction to recommender systems, followed by a systematic literature review on the subject. The next part describes the course of the research and its implementation of two deep learning approaches: (1) the entity embeddings of organisations and (2) the embedding space of keywords. The paper ends with a summary of the entity embedding-based recommendation characterized by coverage, average accuracy, low Gini index and the entropy measure.

Keywords: Convolutional neural network · Deep learning ·
Knowledge discovery · Performance evaluation metrics

1 Introduction

The constantly increasing capacity to generate, collect and process data has led to an information overload. Recommender systems are solutions that address this issue. They limit the data available to the user providing a personalized recommendation, often based on their previous choices. Recommender systems have been adopted in many fields such as e-commerce or entertainment in order to facilitate users' decision-making. The interest in this area remains high as researchers continually find that recommendation-based solutions might be applied to other data-abundant fields.

The term business matchmaking refers to the structured process of identifying and connecting business entities that have similar business goals. It can also be used in the context of scientific partners and their common objectives. During events such as conferences or fairs, partner matchmaking is often a part of the agenda. The process is usually supported with software, which automates and

N. T. Nguyen et al. (Eds.): ACIIDS 2022, LNAI 13757, pp. 701–715, 2022.
https://doi.org/10.1007/978-3-031-21743-2_56

integrates matchmaking into the workflow of the event. Accurate matchmaking can save collaborators' time and let them focus on discussing the details, instead of searching for an appropriate institution. The process begins with an organisation registering in the system and providing a detailed description of its goals. After a successful registration of all the participants, the system can determine what entities might be interested in a collaboration. The more registered organisations, the more difficult the matchmaking problem. Here is where the recommender systems may be of use. Not only are they able to process large amounts of data but they can also determine which characteristics of an organisation best determine the probability of collaboration.

The subject of the work lies within the field of machine learning, in particular deep learning, determining the closeness of business and scientific relationships, as well as defining a measure of matchmaking success. The final result of the research will be an innovative method of matching partners, which will assess the compatibility of relationships using neural networks. The work focuses on the construction of a learned classifier that performs the task of selecting the most likely partnership and evaluating its effectiveness. The learned model will be tested experimentally, taking into consideration measures of recommender system evaluation presented in literature.

The following research was conducted as a part of the Mozart project. Mozart is a project financed by the City Council of Wrocław, Poland, to support scientific and business partnerships for the implementation of their proposed projects.

The result of the project will be the creation of a prototype recommending a ready-made solution, i.e. it will identify partners who have similar business/scientific goals and thus their relationship will potentially show the greatest similarity. Potential recipients may be organisations looking for institutions to collaborate with as well as the organisers of conferences or fairs.

2 Systematic Literature Review

In order to properly summarize the current state of knowledge, a systematic literature review was conducted.

There are four document databases selected for this review: IEEE Explore, ProQuest, Science Direct, ACM and Springer Link. The inclusion criteria of the selected documents are the following: full text articles, papers written in English, papers written after 2001, and studies covering the application of recommender systems. The exclusion criteria of the selected documents were the following: articles without the full-text access, papers not written in English, studies that were out of defined time span, and studies not relevant to the research questions.

The following phrases were used to extract the initial set of related documents: *recommender system, recommendation system,recommendation list, matchmaking* and *deep recommendation*. The search was limited to the terms describing the problem, limiting recommendation-related documents to the ones concerning the matchmaking problem. The research follows the PRISMA [11] guideline for systematic reviews. This work aims to address the following research questions:

- What are the best recommendation methods and techniques?
- What are the best matchmaking techniques?
- Are recommender systems often used as a solution to the partner matchmaking problem?

The first search of the terms in the full texts and in the meta data resulted in 185 positions found. After removing duplicates there were 177 documents. After screening of the titles we excluded 87 documents which did not cover the application of recommender systems. After screening of abstracts there were found 28 positions eligible for the full text review. Figure 1 presents the flow of articles' screening.

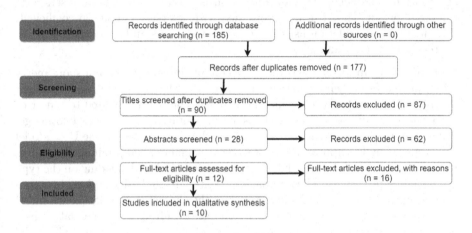

Fig. 1. PRISMA screening flow.

The reviewed papers revealed that the recommender systems can be divided into three categories.

Content-based recommender systems - attributes of items are compared to find similar entities. The main advantage of content-based recommendation is that the items can be recommended even though they are not yet rated by any user. However the method may lack novelty as it has a high probability of recommending items from the same category. In paper [5] we can read about a realisation of this type with a use of matrix factorization algorithm. In the problem of matching Twitter users [8] the content base approach lies within building a good representation of a user profile. It is done based on: their own tweets, tweets of their followees, tweets of their followers. Document [17] presents matchmaking recommendation with the use of ontology-based user description and Naive Bayes classification method. The solution of the recruitment problem described in [1] defines the vacant position with a set of characteristics/

skills of a desired candidate and then calculates Minkowski distance between the representations of the candidate's skill and the vacancy.

Collaborative recommender systems - recommendation is based on the preferences of other users (user-based) or based on the rating a user has given to similar items (item-based). The first subtype bases on the assumption that users, who made similar decisions in the past, are likely to still act in a similar manner. Among the techniques used within that type is Vector Space Model and Term Frequency-Inverse Document Frequency (TF-IDF) [5]. The method demonstrated in [7] requires a User Preference Model based on the user's actions. Having the model allows for calculating the correlation similarity between the models. The K-Nearest Neighbours algorithm is later used to determine the most similar users. The paper [14] states the problem of determining the recommended items for a user as neighborhood formation, pairwise prediction, and prediction aggregation. The neighbourhood formation is a process of defining users with similar interests. They can be determined with Pearson correlation coefficient. The final recommendation list can be calculated as a standard deviation of the nearest neighbours' rankings.

Hybrid recommender systems - a technique integrating two or more methods to benefit from their advantages and avoid their flaws. In paper [20] a concept of User Model is introduced, which is the collection of data related to a user or a specific group of users. User Model is a representation of the user's knowledge, skills, interests and how the user interacts with others. Based on the User Model the proposed hybrid approach uses statistical, semantic, and profiling algorithms to recommend items. The results of algorithms are ranked depending on the type of task being performed. The approach presented in [16] besides taking into consideration the preferences of other users and the characteristics of recommended items, is also aware of the geolocation context. Resulting recommendation lists are therefore re-ranked based on the closeness of recommended items. In HYREC [18] hybrid system collaborative filtering and content-based approaches aim to solve the problem of temporal dependency of recommended items. It is achieved with a sequence recognition algorithm. The approach described in [21] solves a problem of web services recommendation. It uses content-based methods and collaborative filtering and then aggregates the results using a three-way aspect model.

Table 1. Classification of papers based on the recommender system type.

Recommender system types	Reference
Content-based	[1,5,8,17]
Collaborative	[5,7,8,14]
Hybrid	[16,18,20,21]

The conducted research showed that there are multiple techniques of handling a recommendation problem. It is dependent on the type of items we want to make recommendation for and the type of data that we are dealing with. The problem of recommendation in the partner matchmaking process is neither pure content-based nor collaborative in nature. Therefore the approach that needs to be followed must be of a hybrid nature. What is more, to evaluate the effectiveness of the recommender system the following evaluation metric ought to be used: accuracy, coverage [10,13], entropy [4,10,13] and the Gini index [2,13].

3 Data and Methodology

The data source selected for the project is an open-source database of projects funded by European Union (EU). It is referred to as The Community Research and Development Information Service (CORDIS). The CORDIS database consists of data about participants, reports, deliverables and links to open access publications. The data is available in the form of either .csv, .xlsx or .xml files or accessible via an API [6]. Because some data is presented in the form of static files those responsible for maintaining the website inform about possible inconsistencies between what is presented on the CORDIS website and the data sets. There are a few different data sets, each focusing on a different aspect of the data:

- H2020 Organisations
- H2020 Project deliverables
- H2020 Project publications
- H2020 Projects
- H2020 Report summaries
- Principal Investigators in Horizon 2020

For the implementation part of the research Python 3.9 was selected. It is commonly used for the data related tasks. Pandas library was used in the data extraction and preparation process while the Keras library was used for the deep learning part. Keras provides an easy way to create and run neural network experiments and therefore is widely used among researchers. For data preparation and evaluation purposes several additional Python libraries were used. Numpy was used for managing and structuring of data, Natural Language Kit [3] for tokenization and stop words removal, Matplotlib [9] was used to plot the results and SciPy to calculate evaluation measures. The research was conducted with a use of the Jupyter environment.

Before the data can be fitted into neural network it must undergo several preparation procedures. First, it has to be fetched from the data source, which is not always straightforward. Secondly, the data has to be analysed and cleaned. We have to make sure the data has the most accurate entries. The next step is the preparation of descriptions. Descriptions have to be tokenized, we have to remove stop words and the resulting set must be simplified to its lemmatized form. This can be achieved with Python based tools and libraries. After the data preparation we were left with **2384** organisations with lemmatized meaningful words that could be further processed to eventually be fitted into the neural network model.

4 Results

Two neural network models and four different recommendation list creation methods were proposed. The proposed neural network models base on the connections between the organizations that took part in the same projects in the past. If a new organization appears in the database or the description of the organization is updated both of the models would have to be trained again. The small portion of the data that we experimented on makes training the model time consuming only to a small extent but bigger amounts could slow down remarkably the recommendation process. On the other hand, the proposed method takes into consideration the attributes of organization and their relation to the others which makes it more resistant to scarcity of data in one of the fields.

This section details the recommendation method discovery process, the result of applying the methodology to the prepared data. It spans the outcome of models' training, the four created recommendation lists and the evaluation of those lists. There are 4 different measures of evaluation: accuracy, coverage, Gini index and entropy. They serve as a verification of proposed methods.

4.1 Entity Embedding-Based Model

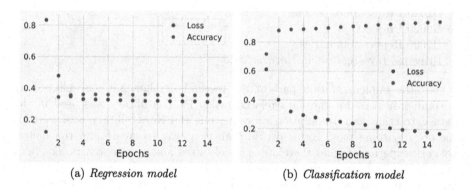

(a) *Regression model*   (b) *Classification model*

Fig. 2. Model loss and accuracy.

In the entity embedding-based model the main focus lies in the weights on the embedding layer. Although there are two embedding layers in the model, each representing a party in a given collaboration, only the first one is taken into consideration in further research. In the training process, the models adjust weights on their layers to place the organisations, which took part in common projects, closer in the embedding space. The models will try to decrease the value of loss and increase accuracy. Because we are interested in training of the embedding layer there can be two supervised learning problems that the model will try to solve in the training process. Based on the trainings' results only one approach will be taken into consideration in further discussion.

Figure 2(a) shows the results of regression model's training process. The loss function is mean squared error. It can be noted that the value of loss decreases significantly in the first two epochs of learning, but after that remains at the same level of about 0.3. Something similar happens to the value of accuracy. It increases in the beginning but stops at a certain level and for the next epochs stays at the same level. The resulting accuracy is relatively low and definitely does not meet the expectations.

Figure 2(b) presents the results of training of classification problem. The loss function is binary cross-entropy. The loss starts with the value of 0.6 and quickly decreases down to about 0.1. The accuracy is high from the first epoch on and continues to steadily grow up to 0.9. Although we expected to obtain similar results in the two machine learning approaches the classification results are undoubtedly better and as such it will be taken into consideration in further research.

Although the embedding space has 50 dimension there are techniques that allow us to visualize the embedding representations of organisations: t-SNE: t-Stochastic Distributed Networks Embeddings [12] and umap: Uniform Manifold Approximation and Projection [19]. Both techniques allow to reduce dimensionality of the embedding space to 2, still capturing the variation between groups of organisations. Figure 3(a) presents the results of dimensionality reduction with t-SNE technique and 3(b) shows the same process with the use of UMAP. On both diagrams we can see small, distinguishable clusters of organisations, but no bigger group spanning more examples. We can interpret it in such a way that the collaborations present in the data set are relatively uniformly distributed. There are no big groups of organisations working closely together without partners from outside of the group.

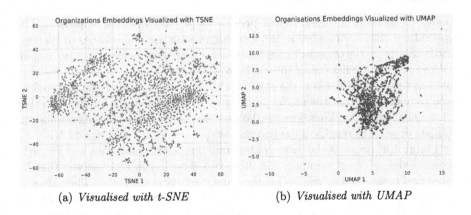

(a) *Visualised with t-SNE*   (b) *Visualised with UMAP*

Fig. 3. Organisations embeddings.

After the training of classification model the resulting weights on the embedding layer were taken to calculate the cosine distances between organisations. The resulting distances were taken to calculate the cosine similarity. The measure of cosine similarity is a basis for the creation of a recommendation list. As an example an organisation with id 937069277 and name *IMP'ROVE - EUROPEAN INNOVATION MANAGEMENT ACADEMY* was chosen. The resulting recommendation list is presented in Table 2.

Table 2. Entity recommendation list of org. 937069277.

Nr	ID	Organisation name	Similarity
1	937069277	IMP'ROVE - EUROPEAN INNOVATION MANAGEMENT...	1.0
2	909258310	909258310	0.8448
3	955606365	955606365	0.8307
4	921448494	NARODNE CENTRUM ROBOTIKY	0.8265
5	949834186	UNIVERSITA DEGLI STUDI NICCOLO CUSANO TELE...	0.8239
6	951881080	UPPSALA KOMMUN	0.8198
7	935540945	MIRANTIS POLAND SPOLKA Z OGRANICZONA...	0.8192
8	929254084	FEDERACION ESPANOLA DE EMPRESAS DE TECN...	0.8155

Each organisation is listed with its unique id, its name, if it is available in the data set, and the measure of similarity. If the data set does not include the organisation's name, in the name field the ID is repeated. It is important to note that the first organisation that appears on the list is the same organisation for which the recommendation is requested. The value of similarity is equal to 1. In the embedding space, the same organisation must be closest to itself.

4.2 Keywords Embedding-Based Model

Keywords embedding-based model presents a slightly different approach. The embeddings that are created in the learning process are embeddings of the keywords describing the organisations. The value of similarity between the two organisations are the predictions of the model whether the organisations have collaborated in the past. Unlike it is the case with the entity embedding-based model we need to make sure that the model predicts correctly the output for the data that it has not seen before. Therefore the data set was divided into training set and the validation set. The results of the training are presented in Fig. 4(a) and 4(b).

Figure 4(a) presents the values of loss function for training data and Fig. 4(b) presents the values of accuracy for those two sets. It can be observed that for those two measures the results for the validation data are worse. The values of loss are between 0.23 and 0.15 for the validation data and between 0.23 and 0.015 for the training data. The values of accuracy for the validation data reach maximally 0.78 and for the training data maximally 0.89. Although the validation measures are not as good as the training measures, they still stay at a fairly high level.

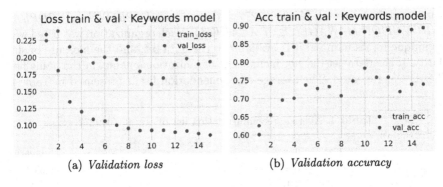

(a) *Validation loss*   (b) *Validation accuracy*

Fig. 4. Keywords model training.

As described in the previous section to create a recommendation list for the keywords embedding-based model we had to generate data set consisting of all possible pairs of keywords sets. Such data set can be verified by the trained model to give us the prediction about the collaboration of the given organizations pair. As an example an organization with id 937069277 and name IMP'ROVE - EUROPEAN INNOVATION MANAGEMENT ACADEMY was chosen. The resulting recommendation list is presented in Table 3.

Table 3. Keywords recommendation list of org. 937069277.

Nr	ID	Organisation name	Similarity
1	999975426	UNIVERSITY OF LEEDS	0.9999
2	999879881	POLITECNICO DI MILANO	0.9995
3	999991722	AGENCIA ESTATAL CONSEJO SUPERIOR DE INVESTG	0.9994
4	999974844	UNIVERSIDAD POLITECNICA DE MADRID	0.9981
5	999988909	NEDERLANDSE ORGANISATIE VOOR TOEGEPAST	0.9975
6	999992401	COMMISSARIAT A L ENERGIE ATOMIQUE	0.9961
7	999865331	UNIVERSIDADE DE AVEIRO	0.9941
8	999979500	CONSIGLIO NAZIONALE DELLE RICERCHE	0.9939

4.3   Final Recommendation List

As described, in the final step of developing a recommendation method we decided to incorporate two recommendation lists based on the models into two additional lists.

The first aggregate list was based on the average of similarities. Given two recommendation lists based on models we took the same organization from both lists, along with its measure of similarity and we calculated the average of the similarities. We performed the same for all items in the lists. The average of similarities served as a new measure to order the elements. As an example we took the same organization with id 937069277 and the name *IMPŔOVE - EUROPEAN INNOVATION MANAGEMENT ACADEMY*. Each organization was listed with its unique id, its name, if it is available in the data set, and the measure of similarity. If the data set did not include the organization's name, in the name field the ID was repeated. The average recommendation list is presented in Table 4.

Table 4. Average recommendation list of org. 937069277.

Nr	ID	Organisation name	Similarity
1	999847289	VAASAN YLIOPISTO	0.6923
2	953824475	CYBRES GMBH	0.6771
3	999428928	999428928	0.6104
4	945195840	GRADO ZERO INNOVATION SRL	0.605
5	996317362	G.E. PUKHOV INSTITUTE FOR MODELINGIN ENERGY...	0.6032
6	997732592	ECOLE NATIONALE SUPERIEURE DE CHIMIE DE PARIS	0.6021
7	903022277	FOLLOWHEALTH SL	0.5982
8	951881080	UPPSALA KOMMUN	0.5978

The second aggregate was created based on the principle of ranks [15]. The position on the list was taken as a rank, added to the second rank and normalized to give a value of similarity. Each organization is listed with its unique id, its name, if it is available in the data set, and the measure of similarity. If the data set did not include the organization's name, in the name field the ID was repeated. The rank recommendation list is presented in Table 5.

Table 5. Rank recommendation list of org. 937069277.

Nr	ID	Organisation name	Similarity
1	953824475	CYBRES GMBH	0.9373
2	997732592	ECOLE NATIONALE SUPERIEURE DE CHIMIE DE PARIS	0.9029
3	951881080	UPPSALA KOMMUN	0.9023
4	910218028	THE SHADOW ROBOT COMPANY ESPANA SL	0.8979
5	948446795	CLEAR VILLAGE TRUSTEE LIMITED	0.8979
6	997224700	STARHOME	0.8968
7	999428928	999428928	0.8849
8	937069277	IMP'ROVE - EUROPEAN INNOVATION MANAGEMENT...	0.8836

4.4   Measures of Recommender System Evaluation

In other recommendation problems researchers are often able to use the help of an expert or users feedback to check the proposed recommendation correspond to their preferences or not in order to evaluate their recommendation method. In our case we need to choose evaluation metrics to determine which recommendation list among those created is the best. There are 4 metrics proposed: *accuracy*, *coverage*, *Gini index* and *entropy*. Each metric determines a characteristic of a distribution and we asses obtained lists in terms of the desired value of a given metric.

As mentioned in the introduction of this section we did not have access to the information on what good recommendations were for a specific organization provided by a user (or in this case e.g. organization's authorities). Therefore, we chose another approach to determine the accuracy of our recommendation lists. We defined a relevancy threshold to be a median of all similarities. The organizations below the threshold of relevancy were treated as less relevant to us and therefore its recommendation as less accurate. We calculated accuracy as the percentage of the elements in the list that are above the median of similarities of each recommendation list. The results can be seen in Table 6.

The best accuracy is achieved in the case of a recommendation list based on ranks. Entity and keywords lists reach similar values (0.5) for all the chosen lengths. The worst in terms of accuracy is the list based on averages. Its values do not exceed 0.2.

Table 6. Accuracy values for all lists.

List length	Entity	Keywords	Average	Rank
1	0.500000	0.515101	**0.505453**	0.344379
5	0.500000	0.489010	0.504027	**0.689178**
10	0.500000	0.483347	0.491023	**0.592492**
20	0.500000	0.470680	0.484857	**0.515520**
50	0.500000	0.476779	0.490948	**0.552005**
100	0.500000	0.488578	0.493440	**0.532626**

The second metric that we chose to evaluate the recommendation lists was coverage. Coverage is the percentage of items which appear at least once in the recommendation lists [13]. It is important to mention that in case of the entity recommendation list the first value had to be ignored as it is always 1 because it is a cosine similarity between the same two vectors of organisation's embedding. The values of coverage for the lists are presented in Table 7. The highest values of coverage are obtained for the entity and average list. For all of the methods, the values increased along the increasing length of the lists. We obtained the worst results for the keywords recommendation list where the values of coverage did not exceed 0.2. It means that only 20 percent of all of the organisations were recommended in any of the recommendation lists.

Besides the accuracy of recommended items and the coverage of the organisations in the lists we decided to check fairness of the distribution of recommended items. *Gini index* is a measure that takes into account how uniformly items appear in the recommendation lists [13]. Given all the recommendation lists for users, L, and $p(i_k|L)$ as the probability of k-th least recommended item

712   D. Król et al.

Table 7. Coverage values for all lists.

List length	Entity	Keywords	Average	Rank
1	0.000839	0.011326	**0.348993**	0.088507
5	**0.911913**	0.033977	0.731544	0.204279
10	**0.990352**	0.054530	0.872903	0.292785
20	**0.999161**	0.089346	0.946309	0.404362
50	**1.000000**	0.162332	0.990352	0.578020
100	**1.000000**	0.269715	0.999581	0.710990

being drawn from L, where L_u is the recommendation list for user u, Gini index can be calculated as 1:

$$Gini(L) = \frac{\sum_{k=1}^{|I|}(2k - |I| - 1)p(i_k|L)}{|I| - 1} \tag{1}$$

The desired value of Gini index is 0 for the ideally uniformly distributed items. Table 8 shows the results of calculating Gini index for the recommendation lists of different sizes. The results can be ordered from the list of the highest Gini index (the least successful) to the lowest Gini index (the most) in such a way: keywords, embedding-based model list, average of lists, entity embedding-based list and rank based list.

Table 8. Gini index values for all lists.

List length	Entity	Keywords	Average	Rank
1	0.000000	0.018246	0.279548	**0.375745**
5	0.348321	0.053628	0.276863	**0.379760**
10	0.345713	0.064063	0.278350	**0.378293**
20	0.345188	0.071454	0.279611	**0.372790**
50	0.347549	0.079096	0.287044	**0.365775**
100	0.351002	0.085931	0.295009	**0.360678**

Entropy is the measure of the uniformity of a distribution. Uniform distributions have higher information gain and therefore are more desired when higher diversity of recommended items is needed. Entropy can be calculated with Formula 2:

$$Entropy(L) = -\sum_{i\in I} p(i|L) \log p(i|L) \tag{2}$$

where $p(i|L)$ is the probability of item i being in recommendation list L.

The results that we obtained are presented in Table 9. Entity list characterized the biggest values of entropy for any given length of a list. Next are: average-based, rank-based and keyword-based list in a mentioned order. For all recommendation lists the value of entropy increases with the increase of the list size.

The results of training and the evaluation of obtained recommendation lists do not lead to a strong conclusion that any of the proposed list creation methods

Table 9. Entropy values for all lists.

List length	Entity	Keywords	Average	Rank
1	0.003681	1.872468	**5.654874**	4.148033
5	**7.459555**	2.963007	6.139525	4.918314
10	**7.490582**	3.483342	6.346608	5.282652
20	**7.463683**	3.989657	6.549774	5.697119
50	**7.400253**	4.692202	6.799791	6.200449
100	**7.364463**	5.274451	6.968509	6.547615

is significantly better than the others. One recommendation list is always significantly better than the other but it differs depending on which metric is taken into consideration. However, for a majority of metrics, the entity embedding-based list gives good results. It is characterized by high coverage, average accuracy, low Gini index and the highest entropy. If any method were to be chosen to be proposed, the method based purely on the entity embedding-based is the first to recommend.

5   Concluding Remarks

The main purpose of this work was developing a recommendation method for matching scientific and business partners. The main goal of the research was to perform experiments on the available data with a use of different deep learning techniques. After a detailed systematic literature review there were two deep learning approaches proposed. The first focuses on the entity embeddings of organisations. During the learning process it tries to place similar organisations closer in the embedding space and thus provides a possibility to determine which partners may be interested in a collaboration. The second, on the other hand, focuses on the keywords describing each organisation and creates an embedding space of those keywords. Based on the keywords' proximity the model tries to predict whether two organisations characterized with a given set of keywords may have collaborated. The two models give us a measure of similarity which in the end lets us create four recommendation lists. Using common evaluation measures we evaluated which of the created lists gives the best recommendation. In the course of the research it became clear that recommendation problems are complex and require a lot of experimentation to adjust all the parameters to fit the given data. The evaluation part showed that even effective prediction models may not be the most effective when it comes to create a diverse, uniformly distributed and accurate recommendation lists. Because the problem is so complex there is no simple solution to it problem and the answer differs depending for example on the length of the list that we want to obtain or whether we want to create more accurate or more diverse lists. This research has shown that there is still a lot to discover in that field. Several opportunities for further development are presented in the last section.

There are several directions in which the proposed method can be further developed. The CORDIS database is a source of huge amounts of data and there are many attributes of organisations that could be taken into consideration in a deep learning model. For the purposes of this work only a small portion of the data has been used. It underwent several data cleaning and transformation procedures but there is still a lot that can be done. The complexity of the CORDIS API raises a problem of data consistency. Perhaps the accuracy of the method could be improved if the data was more consistent. In further development it would be important to determine the reason of so big inconsistencies. The static nature of the model makes it difficult to handle the updates of the database. The data was fetched, cleaned, transformed on the basis of the database state at a given point of time. If the new data appears in the database the model will have to be trained again. The addition of a new organisation in the data set is something worth considering. The number of layers and their parameters that were chosen for the experiments correspond to the examples seen in the literature. As part of the research some of the parameters were changed, and adjusted, but it could be done in a systematic way, reporting and comparing the learning outcomes. The approach presented in this work uses two separate deep learning models to create a recommendation method. Another approach that could be explored is incorporating the two proposed models into one containing both entity and keyword embeddings. In many machine learning studies related to natural language processing pre-trained word embeddings are used. Such embeddings are vectors representing words in space, where the words which have similar meaning are closer together. There is a chance that the pre-trained word embeddings could be adjusted based on the collaborations data, similarly to what have been done with the keywords embedding-based model.

Acknowledgments.. The publication has been prepared as a part of the Support Programme of the Partnership between Higher Education and Science and Business Activity Sector financed by City of Wroclaw.

References

1. Almalis, N.D., Tsihrintzis, G.A., Karagiannis, N.: A content based approach for recommending personnel for job positions. In: IISA 2014, the 5th International Conference on Information, Intelligence, Systems and Applications, pp. 45–49 (2014)
2. Antikacioglu, A., Bajpai, T., Ravi, R.: A new system-wide diversity measure for recommendations with efficient algorithms. SIAM J. Math. Data Sci. **1**, 759–779 (2019)
3. Bird, S., Klein, E., Loper, E.: Natural Language Processing with Python. O'Reilly Media, Inc. (2009)
4. Chandrashekhar, H., Bhasker, B.: Personalized recommender system using entropy based collaborative filtering technique. J. Electron. Commer. Res. **12**(3), 214 (2011)
5. Chinchanachokchai, S., Thontirawong, P., Chinchanachokchai, P.: A tale of two recommender systems: the moderating role of consumer expertise on artificial intelligence based product recommendations. J. Retail. Consum. Serv. **61** (2021)
6. CORDIS API. https://ec.europa.eu/info/funding-tenders/opportunities/portal/screen/support/apis

7. Guangyao, C.: Research on the recommending method used in c2c online trading. In: Proceedings of the 2007 IEEE/WIC/ACM International Conferences on Web Intelligence and Intelligent Agent Technology - Workshops, pp. 103–106. WI-IATW 2007, IEEE Computer Society, USA (2007)

8. Hannon, J., Bennett, M., Smyth, B.: Recommending twitter users to follow using content and collaborative filtering approaches. In: Proceedings of the Fourth ACM Conference on Recommender Systems, pp. 199–206. RecSys 2010, Association for Computing Machinery, New York, NY, USA (2010)

9. Hunter, J.D.: Matplotlib: a 2D graphics environment. Comput. Sci. Eng. 9(3), 90–95 (2007)

10. Kouki, A.B.: Recommender system performance evaluation and prediction an information retrieval perspective (2014)

11. Liberati, A., et al.: The PRISMA statement for reporting systematic reviews and meta-analyses of studies that evaluate health care interventions: explanation and elaboration. J. Clin. Epidemiol. 62(10), e1–e34. PubMed (2009)

12. Maaten, L.V.D., Hinton, G.E.: Visualizing data using t-sne. J. Mach. Learn. Res. 9, 2579–2605 (2008)

13. Mansoury, M., Abdollahpouri, H., Pechenizkiy, M., Mobasher, B., Burke, R.: Fairmatch: a graph-based approach for improving aggregate diversity in recommender systems. In: Proceedings of the 28th ACM Conference on User Modeling, Adaptation and Personalization, pp. 154–162 (2020)

14. Miller, B.N., Konstan, J.A., Riedl, J.: Pocketlens: toward a personal recommender system. ACM Trans. Inf. Syst. 22(3), 437–476 (2004)

15. Moskovkin, V., Golikov, N., Peresypkin, A., Serkina, O.: Aggregate ranking of the world's leading universities. Webology 12 (2015)

16. Noguera, J.M., Barranco, M.J., Segura, R.J., Martínez, L.: A mobile 3D-gis hybrid recommender system for tourism. Inf. Sci. 215, 37–52 (2012)

17. Ozbal, G., Karaman, H.: Matchbook a web based recommendation system for matchmaking. In: 2008 23rd International Symposium on Computer and Information Sciences, pp. 1–6 (2008)

18. Prasad, B.: Hyrec: a hybrid recommendation system for e-commerce. In: Muñoz-Ávila, H., Ricci, F. (eds.) Case-Based Reasoning Research and Development, pp. 408–420. Springer (2005)

19. Sainburg, T., McInnes, L., Gentner, T.Q.: Parametric umap: learning embeddings with deep neural networks for representation and semi-supervised learning. ArXiv e-prints (2020)

20. Strassner, J.: The design of a hybrid semantic recommender system. In: 2011 IEEE Consumer Communications and Networking Conference (CCNC), pp. 147–152 (2011)

21. Yao, L., Sheng, Q.Z., Segev, A., Yu, J.: Recommending web services via combining collaborative filtering with content-based features. In: 2013 IEEE 20th International Conference on Web Services, pp. 42–49 (2013)

Author Index